I0881918

# Chemical contaminants in foods

## Understanding and managing risks

It is widely recognised that agriculture is a significant contributor to global warming and climate change. Agriculture needs to reduce its environmental impact and adapt to current climate change whilst still feeding a growing population, i.e. become more 'climate-smart'. Burleigh Dodds Science Publishing is playing its part in achieving this by bringing together key research on making the production of the world's most important crops and livestock products more sustainable.

Based on extensive research, our publications specifically target the challenge of climate-smart agriculture. In this way we are using 'smart publishing' to help achieve climate-smart agriculture.

Burleigh Dodds Science Publishing is an independent and innovative publisher delivering high quality customer-focused agricultural science content in both print and online formats for the academic and research communities. Our aim is to build a foundation of knowledge on which researchers can build to meet the challenge of climate-smart agriculture.

For more information about Burleigh Dodds Science Publishing simply call us on +44 (0) 1223 839365, email info@bdspublishing.com or alternatively please visit our website at www.bdspublishing.com.

**Related titles:**

Advances in ensuring the microbiological safety of fresh produce
Print (ISBN 978-1-80146-268-6); Online (ISBN 978-1-80146-269-3, 978-1-80146-270-9)

Advances in postharvest management of cereals and grains
Print (ISBN 978-1-78676-352-5); Online (ISBN 978-1-78676-354-9, 978-1-78676-355-6)

Advances in postharvest management of horticultural produce
Print (ISBN 978-1-78676-288-7); Online (ISBN 978-1-78676-290-0, 978-1-78676-291-7)

Developing smart agri-food supply chains
Print (ISBN 978-1-78676-749-3); Online (ISBN 978-1-78676-751-6, 978-1-78676-752-3)

Improving standards and certification in agri-food supply chains
Print (ISBN 978-1-80146-451-2); Online (ISBN 978-1-80146-452-9, 978-1-80146-453-6)

Chapters are available individually from our online bookshop:
https://www.bdspublishing.com/webshop/

climate•SMART•publishing
IN AGRICULTURAL SCIENCE

BURLEIGH DODDS SERIES IN AGRICULTURAL SCIENCE
NUMBER 172

# Chemical contaminants in foods

## Understanding and managing risks

Edited by Dr Rob Theelen

burleigh dodds
SCIENCE PUBLISHING

Published by Burleigh Dodds Science Publishing Limited
82 High Street, Sawston, Cambridge CB22 3HJ, UK
www.bdspublishing.com

Burleigh Dodds Science Publishing, 1518 Walnut Street, Suite 900, Philadelphia, PA 19102-3406, USA

First published 2025 by Burleigh Dodds Science Publishing Limited
© Burleigh Dodds Science Publishing, 2025, except the following: Chapters 1 and 2 are © of the Food and Agriculture Organization of the United Nations (FAO). All rights reserved.

This book contains information obtained from authentic and highly regarded sources. Reprinted material is quoted with permission and sources are indicated. Reasonable efforts have been made to publish reliable data and information but the authors and the publisher cannot assume responsibility for the validity of all materials. Neither the authors nor the publisher, nor anyone else associated with this publication shall be liable for any loss, damage or liability directly or indirectly caused or alleged to be caused by this book.

No part of this publication may be reproduced, stored in a retrieval system or transmitted in any form or by any means electronic, mechanical, photocopying, recording or otherwise without the prior written permission of the publisher.

The consent of Burleigh Dodds Science Publishing Limited does not extend to copying for general distribution, for promotion, for creating new works, or for resale. Specific permission must be obtained in writing from Burleigh Dodds Science Publishing Limited for such copying.

Permissions may be sought directly from Burleigh Dodds Science Publishing at the above address. Alternatively, please email: info@bdspublishing.com or telephone (+44) (0) 1223 839365.

Trademark notice: Product or corporate names may be trademarks or registered trademarks and are used only for identification and explanation, without intent to infringe.

**Notice**
No responsibility is assumed by the publisher for any injury and/or damage to persons or property as a matter of product liability, negligence or otherwise, or from any use or operation of any methods, products, instructions or ideas contained in the material herein.

Library of Congress Control Number: 2025946836

British Library Cataloguing in Publication Data
A catalogue record for this book is available from the British Library

ISBN 978-1-83545-050-5 (print)
ISBN 978-1-83545-051-2 (ePub)
ISBN 978-1-83545-052-9 (PDF)
ISSN 2059-6936 (print)
ISSN 2059-6944 (online)

DOI: 10.19103/AS.2025.0158

Typeset by Deanta Global Publishing Services, Dublin, Ireland

# Contents



© Burleigh Dodds Science Publishing Limited, 2025. All rights reserved.

© Burleigh Dodds Science Publishing Limited, 2025. All rights reserved.

© Burleigh Dodds Science Publishing Limited, 2025. All rights reserved.

© Burleigh Dodds Science Publishing Limited, 2025. All rights reserved.

© Burleigh Dodds Science Publishing Limited, 2025. All rights reserved.

## Part 4 Environmental and processing contaminants and natural toxins



© Burleigh Dodds Science Publishing Limited, 2025. All rights reserved.

© Burleigh Dodds Science Publishing Limited, 2025. All rights reserved.

© Burleigh Dodds Science Publishing Limited, 2025. All rights reserved.

# Series list

| Title | Series number |
|---|---|
| **Achieving sustainable cultivation of maize - Vol 1**<br>From improved varieties to local applications<br>Edited by: Dr Dave Watson, CGIAR Maize Research Program Manager, CIMMYT, Mexico | 001 |
| **Achieving sustainable cultivation of maize - Vol 2**<br>Cultivation techniques, pest and disease control<br>Edited by: Dr Dave Watson, CGIAR Maize Research Program Manager, CIMMYT, Mexico | 002 |
| **Achieving sustainable cultivation of rice - Vol 1**<br>Breeding for higher yield and quality<br>Edited by: Prof. Takuji Sasaki, Tokyo University of Agriculture, Japan | 003 |
| **Achieving sustainable cultivation of rice - Vol 2**<br>Cultivation, pest and disease management<br>Edited by: Prof. Takuji Sasaki, Tokyo University of Agriculture, Japan | 004 |
| **Achieving sustainable cultivation of wheat - Vol 1**<br>Breeding, quality traits, pests and diseases<br>Edited by: Prof. Peter Langridge, The University of Adelaide, Australia | 005 |
| **Achieving sustainable cultivation of wheat - Vol 2**<br>Cultivation techniques<br>Edited by: Prof. Peter Langridge, The University of Adelaide, Australia | 006 |
| **Achieving sustainable cultivation of tomatoes**<br>Edited by: Dr Autar Mattoo, USDA-ARS, USA and Prof. Avtar Handa, Purdue University, USA | 007 |
| **Achieving sustainable production of milk - Vol 1**<br>Milk composition, genetics and breeding<br>Edited by: Dr Nico van Belzen, International Dairy Federation (IDF), Belgium | 008 |
| **Achieving sustainable production of milk - Vol 2**<br>Safety, quality and sustainability<br>Edited by: Dr Nico van Belzen, International Dairy Federation (IDF), Belgium | 009 |
| **Achieving sustainable production of milk - Vol 3**<br>Dairy herd management and welfare<br>Edited by: Prof. John Webster, University of Bristol, UK | 010 |
| **Ensuring safety and quality in the production of beef - Vol 1**<br>Safety<br>Edited by: Prof. Gary Acuff, Texas A&M University, USA and Prof. James Dickson, Iowa State University, USA | 011 |
| **Ensuring safety and quality in the production of beef - Vol 2**<br>Quality<br>Edited by: Prof. Michael Dikeman, Kansas State University, USA | 012 |
| **Achieving sustainable production of poultry meat - Vol 1**<br>Safety, quality and sustainability<br>Edited by: Prof. Steven C. Ricke, University of Arkansas, USA | 013 |
| **Achieving sustainable production of poultry meat - Vol 2**<br>Breeding and nutrition<br>Edited by: Prof. Todd Applegate, University of Georgia, USA | 014 |

© Burleigh Dodds Science Publishing Limited, 2025. All rights reserved.

© Burleigh Dodds Science Publishing Limited, 2025. All rights reserved.

© Burleigh Dodds Science Publishing Limited, 2025. All rights reserved.

© Burleigh Dodds Science Publishing Limited, 2025. All rights reserved.

© Burleigh Dodds Science Publishing Limited, 2025. All rights reserved.

© Burleigh Dodds Science Publishing Limited, 2025. All rights reserved.

© Burleigh Dodds Science Publishing Limited, 2025. All rights reserved.

© Burleigh Dodds Science Publishing Limited, 2025. All rights reserved.

© Burleigh Dodds Science Publishing Limited, 2025. All rights reserved.

© Burleigh Dodds Science Publishing Limited, 2025. All rights reserved.

© Burleigh Dodds Science Publishing Limited, 2025. All rights reserved.

# Introduction

There is growing awareness of the degree to which human health is influenced by the exposome: the overall array of dietary, lifestyle and other environmental factors affecting individuals over their lifetime. Diet has consistently been shown to the dominant source of exposure to the majority of contaminants of health concern, often via low-level exposure to multiple chemicals over time with varying effects on different groups of consumers.

There is a growing awareness that current methods for detection, analysis and risk assessment of chemical residues in food (the latter traditionally based on *in-vivo* animal testing of the effects of single chemicals) may be not sufficiently accurate or flexible to fully reflect this more complex and evolving picture.

There is a growing interest in what have come to be known as new approach methodologies (NAMs). Emerging technologies include human biomonitoring (HBM), the development of adverse outcome pathways (AOPs) to link molecular-level changes to negative/adverse health outcomes/effects that can then inform more sophisticated next generation risk assessments, as well as use of in-silico physiologically-based pharmacokinetic (PBPK) models to improve risk assessment. NAMs have the potential to provide a more flexible approach better equipped to identify and deal with emerging chemicals/ combinations of chemicals (whether from food or elsewhere) than current methods and national regulations.

This collection reviews and synthesises the wide range of recent research in this exciting new area which is redefining how we can measure levels of and potential health risks from chemical contaminants in food.

## Part 1 Current principles and systems

Part 1 looks at the current policy context relating to managing chemical contaminant risks in foods. Chapter 1 sets the scene by looking at the global framework for food chemical risk assessment. It outlines the key role of the Codex Alimentarius, established by the Food and Agriculture Organization of the United Nations (FAO) and the World Health Organization (WHO) as an international collection of science-based international standards, guidelines and codes of practice (CoPs) relating to food safety and quality. The chapter outlines the coverage in Codex of food chemical risk management with a focus on contaminants in food, how its standards relate to food trade and the agreements of the World Trade Organization (WTO), the scientific basis for its standards and the work of the Codex Committee on Contaminants in Food (CCCF).

© Burleigh Dodds Science Publishing Limited, 2025. All rights reserved.

Building on the previous chapter, Chapter 2 explores the role of the Joint FAO/WHO Expert Committee on Food Additives (JECFA) as an international scientific expert committee in ensuring Codex is kept up to date in terms of new contaminants to assess and broader developments in scientific analysis and understanding of contaminant risks. This work is illustrated by a case study analyzing the way JECFA has assessed co-exposure to fumonisins and aflatoxins as food contaminants in establishing robust safe limits in food.

Completing Part 1, Chapter 3 addresses key issues and challenges in exposure assessment and risk characterization. As the chapter points out, exposure assessment is a key step in food safety risk assessment, in which the dietary intake of chemical substances is estimated. The health risk of that intake is then evaluated in a risk characterization, where intake is compared with a Health Based Guidance Value (HBGV). To estimate dietary intake, the exposure assessor needs data on consumption quantities of foods and the concentration of substances in these foods. The chapter explores the range of consumption and other data needed for exposure assessment. It then explores challenges arising from both limitations in available data and in how best to interpret such data, as well as offering solutions to optimize the reliability of exposure assessments and risk characterization.

## Part 2 Developments in risk assessment: hazard identification/characterisation and exposure assessment

Building on some of the issues in Chapter 3, chapters in Part 2 discuss developments in hazard identification and characterization as well as exposure assessment. Chapter 4 begins by highlighting common challenges in the analysis of food contaminants and residues such as the presence of multiple residues in low concentrations within a complex food matrix. It then explores the role of chromatographic methods such as gas chromatography (GC), liquid chromatology (LC), supercritical fluid chromatography (SFC) and multi-dimensional chromatography (MDC) in addressing these challenges.

Complementing the previous chapter, Chapter 5 reviews the role of electrochemical biosensors in providing real-time, online monitoring of contaminant risks along agri-food supply chains. Electrochemical sensors are devices designed to measure current and voltage between electrodes in electrolyte to sense different bio-analytes dissolved in the electrolyte. They are typically combined with biosensors based on a biological sensing element such as an enzyme-substrate, antibody-antigen and nucleic acids-complementary sequences. As the chapter shows, this allows rapid and accurate identification e.g. of antibiotic and veterinary residues. The chapter explains the principles, technologies and applications of this class of sensor, as well challenges such

© Burleigh Dodds Science Publishing Limited, 2025. All rights reserved.

as the ability to identify multiple residues in a sample and how these might be addressed.

As Chapter 6 discusses, the concept of effect biomarkers has emerged as a promising alternative to traditional toxicity tests due to the need for more sensitive, specific, and human-relevant methods to assess chemical exposure effects. An effect biomarker (or biomarker of effect) is defined as a biological indicator that provides evidence of a biological response to exposure to a contaminant. It reflects changes in biochemical, physiological, behavioral, or other alterations in an organism that can be associated to adverse or potentially adverse health effects. Alterations in gene expression, protein levels and metabolic profiles can e.g. serve as indicators of biological effects. The chapter discusses the use of effect biomarkers in food safety and environmental health, examining types, relevance and advances such as the use of metabolomics. Two case studies relating to veterinary residues and endocrine-disrupting compounds are explored, highlighting varied applications in using effect biomarkers in food safety and environmental health. Finally, the chapter addresses challenges and future directions for integrating effect biomarkers into regulatory science to improve chemical risk assessment and management.

As noted earlier, diet represents the main source of exposure to many chemicals for much of the population. However, tracking chronic, low-level exposure to combinations of chemical contaminants in food is hugely challenging. As discussed in Chapter 3, dietary exposure assessments provide valuable information on levels of risk, exposure trends, the effectiveness of control measures, and the impact of new agricultural, safety or food production practices. Building on some of the issues highlighted in Chapter 3, Chapter 7 reviews the complex challenges faced by dietary exposure studies and the range of advances addressing those challenges. It looks at challenges and advances in food consumption data collection, determination of food contaminant levels and in modeling to establish dietary exposure.

Human biomonitoring (HBM) has emerged as a pivotal tool in environmental health research, providing critical insights into the levels of potentially harmful chemicals in the human body. Building on Chapter 6, Chapter 8 reviews recent advances in HBM methodologies. Using different European case studies, the chapter illustrates the practical applications of biomonitoring data to study exposure ro contaminants from food in different populations. It shows this by comparing exposure models to estimate intake (using cadmium and acrylamide as examples), discusses the revision of hazard assessment (bisphenols and pyrethroids), reviews risk communication issues and summarizes progress in mixture risk assessment based on HBM data. The chapter also discusses ongoing challenges and ways they can be addressed going forward.

© Burleigh Dodds Science Publishing Limited, 2025. All rights reserved.

## Part 3 Developments in risk assessment: risk characterisation

Part 3 reviews key developments in risk assessment. As Chapter 9 highlights, traditional exposure assessments based solely on dietary intake levels do not account for how these chemicals are absorbed, distributed, metabolized and excreted in the human body. Chapter 9 reviews recent advancements in toxicokinetic modeling, emphasizing physiologically based toxicokinetic (PBTK) models that predict impact of internal doses more realistically. Innovations include generic PBTK models adaptable to numerous chemicals and sophisticated computational tools, such as quantitative structure-activity relationship (QSAR) models, machine learning, and Bayesian inference, which are able to address population variability, aggregate and cumulative exposures, interactions among multiple chemicals and exposure routes, thus refining the accuracy of health risk assessments. Integrating toxicokinetic models with biomonitoring data also allows for reverse dosimetry, enhancing interpretation of real-world exposure studies. The chapter describes the shift toward Next-Generation Risk Assessment (NGRA), utilizing high-throughput *in vitro* data and *in vitro*-to-*in vivo* extrapolation (IVIVE) methods to minimize reliance on animal testing. Incorporating advanced toxicokinetic modeling approaches into regulatory practices provides robust, human-centric risk evaluations, enabling more effective public health protection measures against foodborne chemical exposure.

As discussed earlier in the book, chemical risk assessment is moving more towards non-animal approaches that rely on the mechanistic understanding of toxicity. As Chapter 10 explains, the adverse outcome pathway (AOP) framework is a promising tool providing a mechanistic understanding of toxicological processes. AOPs link a molecular initiating event to a specific adverse outcome via a series of key events and key event relationships. Sharing common key features to the AOP framework, the aggregate exposure pathway (AEP) framework is a complementary tool to systematically collect chemical-specific exposure information from the source to the site of action via key exposure states and key transitional relationships. Integrating exposure and toxicological knowledge into combined AEP-AOP constructs can facilitate prediction of chemical-induced toxicity and improve risk-based decision making. Chapter 10 summarises the structure, development, assessment and applications of AOP and AEP frameworks, and illustrates their value with a case study on the hepatotoxicity of pesticide residues.

Chapter 11 reviews developments in the use of *in silico* computational toxicology tools for safety and hazard assessment of chemicals in food. Regulatory agencies worldwide are increasingly incorporating computational toxicology tools into their risk assessment frameworks to enhance food safety and protect public health from dietary exposure to contaminants. Chapter 11

© Burleigh Dodds Science Publishing Limited, 2025. All rights reserved.

reviews a broad range of publicly, commercially-available and emerging in silico tools and sets a framework for comparison and reliability assessment. It also discusses Consensus Modeling for Endpoint Prediction (CoMPARA), an approach which involves combining multiple computational models to predict toxicological endpoints or properties of chemical compounds. This approach is used in a case study assessing potential risks from the brominated flame retardant (BFR) DBE-DBCH, a potential endocrine disruptor. However, these tools rely on available data and models, which can introduce uncertainties and limitations in model predictions, the reliability and relevance of the underlying data, and the need for validation with experimental data. The chapter advocates for a synergistic approach that combines computational predictions with experimental data to bolster safety and hazard assessments.

## Part 4 Environmental and processing contaminants and natural toxins

Part four addresses advances in detecting particular environmental and processing contaminants as well as natural toxins.

Chapter 12 provides a comprehensive overview of assessing risks from heavy metal (HM) residual contentrations in food, including lead (Pb), mercury (Hg), cadmium (Cd), and arsenic (As), which pose significant health risks even at low concentrations. The chapter explores the mechanisms of HM uptake and bioaccumulation, particularly in fish and fresh produce. A case study on turmeric adulteration with lead chromate illustrates the public health implications of economically-motivated fraud. The chapter concludes by emphasizing the need for rapid, cost-effective screening tools, such as electrochemical sensors and lateral flow assays, and the integration of artificial intelligence for enhanced detection, data analysis, and regulatory compliance.

Chapter 13 highlights the advances that have been made in the analysis of flame retardant residues in food, such as polybrominated diphenyl ethers (PBDEs), chlorinated flame retardants (CFRs), organophosphorus flame retardants (OPFRs) and nano-based flame retardants. The chapter reviews contamination pathways, sampling and extraction methods as well as analytical techniques, including separation (e.g. using gas, liquid or supercritical fluid (SCF) chromatography) and detection (e.g. using mass spectrometry). The chapter also reviews other detection techniques to detect flame retardants such as infrared spectroscopy (IR), Fourier transform infrared spectroscopy (FTIR), immunoassays and biosensors

Per- and polyfluoroalkyl substances (PFAS) are persistent organic pollutants (POP) that can accumulate in food systems. Analyzing these substances is complex due to the factors such as the diversity of PFAS and complexity of food matrices as well as analytical limitations. Chapter 14 begins by examining

© Burleigh Dodds Science Publishing Limited, 2025. All rights reserved.

sources of PFAS in crops and livestock. It then reviews key issues in analysing PFAS in food including sampling, extraction and clean-up, as well as analytical techniques (e.g. high and ultra performance liquid-chromatography tandem mass spectrometry (HPLC-MS/MS or UPLC-MS/MS)), and quality control. The chapter also discusses mitigation strategies including phytomanagement. To illustrate challenges and advances in the field, the chapter provides two case studies reassessing PFAS levels in Belgian food products and in processed seafood products.

An area of growing concern is the presence and potential adverse health effects of microplastics and nanoplastics in the environment and in food. Chapter 15 reviews what we know about types of micro/nanoplastics and levels of occurrence in the environment. It also assesses their potential impact as contaminants in food, including oral bioavailability and toxicology, providing the basis for improved risk assessments going forward.

Chapter 16 assesses contaminants related to high-temperature processing of food. The chapter examines the formation and occurrence of acrylamide (AA), heterocyclic amines (HCAs), and polycyclic aromatic hydrocarbons (PAHs) in foods. It focuses on precursors and the conditions under which they form as a foundation to help identify effective mitigation strategies. The chapter analyzes dietary exposure levels to AA, HCAs and PAHs in different countries, with a special focus on the Republic of Korea as a case study. Finally, it summarizes management strategies to reduce these contaminants in food.

Chapter 17 assesses the role of pyrrolizidine alkaloids (PAs) in food. It reviews their contamination pathways, current regulation and recent reports on levels of contamination. The chapter also surveys analytical methods and strategies for the assessment of PAs in food. It shows how foods, including teas, herbs, spices, cereals, vegetables and animal products, can influence the content and stability of these alkaloids. It also discusses the stability and degradation products of pyrrolizidine alkaloids in food processing.

Tropane alkaloids (TAs) found in plants from the Solanaceae family can contaminate crops entering the food chain, leading to severe toxic effects in humans. Chapter 18 summarises what we know about the health hazards posed by TAs (including Atropine and Scopolamine). It also reviews existing studies documenting on levels of food contamination, detection methods and potential decontamination techniques.

© Burleigh Dodds Science Publishing Limited, 2025. All rights reserved.

# Part 1

## Current principles and systems

# Chapter 1

## The global framework for food chemical risk assessment*

*Gracia Brisco and Verna Carolissen-Mackay, Joint FAO/WHO Food Standards Programme Secretariat - Codex Alimentarius Commission, Italy; and Annamaria Bruno, formerly Joint FAO/WHO Food Standards Programme Secretariat - Codex Alimentarius Commission, Italy*



### 1 Introduction

The Codex Alimentarius – or 'food code' – is the collection of science-based international standards, guidelines, codes of practice (CoPs), and miscellaneous texts (collectively, 'standards') on food safety and quality adopted by the Codex Alimentarius Commission (CAC), with the statutory purpose of protecting consumer health and facilitating fair practices in the food trade. Known as 'Codex', this intergovernmental body is also responsible for promoting coordination of all food standards work undertaken by international governmental and non-governmental organizations to avoid duplication of work and contradictory standards that may negatively impact trade [1]. The compilation of Codex standards adopted by CAC is publicly available on the Codex website (https://www.fao.org/fao-who-codexalimentarius/home/en/).

Codex was established by the Food and Agriculture Organization of the United Nations (FAO) and the World Health Organization (WHO) and met for the first time in 1963. Currently, CAC has 189 members, of which 188 are

* The views expressed in this chapter are those of the authors and do not necessarily the views or policies of the Food and Agriculture Organization of the United Nations.

http://dx.doi.org/10.19103/AS.2025.0158.06
Published by Burleigh Dodds Science Publishing Limited, 2025.

member countries and one is a member organization (the European Union). There are 240 Codex Observer Organizations, which are governmental and non-governmental, public and private. These organizations, which represent food producers, industry, consumer associations and academia, play an important role in the technical discussion of Codex and in ensuring Codex texts are of the highest quality and based on sound science. United Nations' technical agencies like the International Atomic Energy Agency (IAEA) can participate in Codex work as observers. Observers can, therefore, participate in the standards-setting process but not in the decision-making process, which rests with Codex members.

Codex works to the core values of inclusivity, transparency, collaboration, and consensus [2].

## 2 The background to Codex standards

### *2.1 Codex and fair practices in the food trade*

The annual value of trade in agricultural products has continued to grow over the past decades, largely in emerging economies and developing countries; the reduction in tariffs through global and regional trade agreements has provided greater opportunities for expanding global food trade [3].

However, to trade internationally and access markets for high-value products, producers must be able to meet food standards. Governments apply food standards to ensure that food is suitable for human consumption and meets the hygiene, labelling and other relevant requirements to ensure the overall safety, quality and identity of the product. The use of international food standards worldwide helps reduce trade costs by making trade more transparent and efficient, allowing food to move more smoothly between markets [3].

### *2.2 Legal status of Codex standards*

Codex standards are voluntary and become compulsory only when incorporated into national-level regulations. Governments are not obligated to accept or apply Codex standards under the Codex rules or procedures.

Therefore, the added value that may be attributed to Codex standards comes from the agreements of the WTO. In part, the value of Codex standards is reflected in governments' commitment to enter into negotiations on the content of each standard and to base these negotiations on science.

While there is no obligation for countries to accept Codex standards at the national level, they are significant at the international level because they are the benchmark standards for food under the World Trade Organization's (WTO's) agreement on the Application of Sanitary and Phytosanitary Measures

Published by Burleigh Dodds Science Publishing Limited, 2025.

(SPS agreement), which calls upon WTO members to align their national technical regulations to protect human, animal and plant life with standards established by relevant international standards-setting bodies [3]. The CAC is the relevant body concerned with human health protection. The recognition in the SPS agreement is based on the fact that the Codex standards satisfy a fundamental principle of the agreement, i.e. that any sanitary measures must be based on a scientific risk assessment. When a food-related trade dispute is taken to the dispute panel at the WTO, or the SPS committee when addressing specific trade concerns, Codex standards are primarily referenced in an effort to resolve the dispute or concern. This adds legal value to the standards (Article 3.1 of the SPS agreement) as well as to the risk assessment techniques used by the FAO/WHO expert bodies (Article 5.1 of the SPS agreement).

Similarly, although this is not explicitly stated, Codex texts are an international benchmark for the WTO agreement on Technical Barriers to Trade (TBT Agreement) and have been taken as the 'reference standards' in dispute trade settlements, which, for example, can involve the misalignment of national food regulations with Codex standards. They are also referenced regularly in meetings of the WTO TBT committee to address specific trade concerns related to food.

## 2.3 Codex working procedures

Codex's work on food safety conforms to its 'risk-analysis' paradigm. This working mode requires a functional separation between 'risk assessment' and 'risk management'. Risk assessment provides a science-based evaluation of risk, whereas risk managers determine what food safety decisions may or may not be needed based on the assessments provided. The separation of the two is fundamental to ensuring the integrity, independence and transparency of the scientific advice that informs the development of Codex texts. This safeguards the scientific credibility of Codex's work. In Codex, risk analysis additionally involves 'risk communication'.

CAC is Codex's risk management body and the joint FAO/WHO scientific advice programme, such as the Joint Expert Meeting on Food Additives (JECFA), which provides scientific risk assessment to help inform development of safety standards. CAC meets annually to decide on the adoption of new or revised standards. The commission also discusses its work programme, including its strategic plan, approvals for new work and other matters of general relevance to Codex work.

It is in CAC's subsidiary bodies that normative work is carried out. These bodies can be horizontal, or 'general subject' committees concerned with issues related to all food types and mainly related to food safety, vertical or 'commodity' committees related to specific food categories or ad hoc task

Published by Burleigh Dodds Science Publishing Limited, 2025.

forces, which carry out specific work within a limited timeframe. The committee that is involved with contaminants the Codex Committee on Contaminants in Food (CCCF) is a general subject committee.

## 2.4 Codex and science

Codex standards are developed through risk management measures based on sound scientific evidence derived from risk assessments using the risk analysis paradigm. This paradigm allows the systematic documentation of how science has been considered in the standards-setting process, including the degree of uncertainty and variability in the available scientific information for risk assessment (i.e. precaution).

Therefore, precaution is an inherent element of risk analysis vis-à-vis the uncertainties associated with the assessment and management processes that will impact any risk management options selected to protect human health [1]. In the case of contaminants, for instance, this may lead to the development of CoPS instead of maximum levels (MLs) until the data and information available suffice to establish an ML.

The *Working Principles for Risk Analysis for Application in the Framework of the Codex Alimentarius* is the umbrella text for applying risk analysis in Codex work. Other texts addressing risk analysis on specific issues have also been developed, such as the *Risk Analysis Principles applied by the Codex Committee on Contaminants in Foods*. Codex also advises governments on how to apply risk analysis at the domestic level, to enable international comparability of risk analysis and consequently to promote fair practices in the food trade. The compendium of risk analysis texts, except the one directed to governments, can be found in the *Procedural Manual of the Codex Alimentarius Commission*. All risk analysis texts, including the Procedural Manual, are available on the Codex website (https://doi.org/10.4060/cd2280en).

According to the *Working Principles for Risk Analysis for Application in the Framework of the Codex Alimentarius*, the responsibility for risk assessment lies primarily with the joint FAO/WHO expert bodies and consultations (risk assessors), which establish the scientific basis for Codex standards. FAO/WHO experts provide scientific advice to Codex and its subsidiary bodies through two main mechanisms (see Fig. 1).

The competence, integrity and neutrality of the FAO/WHO scientific advice are critical to ensure that Codex food safety standards are based on robust science. Any conclusions and recommendations elaborated depend on the application of strict protocols and procedures developed by these expert bodies, which align with the latest scientific developments. In addition, the objectivity, scientific knowledge and competence of the experts who apply the

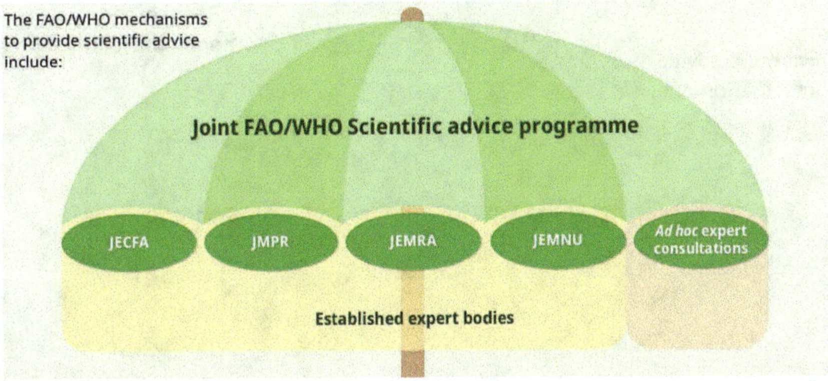


**Figure 1** Joint FAO/WHO scientific advice programme.

procedures and their own 'expert judgement' are essential when developing food safety recommendations for consideration by Codex [4].

FAO and WHO take great care in evaluating the accomplishments of the experts and apply stringent policies to identify and prevent any potential conflict of interest during the selection of the experts who are invited to participate. These procedures thereby seek to ensure the excellence, independence, integrity and transparency of the advice provided. Experts must be pre-eminent field specialists and indisputably objective in their judgement. They are appointed in their personal capacity, not as representatives of a government, organization or institution, and the input they provide is theirs alone [5].

## 2.5 When risk managers also consider factors other than science

Codex's risk managers use independent, transparent and current scientific advice to inform their decision-making. However, according to the Codex Statements of Principle, 'other legitimate factors' relevant to health protection and fair-trade practices may be identified in the risk management process. Consensus is not always possible within Codex. Points of contention often revolve around whether other considerations – including, for instance, those of animal welfare, consumer preferences or environmental considerations such as concern regarding the decline of pollinators – may enter the debate when considering the establishment of safety standards such as maximum residue limits (MRLs) for veterinary drugs or pesticides. However, Codex's statutory purpose is to protect consumer health and facilitate fair practices in the food trade [6]. This is fixed and explicitly does not include considerations other than human health and fair-trade practices, which avoids unnecessarily restrictive

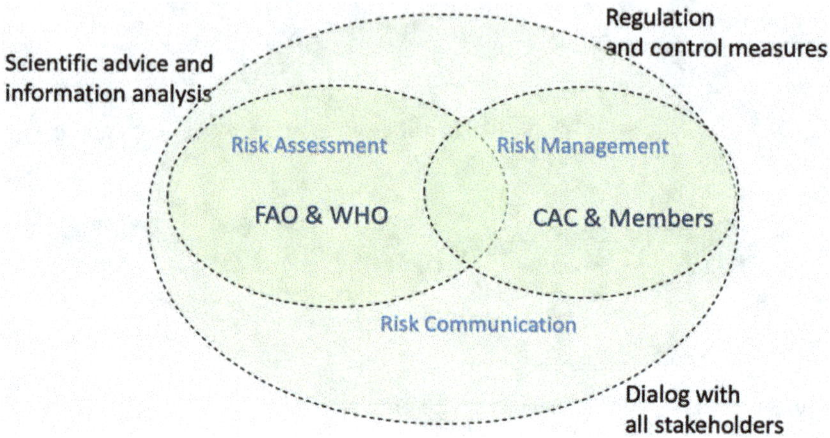


**Figure 2** Codex risk analysis paradigm.

measures but ensures consumer health protection and helps protect the true nature of products in trade.

### 2.6 When risk assessment and risk management coincide

While the separation of risk assessment and risk management is integral to Codex's transparent processes, the interaction between risk assessors and risk managers is nonetheless an essential and integral part of the risk analysis process (see Fig. 2). Frequent interaction between risk managers and risk assessors, particularly during the problem formulation, planning and scoping phases of the risk analysis process, and when communicating the outcomes of the risk assessment to the risk managers, is critical. Risk assessors need to consider the appropriate risk management issue(s) and question(s) and risk managers must determine what food safety decisions may or may not be needed.

## 3 Codex definitions of contaminants, additives and residues in food and feed

### 3.1 Contaminants in food and feed

Codex defines 'contaminants' as 'any substance not intentionally added to food or feed for food-producing animals, which is present in such food or feed as a result of the production (including operations carried out in crop husbandry, animal husbandry, and veterinary medicine), manufacture, processing, preparation, treatment, packing, packaging, transport or holding of such food or feed, or as a result of environmental contamination. The term does not include insect fragments, rodent hairs, and other extraneous matter' [3].

This rather ample definition of contaminants implicitly includes a wide range of substances, such as toxins (e.g. biotoxins, mycotoxins, plant toxins, etc.) as well as industrial, environmental and naturally occurring toxicants (e.g. heavy metals, acrylamide, chloropropanols, dioxins, dioxin-like polychlorinated biphenyls (PCBs), non-dioxin-like PCBs, cyanogenic glycosides in cassava, etc.).

Pesticides no longer used in agriculture but that persist in the environment may find their way into food, and their residues are considered contaminants. The Codex Committee on Pesticide Residues (CCPRs) address the residues of these obsolete pesticides by establishing extraneous maximum residue limits (EMRLs) to differentiate them from MRLs, which usually apply to residues of registered pesticides found in food and feed that have been used in agriculture in line with good agricultural practices. CAC has adopted EMRLs for some of these compounds, e.g. chlordane, DDT, lindane, etc.

Biological hazards, such as microorganisms and their toxins (e.g. botulinum toxin or *Staphylococcus* enterotoxin) present in food that can threaten human health, are considered part of Codex's work on good hygiene practices and are addressed by the Codex Committee on Food Hygiene (CCFH).

CCCF also addresses contaminants in feed, where these can be transferred to food of animal origin and can be relevant to public health.

CCCF has addressed most of the key contaminants of concern for public health and international trade by establishing MLs and CoPs.

However, national or regional competent authorities may include additional MLs for contaminants other than those adopted by CAC, i.e. there is no need to wait for Codex to establish an international standard.

They may also deviate from the Codex MLs to achieve their appropriate level of protection (ALOP) from the identified risk when international standards do not provide the required level of sanitary or phytosanitary protection as long as such measures are still consistent with the provisions of the WTO/SPS agreement in accordance with Articles 3.3 and 5 of the agreement, which requires a robust science-/risk-based justification for the establishment and maintenance of such measures that in any case should not be more trade-restrictive than required to achieve their ALOP.

Codex practices, as opposed to MLs, are not usually subject to inclusion in national regulations. They are normally used in such a way that they are built upon by including specific practices that may be more or less stringent than those described in the CoP to address the particular geographic, climatic, growing conditions, etc. of the country.

Published by Burleigh Dodds Science Publishing Limited, 2025.

### *3.2 Additives and residues of chemicals in food and feed*

For chemicals that are intentionally added to foods and whose residues are present in food as a result of production processes, including operations carried out in crop protection and animal husbandry, Codex establishes MRLs. MRLs for veterinary drugs in food and for pesticides in food and feed are elaborated by the Codex Committee on Residues of Veterinary Drugs in Food (CCRVDF) and CCPR, respectively. The scientific advice provided to CCPR comes from the Joint FAO/WHO Expert Meetings on Pesticide Residues (JMPR), and the scientific advice given to CCRVDF comes from the Joint FAO/WHO Expert Committee on Food Additives (JECFA).

For food additives and processing aids added to foods for a technological purpose, Codex establishes maximum use levels (MLs) for food additives or provides overarching guidance on the safe use of processing aids in the preparation of food and food ingredients. The Codex Committee on Food Additives performs this work based on the provision of scientific advice provided by JECFA, which is a joint FAO/WHO risk assessment body which also provides scientific advice to CCCF for the establishment of MLs and CoPs.

Codex does not establish MLs or MRLs for processing aids as these substances, as opposed to food additives, do not perform any technological function in the final product, nor do they alter their organoleptic properties; they are used at certain stages in the process to aid such process and residues in the final product are likely to be very low or non-expected.

## 4 The Codex Committee on Contaminants in Food

The CCCF carries out Codex risk management work on contaminants in food and feed. The mandate of CCCF includes:

- the establishment of MLs or guidelines levels (GLs) for contaminants in food and feed;
- the elaboration of standards or CoPs or other relevant risk management approaches;
- the consideration of methods of analysis and sampling to support implementation of the ML; and
- the preparation of the priority list of contaminants for JECFA risk assessment.

### *4.1 Some key definitions*

Codex defines the ML for a contaminant in a food or feed commodity as the maximum concentration of that substance recommended by CAC to be legally permitted in that commodity [1, 6].

Published by Burleigh Dodds Science Publishing Limited, 2025.

For the contaminants radionuclides, acrylonitrile and vinyl chloride monomer, Codex has established GLs, which are defined as the ML of a substance in a food or feed commodity which CAC recommends as acceptable for commodities moving in international trade. When the GL is exceeded, governments should decide whether and under what circumstances the food should be distributed within their territory or jurisdiction [1].

Codex has established MLs and GLs for several contaminants, including mycotoxins, metals, radionuclides and other substances (see Table 1).

All MLs and GLs established by CCCF and adopted by CAC are compiled in the *General Standard for Contaminants and Toxins in Food and Feed* (CXS 193-1995). MLs with related CoPs to support their implementation are cross-referenced in the General Standard. Although not limited to this set of contaminants, sampling plans for MLs for different mycotoxins are described in the General Standard as essential for the establishment and implementation of these MLs.

Because of the Codex decision that an ML is the preferred format for a contaminant standard, existing or proposed GLs should be reviewed for possible conversion to an ML after a risk assessment performed by JECFA, if appropriate. This has been the case for GLs for methylmercury in fish, which have been converted into MLs following a full re-evaluation by JECFA, and additional corresponding sampling plans have been developed to facilitate the application of the MLs [7].

## 4.2 Contaminants for which Codex/Codex Committee on Contaminants in Food has established maximum levels

See Table 1.

## 4.3 Other Codex Committee on Contaminants in Food outputs

Besides the establishment of MLs, which can be considered the core work of the committee as described above, CCCF also develops CoPs for the prevention and/or reduction of contamination for a given contaminant in a given food or food group, which are usually associated with the related ML(s). CoPs are based on good agricultural and manufacturing practices along the food chain and as such define the production, processing, manufacturing, transport and storage practices for individual foods or food groups that are considered essential to ensure the safety of the food for consumption.

CoPs can provide useful measures to reduce occurrence levels and, therefore, exposure. They have been developed to facilitate compliance with a (future) ML or where establishing an ML is not feasible. The application of the CoP measures assists countries in reducing the contaminants in a

**Table 1** List of contaminants for which Codex has established MLs or GLs (as per 2023/2024)

| Name of the contaminant(s) |
|---|
| **Mycotoxins** |
| Aflatoxins, total |
| Aflatoxin M1 |
| Deoxynivalenol (DON) |
| Fumonisins |
| Ochratoxin A |
| Patulin |
| **Heavy Metals** |
| Arsenic |
| Cadmium |
| Lead |
| Mercury |
| Methylmercury |
| Tin |
| **Radionuclides** |
| **Others** |
| Acrylonitrile |
| Chloropropanols |
| Hydrocyanic acid |
| Melamine |
| Vinylchloride monomer |

given commodity or group of commodities and in considering lowering or establishing new MLs.

CCCF has also invested in the development of guidance other than MLs and CoPs, such as providing an approach to assist governments in the rapid risk analysis of instances of detection of chemical contaminants in food where there is no regulatory level, in order to adequately protect public health without introducing unnecessary trade barriers (i.e. *Guidelines for rapid risk analysis following instances of detection of contaminants in food where there is no regulatory level* (CXG 92-2019)).

## 5 The process for establishing maximum limits for contaminants

To establish a Codex ML, it is necessary that JECFA or an ad hoc FAO/WHO expert meeting has completed a safety assessment or has performed a quantitative risk assessment, and that appropriate sampling plans and methods of analysis are available to determine their presence in food.

CCCF is guided by several principles for the establishment of MLs and generally considers the establishment of an ML for a contaminant that presents both a significant risk to public health and a known or expected problem in international trade and should be set only for foods, which make a significant contribution to total exposure of the consumer to the contaminant [1].

MLs should be set as low as reasonably achievable (ALARA) and at levels necessary to protect consumers, through best practices such as prevention of food and feed contamination at the source. In applying the ALARA principle, Codex must balance the need for the protection of consumer health, the avoidance of unnecessary trade disruption and food security. Trade is inextricably linked to food security, nutrition and food safety. The fair and safe international trade in food is considered to be one of the most important elements in achieving world food security.

When establishing MLs, Codex strives to ensure they are based on data from various countries and sources and reflect the main and relevant production areas and processes.

When considering safety standards and the dual mandate of Codex, consumer health protection should, in many ways, 'have greater importance than' the issue of trade. However, when taking risk management measures, those measures should have the least trade disruptive effect possible while still guaranteeing that public health is not unduly affected [4]. To balance food safety and trade/food security, CCCF has, over the years, followed an approach that not only considers the scientific output of JECFA when establishing an ML or looking at the data from a health perspective but also considers the ML's impact on trade. To do so, CCCF considers a range of MLs that will provide sufficient consumer health protection against trade rejection (violation) rates at an international level. CCCF generally considers a cutoff rejection (violation) rate of up to 5% as suitable, i.e. that a proposed ML will not lead to more than 5% rejections in trade or a rejection rate close to that.

Balancing consumer health and food security becomes especially important when establishing MLs for carcinogenic and teratogenic contaminants in staple foods, such as mycotoxins in cereal grains (e.g. maize) or rice, and considering that there is no safe level for mycotoxins, the ALARA principle comes strongly into play. In such cases, MLs are set ALARA, considering the most suitable or lowest possible rejection (violation) rates. In cases where a decision cannot easily be taken on an ML, CCCF might also request JECFA to perform an impact assessment of different hypothetical MLs on both consumer health and trade to determine if there would be any definable difference of those MLs in health risks to consumers and then in trade. Generally, decisions are taken to ensure that countries from all regions will be able to comply and with the intent to review the data after a period to give countries sufficient time to implement CoP to prevent or reduce contamination, in the expectation that lower levels will become achievable in the future.

On occasion, CCCF has taken a two-tiered approach to establishing separate, less stringent, MLs for foods supplied in food-aid programmes, as opposed to similar commodities for long-term use. This is in order to ensure a stable supply of such foods and bearing in mind that foods are made available for emergency use in food-aid programmes for a limited period of time. For example, Codex has established a ML of 5 μg/kg for total aflatoxins in cereal-based foods for infants and young children and a ML of 10 μg/kg for cereal-based foods for infants and young children destined for food-aid programmes [1].

In the above scenarios, the Committee has to weigh the constraints that a more stringent ML would have on trade vis-à-vis food security or humanitarian responses and their supply chains, and still ensure limiting exposure to the given contaminant considering that such foods are given for a limited period of time.

There are, however, instances where MLs have been established to address a specific trade concern rather than a health concern. In these instances, the food in question doesn't contribute significantly to total exposure to the contaminant. Still, it is widely traded and can be an important commodity, such as chocolates or spices. These cases are rare for Codex, but they help address its strategic objective of ensuring fair practices in the food trade by providing a forum for harmonization of national requirements, so that the ML is not more stringent than necessary to protect consumer health and enable trade. In the case of cadmium, e.g., JECFA had concluded that total cadmium exposure, including for high consumers of cocoa and cocoa products, was not considered a health concern; however, CCCF considered it necessary to harmonize MLs for cadmium in cocoa products due to an increased number of diverse regulations concerning MLs for cadmium, especially in importing countries. This approach was to avoid overly restrictive MLs that could impact exporting countries, mainly developing countries, especially from certain regions where the cadmium concentration in the soil was higher due to its volcanic nature. These restrictive MLs could have a significant impact on the economies of these countries, as cacao beans are one of the key commodities for economic growth/foreign exchange.

The issue of cadmium in cocoa and cocoa products led to several years of negotiations that ended in establishing MLs for cadmium in different chocolate categories and cocoa powder. Despite the absence of safety concerns at the levels of exposure, cadmium is, nonetheless, a heavy metal, and the ALARA principle was followed. Thus, the MLs for cadmium in chocolate and cocoa powder were derived based on achievability, with a minimum negative impact on trade [1].

In conclusion, the issue of cadmium in chocolates was a trade issue rather than a public health issue, and the work undertaken by CCCF has contributed to minimizing a negative effect on exports by providing internationally harmonized MLs as benchmark standards for international trade in chocolate while still contributing to lowering consumers' exposure to cadmium.

## 6 Conclusion and future trends

Increasingly Codex is confronted with other challenges: how to contribute to sustainable food systems, new food sources and the need to respond to the impacts of climate change on food safety, amongst other issues, and it needs to respond in a timely manner. In April 2024, CCCF demonstrated its responsiveness to new challenges and completed work on a CoP to reduce or prevent ciguatera poisoning. Ciguatera is an increasingly important global health issue due to rising temperatures and extensive seafood trade, and it can impact human health when consuming contaminated fish [8].

CCCF continues to refine its work procedures, but the basic principles for establishing MLs are as valid today as they were when they were originally established. The work of CCCF, however, demonstrates that when balancing consumer health and trade/food security, regional differences and needs will continue to be important factors in undertaking its work. That is why part of the current work in CCCF is to review the MLs and CoP to determine if they are still fit-for-purpose, or whether they need to be updated to address new developments in science and technology and to the extent the mandate of Codex allows, some of the emerging issues mentioned above.

## 7 References

1. FAO and WHO. 2023. *General Standard for Contaminants and Toxins in Food and Feed, No. CXS 193-1995*. Codex Alimentarius Commission, Rome.
2. FAO and WHO. 2020. *Codex Strategic Plan 2020–2025*. Codex Alimentarius Commission, Rome.
3. FAO and WHO. 2024. *Report of the 17th Session of the Codex Committee on Contaminants in Foods*. Codex Alimentarius Commission, Rome.
4. FAO and WTO. 2017. *Trade and Food Standards*. Codex Alimentarius Commission, Rome.
5. FAO and WHO. 2021. *Report of the 14th Session of the Codex Committee on Contaminants in Foods (REP21/CF)*. Codex Alimentarius Commission, Rome.
6. FAO and WHO. 2018. *FAO/WHO Framework for the Provision of Scientific Advice on Food Safety and Nutrition*. Codex Alimentarius Commission, Rome.

7. FAO and WHO. 2024. *Codex Alimentarius Commission Procedural Manual 29th Edition*. Codex Alimentarius Commission, Rome.
8. FAO and WHO. 2018. *Report of the 41st Session of the Codex Alimentarius Commission*. Codex Alimentarius Commission, Rome.

Published by Burleigh Dodds Science Publishing Limited, 2025.

# Chapter 2

## The role of the Joint Food and Agriculture Organization/World Health Organization Expert Committee on Food Additives (JECFA) in food chemical risk assessment

*Angeliki Vlachou, Vittorio Fattori and Markus Lipp, Food and Agriculture Organisation (FAO), Italy*



## 1 Introduction

The Joint Food and Agriculture Organization/World Health Organization Expert Committee on Food Additives (JECFA) is an international scientific expert committee administered jointly by the Food and Agriculture Organization of the United Nations (FAO) and the World Health Organization (WHO) (FAO, 2024a). It has been meeting since 1956, initially to evaluate the safety of food additives. Currently, the Committee evaluates the safety of chemicals in foods including food additives, contaminants, naturally occurring toxicants, and residues of veterinary drugs in food (FAO, 2024a).

http://dx.doi.org/10.19103/AS.2025.0158.07
Published by Burleigh Dodds Science Publishing Limited, 2025.

The evaluations of contaminants can result in Health-Based Guidance Values (HBGVs), such as tolerable daily intake (TDI) estimates and estimates of exposure to contaminants and toxins for various populations worldwide.

Results from assessments by JECFA are usually reported to the Codex Committee on Contaminants in Foods (CCCF). The CCCF uses that information to make decisions about food safety standards for contaminants in food to be included in the General Standard for Contaminants and Toxins in Food and Feed (CODEX STAN CXS 193/1995).

## 2 The role of the Food and Agriculture Organization (FAO) in food safety

The Food and Agriculture Organization (FAO) is a specialized agency of the United Nations that leads international efforts to defeat hunger. FAO is the only international organization overseeing all aspects of the food chain, such as agricultural production, food safety, animal health, and markets and trade, thereby providing a unique, 360° vision on food safety (FAO, 2024b).

FAO's goal is to achieve food security for all and make sure that people have regular access to enough high-quality food to lead active, healthy lives (FAO, 2024c). FAO's strategic commitment is dedicated to the United Nations Sustainable Development Goals (SDGs) that form its strategic framework for 2022-2031 (FAO, 2023). This framework is designed to support the transformation of agrifood systems into more efficient, inclusive, resilient, and sustainable entities. Central to this transformative agenda are the 'Four Betters': better production, better nutrition, a better environment, and a better life. These guiding principles reflect FAO's holistic approach to addressing the interconnected economic, social, and environmental dimensions of agrifood systems (FAO, 2023):

- **Better production.** Objectives are to ensure sustainable consumption and production patterns through efficient and inclusive food and agriculture supply chains at local, regional, and global levels and to promote resilient and sustainable agrifood systems in the face of changing climate and environmental conditions.
- **Better nutrition.** Objectives are to end hunger, achieve food security, and improve nutrition in all its forms and prioritize nutritious food and increase access to healthy diets.
- **Better environment.** Objectives are to protect, restore, and promote the sustainable use of terrestrial and marine ecosystems and to combat climate change through efficient, inclusive, resilient, and sustainable agrifood systems.

- **Better life.** Objectives are to promote inclusive economic growth by reducing inequalities between urban and rural areas, rich and poor countries, and men and women.

The 'Four Betters' align closely with key SDGs, such as SDG1 (No Poverty), SDG2 (Zero Hunger), and SDG10 (Reduced Inequalities) (FAO, 2023).

The Strategic Priorities for Food Safety for 2022–2031 are articulated around four Strategic Outcomes that result from an iterative consultative process led by FAO with its members and international partner organizations, including, notably, WHO. The four Strategic Outcomes of the Strategic Priorities for Food Safety are shown in Fig. 1.

- Strategic Outcome 1 – Intergovernmental and multi-stakeholder engagement in intersectoral coordination of food safety governance is reinforced at all levels.
- Strategic Outcome 2 – Sound scientific advice and evidence are provided as the foundation for food safety decision-making.
- Strategic Outcome 3 – National food control systems are further strengthened and are continuously improved.
- Strategic Outcome 4 – Public and private stakeholder collaboration is promoted to ensure food safety management and controls throughout agrifood systems. (FAO, 2023)

FAO is recognized as a leading organization in developing international food safety initiatives that can be converted into country-level action. The Food Safety program endorses an integrated and multidisciplinary approach to managing food safety issues. It offers holistic and practical 'food chain' solutions to specific food safety issues, as outlined in FAO's Strategy for Improving Food Safety Globally. The foundations for this approach are based on science. FAO's Food Safety Unit frequently partners with national and international bodies and organizations in mutually beneficial collaborations that align with shared mandates and guiding principles.

The provision of scientific advice is a foundational element of global food safety governance. For over 60 years FAO, in collaboration with WHO, has been providing neutral and independent scientific advice as the basis for the international food safety standards, guidelines, and codes of practice established by the Codex Alimentarius Commission (CAC) and for supporting the development of modern food control systems by national authorities (FAO and WHO, 2007). FAO and WHO have complementary mandates that address a broad spectrum of issues to promote global food safety and safeguard consumer health. FAO generally addresses food safety issues along the food

Published by Burleigh Dodds Science Publishing Limited, 2025.

**Figure 1** Food and Agriculture Organisation strategic priorities for food safety, within the Food and Agriculture Organisation strategic framework 2022–2031, in support of Sustainable Development Goals. (FAO, 2023; https://doi.org/10.4060/cc4040en)

production chain, while WHO primarily focuses on public health. FAO and WHO have a longstanding tradition of offering scientific advice on food safety and nutrition issues to the CAC and its subsidiary bodies, member countries, and other interested parties (such as industry, consumer groups, and academic and research institutes).

Joint FAO/WHO scientific advice is provided through the convening of established expert committees (known as expert bodies) and other expert meetings and ad hoc consultations on issues related to (FAO and WHO, 2007)

- the safety assessment of chemicals in food (e.g. food additives, veterinary drug residues, pesticide residues, contaminants, and natural toxins);
- the safety assessment of biological agents in food (e.g. microorganisms, fungi, and parasites and prions);
- the assessment of practices and technologies used to produce foods (e.g. the safety assessment of foods derived from biotechnology); and
- human nutrition.

Scientific advice plays an important role in the effective planning and implementation of food control and nutrition programs at both international and national levels (Fig. 2). Science has an important role in informing policy and decision-making processes related to food safety and nutrition. CAC and its members have agreed on the importance of ensuring that 'the food standards, guidelines and other recommendations of Codex Alimentarius shall be based on sound scientific analysis and evidence, involving a thorough review of all relevant information'.

The relevance of scientific advice for global trade in food and agricultural products is reaffirmed in the Agreement on the Application of Sanitary and Phytosanitary Measures (SPS Agreement) of the World Trade Organization (WTO). The Agreement underscores the importance of scientific and technical

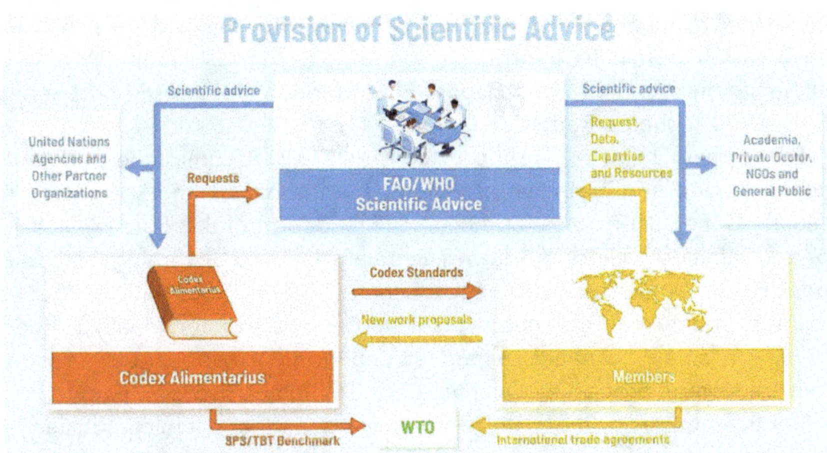


**Figure 2** Provision of scientific advice by Food and Agriculture Organisation and World Health Organization.

Published by Burleigh Dodds Science Publishing Limited, 2025.

guidance and requires that SSPS measures be established on scientific risk assessment.

All FAO and WHO activities involving the provision of scientific advice are conducted in accordance with the following core principles (FAO and WHO, 2007):

- **Soundness** is the need for scientific excellence and applies to both the experts and the process.
- **Responsibility** encompasses the various aspects of accountability and applies both to the need to safeguard the integrity of the process and to consider experts answerable for their views.
- **Objectivity** is considered to include neutrality and applies both to the experts and to the advice provided.
- **Fairness** applies to the conduct of the scientific advice process and requires respect from all participants for each other and for their scientific views.
- **Transparency** may also involve review of both the advice and the procedures involved in providing advice.
- **Inclusiveness** is considered to include group balance.

# 3 The role of the Joint Food and Agriculture Organization/World Health Organization Expert Committee on Food Additives (JECFA)

## *3.1 Background*

The JECFA was established in 1956 following the recommendation of the first Joint FAO/WHO Conference on Food Additives that the two organizations collect and disseminate information on food additives (Joint FAO/WHO Conference on Food Additives, Report. FAO Nutrition Meetings Report Series, No. 11; WHO Technical Report Series, No. 107, 1956).

The primary objective was to establish an expert committee dedicated to technical and administrative aspects of food additives and specifically to formulate General Principles for their use (First Report of the Joint FAO/WHO Expert Committee on Food Additives, General Principles Governing the Use of Food Additives, WHO Technical Report Series, No. 129, 1956). Initially, JECFA's scope was confined to non-nutritive substances added intentionally to food, generally in small quantities, aimed at enhancing its appearance, flavour, texture, or storage properties. Over time, the Committee's scope was extended to cover the inquires by the members pertaining to other chemical substances in food beyond food additives. For instance, in 1967, the Committee held the first meeting on food contaminants.

JECFA serves as an independent scientific expert committee, which performs risk assessments on chemical hazards in food and provides advice to FAO, WHO, and the member countries of both organizations, as well as to the CAC. JECFA uses as a reference document for conducting safety evaluations the principles and methods for the risk assessment of chemicals in food as published in the Environmental Health Criteria 240 (FAO and WHO, 2020).

JECFA members are distinguished scientists from around the globe recognized for their scientific excellence and expertise in various disciplines within JECFA's scope:

- Risk assessments/safety evaluations: hazard identification, hazard characterization, and exposure assessment of chemicals in food.
- Specifications and analytical methods, residue definition, proposals of maximum residue limits for veterinary drugs.
- Guidelines for the safety assessment of chemicals in foods consistent with current thinking on risk assessment in toxicology and other relevant sciences.

JECFA normally meets twice a year with individual agendas covering either:

- food additives, contaminants, and naturally occurring toxicants in food; or
- residues of veterinary drugs in food.

The composition of the experts at the meetings varies, with different sets of experts being called on depending on the subject matter.

FAO and WHO can also organize expert meetings to offer scientific advice on chemical-related food safety issues beyond JECFA's scope. These meetings may be convened in response to specific requests from Codex and/or to assist national authorities in addressing specific risks or incidents that impact on consumers' health and/or have significant economic and trade consequences.

## *3.2 Selection of experts*

Both Organizations are responsible for selecting experts for establishing rosters. In order to establish the rosters, FAO and WHO issue calls for applications. The rosters are renewed every 5 years. The calls describe the essential qualifications of the applicants, the selection procedures for the roster and other relevant information. The selection of members is made after careful consideration of the various candidates' scientific credentials, and a balance of scientific expertise and other experience is considered essential.

FAO places great value on the technical qualifications and independence of the participating experts, as well as to the transparency of its selection process. Therefore, FAO has in place procedures for selecting experts. Those procedures safeguard transparency, excellence, and independence of opinions. The procedures for the work of JECFA and the process for the selection of experts are available in the FAO/WHO framework for the provision of scientific advice on food safety and nutrition (https://www.sciencedirect.com/science/article/pii/B9780128225219001349?via%3Dihub).

FAO and WHO have complementary functions in selecting experts. Experts are selected from the rosters to serve at a specific JECFA meeting. FAO is responsible for selecting experts to develop analytical methods and specifications for the identity and purity of food additives and to perform assessments of residue of veterinary drugs in food. WHO is responsible for selecting members to perform the toxicological evaluation of the substances under consideration. Both FAO and WHO invite members who are responsible for assessing dietary exposure.

Individuals serving in the JECFA meetings are experts with specializations in a range of scientific subjects. They come from many different countries and organizations but act in their individual capacity and not as representatives of any institution or country. To secure the independent character of these experts, FAO and WHO budget covers their expenses like travel, hotel, and daily allowance (FAO and WHO, 2007).

### *3.3 Communication*

Communication materials and channels utilized for disseminating the outcomes of the JECFA meetings primarily consist of written publications. The publications aim at documenting the results of the meetings and facilitating the dissemination of the formulated scientific advice.

Specifically, in the context of the JECFA meetings for contaminants and natural toxins in food, three distinct publications are developed:

- a summary report,
- a technical report series, and
- JECFA Monographs.

The summary report serves the immediate communication needs of the main results of the assessments. It is a condensed version of the main report focusing on the main findings. Typically, it is published shortly after the meeting has concluded in 2 weeks' time.

The technical report series encompasses the full assessment of the chemical substance. The safety evaluation includes the description of the inquiries to be addressed. Usually, there is a request from the CCCF for a risk assessment of a specific chemical compound or group of substances.

Further to that, the evaluation and, consequently, the report include the toxicological evaluation of the substance, analytical methods to be used for its determination, sampling protocols, effects of processing, prevention, and control, as well as the levels and patterns of contaminants in food commodities. Furthermore, these assessments are extended to estimating dietary exposure through the consideration of both chronic and acute exposure scenarios. The conclusions of the Committee on the substance evaluation and recommendations for future work, if deemed appropriate, would also be addressed in the report.

The monographs produced by JECFA provide comprehensive and detailed descriptions of the biological and toxicological data that are considered during its evaluations. These documents also include thorough assessments of intake levels, ensuring a robust analysis of the safety and potential health impacts of various food chemicals considered.

In conjunction with the primary scientific findings, reports encompass additional pertinent information for the experts' evaluation. They include a summary of the most relevant available data and sources, along with a summary of data used to answer the question. Explanations are provided for the exclusion of certain datasets. The report also documents the reliability of the data and any assumptions made, including their impact on uncertainty.

In certain cases, interpretative and executive summaries may be added to reports or published separately to make the advice more useful for target audiences (FAO and WHO, 2007).

The reports are available in the FAO Resources webpage (https://www.fao.org/food-safety/resources/publications/en/) and the WHO webpage (https://www.who.int/groups/joint-fao-who-expert-committee-on-food-additives-(jecfa)/publications/reports).

## 4 JECFA: risk assessment of contaminants and natural toxins in food

The principal product of scientific advice provided by JECFA associated with the risk assessment is, in most cases, the establishment of a TDI and the recommendation of Maximum Levels (MLs) for the contaminants in relevant food commodities. Advice is also produced on exposure assessment at both national and international levels. Additionally, the JECFA contributes to the development of guidelines and resource documents on various pertinent topics, such as in the case of acrylamide.

Moreover, the Committee developed the principles for conducting risk assessment and continues to refine existing risk assessment methodologies, ensuring their adaptation and evolution to keep pace with emerging scientific knowledge (FAO and WHO, 2020). FAO actively engages with international networks dedicated to discussing risk assessment methodologies and approaches as well as advancing method harmonization and developmental initiatives.

## 5 Case study: JECFA assessment of co-exposure to fumonisins and aflatoxins

Further to a request from the CCCF, at its 83rd meeting in 2016 JECFA evaluated and updated toxicological and exposure data for fumonisins and aflatoxins separately, as well as evaluated co-exposure to these mycotoxins.

Fumonisins and aflatoxins are two groups of mycotoxins produced by fungi of *Fusarium* and *Aspergillus* species that are frequently found in cereals (especially maize, rice, sorghum, and wheat) and cereal-based foods. Aflatoxins are also common contaminants in ground nuts and tree nuts. As these mycotoxins can affect the same commodities, co-exposure to both mycotoxins can happen in regions where these foods are part of the routine diet (WHO, 2018).

JECFA first assessed the co-occurrence of fumonisins and aflatoxins in food. In doing so, the Committee noted that numerous factors, such as the type of commodities, the geographical area, sampling time, storage conditions, and methods of food preparation/processing, can affect the degree of co-occurrence of aflatoxins and fumonisins. Additionally, while also noting that data related to the various aflatoxins and fumonisins found in foods (either separately or combined) are available, the co-occurrence of Aflatoxin B1 (AFB1) and Fumonisin B1 (FB1) was the focus of the JECFA evaluation.

As per standard practice, prior to a JECFA meeting, FAO and WHO publish an open call for data, inviting governments, interested organizations, research institutes, etc. to submit relevant data on the contaminants to be assessed by JECFA. The data submitted in response to such calls may be published or unpublished and should contain detailed reports of laboratory studies, including individual animal data, epidemiological evidence of human exposures and associated health endpoints, or examination of the occurrence of chemicals in the food system. Submissions of regulatory reviews or comprehensive literature reviews with supporting publications are also encouraged.

In the case of the JECFA assessment of the co-exposure to fumonisins and aflatoxins, data submitted to the Global Environment Monitoring System (GEMS)/Food Contamination database[1] and the studies from the literature that considered both mycotoxins were the two main sources of information used by the committee, and one of the key part of the assessment was to identify

1 Global Environment Monitoring System (GEMS)/Food Contamination Monitoring and Assessment Programme.

the food categories that contain both mycotoxins. Most data available in the GEMS/food database for both AFB1 and FB1 were in the cereals and cereal-based products group (including foods like maize, sorghum, rice, barley, millet, etc.). At the same time, the review of the literature showed that AFB1 and FB1 co-occurrence in food samples from the same study is largely unknown. Most of the studies in the literature were conducted in the African region, and the primary food groups evaluated were maize and products made from maize. Coffee, smoked meat, black pepper, and various cereals (such as sorghum, millet, rice, and wheat) were among the other foods where co-occurrence was noted.

In addition to the analysis of the specific co-occurrence in foods, an assessment of the co-exposure to AFB1 and FB1, considering the overall diet, was conducted to identify populations that can be exposed to both mycotoxins. To achieve this, JECFA evaluated foods that have higher AFB1 and FB1 levels as well as populations that may consume these foods more frequently. Clusters of countries which include, for example the United Republic of Tanzania and Guatemala, were those with high dietary exposure to both AFB1 and FB1, and where indeed the co-exposure was confirmed through urinary or plasma exposure biomarkers of FB1 and AFB1.

Among other conclusions, JECFA also noted the concern related to the interaction between AFB1, a potent genotoxic substance, and fumonisins, which can induce regenerative cell proliferation. As part of the assessment, JECFA also made the following recommendations:

- There is a need to reduce human exposure to aflatoxins and fumonisins, alone or in combination, in particular in developing countries.
- With regards to human studies, the emphasis should be on biomarker-based approaches. Biomarker-based studies in high-risk areas should include attempts to characterize the health issues common in individuals within communities where exposure is high, which can be compared with similar communities where exposure is low.
- Experimental animal feeding studies should also use biomarker-based approaches and should be designed with multiple dose levels that reflect the levels of contamination seen in areas at high risk for co-exposure. (WHO, 2018)

## 6 Risk management and Codex Alimentarius

The CAC was established in 1961 during the 11th FAO Conference. Within the framework of the CAC and its procedures, the responsibility for providing advice on risk management lies with the CAC and its subsidiary bodies (risk managers), while the responsibility for risk assessment lies primarily with the joint FAO/WHO expert bodies and consultations (risk assessors) (FAO and WHO, 2024a).

The foundation of Codex Standards is sound scientific evidence. From the very beginning, the Codex Alimentarius has been a science-based activity. In line with the *Codex Procedural Manual,* 'The food standards, guidelines and other recommendations of Codex Alimentarius shall be based on the principle of sound scientific analysis and evidence, involving a thorough review of all relevant information, in order that the standards assure the quality and safety of the food supply.'

Therefore, risk management decisions should be based on risk assessment, and when appropriate, take into account other legitimate factors crucial to protecting consumer health and promoting fair practices in food trade (FAO and WHO, 2023). The safety assessments and evaluations performed by JECFA are based on the best scientific information available, incorporating contributions from many authoritative sources, and resulting in publications regarded as international reference works (FAO and WHO, 2024)

## *6.1 Codex Standards*

It is important to note that Codex Standards and related texts are voluntary. They do not replace national legislation, and they do not have binding effect on national food legislation. The WTO Agreements on Sanitary and Phytosanitary Measures (SPS Agreement) and on Technical Barriers to Trade (TBT Agreement) encourage member countries to align their regulations with the international standards. The SPS Agreement specifically designates Codex Standards, guidelines, and recommendations as the international benchmark for food safety. Under the SPS Agreement, WTO members may adopt stricter national standards than those established by Codex. In that case they may be requested to provide scientific justification for the need for these measures to achieve their desired level of protection. Members may also be requested to demonstrate that the measure taken is based on a risk assessment. Additionally, Codex Standards, guidelines, and other recommendations can serve as a reference point in case food trade disputes (FAO and WHO, 2024).

## *6.2 Standard for contaminants*

The General Standard for Contaminants and Toxins in Food and Feed CXS 1993–1995 (GSCTFF) (Codex Alimentarius, 2023) contains the main principles which are recommended by the Codex Alimentarius in dealing with contaminants and toxins in food and feed.

Published by Burleigh Dodds Science Publishing Limited, 2025.

The Codex Alimentarius defines a contaminant as 'Any substance not intentionally added to food or feed for food producing animals, which is present in such food or feed as a result of the production (including operations carried out in crop husbandry, animal husbandry and veterinary medicine), manufacture, processing, preparation, treatment, packing, packaging, transport or holding of such food or feed, or as a result of environmental contamination. The term does not include insect fragments, rodent hairs and other extraneous matter'. The definition includes naturally occurring toxicants including toxic metabolites of certain microfungi that are not intentionally added to food and feed (mycotoxins). Toxins that are produced by algae and that may be accumulated in edible aquatic organisms, such as shellfish (phycotoxins), are also included in this standard. However, endogenous toxins (like solanine in potatoes) are addressed on a case-by-case basis by the CCCF (Codex Alimentarius, 2023).

Contamination in food and feed may pose risks to human and animal health as well as compromise food quality. Contaminants can occur from various sources and processes. The guiding principle is to ensure that contaminant levels in food are kept as low as reasonably achievable. This can be achieved through the implementation of best practices, such as good agricultural practice and good manufacturing practice, following an appropriate risk assessment and a code of practice. MLs should be only established for food that contribute significantly to the total exposure, considering the *Policy of the Committee on Contaminants in Foods for Exposure Assessment of Contaminants and Toxins in Foods or Food Groups*, as written in the Procedural Manual (FAO and WHO, 2023). Information that should be considered includes toxicological information, analytical data, intake data, technological information, and risk assessment and management options.

The Codex Standard for contaminants (Codex Alimentarius, 2023) outlines the principles for managing contaminants and establishing MLs in food and feed. It elaborates on specific criteria to be considered when developing MLs and other measures in connection with the Standard. The Standard also foresees monitoring and survey programs as well as more specialized research programs, where necessary. It lists the MLs and associated sampling plans of contaminants and natural toxicants in food and feed which are recommended by the CAC to be applied to commodities in international trade. This standard is periodically updated, based on the outcome of the meetings of the CCCF with the support of JECFA. The latest version of the document is available from the Codex Alimentarius website at https://www.fao.org/fao-who-codexalimentarius/codex-texts/list-standards/en/ (Codex Alimentarius, 2024).

# 7 The role of JECFA in the Codex Alimentarius decision-making process

JECFA has the mandate to carry out the risk assessments upon which the Codex Committee of Food Additives (CCFA), the CCCF, the Codex Committee on Residues of Veterinary Drugs in Food (CCRVDF), and ultimately the CAC base their risk management decisions. While this section focuses on the interaction between JECFA and CCCF, similar interactions occur in other committees as well.

## *7.1 Key considerations in risk assessment and decision-making*

In the area of contaminants, JECFA determines, to the extent possible, the risks associated with various levels of dietary exposure to contaminants and toxins. In doing so, JECFA uses worldwide data, including data from developing countries, as the foundation for its risk evaluations. During the risk assessment, JECFA considers variations in food consumption habits across different regions, and a key consideration in this process is to provide CCCF with details on any limitations, uncertainties, and assumptions of the assessment; in particular, JECFA should inform CCCF about potential risks to specific groups of populations with higher vulnerability (such as children women of childbearing age and the elderly). JECFA also informs CCCF regarding the reliability and distribution of the existing data on contaminants and toxins in food and feed for exposure assessments and outlines the extent to which certain foods and feeds contribute to exposure. This information could be very useful for CCCF's risk management recommendations.

Consensus is used by JECFA to make decisions on interpreting key data, evaluations, and final conclusions, rather than resorting to voting. If there is no agreement, a minority viewpoint can be expressed and will be documented in the meeting summary and published separately, explaining the reasons for the differing perspective.

While JECFA's risk assessment principles are internationally recognized and accepted, their application (including specific methodologies like for example the benchmark dose - BMD), or even the database underlying the assessment might differ in other risk assessment bodies at national/regional level. This sometimes may lead to different conclusions among the various national, regional and international assessment bodies.

## *7.2 The relation between JECFA and Codex Committee on Contaminants in Foods*

CCCF provides risk management recommendations to the CAC regarding contaminants and toxins based on the principles outlined in the Codex General

Standard for Contaminants and Toxins in Food and Feed (CXS 193-1995). The recommendations to CAC should be derived from JECFA's risk assessments and should consider the uncertainties and safety factors outlined in JECFA's assessments. It is important to note that CCCF should clearly indicate in its standards when considering other legitimate factors besides JECFA's risk assessment and explain why/how they are taken into consideration. It is also worth important to mention that international MLs will only be established by Codex for contaminants that have undergone a quantitative risk assessment by JECFA (or other FAO/WHO expert consultations), are considered a significant contributor to total dietary exposure, and can be accurately determined in food or feed using approved sampling plans and analytical methods. In the risk management process to set MLs for food contaminants, it is also key to consider variations in regional/national food consumption habits and levels of dietary exposure evaluated by JECFA.

## 8 Conclusion

The JECFA plays a significant role in the global food safety setting by providing independent, science-based risk assessments of chemical contaminants in food. Through its rigorous evaluations and transparent processes, JECFA supports the development of international food safety standards and informs national regulatory frameworks. The scientific advice provided by JECFA ensures that risk management decisions by Codex Alimentarius Commission are based on robust scientific evidence.

## 9 Where to look for further information

The following provide further information on the subject:

- Food safety webpage of Food and Agriculture Organization of the United Nations: https://www.fao.org/food-safety/en/
- Codex Alimentarius International Food Standards : https://www.fao.org/fao-who-codexalimentarius/en/
- Contributions of the joint FAO/WHO expert committee on food additives to international food safety: celebrating the 100th meeting of the committee. Diane Benford, Alan Boobis, Richard Cantrill, Peter Cressey, Eugenia Dessipri, Shruti V. Kabadi, Suzanne Jeurissen, Utz Mueller, Susan Barlow. Regulatory Toxicology and Pharmacology. Volume 160, 2025: https://doi.org/10.1016/j.yrtph.2025.105833

## 10 References

Codex Alimentarius. 2023. General Standard for contaminants and toxins in food and feed (CXS 193-1995). https://www.fao.org/fao-who-codexalimentarius/sh-proxy/en/?lnk=1&url=https%253A%252F%252Fworkspace.fao.org%252Fsites%252Fcodex%252FStandards%252FCXS%2B193-1995%252FCXS_193e.pdf

Codex Alimentarius. 2024. Codex Alimentarius standards. https://www.fao.org/fao-who-codexalimentarius/codex-texts/list-standards/en/

FAO. 2023. FAO Strategic Priorities for Food Safety within the FAO Strategic Framework 2022–2031. Rome. https://doi.org/10.4060/cc4040en

FAO. 2024a. Joint FAO/WHO Expert Committee on Food Additives (JECFA). https://www.fao.org/food-safety/scientific-advice/jecfa/en/

FAO. 2024b. FAO's role. https://www.fao.org/food-safety/en/

FAO. 2024c. About FAO. https://www.fao.org/about/about-fao/en/

FAO and WHO. 1956. Joint FAO/WHO Conference on Food Additives, Report. FAO Nutrition Meetings Report Series, No. 11; WHO Technical Report Series, No. 107, 1956.

FAO and WHO. 2007. *Framework for the Provision of Scientific Advice on Food Safety and Nutrition*, FAO/WHO, Rome. https://www.fao.org/4/a1296e/a1296e00.pdf

FAO and WHO. 2020. Environmental health criteria 240. Principles and methods for the risk assessment of chemicals in food. https://www.who.int/publications/i/item/9789241572408

FAO and WHO. 2023. Codex Alimentarius Commission Procedural Manual. Twenty-eighth edition, revised. Rome. https://doi.org/10.4060/cc5042en

FAO and WHO. 2024. About Codex Alimentarius. https://www.fao.org/fao-who-codexalimentarius/about-codex/en/

First Report of the Joint FAO/WHO Expert Committee on Food Additives, General Principles Governing the Use of Food Additives, WHO Technical Report Series, No. 129, 1956. https://iris.who.int/bitstream/handle/10665/40387/WHO_TRS_129.pdf?sequence=1

https://www.sciencedirect.com/science/article/pii/B9780128225219001349?via%3Dihub

WHO. 2018. Safety evaluation of certain contaminants in food: prepared by the eighty-third meeting of the Joint FAO/WHO Expert Committee on Food Additives (JECFA). WHO Food Additives Series, No. 74; FAO JECFA Monographs 19 bis. https://iris.who.int/bitstream/handle/10665/276868/9789241660747-eng.pdf?ua=1#page=423

# Chapter 3

# Key issues in exposure assessment and risk characterization of chemical contaminants in food

*Rob Theelen, Food Safety Portal, The Netherlands*



## 1 Introduction: key steps in risk assessment

Risk analysis has been defined within the Codex Alimentarius as a series of three activities: risk assessment being a scientific process, risk management being the result of policy interactions, and risk communication about the assessment and management decisions. Regarding the risk assessment of chemical substances in food and feed, the Codex Alimentarius adopted the terms "hazard" and "risk". "Hazard" refers to the potential of agents to cause adverse health effects, and "risk" is the probability of health effects when exposed (FAO and WHO, 2024).

The risk assessment of chemicals in food and feed is divided into four separate steps. The first step is the "hazard identification". Using data from human and animal studies, the toxicity of the chemicals is identified, which

http://dx.doi.org/10.19103/AS.2025.0158.03
© Burleigh Dodds Science Publishing Limited, 2025. All rights reserved.

organs and tissues are affected, and its impact on the subject's health. In the second step, the "hazard characterization", the dose-responses of the hazards are quantified. Then the lowest dose level with adverse health effects is identified and converted into a health-based guidance level (HBGV). Relevant HBGVs are the Acceptable Daily Intake (ADI) and the Tolerable Daily Intake (TDI) (Herman J.L. and Younes M., 1999). Benchmark Dose Levels (BMDLs) can also be derived in the hazard characterization; they refer to specific adverse effects, in contrast to an ADI or TDI.

The third step is the "exposure assessment", in which dietary intake of a substance is estimated for a population or sub-populations such as infants, children, and adults, or consumers with a specific diet (e.g. vegetarians). In the fourth step, the "risk characterization", is the dietary intake compared with the HBGV in a quantitative manner. Based on that comparison, the likelihood of adverse health effects is determined (IPCS, 2009).

## 2 Performing exposure assessments

An exposure assessment does not depend on the toxicological data from the hazard identification and characterization, but on consumption and occurrence of substances. The HBGV is the endpoint of the hazard assessment; many publications present HBGVs, such as EFSA's opinions (EFSA Publications) and JECFA's toxicological monographs (WHO, JECFA). Many exposure assessments deal with the state of affairs of local issues, such as foods sampled in an import control which do not comply with the (EU) standards, or accidental contamination of foods produced in contaminated areas such as heavy metals in vegetables. Most of these cases are evaluated by national food safety authorities, such as Bundesinstitut für Risikobewertung (BfR) in Germany (BfR, 2024), ANSES in France (ANSES, 2024), or the Office for Risk Assessment & Research (BuRO) in the Netherlands (NVWA, 2024). Results of such national evaluations might be published in public sources and/or discussed with the European Commission and the Member States.

According to the risk analysis paradigm of the Codex Alimentarius, risk assessment is science-based, whereas risk management is policy-based. This makes risk assessors independent from the food safety policy framework, with no apparent restrictions on how to perform an exposure assessment. As a result, there are only a few general guidelines about how to perform an exposure assessment. Consequently, there are many opportunities to discuss the results. For the risk manager, this situation is unsatisfactory. A risk manager has to decide on the management options and prefers one conclusion: "*safe*" or "*not safe*"; this is to be interpreted as: "actions not needed" or "actions needed".

The dietary intake of a chemical substance is equal to the consumption quantity of the food multiplied with the concentration of the chemical in the

© Burleigh Dodds Science Publishing Limited, 2025. All rights reserved.

food. To evaluate the intake, it is to be divided by the body weight of the consumer (Annex, 1). In the current approach, a chemical substance in food does not pose a risk for the consumer when the dietary intake is less than the HBGV of that chemical substance (Annex, 2). In the European Union, the term "*safe*" is preferred, as it fits with article 14 of Regulation 178 of 2002, stating that unsafe food cannot be traded (Regulation (EC) No 178/2002).

## 3 Food consumption data

### 3.1 Surveys

Many data on consumers and their consumption quantities come from food consumption surveys. The drivers for such studies are diverse, such as consumer behavior, impact of cultural differences, diets and health, and economy (Wherry and Woodward, 2019). With regard to food safety, the focus is to obtain data about what foods are consumed, by whom, in what quantities, and when and where. Various techniques are used, including interviews and questionnaires, 24-hour recalls, and duplicate diets. For the evaluation of food consumption within the EU, EFSA has developed a Guidance document on sampling, assessment methodologies, tools, and quality control (EFSA, 2009, 2014). Most EU member countries follow these recommendations and have collected consumption data accordingly in the last decades. An overview of the available European surveys can be found on the EFSA website (EFSA, 2024a). There are also surveys available from other parts of the world; many of them are published in the GEMS/Food database of the WHO (WHO, GEMS/Food), or the Global Individual Food Consumption Data Tool (GIFT) databases (FAO/WHO GIFT, 2024). Other examples of food consumption surveys can be found on the Internet, such as the Canadian Food Consumption Table (Health Canada, 2017) and the USA-NHANES data set (CDC, 2024).

The data sets have different formats, e.g. due to cultural differences, different staple foods, or foods controlled by religious rules ("halal", "kosher"). The surveys are never complete; consumption of rare foods are most likely not included. And, there is not a universal list of foods noted in all surveys. Different descriptors for food items are used. Consequently, an issue in the exposure assessment is the selection of the food items to be evaluated. There are efforts to harmonize the coding system of the list of food items, such as the FoodEx2 system of EFSA (EFSA, 2024b) but most older surveys do not refer to this coding system. Most exposure assessors still prefer to search foods by their literal name.

### 3.2 Databases

EFSA's Comprehensive European Food Consumption Database: Many data sets of the EU countries are available through EFSA's Comprehensive

© Burleigh Dodds Science Publishing Limited, 2025. All rights reserved.

European Food Consumption Database website (EFSA, 2022). This database can be approached online and contains groups (e.g. toddler, child, adult) and food items. The food items are presented at different levels, with an increasing level of detail; for example: (1) Vegetables and vegetable products, (2) Leafy vegetables, (3) Spinach type leaves, (4) Spinaches and similar. Based on the selection, the database produces a table that can be downloaded as spreadsheet (xlsx) or raw data file (csv). The tables show values for various percentiles, for all subjects and for consumers only, for a consumption quantity in grams per day or in grams per kg body weight per day, for chronic and acute exposures.

PRIMo: Another public data set from EFSA is the pesticide residue intake model (PRIMo). It is developed for the exposure assessment of pesticides and needs the user's input of a HBGV and the concentrations of a substance. Currently, versions 2, 3, and 3.1 can be downloaded as spreadsheets. In the sheets, the consumption quantities of various EU member states are included (EFSA, 2024c).

How do these consumption quantities relate to the data of EFSA's Comprehensive European Food Consumption Database? According to PRIMo_3.1, the consumption of spinach for the general Dutch population is 0.156 gram per kg body weight per day (*"chronic_consumption", cell X169*). However, using the Dutch chronic data set of 2012 in the Comprehensive database, the consumption for adults is 1.2 ± 0.91 gram per kg body weight per day for consumers only, and 0.10 ± 0.090 gram per kg body weight per day for all adults. So, the average consumption according to PRIMo falls within the range of all adults as presented in the Comprehensive database, but the numbers do not exactly match. There is no publication of how the values in PRIMo and those in the Comprehensive European Food Consumption Database were derived from the initial data of the individual consumers. Consequently, it is not quite clear how the values in PRIMo and the Comprehensive database came to be.

WHO GEMS/Food and Food Safety Collaborative Platform (FOSCOLLAB) The WHO GEMS/Food database shows a distribution of different countries of the world, divided into 17 groups. For each group, a "GEMS/Food cluster diet" is given. Values "per capita" per food category are provided. It is noted that "these diets are to be used for the assessment of chronic dietary exposure to chemicals in food, but are suitable only for estimating mean dietary exposure for the general population" (WHO, Food Cluster Diets). The website shows overviews that can be exported.

The FOSCOLLAB presents values for a series of countries, over different years, filtered by gender and age of the consumers (WHO, 2018). Its output is percentiles, quite similar to the format of EFSA's Comprehensive European Food Consumption Database. A comparison of Dutch data for spinach in the

© Burleigh Dodds Science Publishing Limited, 2025. All rights reserved.

WHO Collaborative database and in EFSA's Comprehensive database showed that the data are identical (results not shown). So, the data in both databases refer to the same surveys and were derived in a similar manner.

FAO/WHO GIFT: Another source of consumption data is the FAO/WHO GIFT online platform. It is a growing repository with access to individual consumption quantities. A series of publications is available, e.g. about the data set and the underlying surveys, and some reports based on analysis of the data. The current platform contains 62 data sets of countries in Central and South America, Africa, and the Far East that are available for download. For Europe, only data from Italy, Romania, and Bulgaria are included (FAO/WHO GIFT, 2024).

## 4 Establishing concentrations of chemicals in food

### *4.1 Sampling*

The method of sampling is relevant in relation to the concentrations of chemical substances in food. Samples can be selected in various ways. For example, to perform an import control or an official control programme, samples must be taken in a random manner according to the EU legislation; only then are the results considered representative. In case of a contamination, however, "suspected" samples can be taken e.g. to look for the sources. It can be understood that concentrations in suspected samples are higher than those in randomly taken samples. So, it is important that the exposure assessor is aware of the sampling strategy that was followed, and to consider additional samples to be taken when the sampling procedure does not fit the intentions of the exposure assessment.

### *4.2 Detection limits*

The detection limit of the chemical analysis is also a key issue in exposure assessment. If a concentration in a sample is below the detection limit, it is usually reported as "*n.d.*" (non-detectable). They are often called "left-censored" data (Wikipedia, Censoring(statistics)). Its actual concentration will be somewhere between zero and the detection limit. The common approach then is the use of the "lower/upper bound" method. Here it is assumed that the concentration can be 0 (lower bound), and the dietary intake will also be 0. Or it is assumed that the concentration is equal to the detection limit (upper bound). One can also use a "medium bound" scenario (half of the detection limit). For official control of some contaminants, the approach that is to be used is described in the EU legislation. For example, according to Regulation (EU) 2023/915, the concentrations of PAHs are to be expressed as "lower bound" concentrations,

© Burleigh Dodds Science Publishing Limited, 2025. All rights reserved.

whereas maximum levels for dioxins and dioxin-like polychlorinated biphenyls (PCBs) refer to "upper bound" concentrations. (Regulation (EU) 2023/915). The exposure assessor, however, can deviate from these obligations, as the exposure assessment is not regulated by EU law.

The method of lower/upper bounds may lead to problems in the risk characterization. As an example: a person with a body weight of 50 kg drinks two bottles of beer per day, being a total of 600 grams. In this beer, a substance is found that is assigned an ADI of 1 ng per kg body weight per day. The concentration was below the detection limit of 100 ng per kg. Using the lower bound, intake is 0, and the beer is "safe". Following the upper bound, the intake is 0.6 (kg beer) × 100 (ng/kg beer) ÷ 50 (kg body weight), equal to 1.2 ng per kg body weight per day. Then the ADI is exceeded, and the beer is "not safe". It can be calculated (Annex 3) that the critical concentrations to discriminate between safe and not safe in this example are 83 ng/kg spinach (1 ng/kg/day × 50 kg bw/0.6 kg spinach) (Foodsafetyportal, Limit of Rejection). Consequently, the detection limit for the substance in the beer should be lower than 83 ng/kg for the exposure assessment to decide about its safety.

## *4.3 Series*

In the food safety system, control authorities and food industries do not take one sample, but usually a series of samples. As only one number for the intake is to be compared with the HBGV, one value for the concentration has to be selected from the series. Most assessors select the highest value of the series, thus protecting all consumers. However, high values might be a statistical "outlier" (Wikipedia, Outlier) caused by errors during sampling or analysis. There are different ways to identify outliers, for example based on the values of the quartiles (Soetewey, 2020; Bobbitt, 2023), or the use of QQ-plots (Wikipedia, Q-Q plot) where a visual observation helps to find numbers that deviate from the distribution. If there is proof of errors, one has to exclude these numbers from the calculations. Such errors are usually hard to find; therefore, it is better to select high percentile values, such as the 95th or 99th percentile of the series, instead of the highest value to avoid the selection of outliers or otherwise biased values.

There are different ways to calculate percentiles; in more recent versions of MS Excel, for example one will find various pre-programmed functions. There are different acceptable methods (Rumsey, 2021), so it is possible that values of a higher percentile differ between experts. In this regard, it is important to understand that most series of concentrations are log-normally distributed, whereas many statistical functions are based on normal distributions. A transformation might then be needed. And when a series shows a different

© Burleigh Dodds Science Publishing Limited, 2025. All rights reserved.

distribution than log-normal, it could be a sign that one is dealing with a unconventional situation that needs further investigation.

Outlier example: As an example of outlier detection, a data set was taken with a series of concentrations of fipronil in eggs sampled in the period of September to November 2017 in EU member states. Details of this food safety issue are published (Miko, L., 2018; EFSA, 2018). Analysis of the data was done using the statistical package R for Windows (Version 4.2.2 of 2022) (CRAN). The data set contains 1954 samples. The series shows 707 samples above the detection limits, ranging from 0.002 mg to 2.8 mg per kg. A histogram of the data after logarithmic transformation is shown in Fig. 1. The higher percentiles based on rank are presented in Table 1.

To check for outliers, a Q-Q plot is presented for the log-transformed data. It is presented in Fig. 2. It shows that the highest value of 2.8 mg fipronil per kg (pointed to by the black arrow) deviates substantially from the log-normal

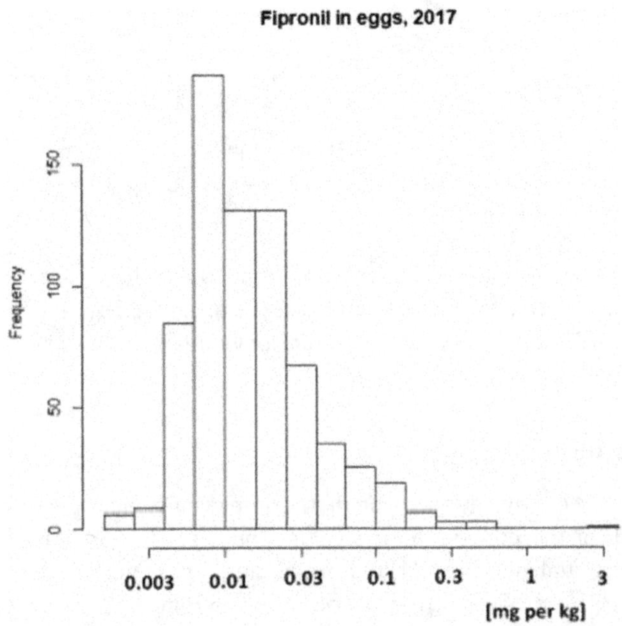


**Figure 1** Histogram of concentrations of fipronil in eggs, positive samples only ($n$ = 707).

**Table 1** Percentiles of concentrations of fipronil in eggs, positive samples only ($n$ = 707), based on their rank

| [mg fipronil per kg egg] | | | | |
|---|---|---|---|---|
| 50 percentile | 75 percentile | 90 percentile | 95 percentile | 99 percentile |
| 0.013 | 0.023 | 0.050 | 0.10 | 0.19 |

© Burleigh Dodds Science Publishing Limited, 2025. All rights reserved.

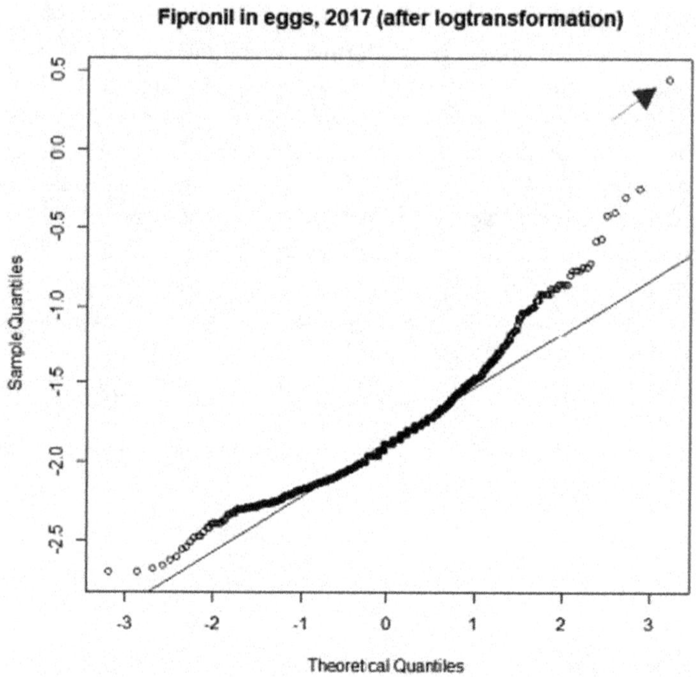


**Figure 2** QQ plot of concentrations of fipronil in eggs (707 positive samples). The black arrow points to a possible outlier.

distribution and this value might be a potential outlier. Besides, it can be noticed from this figure that the values after a log transformation are still not really normally distributed, as the lower and higher values deviate systematically from the normal line.

## *4.4 Corrections*

Before using a concentration for the dietary intake calculation, a correction might be needed for the concentration's value. There are two major items that can have a substantial influence: (1) the processing factor, and (2) the composition of the food. The processing factor refers to the differences between samples as taken and as being consumed. For example, many vegetables and fruits are sampled fresh, whereas they might be traded in cans or in a dried form. The concentrations will change during processing, e.g. by dilution in a fluid, or by mass concentration during heating and drying. Processing factors are mostly derived and published by food producers (RIVM, 2024). Unfortunately, processing factors for household and restaurants' preparations are not frequently looked for. When relevant data on processing are lacking, it should be considered to analyze the foods as consumed.

© Burleigh Dodds Science Publishing Limited, 2025. All rights reserved.

Composition might also be important for identifying the appropriate value for a concentration. As an example: the analytical results of dioxins and related compounds have to be reported per gram fat (Regulation (EU) 2023/915). To estimate exposure, it is needed to convert the concentration per gram fat into the concentration in food as consumed (wet weight). So, the concentration of dioxins in cow's milk with 3% fat equals 0.03 times the value as reported. It is also possible to estimate the concentration in a complex food on the basis of the major components. Components can be found on food labels and in food composition tables like those presented by EuroFIR (EuroFIR, 2024). For example: dioxins and related compounds are found in mozzarella cheese with 5 pg TEQ per gram fat, exceeding the EU standard of 4 pg (Regulation (EU) 2023/915). What is the concentration of dioxins in a pizza containing 30% mozzarella? The amount of mozzarella is 300 grams per kg pizza. Other ingredients do not contain relevant amounts of dioxins. The mozzarella contains 18% fat (Eatthismuch, 2024). So, the pizza contains 54 [300 × 0.18] grams of fat per kg pizza, with 5 pg TEQ from dioxins and related compounds per gram fat, equalling 270 [54 × 5] pg TEQ per kg pizza.

## 5 Calculating levels of exposure

### *5.1 Units*

The units of the different variables can pose problems for an exposure assessor. For example, the "microgram" can lead to confusion. The use of its abbreviation is not consistent; many people use *"µg"* (the recommended symbol), but some text editors convert this automatically into "mg". Other people therefore use *"ug"*. In North America and the UK, *"mcg"* is used (Wikipedia, Mikrogram). Another pitfall is the kg, as it can refer to the body weight of the consumer, or to the concentrations in the foods. Consumption quantity of foods is usually expressed in grams, whereas the concentration is per kg.

### *5.2 Acute or chronic intake*

The consumption data in PRIMo are to be used for a chronic evaluation, whereas other data sets show data for both acute and chronic exposure. "Acute" means that health effects are evaluated within a day after a single dose or meal, whereas "chronic" refers to lifetime repeating exposure. The limit value for chronic exposure is the ADI or TDI, but for acute exposure one has to use another limit value: the Acute Reference Dose (ARfD) (European Commission, 2001) (OECD Series on Testing and Assessment, 2010). When an exposure assessment is to be made for a consignment exceeding a MRL or ML, it can be assumed that exposure will be short as the consignment is limited

© Burleigh Dodds Science Publishing Limited, 2025. All rights reserved.

and remaining foods will be rejected. To protect all consumers, it is needed to perform a realistic worst-case calculation: this is a consumer with a low body weight and a high consumption quantity (for example, the 95th percentile), with a high concentration of the chemical substance in the food item (also 95%). The probability of that situation is 0.05 × 0.05, which equals to 25 persons in a population of 10000 people (1 person in 400 consumers exceeding the ADI, being at risk). For most chemical substances, the ARfD is missing. A search in the OOFT database of the Foodsafetyportal (Foodsafetyportal, OOFT) showed a list of 5181 chemicals with an ADI or TDI; only 20 of these had also an ARfD. For the 20 chemical substances assessed, the ARfD ranged 1 to 10 times higher than the corresponding ADI or TDI, with a median ratio of 3. An ADI or TDI can be used when the ARfD is unavailable, but this approach tends to overestimate the risk by an average factor of three. So, an ADI or TDI can be used when the ARfD is missing, but this will overestimate the risk on average three times.

To evaluate chronic exposure, one should be aware that many foods are not consumed every day. Staple foods such as rice and bread are consumed frequently, but more rare items only a few times per year. The consumption frequency can be defined by the ratio between consumers vs. all subjects in the databases. As an example: according to the Dutch survey of 2012 (taken from EFSA's Comprehensive European Food Consumption Database), spinach was consumed in 2012 by 48 Dutch adolescents in a population of 870 adolescents (5.5%). So, the consumption for adolescents of spinach is on average 20 times per year. The median consumption is 81 grams per meal, equalling consumption of 3888 [48 × 81] grams for all adolescent consumers. For the whole population, on average this is 4.5 [3888 ÷ 870] grams per meal. So, to evaluate chronic dietary intake it is best to use median or average consumption quantities for the whole population or to correct the consumption quantity of consumers on the basis of the consumption frequency, whereas for acute exposure one should select a high percentile of consumers only.

## 5.3 Calculation tools

EFSA's tools for intake calculations are described by Ioannidou et al. (2021). The Food Additives Intake Model (FAIM) is developed for exposure to food additives. The tool Rapid Assessment of Contaminant Exposure (RACE) is provided for a simplified risk evaluation of chemical contaminants in food, and the Feed Additive Consumer Exposure (FACE) calculator estimates exposure to residues of feed additives. The Food Enzyme Intake Model (FEIM) model estimates dietary exposure to food enzymes. According to the paper, they all use the Comprehensive database *"for each individual"*. The latter statement is remarkable, as the databases present summarized consumption data only.

© Burleigh Dodds Science Publishing Limited, 2025. All rights reserved.

The data from PRIMO_3.1 were copied into the online calculation tool EAST2 of the Foodsafetyportal (https://foodsafetyportal.eu/) for the calculation of the chronic dietary intake of chemical substances in food, for the general population. The data from the Comprehensive European Food Consumption Database are copied into the tool EAST3 (EAST, 2024). The user interfaces in both EAST tools filter the data (based on a selection of country, survey, consumer, food item) and use the numbers from the databases for calculation of dietary intake. Both tools then compare the intake with an HBGV; these are copied from the OpenFoodTox database (EFSA, 2021) and the IRIS database (US-EPA, 2025).

## 5.4 Monte Carlo

The calculation of the dietary intake is a "point estimate", based on one number for the consumption and one for the concentration in the food, for an average or extreme consumer. This estimate is compared with a single number for the HBGV. All values come from a range of data, and there are no fixed rules about what number to select. The tendency is to use summarized consumption quantities, and it is advised to select the higher concentrations (95th or 99th percentile) for acute exposure, with the most sensitive consumer (low body weight, most often children). One should be aware that the results will have a higher chance of exceeding the HBGV than the average situation. This can lead to the destruction of a consignment although only a small percentage of the samples will actually show the concentration that was used for the intake calculations. So, using higher percentiles gives higher intake numbers with a lower probability of happening, and a higher chance of rejection of foods that are actually safe.

To overcome the differences between a point estimate and the exposure of a population one can use "probabilistic modelling" (Wikipedia, Predictive intake modelling). It is based on a calculation where all distributions are multiplied with each other, instead of one value taken from each distribution. This is also called a "Monte Carlo" calculation. The Monte Carlo method is a class of computational algorithms that uses randomness to reach a result (Wikipedia, Monte Carlo method). For such calculations a computer is needed that reads the consumption quantities of all consumers and the concentrations of all samples. Then the computer selects a number randomly from the data, calculates the intake, giving one result. Then it selects another one, and so on, leading to a series of intakes. According to US-EPA "*a Monte Carlo analysis may be useful when … calculations using conservative point estimates fall above the levels of concern. Or … to disclose the degree of bias associated with point estimates … ; or when it is necessary to rank exposure pathways, …*". It pleads for a "tiered approach" (EPA, 1997).

© Burleigh Dodds Science Publishing Limited, 2025. All rights reserved.

As the output of a Monte Carlo exposure calculation is a series of numbers of intakes of individual consumers, it is well suited for statistical analysis, e.g. to determine various percentiles and their probability. The exact percentage of a population exceeding the HBGV can also be identified. It is possible to relate the intake to consumers' parameters such as gender and age. Doing so, one can estimate lifelong intakes. And, by comparison of consumption data over different time periods, changes in exposure to substances can be identified, such as reported by Liem and Theelen (1997) about exposure to dioxins in the Netherlands.

Monte Carlo calculation tools on the Internet are MCRA (MCRA), XI (Foodsafetyportal, XI), and ImproRisk (State General Laboratory, 2022). Most of these tools do not include the data set with consumption quantities of individual consumers; the user has to provide this data by himself. XI is the exception; however, it contains only the consumption data of the Netherlands for the periods 2012–2016 and 2016–2019. The FAO/WHO GIFT data sets might be used in the tools, but the surveys of most EU countries are still missing in GIFT. At the moment of writing, the use of data sets from the GIFT database for Monte Carlo calculations, next to the Dutch data in XI in the Foodsafetyportal, is being explored. So it can be expected that additional tools to perform Monte Carlo calculations on the dietary intake of chemical substances in the Foodsafetyportal will become available.

## 6 Risk characterization: limit values

To evaluate the dietary intake, it has to be compared with a limit value for the maximal safe exposure. There are two types of such limit values: (1) HBGVs such as ADI, TDI, and ARfD and (2) BMDLs. Both types of values are derived by experts in the hazard characterization (step two of the risk assessment), and the results can be found in different publications and databases. The ADI is used to describe the maximal acceptable intake of "residues" of, for example, pesticides, additives, and veterinary drugs. For "contaminants" such as mycotoxins and environmental and process contaminants, the limit value is called a TDI. The ARfD can be used for acute exposure to both residues and contaminants.

The method of deriving HBGVs differs from that or BMDLs (Foodsafetyportal, HBGVs and BMDLs). A HBGV is based on an evaluation of all known effects, leading to a single value; usually, the lowest NOAEL is chosen (ChemSafetyPro, 2018). And, a HBGV is derived using "Safety Factors". These factors are actually extrapolation factors, which are intended to transform the NOAEL of a research study into a NOAEL for the general human population (European Medicines Agency, 2014). Doing so, the dietary intake can directly be compared with an ADI or TDI or ARfD, as it can be considered to be the

© Burleigh Dodds Science Publishing Limited, 2025. All rights reserved.

NOAEL for the most sensitive humans. On the other hand, a BMDL refers to a specific effect, such as a BMDL01 of 0.50 μg/kg bw per day for developmental neurotoxicity of lead in children, and a BMDL01 of 1.5 μg/kg bw per day for clinical effects of lead in adults (EFSA, 2010). So, BMDLs describe a specific effect in an animal or human study, which is not necessarily the most sensitive effect, without an extrapolation to the most sensitive humans. For that extrapolation, the risk characterization has to be made using a "Margin of Exposure" (MoE).

## *6.1 Tolerable weekly intakes and provisional tolerable monthly intakes*

A TDI is set and used in the same manner as the ADI. Instead of a TDI, some contaminants have a tolerable weekly intake (TWI) or a provisional tolerable monthly intake (PTMI). When using a TWI or PTMI, it is necessary to evaluate exposure over a time period. It is recommended here not to use a "dummy TDI" (a TWI divided by 7) as some people do. As an example: a person of 60 kg body weight (bw) consumes 145 grams of spinach on a day, with 0.18 mg cadmium per kg (equal to 0.18 μg/g) of spinach. Then the intake is 26 μg [145 (g) × 0.18 (μg/g)] cadmium. This is 0.43 μg [26 μg ÷ 60 kg] per kg bw per day. The TWI for cadmium is 2.5 μg per kg bw per week (EFSA, 2013) which is equal to a "dummy TDI" of 0.36 [2.5 ÷ 7] μg per kg bw per day. The dummy TDI is exceeded. However, if the consumer eats on average three portions of spinach per week, then the intake is 78 [3 × 26] μg per week, which is 1.3 μg cadmium per kg bw per week. It can be calculated that the TWI is not exceeded as long as the consumer eats less than 6 meals (actually 5.7 meals of 145 grams) with 0.18 mg cadmium per kg spinach per week.

## *6.2 Selection of a limit value*

In recent times deriving HBGVs is becoming rare, and more BMDLs are set. The BMDLs are limit values from a curve fit of dose-effect relations of health effects in experimental animal or human studies. In these fits one can statically calculate a minimal response of 1%, 5%, or 10%, giving a BMDL01, resp. BMDL05, or BMDL10. Consequently, one will find series of ADIs, TDIs, and ARfDs, and different BMDLs today in more recent EFSA opinions (EFSA Publications) and JECFA toxicological monographs (WHO, JECFA). These reports can contain one HBGV and/or one or more BMDLs. See for example the limit values of EFSA for total lead in Table 2.

How to select the appropriate limit value from such series? Most experts will select the most recent study, but how must one deal with both a HBGV and various BMDLs in the same opinion? It is proposed by Theelen (2024) to use

© Burleigh Dodds Science Publishing Limited, 2025. All rights reserved.

**Table 2** Limit values for total lead, as taken from EFSA opinions

| Opinion | Type | Value | Note |
|---|---|---|---|
| EFSA (2005) | TWI (provisional) | 25 μg per kg bw per week | JECFA has established a PTWI for lead in 1986. It is set to 25 ug/kg b.w./week for infants and children on the basis that lead is accumulating in the body and an increase of the body burden of lead should be avoided. In 1993 and 2000, JECFA reconfirmed this PTWI and extended it to all age groups |
| EFSA (2010) | BMDL01 (Human) | 1.5 μg per kg bw per day | clinical signs |
| | BMDL01 (Human) | 0.5 μg per kg bw per day | neurology |
| | BMDL10 (Human) | 0.63 μg per kg bw per day | clinical chemistry |

**Table 3** Selection of a HBGV or BMDL, when different studies and values are available

| Step | |
|---|---|
| 1 | Select the most recent study |
| 2 | Use the HBGV when available, otherwise select the BMDLs from that study |
| 3 | Prefer BMDLs from human studies above animal studies |
| 4 | Prefer a BMDL01 over a BMDL05, and a BMDL05 over a BMDL10 |
| 5 | Select the lowest value of BMDLs of the same type |

distinct rules for the selection, as presented in Table 3. When following these rules, the BMDL01 of 0.5 μg per kg bw per day will be selected for the risk characterization of dietary intake of lead.

### 6.3 Margin of Exposure

The EFSA made the statement that the MoE for genotoxic carcinogens should be at least 10 000 lower than a BMDL10 from an animal study (EFSA, 2012). A similar type of statement for non-genotoxic substances was not made. Various EFSA Opinions demonstrate, however, that EFSA is using the MoE concept for non-genotoxic chemicals with a critical level of 100 for a BMDL10 from animal studies. Existing EFSA evaluations do, however, not make clear what the MoE should be for a BMDL01 or BMDL05 from animal studies or for BMDLs from human studies. Other critical levels for the MoE should then be used, but it is not noted what the MoE should be for the different BMDLs (EFSA, 2023). Theelen (2024) presents MoEs for the various BMDLs, shown in Table 4. These values are derived from the safety factors of HBGVs; this list is included in the

© Burleigh Dodds Science Publishing Limited, 2025. All rights reserved.

**Table 4** Margin of exposure of different BMDLs

| Chemical | BMDL | Study | MoE |
|---|---|---|---|
| Genotoxic | All BMDLs | Animal and human | 10000 |
| Non-genotoxic | BMDL10 | Animal | 100 |
| | | Human | 10 |
| | BMDL05 | Animal | 30 |
| | | Human | 3 |
| | BMDL01 | Animal | 10 |
| | | Human | 1 |

calculation tools of the Foodsafetyportal, leaving the user the possibility to deviate from these values.

## 7 Risk characterization: setting 'safe' levels

Foods are to be considered not safe for consumption when the dietary intake is above the HBGV or smaller than the MoE of a BMDL. This is not similar to the statement that adverse health effects will occur. Most toxicological experts are of the opinion that health effects cannot be excluded anymore; there is always a certain risk. That risk will usually be very small as the safety factors and the MoE leave room between the intake by the consumer and the doses for which effects are actually demonstrated in humans or animals. So, the appearance of adverse health effects is not very likely when the permissible exposure limit value is exceeded a few times.

Some toxicological experts state that there is no such thing as a "safe" level, as one cannot derive a threshold value for some adverse health effects, e.g. for carcinogenicity. For these effects limit values are set in another way; most often by a low dose linear extrapolation (Foodsafetyportal, Genotoxic carcinogens). Other experts criticize this approach, as it sometimes leads to "unreasonable conclusions" (Bolt et al., 2009). Another comment is that a NOAEL cannot be considered a real threshold between "safe" and "not safe", as hormetic dose response can lead to effects below the NOAEL (Calabrese, 2008). Although these comments are seriously debated within the scientific toxicological community, it should be concluded that the HBGVs and BMDLs are still considered the boundary values between safe and not safe exposure.

Another comment refers to the fact that a person is exposed to mixtures of chemical substances and not to a single substance. This is correct, but there are not yet unified methods on how to evaluate these mixtures. The issue here is that the effects of different substances in mixtures can be additive (equal to the sum of all substances), or synergistic (the mixture is more toxic than its components), or antagonistic (the mixture becomes less toxic) (Roell et al.,

© Burleigh Dodds Science Publishing Limited, 2025. All rights reserved.

2017). In the current situation, only additive action is taken into account for mixtures of dioxins and dioxin-like PCBs, and for a few PAHs. So, exposure to mixtures is taken care of, albeit for only a few well-defined combinations of substances.

The risk characterization deals with deterministic calculations. Many assessors use a worst-case scenario by selection of the highest concentration and largest consumption quantities. If that is "safe", all other intake levels are considered safe. The real situation is that there will always be a distribution with safe and unsafe intakes. These can only be determined using a probabilistic method such as a Monte Carlo calculation. Then the question arises, what percentage of the actual population defines a "safe" situation. For example: one consumer in 1 000 000 consumers? If so, how to evaluate an exposed population of 100 persons?

## 8 Conclusion

Data on food consumption of various countries in the world are available on the Internet in public databases, but these lack a unified, harmonized structure. Data on the consumption of rare foods might be missing. The description of food commodities is not yet harmonized, making the selection of the appropriate food items a challenge. EFSA's FoodEx2 is a good step in the harmonization, but its success depends on its use in the food consumption surveys of the EU member states and its ability to describe the foods in the appropriate way.

The comprehensive databases from EFSA and WHO-FOSCOLLAB contain summarized data. They present average consumption data, with a standard deviation and percentiles for population groups of different countries for the whole population and for consumers only. Doing so, the actual variation and extreme values are lost. PRIMo is based on averages of the whole population. It is not possible to compare the data of the same surveys in the different data sets, as the average values and higher percentiles are presented with little or no explanation of how these are derived.

Most data sets follow log-normal distributions; these might show other percentiles than those estimated statistically for a normal distribution (Sokal and Rohlf, 1981; Thisvsthat, 2023). Whether or not the appropriate method is used is often not clear.

An exposure assessor should be aware of the background of the occurrence data that he is using, as samples might be taken in a risk-based manner or "at random". Another issue about concentrations is the interpretation of non-detectable values. The use of "lower/upper bound" values can be helpful, but it should be checked if the LOD or LOQ is sufficiently low to discriminate between "safe" and "unsafe" concentrations. If not, an alternative analytical method should be considered.

© Burleigh Dodds Science Publishing Limited, 2025. All rights reserved.

Misunderstanding the meaning of units in an exposure calculation is a common source of calculation errors. Special attention is needed for the different notations of "*micrograms*". Series of consumption quantities and concentrations usually follow a log-normal distribution. Outliers should be excluded when there are reasons to conclude about possible bias; the use of higher percentiles is preferred above the single highest value. If the distribution deviates from log-normal, it could be a sign that one is dealing with a unconventional situation that may need further investigation.

Many data sets discriminate between "acute" and "chronic" consumption. For the evaluation of a short-term situation, the selection of worst-case conditions is preferred, such as higher percentiles for consumption and concentration, to protect all consumers. The dietary intake is then to be compared with the ARfD. In many cases, however, ARfDs are missing. Use of an ADI or TDI instead can be considered, but it might slightly overestimate the risk. For an estimate of long-term exposure, average data are preferred. To obtain the most reliable results for chronic dietary intake, the consumption frequency should be taken into account. If not, another overestimation is likely to occur.

The calculation with one value for consumption and one for concentration is a point estimate. Using a worst-case scenario has a higher probability of establishing safe results, but will lead to a higher chance of rejecting (and destroy) safe foods.

When the impact of the actual variation is to be known, a Monte Carlo calculation is to be considered with data of individual consumers. Monte Carlo tools that are accessible through the Internet ask the users to provide the data of individual consumers, with the exception of the Dutch data in XI (Foodsafetyportal, XI). It is very difficult to obtain this data, as no public sources are available, with the exception of the WHO-GIFT data sets.

In the hazard identification, the use of HBGVs is preferred over BMDLs, as they already include the evaluation of the relevant effects and the conversion into a human NOAEL by expert judgment. The use of BMDLs forces the assessor to select a Margin of Exposure; guidance for the appropriate MoE and on the selection of the HBGV or BMDLs from a series of values is lacking, except for the methods developed by Theelen (2024) as implemented in the Foodsafetyportal (https://foodsafetyportal.eu/).

When dealing with a TWI or PTMI, it is needed to define intake per week or month. The calculation of the intake should be based on the consumption frequency of the foods. For an acute assessment, however, one is dealing with a quantity per portion, so then the consumption frequency can be ignored.

The risk characterization is based on the concept of "safe" foods. A known issue here is that many people do not believe that there is such a thing as "safe" food. The exposure to mixtures and unusual dose-response relationships are

© Burleigh Dodds Science Publishing Limited, 2025. All rights reserved.

cited by (sometimes self-proclaimed) experts as reasons why a HBGV is not the limit value for the evaluation of safety of foods. Toxicological research and discussions on these topics between experts are still needed.

The issue with Monte Carlo calculations is the evaluation of an output of a series instead of a single value. In the current system of food safety exposure assessment, it is yet not defined how to evaluate the risk of a range of intakes and what number of consumers exceeding the HBGV can be considered safe.

## 9 The future of exposure assessment

According to the risk analysis concept of the Codex Alimentarius, exposure assessment is independent of risk management. This has led to different implementations of the calculations of dietary intake. Besides, one can notice different formats for the storage of relevant data in databases. The latter makes searching for the appropriate numbers more complicated. Consequently, results might differ, leading to discussions between experts and stakeholders. To solidify the quality of the results and to improve the transparency of the evaluations, it is recommended to better harmonize the method of exposure assessment and the selection of data. Special attention should be given to updates of data sources to guarantee consistencies between previous and present evaluations.

Most current evaluations of chemical substances in foods are point estimates, based on a limited number of data with worst-case conditions. Monte Carlo calculations are preferred for more realistic assessments; to do so, existing data sets of different countries containing the consumption quantities of individual consumers should become public. In this regard, discussions are also needed about the method of evaluation of results of Monte Carlo calculations, to define "safe" situations.

The discussions about the impact of mixtures and about how to deal with dose-response thresholds must continue, but academics should understand that clear-cut boundaries and agreements are still needed for consistent conclusions in the current global food safety system.

According to the risk analysis approach, discussion on changes and updates of the current system of exposure assessment is the responsibility of the scientific community. Risk managers should, however, join these discussions as observers to create social and legal acceptance of new developments and concepts.

## 10 Abbreviations

ADI Acceptable Daily Intake
ARfD Acute Reference Dose

© Burleigh Dodds Science Publishing Limited, 2025. All rights reserved.

| | |
|---|---|
| BMDL | Benchmark Dose Level |
| CSV | comma separated values |
| EFSA | European Food Safety Authority |
| FAO | Food and Agricultural Organization of the United Nations |
| HBGV | Health Based Guidance Value |
| LOD | Limit of Detection |
| LOQ | Limit of Quantification |
| ML | Maximum Level |
| MoE | Margin of Exposure |
| MRL | Maximum Residue Level |
| n.d. | non-detectable |
| NOAEL | No Observed Adverse Effect Level |
| pg | picogram ($10^{-12}$ gram) |
| PTMI | Provisional Tolerable Monthly Intake |
| TDI | Tolerable Daily Intake |
| TWI | Tolerable Weekly Intake |
| WHO | World Health Organization |

## 11 Where to look for further information

Sources of information on exposure assessment and risk characterization of chemical substances in food are:

- Environmental health criteria, 240 (WHO 2008) Principles and methods for the risk assessment of chemicals in food. Chapter 6 Dietary exposure assessment for chemicals in food. Second edition (2020)
- Environmental health criteria, 240 (WHO 2009) Principles and methods for the risk assessment of chemicals in food. Chapter 7 Risk characterization
- FAO and WHO. 2023. Codex Alimentarius Commission Procedural Manual. Section 4 Twenty-eighth edition. Rome.
- For practical support with exposure assessment and risk characterization consulting the Food Safety Portal (https://foodsafetyportal.eu) is advised.

## 12 Annex

$I = (C \times Q) \div bw$ where: $(C = C0 \times cC)$ and $(Q = Q0 \times cQ)$ [1]

$\text{"safe"} = (I \leq HBGV)$ [2]

$HBGV \times bw / Q$ [3]

| | |
|---|---|
| C | Final concentration of substance in food as consumed [mg per kg] |
| Q | Final consumption quantity [kg per day] |

© Burleigh Dodds Science Publishing Limited, 2025. All rights reserved.

C0 Initial concentration of substance in food [mg per kg]
cC Correction factor for initial concentration [-]
Q0 Initial consumption quantity [kg per day]
cQ Correction factor for consumption quantity [-]
bw Body weight of the consumer [kg]
"safe" No adverse health effects for the consumer
I Dietary intake of the consumer [mg chemical per kg body weight per day]
HBGV Health Based Guidance Value of the chemical substance [mg chemical per kg body weight per day]

## 13 References

ANSES, 2024. République Française. Anses. https://www.anses.fr/en

BfR, 2024. Bundesinstitut für Risikobewertung (BfR). https://www.bfr.bund.de/en/home.html

Bobbitt, Z., 2023. How to Find Outliers in R. https://www.statology.org/find-outliers-in-r/

Bolt, H.M., et al., 2009. Low-dose extrapolation in toxicology: an old controversy revisited. *Arch Toxicol* 83: 197–198. https://doi.org/10.1007/s00204-009-0413-3

Calabrese, E.J. 2008. Hormesis: why it is important to toxicology and toxicologists. *Environ Toxicol Chem* 27(7): 1451–1474. https://doi.org/10.1897/07-541

CDC, 2024. National Center for Health Statistics, NHANES Tutorials. https://wwwn.cdc.gov/nchs/nhanes/tutorials/default.aspx?CDC_AA_refVal=https%3A%2F%2Fwww.cdc.gov%2Fnchs%2Fnhanes%2Fmodules%2Fdata-analysis-tutorial.html

ChemSafetyPro, 2018. What Is Acceptable Daily Intake (ADI) and How to Calculate It. https://www.chemsafetypro.com/Topics/CRA/What_Is_Acceptable_Daily_Intake_(ADI)_and_How_to_Calculate_It.html

CRAN. The R Project for Statistical Computing. https://www.r-project.org/

EAST, 2024. Differences between EAST, EAST2, and EAST3. https://foodsafetyportal.eu/east/ea_versions.html

Eatthismuch, 2024. Mozzarella 365, Whole Foods Market. https://www.eatthismuch.com/calories/mozzarella-3300651

EFSA Publications. https://www.efsa.europa.eu/en/publications

EFSA, 2005. Opinion of the Scientific Panel on Contaminants in the Food Chain CONTAM Related to the Safety Assessment of Wild and Farmed Fish. https://doi.org/10.2903/j.efsa.2005.236

EFSA, 2009. General principles for the collection of national food consumption data in the view of a pan-European dietary survey. *EFSA J* 7(12): 1435. https://doi.org/10.2903/j.efsa.2009.1435

EFSA, 2010. Scientific Opinion on Lead in Food EFSA Panel on Contaminants in the Food Chain (CONTAM) 20 April 2010. https://doi.org/10.2903/j.efsa.2010.1570

EFSA, 2012. Statement on the Applicability of the Margin of Exposure Approach for the Safety Assessment of Impurities Which Are Both Genotoxic and Carcinogenic in Substances Added to Food/Feed. https://doi.org/10.2903/j.efsa.2012.2578

EFSA, 2013. Scientific Opinion on the Public Health Hazards to Be Covered by Inspection of Meat (Solipeds). https://doi.org/10.2903/j.efsa.2013.3263

© Burleigh Dodds Science Publishing Limited, 2025. All rights reserved.

EFSA, 2014. Guidance on the EU Menu Methodology. https://doi.org/10.2903/j.efsa.2014.3944

EFSA, 2018. Fipronil: Results of Follow-Up Monitoring Published. https://www.efsa.europa.eu/en/press/news/180503

EFSA, 2021. Chemical Hazards Database (OpenFoodTox). https://www.efsa.europa.eu/en/data-report/chemical-hazards-database-openfoodtox

EFSA, 2022. EFSA Food Consumption Data. https://www.efsa.europa.eu/en/data-report/food-consumption-data

EFSA, 2023. Margin of Exposure. https://www.efsa.europa.eu/en/topics/topic/margin-exposure

EFSA, 2024a. Food Consumption- Survey Details. https://www.efsa.europa.eu/en/microstrategy/food-consumption-survey

EFSA, 2024b. Food Classification Standardisation – The FoodEx2 System. https://www.efsa.europa.eu/en/data/data-standardisation

EFSA, 2024c. Pesticide Evaluation: Tools. PRIMo - Pesticide Residue Intake Model. https://www.efsa.europa.eu/en/applications/pesticides/tools

EPA, 1997. Guiding Principles for Monte Carlo Analysis, EPA/630/R-97/001, p. 5. https://www.epa.gov/sites/default/files/2014-11/documents/montecar.pdf

EuroFIR, 2024. European Food Information Resource. List of EuroFIR Databases. https://www.eurofir.org/food-information/food-composition-databases/

European Commission, 2001. Guidance for the Setting of an Acute Reference Dose (ARfD). https://food.ec.europa.eu/system/files/2016-10/pesticides_ppp_app-proc_guide_tox_acute-ref-dose.pdf

European Medicines Agency, 2014. Guideline on Setting Health Based Exposure Limits, pp. 5–6. https://www.ema.europa.eu/en/documents/scientific-guideline/guideline-setting-health-based-exposure-limits-use-risk-identification-manufacture-different-medicinal-products-shared-facilities_en.pdf

FAO and WHO, 2024. Codex Alimentarius Commission. Procedural Manual – Twenty-Ninth Edition. Rome. https://openknowledge.fao.org/handle/20.500.14283/cd2280en

FAO/WHO GIFT, 2024. Global Individual Food Consumption Data Tool. https://www.fao.org/gift-individual-food-consumption/data/en

Foodsafetyportal, Genotoxic Carcinogens. https://foodsafetyportal.eu/cancer/calculation.html

Foodsafetyportal, HBGVs and BMDLs. https://foodsafetyportal.eu/hbgv/hbgv.html

Foodsafetyportal, Limit of Rejection. https://foodsafetyportal.eu/ra/lor.html

Foodsafetyportal, OOFT. https://foodsafetyportal.eu/oft/oft_ooft.html

Foodsafetyportal, XI. https://foodsafetyportal.eu/xi/xi.html

Health Canada, 2017. Food Consumption Table (2015 CCHS Nutrition). https://open.canada.ca/data/en/dataset/a9c18c37-b7b9-48ce-ac1d-b40a65bafa9e

Herman, J.L. and Younes, M. 1999. Background to the ADI/TDI/PTWI. *Regul Toxicol Pharmacol.* 30(2 Pt 2): S109–S113. https://doi.org/10.1006/rtph.1999.1335

IPCS, 2009. *Environ. Health Crit.* 240, Principles and Methods for the Risk Assessment of Chemicals in Food, Chapter 2: Risk Assessment and its Role in Risk Analysis, IPCS WHO 2009.

Ioannidou, S., et al., 2021 European Food Safety Authority Open Access Tools to Estimate Dietary Exposure to Food Chemicals. *Environment International* 149: 106357. https://pubmed.ncbi.nlm.nih.gov/33556818/

© Burleigh Dodds Science Publishing Limited, 2025. All rights reserved.

Liem, A.K.D. and Theelen, R.M.C., 1997. *Dioxins: Chemical Analysis, Exposure and Risk Assessment*. Thesis University Utrecht.

MCRA. Exposure, Hazard & Risk Assessment. https://mcra.rivm.nl/mcra/#/

Miko, L., 2018. Fipronil in Eggs. https://food.ec.europa.eu/system/files/2018-01/adv-grp_plenary_20171122_pres_01.pdf

NVWA, 2024. Office for Risk Assessment & Research. https://english.nvwa.nl/about-us/office-for-risk-assessment--research

OECD Series on Testing and Assessment, 2010. OECD Series on Testing and Assessment, 2010, Guidance for the Derivation of an Acute Reference Dose. https://doi.org/10.1787/916afb91-en

Regulation (EC) No 178/2002. Regulation (EC) No 178/2002 of the European Parliament and of the Council of 28 January 2002 Laying Down the General Principles and Requirements of Food Law, Establishing the European Food Safety Authority and Laying Down Procedures in Matters of Food Safety.

Regulation 2023/915 (EU). COMMISSION REGULATION (EU) 2023/915 of 25 April 2023 on Maximum Levels for Certain Contaminants in Food and Repealing Regulation (EC) No 1881/2006.

RIVM, 2024. Processing Factors, Update 11 June 2020. https://www.rivm.nl/en/chemkap/processing-factors

Roell. K.R., et al., 2017. An introduction to terminology and methodology of chemical synergy-perspectives from across disciplines. *Front Pharmacol* 8: 158. https://doi.org/10.3389/fphar.2017.00158

Rumsey, D.J., 2021. How to Calculate Percentiles in Statistics, From The Book: Statistics For Dummies. https://www.dummies.com/article/academics-the-arts/math/statistics/how-to-calculate-percentiles-in-statistics-169783/

Soetewey, A., 2020. Outliers Detection in R. https://statsandr.com/blog/outliers-detection-in-r/

Sokal, R.R. and Rohlf, F.J., 1981. *Biometry*, 2nd edition. W.H.Freeman and Company, New York, p. 419.

State General Laboratory, 2022. ImproRisk. http://www.improrisk.com/

Theelen, R., 2024. The Foodsafetyportal. Online Expert Support for Enforcement and Risk Assessment. *Food Control* 158: 110227. https://foodsafetyportal.eu/portal/doi_2023_110227.pdf

Thisvsthat, 2023. Lognormal Distribution vs. Normal Distribution What's the Difference? https://thisvsthat.io/lognormal-distribution-vs-normal-distribution

USDA, 2021. Food Consumption. https://www.ers.usda.gov/topics/food-choices-health/food-consumption-demand/food-consumption/

US-EPA, 2024. IRIS Assessments. https://iris.epa.gov/AtoZ/?list_type=alpha

Wherry, F.F. and Woodward, I., 2019. The Oxford Handbook of Consumption. https://doi.org/10.1093/oxfordhb/9780190695583.001.0001

WHO, 2018. WHO/FAO. Food Safety Collaborative Platform (FOSCOLLAB). https://apps.who.int/foscollab/

WHO, Food Cluster Diets. WHO, Food Cluster Diets. https://www.who.int/data/gho/samples/food-cluster-diets

WHO, GEMS/Food. GEMS/Food. https://extranet.who.int/gemsfood/

WHO, JECFA. Monographs. https://www.who.int/groups/joint-fao-who-expert-committee-on-food-additives-(jecfa)/publications/toxicological-monographs

© Burleigh Dodds Science Publishing Limited, 2025. All rights reserved.

Wikipedia, Censoring(statistics). https://en.wikipedia.org/wiki/Censoring_(statistics)
Wikipedia, Mikrogram. https://en.wikipedia.org/wiki/Microgram
Wikipedia, Monte Carlo Method. https://en.wikipedia.org/wiki/Monte_Carlo_method
Wikipedia, Outlier. https://en.wikipedia.org/wiki/Outlier
Wikipedia, Predictive Intake Modelling. https://en.wikipedia.org/wiki/Predictive_intake_modelling
Wikipedia, Q.-Q. Plot. https://en.wikipedia.org/wiki/Q%E2%80%93Q_plot

© Burleigh Dodds Science Publishing Limited, 2025. All rights reserved.

# Part 2

## Developments in risk assessment: hazard identification/characterisation and exposure assessment

# Chapter 4

## Advances in techniques for monitoring/detecting chemical contaminants in food: chromatography

*Merrik Kobarfard, Alexander Sweett and Tadeusz Górecki, University of Waterloo, Canada*



### 1 Introduction

Chromatography is a technique used for the separation of components within a mixture. This separation process occurs as the mobile phase transports various components through the stationary phase. Due to their differing affinities for the stationary phase, the components are carried at varying rates. On a given stationary phase, this affinity is directly influenced by the physical and chemical properties of the compounds, and different chromatographic techniques leverage one or more of these properties to achieve effective separation.[1,2] Chromatographic methods serve both analytical and preparative purposes, with preparative chromatography being particularly valuable for purifying chemical components. This chapter focuses on the analytical applications of chromatography within the domain of food contaminant and residue analysis.

It is important to note that, for official purposes, chemicals present in food are categorized based on their source of contamination. Contaminants

http://dx.doi.org/10.19103/AS.2025.0158.10
© Burleigh Dodds Science Publishing Limited, 2025. All rights reserved.

**Table 1** Overview of the chromatographic techniques frequently used for food chemical analysis

| Analytical technique | Analyte coverage | Method development complexity | Analysis time | Compatibility with MS | Robustness | Expense | Section of this chapter |
|---|---|---|---|---|---|---|---|
| GC | Volatile & semi-volatile compounds | Low to moderate | Fast | Excellent | High | Moderate/high* | Section 3 |
| HPLC/UHPLC | Very wide, including volatile and non-volatile compounds | Moderate/moderate to high | Moderate/very fast | Good/excellent | High | Moderate/high/very high* | Section 4 |
| SFC | Similar to HPLC/UHPLC | High | Fast to very fast | Good | Moderate to high | High/very high* | Section 5 |
| TLC | Non- and semi-volatile compounds | Low | Fast | Limited | Low | Very low | Section 4.2 |
| 2D-GC | Similar to GC, with improved peak capacity and selectivity | High | Longer | Excellent | Moderate to high | High/very high* | Section 6.1 |
| 2D-LC | Similar to HPLC/UHPLC, with improved peak capacity and selectivity | Very high | Longer | Excellent | Moderate to low | High/very high* | Section 6.2 |
| LC-GC | Similar to GC | High | Moderate to long | Excellent | Moderate | High/very high* | Section 6.3 |

Note: GC = gas chromatography; HPLC = high performance liquid chromatography; UHPLC = ultra-high performance liquid chromatography; SFC = supercritical fluid chromatography; TLC = thin layer chromatography; 2D-GC = two-dimensional gas chromatography; 2D-LC = two-dimensional liquid chromatography; LC-GC = liquid chromatography-gas chromatography.
*In combination with MS.

© Burleigh Dodds Science Publishing Limited, 2025. All rights reserved.

and toxins are defined as chemicals that are not intentionally added to food; rather, they are present in food as a result of various stages of production, packaging, transportation, and storage or due to environmental contamination. Examples of such contaminants include mycotoxins and polychlorinated biphenyls (PCBs).[3] In contrast, residues refer to the remnants of chemicals that are intentionally applied to food, such as pesticides.[4] Although these two groups of compounds differ in their required risk mitigation strategies, they exhibit similarities in terms of analysis. Therefore, in this chapter, they are not extensively differentiated.

In analytical contexts, chromatography can be employed to separate components within a sample. The unique retardation of different components, due to their distinct physical and chemical properties, facilitates their identification. The analytical efficacy of chromatography can be further enhanced when coupled with other techniques, particularly mass spectrometry (MS). While MS is very effective in identifying and quantifying compounds, it necessitates an interface for analyte introduction. Analyzing multiple compounds simultaneously (with no pre-separation) can compromise the sensitivity and selectivity of the mass spectrometer. Chromatography addresses this by separating sample components and sequentially introducing them into the mass spectrometer, thereby providing optimal conditions for MS operation.

Chromatographic techniques are generally categorized based on the state of the mobile phase into three primary types: gas chromatography (GC), liquid chromatography (LC), and supercritical fluid chromatography (SFC).[1] The following sections will briefly explore some of the challenges frequently encountered in food analysis, offer a concise overview of various chromatographic techniques, discuss recent advancements in the field, and examine how these innovations help address the analytical difficulties associated with food samples. A brief overview of the analytical techniques covered in this chapter can be found in Table 1.

## 2 Common challenges in the analysis of contaminants and residues in food

### 2.1 Low concentrations of contaminants and residues

The adverse effects of many food contaminants and residues can manifest at very low exposure concentrations. This phenomenon may be attributed to the high potency of these contaminants and their accumulation resulting from consistent exposure through food and beverages.[5] Consequently, this poses a significant challenge for the analysis, as the lower limits of detection or quantification might necessitate extensive sample preparation and advanced analytical techniques.

© Burleigh Dodds Science Publishing Limited, 2025. All rights reserved.

## 2.2 Matrix effect

One of the primary challenges in food analysis is the presence of complex matrices in the samples. The matrix refers to all components of the sample excluding the analyte of interest. The influence of these matrix components on the analytical results, compared to a clean standard, is known as the matrix effect. This effect can either enhance or suppress the analyte's signal, potentially leading to inaccuracies in contamination level estimations.[6]

## 2.3 Presence of multiple contaminants and residues in the sample

Another challenge with food samples is the presence of multiple contaminants and residues simultaneously. These chemicals can be very similar in structure, making them difficult to distinguish using analytical methods. As a result, the analytical method must exhibit high selectivity to accurately differentiate and identify the specific compounds of interest.

## 2.4 Complex and time-consuming sample preparation

Effective sample preparation is crucial in food analysis, as it aims to minimize the effect of matrix components and interfering substances while preserving the analyte. Despite many advancements, some sophisticated techniques can be time-consuming, applicable only to a narrow range of analytes, and may result in suboptimal analyte recovery. Newer sample preparation procedures have been developed to address these problems. QuEChERS sample preparation procedure is a well recognized example of such procedures.

QuEChERS, which stands for Quick, Easy, Cheap, Effective, Rugged, and Safe, is a relatively new sample preparation method for contaminants and residues that is valued for its simplicity and minimal solvent usage.[7] The process typically involves two main steps: since most contaminants and residues are non-polar, they can be extracted using organic solvents. Initially, samples are homogenized, and the analytes are extracted into an organic solvent, usually acetonitrile. A buffering salt is then added to the extract, enhancing the tendency of non-polar analytes to migrate into the extraction phase.[8]

In the subsequent step, particulate matter is separated from both the aqueous and the organic phases, with the organic phase undergoing a clean-up process. Dispersive Solid-Phase Extraction (D-SPE) is the standard clean-up method used in QuEChERS, which involves adding particles coated with a specific sorbent material to the extract.[8] Ideally, these particles selectively adsorb matrix components without affecting the analytes. However, in practice, the sorbent may also adsorb some analytes, resulting in lower

© Burleigh Dodds Science Publishing Limited, 2025. All rights reserved.

analyte recoveries. Nonetheless, due to the sorbent's greater affinity for matrix components, the overall effect is an increased signal for the analytes, attributed to a reduced background interference from the matrix. QuEChERS is considered advantageous due to its rapid and straightforward approach to sample preparation. More importantly, it is considered a generic extraction method for pesticides, which can extract many pesticides simultaneously with acceptable recoveries.[7]

Similar to other domains within chemical analysis, numerous sample preparation and extraction procedures have been developed over the years.[9] While these methods may not be as widely recognized as the QuEChERS technique, they have the potential to broaden the applicability of specific analytical techniques, reduce analytical interferences, and lower both the time and costs associated with analysis. Therefore, it is essential to consider various facets of analysis beyond the final determination to optimize overall analytical performance.

## 3 Gas chromatography

GC is a chromatographic technique in which the mobile phase is a gas. It is highly effective for analyzing a wide range of food contaminants and residues. As illustrated in Fig. 1, the GC system consists of several key components: an injection system, a column coated with a stationary phase, a mobile phase, an oven, and a detector. The injection system is responsible for introducing the sample into the column for separation.[10] The sample components are then carried through the column by the mobile phase, which is typically an inert gas such as helium due to its safety and chromatographic efficiency.[11]

In GC, the stationary phase can be either solid or liquid.[1] Analytes interact with such stationary phases through adsorption or partitioning, respectively. Solid stationary phases are generally less suitable for food chemical analysis due to typically longer analysis times for common contaminants and residues compared to liquid stationary phases, as well as overall poorer chromatography (e.g. asymmetrical peaks). Consequently, liquid stationary phases are more commonly used.[11] In partition GC, when the sample reaches the column, its components repeatedly partition into and out of the stationary phase. As a result, they move slower than the mobile phase, with the net speed of migration dependent on their partition coefficients. The differential partitioning causes each component to be delayed to varying degrees, leading to their separation.[2] Various stationary phases are available, differing in polarity and selectivity, and can be selected based on the specific analytes of interest.

GC can be operated under isothermal or temperature-programmed conditions. In isothermal separation, the oven temperature is maintained constant throughout the run, which provides maximum chromatographic

© Burleigh Dodds Science Publishing Limited, 2025. All rights reserved.

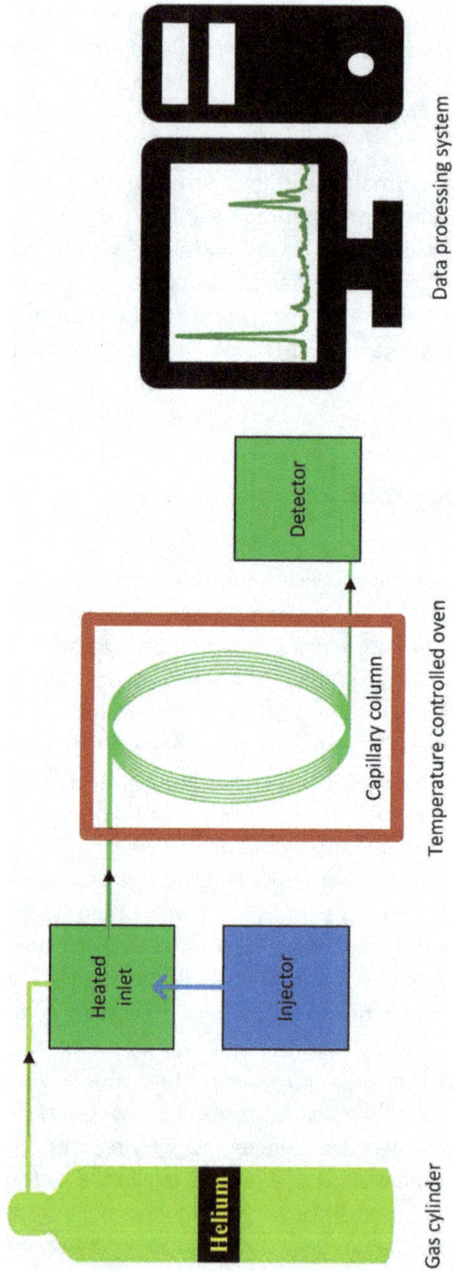


**Figure 1.** Schematic diagram of a typical gas chromatography system.

© Burleigh Dodds Science Publishing Limited, 2025. All rights reserved.

resolution. However, this approach leads to the so-called general elution problem when the sample contains analytes with widely ranging retention properties. At low temperatures, volatile analytes may be well separated, but the less volatile ones take a very long time to elute from the column and might not be detectable as a result of peak broadening. At higher temperatures, less volatile analytes might be resolved, but the more volatile ones coelute. Thus, no single temperature can be found to effectively separate all sample components. Temperature programming, on the other hand, involves starting the analysis at a lower temperature and gradually increasing it, which eliminates the general elution problem and allows the separation of a wide range of compounds.[1]

Compounds are separated in GC based on their volatility when their polarity aligns with that of the stationary phase, with the most volatile compounds eluting first and the least volatile ones eluting later. When the polarity of the analyte does not match the stationary phase, however, the elution order becomes more complex and is influenced by both polarity and volatility. Following elution from the column, the separated compounds are detected by a detector. The types of GC detectors commonly used in food analysis are discussed in the subsequent section.

## 3.1 Gas chromatography detectors

In GC, the detector's primary function is to detect and quantify the analytes eluting from the column. An ideal detector should be simple to use and robust, while demonstrating high selectivity and sensitivity for the analyte(s) of interest. Sensitivity is defined as the slope of the calibration curve produced by the detector for a certain analyte, and selectivity refers to the detector's ability to respond exclusively to the target analyte, distinguishing it from other components in the sample. The detector should provide appropriate sensitivity and selectivity for the specific application at hand.

Another critical aspect to consider is the detector's linear dynamic range, which is the range over which changes in analyte concentration lead to a proportional changes in detector response. Detectors with a wider dynamic range allow the development of methods capable of measuring both low and high concentrations of analytes in samples without the need for additional steps prior to analysis. Additionally, the data acquisition rate of the detector is very important, particularly in multidimensional chromatography (MDC) and fast chromatography, where the eluting peaks may be very narrow and require rapid detection to accurately capture the data. In the following sections, the GC detectors commonly utilized for chemical analysis of food and feed, and their advancements are briefly discussed.

© Burleigh Dodds Science Publishing Limited, 2025. All rights reserved.

### 3.1.1 Flame ionization detector

The flame ionization detector (FID) is one of the most widely utilized detectors in gas chromatography. It is a hydrocarbon-selective detector, meaning it responds to all compounds containing C-H bonds. Given that carbon is present in most food contaminants and residues, the FID is often considered a universal detector. Its advantages include reasonably high sensitivity, the widest linear dynamic range of all GC detectors (7 orders of magnitude), robustness, straightforward operation, and rapid data acquisition rate.[1] However, as the FID is not selective, it depends almost entirely on chromatographic separation for the identification of compounds. One of its notable advantages is its consistent response to chemical compounds based on the total carbon content. This uniformity makes the FID the detector of choice for analyses where a variety of compounds are targeted and obtaining standards for each compound is not practical. This is particularly useful in the analysis of polycyclic aromatic hydrocarbons (PAHs) and mineral oils in food samples, where a range of compounds is typically encountered rather than a single compound.[12,13]

### 3.1.2 Flame photometric detector

Flame photometric detectors (FPDs) show selectivity for sulfur and phosphorus compounds. These elements are prevalent in pesticides and various other pollutants, making FPDs particularly valuable in food analysis.[1] Their high sensitivity and selectivity have led to widespread application in the detection of organophosphorus pesticides, as extensively documented in the literature.[14,15]

### 3.1.3 Nitrogen-phosphorus detector

Thermionic detectors, commonly referred to as nitrogen-phosphorus detectors (NPD), selectively detect compounds containing nitrogen and phosphorus. Similar to FPDs, NPDs are employed in the analysis of pesticides and other phosphorus-containing chemicals, such as organophosphorus flame retardants.[15,16] They offer several advantages, including high sensitivity and selectivity, straightforward operation, and cost-effectiveness.

### 3.1.4 Electron capture detector

Electron capture detectors (ECDs) are highly selective for compounds that contain electronegative substituents, such as halogens.[1] Many environmental contaminants, including PCBs and certain pesticides, contain these electronegative elements, making ECDs particularly effective for their detection.[17–19] ECDs are known for their high sensitivity and robustness,

© Burleigh Dodds Science Publishing Limited, 2025. All rights reserved.

which allows them to detect compounds at trace levels. Furthermore, due to their high sensitivity, ECDs have been employed for compounds that are not naturally detectable by the detector, such as acrylamides, through the use of derivatization techniques (for more details regarding derivatization refer to the Section 3.4 of this chapter).[20] However, ECDs have a limited linear dynamic range, making them suitable primarily for analyzing low concentrations of target compounds.[1]

### 3.1.5 Mass spectrometry

MS has become a preferred analytical detector for a wide range of GC applications, including the analysis of food contaminants and residues, due to its high sensitivity and exceptional selectivity. The extensive use of MS in both official and research methodologies indicates its significant advantages.[1] Despite its widespread adoption, MS is not as robust or user-friendly as some other GC detectors and is considerably more expensive to acquire and maintain. Moreover, MS can be more prone to matrix effects. It is also not the detector of choice when a range of compounds without the availability of standards is the target of the analysis.[12,16] The sensitivity and selectivity of MS can also vary depending on the specific compound being analyzed. Nevertheless, the remarkable sensitivity and selectivity of MS have contributed to its popularity, and recent advancements have further enhanced its capabilities.[21]

One notable advancement is Tandem Mass Spectrometry (MS/MS), which greatly improves the selectivity and sensitivity of traditional MS. Tandem MS, when coupled with GC systems, offers enhanced performance in analyzing complex food samples and effectively addresses potential matrix effects. As a result, Tandem MS has become a highly effective and popular choice for quantifying many types of contaminants and residues in food.[22–25]

High-resolution mass spectrometry (HRMS) represents another significant advancement in MS technology. Coupled with GC systems, HRMS provides high-resolution capabilities that enable the identification of molecular formulas of analytes without prior knowledge of their structures. This feature makes HRMS particularly valuable for non-targeted analysis.[22] Although LC is more commonly used for non-targeted analysis due to its broader compound coverage, there are documented instances of HRMS being used with GC systems for such purposes. For example, a study by Li et al. demonstrated the use of GC-HRMS to identify compounds in plastic tableware, successfully identifying 85 substances and assessing 14 as high-risk based on their toxicity, and detection frequency.[26] This study illustrates the complementary strengths of GC and HRMS, as GC enhances selectivity and HRMS provides detailed

© Burleigh Dodds Science Publishing Limited, 2025. All rights reserved.

compound identification, which neither technique could achieve independently in the presence of a complex matrix.

### 3.2 Ionic liquid stationary phases

Non-polar stationary phases in GC have traditionally been able to withstand high temperatures up to 400°C for extended periods. In contrast, polar stationary phases have been limited to much lower temperatures, around 280°C at the most.[27] This temperature constraint has been particularly problematic in applications such as multidimensional gas chromatography (MDGC), where higher temperatures are necessary for the rapid elution of compounds from the second dimension. Ionic liquid stationary phases have been introduced as a natural solution to this limitation.

Ionic liquids are organic salts with melting points below 100°C. They exhibit stability at elevated temperatures and can be tailored for specific applications by varying their cations or anions, allowing manufacturers to adjust their selectivity.[28] Unlike conventional polar stationary phases, ionic liquids do not suffer from excessive bleeding at high temperatures.[29] However, despite their advantages, some ionic liquids have been reported as non-biodegradable and potential endocrine disruptors.[30] Further research is needed to better assess the potential environmental and long-term health risks associated with their use in chromatography.

As previously noted, ionic liquid stationary phases are particularly advantageous for MDGC applications in contaminant and residue analysis. For instance, Tranchida et al. employed GC×GC with a triple quadrupole mass spectrometer to analyze 22 herbicides in spearmint essential oils.[31] For additional details on MDGC, refer to Section 6 of this chapter.

### 3.3 Fast gas chromatography to increase analysis throughput

Over the years, various approaches have been developed to enhance the analysis speed of GC systems and thereby increase their throughput. Despite these advancements, the effectiveness of these approaches has been variable, with each method presenting its own set of advantages and limitations. The following section reviews several of these strategies and their applications in the analysis of chemicals in food.

#### 3.3.1 Microbore columns

Conventional GC capillary columns typically have diameters ranging from 0.53 mm to 0.2 mm. To enhance analysis speed, one approach is to increase the mobile phase flow rates. However, exceeding optimal flow rates can

© Burleigh Dodds Science Publishing Limited, 2025. All rights reserved.

significantly impair separation efficiency.[32] Furthermore, higher flow rates may be incompatible with MS due to its limited capacity to handle large gas volumes.

Advancements in capillary column technology have introduced microbore columns, which have diameters of 0.1-0.2 mm. These microbore columns have higher optimum linear velocities of the carrier gas, resulting in shorter analysis times. The reduced diameter and thinner stationary phase coating of microbore columns also lead to less band broadening and increased separation efficiency. Additionally, these columns produce taller and narrower peaks, which can enhance the system's sensitivity.[33]

The benefits of microbore columns are demonstrated in a study by Lee et al., where they analyzed 356 pesticides using a multi-residue method and compared conventional and microbore GC capillary columns.[34] The use of microbore columns reduced the analysis time by half to 18 minutes and resulted in narrower and taller peaks. The taller peaks improved the signal-to-noise ratio, thus increasing sensitivity.

Despite these advantages, microbore columns have notable limitations. Their reduced sample capacity due to the thinner stationary phase means they cannot handle large sample volumes or high concentrations of sample components. Although the taller peaks may suggest increased sensitivity, the overall sensitivity of the method may be compromised by the lower injection volumes.[35] Additionally, methods utilizing microbore columns are generally less rugged and more sensitive to instrumental variations, leading to increased costs and reduced reliability of the overall analytical method.[32]

### 3.3.2 Mid-column backflushing

Mid-column backflushing is an advanced technique in GC designed to prevent system contamination with highly retained components and improve analytical efficiency. Under standard GC conditions, the mobile phase carries compounds through the column to the detector. The next analysis can only be started once all sample components (including the matrix) leave the column. Backflushing involves reversing the flow of the mobile phase to remove compounds still in the column back to the injector, preventing column contamination by low- or non-volatile components. Initially, backflushing was implemented by introducing the reverse gas flow just before the detector, with the reversal initiated after each run to clear non-volatile contaminants.[36] Mid-column backflushing enhances this approach by initiating the flow reversal closer to the injection site, typically partway through the run.[36] This allows the removal of non-target compounds before they travel the entire length of the column. Consequently, mid-column backflushing not only shortens analysis time by eliminating the wait time

© Burleigh Dodds Science Publishing Limited, 2025. All rights reserved.

for these compounds to elute, but also prevents them from reaching the detector, hence prevents its contamination.

The advantages of mid-column backflushing for food contamination and residue analysis are well illustrated by a study conducted by Mastovska et al.[37] They applied mid-column backflushing in the analysis of over 100 pesticides in dietary supplements using GC-MS/MS. This technique resulted in a 50% increase in throughput. They demonstrated that mid-column backflushing not only accelerates the analysis but also enhances the robustness and durability of the method while protecting both the column and the detector from potential contamination.

### *3.3.3 Low-pressure gas chromatography*

Low-pressure gas chromatography (LPGC) is an advanced GC technique in which the major parts of the capillary column are maintained under vacuum or near-vacuum conditions.[38] Reducing the column pressure shifts the optimal flow rates to higher values due to the increased diffusivity of the analytes at lower pressures. This results in faster GC separations. Additionally, compounds can be eluted from the column at lower temperatures, reducing the risk of their thermal degradation. Furthermore, reduced band broadening, and consequently sharper peaks, lead to higher signal-to-noise ratios and improved sensitivity. However, these advantages come with trade-offs, including reduced overall separation efficiency and decreased robustness due to the complexity of the instrumental setup.[39]

A notable benefit of LPGC is its excellent compatibility with MS detectors. While most ionization techniques for GC-MS systems require vacuum conditions, pneumatic resistance of conventional capillary columns does not allow the vacuum to extend deep into the column. In LPGC, a short restrictor column without a stationary phase is installed at the inlet and connected to a longer analytical column with a larger internal diameter. This setup enables the analytical column to operate under near-vacuum conditions. Unlike microbore columns used in fast GC, LPGC allows the use of thicker film stationary phases, which increases the sample loadability of the column and enhances the overall sensitivity of the analysis.[39]

The advantages of LPGC are well-documented in studies focused on food contaminant analysis. For example, Sapozhnikova et al. successfully analyzed 207 types of contaminants and residues, including pesticides, PAHs, and PCBs, in fish meat within 10 minutes by coupling a 5 m restrictor column with an inner diameter of 0.18 mm to a 15 m analytical column with an inner diameter of 0.53 mm.[40] Additionally, Poynter et al. compared LPGC with microbore column approaches and found that, despite a reduction in separation efficiency, LPGC

© Burleigh Dodds Science Publishing Limited, 2025. All rights reserved.

increased sensitivity by a factor of three while halving the analysis time.[41] They concluded that the loss in separation efficiency was worth the analysis time and sensitivity gain by LPGC.

## 3.4 Derivatization in gas chromatography

Polar and low-volatility compounds pose significant challenges for GC analysis. These compounds either fail to reach the GC column, or elute at temperatures exceeding the maximum permissible temperature for GC stationary phases.[42] Historically, due to the limited robustness of alternative techniques and the simplicity and reliability of GC, chemists have addressed this issue by chemically modifying these compounds prior to analysis through a process known as derivatization.

Traditionally, derivatization is conducted in a separate vessel from the GC instrument, a method referred to as off-line derivatization. In this approach, derivatizing reagents are added to the sample, which is then usually heated to facilitate the reaction. Off-line derivatization has the advantage of not requiring modifications to the GC instrument, enabling the analysis of compounds not naturally amenable to GC. However, this method is time-consuming, with potential issues such as analyte loss, contamination during sample processing, and higher consumption of organic solvents. Additionally, derivatized analytes are sometimes highly sensitive to moisture and require careful handling.[43]

Online derivatization represents an advancement aimed at addressing these limitations by integrating the derivatization process into the GC analysis itself. In online or in-port derivatization, the derivatization reaction takes place within the inlet after injection. This approach offers several benefits, including improved reproducibility, reduced analyte loss, lower solvent consumption, and time savings. Various strategies have been developed over the years with distinct designs for GC inlets and sample preparation procedures, each with its own advantages and drawbacks. For more information regarding in-port derivatization strategies refer to the review article by Wang et al.[44]

A notable example of online derivatization is demonstrated by Tzing and Ding et al. in their study of melamine and cyanuric acid – two polar analytes not typically amenable to GC – in milk powder.[45] Although sample preparation to remove matrix interferences and clean up of the samples were still necessary, derivatization was performed directly within the GC injection port. The procedure involved initially raising the temperature to the derivatization temperature, followed by a rapid increase to transfer the derivatized analytes from the inlet to the column.

© Burleigh Dodds Science Publishing Limited, 2025. All rights reserved.

# 4 Liquid chromatography

Liquid chromatography (LC) is a chromatographic technique where the mobile phase is a liquid.[1] LC can be categorized based on the form and type of the stationary phase used. A brief description as well as the major advances regarding different types of LC employed in food chemical analysis are discussed in the following sections.

## 4.1 High-performance liquid chromatography

High-performance liquid chromatography (HPLC) is an advanced analytical technique in which a liquid mobile phase is driven through a column by high-pressure pumps. The sample, introduced as a solution onto the column packed with the stationary phase, interacts with this stationary phase as it passes through the column. This interaction causes the sample components to migrate at different rates relative to the mobile phase, leading to their separation. The separation occurs because different components experience varying degrees of retardation due to their interactions with the stationary phase.[1] A schematic diagram of a typical HPLC system is illustrated in Fig. 2.

In HPLC, there are two primary types of elution based on the composition of the mobile phase. Isocratic elution maintains a constant mobile phase composition throughout the analysis, offering the best resolution of the sample components but inevitably creating the general elution problem discussed earlier. Conversely, gradient elution involves varying the mobile phase composition during the run. Gradient elution may provide a shorter analysis time by optimization to achieve adequate resolution of the components within a reduced timeframe.[2]

### 4.1.1 Instrumentation and column technology

#### 4.1.1.1 Ultra-high-performance liquid chromatography and sub-2 μm particles packings

Since the 1970s, HPLC systems have been capable of reaching pressures up to 400 bar. Typically, HPLC columns had inner diameters of 4.6 mm, and the packing particles ranged from 3 μm to 5 μm in diameter.[46] To enhance separation efficiency, one approach is to reduce particle size. However, smaller particles lead to increased column back pressure, necessitating higher pressures to maintain the same linear flow rates.[47]

Recognizing that higher pressure and smaller packing sizes are crucial for improved performance, new instrumentation with higher pressure capabilities was developed.[48] These instruments, known as UHPLC systems, were introduced commercially in 2004.[49]

© Burleigh Dodds Science Publishing Limited, 2025. All rights reserved.

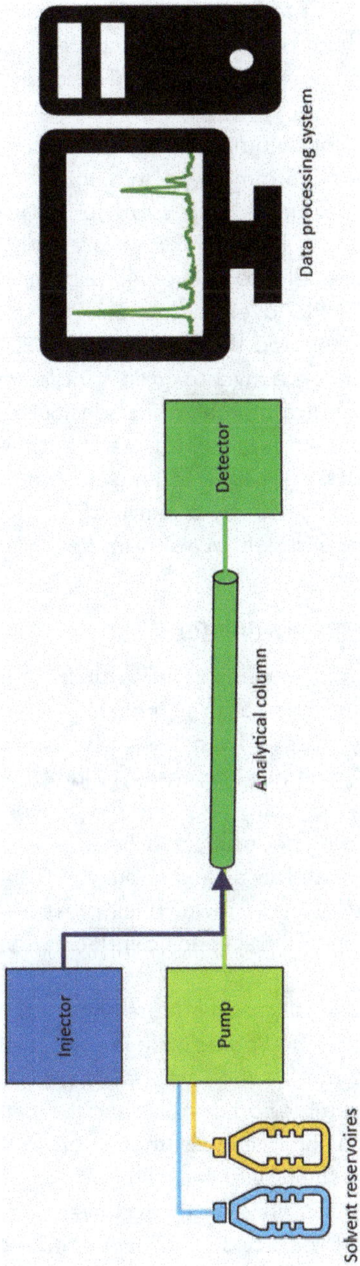


**Figure 2.** Schematic diagram of a typical high-performance liquid chromatography system.

© Burleigh Dodds Science Publishing Limited, 2025. All rights reserved.

Alongside UHPLC, columns packed with sub-2 µm particles were also introduced. These columns typically had a diameter of 2.1 mm and lengths of 50 mm to 100 mm.[49] The reduction in column diameter was mandated by the necessity of better heat dissipation, which decreases radial temperature variability and enhances column efficiency.[50] These columns were packed with 1.7 µm particles.[50]

Over time, a variety of column sizes have become available, with inner diameters ranging from 0.075 mm to 4.6 mm. Initially designed for reversed-phase chromatography, columns with sub-2 µm packings have also been developed for other types of liquid chromatography.[47] For further details on the chromatography types, refer to Section 4.1.2 of this chapter.

The advantages of UHPLC over traditional HPLC are well documented. For instance, Ok et al. demonstrated the benefits of UHPLC in their study on the determination of zearalenone, a mycotoxin, in noodles, cereal snacks and infant formulas.[51] Their results indicated that, while sensitivity remained comparable, UHPLC significantly improved peak width, peak symmetry, and analysis time. Specifically, UHPLC reduced the analysis time by more than half compared to HPLC and decreased peak width by a factor of 7, allowing the analysis of a greater number of compounds within a shorter time frame.

#### *4.1.1.2 Solid core particle packing*

Traditionally, HPLC columns, including those utilizing sub-2 µm packings for UHPLC, have used fully porous particles (FPPs). These particles are irregularly shaped and completely porous from the outer shell to the core.[52] While reducing particle size generally increases column efficiency, a significant challenge is the increased pressure required for smaller particles.

Manufacturers have long been able to produce non-porous, homogeneous, spherical silica particles with low size variability.[52] The spherical and regular shape of these particles results in lower back pressure.[47] However, their use has been limited due to their small specific surface area, which results in lower column efficiency.[52]

The concept of superficially porous (core-shell) particles re-emerged more than a decade ago.[46] Core-shell particles feature a solid, non-porous core and a porous outer shell. The porous shell increases the surface area and facilitates interaction between the sample components and the stationary phase, while the solid core minimizes diffusion distances, thus improving separation efficiency. As a result, larger core-shell particles can produce separations similar to smaller, fully porous particles while requiring lower back pressure.[47] However, the benefits of core-shell particles come with trade-offs, such as reduced stationary phase volume and lower saturation capacity. The thinner the porous shell relative to the total particle diameter, the lower the saturation capacity.[52]

© Burleigh Dodds Science Publishing Limited, 2025. All rights reserved.

Columns packed with core-shell particles can match the separation efficiency of UHPLC systems on standard HPLC systems. When using sub-2 µm core-shell particles with UHPLC systems, even higher efficiencies can be achieved, leading to improved separation performance within a shorter analysis time.[47]

### 4.1.2 Modes of high-performance liquid chromatography

HPLC is renowned for its versatility and widespread applicability in chemical analysis. This versatility arises from the various separation mechanisms it employs, which are categorized based on the types of interactions used to separate different groups of compounds.[1] The primary modes of HPLC used in food analysis and their advances in recent years are outlined in the following sections.

#### 4.1.2.1 Reversed phase (partition) liquid chromatography

Partition liquid chromatography, usually referred to as reversed phase liquid chromatography (RPLC), is a type of HPLC in which the stationary phase is a non-polar liquid and the mobile phase is a polar liquid.[1] Despite the broad range of chromatographic techniques available, this approach is sometimes referred to simply as HPLC in the literature. In reversed phase chromatography, the sample solution is introduced onto the column and separation occurs based on the differential partitioning of components between the stationary and the mobile phases. Components partition to different extent into the stationary phase (which means that they spend different times associated with it), leading to their separation as they elute at different times.[1]

Reversed phase chromatography is favored for its ability to analyze a wide variety of contaminants and residues due to the chemical nature of common analytes.[1] This technique is particularly effective for analyzing diverse contaminants and residues in food, including (but not limited to) pesticides, anatoxin-A, acrylamides, polyfluoroalkyl substances (PFAS), mycotoxins, and anabolic steroids.[53–57] Although other techniques like GC may be preferred for certain analytes (e.g. chloropropanols, PAHs and PCBs), HPLC is also employed for these analyses.[24,58,59] RPLC allows detection of multiple contaminants and residues within a single analytical run and integrates effectively with MS, enhancing the selectivity of the LC methods.[60] Furthermore, when coupled with HRMS, RPLC can support non-targeted analyses.[61] Last but not least, the mobile phase typically consists of water mixed with methanol or acetonitrile, making this technique less environmentally detrimental compared to normal phase chromatography.[1,62]

© Burleigh Dodds Science Publishing Limited, 2025. All rights reserved.

Despite its advantages, RPLC has several drawbacks, including the need for high-purity and costly mobile phases, the requirement for skilled personnel, and the use of sophisticated and expensive instrumentation. Consequently, RPLC is generally not suitable for field applications and is predominantly employed in academic or commercial laboratories staffed by highly trained personnel.

In comparison to GC, RPLC offers a broader range of analytical capabilities. GC often requires derivatization of polar and/or high-boiling-point compounds to facilitate their analysis, whereas RPLC typically accommodates these compounds without such modifications. Derivatization in RPLC is usually performed to address detector limitations or to enhance method sensitivity.[63] While both LC and GC can analyze multiple compounds rapidly, RPLC's ability to handle a broader range of analytes make it the preferred choice for many multi-contaminant and multi-residue analytical methods.[64]

### 4.1.2.2 Normal phase chromatography

Normal phase chromatography, introduced by Tswett in the early twentieth century, was the first chromatographic technique developed. Initially, Tswett's method relied on gravity to move the mobile phase through the column. In contrast, modern HPLC systems utilize high-pressure pumps to propel the mobile phase, significantly improving separation efficiency and speed. In normal phase chromatography, the mobile phase is non-polar, while the stationary phase is polar. The primary mechanism of analyte retention in this mode is adsorption. Normal phase chromatography is particularly suited for separating mixtures with polar components.[1]

Normal phase chromatography is far less popular than reversed phase chromatography.[1] However, a type of normal phase chromatography, known as hydrophilic interaction liquid chromatography (HILIC) has been the subject of a lot of recent advances in chromatography.

#### 4.1.2.2.1 Hydrophilic interaction liquid chromatography

HILIC is a subtype of normal phase chromatography where the stationary phase is polar, and the mobile phase consists of a mixture of water and a significant proportion of a water-miscible organic solvent, such as acetonitrile. Although there is no universal consensus on the exact mechanism of separation, the most widely accepted model suggests that water molecules in the mobile phase create an aqueous layer around the polar stationary phase. The separation process primarily results from the different degrees of partitioning of the analytes between this aqueous layer and the organic solvent-rich mobile phase.[65,66]

© Burleigh Dodds Science Publishing Limited, 2025. All rights reserved.

HILIC is particularly effective for compounds that are poorly retained by RPLC due to their high polarity.[67] The complementary nature of HILIC is very valuable. Additionally the solvents used in HILIC are less problematic compared to those in conventional normal phase chromatography.[62]

The efficacy of HILIC has been demonstrated in various studies. For instance, Francesquett et al. developed a UHPLC-HILIC-MS method for analyzing quaternary ammonium pesticides (such as paraquat, diquat, mepiquat, and chlormequat) in just 4 minutes.[68] Although these pesticides can be analyzed using a normal phase system with modifications to the mobile phase, these modifications are often incompatible with MS.

Additionally, HILIC offers the advantage of mobile phase compatibility with reversed phase chromatography, while providing orthogonal selectivity. This orthogonality is particularly useful in multidimensional liquid chromatography (MDLC), where combining different separation modes can enhance overall analytical performance.[69]

### 4.1.2.3 Ion chromatography

Ion chromatography (IC) is a specialized type of HPLC designed to separate and analyze ionic species.[1] In IC, the column is packed with anion-exchange or cation-exchange resins, enabling the separation of both inorganic ions and small organic ionic species, such as amino acids.[70] Detection methods commonly used with IC include conductivity detectors, electrochemical detectors, and MS.[1,70]

Recent advancements in IC instrumentation have focused on increasing automation and sensitivity. Modern IC systems now feature online eluent generation, which produces the eluent through electrochemical reactions.[71] Additionally, significant progress has been made in suppressor technology.[72] Suppressors are used to remove interfering ions from the mobile phase, thereby enhancing the sensitivity of ion detection.[1]

Suppressors operate via various mechanisms but all have limited capacity and require regeneration after a certain number of runs.[72] There are two primary types of suppressors: chemically regenerated and electronically regenerated. Chemically regenerated suppressors rely on an external chemical regeneration solution that must be supplied to the system.[73] In contrast, electronically regenerated suppressors utilize electrochemical reactions to produce the regeneration solution internally. Recent technological advancements have integrated eluent generation and regeneration processes, resulting in advanced IC systems that minimize the need for external consumables.[74]

Despite these advancements, IC systems are not yet capable of achieving the high pressures seen in UHPLC systems for reversed phase chromatography. This limitation is partly due to the use of alkaline or acidic mobile phases in IC

© Burleigh Dodds Science Publishing Limited, 2025. All rights reserved.

systems, which restricts the choice of materials that can withstand these harsh chemicals.[72] Another reason is the poor mechanical stability of typical ion-exchange resin particles.

### 4.1.2.4 Mixed mode liquid chromatography

Mixed-mode liquid chromatography was developed to address the limitations of traditional chromatography techniques, particularly reversed phase chromatography, in separating highly polar components within complex mixtures that also contain non-polar and mid-polar components. Mixed-mode stationary phases incorporate both reversed phase and ion-exchange functionalities. This dual functionality allows the separation of highly polar and ionic compounds, which are typically challenging to retain with reversed phase columns alone.[75]

By employing two separation mechanisms within a single stationary phase, mixed-mode chromatography enables the analysis of complex mixtures with a broad range of polarities in a single run.[76] The effectiveness of this approach was demonstrated in a study by Montes et al., which involved separation and quantification of a range of persistent and mobile organic contaminants in water.[77] Despite the contaminants' broad polarity range and the presence of negatively charged species, the study successfully analyzed all contaminants in a single analytical run. This outcome highlights that highly polar and charged species, which are not retained on traditional reversed phase columns, can be effectively analyzed using mixed-mode chromatography.

### 4.1.2.5 Affinity chromatography

Affinity chromatography is a technique in which the stationary phase contains specific binding sites such as antibodies that selectively and reversibly interact with a target compound or a group of compounds. Following the isolation of the target analyte within the affinity column, other sample components are removed. The analyte is then eluted from the column by altering the mobile phase composition, such as by adjusting the pH. This method offers high selectivity and can almost completely eliminate matrix components, thereby enhancing the sensitivity of the analysis.[78] However, it is relatively expensive, and its applicability depends on the availability of a suitable affinity chromatography column.[1]

In addition to its efficiency in sample cleanup, AC is particularly valuable in isolating compounds with specific biological functions. This is achieved by utilizing biological receptors as the stationary phase. For example, a study by Jondeau-Cabaton et al. employed affinity chromatography columns with

© Burleigh Dodds Science Publishing Limited, 2025. All rights reserved.

estrogen receptors to selectively bind compounds in samples that interact with these receptors, identifying potential endocrine disruptors. These molecules were subsequently characterized using HRMS.[79]

### 4.1.3 High-performance liquid chromatography detectors

The choice of detector in HPLC is primarily influenced by the nature of the analyte. While detectors used in UHPLC and standard HPLC are largely similar, UHPLC systems require high-speed detectors with low internal volume due to their high efficiency. Commonly used detectors in liquid chromatography include ultraviolet-visible (UV-Vis) detectors, evaporative light scattering detectors (ELSD), fluorescence detectors (FLD), electrochemical detectors, and MS. Each of these detectors has unique characteristics and applications, which are briefly discussed below.

#### 4.1.3.1 Ultraviolet-visible and fluorescence detectors

For UV-Vis and FLD, the analyte must have a chromophore capable of absorbing light. UV-Vis detectors are widely utilized due to their ability to detect many compounds that absorb UV light. However, they are often limited by lower sensitivity, necessitating extensive sample preparation for trace analysis.[1]

In contrast, FLDs are among the most sensitive options available for LC, with several studies demonstrating detection limits in the sub ng per g range.[80–82] The challenge with FLD is that intrinsic fluorescence in analytes is relatively rare. Compounds such as some mycotoxins, including aflatoxins and ochratoxin A, naturally exhibit fluorescence and can be effectively analyzed using FLD.[83] For other analytes, derivatization can be used to introduce or enhance fluorescence. Although derivatization can improve sensitivity and selectivity, it complicates sample preparation and may introduce additional sources of error. Traditional derivatization occurs before analysis (pre-column), but recent advancements introduced post-column derivatization, which enhances both sensitivity and selectivity and is well-suited to LC systems.[63,84]

#### 4.1.3.2 Evaporative light scattering detectors and condensation nucleation light scattering detector

In evaporative light scattering detection (ELSDs), the eluate from the chromatographic column is nebulized to create a mist, followed by the evaporation of the solvent. The remaining non-volatile components then pass through a laser beam, and the scattering of this beam by the non volatile particles is measured as the detector response. ELSD offers a universal and relatively uniform response to all non-volatile compounds, which is a significant

© Burleigh Dodds Science Publishing Limited, 2025. All rights reserved.

advantage over the UV-Vis detector, a commonly used detector that lacks this universality. However, it is generally less sensitive compared to other detection methods, and its response is non-linear.[1]

Condensation nucleation light scattering detectors (CNLSD) are similar to ELSDs but with enhanced sensitivity. In CNLSD, after solvent evaporation, a stream of gas saturated with a solvent is introduced to the particles, causing them to grow in size.[85] This increase in size results in a sensitivity enhancement of one or two orders of magnitude compared to ELSD.[86]

#### 4.1.3.3 Corona charged aerosol detector

In a corona charged aerosol detector (CAD), the eluate is nebulized, and after the evaporation of the mobile phase, the resulting particles are exposed to a stream of positively charged nitrogen gas generated by passing nitrogen over a high-voltage platinum wire generating a corona discharge. As the particles pass through the stream of charged nitrogen, they acquire some of the charge, leading to a loss of charge in the nitrogen stream. This loss is recorded as the detector response.[86] CAD is highly sensitive and provides a universal response to non-volatile compounds, with a uniform response across different non-volatile substances. Additionally, CAD offers a wide dynamic range, making it a versatile choice for various analytical applications.[87] In the context of food contamination analysis, CAD has been effectively employed for detecting bisphenol analogues. A study by Wang et al. compared UV-Vis, CAD, and ELSD and found that, although the UV-Vis detector exhibited higher sensitivity, CAD offered a significant advantage by providing a uniform response irrespective of the compound's structure.[88] Moreover, CAD demonstrated a sensitivity five times greater than that of ELSD.

#### 4.1.3.4 Electrochemical detectors

Electrochemical detectors operate based on electroanalytical principles, detecting compounds with oxidizable or reducible functional groups. They offer high sensitivity, selectivity, and a broad range of applications but are not compatible with gradient elution, which limits their versatility.[1] Despite their advantages, EDs are more commonly used in sensors for food contaminants detection rather than as HPLC detectors.[89,90]

#### 4.1.3.5 Mass spectrometry

MS coupled with LC is a powerful detection method. Due to the limited capacity of mass spectrometers to handle large volumes of gas formed from liquid mobile phases, UHPLC systems with smaller diameter columns and smaller

© Burleigh Dodds Science Publishing Limited, 2025. All rights reserved.

packing particles are typically used. The combination of LC with MS, especially LC-MS-MS, leverages the high selectivity and sensitivity of MS alongside the broad analytical range offered by LC for various analytes.[60] HRMS further enhances LC capabilities by offering precise compound identification, making it invaluable for non-targeted analyses.[61] This combination may be more effective than GC coupled with HRMS due to the superior analyte coverage provided by LC. Furthermore, certain LC-HRMS instruments, such as LC-ESI-QOrbitrap®-MS, can simultaneously quantify a targeted group of compounds while also collecting 'full-scan data', enabling the screening of non-targeted compounds.[91,92] For more detailed information regarding MS systems refer to the ninth chapter of this book, 'Advances in techniques for detecting chemical contaminants: spectrometry'.

## *4.2 Thin layer chromatography*

Thin layer chromatography (TLC) is a technique used to separate the components of a mixture on a thin layer of stationary phase, which is typically adhered to a support material such as glass or metal. In this process, the sample is dissolved in a volatile solvent and applied to the stationary phase. Following the evaporation of the solvent, one end of the TLC plate is placed in a mobile phase composed of an organic solvent or a mixture of solvents. Capillary action causes the mobile phase to migrate toward the opposite end of the plate. The components of the sample move at different rates depending on their affinity for the mobile phase and the stationary phase, resulting in their separation.[1]

After separation, the individual bands can be scraped off of the stationary phase and extracted for further analysis using other techniques.[93] Additionally, the distance travelled by a compound in comparison to the mobile phase is unique to different compounds and can be used to identify compounds with relative selectivity (e.g. by using standard compounds as a reference).[1] After the separation, different techniques are used to reveal the separated chemicals on the TLC plate. Chemical reagents such as iodine can be used to reveal the chemicals on the plate.[1] Additionally, if a compound absorbs UV or visible light, spectroscopic methods can be employed for identification and quantification.[94]

Mycotoxins, including aflatoxins and ochratoxins, are commonly analyzed using TLC.[95–103] Notably, one of the official methods recommended by AOAC INTERNATIONAL for quantifying aflatoxins is based on TLC.[94] This method is valued for its simplicity, speed, and cost-effectiveness. Its straightforward procedure and minimal equipment requirements also make it suitable for field applications. However, TLC typically achieves detection limits in the low parts per billion (ppb) range.[53] While this level of sensitivity may suffice for

© Burleigh Dodds Science Publishing Limited, 2025. All rights reserved.

certain products, it may fall short of the requirements established by regulatory bodies for products with lower maximum permissible levels, thereby limiting its application in quantifying aflatoxins in such products.[53,104]

### 4.2.1 High-performance thin layer chromatography

High-performance thin layer chromatography (HPTLC) is an advanced, automated variant of TLC designed to enhance reproducibility and sensitivity. HPTLC offers improved performance over conventional TLC, including higher sensitivity and lower sample volume requirements.[105] Nevertheless, the technique's higher cost, more complex sample preparation, and need for specialized equipment and trained personnel make it less practical for field use.

## 5 Supercritical fluid chromatography

Supercritical fluid chromatography (SFC) is a chromatographic technique where the mobile phase is a supercritical fluid (SF). A supercritical fluid is defined as a substance at a temperature and pressure above its critical point, where it cannot exist as a liquid regardless of pressure. Supercritical fluids exhibit unique properties that combine characteristics of both gases and liquids. They possess densities significantly higher than gases, but lower viscosities compared to liquids. Additionally, their diffusivity is much higher than that of liquids.[1]

Modern SFC instrumentation mirrors that of HPLC systems. The most commonly used mobile phase in SFC is supercritical carbon dioxide (SC-$CO_2$). SC-$CO_2$ is both environmentally friendly and cost-effective compared to traditional HPLC solvents. By adjusting the pressure of SC-$CO_2$, the elution process can be accelerated. Unlike GC, where the gas serves primarily as a medium to move analytes, SFC utilizes the supercritical fluid as an active separation medium. This allows modifications of the mobile phase by adding small amounts of organic solvents to adjust the polarity and selectivity, thereby expanding the types of compounds that can be effectively separated.[106]

Moreover, the high diffusivity of the supercritical state leads to more efficient separations compared to HPLC. The lower viscosity of the mobile phase allows higher flow rates at significantly reduced pressures, resulting in faster analyses. Recent advancements include the development of high-performance supercritical fluid chromatography (HPSFC) columns packed with sub-2 μm particles, which further enhance the efficiency of these systems.[107]

SFC with SC-$CO_2$ closely resembles normal-phase HPLC, making it less suitable for non-polar analytes. However, by adjusting the mobile phase with

© Burleigh Dodds Science Publishing Limited, 2025. All rights reserved.

small amounts of organic solvents, the scope of SFC can be broadened to include both reversed-phase and normal-phase LC compounds. Additionally, SFC can be coupled with a range of detectors similar to those used in HPLC, including UV, MS, and ELSD detectors. This versatility allows for the analysis of a wide variety of compounds.[108]

Recent studies have highlighted the effectiveness of SFC in chemical analysis within food matrices. For instance, Cutillas et al. evaluated SFC coupled with MS for analyzing samples prepared by the QuEChERS method. They identified and quantified 164 pesticides in three types of vegetables, achieving high sensitivity for over 90% of the pesticides with minimal matrix effects. The total runtime for the analysis was under 8 minutes.[109] Similarly, Ishibashi et al. demonstrated that coupling of SFC with HRMS allowed the screening of 373 pesticides with a wide range of polarities and molecular weights in QuEChERS extracts without further clean-up, with a total run time of just 20 minutes.[110] These studies underscore the advantages of SFC in providing rapid, sensitive and efficient analysis of complex samples, comparable to or surpassing traditional LC and GC methods.

Another area where SFC has demonstrated significant advantages is in chiral chromatography. Chiral separation is crucial for analyzing pesticides and other biologically active contaminants and residues, because different enantiomers can have varying effects and potencies in biological systems.[111] Traditional LC methods for chiral chromatography are generally slow. However, SFC's high efficiency and rapid separation capabilities make it a superior choice for chiral separations. This is illustrated by a study conducted by Chen et al., where they successfully separated the enantiomers of isofenphos-methyl, a pesticide, in just 3.5 minutes, achieving baseline separation of the peaks.[112] In contrast, the same separation would have taken over 70 minutes with much lower resolution on a traditional normal-phase chiral column. This efficiency has been consistently demonstrated in other studies as well.

Despite these clear advantages in terms of chromatographic performance, resource conservation and environmental benefits, SFC is not as widely established as GC or LC. Its limited adoption can be attributed to its relatively recent successful implementation and the need for further optimization of analytical procedures. Furthermore, the additional complexities and numerous optimizable parameters in SFC require more extensive and time-consuming optimization.[113] However, as SFC technology evolves and becomes more comparable to LC analyses in terms of sample preparation and detection technologies, its use is likely to expand, offering even more applications in various analytical fields.

© Burleigh Dodds Science Publishing Limited, 2025. All rights reserved.

## 6 Multidimensional chromatography

While chromatography techniques are highly effective, and the availability and progress of selective detectors such as MS/MS extend their capability for complex samples, some samples such as food often contain too many components to be fully separated in a single dimension. To obtain more detailed information about sample components, MDC can be employed.

MDC is a technique in which the eluate from an initial separation undergoes one or more additional separations using a method with a different retention mechanism. The degree of difference in separation mechanisms between these dimensions is referred to as orthogonality. Greater orthogonality enhances the ability of the MDC system to differentiate and separate compounds.[114]

MDC can be approached in different ways. In 'heart-cutting,' specific fractions of the eluate from the first dimension ($^{1}$D) are collected and then subjected to a subsequent separation. In contrast, comprehensive MDC involves continuously sampling the eluate from $^{1}$D at regular intervals and injecting these samples into a second dimension ($^{2}$D) for further separation. While comprehensive MDC offers a more detailed analysis of the entire sample and its components, heart-cutting allows for more focused information to be obtained from specific fractions of $^{1}$D. Furthermore, comprehensive MDC is more complex than heart-cutting due to the necessity of real-time collection and injection of eluate from $^{1}$D into $^{2}$D, which imposes limitations on flow rates and analysis times.[115]

MDC can be performed offline, where the second separation is conducted separately after collecting the eluate from $^{1}$D. Recent advancements in chromatography systems have enabled online MDC, where fractions of interest are automatically collected and injected into the second dimension for immediate separation. The modulator, a crucial component placed between the columns, is responsible for transferring the eluate from $^{1}$D to $^{2}$D. While both heart-cutting and comprehensive MDC can be performed offline, advancements in modulator technology have enabled the practical execution of online heart-cutting and, more importantly, comprehensive MDC, significantly improving efficiency and analysis time.[116]

MDC can be categorized based on the type of chromatography and its dimensional configurations. This classification includes MDGC, MDLC, and hybrid systems that combine both GC and LC. The following sections provide a brief overview of these techniques and their applications in the analysis of food contaminants and residues.

© Burleigh Dodds Science Publishing Limited, 2025. All rights reserved.

## 6.1 Multidimensional gas chromatography

The introduction of MDGC represents a significant advancement in GC for food analysis. Among the various types of MDGC, two-dimensional gas chromatography (2DGC) is the most commonly employed. As previously discussed, 2DGC can be implemented as either heart-cutting (GC-GC) or comprehensive two-dimensional gas chromatography (GC×GC).

In GC-GC, specific fractions of the eluate from $^1$D are collected and subsequently injected into the $^2$D. GC-GC is particularly useful for the targeted analysis of compounds that cannot be effectively separated using a single dimension. For instance, it has proven advantageous for analyzing enantiomers in complex matrices, where initial separation from the matrix is required before further separation on a chiral column.[117] Despite these advantages, the early availability and feasibility of GC×GC have led to its more widespread adoption in studies related to food chemical analysis.

GC×GC utilizes two capillary columns with orthogonal selectivities, connected by a modulator. The modulator periodically collects, refocuses, and injects the analytes from the $^1$D column into the $^2$D column.[118] This setup enhances separation and peak capacity, but these benefits come with trade-offs, including typically longer analysis times, reduced robustness, and the need for skilled operators. Additionally, GC×GC involves more complex instrumentation and requires high data acquisition rate detectors, which can increase the overall cost of analysis.[118]

GC×GC in food chemical analysis not only mitigates matrix effects but also enhances both selectivity and sensitivity of the method. Given the complexity and diverse range of compounds classified as pesticides, GC×GC is particularly effective for pesticide analysis. For example, a study by Liu et al. demonstrated that GC×GC could separate pesticides that were unresolved in the $^1$D and achieve significant improvements in sensitivity.[119] Although the FPD used in the study is not the most selective by itself, the use of GC×GC enhanced selectivity, thereby improving compound identification accuracy. In another study, Dallüge et al. developed a GC×GC method tailored for analyzing vegetable extracts.[120] This approach enabled the identification and quantification of 58 pesticide residues, something that single-dimension GC could not achieve due to the sample's complexity and matrix interference.

GC×GC is also frequently applied to analyze persistent organic pollutants, such as PCBs and PAHs, in food samples. For instance, Kalachova et al. successfully analyzed 16 PAH compounds and 18 PCB compounds in fish tissue using GC×GC coupled with MS.[121] Additionally, this method significantly improved the detection of non-target compounds due to the enhanced separation efficiency offered by GC×GC.

© Burleigh Dodds Science Publishing Limited, 2025. All rights reserved.

## 6.2 Multidimensional liquid chromatography

As in MDGC, MDLC involves subjecting the eluate from an initial LC separation to one or more additional LC separations utilizing different separation mechanisms. Two-dimensional liquid chromatography (2DLC) is the most common form of MDLC.[122]

A basic form of two-dimensional chromatography can be observed in TLC. In two-dimensional TLC, after initial separation, the TLC plate is rotated by 90° and developed with a second solvent system for further separation. Although this method complicates the procedure, it extends the applicability of TLC to more complex samples.[123] The advantages of MDLC become more pronounced in HPLC systems.

Due to the complexity of comprehensive two-dimensional liquid chromatography (LC×LC), heart-cutting liquid chromatography (LC-LC) is predominantly used in food contaminant and residue analysis. However, LC×LC has proven beneficial in food verification and authenticity, with extensive literature supporting its use.[124–129] Despite the limited usage for food contaminant analysis, LC×LC has been proven to be beneficial for non-target analysis in other matrices. For example, a study by Saint Germain et al. demonstrated that LC×LC coupled with HRMS significantly improved non-target analysis of pharmaceutical contaminants in water, increasing the number of identified compounds from 130 to 240.[130] As LC×LC systems become more accessible and user-friendly, applications of this technique are expected to expand.

In the analysis of food contaminants and residues, LC-LC is frequently employed to simplify or eliminate the need for sample cleanup. Additionally, this technique can separate structurally similar compounds, allowing further separation in the second dimension. When intended to eliminate sample matrix, the $^{1}$D serves as a cleanup stage, selectively retaining matrix components and other non-target compounds. The eluate from the $^{1}$D is then analyzed in the $^{2}$D. An early study by Pascoe et al. illustrated the potential of LC-LC to analyze multiple compounds with minimal sample cleanup and reduced matrix effects.[131] Subsequent studies have confirmed the effectiveness of LC-LC in achieving quantitative results with reduced sample preparation.

One notable application of LC-LC is in pesticide analysis. For instance, Kittlaus et al. coupled a HILIC column as $^{1}$D with a reversed-phase column as $^{2}$D. This column combination resulted in the elution of all pesticides in the study in a single fraction at the beginning of the run from $^{1}$D, enabling the analysis of over 300 pesticides in acetonitrile extracts from various food matrices without the need for additional cleanup steps.[132] This method outperformed the QuEChERS sample clean-up method in reducing matrix effects, though neither method fully eliminated them.[133]

© Burleigh Dodds Science Publishing Limited, 2025. All rights reserved.

Mycotoxin analysis is another area where LC-LC has been beneficial. AC has been the method of choice for sample cleanup due to the low concentrations of mycotoxins and the complexity of the sample matrix.[53] However, AC is both costly and specific, limiting its use to a narrow range of mycotoxins and requiring the availability and replacement of the AC column to expand the method's scope.[134] As an alternative, several studies have demonstrated that LC-LC, employing two reversed phase columns in a two-dimensional configuration, can effectively analyze mycotoxins. In this setup, a C8 or C18 column is used as the $^{1}$D, while a phenyl-containing stationary phase is employed as the $^{2}$D. Despite both columns being reversed phase, they offer distinct selectivities: the C8 or C18 column differentiates analytes based on polarity and lipophilicity, whereas the phenyl column separates compounds through $\pi$–$\pi$ interactions. This combination allows the successful quantification of multiple mycotoxins with high sensitivity in a single analytical run, eliminating the need for extensive sample cleanup procedures.

Additionally, LC-LC has been employed for multi-class contaminant and residue analysis, significantly enhancing the number of analytes that can be analyzed in a single run. This approach not only increases the number of detectable compounds to several hundred per run, but also reduces analysis time by minimizing or even eliminating the sample cleanup step.[132,135]

Although not widely used, LC×LC has been employed in studies for pesticide residue analysis in food. One such study by Donato et al. utilized two reversed-phase columns with distinct selectivities for the $^{1}$D and $^{2}$D.[127] This approach enabled the separation and identification of 43 polyphenols, which serve as indicators of wine quality and origin. Additionally, the study quantified monuron, a herbicide commonly used on grapes. Notably, the analysis did not include a sample cleanup step; instead, matrix separation from the analyte was achieved using the 2D system. While this study did not fully exploit the resolving power of LC×LC for residue analysis, it demonstrated the potential of this technique for analyzing contaminants and residues in complex food matrices.

## 6.3 Multidimensional liquid chromatography - gas chromatography

A notable advancement in chromatographic techniques is the utilization of GC following an LC separation. In this multidimensional approach, LC as the first dimension effectively separates compounds based on various selectivity mechanisms only available to LC systems. Subsequently, volatile compounds are analyzed using the high efficiency and robustness of GC. This combination leverages the versatility and large sample capacity of LC systems with the efficiency and reliability of GC systems.[136]

© Burleigh Dodds Science Publishing Limited, 2025. All rights reserved.

Typically, this technique is employed in heart-cutting mode rather than comprehensive mode, largely due to the technical challenges associated with comprehensive liquid chromatography–gas chromatography (LC×GC). These challenges include extended analysis times for GC, lengthy cooling periods after each run, and overall prolonged analytical durations.[137] Furthermore, for comprehensive analysis, alternative techniques such as LC×LC and GC×GC generally provide superior separation and more detailed information.

In the context of food analysis, the development of this technique has been particularly focused on the determination of mineral oil hydrocarbons (MOHs) in food samples.[138] The European Union currently recommends LC-GC as a method for monitoring MOHs in food products.[139] MOHs, being lipophilic molecules, pose a challenge in separating them from other abundant naturally occurring lipophilic compounds, such as triglycerides, or naturally occurring n-alkanes in plants.[137] In this application, the first dimension of LC is employed to isolate MOHs from other lipophilic substances in the matrix, followed by GC separation for further analysis of the isolated aliquot. This exemplifies offline MDC, which is advantageous because it requires no modifications to the LC or GC systems, making implementation in laboratories more straightforward.

Recently, more robust, commercially available systems have emerged that facilitate the online transfer of aliquots from LC to GC for further separation.[137] These systems enhance throughput and improve reproducibility, making them more suitable for routine analysis.

## 7 An overview of chromatographic methods of analysis from the perspective of regulatory bodies

The United States Food and Drug Administration (FDA) publishes a comprehensive set of manuals, detailing the analysis of contaminants in various food samples.[140] These manuals outline guidelines for each step of the analytical process, from sampling to the final determination of chemical residues. Additionally, the FDA provides general guidelines for the validation of chemical analytical methods, which can be employed to validate novel analytical techniques.[141] In contrast, the European Union typically does not publish official methods of analysis; instead, it establishes criteria for groups of analytes.[142] While recommendations for analytical methods for various compounds are provided from European Union agencies such as European Food Safety Authority (EFSA), any method that meets these criteria can be utilized.[143]

Established chromatographic techniques, such as GC and LC, are widely accepted by both regulatory agencies and are regarded as standard methods for analyzing various contaminants and residues. Conversely, newer chromatographic methods, such as SFC and MDC, are not recognized as

© Burleigh Dodds Science Publishing Limited, 2025. All rights reserved.

standard analytical methods for any contaminants or residues, with no mention of SFC or MDC in FDA manuals.[140]

While the European Union does not explicitly endorse these newer techniques, its performance-based approach to official methods does not preclude their use. The recommendation of LC-GC, and GC×GC by EFSA as the preferred methods of analysis for MOHs in food suggests that these emerging chromatographic techniques can be effectively implemented.[143] This is particularly relevant for MDC, given the recent advancements and current availability of robust commercial systems suitable for routine analysis. It is anticipated that the utilization of these innovative techniques will increase as their availability and reliability improve, owing to their advantages over traditional chromatographic methods.

## 8 Conclusion

As previously discussed, sample preparation is often critical for analyzing food contaminants and residues due to the inherent complexity of these samples. Recent advancements in sample preparation techniques have significantly enhanced the sensitivity and reliability of analytical methods. Nevertheless, advancements in chromatography technologies have reduced the reliance on extensive sample preparation while maintaining high-quality analysis. These improvements can lead to higher throughput, more economical, more eco-friendly, and more robust analysis.

GC is a dependable analytical technique for detecting food contaminants and residues, particularly when analytes are volatile or semi-volatile. Recent innovations have made GC faster, more accurate, and more robust. Key improvements in this technology have focused on mitigating the effects of complex matrices, such as those found in food, and addressing matrix interference.

One significant development is multidimensional GC, which utilizes multiple columns and online injection with modulation hardware to achieve further separation of analytes. By employing columns with varying polarities, multidimensional GC enhances separation and peak capacity, particularly for compounds that elute within similar temperature ranges.

While GC remains the preferred method for volatile compounds, it is less effective for non-volatile and polar compounds, which typically require derivatization to be analyzed by this technique. Despite advancements in derivatization techniques, this process can complicate sample preparation and analysis. In such cases, LC offers a versatile alternative, capable of analyzing both volatile and non-volatile compounds, as well as large biomolecules.

In recent years, advancements in column and instrumentation technology have significantly enhanced LC. The introduction of UHPLC systems, featuring

© Burleigh Dodds Science Publishing Limited, 2025. All rights reserved.

sub-2 μm particle column packings, has led to increased separation efficiency and reduced analysis time. Additionally, the development of columns with SCP packings has enabled conventional HPLC systems to achieve similar separation efficiencies.

Despite these advancements in stationary phases and HPLC instrumentation, some complex samples cannot be effectively separated using a single stationary phase. MDLC addresses this challenge by providing enhanced peak capacity and further separation capabilities, making the analysis of highly complex samples feasible. MDLC also reduces the need for extensive sample preparation to remove interfering compounds. Although there is still potential for further development in the application of these techniques for food analysis, the increasing availability of these systems is likely to lead to their more widespread use in analyzing food contaminants and residues.

The integration of advanced mass spectrometric techniques (e.g. tandem MS) and high-resolution (HR) instruments like Orbitrap or quadrupole-time-of-flight (QTOF) with chromatographic techniques has yielded significant benefits for both technologies. MS benefits from the gradual introduction of the components of complex samples via chromatography, which helps mitigate the matrix effects that can compromise data reliability. Similarly, MS enhances the sensitivity and selectivity of chromatographic methods. The combined use of LC and HRMS, for example, exemplifies the synergistic benefits of these techniques, enabling non-targeted analysis of chemicals in food samples, an approach that is not achievable with either technique alone.

The continued evolution of chromatographic and mass spectrometric technologies has revolutionized the analysis of food contaminants and residues. These advancements have significantly improved the efficiency, sensitivity, and reliability of analytical methods. These innovations not only enhance the capability to analyze complex food matrices, but also reduce the need for extensive sample preparation, enabling more accurate and comprehensive contaminant detection. As these technologies advance and become more accessible, they promise to further elevate the standards of food safety and quality analysis, addressing emerging challenges in the field with greater precision and efficiency.

## 9 Where to look for further information

**The following review articles provide a good overview of the subjects:**

- Brandi, J., et al. (2024). Analysis of veterinary drugs and pesticides in food using liquid chromatography-mass spectrometry. *TrAC Trends in Analytical Chemistry*, *179*, 117888. https://doi.org/10.1016/j.trac.2024.117888.[144]

© Burleigh Dodds Science Publishing Limited, 2025. All rights reserved.

- Liu, L., et al. (2020). The application of supercritical fluid chromatography in food quality and food safety: an overview. *Critical Reviews in Analytical Chemistry, 50*(2), 136–160. https://doi.org/10.1080/10408347.2019.1586520.[145]
- Nolvachai, Y., et al. (2023). Foods and contaminants analysis using multidimensional gas chromatography: an update of recent studies, technology, and applications. *Analytical Chemistry, 95*(1), 238–263. https://doi.org/10.1021/acs.analchem.2c04680.[115]

**The following books and book chapters provide comprehensive information about the analytical techniques discussed:**

- Muntean, E., Chapter 17 - Applications of ion chromatography in food analysis. In P. Nesterenko, et al. (Ed.), *Ion-exchange chromatography and related techniques (part of the handbooks in separation science)* Elsevier, 2024, pp. 351–369. https://doi.org/10.1016/B978-0-443-15369-3.00013-4.[146]
- Mostafa, A., et al., Multidimensional liquid chromatography applied to the analysis of pesticides. In T. Tuzimski & J. Sherma (Eds.), *High performance liquid chromatography in pesticide residue analysis (part of the chromatographic science series)* CRC Press, Taylor & Francis Group, 2015, pp. 393–410.[147]
- Mondello, L. (Ed.), *Comprehensive chromatography in combination with mass spectrometry (Part of Wiley-Interscience Series in Mass Spectrometry.* Wiley, 2011.[122]

**The following organizations provide useful information regarding the subjects:**

- European Commission – Food chemical safety (https://food.ec.europa.eu/food-safety/chemical-safety_en).
- U.S. Food & Drug Administration – Food chemical contaminants and pesticides (https://www.fda.gov/food/chemical-contaminants-pesticides).

## 10 References


1. Skoog, D. A., et al. (2018). *Principles of instrumental analysis* (Seventh edition). Cengage Learning.
2. Harris, D. C. & Lucy, C. A. (2020). *Quantitative chemical analysis* (Tenth edition). Macmillan Learning.
3. *Contaminants–European Commission*. (n.d.). Retrieved October 28, 2024, from https://food.ec.europa.eu/food-safety/chemical-safety/contaminants_en


© Burleigh Dodds Science Publishing Limited, 2025. All rights reserved.

4. *Maximum Residue Levels–European Commission*. (n.d.). Retrieved October 28, 2024, from https://food.ec.europa.eu/plants/pesticides/maximum-residue-levels_en
5. Mostafalou, S. & Abdollahi, M. (2017). Pesticides: an update of human exposure and toxicity. *Archives of Toxicology*, *91*(2), 549–599. https://doi.org/10.1007/s00204-016-1849-x
6. Belarbi, S., et al. (2021). Comparison of new approach of GC-HRMS (Q-Orbitrap) to GC–MS/MS (triple-quadrupole) in analyzing the pesticide residues and contaminants in complex food matrices. *Food Chemistry*, *359*, 129932. https://doi.org/10.1016/j.foodchem.2021.129932
7. Santana-Mayor, A., et al. (2023). Updated overview of QuEChERS applications in food, environmental and biological analysis (2020–2023). *TrAC Trends in Analytical Chemistry*, *169*, 117375. https://doi.org/10.1016/j.trac.2023.117375
8. Anastassiades, M., et al. (2003). Fast and easy multiresidue method employing acetonitrile extraction/partitioning and "dispersive solid-phase extraction" for the determination of pesticide residues in produce. *Journal of AOAC International*, *86*(2), 412–431. https://doi.org/10.1093/jaoac/86.2.412
9. Eyring, P., et al. (2021). Development of a new generic extraction method for the analysis of pesticides, mycotoxins, and polycyclic aromatic hydrocarbons in representative animal feed and food samples. *Food Chemistry*, *356*, 129653. https://doi.org/10.1016/j.foodchem.2021.129653
10. Khan, M. S. & Shafiur Rahman, M. (Eds.). (2021). *Techniques to measure food safety and quality: microbial, chemical, and sensory*. Springer International Publishing. https://doi.org/10.1007/978-3-030-68636-9
11. Lehotay, S. J. & Hajšlová, J. (2002). Application of gas chromatography in food analysis. *TrAC Trends in Analytical Chemistry*, *21*(9–10), 686–697. https://doi.org/10.1016/S0165-9936(02)00805-1
12. Bevan, R., et al. (2020). Evaluating the risk to humans from mineral oils in foods: Current state of the evidence. *Food and Chemical Toxicology*, *136*, 110966. https://doi.org/10.1016/j.fct.2019.110966
13. Fatoki, O. S., et al. (2011). Polycyclic Aromatic Hydrocarbons (PAHs) in food and environmental samples: an overview. *Fresenius Environmental Bulletin*, *20*(8A), 2012–2020.
14. Zhang, W., et al. (2019). A method for determining glyphosate and its metabolite aminomethyl phosphonic acid by gas chromatography-flame photometric detection. *Journal of Chromatography A*, *1589*, 116–121. https://doi.org/10.1016/j.chroma.2018.12.039
15. Masci, M., et al. (2024). Chromatographic methods for the determination of glyphosate in cereals together with a discussion of its occurrence, accumulation, fate, degradation, and regulatory status. *Methods and Protocols*, *7*(3), Article 3. https://doi.org/10.3390/mps7030038
16. Quintana, J. B., et al. (2008). Organophosphorus flame retardants and plasticizers in water and air II. Analytical methodology. *TrAC Trends in Analytical Chemistry*, *27*(10), 904–915. https://doi.org/10.1016/j.trac.2008.08.004
17. Soceanu, A., et al. (2009). Pesticide residues determination in vegetables from Romania by Gc-Ecd. In L. I. Simeonov & M. A. Hassanien (Eds.), *Exposure and risk assessment of chemical pollution - contemporary methodology* (p. 423). Springer. https://doi.org/10.1007/978-90-481-2335-3_33


© Burleigh Dodds Science Publishing Limited, 2025. All rights reserved.

18. Li, Z., et al. (2012). Optimization and application of accelerated solvent extraction for rapid quantification of PCBs in food packaging materials using GC-ECD. *Food Control*, *27*(2), 300–306. https://doi.org/10.1016/j.foodcont.2012.04.006
19. Kawasaki, H., et al. (2009). Analysis of hexachlorobenzene in food red nos. 104 (Phloxine) and 105 (Rose Bengal) by GC-ECD. *Journal of the Food Hygienic Society of Japan*, *50*(1), 6–9. https://doi.org/10.3358/shokueishi.50.6
20. Notardonato, I., et al. (2013). Validation of a novel derivatization method for GC-ECD determination of acrylamide in food. *Analytical and Bioanalytical Chemistry*, *405*(18), 6137–6141. https://doi.org/10.1007/s00216-013-7001-3
21. Sun, Q., et al. (2023). A review on recent advances in mass spectrometry analysis of harmful contaminants in food. *Frontiers in Nutrition*, *10*, 1244459. https://doi.org/10.3389/fnut.2023.1244459
22. Picó, Y. & Campo, J. (Eds.). (2023). *Mass spectrometry in food and environmental chemistry* (Vol. 119). Springer International Publishing. https://doi.org/10.1007/978-3-031-19093-3
23. Peng, P. L. & Lim, L. H. (2022). Polycyclic Aromatic Hydrocarbons (PAHs) sample preparation and analysis in beverages: a review. *Food Analytical Methods*, *15*(4), 1042–1061. https://doi.org/10.1007/s12161-021-02178-y
24. Zhao, J., et al. (2024). A comprehensive review of chloropropanol analytical method in the context of food safety. *Food Chemistry*, *446*. https://doi.org/10.1016/j.foodchem.2024.138776
25. Rivera-Pérez, A., et al. (2022). Persistent organic pollutants (PCBs and PCDD/Fs), PAHs, and plasticizers in spices, herbs, and tea–A review of chromatographic methods from the last decade (2010–2020). *Critical Reviews in Food Science and Nutrition*, *62*(19), 5224–5244. https://doi.org/10.1080/10408398.2021.1883546
26. Li, M., et al. (2024). Non-targeted identification and risk assessment of unknown substances in disposable plastic tableware by GC-Orbitrap HRMS. *Food Chemistry*, *454*, 139837. https://doi.org/10.1016/j.foodchem.2024.139837
27. Shashkov, M. V., et al. (2023). Ionic liquids–new gas chromatographic phases with unique properties: a review. *Doklady Physical Chemistry*, *508*(1), 1–16. https://doi.org/10.1134/S0012501623600018
28. *Ionic Liquids in Analytical Chemistry: Applications and Recent Trends-Web of Science Core Collection*. (n.d.). Retrieved August 29, 2024, from https://www.webofscience.com/wos/woscc/full-record/WOS:000727726300010
29. Ros, M., et al. (2018). Evaluation of ionic liquid gas chromatography stationary phases for the separation of polychlorinated biphenyls. *Journal of Chromatography A*, *1559*, 156–163. https://doi.org/10.1016/j.chroma.2017.12.029
30. Abdelghany, T. M., et al. (2024). Methylimidazolium ionic liquids–A new class of forever chemicals with endocrine disrupting potential. *Chemosphere*, *363*, 142827. https://doi.org/10.1016/j.chemosphere.2024.142827
31. Tranchida, P. Q., et al. (2013). Untargeted and targeted comprehensive two-dimensional GC analysis using a novel unified high-speed triple quadrupole mass spectrometer. *Journal of Chromatography A*, *1278*, 153–159. https://doi.org/10.1016/j.chroma.2012.12.066
32. Maštovská, K. & Lehotay, S. J. (2003). Practical approaches to fast gas chromatography–mass spectrometry. *Journal of Chromatography A*, *1000*(1), 153–180. https://doi.org/10.1016/S0021-9673(03)00448-5

© Burleigh Dodds Science Publishing Limited, 2025. All rights reserved.

33. Zoccali, M., et al. (2019). Fast gas chromatography-mass spectrometry: a review of the last decade. *TrAC Trends in Analytical Chemistry*, *118*, 444–452. https://doi.org/10.1016/j.trac.2019.06.006
34. Lee, J., et al. (2017). Sensitivity enhancement using a microbore column and pulsed pressure injection in the simultaneous analysis of 356 pesticide multiresidues by gas chromatography–tandem mass spectrometry. *Applied Biological Chemistry*, *60*(4), 385–390. https://doi.org/10.1007/s13765-017-0290-z
35. Korytár, P., et al. (2002). Practical fast gas chromatography: methods, instrumentation and applications. *TrAC Trends in Analytical Chemistry*, *21*(9), 558–572. https://doi.org/10.1016/S0165-9936(02)00811-7
36. Klee, M. S. (2009). Optimizing capillary column backflush to improve cycle time and reduce column contamination. *Journal of Separation Science*, *32*(1), Article 1. https://doi.org/10.1002/jssc.200800434
37. Mastovska, K. & Wylie, P. L. (2012). Evaluation of a new column backflushing set-up in the gas chromatographic–tandem mass spectrometric analysis of pesticide residues in dietary supplements. *Journal of Chromatography A*, *1265*, 155–164. https://doi.org/10.1016/j.chroma.2012.09.094
38. Ravindra, K., et al. (2008). Low-pressure gas chromatography: recent trends and developments. *TrAC Trends in Analytical Chemistry*, *27*(4), 291–303. https://doi.org/10.1016/j.trac.2008.02.003
39. Sapozhnikova, Y. & Lehotay, S. J. (2015). Review of recent developments and applications in low-pressure (vacuum outlet) gas chromatography. *Analytica Chimica Acta*, *899*, 13–22. https://doi.org/10.1016/j.aca.2015.10.003
40. Sapozhnikova, Y. & Lehotay, S. J. (2015). Evaluation of different parameters in the extraction of incurred pesticides and environmental contaminants in Fish. *Journal of Agricultural and Food Chemistry*, *63*(21), 5163–5168. https://doi.org/10.1021/jf506256q
41. Poynter, S. D. H. & Shellie, R. A. (2008). High-speed, low-pressure gas chromatography–mass spectrometry for essential oil analysis. *Journal of Chromatography A*, *1200*(1), 28–33. https://doi.org/10.1016/j.chroma.2008.03.069
42. Arkan, T. & Molnár-Perl, I. (2015). The role of derivatization techniques in the analysis of glyphosate and aminomethyl-phosphonic acid by chromatography. *Microchemical Journal*, *121*, 99–106. https://doi.org/10.1016/j.microc.2015.02.007
43. Bizkarguenaga, E., et al. (2013). In-port derivatization after sorptive extractions. *Journal of Chromatography A*, *1296*, 36–46. https://doi.org/10.1016/j.chroma.2013.03.058
44. Wang, Q., et al. (2013). Developments in injection port derivatization. *Journal of Chromatography A*, *1296*, 25–35. https://doi.org/10.1016/j.chroma.2013.04.036
45. Tzing, S.-H. & Ding, W.-H. (2010). Determination of melamine and cyanuric acid in powdered milk using injection-port derivatization and gas chromatography–tandem mass spectrometry with furan chemical ionization. *Journal of Chromatography A*, *1217*(40), 6267–6273. https://doi.org/10.1016/j.chroma.2010.07.081
46. De Vos, J., et al. (2021). Advances in ultra-high-pressure and multi-dimensional liquid chromatography instrumentation and workflows. *Analytical Science Advances*, *2*(3–4), 171–192. https://doi.org/10.1002/ansa.202100007
47. Walter, T. H. & Andrews, R. W. (2014). Recent innovations in UHPLC columns and instrumentation. *TrAC Trends in Analytical Chemistry*, *63*, 14–20. https://doi.org/10.1016/j.trac.2014.07.016

© Burleigh Dodds Science Publishing Limited, 2025. All rights reserved.

48. MacNair, J. E., et al. (1999). Ultrahigh-pressure reversed-phase capillary liquid chromatography: isocratic and gradient elution using columns packed with 1.0-μm particles. *Analytical Chemistry*, *71*(3), 700–708. https://doi.org/10.1021/ac9807013
49. Advancing LC performance with smaller particles and higher pressure. (2005). *Analytical Chemistry*, *77*(23), 460 A-467 A. https://doi.org/10.1021/ac053516f
50. De Vos, J., et al. (2016). Advances in ultrahigh-pressure liquid chromatography technology and system design. *Analytical Chemistry*, *88*(1), 262–278. https://doi.org/10.1021/acs.analchem.5b04381
51. Ok, H. E., et al. (2014). HPLC and UPLC methods for the determination of zearalenone in noodles, cereal snacks and infant formula. *Food Chemistry*, *163*, 252–257. https://doi.org/10.1016/j.foodchem.2014.04.111
52. Guiochon, G. & Gritti, F. (2011). Shell particles, trials, tribulations and triumphs. *Journal of Chromatography A*, *1218*(15), 1915–1938. https://doi.org/10.1016/j.chroma.2011.01.080
53. Dib, A. A., et al. (2023). A comparative review on methods of detection and quantification of mycotoxins in solid food and feed: a focus on cereals and nuts. *Mycotoxin Research*, *39*(4), 319–345. https://doi.org/10.1007/s12550-023-00501-6
54. Jia, Q., et alJ. (2024). Pesticide residues in animal-derived food: current state and perspectives. *Food Chemistry*, *438*. https://doi.org/10.1016/j.foodchem.2023.137974
55. Plata-Calzado, C., et al. (2024). Analytical methods for anatoxin-a determination: a review. *Toxins*, *16*(4). https://doi.org/10.3390/toxins16040198
56. Cui, Y., et al. (2024). Advancements in detection techniques for per- and polyfluoroalkyl substances: a comprehensive review. *TrAC - Trends in Analytical Chemistry*, *176*. https://doi.org/10.1016/j.trac.2024.117754
57. Chen, T., et al. (2024). Anabolic steroids in livestock production: background and implications for chemical food safety. *Steroids*, *206*. https://doi.org/10.1016/j.steroids.2024.109420
58. Chen, B.-H., et al. (2022). Recent advances in the analysis of polycyclic aromatic hydrocarbons in food and water. *Journal of Food and Drug Analysis*, *30*(4), 494–522. https://doi.org/10.38212/2224-6614.3429
59. Liu, J., et al. (2023). Recent progress on toxicity and detection methods of polychlorinated biphenyls in environment and foodstuffs. *Critical Reviews in Analytical Chemistry*, *53*(4), 928–953. https://doi.org/10.1080/10408347.2021.1997570
60. Sapozhnikova, Y. (2023). Liquid and gas chromatography–mass spectrometry methods in food and environmental safety. *Handbook of Environmental Chemistry*, *119*, 127–147. https://doi.org/10.1007/698_2022_883
61. Feng, Y.-L. (2023). Chromatography high-resolution mass spectrometry in food and environmental chemistry. *Handbook of Environmental Chemistry*, *119*, 149–185. https://doi.org/10.1007/698_2022_892
62. Tobiszewski, M., et al. (2017). Environmental risk-based ranking of solvents using the combination of a multimedia model and multi-criteria decision analysis. *Green Chemistry*, *19*(4), 1034–1042. https://doi.org/10.1039/C6GC03424A
63. Shi, E., et al. (2024). Recent advances in post column derivatization: enhancing sensitivity and selectivity in liquid chromatography analysis. *TrAC Trends in Analytical Chemistry*, *177*, 117799. https://doi.org/10.1016/j.trac.2024.117799

© Burleigh Dodds Science Publishing Limited, 2025. All rights reserved.

64. Pareja, L., et al. (2022). Determination of multiclass emerging contaminants using QuEChERS method. In D. C. Dominguez, H. Sarma, W.-Y. Lee (Eds.), *Emerging contaminants in the environment: challenges and sustainable practices* (pp. 335–380). Elsevier. https://doi.org/10.1016/B978-0-323-85160-2.00003-2
65. Jandera, P. & Hájek, T. (2018). Mobile phase effects on the retention on polar columns with special attention to the dual hydrophilic interaction–reversed-phase liquid chromatography mechanism, a review. *Journal of Separation Science*, *41*(1), 145–162. https://doi.org/10.1002/jssc.201701010
66. Jandera, P. & Janás, P. (2017). Recent advances in stationary phases and understanding of retention in hydrophilic interaction chromatography: a review. *Analytica Chimica Acta*, *967*, 12–32. https://doi.org/10.1016/j.aca.2017.01.060
67. Marrubini, G., et al. (2018). Hydrophilic interaction chromatography in food matrices analysis: an updated review. *Food Chemistry*, *257*, 53–66. https://doi.org/10.1016/j.foodchem.2018.03.008
68. Francesquett, J. Z., et al. (2019). Simultaneous determination of the quaternary ammonium pesticides paraquat, diquat, chlormequat, and mepiquat in barley and wheat using a modified quick polar pesticides method, diluted standard addition calibration and hydrophilic interaction liquid chromatography coupled to tandem mass spectrometry. *Journal of Chromatography A*, *1592*, 101–111. https://doi.org/10.1016/j.chroma.2018.12.060
69. de Villiers, A. & Kalili, K. M. (2016). Comprehensive two-dimensional hydrophilic interaction chromatography x reversed-phase liquid chromatography (HILIC x RP-LC) *Theory, Practice, and Applications*. In E. Grushka & N. Grinberg (Eds.), *Advances in chromatography* (Vol. 53, pp. 217–299). CRC Press-Taylor & Francis Group. https://www.webofscience.com/wos/woscc/full-record/WOS:000394569500006
70. Yates, H. S. A., et al. (2023). Ion chromatography and ion chromatography / mass spectrometry as a complementary analysis technique for amino acid analysis in food, a review. *Food Chemistry Advances*, *3*, 100415. https://doi.org/10.1016/j.focha.2023.100415
71. *Semiautomated pH Gradient Ion-Exchange Chromatography of Monoclonal Antibody Charge Variants | Analytical Chemistry*. (n.d.). Retrieved August 14, 2024, from https://pubs.acs.org/doi/10.1021/ac502372r
72. Wouters, S., et al. (2017). System design and emerging hardware technology for ion chromatography. *Chromatographia*, *80*(5), 689–704. https://doi.org/10.1007/s10337-016-3184-z
73. Hadda, P. R., et al. (2003). Developments in suppressor technology for inorganic ion analysis by ion chromatography using conductivity detection. *Journal of Chromatography. A*, *1000*(1–2), 725–742. https://doi.org/10.1016/s0021-9673(02)01999-4
74. Elkin, K., et al. (2015). Improvements in ion reflux: an electrodialytic eluent generation and suppression device for ion chromatography. *Journal of Chromatography A*, *1403*, 63–69. https://doi.org/10.1016/j.chroma.2015.05.027
75. Ordoñez, E. Y., et al. (2012). Computer assisted optimization of liquid chromatographic separations of small molecules using mixed-mode stationary phases. *Journal of Chromatography A*, *1238*, 91–104. https://doi.org/10.1016/j.chroma.2012.03.055
76. Kazarian, A. A., et al. (2014). Comprehensive analysis of pharmaceutical products using simultaneous mixed-mode (ion-exchange/reversed-phase) and hydrophilic

© Burleigh Dodds Science Publishing Limited, 2025. All rights reserved.

interaction liquid chromatography: liquid chromatography. *Journal of Separation Science, 37*(16), 2138–2144. https://doi.org/10.1002/jssc.201400411

77. Montes, R., et al. (2019). Determination of Persistent and Mobile Organic Contaminants (PMOCs) in water by mixed-mode liquid chromatography–tandem mass spectrometry. *Analytical Chemistry, 91*(8), 5176–5183. https://doi.org/10.1021/acs.analchem.8b05792
78. Rodriguez, E. L., et al. (2020). Affinity chromatography: a review of trends and developments over the past 50 years. *Journal of Chromatography. B, Analytical Technologies in the Biomedical and Life Sciences, 1157*, 122332. https://doi.org/10.1016/j.jchromb.2020.122332
79. Jondeau-Cabaton, A., et al. (2013). Characterization of endocrine disruptors from a complex matrix using estrogen receptor affinity columns and high performance liquid chromatography–high resolution mass spectrometry. *Environmental Science and Pollution Research, 20*(5), 2705–2720. https://doi.org/10.1007/s11356-012-1458-z
80. Du, Y., et al. (2021). A sensitive HPLC-FLD method for the quantitative determination of 3-chloro-1,2-propanediol by pre-column fluorescence derivatization with 9-(2-Hydroxypropyl)adenine. *Journal of Liquid Chromatography & Related Technologies, 44*(9–10), 445–456. https://doi.org/10.1080/10826076.2021.1952427
81. Daou, R., et al. (2020). Occurrence of aflatoxin M1 in raw, pasteurized, UHT cows' milk, and dairy products in Lebanon. *Food Control, 111*, 107055. https://doi.org/10.1016/j.foodcont.2019.107055
82. Molle, D. R. D., et al. (2017). Polycyclic aromatic hydrocarbons in canola, sunflower and corn oils and estimated daily intake. *Food Control, 81*, 96–100. https://doi.org/10.1016/j.foodcont.2017.05.045
83. Janik, E., et al. (2021). The existing methods and novel approaches in mycotoxins' detection. *Molecules, 26*(13), 3981. https://doi.org/10.3390/molecules26133981
84. Tsiasioti, A. & Tzanavaras, P. D. (2024). High performance liquid chromatography coupled with post – Column derivatization methods in food analysis: chemistries and applications in the last two decades. *Food Chemistry, 443*, 138577. https://doi.org/10.1016/j.foodchem.2024.138577
85. Allen, L. B., et al. (1995). Condensation nucleation light scattering detection for conventional reversed-phase liquid chromatography. *Analytical Chemistry, 67*(3), 659–666. https://doi.org/10.1021/ac00099a026
86. Swartz, M. (2010). Hplc detectors: a brief review. *Journal of Liquid Chromatography & Related Technologies, 33*(9–12), 1130–1150. https://doi.org/10.1080/10826076.2010.484356
87. Górecki, T., et al. (2006). Universal response in liquid chromatography using charged aerosol detection. *Analytical Chemistry, 78*(9), 3186–3192. https://doi.org/10.1021/ac060078j
88. Wang, H., et al. (2019). Determination of bisphenol analogues in food-contact plastics using diode array detector, charged aerosol detector and evaporative light-scattering detector. *Ecotoxicology and Environmental Safety, 186*, 109778. https://doi.org/10.1016/j.ecoenv.2019.109778
89. Perez-Fernandez, B. & de la Escosura-Muniz, A. (2022). Electrochemical biosensors based on nanomaterials for aflatoxins detection: a review (2015–2021). *Analytica Chimica Acta, 1212*, 339658. https://doi.org/10.1016/j.aca.2022.339658

© Burleigh Dodds Science Publishing Limited, 2025. All rights reserved.

90. Pérez-Fernández, B., et al. (2020). Electrochemical (bio)sensors for pesticides detection using screen-printed electrodes. *Biosensors*, *10*(4), 32. https://doi.org/10.3390/bios10040032
91. Zomer, P. & Mol, H. G. J. (2015). Simultaneous quantitative determination, identification and qualitative screening of pesticides in fruits and vegetables using LC-Q-Orbitrap™-MS. *Food Additives & Contaminants: Part A*, *32*(10), 1628–1636. https://doi.org/10.1080/19440049.2015.1085652
92. dos Santos, I. D., et al. (2024). Multi-residue determination of biocides in dairy products and slurry feed using QuEChERS extraction and liquid chromatography combined with high resolution mass spectrometry (LC-ESI-QOrbitrap™-MS). *Food Chemistry*, *457*, 140117. https://doi.org/10.1016/j.foodchem.2024.140117
93. Boukaew, S., et al. (2023). Efficacy of the antifungal metabolites of *Streptomyces philanthi* RL-1-178 on aflatoxin degradation with its application to prevent aflatoxigenic fungi in stored maize grains and identification of the bioactive compound. *World Journal of Microbiology & Biotechnology*, *39*(1), 24. https://doi.org/10.1007/s11274-022-03470-7
94. Latimer, G. W. (Ed.). (2023). AOAC official method 971.22 standards for aflatoxins: thin-layer chromatographic method. In *Official methods of analysis of AOAC international* (22nd ed.). Oxford University Press. https://doi.org/10.1093/9780197610145.003.3957
95. ul Haq, I., et al. (2024). Mycotoxin detection in selected medicinal plants using chromatographic techniques. *Biomedical Chromatography*, *38*(4). https://doi.org/10.1002/bmc.5831
96. Rahim, M., et al. (2024). Assessment of total aflatoxin and ochratoxin A in poultry feed ingredients by thin-layer chromatography and enzyme-linked immunosorbent assay. *Journal of Advanced Veterinary and Animal Research*, *11*(1), 107–113. https://doi.org/10.5455/javar.2024.k754
97. Ismail, A. M., et al. (2024). Aflatoxins in wheat grains: detection and detoxification through chemical, physical, and biological means. *Life-Basel*, *14*(4), 535. https://doi.org/10.3390/life14040535
98. Bata-Vidacs, I., et al. (2024). Molecular and chemical evaluation of patulin production of *Aspergillus* and *Penicillium*-like species isolated from Hungarian apples. *Food Additives and Contaminants Part A-chemistry Analysis Control Exposure & Risk Assessment*. https://doi.org/10.1080/19440049.2024.2364364
99. Akbar, M., et al. (2024). Spherical Fe2O3 nanoparticles inhibit the production of aflatoxins (B1 and B2) and regulate total soluble solids and titratable acidity of peach fruit. *International Journal of Food Microbiology*, *410*, 110508. https://doi.org/10.1016/j.ijfoodmicro.2023.110508
100. Ullah, R., et al. (2023). Exploitation of selected plant extracts as bio-control against fungal contaminants in animal feed. *Journal of King Saud University Science*, *35*(5), 102685. https://doi.org/10.1016/j.jksus.2023.102685
101. Naz, G., et al. (2023). Molecular characterization of ochratoxin a producing indigenous *Aspergillus* strains from poultry feed in Pakistan. *Sains Malaysiana*, *52*(3), 821–835. https://doi.org/10.17576/jsm-2023-5203-11
102. Naqvi, S. I. Z., et al. (2023). Antifungal activity of *Juglans-regia*-mediated silver nanoparticles (AgNPs) against *Aspergillus-ochraceus*-induced toxicity in in vitro and in vivo settings. *Journal of Functional Biomaterials*, *14*(4), 221. https://doi.org/10.3390/jfb14040221


© Burleigh Dodds Science Publishing Limited, 2025. All rights reserved.

103. Barborakova, Z., et al. (2023). Toxin producing micromycetes of the genus penicillium and aspergillus on berries, grapes, and tomato fruits in slovak stores. *Journal of Microbiology Biotechnology and Food Sciences, 13*(1). https://doi.org/10.55251/jmbfs.9927
104. Commission Regulation (EU) 2023/915 of 25 April 2023 on maximum levels for certain contaminants in food and repealing regulation (EC) No 1881/2006 (Text with EEA Relevance), 119 OJ L (2023). http://data.europa.eu/eli/reg/2023/915/oj/eng
105. Ramesh, J., et al. (2013). Analysis of feed samples for aflatoxin B1 contamination by HPTLC – a validated method. *International Journal of Current Microbiology and Applied Sciences, 2*(5), 373–377.
106. Saito, M. (2013). History of supercritical fluid chromatography: Instrumental development. *Journal of Bioscience and Bioengineering, 115*(6), 590–599. https://doi.org/10.1016/j.jbiosc.2012.12.008
107. Breitenbach, S., et al. (2016). Assessment of ultra high performance supercritical fluid chromatography as a separation technique for the analysis of seized drugs: applicability to synthetic cannabinoids. *Journal of Chromatography A, 1440*, 201–211. https://doi.org/10.1016/j.chroma.2016.02.047
108. Bernal, J. L., et al. (2013). Supercritical fluid chromatography in food analysis. *Journal of Chromatography A, 1313*, 24–36. https://doi.org/10.1016/j.chroma.2013.07.022
109. Cutillas, V., et al. (2018). Evaluation of supercritical fluid chromatography coupled to tandem mass spectrometry for pesticide residues in food. *Journal of Chromatography A, 1545*, 67–74. https://doi.org/10.1016/j.chroma.2018.02.048
110. *High-throughput simultaneous analysis of pesticides by supercritical fluid chromatography coupled with high-resolution mass spectrometry | Journal of Agricultural and Food Chemistry*. (n.d.). Retrieved August 13, 2024, from https://pubs.acs.org/doi/10.1021/jf5056248
111. Nillos, M. G., et al. (2007). Enantioselective acetylcholinesterase inhibition of the organophosphorous insecticides profenofos, fonofos, and crotoxyphos. *Environmental Toxicology and Chemistry, 26*(9), 1949–1954. https://doi.org/10.1897/07-001R.1
112. Chen, X., et al. (2016). Enantioseparation and determination of isofenphos-methyl enantiomers in wheat, corn, peanut and soil with Supercritical fluid chromatography/tandem mass spectrometric method. *Journal of Chromatography B, 1015–1016*, 13–21. https://doi.org/10.1016/j.jchromb.2016.02.003
113. *On the discrepancy between crossovers of solubility in supercritical carbon dioxide and retention in supercritical fluid chromatography–ScienceDirect*. (n.d.). Retrieved August 29, 2024, from https://www.sciencedirect.com/science/article/pii/S0021967324005843
114. Herrero, M., et al. (2009). Multidimensional chromatography in food analysis. *Journal of Chromatography A, 1216*(43), Article 43. https://doi.org/10.1016/j.chroma.2009.08.014
115. Nolvachai, Y., et al. (2023). Foods and contaminants analysis using multidimensional gas chromatography: an update of recent studies, technology, and applications. *Analytical Chemistry, 95*(1), 238–263. https://doi.org/10.1021/acs.analchem.2c04680
116. Caño-Carrillo, I., et al. (2023). Comprehensive and heart-cutting multidimensional liquid chromatography-mass spectrometry and its applications in food analysis. *Mass Spectrometry Reviews*. https://doi.org/10.1002/mas.21843


© Burleigh Dodds Science Publishing Limited, 2025. All rights reserved.

117. Pons, A., et al. (2016). Identification and analysis of piperitone in red wines. *Food Chemistry*, *206*, 191–196. https://doi.org/10.1016/j.foodchem.2016.03.064
118. Mostafa, A., et al. (2012). Optimization aspects of comprehensive two-dimensional gas chromatography. *Journal of Chromatography A*, *1255*, 38–55. https://doi.org/10.1016/j.chroma.2012.02.064
119. Liu, X., et al. (2013). Comprehensive two-dimensional gas chromatography with flame photometric detection applied to organophosphorus pesticides in food matrices. *Microchemical Journal*, *111*, 25–31. https://doi.org/10.1016/j.microc.2012.07.013
120. Dallüge, J., et al. (2002). Comprehensive two-dimensional gas chromatography with time-of-flight mass spectrometric detection applied to the determination of pesticides in food extracts. *Journal of Chromatography A*, *965*(1), 207–217. https://doi.org/10.1016/S0021-9673(01)01324-3
121. Kalachova, K., et al. (2012). Implementation of comprehensive two-dimensional gas chromatography–time-of-flight mass spectrometry for the simultaneous determination of halogenated contaminants and polycyclic aromatic hydrocarbons in fish. *Analytical and Bioanalytical Chemistry*, *403*(10), 2813–2824. https://doi.org/10.1007/s00216-012-6095-3
122. Mondello, L. (Ed.). (2011). *Comprehensive chromatography in combination with mass spectrometry*. Wiley.
123. Jain, A. V. & Hatch, R. C. (1980). Determination of aflatoxins in compound feedstuffs containing citrus pulp. *Journal of AOAC International*, *63*(3), 626–630. https://doi.org/10.1093/jaoac/63.3.626
124. Kalili, K. M., et al. (2013). Toward unraveling grape tannin composition: Application of online hydrophilic interaction chromatography × reversed-phase liquid chromatography-time-of-flight mass spectrometry for grape seed analysis. *Analytical Chemistry*, *85*(19), 9107–9115. https://doi.org/10.1021/ac401896r
125. Wei, F., et al. (2015). Quantitation of triacylglycerols in edible oils by off-line comprehensive two-dimensional liquid chromatography-atmospheric pressure chemical ionization mass spectrometry using a single column. *Journal of Chromatography. A*, *1404*, 60–71. https://doi.org/10.1016/j.chroma.2015.05.058
126. Arena, P., et al. (2021). Pattern-type separation of triacylglycerols by silver thiolate×non-aqueous reversed phase comprehensive liquid chromatography. *Separations*, *8*(6), Article 6. https://doi.org/10.3390/separations8060088
127. Donato, P., et al. (2016). Comprehensive two-dimensional liquid chromatography–tandem mass spectrometry for the simultaneous determination of wine polyphenols and target contaminants. *Journal of Chromatography A*, *1458*, 54–62. https://doi.org/10.1016/j.chroma.2016.06.042
128. Vergara-Barberán, M., et al. (2020). Classification of olive leaves and pulp extracts by comprehensive two-dimensional liquid chromatography of polyphenolic fingerprints. *Food Chemistry*, *320*, 126630. https://doi.org/10.1016/j.foodchem.2020.126630
129. Montero, L., et al. (2013). Characterization of grape seed procyanidins by comprehensive two-dimensional hydrophilic interaction × reversed phase liquid chromatography coupled to diode array detection and tandem mass spectrometry. *Analytical and Bioanalytical Chemistry*, *405*(13), 4627–4638. https://doi.org/10.1007/s00216-012-6567-5


© Burleigh Dodds Science Publishing Limited, 2025. All rights reserved.

130. Saint Germain, F. M., et al. (2022). On-line 2D-RPLC x RPLC – HRMS to assess wastewater treatment in a pharmaceutical plant. *Journal of Pharmaceutical and Biomedical Analysis, 208*, 114465. https://doi.org/10.1016/j.jpba.2021.114465
131. Pascoe, R., et al. (2001). Reduction in matrix-related signal suppression effects in electrospray ionization mass spectrometry using on-line two-dimensional liquid chromatography. *Analytical Chemistry, 73*(24), 6014–6023. https://doi.org/10.1021/ac0106694
132. Kittlaus, S., et al. (2013). Development and validation of an efficient automated method for the analysis of 300 pesticides in foods using two-dimensional liquid chromatography–tandem mass spectrometry. *Journal of Chromatography A, 1283*, 98–109. https://doi.org/10.1016/j.chroma.2013.01.106
133. Kittlaus, S., et al. (2013). Evaluation of matrix effects in different multipesticide residue analysis methods using liquid chromatography–tandem mass spectrometry, including an automated two-dimensional cleanup approach. *Journal of Separation Science, 36*(13), 2185–2195. https://doi.org/10.1002/jssc.201300044
134. Wilcox, J., et al. (2015). The use of immunoaffinity columns connected in tandem for selective and cost-effective mycotoxin clean-up prior to multi-mycotoxin liquid chromatographic–tandem mass spectrometric analysis in food matrices. *Journal of Chromatography A, 1400*, 91–97. https://doi.org/10.1016/j.chroma.2015.04.053
135. Kresse, M., et al. (2019). Simultaneous determination of pesticides, mycotoxins, and metabolites as well as other contaminants in cereals by LC-LC-MS/MS. *Journal of Chromatography B, 1117*, 86–102. https://doi.org/10.1016/j.jchromb.2019.04.013
136. *Advances in Multidimensional LC–GC for Food Analysis*. (n.d.). Retrieved October 24, 2024, from https://www.chromatographyonline.com/view/advances-multidimensional-lc-gc-food-analysis
137. Purcaro, G., et al. (2012). Hyphenated liquid chromatography–gas chromatography technique: recent evolution and applications. *Journal of Chromatography A, 1255*, 100–111. https://doi.org/10.1016/j.chroma.2012.02.018
138. Sdrigotti, N., et al. (2021). Evolution of hyphenated techniques for mineral oil analysis in food. *Journal of Separation Science, 44*(1), 464–482. https://doi.org/10.1002/jssc.202000901
139. Bratinova, S., et al. (2023). *Guidance on sampling, analysis and data reporting for the monitoring of mineral oil hydrocarbons in food and food contact materials–2nd edition: in the frame of Commission Recommendation (EU) 2017/84*. Publications Office. https://data.europa.eu/doi/10.2760/963728
140. Program, H. F. (2024, September 6). *Laboratory methods (food)*. US Food and Drug Administration; FDA. https://www.fda.gov/food/science-research-food/laboratory-methods-food
141. Stromgren, S. (2019, October). *Guidelines for the validation of chemical methods in food, feed, cosmetics, and veterinary products*. US Food and Drug Administration. https://www.fda.gov/media/81810/download
142. *Sampling and analysis of contaminants–European Commission*. (n.d.). Retrieved October 24, 2024, from https://food.ec.europa.eu/food-safety/chemical-safety/contaminants/sampling-and-analysis_en
143. EFSA Panel on Contaminants in the Food Chain (CONTAM), et al. (2023). Update of the risk assessment of mineral oil hydrocarbons in food. *EFSA Journal, 21*(9), e08215. https://doi.org/10.2903/j.efsa.2023.8215

© Burleigh Dodds Science Publishing Limited, 2025. All rights reserved.

144. Brandi, J., et al. (2024). Analysis of veterinary drugs and pesticides in food using liquid chromatography-mass spectrometry. *TrAC Trends in Analytical Chemistry, 179*, 117888. https://doi.org/10.1016/j.trac.2024.117888
145. Liu, L., et al. (2020). The application of supercritical fluid chromatography in food quality and food safety: an overview. *Critical Reviews in Analytical Chemistry*, *50*(2), 136–160. https://doi.org/10.1080/10408347.2019.1586520
146. Muntean, E. (2023). Applications of ion chromatography in food analysis. In P. N. Nesterenko, C. F. Poole, & Y. Sun (Eds.), *Ion-exchange chromatography and related techniques* (pp. 351–369). Elsevier. https://doi.org/10.1016/B978-0-443-15369-3.00013-4
147. Mostafa, A., et al. (2015). Multidimensional liquid chromatography applied to the analysis of pesticides. In T. Tuzimski & J. Sherma (Eds.), *High performance liquid chromatography in pesticide residue analysis*. CRC Press.

© Burleigh Dodds Science Publishing Limited, 2025. All rights reserved.

# Chapter 5

# Advances in techniques for detecting chemical contaminants in food: electrochemical biosensors

*Daniel Cozzolino, The University of Queensland, Australia*



## 1 Introduction

Food safety and security are of primary interest for both the food manufacturing industry and consumers. Food fraud, food contamination, and food miss-labelling are contemporary issues that disrupt the food supply and value chains around the world (Manning and Soon, 2016; Rong et al., 2021; Shen et al., 2022). These issues not only affect the transparency and trust of the food manufacturing industry but also have significant public concerns linked with human distress, well-being, and mortality (Carcea et al., 2009; Mannin and Soon, 2016; Rong et al., 2021; Shen et al., 2022).

As stated above, different practices associated with food safety and security, and more specifically with food contamination, could have an impact on human nutrition and ultimately human health (Carcea et al., 2009; Manning and Soon, 2016; Miranda et al., 2019). Food contamination occurs at different steps of the supply chain, from farm to fork, along the production line, processing, packaging, distribution, storage and retail, resulting in the occurrence of a wide range of contaminants entering the food (Miranda et al.,

http://dx.doi.org/10.19103/AS.2025.0158.11
© Burleigh Dodds Science Publishing Limited, 2025. All rights reserved.

2019). These contaminants might include the presence of hazardous elements and metals, antibiotics, chemical residues, foodborne pathogens, and biotoxins (Rong et al., 2021; Shen et al., 2022). The contaminated food ingredients and food products navigate along the global trade networks and might pose health risks to consumers (Duan et al., 2023; Gao, et al., 2022). For example, an 'adulterated or contaminated food' can be highly toxic, leading to several health problems, such as nutrition deficiencies and health disorders, which could in turn result in the failure of different organs (e.g. heart, kidney, and liver), where symptoms can be varied from vomiting, diarrhoea, fever, intestinal inflammation, to even fatal (Miranda et al., 2019; Spink, 2019). One important human health aspect intimately linked with food contamination is the potential for undeclared allergens in foods reaching the supply and value chains and ultimately causing serious health problems to the consumer (Aung and Chang, 2014; Spink, 2019; Cozzolino, 2015; 2019; Callao and Ruisánchez 2018; Astill et al., 2019; Visciano and Schirone, 2021). Although allergens as such only pose health risks for a sensitive subpopulation, and therefore not really considered as 'contaminants', food ingredients and products are considered of high risk. However, independent of the adoption and implementation of strict safety regulations and standard protocols by the different stakeholders, regulatory authorities, and governments, both food safety and security remain of concern for the food manufacturing industry worldwide (Carcea et al., 2009; Spink, 2019; Cozzolino, 2015; Callao and Ruisánchez, 2018).

Several researchers and industry reports have highlighted and demonstrated the current state of the art on the different technologies available to address issues associated with food contamination (Spink, 2019; Cozzolino, 2015; Callao and Ruisánchez, 2018; McGrath et al., 2018). Different methods and techniques, including the incorporation of the internet of things, machine learning, artificial intelligence, three-dimensional printing, and robots are considered as disruptors as they have an impact on the production, storage, and transport of foods and food systems (e.g. automation, implementation novel management tools). These technologies have been utilised alone or in combination with other tools to target food contamination in different food ingredients and products (Miranda et al., 2019; Kachouei et al., 2023).

Most of the analytical techniques utilised to assure and monitor food safety and security are considered non-target in nature. However, the vast amount of research and the wide range of practical and successful examples of analytical methods in this field have indicated their potential to be utilised as routine quality control methods targeting food contamination (Cozzolino, 2019; Miranda et al., 2019). These analytical methods include the use of liquid chromatography, gas chromatography (GC), mass spectrometry (MS), high liquid performance chromatography (HPLC), and vibrational spectroscopy (e.g. mid- and near-infrared, hyperspectral imaging, Raman spectroscopy) (Cozzolino, 2015; Adley,

© Burleigh Dodds Science Publishing Limited, 2025. All rights reserved.

2014). Although these techniques provide with high-resolution analysis (e.g. detection and identification of molecules, compounds), they are considered time consuming, costly, require highly specialised instrumentation and skilled operators, and slow for the demands imposed by the quality control systems in the agile food supply and values chains (McGrath et al., 2012; Manning and Soon, 2016). However, these techniques are still considered as the gold standard by official, regulatory, and government agencies worldwide. Nevertheless, some of these issues have prompted the development and incorporation of sensing technologies, where less consumables and less time in sample preparation are required. These sensors have become more attractive as analytical techniques, as they can be time consuming and cost-effective to the traditional analytical methods used in the laboratory (Cozzolino, 2015; Adley, 2014).

The utilisation of sensing technologies to analyse or monitor foods from the production along the distribution centres has become important in testing for compliance within the legal thresholds and contaminant levels (Adley, 2014; Bec and Huck, 2019; Bec et al., 2020). For example, testing at the sample collection point (e.g. distribution centres, supermarket) minimises possible deterioration and damage to the samples during transportation to the more traditional analytical labs. The same sensing technologies are being tested in food forensic cases, to assess the decrease in quality in high value food ingredients due to breaches in the custody chain (Adley, 2014; Bhalla et al., 2016; Bettazzi et al., 2017; Bec and Huck, 2019; Bec et al., 2020). Moreover, the advent of sensors will eliminate the need for various expensive and time-consuming analyses along the food supply and value chains (Zaukuu et al., 2020). More importantly, the utilisation of different sensing techniques has demonstrated their capability to assess and monitor complex samples such as agricultural commodities, food ingredients, and products rather than to depend on the subjective analysis by humans (Adley, 2014; Bhalla et al., 2016; Bettazzi et al., 2017; Rai et al., 2015; Velasco-Garcia and Mottram, 2003; Zaukuu et al., 2020).

## 2 Sensor definition

The International Union of Pure and Applied Chemistry defines that '*a chemical sensor is a device that transforms chemical information, ranging from the concentration of a specific sample component to total composition analysis, into an analytical useful signal. These sensors usually comprise of two components connected in series the chemical (molecular) recognition system (receptor) and a physico-chemical transducer*' (Snopok and Kruglenko, 2002; Ruiz-Altisent et al., 2010).

The available sensors can be classified according to their specificity, while biosensors belong to chemical sensors that have incorporated a biological

© Burleigh Dodds Science Publishing Limited, 2025. All rights reserved.

sensing element (Snopok and Kruglenko, 2002; Ruiz-Altisent et al., 2010; Justino et al., 2015). This biological sensing element can be made of the combination of enzyme–substrate, antibody–antigen and nucleic acids–complementary sequences (Snopok and Kruglenko, 2002; Ruiz-Altisent et al., 2010; Justino et al., 2015). Consequently, the selectivity of the biological sensing element offers the opportunity to develop devices that are highly specific for real-time analysis of complex mixtures such as food (Snopok and Kruglenko, 2002; Ruiz-Altisent et al., 2010). An example of biological sensing is the *Calux* bioassay technique (Hoogenboom et al., 2006; Sakthivel et al., 2016).

It has been proposed that the ideal sensors must be objective, cost-effective, and provide rapid, reproducible results, allowing for continuous operation (Adley, 2014; Ali et al., 2017; Korotkaya, 2014; Justino et al., 2015). As stated above, a sensor refers to a single device that is based on fundamental knowledge about chemistry and physics principles. They include electrochemistry, fluorescence, thermal, surface plasmon resonance, piezo, reflectometry, chemo or bioluminescence, and optics (Adley, 2014; Ali et al., 2017; Korotkaya, 2014; Justino et al., 2015; Zaukuu et al., 2020). For example, a chemical sensor will be defined or described as a small device that can deliver rapid information about specific molecules or compounds present in a food sample (Adley, 2014; Ali et al., 2017; Korotkaya, 2014).

The main advantage of using sensor devices is that it can be utilised throughout the different steps of the food supply and value chains (e.g. at harvest, processing, storage, and market) (Arora et al., 2011; McGrath et al., 2012; Manning and Soon, 2016; Justino et al., 2015; Zaukuu et al., 2020).

Recent advances in electronics and hardware have facilitated the utilisation of a wide range of sensors in food assessment and monitoring, where instrument portability and miniaturisation (Mahato et al., 2022; Derkus, 2016; Wilson et al., 2001), the ability of analysing low sample volumes, as well as improvements in sampling, concomitantly with advances in chemical and pattern recognition tools have made these tools to be easily implemented by the food manufacturing industry (Adley, 2014; Ali et al., 2017; Korotkaya, 2014). In addition to the ease of manufacture, using recognition methods (e.g. molecularly imprinted polymers), other characteristics such as automation and the inclusion during process (e.g. at/in/on-line) have permitted to move from the laboratory into the processing plant (Adley, 2014; Ali et al., 2017; Korotkaya, 2014). These developments have cemented the increased number of low costs and autonomous sensors that can be economically produced and implemented in a wide range of food applications (Adley, 2014; Ali et al., 2017; Korotkaya, 2014).

Advantages on the utilisation of sensing technologies also allowed for the continuous collection of data and quick response providing with efficient means to gather information and knowledge about the sample (e.g. chemical

© Burleigh Dodds Science Publishing Limited, 2025. All rights reserved.

changes during manufacturing, spoilage) or the system (Adley, 2014; Ali et al., 2017; Korotkaya, 2014; Zaukuu et al., 2020). The implementation of sensing technologies has resulted in an unparalleled collection of real-time measurements that allowed to better assess, control, and monitor all the steps in the food processing, supply, and value chains. Gathering data, and consequently information, is allowing for a better understanding of the food production systems (Adley, 2014; Ali et al., 2017; Korotkaya, 2014). However, the use and implementation of sensing technologies is facing with new challenges, like how to extract and interpret the information generated from data (e.g. big data) to improve or maintain the consistency of the food (Adley, 2014; Ali et al., 2017; Korotkaya, 2014; Zaukuu et al., 2020). The challenges imposed by the utilisation of these sensors through the generation of big data can be overcome with application of advanced data analytical techniques (e.g. ML, chemometrics) that provide with the means to accurately interpret the data collected (Adley, 2014; Ali et al., 2017; Korotkaya, 2014).

Overall, the assessment of food contamination demands rapid, instantaneous, accurate, and selective identification of a wide range of analytes and contaminants present within the food matrix (e.g. pathogenic bacteria, presence of pesticides and toxins) (Gaudin, 2017; Gao et al., 2022; Pilolli et al., 2013; Perumal and Hashin, 2014; Wu et al., 2017; Yeni et al., 2014; Zaukuu et al., 2020). Consequently, developments in sensors have enabled the quantitative detection and screening of different and wide range of microorganisms and molecules in agricultural commodities, food ingredients, and products. A schematic representation of a food sensor illustrating the link between the food, contaminant (e.g. analyte, molecule), the recognition component, and transducer, as well as the signal and the signal-processing element is shown in Fig. 1., adapted from Weston et al. (2021).

## 3 Type of sensors

Advances in chemistry and physics, electronics, and computing science have determined an increase in the number of types of sensors available including biosensors, electrochemical, and nano sensors, as well as optical sensors (Hoogenboom et al., 2006; Sakthivel et al., 2016; Adley, 2014; Poltronieri et al., 2014; Thakur and Ragavan, 2013).

A biosensor is considered as an analytical device that combines the biological sensitive recognition ability of a component (e.g. antibodies, nucleic acids, enzymes, whole cells, and aptamers) connected with a physicochemical transducer, combined with a detector that allows to identify (e.g. what is it?) and quantify (e.g. how much is it?) either a single or several analytes, molecules, or compounds present in food ingredients and products (Hoogenboom et al., 2006; Sakthivel et al., 2016; Arora et al., 2011; Adley, 2014; Velusamy et al.,

© Burleigh Dodds Science Publishing Limited, 2025. All rights reserved.

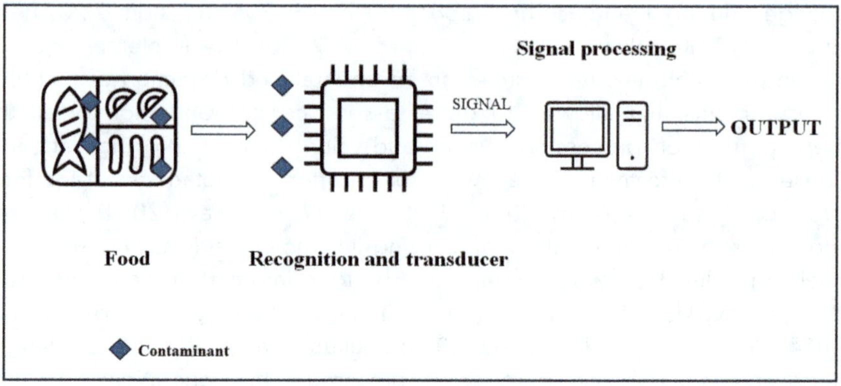


**Figure 1** A schematic representation of a food sensor including the contaminant (e.g. analyte), the recognition component and transducer, the signal and the signal-processing element. Source: adapted from Weston et al. (2021).

2010; Turner, 2013). The main advantage of the utilisation of biosensors is that they can identify biological substances providing a superior level of specificity and binding affinity with the target molecule (Arora et al., 2011; Adley, 2014). These binding characteristics are called *specific binding* or *coupling*, allowing to determine if an interaction occurs, by creating an electrical signal that is recorded and amplified by the sensing device during the analysis of food (Arora et al., 2011; Adley, 2014; Velusamy et al., 2010; Turner, 2013). Additionally, the great ability of molecular recognition that the biosensors have allowed for high selectivity (Hoogenboom et al., 2006; Sakthivel et al., 2016). This molecular selectivity is accomplished which results in signals generated only by the specific interactions between the molecules and the sensors, irrespective of the matrix complexity (Arora et al., 2011; Adley, 2014; Velusamy et al., 2010; Turner, 2013).

Specificity is another important property of biosensors (Turner, 2013). This property has been proven to overcome the high analytical accuracy required during the analysis of complex food matrices, overpowering issues such as differences in density, pH, and temperature (Thakur and Ragavan, 2013; Turner, 2013). Another benefit that makes the utilisation of biosensors very attractive to use in food applications is the possibility to miniaturisation. This can be accomplished using small sample sizes or volumes during the analysis of the sample (Adley, 2014; Thakur and Ragavan, 2013). Unfortunately, miniaturisation has only partially been achieved because of the limitations in the structural integrity of very fine electrode tips associated with microelectrodes (Thakur and Ragavan, 2013).

Developments in nanomaterials, such as nanoparticles and nanofibres, have provided new alternatives for the miniaturisation of this type of sensors (Pérez-López and Merkoçi, 2011; Zhu et al., 2015; Zeng et al., 2016; Pashazadeh

© Burleigh Dodds Science Publishing Limited, 2025. All rights reserved.

et al., 2017). Nanomaterials have enhanced the electrochemical biosensors by filtering the response of the electrode by increasing the surface area as well as supporting the robust attachment of the bioreceptor or recognition unit (Adley, 2014; Pérez-López and Merkoçi, 2011; Zhu et al., 2015; Zeng et al., 2016; Pashazadeh et al., 2017). The possibility of providing a greater surface-to-area volume ratio, nanomaterials give greater catalytic ability, warrant biocompatibility, and achieve lower mass transfer resistance. This characteristic has improved the selectivity, sensitivity, time efficiency, and cost effectiveness of most of the current biosensors available in the market (Pérez-López and Merkoçi, 2011; Zhu et al., 2015; Zeng et al., 2016). In the same way, the increase in transducer surface area delivers greater conductivity and sensitivity, promotes greater interaction capacity, and lowers detection limits (Adley, 2014; Pérez-López and Merkoçi, 2011; Pashazadeh et al., 2017; Eivazzadeh-Keihan et al., 2017; Verma, 2017).

Nano-biosensors have been developed and evaluated in the agriculture and food manufacturing industries to identify and quantify pesticides, herbicides, pathogenic microorganisms, and other microbial contaminant or unwanted microorganisms including viruses and bacteria, insects, fungi, hormones, and other foreign substances (Heo and Hua, 2009; Adley, 2014; Pérez-López and Merkoçi, 2011; Pashazadeh et al., 2017; Eivazzadeh-Keihan et al., 2017; Verma, 2017). For example, the utilisation of nano-biosensor allowed to detect pesticide residues at low concentrations (e.g. below 0.4 pM) (Heo and Hua, 2009; Pérez-López and Merkoçi, 2011). The incorporation of other nanomaterials at the transducer level (e.g. carbon nanotubes) has been demonstrated to increase both the electron transfer and transducer activity (Heo and Hua, 2009; Pérez-López and Merkoçi, 2011). Evidence of these improvements is in the slow but gradual replacement of traditional enzyme-substrate biosensors by nano-biosensor technology (Adley, 2014; Pérez-López and Merkoçi, 2011).

Developments in microfluidics and their incorporation into biosensors has provided with some analytical advantages (Kim et al., 2015; Luka et al., 2015). For example, microfluidics has determined a reduction in both sample and reagents volume (e.g. nanolitre), enhancing the sensitivity, as well as prompted for the development of a single platform for both sample preparation and detection (Heo and Hua, 2009; Adley, 2014; Mairhofer et al., 2009; Kim et al., 2015; Luka et al., 2015). However, the main advantage of using microfluidics strives in their portability allowing for the development of reusable or disposable as well as real-time recognition devices. This allows for the instantaneous analysis of the analyte (e.g. toxins, molecules, compounds) in a single device with high accuracy (Heo and Hua, 2009; Mairhofer et al., 2009). Recent examples of microfluidics are related with the detection of pathogenic bacterial species in food such as Salmonella (Kim et al., 2015).

© Burleigh Dodds Science Publishing Limited, 2025. All rights reserved.

Sensors used to measure or monitor aroma, flavour, and volatiles in foods are the so-called electronic noses (E_nose) and in some extent the electronic tongues (E_tongue). The E_nose is a device that mimics the perception of the mammalian olfactory system (Bartlett et al., 1997; Ampuero and Bosset, 2003; Deisingh et al., 2004; Loutfi et al., 2015; Hu et al., 2019). The principle of an E_nose is associated with the recognition of odours (e.g. volatile compounds) where E_nose instruments offer the ability to detect different volatiles or gases with no odour activity (Bartlett et al., 1997; Ampuero and Bosset, 2003; Deisingh et al., 2004; Loutfi et al., 2015; Hu et al., 2019). The E_nose can be improved to assess or monitor specific molecules or compounds of importance to humans and animals such as the scent of other animals, food ingredients, or spoilage (Bartlett et al., 1997; Ampuero and Bosset, 2003; Deisingh et al., 2004; Loutfi et al., 2015; Sanaeifar et al., 2017). A typical E_nose instrument consists of a series of array of sensors, where metal oxide sensors, MS, ion mobility spectrometry, GC, and conducting organic polymers are the most utilised in food applications (Bartlett et al., 1997; Ampuero and Bosset, 2003; Deisingh et al., 2004; Loutfi et al., 2015; Hu et al., 2019; Sanaeifar et al., 2017). However, these types of sensors are well known to lack in both sensitivity (ppm/ppb concentration) and selectivity and regrettably, this type of devices cannot respond in a manner comparable with that of the human olfactory system (Bartlett et al., 1997; Ampuero and Bosset, 2003; Deisingh et al., 2004; Loutfi et al., 2015; Hu et al., 2019). The use of E_noses has been extensively evaluated in different food applications (Gliszczy, 2017), such as quality assurance in bakery products (Ghasemi-Varnamkhasti and Lozano, 2016), the quality and composition of alcoholic beverages (Marti et al., 2005), and the quality control of edible oils (Majchrzak et al., 2018).

## 4 Electrochemical sensors

Electrochemical sensors are devices designed to measure current and voltage between electrodes in the electrolyte to sense the occurrence of molecules and compounds (e.g. different bio-analytes dissolved in the electrolyte) (Adley, 2014; Vidal et al., 2013; Rotariu et al., 2016). Typically, there is a working, counter, and reference electrode in an electrochemical cell. An example of electrochemical sensors in food applications is the electronic tongue (Adley, 2014; Vidal et al., 2013; Rotariu et al., 2016). This type of sensors is usually combined with biosensors that can be coupled with electrochemical transducers (Adley, 2014; Vidal et al., 2013; Rotariu et al., 2016). A wide range of chemical sensors have been used or designed as E_tongue devices including electrochemical, potentiometric, voltammetric (Winquist, 2008), amperometric, impedimetric, conductimetric, optical, mass, and enzymatic biosensors (Wilson and Baietto, 2009; Adley, 2014;

© Burleigh Dodds Science Publishing Limited, 2025. All rights reserved.

Vidal et al., 2013; Rotariu et al., 2016; Rackus et al., 2015). Generally, an E_tongue is defined as an advanced analytical device made up of an array of sensors combined with pattern recognition technologies that are able toexecute quantitative and qualitative analyses (Adley, 2014; Vidal et al., 2013). Voltammetric E_tongues are restricted to redox-active substances and are sometimes associated with low detection limits (Adley, 2014; Vidal et al., 2013). The potentiometric E_tongues are made up of an array of nonselective chemical sensors with partial specificity to different constituents in solution (Adley, 2014; Vidal et al., 2013). This type of E_tongue is most frequently used in food applications (Rotariu et al., 2016). The best example of an electrochemical detectors is the E_tongue. Several examples can be found in the literature where E_tongue has been reported to monitor different biological systems (Valle et al., 2010) and food applications (Escuder-Gilabert and Peris, 2010; Ciosek and Wroblewski, 2011).

## 5 Case studies

In recent years, the utilisation of sensing technologies contributed with the development of devices or platforms that have allowed to implement simple, rapid, cost-effective, and multiparametric response to the presence of contaminants, specific markers of various types of diseases, and/or indicators of safety levels in food ingredients and products (Ruiz-Altisent et al., 2010; Weston et al., 2021). In recent years, different applications have been developed and evaluated in the agriculture and food manufacturing industries to identify and quantify pesticides, herbicides, pathogenic microorganisms (e.g. viruses and bacteria), insects, and fungi, as well as hormones and other contaminants (e.g. toxins) (Ruiz-Altisent et al., 2010; Weston et al., 2021).

The application of sensors in the field of food analysis has exponentially grown in the last two decades as results in improved accuracy in the type and concentration of detection, increased interest from consumers about quality and safety of food ingredients and products, and the demand for reduction in analysis times associated with electrochemical detection (Ruiz-Altisent et al., 2010; Weston et al., 2021). Pathogen (e.g. presence of microorganism) and toxins (e.g. mycotoxins) detection are the two main areas of which electrochemical and other sensors have been applied the most (Ruiz-Altisent et al., 2010; Weston et al., 2021). Some specific applications on the application of biosensors to food ingredients and agriculture commodities are discussed below.

The simultaneous detection of antibiotics in milk using an automated microarray system was reported by Knecht et al. (2004). The proposed microarray system is based on the utilisation of an indirect enzyme-linked immunosorbent assay (ELISA) biosensor. Mishra et al. (2012) demonstrated that

© Burleigh Dodds Science Publishing Limited, 2025. All rights reserved.

a flow-based biosensor was able to detect organophosphate pesticides in milk, through the utilisation of a genetically modified acetylcholinesterase enzyme. Detection of different food pathogens such as *Salmonella* is essential to assure food quality and ensuring compliance with the regulations (Ehuwa et al., 2021). The detection of *Salmonella* in food using traditional methods requires between 4 and 5 days (Bell et al., 2016; Ehuwa et al., 2021). Methods such as ELISA and polymerase chain reaction involve complex sample pretreatment protocols requiring long analysis times, ranging from 12 to 24 h (Nowak et al., 2007). It has been reported that using microfluidic biosensors, the analysis time required to detect *Salmonella* in foods can be reduced (Qi et al., 2021).

## 6 Advantages and limitations on the use of sensors

The different reports available demonstrated the ability of electrochemical and other types of sensors applied to detect contamination in foods (Sharma et al., 2017). Most of these applications are based on the identification of a specific molecule or compound (target analysis), where others are based on the development of untargeted approaches where the use of classification tools (e.g. discriminant and cluster analysis) or regression models (e.g. calibration models) to predict those compounds in the sample.

The use of sensors brings a new set of challenges to the analysis of foods. For example, the determination of more than one metabolite using a single measurement in real time in complex matrices such as foods are still one of the main challenges of the application of biosensors. Although, some of these issues can be addressed by the incorporation of aptamers and antibodies as specific recognition elements.

The different studies and commercial applications of these devices also concluded that the use of sensing technologies allows for the development of efficient and robust systems to assess and monitor the contamination in a wide range of fresh produces and food ingredients. These methods have shown the potential of these non-invasive, inexpensive, and rapid technologies to aid in the development of management decision tools that would allow for the evaluation and monitoring food contamination along the food supply and value chains. The utilisation of sensing devices is providing the food industry with a robust analytical tool that can be incorporated into decision management tools. The development of these type of systems will provide with easy to use and rapid and less expensive options to monitor food contamination.

Despite the advantages that these technologies provide the implementation of these technologies is not fully embraced by the food manufacturing industry. Traditional analytical methods, such as HPLC and MS spectrometry, are still preferred by the analyst over the use of sensing

© Burleigh Dodds Science Publishing Limited, 2025. All rights reserved.

**Table 1** Advantages and challenges of the use of electrochemical sensors in food applications

| Advantages | Challenges |
|---|---|
| Real-time measurement | Food mixture complexity and natural variation |
| Miniaturisation and portability | Type of contaminant (e.g. antibiotic and toxin) |
| Reusable or disposable devices | Food sensor interface |
| Inexpensive | Signal processing |
| Robust devices | Specificity |

technologies. In addition, it is important to highlight that these traditional analytical methods are described in the official food safety systems and 'fully embraced' by government and authorities. The inclusion of sensing technologies is often exacerbated by the lack of training and the preference of classical food safety methods by the official food safety systems over the new technologies, where the lack of understanding of the principles underlying the use of sensing technologies is overwhelming. In some cases, the utilisation of sensing devices is still considered by researchers and practitioners in the field of food science and food safety authorities as '*black box*' approach due to many reasons, some of them described earlier. Table 1 summarises the advantages and challenges of the use of electrochemical sensors in food applications.

## 7 Final considerations

The available scientific literature and commercial examples on the application of sensing technologies in real industrial settings have proven that these techniques are a valuable tool to monitor food contamination. However, knowledge about the limitations of these techniques, as well as a proper understanding of the data analytics used, will avoid doubtful conclusions to be made from the application of these techniques. Continuous improvements on instrumentation (e.g. miniaturisation, portability, and robustness) and proper training on the utilisation and interpretation of the models will increase its trustworthiness, contributing to the standardisation of methods and systems along the food supply and value chain.

## 8 Conclusion and future trends

Future developments and trends in electrochemical sensors are focused in the inclusion of aptasensors that will allow the combination of high sensitivity and versatility of electrochemical detection systems with the specificity of the aptamers (Sharma et al., 2015). This combination makes this type of sensors highly capable and effective probes which undoubtedly will offer

© Burleigh Dodds Science Publishing Limited, 2025. All rights reserved.

a wide range of possibilities for the simultaneous detection of multiple analytes. They constitute a rapidly developing area of research, particularly in the development of labels that allow for a one-to-one relationship with the analyte and signal amplification. In this regard, one of the areas that has seen intensified research is the field of new materials, especially those at the nanoscale.

The use of nanomaterials as multiple labels presents with a new approach, with the inclusion of metallic quantum dots, metal–organic frameworks, and silica particles (Barsan et al., 2015). These materials exhibit high porosity, providing with a large specific surface area and a high density of molecular loading. These materials are complemented by more traditional labels such as redox and enzymatic labels. Despite intensive research efforts in the development of such multisensing platforms in recent years, their application is scarce and limited to a maximum of three analytes.

In summary, pathogen detection is the main trend and focus on the application and implementation of sensor devices by the food industry, particularly in the detection of multiple pathogens or derived toxins. More recently, biotechnology has shifted into ever smaller systems to allow for portability, cost reduction, analysis time reduction, and commercial viability of this type of sensors. Improvements in microfabrication systems have similarly aided in advancing biosensor technology and utility.

## 9 Where to look for further information

### The following articles provide a good overview of the subject:

- Astill, J., et al. (2019). Transparency in food supply chains: a review of enabling technology solutions. *Trends in Food Science and Technology*, 91, 240–247.
- Gaudin, V. (2017). Advances in biosensor development for the screening of antibiotic residues in food products of animal origin – a comprehensive review. *Biosensors and Bioelectronics*, 90, 363–377.
- Kachouei, A., et al. (2023). Internet of things-enabled food and plant sensors to empower sustainability. *Advanced Intelligent Systems*, 5, 2300321.

### *Key research in this area can be found at the following organisations:*

- United States Department of Agriculture (USDA) (www.usda.gov).
- School of Chemical Engineering and Australian Centre for NanoMedicine (ACN), The University of New South Wales.
- A wide range of private companies and laboratories.

© Burleigh Dodds Science Publishing Limited, 2025. All rights reserved.

## 10 References


Ali, J., et al. (2017). Biosensors: Their fundamentals, designs, types and most recent impactful applications: a review. *Journal of Biosensors & Bioelectronics,* 8(1), 1–9.

Adley, C. C. (2014). Past, present and future of sensors in food production. *Foods,* 3(3), 491–510.

Ampuero, S. and Bosset, J. (2003). The electronic nose applied to dairy products: a review. *Sensors and Actuators B: Chemical*, 94(1), 1–12.

Arora, P., et al. (2011). Biosensors as innovative tools for the detection of food borne pathogens. *Biosensors and Bioelectronics*, 28, 1–12.

Astill, J., et al. (2019). Transparency in food supply chains: a review of enabling technology solutions. *Trends in Food Science and Technology*, 91, 240–247.

Aung, M. M. and Chang, Y. S. (2014). Traceability in a food supply chain: safety and quality perspectives. *Food Control*, 39, 172–184.

Barsan, M. M., et al. (2015). Electrochemical sensors and biosensors based on redox polymer/carbon nanotube modified electrodes: a review. *Analytica Chimica Acta*, 8811–8823. https://doi.org/10.1016/j.aca.2015.02.059

Bartlett, P. N., et al. (1997). Electronic noses and their application in the food industry. *Food Technology*, 51 (12), 44–48.

Bec, K. B., et al. (2020). Review near-infrared spectroscopy in bio-applications. *Molecules*, 25, 2948. https://doi.org/10.3390/molecules25122948

Bec, K. B. and Huck, C. W. (2019). Breakthrough potential in near-infrared spectroscopy: spectra simulation. A review of recent developments. *Frontiers in Chemistry*. https://doi.org/10.3389/fchem.2019.00048

Bell, R. L., et al. (2016). Recent and emerging innovations in salmonella detection: a food and environmental perspective. *Microbial Biotechnology*, 9, 279–292.

Bettazzi, F., et al. (2017). Biosensors and related bioanalytical tools. *Comprehensive Analytical Chemistry*, 77, 1–33.

Bhalla, N., et al. (2016). Introduction to biosensors. *Essays in Biochemistry*, 60(1), 1–8.

Callao, M. P. and Ruisánchez, I. (2018). An overview of multivariate qualitative methods for food fraud detection. *Food Control*, 86, 283–293.

Carcea, M., et al. (2009). Food authenticity assessment: ensuring compliance with food legislation and traceability requirements. *Quality Assurance and Safety of Crops & Foods*, 1(2), 93–100.

Ciosek, P. and Wroblewski, W. (2011). Potentiometric electronic tongues for foodstuff and biosample recognition: an overview. *Sensors*, 11(5), 4688–701. https://doi.org/10.3390/s110504688

Cozzolino, D. (2015). Foodomics and infrared spectroscopy: from compounds to functionality. *Current Opinion in Food Science*, 4, 39–43.

Cozzolino, D. (2019). Food for thought: the digital disruption and the future of food production. *Current Research in Nutrition and Food Sciences*, 7, 607–609.

Deisingh, A. K., et al. (2004). Applications of electronic noses and tongues in food analysis. *International Journal of Food Science Technology*, 39(6), 587–604.

Derkus, B. (2016). Applying the miniaturization technologies for biosensor design. *Biosensors and Bioelectronics*, 79, 901–913.

Duan, M., et al. (2023). A CRISPR/Cas12a-mediated, DNA extraction and amplification-free, highly direct and rapid biosensor for *Salmonella typhimurium*. *Biosensors and Bioelectronics*, 219, 114823.


© Burleigh Dodds Science Publishing Limited, 2025. All rights reserved.

Ehuwa, O., et al. (2021). Salmonella, food safety and food handling practices. *Foods*, 10, 907.

Eivazzadeh-Keihan, R., et al. (2017). Recent advances in nanomaterial-mediated bio and immune sensors for detection of aflatoxin in food products. *TRAC Trends in Analytical Chemistry*, 87, 112–128.

Escuder-Gilabert, L. and Peris, M. (2010). Review: highlights in recent applications of electronic tongues in food analysis. *Analytica Chimica Acta*, 665(1), 15–25. https://doi.org/10.1016/j.aca.2010.03.017

Gao, D., et al. (2022). Recent advances in microfluidic devices for foodborne pathogens detection. *TrAC Trends in Analytical Chemistry*, 157, 116788.

Gaudin, V. (2017). Advances in biosensor development for the screening of antibiotic residues in food products of animal origin: a comprehensive review. *Biosensors and Bioelectronics*, 90, 363–377.

Ghasemi-Varnamkhasti, M. and Lozano, J. (2016). Electronic nose as an innovative measurement system for the quality assurance and control of bakery products: a review. *Engineering in Agriculture, Environment Food*, 9(4), 365–374.

Gliszczy, A. (2017). Electronic nose as a tool for monitoring the authenticity of food: a review. *Food Analytical Methods*, 10, 1800–1816. https://doi.org/10.1007/s12161-016-0739-4

Heo, J. and Hua, S. Z. (2009). An overview of recent strategies in pathogen sensing. *Sensors*, 9, 4483–4502.

Hoogenboom, L., et al. (2006). The CALUX bioassay: current status of its application to screening food and feed. *TrAC Trends in Analytical Chemistry*, 25, 410–420.

Hu, W., et al. (2019). Electronic noses: from advanced materials to sensors aided with data processing. *Advanced Materials Technologies*, 4, 1800488.

Justino, C. I. L., et al. (2015). Recent developments in recognition elements for chemical sensors and biosensors. *TRAC Trends in Analytical Chemistry*, 68, 2–17.

Kachouei, A., et al. (2023). Internet of things-enabled food and plant sensors to empower sustainability. *Advanced Intelligent Systems*, 5, 2300321.

Kim, G., et al. (2015). A microfluidic nano-biosensor for the detection of pathogenic salmonella. *Biosensors and Bioelectronics*, 67(Supplement C), 243–247.

Knecht, B. G., et al. (2004). Automated microarray system for the simultaneous detection of antibiotics in milk. *Analytical Chemistry*, 76, 646–654.

Korotkaya, E. (2014). Biosensors: design, classification, and applications in the food industry. *Foods and Raw Materials*, 2(2), 161–171.

Kukla, A., et al. (2009). Application of sensor arrays based on thin films of conducting polymers for chemical recognition of volatile organic solvents. *Sensors and Actuators B: Chemical*, 135(2), 541–551.

Loutfi, A., et al. (2015). Electronic noses for food quality: a review. *Journal of Food Engineering*, 144, 103–111.

Luka, G., et al. (2015). Microfluidics integrated biosensors: a leading technology towards lab-on-a-chip and sensing applications. *Sensors*, 15(12), 30011.

Mahato, K., et al. (2022). Onsite quality controls for food safety based on miniaturized biosensing. In *Miniaturized Biosensing Devices* (pp. 251–272). https://doi.org/10.1007/978-981-16-9897-2_11

Mairhofer, J., et al. (2009). Microfluidic systems for pathogen sensing: a review. *Sensors*, 9, 4804–4823.

© Burleigh Dodds Science Publishing Limited, 2025. All rights reserved.

Manning, L. and Soon, J. M. (2016). Food safety, food fraud, and food defense: a fast evolving literature. *Journal of Food Science*, 81, 823–834.

Martí, M. P., et al. (2005). Electronic noses in the quality control of alcoholic beverages. *TrAC Trends in Analytical Chemistry*, 24(1), 57–66.

Majchrzak, T., et al. (2018). Electronic noses in classification and quality control of edible oils: a review. *Food Chemistry*, 246, 192–201.

Miranda, J., et al. (2019). Sensing, smart and sustainable technologies for Agri-Food 4.0. *Computers in Industry*, 108, 21–36.

Mishra, R. K., et al. (2012). A novel automated flow-based biosensor for the determination of organophosphate pesticides in milk. *Biosensors and Bioelectronics*, 32, 56–61.

McGrath, T. F., et al. (2012). Biosensors for the analysis of microbiological and chemical contaminants in food. *Analytical and Bioanalytical Chemistry*, 403(1), 75–92.

McGrath, T. F., et al. (2018). What are the scientific challenges in moving from targeted to non-targeted methods for food fraud testing and how can they be addressed? – Spectroscopy case study. *Trends in Food Science and Technology*, 76, 38–55.

Nowak, B., et al. (2007). Salmonella contamination in pigs at slaughter and on the farm: a field study using an antibody ELISA test and a PCR technique. *International Journal of Food Microbiology*, 115, 259–267.

Nychas, G.-J. E., et al. Novel approaches for food safety management and communication. *Current Opinion in Food Science*, 12, 13–20.

Pashazadeh, P., et al. (2017). Nano-materials for use in sensing of salmonella infections: recent advances. *Biosensors and Bioelectronics*, 87, 1050–1064.

Pérez-López, B. and Merkoçi, A. (2011). Nanomaterials based biosensors for food analysis applications. *Trends in Food Science and Technology*, 22, 625–639.

Perumal, V. and Hashim, U. (2014). Advances in biosensors: principle, architecture and applications. *Journal of Applied Biomedicine*, 12(1), 1–15.

Pilolli, R., et al. (2013). Advances in biosensor development based on integrating nanotechnology and applied to food-allergen management. *TRAC Trends in Analytical Chemistry*, 47, 12–26.

Poltronieri, P., et al. (2014). Biosensors for the detection of food pathogens. *Foods*, 3(3), 511–526.

Qi, W., et al. (2021). A microfluidic biosensor for rapid and automatic detection of salmonella using metal-organic framework and raspberry Pi. *Biosensors and Bioelectronics*, 178, 113020.

Rai, M., et al. (2015). Emerging nanotechnology for detection of mycotoxins in food and feed. *International Journal of Food Sciences and Nutrition*, 66(4), 363–370.

Rackus, D. G., et al. (2015). Electrochemistry, biosensors and microfluidics: a convergence of fields. *Chemical Society Reviews*, 44(15), 5320–5340.

Rong, Y., et al. (2021). Lanthanide ion (Ln3+)-based upconversion sensor for quantification of food contaminants: a review. *Comprehensive Reviews in Food Science and Food Safety*, 20(4), 3531–3578.

Rotariu, L., et al. (2016). Electrochemical biosensors for fast detection of food contaminants – trends and perspective. *TRAC Trends in Analytical Chemistry*, 79, 80–87.

Ruiz-Altisent, M., et al. (2010). García-Ramos, sensors for product characterization and quality of specialty crops: a review. *Computers and Electronics in Agriculture*, 74(2), 176–194.

© Burleigh Dodds Science Publishing Limited, 2025. All rights reserved.

Sakthivel, S. (2016). CALUX bioassay: a cost-effective rapid screening technique for screening dioxins like compounds. *Reviews on Environmental Health*, 31(1), 149–152.

Sanaeifar, A., et al. (2017). Early detection of contamination and defect in foodstuffs by electronic nose: a review. *TrAC Trends in Analytical Chemistry*, 97, 257–271.

Sharma, A., et al. (2017). Recent advances in electrochemical-based sensing platforms for aflatoxins detection. *Chemosensors*, 5(1), 1.

Sharma, R., et al. (2015). Recent advances in nanoparticle based aptasensors for food contaminants. *Biosensors and Bioelectronics*, 74, 612–627.

Shen, Y., et al. (2022). Ratiometric fluorescent signals-driven smartphone-based portable sensors for onsite visual detection of food contaminants. *Coordination Chemistry Reviews*, 458, 214442.

Snopok, B. A. and Kruglenko, I. V. (2002). Multisensor systems for chemical analysis: state-of-the-art in electronic nose technology and new trends in machine olfaction. *Thin Solid Films*, 418(1), 21–41.

Spink, J. (2019). The current state of food fraud prevention: overview and requirements to address how to start? and How much is enough? *Current Opinion in Food Science*, 27, 130–138.

Thakur, M. S. and Ragavan, K. V. (2013). Biosensors in food processing. *Journal of Food Science and Technology*, 50(4), 625–641.

Turner, A. P. F. (2013). Biosensors: sense and sensibility. *Chemical Society Reviews*, 42(8), 3184–3196.

Valle, M., et al. (2010). A review of the use of the potentiometric electronic tongue in the monitoring of environmental systems. *Environmental Modelling & Software*, 25, 1023–1030. https://doi.org/10.1016/j.envsoft.2009.12.003

Velasco-Garcia, M. N. and Mottram, T. (2003). Biosensor technology addressing agricultural problems. *Biosystems Engineering*, 84(1), 1–12.

Velusamy, V., et al. (2010). An overview of foodborne pathogen detection: in the perspective of biosensors. *Biotechnology Advances*, 28(2), 232–254.

Verma, M. L. (2017). Nanobiotechnology advances in enzymatic biosensors for the agri-food industry. *Environmental Chemistry Letters*, 15(8), 1–6.

Vidal, J. C., et al. (2013). Electrochemical affinity biosensors for detection of mycotoxins: a review. *Biosensors and Bioelectronics*, 49, 146–158.

Visciano, P. and Schirone, M. (2021). Food frauds: global incidents and misleading situations. *Trends in Food Science and Technology*, 114, 424–442.

Weston, M., et al. (2021). Food sensors: challenges and opportunities. *Advanced Materials Technologies*, 6, 2001242.

Wilson, A. and Baietto, M. (2009). Applications and advances in electronic-nose technologies. *Sensors*, 9(7), 5099–5148.

Wilson, D. M., et al. (2001). Chemical sensors for portable, handheld field instruments. *IEEE Sensors Journal*, 1(4), 256–274.

Winquist, F. (2008). Voltammetric electronic tongues – Basic principles and applications. *Microchimica Acta*, 163, 3–10.

Wu, M. Y.-C., et al. (2017). Point-of-care detection devices for food safety monitoring: proactive disease prevention. *Trends in Biotechnology*, 35(4), 288–300.

Yeni, F., et al. (2014). Rapid and standardized methods for detection of foodborne pathogens and mycotoxins on fresh produce. *Food Control*, 40, 359–367.

© Burleigh Dodds Science Publishing Limited, 2025. All rights reserved.

Zaukuu, J. L. Z., et al. (2020). Emerging trends of advanced sensor based instruments for meat, poultry and fish quality: a review. *Critical Reviews in Food Science and Nutrition*, 60(20), 3443–3460.

Zeng, Y., et al. (2016). Nanomaterial-based electrochemical biosensors for food safety. *Journal of Electroanalytical Chemistry*, 78, 1147–1154.

Zhu, C., et al. (2015). Electrochemical sensors and biosensors based on nanomaterials and nanostructures. *Analytical Chemistry*, 87(1), 230–249.

© Burleigh Dodds Science Publishing Limited, 2025. All rights reserved.

# Chapter 6

# The use of effect biomarkers to assess potential effects of exposure to chemical residues from food and other sources

*G. Dervilly, M. Hernández-Mesa and B. Le Bizec, ONIRIS - INRAE - LABERCA, France*



## 1 Introduction: challenges in assessing risks from chemical residues in food

Chemical risk analysis in food safety and public health faces significant challenges due to the vast array of chemicals to which populations are exposed daily. As of 2024, the European Chemicals Agency (ECHA) has approximately 27 000 chemicals registered under the Registration, Evaluation, Authorisation and Restriction of Chemicals (REACH) regulation (ECHA, 2024), while the US Environmental Protection Agency (EPA) lists over 86 000 chemicals on the Toxic Substances Control Act inventory (EPA, 2024). Yet, only a fraction have undergone comprehensive risk assessments, particularly in the US, where less than 5% of the most widely used chemicals are assessed by the EPA. This challenge is even more complicated by the concept of the exposome, which

http://dx.doi.org/10.19103/AS.2025.0158.12
© Burleigh Dodds Science Publishing Limited, 2025. All rights reserved.

encompasses the totality of environmental exposures individuals encounter throughout their lifetime, including chemicals from air, water, diet, and lifestyle factors (Wild, 2012). Chemical hazards arise from various sources–environmental exposure, occupational settings, and habits like smoking–with food being a major exposure route for the general population. The European Food Safety Authority (EFSA) has identified over 5000 chemicals in food as potentially posing health risks (EFSA, 2023). Despite the EU regulating around 8000 food-related chemicals, the true scope of chemical hazards remains uncertain (European Commission, 2024). The complexity of the exposome highlights the need for integrated approaches to assess chemical exposure across different media and over time, significantly influencing risk assessments in food safety.

Risk analysis is an ongoing and dynamic process that not only addresses known hazards–often revisiting regulated chemicals with new data–but also anticipates emerging risks (WHO, 2009). These emerging risks may be tied to new products and technologies, such as nanomaterials, contaminants of emerging concern, or known chemicals with updated risk profiles, including pesticides, their transformation products, and endocrine disruptor compounds (EDCs). Additional factors also contribute to emerging risks, such as climate change, which can increase mycotoxin contamination, and globalized trade, which can lead to the introduction of new hazards by intensive transport (Ng and von Goetz, 2017).

Assessing the toxicity of chemicals is a critical step in understanding their potential risks to human health and the environment. In a risk assessment framework, traditional methods used to assess chemical toxicity primarily rely on *in vivo* (animal) studies, such as acute, sub-chronic, and chronic toxicity tests in rodents (Choudhuri et al., 2018). These studies involve exposing animals (typically rodents like mice and rats) to chemicals over different durations–acute (single or short-term exposure), sub-chronic (up to 90 days), or chronic (more than 6 months). The effects are observed at different levels, including organ toxicity, carcinogenicity, reproductive toxicity, neurotoxicity, etc. They are valuable for identifying toxic doses, mechanisms of action, and potential health risks for humans. In particular, reference toxicological value (VTR), are defined, focusing on relevant toxicological and epidemiological studies based on species, exposure route, duration, and dose. The adverse effect at the lowest dose, known as the critical dose, is selected using one of these three types:

- LOAEL (Lowest Observed Adverse Effect Level), the lowest dose causing a significant effect;
- NOAEL (No Observed Adverse Effect Level), the highest dose with no adverse effect; and

© Burleigh Dodds Science Publishing Limited, 2025. All rights reserved.

- BMD (Benchmark Dose), a modeled dose that produces a specific effect compared to a control group, with BMDL (Benchmark Dose Lower Bound) often used.

However, these methods have several limitations, including:

- ethical concerns over animal welfare;
- species differences that limit the extrapolation of results to humans due to toxicokinetics and toxicodynamics differences between species; and
- high costs and time requirements.

Additionally, these tests often involve a limited number of chemicals, due to the complexity and cost, which means only a limited number of chemicals can be tested comprehensively, not to mention the fact that regulations require a limited set of data for chemicals produced in small quantities. Traditional methods also struggle with assessing the toxicity of low-dose exposures and chemical mixtures, which are more reflective of real-world scenarios. The drive to satisfy the ethical expectations (3R principle) surrounding animal experimentation while improving human relevance has led to the development of alternative methods, such as *in vitro* and computational models, to address these limitations.

Toxicity may also be assessed using *in vitro* models of cultured cells, derived from humans or animals, to study the cytotoxicity, genotoxicity, and specific organ toxicities (Choudhuri et al., 2018). However, traditional 2D cell cultures may not accurately mimic the complexity of whole organisms, limiting their predictive value for *in vivo* outcomes. Such models further face limited metabolic capacity as some cell lines do not replicate the metabolic capacity of an entire organ, affecting the relevance of data, especially for chemicals that require metabolic activation. Finally, variability in the cell model is a recognized issue as different cell lines can yield different results, leading to inconsistencies. More recently, high-throughput screening (HTS) appeared as a method to quickly evaluate the toxicity of large libraries of chemicals or drugs. It employs automated robotic systems, miniaturized assays, and advanced data analysis to screen thousands of compounds simultaneously. It is widely used to identify toxicants by measuring biological responses like cell viability or enzyme activity. While HTS is efficient and cost-effective, it often uses simplified models that may not fully replicate complex biological systems, requiring further validation to ensure the relevance of findings in real-world scenarios.

Quantitative structure-activity relationship (QSAR) models are computational tools used to predict the toxicity and biological activity of chemicals based on their molecular structure (Dearden, 2017, Roy et al., 2015). By analyzing the relationship between chemical structures and their known biological effects,

© Burleigh Dodds Science Publishing Limited, 2025. All rights reserved.

QSAR models provide rapid, cost-effective screening for potentially toxic substances, helping to prioritize chemicals for further testing. These models utilize mathematical and statistical techniques to correlate chemical properties, such as molecular weight or electronegativity, with biological activity, allowing for predictions without requiring extensive experimental data. However, QSAR models have limitations. They depend on the quality and diversity of the training datasets, which may not cover all chemical structures, reducing their applicability to novel or structurally distinct compounds. Moreover, QSAR predictions are often based on simplified assumptions, lacking detailed mechanistic insights into toxicity pathways. Therefore, while valuable for initial screening, QSAR models require validation; they are typically used in combination with other methods for comprehensive hazard characterization.

In parallel, human-based models may also be relevant for assessing chemical toxicity (Choudhuri et al., 2018). Besides *in vitro* models using human cells and tissues, they include epidemiological studies and human volunteer studies. Epidemiological studies examine the associations between chemical exposures and health outcomes in real-world populations, providing valuable insights into long-term effects and dose-response relationships. However, these studies can be limited by confounding factors and difficulties in establishing causality. Reliable exposure and outcome data can indeed be difficult to obtain, and studies are often retrospective. Human volunteer studies, such as microdosing trials, offer data on pharmacokinetics and early biological responses to low, non-toxic doses but are restricted by ethical and safety considerations, limiting their use for comprehensive toxicity assessments. Further, these studies do not provide comprehensive toxicity data, especially for effects that require higher exposures or longer durations.

It is in this context that the concept of biomarkers of effect has emerged as a promising alternative to traditional toxicity tests due to the need for more sensitive, specific, and human-relevant methods to assess chemical exposure effects (Gupta, 2014, Campion et al., 2013). Unlike most traditional tests, which rely on animal models and are time-consuming, costly, and ethically sensitive, biomarkers of effects provide a direct measure of biological changes in response to chemical exposure. They enable a more global approach, without any preconceived ideas about the expected effects, unlike traditional approaches that examine effects that have already been described. They can also be used to assess the effects of mixtures.

## 2 Defining effect biomarkers for chemical risk assessment

The concept of 'effect biomarkers,' also known as 'biomarkers of effect,' began to take shape in the late 1980s and early 1990s as part of the evolving field of

© Burleigh Dodds Science Publishing Limited, 2025. All rights reserved.

toxicology and environmental health. The first formal definitions were provided by key regulatory and research organizations. One of the earliest and most influential definitions of effect biomarkers came from the National Research Council (NRC) in its 1989 report titled *'Biologic Markers in Reproductive Toxicology.'* (National Research Council, 1989). This report introduced the broader concept of 'biologic markers' and categorized them into three types:

- Biomarkers of exposure: Indicate the presence of a chemical or its metabolites in the body.
- Biomarkers of effect: Indicate a biological response to an exposure that could lead to a health outcome.
- Biomarkers of susceptibility: Indicate an individual's predisposition to respond to a particular exposure.

The NRC defined biomarkers of effect as measurable biochemical, physiological, behavioral, or other alterations within an organism that, depending on the magnitude, can be recognized as associated with an established or possible health impairment or disease. This foundational definition provided a basis for further research and development in the field of biomarkers, leading to more refined definitions and classifications in later years, especially by organizations such as the World Health Organization (WHO) (WHO, 1993, 2001).

An effect biomarker (or biomarker of effect) is defined as a biological indicator that provides evidence of a biological response to exposure to a chemical, environmental agent, or other stressors. It reflects changes in biochemical, physiological, behavioral, or other alterations in an organism that can be associated with adverse or potentially adverse health effects.

More specifically an effect biomarker is associated with the following key characteristics:

- *Indication of biological response*: Effect biomarkers indicate that a biochemical or physiological process has been altered due to exposure to a specific substance or condition.
- *Mechanistic insight*: They often provide insight into the mechanism of toxicity or the physiopathological process occurring due to exposure.
- *Disease prediction and progression*: Effect biomarkers can help in predicting disease development or monitoring the progression of a disease state resulting from exposure.
- *Measurable changes*: These biomarkers are typically measurable in biological samples such as blood, urine, saliva, tissues, or any accessible fluids or cells.

© Burleigh Dodds Science Publishing Limited, 2025. All rights reserved.

Effect biomarkers are generally categorized into three types:

- *Molecular Biomarkers*: Include DNA adducts, gene expression changes, and protein modifications.
- *Cellular Biomarkers*: Include changes in cell proliferation, apoptosis, or cell signaling pathways.
- *Biochemical Biomarkers*: Involve changes in enzyme activities, hormone levels, or other metabolic processes such as oxidative stress or inflammatory markers.

From a time perspective, the term 'effect biomarker' first appeared in scientific literature in the early 1990s, coinciding with the growing interest in alternative tools for assessing the biological impact of environmental

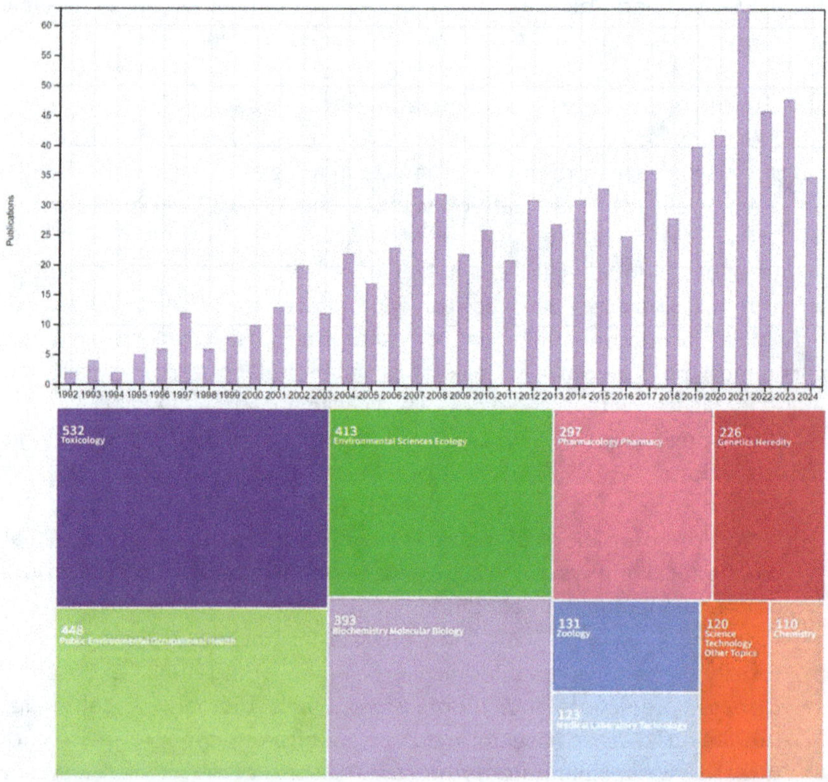


**Figure 1** Query of the Web of Science™ database on the terms 'biomarker of effect,' 'biomarkers of effect,' 'effect biomarker' and 'effect biomarkers,' (all fields) on all available databases, showing (up) the first occurrences of the term in the literature and the evolution in terms of number of publications (n = 781 in total at the date of the query) as well as (down) the main scientific fields concerned. (Accessed 13 September 2024).

© Burleigh Dodds Science Publishing Limited, 2025. All rights reserved.

exposures, including chemical and pharmaceutical agents. Initially, the term was used primarily in toxicology and pharmacology to describe measurable changes in biological systems that indicate a physiological response to exposure. Over time, its use expanded across various domains, including environmental health and occupational medicine (Fig. 1). By the early 2000s, the concept of effect biomarkers had evolved to encompass a broader range of applications, such as understanding disease mechanisms, assessing drug efficacy, and evaluating risk in personalized medicine. The increasing use of high-throughput omics technologies, computational biology, and systems toxicology in the 2010s further accelerated the term's prominence in scientific publications, leading to a more integrative and mechanistic understanding of how chemical exposures impact biological pathways and lead to adverse outcomes (AOs).

## 3 Discovery of effect biomarkers

The discovery of effect biomarkers represents a significant advancement in the fields of toxicology and environmental health and was driven by the limitations of traditional toxicity assessment methods, which primarily relied on animal studies and epidemiological data. These conventional methods were often time-consuming, ethically challenging, and limited in their ability to detect early or subtle effects of exposure. Moreover, the complex nature of chemical exposures in real-life scenarios, which often involve low doses of multiple chemicals, could not be adequately addressed by traditional models.

With advances in molecular biology and biotechnology, researchers began to identify specific biological changes that occur in response to toxicant exposure. These changes included alterations in gene expression, protein levels, and metabolic profiles, which could serve as indicators of biological effects (Choudhuri et al., 2018, Buesen et al., 2017).

The advent of omics technologies–including genomics, proteomics, metabolomics, and epigenomics–marked a significant turning point in the discovery of effect biomarkers (Mortimer et al., 2022, Nguyen et al., 2022, Espín-Pérez et al., 2014). These high-throughput technologies enabled comprehensive analysis of thousands of genes, proteins, and metabolites simultaneously, allowing researchers to detect broad-spectrum changes in biological systems in response to chemical exposures.

Discovering biomarkers of effect in the context of chemical exposure involves designing experiments that can reliably identify biological changes associated with exposure to toxicants. By comparing exposed and control samples, researchers can identify specific molecular changes that serve as potential biomarkers of effect. A well-structured experimental design for discovering biomarkers of effect generally involves multiple steps: selecting

© Burleigh Dodds Science Publishing Limited, 2025. All rights reserved.

the appropriate model system, defining exposure conditions (route, dose, and duration of exposure), choosing relevant endpoints, and employing advanced analytical techniques. Below is a typical experimental design outline used in toxicological studies to identify biomarkers of effect:

- Selection of model system
- Defining exposure conditions
- Selection of endpoints and biomarkers
- Analytical methods and high-throughput technologies
- Quality controls and validation

These stages are discussed in more detail in the coming sections.

### 3.1 Selection of model system

The first step is choosing an appropriate model system, which can include *in vivo*, *in vitro*, or *in silico* approaches. While animal models, such as rodents, are commonly used to study the systemic effects of chemical exposure, ethical concerns and differences in species responses have prompted a shift towards alternative models. Advanced *in vitro* models, including 3D cell cultures, organoids, and organ-on-a-chip systems, are increasingly used to assess the effects of chemical residues (Liu et al., 2019). These models offer more human-relevant data compared to traditional 2D cell cultures and animal models, allowing researchers to study tissue-specific responses and interactions in a controlled environment. For instance, liver and gut organoids are being used to understand how foodborne chemicals influence metabolic pathways and the gut microbiome, respectively. Also, when ethical and safe, human-based studies can also provide direct relevance to human health outcomes. The sampling method is in this context guided by the nature of the biomarker, its biological location, and the target organ or system being studied. Biomarkers related to systemic conditions (e.g. hormones, inflammatory markers) can often be detected in blood, saliva, or urine. These fluids are easily accessible (non invasive) and provide a broad view of the body's physiological state without causing discomfort to the participant. For biomarkers specific to particular tissues, such as skin for dermatological studies or buccal cells for genetic analyses, minimally invasive techniques like swabbing or dermal tape stripping are suitable. When biomarkers originate from less accessible organs (e.g. liver enzymes, or kidney function markers), invasive techniques such as biopsies might be required. These are used sparingly due to ethical and safety concerns.

© Burleigh Dodds Science Publishing Limited, 2025. All rights reserved.

## 3.2 Defining exposure conditions

The next step involves defining the chemical exposure conditions, including the choice of chemical(s), dose levels, duration, and route of exposure. A single chemical or a mixture of chemicals relevant to real-world exposures (e.g. environmental pollutants, food additives, and industrial chemicals) is chosen. Multiple concentrations are selected, including low, medium, and high doses, to capture a range of effects from sub-toxic to toxic levels. Dose selection is guided by existing toxicological data or prior studies. Regarding the duration of exposure: short-term (acute), medium-term (sub-chronic), and long-term (chronic) exposures are considered to capture both immediate and delayed effects. Finally, the route of exposure such as oral, dermal, inhalation, or injection are chosen to reflect realistic exposure scenarios.

## 3.3 Selection of endpoints and biomarkers

A comprehensive range of endpoints are selected to cover different levels of biological organization. Molecular endpoints include changes in gene expression, protein levels, and metabolite profiles. Technologies like RNA sequencing (transcriptomics), mass spectrometry (proteomics and metabolomics), and DNA methylation arrays (epigenomics) are commonly employed. Cellular endpoints measure cellular responses such as oxidative stress, inflammation, cell proliferation, apoptosis, and DNA damage. Physiological and clinical endpoints assess changes at the organ or system level, such as liver function tests, neurological assessments, or immune responses.

## 3.4 Analytical methods and high-throughput technologies

The discovery of effect biomarkers relies heavily on high-throughput technologies and advanced analytical methods to detect and quantify biological changes (Dai and Shen, 2022). Omics technologies such as genomics, proteomics, metabolomics, and epigenomics provide a holistic view of the biological impact of chemical exposure. The choice of measurement method depends on the nature of the biomarker. Genomics and transcriptomics reveal gene mutations or altered expression levels, relying on polymerase chain reaction (PCR), real-time PCR (qPCR), or Next Generation Sequencing techniques; proteomics uncovers changes in protein abundance or modifications, involving immunoassays such as Enzyme-linked immunosorbent assay (ELISA) or Western Blot, or mass spectrometry within specific proteomics workflows; metabolomics detects shifts in metabolite concentrations, relying on cutting-edge analytical platforms involving high-resolution mass spectrometry coupled with chromatographic separation (LC/GC-HRMS) and/or

© Burleigh Dodds Science Publishing Limited, 2025. All rights reserved.

nuclear magnetic resonance systems (Letertre et al., 2020); and epigenomics examines DNA methylation or histone modifications implementing both PCR related tools or liquid chromatography coupled with mass spectrometry. These methods can identify thousands of potential biomarkers in a single experiment. Sophisticated bioinformatics tools are therefore required to process these large datasets and perform statistical analyses (Babin et al., 2023, Kaur et al., 2021, Cai et al., 2020). Data analysis involves statistical and computational approaches to identify significant changes in biological endpoints associated with chemical exposure, mainly relying on differential analysis which identifies genes, proteins, or metabolites that show significant changes in expression or abundance between exposed and control groups. Bioinformatics tools are also implemented to integrate data across multiple omics platforms, within a systems biology approach that enhances the understanding of the causal links between exposure and health outcomes (Santiago-Rodriguez and Hollister, 2021, Chen et al., 2022). Also, machine learning and AI-based approaches are increasingly used to identify patterns and predict potential biomarkers of effect (Reel et al., 2021).

## 3.5 Quality controls and validation

One of the special features of these approaches is the importance of quality control (Brooks et al., 2024, Gonzalez-Dominguez et al., 2024, Dunn et al., 2012). Careful design of controls and replicates is therefore essential for ensuring the validity of the study. Negative controls (untreated or non-exposed groups) are used to establish baseline responses, while positive controls (known exposure to toxicants) help validate the sensitivity of the experimental system. Multiple biological replicates are included to account for biological variability and ensure the robustness of the findings, while for each biological replicate, multiple technical replicates are often analyzed to account for variability in measurement techniques. Finally, identified biomarkers are validated in independent studies to confirm their relevance and reliability. This involves reproducing the findings in different models or populations and under varied conditions.

## 3.6 Improving discovery: the role of metabolomics

While the various omics techniques that have been detailed are useful in the search for and identification of effect markers in environmental health and exposome research, metabolomics occupies a central place (Hernandez-Mesa et al., 2021, Dehghani et al., 2022, Wei et al., 2022, Misra, 2019, Cai et al., 2020, van Ravenzwaay et al., 2007, 2014). The move by toxicologists towards metabolomics represents indeed a major change in toxicological studies, driven by the need to understand complex biological responses to chemical

© Burleigh Dodds Science Publishing Limited, 2025. All rights reserved.

exposure. After an initial phase of reluctance for this science (Robertson, 2005), it is now part of the complete arsenal used in the field of toxicology (Araujo et al., 2021). Specifically referring to the comprehensive study of small molecules or metabolites within cells, tissues, biofluids, or organisms, representing the downstream products of gene expression and protein activity, it can thus provide a real-time snapshot of the physiological state of an organism. Because metabolites are directly linked to the phenotype, metabolomics is particularly useful in identifying biomarkers of effect, especially for complex exposures like those to chemicals in food and the environment. Unlike targeted approaches that focus on specific biomarkers, metabolomics offers an unbiased, global analysis of metabolites, capturing the entire biochemical network affected by chemical exposure. This holistic approach allows for the detection of unexpected and novel biomarkers of effect. It can detect subtle changes in metabolic pathways that may occur even at low levels of chemical exposure. This sensitivity is crucial for identifying early biological effects before clinical symptoms appear. By identifying alterations in metabolic pathways, metabolomics provides insights into the mechanisms of toxicity and adverse effects of chemicals. For example, changes in energy metabolism, oxidative stress markers, or lipid profiles can indicate specific toxicological effects.

# 4 Interpretation of effect biomarkers

## *4.1 Challenges in interpreting results*

Identifying biomarkers using omics approaches presents, however, several challenges that can impact the reliability and interpretation of the results, and have to be addressed very early in the experimental design (Brooks et al., 2021). One important key issue refers to the high dimensionality and complexity of the data. Such data are indeed high-dimensional, with thousands of variables (e.g. genes, proteins, and metabolites) but often limited sample sizes. This can lead to overfitting in statistical models, where the identified biomarkers may not be truly associated with the condition of interest but rather with random noise. Further, when it comes to reproducibility and validation of the workflow, many identified biomarkers lack reproducibility across different studies, populations, or experimental conditions. Batch effects, differences in sample preparation, and variations in analytical platforms can lead to inconsistent findings. Robust validation in independent cohorts is often needed but is not always performed. Biological variability between individuals and technical variability in sample handling, data processing, and acquisition should not be overlooked, as they can mask the true signals. Controlling these variabilities is therefore essential, and is the responsibility of laboratories as far as the technical aspects are concerned (implementation of QA/QC procedures), but more difficult to

© Burleigh Dodds Science Publishing Limited, 2025. All rights reserved.

control in the case of complex biological samples. Omics biomarkers may be influenced by confounding factors such as age, gender, diet, or environmental exposures that are not related to the chemical exposure of interest. Additionally, some biomarkers may lack specificity, being associated with multiple conditions rather than a particular toxicant or disease state. When it comes to combining data from different omics layers (e.g. genomics, proteomics, and metabolomics), issues are encountered due to differences in data types, scales, and quality (Chen et al., 2022). Effective data integration requires sophisticated bioinformatics and computational tools, which can be computationally intensive and require specialized expertise. Finally, omics technologies, while powerful, are often expensive and require substantial infrastructure, expertise, and computational resources. This limits their accessibility, particularly for routine use in toxicological assessments or in low-resource settings.

## 4.2 Mechanistic insights into chemical exposures and health outcomes

Even when potential biomarkers are identified, understanding their biological relevance and linking them to specific mechanisms or pathways remains challenging. Many omics studies generate large lists of potential biomarkers without clear insights into their functional roles in toxicity or disease. To give meaning to the discovered effect biomarkers, innovative approaches help in understanding the mechanistic links between chemical exposures and adverse health effects, thereby improving the identification and validation of the biomarkers and enabling more accurate predictions of toxicity. Critical tools in modern toxicology and risk assessment include:

- Biological pathway analysis;
- Adverse outcome pathways (AOPs); and
- Quantitative biological activity relationship (QBAR) models.

These are briefly reviewed in the coming sections.

## 4.3 Biological pathway analysis

This approach involves examining the complex networks of interactions among genes, proteins, and metabolites to understand how biological systems respond to chemical exposures (Garcia-Campos et al., 2015, Khatri et al., 2012). This approach utilizes data from high-throughput omics technologies to identify key pathways that are altered due to toxicant exposure. Common tools for pathway analysis include Kyoto Encyclopedia of Genes and Genomes, Reactome, Ingenuity Pathway Analysis, and Gene Ontology. Pathway analysis

© Burleigh Dodds Science Publishing Limited, 2025. All rights reserved.

helps identify potential biomarkers by highlighting the key molecular players and processes involved in toxicity. For instance, exposure to a chemical might trigger oxidative stress pathways, DNA damage responses, or inflammatory cascades. By identifying these pathways, researchers can pinpoint specific genes, proteins, or metabolites that serve as biomarkers of effect. Moreover, pathway analysis facilitates the integration of multi-omics data, providing a systems biology perspective that captures the complexity of biological responses to chemical exposures. This comprehensive view is crucial for understanding how disruptions in cellular pathways can lead to adverse health outcomes.

## 4.4 Adverse outcome pathways

Adverse outcome pathways (AOPs) provide a structured framework for linking molecular-level disruptions caused by a chemical to adverse health effects at higher levels of biological organization, such as cellular, tissue, organ, and whole-organism levels (Vinken et al., 2017). An AOP typically starts with a molecular initiating event (MIE)–the initial interaction between a chemical and a biological target, such as a receptor or enzyme (Fig. 2). This MIE triggers a cascade of key events (KEs), which are measurable biological changes that occur sequentially and lead to an AO, such as organ toxicity or disease. By elucidating the series of biological events that lead to an AO, AOPs help identify critical points for intervention and guide the development of more focused

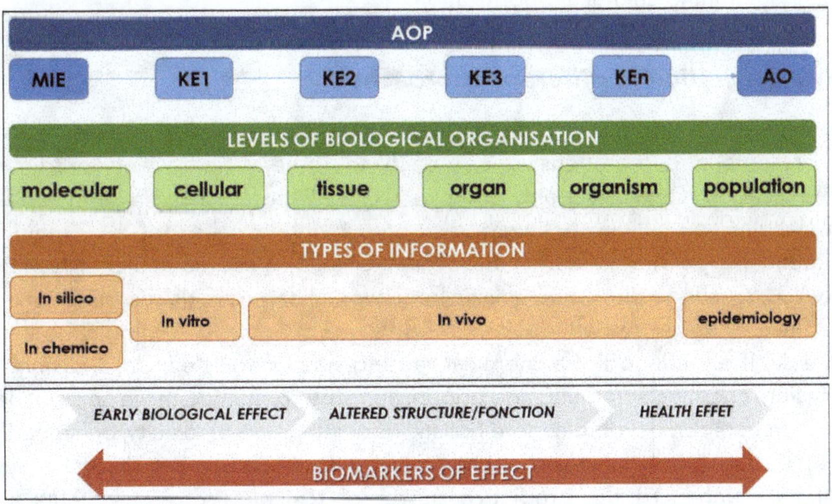


**Figure 2** Simplified representation of AOPs and the role of biomarkers of effect. (AOP: adverse outcome pathways, MIE: molecular initiating event, KE: key events, AO: adverse outcome).

© Burleigh Dodds Science Publishing Limited, 2025. All rights reserved.

and relevant biomarkers of effect. Moreover, AOPs can be used to evaluate the potential toxicity of untested chemicals by comparing their molecular profiles with known AOPs, thereby reducing reliance on animal testing. Launched in 2012 by the Organisation for Economic Co-operation and Development (OECD), a programme on the development of AO pathways has proposed a guidance document on how AOPs are to be developed, reviewed, agreed and published at the OECD level.

## 4.5 Quantitative biological activity relationship models

Quantitative biological activity relationship (QBAR) models extend the concept of QSAR models by incorporating quantitative biological data, such as gene expression profiles, protein levels, or cellular phenotypes, to predict the biological activity or toxicity of chemicals (van Ravenzwaay et al., 2012). Quantitative biological activity relationship models aim to establish a relationship between the biological activity of a chemical and its structure or physicochemical properties, but with a focus on quantitative biological endpoints rather than just chemical properties. Quantitative biological activity relationship models integrate data from omics technologies and pathway analysis to develop robust, quantitative models that can predict toxicity outcomes. By incorporating biological data, QBAR models provide a more mechanistic basis for toxicity predictions compared to traditional QSAR models. These models can be used to predict the likelihood of a chemical triggering a specific pathway or AOP, offering a powerful tool for high-throughput risk assessment and chemical prioritization.

## 4.6 Integration of different approaches

The integration of biological pathway analysis, AOPs, and QBAR models represents a comprehensive approach to understanding and predicting chemical toxicity (Ankley and Edwards, 2018). Pathway analysis identifies key molecular and cellular pathways disrupted by chemical exposure, providing potential targets for biomarker discovery. AOPs Link these molecular disruptions to AOs, creating a clear mechanistic framework that guides risk assessment and regulatory decisions, while QBAR models use quantitative biological data to predict the likelihood of a chemical causing specific toxic effects, allowing for efficient chemical screening and prioritization. By combining these approaches, researchers can develop a more holistic understanding of toxicity, bridging the gap between molecular-level events and organism-level effects. This integrative strategy enhances the predictive power of toxicological assessments and supports the development of safer chemicals and more targeted regulatory policies.

© Burleigh Dodds Science Publishing Limited, 2025. All rights reserved.

## 5 Examples of effect biomarkers associated with specific pathways or outcomes

Biomarkers of effect can span various levels of biological organization, from molecular changes to physiological responses. Catalogs of effect markers, such as AOP-Wiki (Society for the Advancement of Adverse Outcome Pathways, 2024), the OECD's Adverse Outcome Pathway Knowledge Base (AOP-KB) (OECD), and the US EPA's TOXCAST/TOX21 (US EPA, 2024) databases, provide essential resources for linking chemical exposures to health outcomes. These tools document biomarkers and pathways associated with toxicity. Additional insights are offered by initiatives like the Human Toxome Project (Bouhifd et al., 2015) and peer-reviewed literature that catalog markers for oxidative stress, inflammation, and other effects, enhancing our understanding of exposure-health linkages. The aim is not to draw up here an exhaustive catalog of markers already characterized, but to offer a panorama of compounds whose biological variations are now recognized as being associated with specific pathways or outcomes, in relation to chemical exposure (Sun et al., 2022).

Chronic activation of inflammation pathways can lead to negative consequences such as chronic inflammatory diseases, atherosclerosis, and metabolic disorders such as diabetes. Biomarkers of inflammation, such as C-reactive protein and interleukin 6 (IL-6), are upregulated in response to chemical exposures such as bisphenol A (BPA). These exposures can activate pro-inflammatory signaling pathways, including nuclear factor kappa B (NF-κB) and pathogen-activated protein kinase (MAPK).

Typical oxidative stress biomarkers are malondialdehyde (MDA) and glutathione (GSH). Malondialdehyde is a biomarker of lipid peroxidation, while GSH is a critical antioxidant. In particular, low GSH levels and high MDA levels indicate cellular oxidative damage, contributing to AOs such as cardiovascular diseases, neurological disorders, and cancer. Such effects have been observed upon heavy metals exposure.

Liver toxicity can lead to liver disease, including steatosis (fatty liver disease), fibrosis, cirrhosis, or even hepatocellular carcinoma. Alanine aminotransferase (ALT) and Aspartate aminotransferase (AST) are liver enzymes that serve as biomarkers of liver damage. Their elevated blood levels indicate hepatocellular damage caused by exposure to hepatotoxic substances such as mycotoxins, or industrial chemicals. These chemicals often induce toxicity via pathways such as oxidative stress, mitochondrial dysfunction, and lipid peroxidation. MicroRNA-122 (miR-122) is another recognized biomarker for liver toxicity as its expression is highly correlated with liver damage, particularly in cases of hepatotoxicity.

© Burleigh Dodds Science Publishing Limited, 2025. All rights reserved.

Oxidative DNA damage can lead to mutagenesis and carcinogenesis, contributing to the development of cancers. 8-Hydroxy-2′-deoxyguanosine (8-OHdG) is a biomarker of oxidative DNA damage, resulting from the generation of reactive oxygen species due to chemical exposure, such as polycyclic aromatic hydrocarbons or heavy metals. Elevated levels of 8-OHdG are linked to the activation of DNA damage response pathways, which include nucleotide excision repair and base excision repair.

Neurotoxicants like lead, mercury, organophosphates, and certain pesticides trigger pathways that involve oxidative stress and calcium dysregulation leading to neuronal damage and cell death. Persistent exposure and elevated levels of these biomarkers are linked to AOs such as Alzheimer's disease, Parkinson's disease, and other neurodegenerative conditions. Tau protein and glial fibrillary acidic protein are biomarkers associated with neurotoxicity and neurodegeneration.

Endocrine disruption biomarkers such as thyroid hormones (T3, T4) and sex hormone-binding globulin indicate disruption of the hypothalamic-pituitary-thyroid axis and reproductive hormone pathways, respectively. Such disruptions can lead to AOs such as reproductive toxicity, developmental disorders, thyroid dysfunction, and hormone-dependent cancers. Chemicals like bisphenol A (BPA), phthalates, and polychlorinated biphenyls (PCBs) are known endocrine disruptors that interfere with hormone biosynthesis, metabolism, or signaling.

## 6 Regulatory and risk assessment issues

The adoption of new methods and biomarkers into regulatory frameworks represents a critical evolution in modern toxicology and risk assessment. Traditional approaches to assessing the safety of chemicals, such as animal testing and observational studies, are increasingly being supplemented or replaced by more advanced, mechanism-based methods involving the identification of biomarkers of effect. They offer the potential for more accurate, faster, and ethically sound evaluations of chemical toxicity.

### *6.1 Towards next-generation risk assessment*

Omics is being explored as a promising approach for next-generation risk assessment (NGRA) with suggested thresholds such as reference points (RPs); also known as the 'point of departure' (POD), RP is defined as the dose level corresponding to the onset of the compound's effects (Viant et al., 2019). The RP method based on omics data therefore comes under the concept of new approach methods (NAMs) aiming to facilitate risk assessment without generating new *in vivo* data. The Scientific Committee of the EFSA has

© Burleigh Dodds Science Publishing Limited, 2025. All rights reserved.

advised using benchmark doses (BMDs), which are more relevant than the no-observed-adverse-effect level (NOAEL), to derive RPs (EFSA Scientific Committee et al., 2017). For instance, RPs derived from BMDs (*in vivo*) or BMCs (*in vitro*) established from transcriptomics data can predict RPs from conventional 'endpoint' studies, thereby illustrating the benefit of transcriptomics for risk assessment (Black et al., 2022). So far, few studies have demonstrated that metabolomics can also predict RPs; yet applications have been successfully reported to assess the RPs of various chemicals on HepaRG cells (Dubreil et al., 2024, Malinowska et al., 2023) or to derive RPs to substantiate the ecotoxicological risk assessment of chemicals (Larras et al., 2020). In these approaches, the continuous responses of metabolites are modeled over a gradient of concentrations using open-source software such as BMDExpress-2, PROAST (RIVM National Institute for Public Health and the Environment, 2024 ), EFSA software (EFSA), or the DRomics package (Larras et al., 2018).

## 6.2 Challenges in integrating omics-based methods into regulatory frameworks

However, integrating these methods into regulatory frameworks involves several scientific, technical, and policy challenges that are to be detailed.

- *Scientific Validation and Standardization*. For new methods and biomarkers to be accepted into regulatory frameworks, they must undergo rigorous scientific validation to ensure they are reliable, reproducible, and predictive of AOs in humans. Standardized methods ensure consistency and comparability across different studies and regulatory applications. Organizations such as the OECD have established validation guidelines for new test methods (OECD, 2005). These guidelines require demonstrating the method's relevance (biological significance) and reliability (reproducibility across different laboratories and settings). New methods, particularly those involving high-throughput technologies and omics approaches, require standardization of protocols, data analysis methods, and interpretation frameworks. For instance, the OECD has developed guidance documents on the validation of AOPs (OECD, 2018, 2021). With regard more specifically to the validation of biomarkers of effect with a view to their use in a regulatory context, there are few examples of biomarkers being validated with a view to their use in a regulatory context. A typical example is the liver toxicity marker miR-122. Discovered by transcriptomic approaches, its levels are significantly elevated in response to liver damage. It has been proposed as a sensitive biomarker for drug-induced liver injury and has been included in the Food and Drug Administration's (FDA) draft guidance for the assessment of liver toxicity

© Burleigh Dodds Science Publishing Limited, 2025. All rights reserved.

(FDA, 2018). It is considered a promising alternative or complement to traditional liver function tests such as ALT and AST, as it offers greater sensitivity, particularly for the detection of early liver damage.

- *Frameworks.* Adopting new biomarkers and methods requires robust frameworks for data integration and computational modeling. This involves combining diverse data types (e.g. genomics, proteomics, metabolomics, and traditional toxicological data) to provide a holistic view of toxicity. Development of databases and tools are therefore required. Platforms like the AOP Knowledge Base (AOP-KB) and the US EPA's CompTox Chemicals Dashboard that have been listed provide centralized repositories for data on chemical exposures, pathways, and outcomes. These tools help regulatory agencies and researchers access and use integrated data to inform decision-making. Quantitative structure-activity relationship models, QBAR models, and artificial intelligence (AI)-driven tools are increasingly used to predict the potential toxicity of untested chemicals based on their chemical structure and biological activity profiles. However, regulatory acceptance of these models requires comprehensive validation and the demonstration of their predictive power.
- *Regulatory Acceptance and Guidelines Development.* While the assessment of PODs such as BMCs, derived from the concentration-response relationship in omics data, complies with NAMs, there is a need for further knowledge and consensus among the scientific community before it can be useful and fully accepted for use in risk assessment. Genomic and transcriptomic data have been explored for this purpose, and have led to guidelines and recommendations (EFSA Scientific Committee et al., 2017, Farmahin et al., 2017), but application to metabolomics requires further investigation. Regulatory bodies, such as the US Food and Drug Administration (FDA), European Medicines Agency, and International Council for Harmonisation of Technical Requirements for Pharmaceuticals for Human Use (ICH), are increasingly incorporating new methods and biomarkers into their safety evaluation guidelines. However, this integration requires the development of new regulatory frameworks and guidelines; traditional risk assessment relies heavily on dose-response data from animal models. Integrating biomarkers involves shifting towards more mechanistic-based approaches that provide insights into the biological pathways affected by chemical exposure. Frameworks such as NGRA and Integrated Approaches to Testing and Assessment are being developed to incorporate these new approaches. Regulatory guidelines, such as the EPA's ToxCast program (US EPA, 2024) and the European Union's REACH regulation have incorporated HTS data and biomarkers into their chemical safety evaluation processes (Zare Jeddi et al., 2021). In the area of food safety, the EFSA has been quick to address the issue and has studied the

© Burleigh Dodds Science Publishing Limited, 2025. All rights reserved.

potential applications of these modern methodologies and tools in the context of chemical prioritization, integrated testing strategies, and the future of risk assessment (Aguilera et al., 2018, European Food Safety Authority (EFSA), 2014).

The adoption of new methods and biomarkers in toxicology and risk assessment introduces not only scientific and technical challenges but also significant ethical, legal, and social considerations. New approach methodologies, such as *in vitro* systems and omics technologies, support the ethical principle of the 3Rs–replacement, reduction, and refinement of animal use. This shift is essential for aligning with public expectations and advancing humane science. However, the validation and acceptance of these methods still face scrutiny from regulatory bodies and stakeholders who prioritize robustness and predictivity. Many NAMs, including biomarker discovery and validation, involve human-derived data, such as organoids, stem cells, or genomics. Ethical challenges include ensuring proper informed consent, protecting participants' privacy, and addressing concerns over the commodification of human biological materials. With methods like metabolomics and transcriptomics generating massive datasets, regulatory compliance with laws such as the General Data Protection Regulation (European Parliament and Council, 2016) in Europe becomes critical. Ensuring that data is anonymized and secure is paramount. Finally, stakeholders, including the public, industry, and policymakers, may question the reliability of human-relevant methods compared to established animal-based tests. Clear communication and education are necessary to build trust. Efforts like the Tox21 program by the EPA, NIH, and FDA (United States Federal Government Tox21 Collaboration,2024) promote the development and adoption of high-throughput, human-relevant methods. While these methods reduce reliance on animal testing, they also spotlight issues such as the transparency of model predictions and the ethical sourcing of human-derived materials.

Globally, there is a need for harmonization across different regulatory agencies to ensure consistent acceptance of new methods and biomarkers. Organizations like the OECD play a crucial role in this process by developing internationally recognized guidelines and frameworks.

## 7 Use of effect biomarkers in food safety risk analysis: the examples of veterinary residues and endocrine-disrupting compounds

Research work on effect biomarkers had significant implications in the realms of both food safety and environmental health, addressing critical challenges in monitoring and mitigating the impact of toxic chemicals on human and

© Burleigh Dodds Science Publishing Limited, 2025. All rights reserved.

animal health. Such research bridges the gap between exposure to harmful substances and their biological effects by identifying molecular changes, which are crucial in detecting and preventing potential health risks associated with contaminants in food and the environment. This section explores how effect biomarkers have been applied to these two areas, detailing key contributions and methodologies.

Food safety is a global concern, especially as agricultural and industrial practices continue to introduce a wide variety of chemical residues and contaminants into the food chain. These chemicals, which include pesticides, veterinary drug residues, mycotoxins, and persistent organic pollutants (POPs), pose potential risks to human health. The conventional approach to ensuring food safety focuses primarily on identifying and quantifying the presence of these substances, referred to as exposure markers, in food products. This is where effect biomarkers become an interesting alternative. Unlike exposure biomarkers, which only indicate that an organism has come into contact with a potentially harmful agent, effect biomarkers go further by showing the resulting biological response. They provide a more dynamic and insightful means of evaluating the real-world consequences of exposure to food contaminants. Instead of merely detecting the presence of toxicants, they allow researchers and regulatory agencies to observe the physiological and molecular responses triggered by these compounds in living organisms.

## 7.1 Veterinary drug residues and biomarkers of effect

A case study illustrating this aspect is the use of biomarkers of effect to improve the detection of prohibited practices in livestock farming. Veterinary drugs, including antibiotics and growth promoters, may be used in animal farming to enhance productivity and prevent disease. However, the presence of residues from these drugs in meat, milk, and other animal-derived products can be harmful if consumed over time. In Europe in particular, the use of growth promoters has been banned in livestock farming since the late 1980s (Council Directive 96/22/EC, 1996, Council Directive 88/146/EEC, 1988), and controls are carried out to ensure that the regulations are properly applied. Detection of illegal practices classically relies on residue monitoring in a targeted approach and methods based on gas or liquid chromatography coupled to (tandem) mass spectrometry are considered state-of-the-art (De Brabander et al., 2007, Le Bizec et al., 2009). However, these strategies fail when faced with new xenobiotic growth-promoting agents or new ways of application, such as the administration of low-dose cocktails. In this context, screening strategies allowing the detection of the physiological response resulting from the administration of growth promotors are promising approaches to detect misuse (Le Bizec et al., 2023). Profiling biological matrices to reveal biological

© Burleigh Dodds Science Publishing Limited, 2025. All rights reserved.

effects of a drug can be performed by targeting a particular class of compounds (e.g. steroids) (Kaabia et al., 2014, 2018) or in an untargeted way using global strategies, such as transcriptomics, proteomics, or metabolomics (Pinel et al., 2011, Dervilly-Pinel et al., 2012, Antignac et al., 2011, Gallart Ayala et al., 2015, Cloteau et al., 2023b, Benedetto et al., 2021a).

By applying high-throughput techniques like metabolomics, proteomics, or transcriptomics to a lesser extent, researchers have been able to capture the molecular alterations caused by exposure to growth promoter residues. For instance, it has been highlighted how residues from steroids (Dervilly-Pinel et al., 2011, Jacob et al., 2014, 2015, Kouassi Nzoughet et al., 2015a, 2015b, Marchand et al., 2018, Nzoughet et al., 2015, Pinel et al., 2011, Stella et al., 2017), β-agonists (Courant et al., 2009, Marchand et al., 2018, Peng et al., 2017, Stella et al., 2021, Benedetto et al., 2021b), corticosteroids (Stella et al., 2016), or growth hormones (Boyard-Kieken et al., 2011, Kieken et al., 2009, 2011a, 2011b) disrupt normal metabolic pathways in a range of animal species including bovine and porcine. These disruptions can manifest as changes in key metabolites (Cacciatore et al., 2009, Cunningham et al., 2009, Ouzia et al., 2022), lipids (Marchand et al., 2020, Guitton et al., 2018) or proteins (Doue et al., 2015, Kinkead et al., 2018, 2015, McGrath et al., 2013, Mooney et al., 2009a, 2009b), providing a measurable signature of biological stress that can serve as a biomarker of effect. Specific monitoring of these biomarkers means that a robust statistical model can be used to predict treated animals, enabling them to be classified as suspects.

Although many guidelines for validating targeted methods are available and provide actors with performance indicators (e.g. pharmacological drugs Reg 2017/625/EC (European Parliament and Council, 2017), equivalent guidelines for untargeted method validation are not yet available, which certainly explains their limited application for routine or official control purposes. The main difference between non-targeted and targeted approaches is related to the multivariate nature of fingerprints in the former case, compared to single or multi-chemical species analysis in the latter one. In targeted strategies, results and method performance are evaluated compound-by-compound using univariate statistics, while multivariate analysis (which are generally referred to chemometric methods) are required to evaluate relevance of patterns arising from fingerprinting approaches (Alonso et al., 2015, Godzien et al., 2015). While very few examples of metabolomics model performance assessment have been reported (Cloteau et al., 2023a), a model dedicated to the detection of β-agonist administration in cattle has been fully validated: criteria (selectivity, robustness, stability, definition of the threshold of suspicion, false positive and false negative rates) have been proposed in line with EU expectations (European Commission, 2021), enabling performance to be shown to be in line with screening requirements (Dervilly-Pinel et al., 2015). Although some

© Burleigh Dodds Science Publishing Limited, 2025. All rights reserved.

of the biomarkers involved in the predictive model remain to be elucidated, the corresponding method has been accredited to ISO17025, enabling the very first official implementation of a metabolomics model in French national monitoring plans (Dervilly-Pinel et al., 2018). It is worth pointing out that a second barrier to these approaches is the identification of biomarkers in the case of metabolomics strategies, which is not the case for other omics approaches such as proteomics or transcriptomics.

In terms of food safety, this approach allows regulatory authorities to monitor not just the presence of these residues but also their biological effects. By identifying specific biomarkers linked to harmful drug residues, this work supports the development of more robust and efficient monitoring systems, on the one hand, because these approaches extend the detection window, and on the other, because they allow highlighting the use of pharmacologically active substances used at low doses or unknown to control systems. These biomarkers help in distinguishing between safe and hazardous situations in animal products, ultimately ensuring that food products meet safety standards.

## 7.2 Biomarkers of effect of environmental pollutants in the food chain

Another critical area of food chemical safety concerns the presence in foodstuffs of pollutants with endocrine-disrupting (ED) properties, which are substances that interfere with the hormonal systems of animals and humans. These include industrial chemicals that seep into the environment and contaminate foodstuffs. Endocrine-disrupting compounds (EDCs) pose a significant threat to public health due to their ability to mimic, block, or interfere with natural hormones, resulting in reproductive, developmental, and metabolic disorders. The impact of chronic exposure to low levels of EDCs is difficult to assess using traditional toxicological methods. Environmental health studies traditionally rely on measuring external exposure levels, but this approach does not always correlate with the biological consequences of such exposure. Effect biomarkers, which reveal the physiological and molecular responses to environmental stressors, are critical in addressing this gap. This is why research into biomarkers of effects is becoming essential.

By identifying molecular biomarkers that reflect the endocrine-disruptive effects of these chemicals, recent research has contributed to the development of new tools for detecting harmful exposures in the food chain. These molecular changes serve as biomarkers that can indicate early signs of endocrine disruption, even before overt symptoms or diseases develop.

Recent work on PCBs and their effects on pigs focused on understanding how these persistent environmental pollutants impact animal health. Polychlorinated biphenyls are synthetic chemicals once widely used in

© Burleigh Dodds Science Publishing Limited, 2025. All rights reserved.

industrial applications but are now banned due to their toxicity and persistence in the environment. However, due to their POPs properties, they continue to contaminate ecosystems and can enter the food chain, posing health risks to both animals and humans. Adverse effects of PCBs have been documented in humans after accidental and massive exposure. However, little is known about the effect of chronic exposure to low-dose PCB mixtures, and studies regarding scattered lifetime exposures to non-dioxin-like (NDL)-PCBs are especially missing. Research aimed to identify the biological effects of PCB exposure in pigs under a real-case scenario of PCB exposure (chronic exposure through diet), using them as a model organism for studying the impact of these chemicals on livestock and potential implications for human health (Hernandez-Mesa et al., 2022). Pigs share physiological similarities with humans, especially in terms of metabolic and immune systems, making them a suitable model for toxicological studies. This work showed that despite the lack of observable phenotypic effects, exposure to PCBs leads to significant disruptions in metabolic pathways in pigs (Fig. 3). In particular, the combination of several metabolomics platforms covering polar and non-polar metabolites by LC and GC-HRMS has enabled a global and comprehensive view of metabolic changes, allowing to return a more complete picture of the effect of NDL-PCBs on the metabolism of pigs (Narduzzi et al., 2023, 2024a, 2024b). The highlighted trends of the metabolism of the treated pigs seemed to mirror the same tendency towards the several metabolic dysfunctions already reported for NDL-PCB exposures at higher doses, and to be associated with the two main effects reported in the literature, namely hepatotoxicity and neurotoxicity. In this context, metabolic pathways affected by low-dose NDL-PCB exposure include the fatty acid pathway, glycolysis and pentose-phosphate pathway, Krebs cycle, neurotransmitters, purine metabolism, and metabolites derived from the gut microbiota (i.e. indole acids and some secondary bile acids). This research concluded on alterations in pathways related to oxidative stress, immune response, detoxification and energy metabolism, indicating that PCB exposure imposes significant stress on biological systems, leading to potential health risks in exposed animals (Narduzzi et al., 2024c). These findings suggest that the biological effects observed in pigs could be indicative of similar risks in humans exposed to PCBs through diet, especially in vulnerable populations with higher exposure to contaminated food.

## 8 Conclusion

The two case studies presented, while both anchored in the field of food chemical safety and the associated health risks for consumers, highlight the varied applications and nuances in the use of biomarkers of effect in this field. Each case illustrates different challenges and levels of research maturity,

© Burleigh Dodds Science Publishing Limited, 2025. All rights reserved.

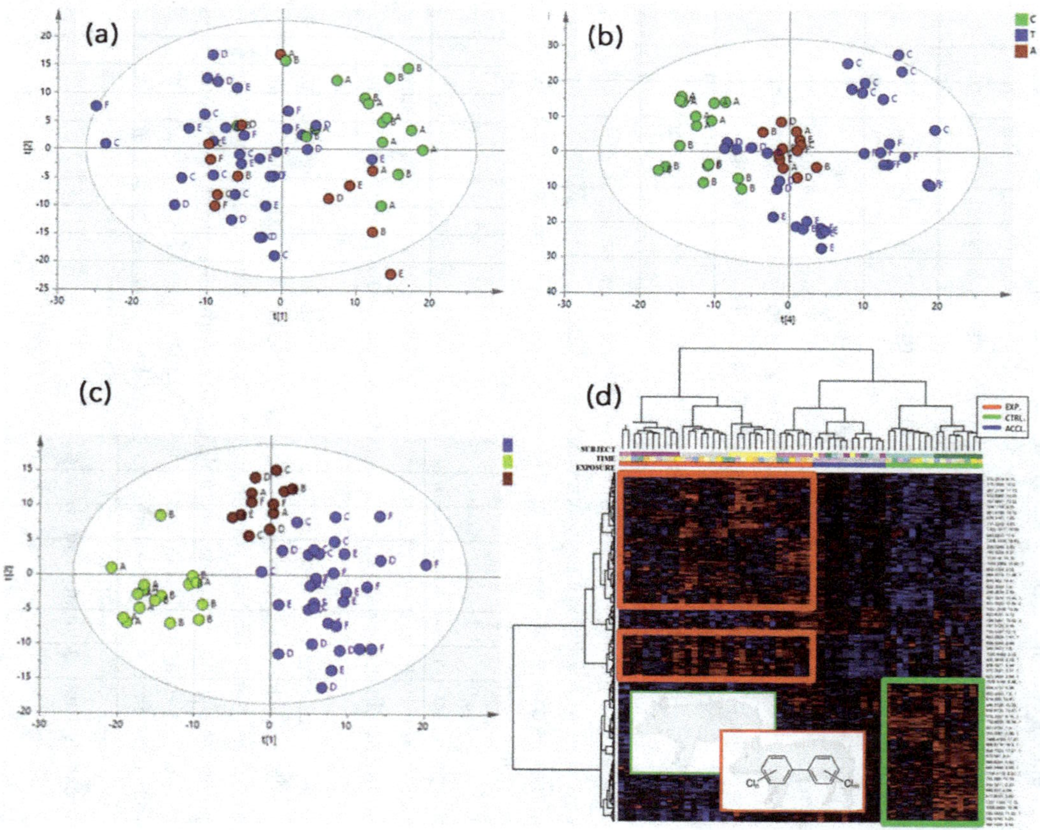


**Figure 3** Illustration of the results of a metabolomics study relating to the characterization of changes in the metabolic profiles of pig serum by LC-HRMS following repeated dietary exposure to Aroclor1260, a mixture of polychlorinated biphenyls (Hernandez-Mesa et al., 2022). (a) Results of the unsupervised PCA analysis (in red the animals before exposure, in green the control animals, in blue the exposed animals). (b) Results of the PCA after application of a T0-centering to address inter-individual differences and highlight differences between groups according to (Narduzzi et al., 2020). (c) Results of the supervised Partial Least Square analysis confirming clustering of samples from 'control', 'exposed' and 'acclimatization' groups, enabling classification accuracy >95%. (d) Heat map showing metabolic disturbances in exposed animals, In red/blue, the group of metabolites with an increase/decrease in their concentration in serum associated with exposure to Aroclor

© Burleigh Dodds Science Publishing Limited, 2025. All rights reserved.

shedding light on the breadth of biomarker utilization in regulatory science and toxicology.

The first example focuses on the use of biomarkers of effect to support strategies aimed at controlling the (mis)use of growth promoters that are banned in livestock farming. In this context, the objective is to develop robust statistical classification tools that can reliably detect illegal practices. These tools are designed to classify animals as compliant or suspect with current regulations, on the basis of their biomarker profiles, even in the absence of direct evidence of the presence of a specific residue of the banned substance. Although the validation of these methods is complicated by the absence of universally accepted guidelines, this does not pose an insurmountable challenge. Indeed, the performance parameters required for regulatory acceptance–such as acceptable levels of false positives and false negatives–are typically well-defined in legal frameworks or regulatory standards. These clear performance benchmarks provide a foundation for moving forward, despite gaps in the standardization of biomarker validation procedures.

One key challenge in this case is the identification of specific biomarkers, particularly when derived from non-targeted metabolomics approaches. In these methods, complex datasets may produce unidentified chemical signals, which could be biomarkers of interest. While it is certainly preferable to assign a chemical identity to each signal, the success of a classification model does not necessarily hinge on this identification. Instead, the priority is to ensure that the biomarker signals are reliable and reproducible. As long as the measurements can be trusted, classification models can be developed to detect patterns that correlate with illegal drug use, even if the exact nature of the chemicals remains unknown. This underscores a pragmatic approach in risk management, where the end goal is to implement regulatory controls that are scientifically sound but not reliant on full chemical characterization.

In contrast, the second example adopts a more fundamental approach to the use of the biomarkers of effect. Here, the focus shifts from practical regulatory applications to advancing the understanding of biological phenomena underlying the effects of chemical exposures. The challenge in this case lies in the identification of biomarkers that not only reflect the presence of a contaminant but also provide insights into biological pathways affected by the exposure. This is why the use of complementary analysis platforms is required here to cover a greater diversity of potential biomarkers. This type of research is crucial for improving our understanding of toxicological mechanisms and contributes directly to risk assessment, specifically during the hazard identification phase. However, identifying biomarkers of effect in this context is significantly more challenging, often requiring substantial human and instrumental resources. The complexity of biological systems, combined with the need for precision in linking specific biomarkers to AOs, makes this

© Burleigh Dodds Science Publishing Limited, 2025. All rights reserved.

a highly resource-intensive endeavor. It should also be stressed here that alterations to biological pathways should not always be considered a 'danger' and that discussions are needed to determine when biological modifications should be considered a 'danger to health.'

These two case studies, while both concerned with chemical risk analysis, represent distinct facets of the field. The first example is closely aligned with risk management, where the goal is to develop operational tools that can be used by regulatory authorities to control food safety risks effectively. The second example, on the other hand, pertains to risk assessment, focusing on understanding the mechanisms by which chemical exposures lead to adverse health effects. These different objectives are reflected in the technology readiness levels (TRLs) of the associated research. The first example, dealing with the detection of banned substances in livestock, operates at a TRL of 7 to 9, indicating that it is practically implemented. The second example, which is more exploratory and centered on biological discovery, falls within TRL levels 2 to 3, indicating that it is still in the early stages of development and is more focused on basic research.

These case studies highlight the diverse applications of biomarkers of effect and the different regulatory needs they fulfill. While both contribute to the broader goal of ensuring chemical safety, they serve distinct purposes within the framework of chemical risk analysis. The first case is primarily concerned with regulatory enforcement and the operationalization of detection methods, while the second aims to advance scientific understanding, laying the groundwork for future regulatory developments. Together, they illustrate how biomarkers can be leveraged in different contexts, with varying degrees of technological maturity and complexity, to address both immediate risk analysis needs and long-term scientific challenges.

## 9 Future trends in research

Despite the promising potential of effect biomarkers in assessing exposure to contaminants and chemicals, numerous challenges remain that must guide and inspire future research efforts.

One major challenge is the standardization and validation of biomarkers to ensure they are both reliable and applicable across different populations, environmental contexts, and experimental conditions. Biomarkers must be rigorously tested in diverse cohorts to establish consistent sensitivity and specificity. Sensitivity is crucial because effect biomarkers should detect subtle changes even at low exposure levels, while specificity ensures that the biomarker accurately reflects the biological effect of a particular exposure without interference from other factors. Comprehensive validation studies are

© Burleigh Dodds Science Publishing Limited, 2025. All rights reserved.

necessary to better address these issues, guaranteeing that the biomarkers are robust, reliable, and applicable in real-world settings.

Moreover, there is a significant need to account for inter-individual variability in biomarker responses. Factors such as genetics, diet, lifestyle, and varying environmental exposures can significantly influence how different individuals respond to contaminants, making the interpretation of biomarker data more complex. To address this, experimental designs must be more inclusive of this variability, incorporating diverse populations and accounting for confounding factors that might skew biomarker effectiveness. Without this, biomarker responses may not be universally applicable, limiting their utility in risk assessments and public health strategies.

In addition to these biological and technical challenges, multi-omics data complexity represents another frontier for research. Effect biomarkers often stem from omics platforms–such as genomics, proteomics, and metabolomics–that generate vast and intricate datasets. The challenge lies not only in generating high-quality data but also in integrating these data streams to produce meaningful and actionable insights into toxicity mechanisms. To do this, advanced bioinformatics tools are essential, as they enable the effective analysis of these complex datasets, uncovering intricate interactions between biological pathways, and identifying how exposure to certain contaminants results in adverse health outcomes. However, data quality assurance and robust computational frameworks for data integration remain underdeveloped, representing an important area for future progress.

Further research is also needed to identify and validate new biomarkers that can cover a broader range of exposures and health effects, particularly in the context of low-level chronic exposures. Additionally, developing non-invasive sampling techniques, such as the use of blood, urine, or saliva, would facilitate broader application, allowing for easier and more frequent monitoring in both clinical and environmental settings. The integration of biomarker data with other datasets, such as epidemiological data and toxicological studies, would create a more comprehensive understanding of how exposures translate into health risks, making biomarker use in risk assessment more robust and reliable.

The interpretation of effect biomarkers, particularly through the use of AOPs, will require research work addressing several challenges. A key issue is the complexity of biological systems, as AOPs aim to map intricate processes within human biology, which is highly complex. Fully understanding the molecular events that lead to AOs requires detailed mechanistic insight, which is often incomplete. The question arises as to whether it is possible to obtain mechanistic information without resorting to 'conventional' *in vivo* toxicological studies. Additionally, data gaps in current AOP frameworks pose a challenge, because while some KEs are well-documented, others lack

© Burleigh Dodds Science Publishing Limited, 2025. All rights reserved.

sufficient experimental evidence or remain poorly understood, hindering the development of comprehensive and reliable pathways.

A critical factor in advancing this field is collaboration among key stakeholders, including regulatory agencies, academic institutions, industries, and non-governmental organizations. These groups must work together to promote the acceptance and application of new biomarker methods in regulatory frameworks. This includes ongoing efforts to refine analytical methods, improve data integration frameworks, and conduct real-world case studies that highlight the relevance of biomarkers in regulatory decision-making. Such studies would demonstrate the practical utility of effect biomarkers in addressing pressing public health concerns, such as chemical exposures through diet or environment.

In summary, overcoming these challenges will require improved research study designs, including larger and more diverse sample populations that reflect the wide range of exposures and genetic variability present in real-world environments. Furthermore, better standardization of protocols for biomarker validation is needed to ensure that results are reproducible across different labs and conditions. The development of more advanced bioinformatics tools to handle large-scale multi-omics data will also be essential. Ultimately, collaborative efforts between scientific communities and regulators will be key to translating biomarker research into practical applications for human health and environmental safety. This collaboration can drive the next wave of innovation in the field, enabling biomarkers to become a central tool in modern toxicology and public health.

## 10 Where to look for further information

The following articles provide a good overview of the subject:

- Gupta, R. Biomarkers in Toxicology, Academic Press, ed Ramesh C. Gupta, 2014, ISBN 9780124046306 (Gupta, 2014).
- Rappaport, S., Smith, M.T. Epidemiology. Environment and disease risks. Science, 2010, 330(6003) :460-1 (Rappaport and Smith, 2010).

Key research in this area has been or is conducted within:

- HBM4EU (Human Biomonitoring for Europe) Initiative. A project that provided data on biomarkers of exposure and effect for various chemicals, including those relevant to environmental health and food safety, and serves as a valuable resource for developing new biomarkers of effect. https://www.hbm4eu.eu.

© Burleigh Dodds Science Publishing Limited, 2025. All rights reserved.

- PARC (Partnership for the Assessment of Risks from Chemicals). A project that aims to develop innovative methodologies and tools to improve chemical risk assessments, including those related to human health. One of its key objectives is to support the transition to NGRAs, which prioritize mechanistic understanding, particularly with omics technologies and biomarkers of effect. https://www.eu-parc.eu.

## 11 References


Aguilera, J., et al. 2018. *EFSA Scientific Colloquium 24 – 'omics in risk assessment: state of the art and next steps*. EFSA Supporting Publications, 15.

Alonso, A., et al. 2015. Analytical methods in untargeted metabolomics: state of the art in 2015. *Front Bioeng Biotechnol*, 3, 23.

Ankley, G. T. & Edwards, S. W. 2018. The adverse outcome pathway: a multifaceted framework supporting 21(st) century toxicology. *Curr Opin Toxicol*, 9, 1–7.

Antignac, J. P., et al. 2011. Mass spectrometry-based metabolomics applied to the chemical safety of food. *Trac-Trends Anal Chem*, 30, 292–301.

Araujo, A. M., et al. 2021. Toxicometabolomics: small molecules to answer big toxicological questions. *Metabolites*, 11, 692.

Babin, E., et al. 2023. A review of statistical strategies to integrate biomarkers of chemical exposure with biomarkers of effect applied in omic-scale environmental epidemiology. *Environ Pollut*, 330, 121741.

Benedetto, A., et al. 2021a. Omics applications in the fight against abuse of anabolic substances in cattle: challenges, perspectives and opportunities. *Curr Opin Food Sci*, 40, 112–120.

Benedetto, A., et al. 2021b. Profiling of transcriptional biomarkers in FFPE liver samples: PLS-DA applications for detection of illicit administration of sex steroids and clenbuterol in veal calves. *Food Control*, 128, 108149.

Black, M. B., et al. 2022. Biological system considerations for application of toxicogenomics in next-generation risk assessment and predictive toxicology. *Toxicol In Vitro*, 80, 105311.

Bouhifd, M., et al. 2015. The human toxome project. *ALTEX*, 32, 112–124.

Boyard-Kieken, F., et al. 2011. Comparison of different liquid chromatography stationary phases in LC-HRMS metabolomics for the detection of recombinant growth hormone doping control. *J Sep Sci*, 34, 3493–501.

Brooks, T. G., et al. 2024. Challenges and best practices in omics benchmarking. *Nat Rev Genet*, 25, 326–339.

Buesen, R., et al. 2017. Applying 'omics technologies in chemicals risk assessment: report of an ECETOC workshop. *Regul Toxicol Pharmacol*, 91(Suppl 1), S3–S13.

Cacciatore, G., et al. 2009. Effect of growth-promoting 17beta-estradiol, 19-nortestosterone and dexamethasone on circulating levels of nine potential biomarker candidates in veal calves. *Anal Chim Acta*, 637, 351–359.

Cai, Y., et al. 2020. Analyzing metabolomics data for environmental health and exposome research. *Methods Mol Biol*, 2104, 447–467.

Campion, S., et al. 2013. The current status of biomarkers for predicting toxicity. *Expert Opin Drug Metab Toxicol*, 9, 1391–1408.


© Burleigh Dodds Science Publishing Limited, 2025. All rights reserved.

Chen, Y., et al. 2022. Guide to metabolomics analysis: a bioinformatics workflow. *Metabolites*, 12, 357.

Choudhuri, S., et al. 2018. From classical toxicology to Tox21: some critical conceptual and technological advances in the molecular understanding of the toxic response beginning from the last quarter of the 20th century. *Toxicol Sci*, 161, 5–22.

Cloteau, C., et al. 2023a. Performance assessment of an equine metabolomics model for screening a range of anabolic agents. *Metabolomics*, 19, 38.

Cloteau, C., et al. 2023b. From targeted methods to metabolomics based strategies to screen for growth promoters misuse in horseracing and livestock: a review. *Food Control*, 148, 109601.

Council Directive 88/146/EEC 1988. Council directive 88/146/EEC prohibiting the use livestock farming of certain substances having a hormonal action. 16–18.

Council Directive 96/22/EC 1996. Council Directive 96/22/EC of 29 April 1996 concerning the prohibition on the use in stockfarming of certain substances having a hormonal or thyrostatic action and of beta-agonists, and repealing Directives 81/602/EEC, 88/146/EEC and 88/299/EEC.

Courant, F., et al. 2009. Development of a metabolomic approach based on liquid chromatography-high resolution mass spectrometry to screen for clenbuterol abuse in calves. *Analyst*, 134, 1637–1646.

Cunningham, R., et al. 2009. Feasibility of a clinical chemical analysis approach to predict misuse of growth promoting hormones in cattle. *Anal Chem*, 81, 977.

Dai, X. & Shen, L. 2022. Advances and trends in omics technology development. *Front Med (Lausanne)*, 9, 911861.

De Brabander, H. F., et al. 2007. Past, present and future of mass spectrometry in the analysis of residues of banned substances in meat-producing animals. *J Mass Spectrom*, 42, 983–998.

Dearden, J. C. 2017. The history and development of quantitative structure-activity relationships (QSARs). *Oncology*, 67–117. https://doi.org/10.4018/978-1-5225-0549-5.ch003

Dehghani, F., et al. 2022. Metabolomics for exposure assessment and toxicity effects of occupational pollutants: current status and future perspectives. *Metabolomics*, 18, 73.

Dervilly-Pinel, G., et al. 2015. LC-HRMS based metabolomics screening model to detect various β-agonists treatments in bovines. *Metabolomics*, 11, 403–411.

Dervilly-Pinel, G., et al. 2012. Metabolomics in food analysis: application to the control of forbidden substances. *Drug Test Anal*, 4, 10.

Dervilly-Pinel, G., et al. 2018. When LC-HRMS metabolomics gets ISO17025 accredited and ready for official controls - application to the screening of forbidden compounds in livestock. *Food Add Cont Part A Chem Anal Control Exp Risk Ass*, 35, 1948–1958.

Dervilly-Pinel, G., et al. 2011. Assessment of two complementary liquid chromatography coupled to high resolution mass spectrometry metabolomics strategies for the screening of anabolic steroid treatment in calves. *Anal Chim Acta*, 700, 144-54.

Doue, M., et al. 2015. Clinical biochemical and hormonal profiling in plasma: a promising strategy to predict growth hormone abuse in cattle. *Anal Bioanal Chem*, 407, 4343–4349.

Dubreil, E., et al. 2024. Modeling HepaRG metabolome responses to pyrrolizidine alkaloid exposure for insight into points of departure and modes of action. *J Hazard Mater*, 474, 134721.

© Burleigh Dodds Science Publishing Limited, 2025. All rights reserved.

Dunn, W. B., et al. 2012. The importance of experimental design and QC samples in large-scale and MS-driven untargeted metabolomic studies of humans. *Bioanalysis*, 4, 2249–2264.

ECHA. 2024. *REACH Registration statistics* [Online]. Available: https://echa.europa.eu/documents/10162/2741157/registration_statistics_en.pdf/58c2d7bd-2173-4cb9-eb3b-a6bc14a6754b

EFSA. 2023. *Chemical Hazards Database (OpenFoodTox)* [Online]. Available: https://www.efsa.europa.eu/en/data-report/chemical-hazards-database-openfoodtox [Accessed 04/10/2024 2024].

EFSA. *OpenAnalytics* [Online]. Available: https://efsa.openanalytics.eu/app/bmd [Accessed 21/11 2024].

EFSA Scientific Committee, et al. 2017. Update: use of the benchmark dose approach in risk assessment. *EFSA J*, 15, e04658.

EPA. 2024. *U.S. Environmental Protection Agency. Toxic Substances Control Act Chemical Substance Inventory (TSCA Inventory)* [Online]. Available: https://www.epa.gov/assessing-and-managing-chemicals-under-tsca/now-available-latest-update-tsca-inventory [Accessed 04/10/2024].

Espín-Pérez, A., et al. 2014. 'OMICS-based' biomarkers for environmental health studies. *Curr Environ Health Rep*, 1, 353–362.

European Commission. 2021. Commission implementing regulation (EU) 2021/808 of 22 march 2021 on the performance of analytical methods for residues of pharmacologically active substances used in food-producing animals and on the interpretation of results as well as on the methods to be used for sampling and repealing decisions 2002/657/ec and 98/179/ec.

European Commission. 2024. *Chemical substances in food* [Online]. Available: https://food.ec.europa.eu/food-safety/chemical-safety_en?prefLang=fr [Accessed 04/10/2024].

European Food Safety Authority (EFSA) 2014. Modern methodologies and tools for human hazard assessment ofchemicals. *EFSA J*, 12, 3638.

European Parliament and Council 2016. REGULATION (EU) 2016/679 on the protection of natural persons with regard to the processing of personal data and on the free movement of such data, and repealing Directive 95/46/EC (General Data Protection Regulation)

European Parliament and Council 2017. REGULATION (EU) 2017/625 on official controls and other official activities performed to ensure the application of food and feed law, rules on animal health and welfare, plant health and plant protection products, amending Regulations (EC) No 999/2001, (EC) No 396/2005, (EC) No 1069/2009, (EC) No 1107/2009, (EU) No 1151/2012, (EU) No 652/2014, (EU) 2016/429 and (EU) 2016/2031 of the European Parliament and of the Council, Council Regulations (EC) No 1/2005 and (EC) No 1099/2009 and Council Directives 98/58/EC, 1999/74/EC, 2007/43/EC, 2008/119/EC and 2008/120/EC, and repealing Regulations (EC) No 854/2004 and (EC) No 882/2004 of the European Parliament and of the Council, Council Directives 89/608/EEC, 89/662/EEC, 90/425/EEC, 91/496/EEC, 96/23/EC, 96/93/EC and 97/78/EC and Council Decision 92/438/EEC (Official Controls Regulation). *OJ L 95, 7.4.2017*, 1–142

Farmahin, R., et al. 2017. Recommended approaches in the application of toxicogenomics to derive points of departure for chemical risk assessment. *Arch Toxicol*, 91, 2045–2065.

© Burleigh Dodds Science Publishing Limited, 2025. All rights reserved.

FDA., U. S. F. A. D. A. 2018. *In Vitro and In Vivo Studies for Biomarker Qualification: Recommendations for Development*. [Online]. Available: https://www.fda.gov. [Accessed 21/11 2024].

Gallart Ayala, H., et al. 2015. Potential of mass spectrometry metabolomics for chemical food safety. *Bioanal Rev*, 7, 133–146.

Garcia-Campos, M. A., et al. 2015. Pathway analysis: state of the art. *Front Physiol*, 6, 383.

Godzien, J., et al. 2015. Controlling the quality of metabolomics data: new strategies to get the best out of the QC sample. *Metabolomics*, 11, 11.

Gonzalez-Dominguez, A., et al. 2024. QComics: recommendations and guidelines for robust, easily implementable and reportable quality control of metabolomics data. *Anal Chem*, 96, 1064–1072.

Guitton, Y., et al. 2018. Rapid evaporative ionisation mass spectrometry and chemometrics for high-throughput screening of growth promoters in meat producing animals. *Food Add Cont Part A Chem Anal Control Exp Risk Ass*, 35, 900–910.

Gupta 2014. *Biomarkers in toxicology*. Biomarkers in Toxicology 1st Edition - January 25, 2014 Imprint: Academic Press Editor: Ramesh C. Gupta Language: English Hardback ISBN: 97801240463069 7 8 - 0 - 1 2 - 4 0 4 6 3 0 - 6eBook ISBN: 9780124046498

Hernandez-Mesa, M., et al. 2021. Metabolomics in chemical risk analysis: a review. *Anal Chim Acta*, 1154, 338298.

Hernandez-Mesa, M., et al. 2022. Metabolomics and lipidomics to identify biomarkers of effect related to exposure to non-dioxin-like polychlorinated biphenyls in pigs. *Chemosphere*, 296, 133957.

Jacob, C., et al. 2015. Global urine fingerprinting by LC-ESI(+)-HRMS for better characterization of metabolic pathway disruption upon anabolic practices in bovine. *Metabolomics*, 11, 184–197.

Jacob, C. C., et al. 2014. Evaluation of specific gravity as normalization strategy for cattle urinary metabolome analysis. *Metabolomics*, 10, 627–637.

Kaabia, Z., et al. 2014. Monitoring the endogenous steroid profile disruption in urine and blood upon nandrolone administration: an efficient and innovative strategy to screen for nandrolone abuse in entire male horses. *Drug Test Anal*, 6, 376–388.

Kaabia, Z., et al. 2018. Comprehensive steroid profiling by liquid chromatography coupled to high it resolution mass spectrometry. *J Steroid Biochem Mol Biol*, 183, 106–115.

Kaur, P., et al. 2021. Computational techniques and tools for omics data analysis: state-of-the-art, challenges, and future directions. *Arch Comput Methods Eng*, 28, 4595–4631.

Khatri, P., et al. 2012. Ten years of pathway analysis: current approaches and outstanding challenges. *PLoS Comput Biol*, 8, e1002375.

Kieken, F., et al. 2011a. Development and application of a metabolomic approach based on liquid chromatography-high resolution mass spectrometry to reveal an illegal administration of recombinant equine growth hormone in horse from urinary and plasmatic biological signatures. *Metabolomics*, 7, 9.

Kieken, F., et al. 2009. Development of a metabonomic approach based on LC-ESI-HRMS measurements for profiling of metabolic changes induced by recombinant equine growth hormone in horse urine. *Anal Bioanal Chem*, 394, 2119–2128.

Kieken, F., et al. 2011b. Generation and processing of urinary and plasmatic metabolomic fingerprints to reveal an illegal administration of recombinant equine growth hormone from LC-HRMS measurements. *Metabolomics*, 7, 84–93.


© Burleigh Dodds Science Publishing Limited, 2025. All rights reserved.

Kinkead, R. A., et al. 2018. Plasma metabolomic profiling based detection of drug specific responses to different bovine growth promoting regimes. *Food Control*, 86, 146–158.

Kinkead, R. A., et al. 2015. Proteomic identification of plasma proteins as markers of growth promoter abuse in cattle. *Anal Bioanal Chem*, 407, 4495–507.

Kouassi Nzoughet, J., et al. 2015a. First insights into serum metabolomics of trenbolone/estradiol implanted bovines; screening model to predict hormone-treated and control animals' status. *Metabolomics*, 11, 1184–1196.

Kouassi Nzoughet, J., et al. 2015b. Original combination of Hydrophilic Interaction (HILIC) and Reverse Phase (RPLC) high resolution LC-MS for characterizing lipids profile disruption in serum of anabolic implanted bovines. *Metabolomics*, 11, 1884–1895.

Larras, F., et al. 2018. DRomics: a turnkey tool to support the use of the dose–response framework for omics data in ecological risk assessment. *Environ Sci Technol*, 52, 14461–14468.

Le Bizec, B., et al. 2009. Options for veterinary drug analysis using mass spectrometry. *J Chromatogr A*, 1216, 8016–8034.

Larras, F., et al. 2020. A multi-omics concentration-response framework uncovers novel understanding of triclosan effects in the chlorophyte Scenedesmus vacuolatus. *J Hazard Mater*, 397, 122727.

Le Bizec, B., et al. 2023. Evolution and revolutions in the control system for residues of veterinary substances in food-producing animals over more than three decades. *Food Control*, 149, 109671.

Letertre, M. P. M., et al. 2020. Combined nuclear magnetic resonance spectroscopy and mass spectrometry approaches for metabolomics. *Anal Chem*, 93, 500–518.

Liu, Z., et al. 2019. Toxicogenomics: A 2020 Vision. *Trends Pharmacol Sci*, 40, 92–103.

Malinowska, J. M., et al. 2023. Derivation of metabolic point of departure using high-throughput in vitro metabolomics: investigating the importance of sampling time points on benchmark concentration values in the HepaRG cell line. *Arch Toxicol*, 97, 721–735.

Marchand, J., et al. 2020. Combining mass spectrometric platforms for lipidome investigation - Application to the characterisation of disruptions in the lipid profile of pig serum upon β agonist treatment. *bioRvix*.

Marchand, J., et al. 2018. A multidimensional H-1 NMR lipidomics workflow to address chemical food safety issues. *Metabolomics*, 14.

Mcgrath, T. F., et al. 2013. Effect-based proteomic detection of growth promoter abuse. *Anal Bioanal Chem*, 405, 1171–1179.

Misra, B. B. 2019. Metabolomics Tools to Study Links Between Pollution and Human Health: an Exposomics Perspective. *Current Pollution Reports*, 5, 93–111.

Mooney, M. H., et al. 2009a. Biosensor-based detection of reduced sex hormone-binding globulin binding capacities in response to growth-promoter administrations. *Anal Chim Acta*, 637, 235–240.

Mooney, M. H., et al. 2009b. Combining biomarker screening and mass-spectrometric analysis to detect hormone abuse in cattle. *Trac-Trends Anal Chem*, 28, 665–675.

Mortimer, M., et al. 2022. Omics approaches in toxicological studies. https://doi.org/10.1007/978-981-16-9116-4_4

Narduzzi, L., et al. 2024a. HILIC-MS and CE-MS as complementary analytical approaches to assess the impact of exposure to polychlorinated biphenyls on the polar serum metabolome of pigs. *Microchem J*, 111839.

© Burleigh Dodds Science Publishing Limited, 2025. All rights reserved.

Narduzzi, L., et al. 2024b. HILIC-MS and CE-MS as complementary analytical approaches to assess the impact of exposure to polychlorinated biphenyls on the polar serum metabolome of pigs. *Microchem J*, 207, 111839.

Narduzzi, L., et al. 2020. Applying metabolomics to detect growth hormone administration in athletes: Proof of concept. *Drug Test Anal*, 12, 887–899.

Narduzzi, L., et al. 2024c. Determination of bile acids in serum of pigs exposed to polychlorinated biphenyls by liquid chromatography-mass spectrometry. *J Chromatogr B Analyt Technol Biomed Life Sci*, 1232, 123976.

Narduzzi, L., et al. 2023. Deeper insights into the effects of low dietary levels of polychlorinated biphenyls on pig metabolism using gas chromatography-high resolution mass spectrometry metabolomics. *Chemosphere*, 341, 140048.

National Research Council. 1989. *Biological markers in reproductive toxicology*. National Research Council.

Ng, C. A. & VON Goetz, N. 2017. The Global Food System as a Transport Pathway for Hazardous Chemicals: The Missing Link between Emissions and Exposure. *Environ Health Perspect*, 125, 1–7.

Nguyen, N., et al. 2022. Omics technologies to understand drug toxicity mechanisms. *Drug Discov Today*, 27, 103348.

Nzoughet, J. K.,et al. 2015. Hydrophilic interaction (HILIC) and reverse phase liquid chromatography (RPLC)-high resolution MS for characterizing lipids profile disruption in serum of anabolic implanted bovines. *Metabolomics*, 11, 1884–1895.

OECD. *Adverse Outcome Patway Knowledge Base (AOP-KB)* [Online]. Available: https://aopkb.oecd.org/index.html [Accessed 04/10 2024].

OECD. 2005. *Guidance document on the validation and international acceptance of new or updated test methods for hazard assessment*. OECD.

OECD. 2018. *Users' handbook supplement to the guidance document for developing and assessing adverse outcome pathways*. OECD.

OECD. 2021. *Guidance document for the scientific review of adverse outcome pathways*. 22. OECD.

Ouzia, S., et al. 2022. Nandrolone and estradiol biomarkers identification in bovine urine applying a liquid chromatography high-resolution mass spectrometry metabolomics approach. *Drug Test Anal*, 14, 879–886.

Peng, T., et al. 2017. Serum-based metabolomics characterization of pigs treated with ractopamine. *Metabolomics*, 13, 77.

Pinel, G., et al. 2010. Targeted and untargeted profiling of biological fluids to screen for anabolic practices in cattle. *Trac-Trends Anal Chem*, 29, 1269–1280.

Pinel, G., et al. 2011. Assessment of two complementary LC-HRMS metabolomics strategies for the screening of anabolic steroid treatment in calves. *Analytica Chimica Acta*, 700, 10.

Rappaport, S. M. & Smith, M. T. 2010. Epidemiology. Environment and disease risks. *Science*, 330, 460–1.

Reel, P. S., et al. 2021. Using machine learning approaches for multi-omics data analysis: a review. *Biotechnol Adv*, 49, 107739.

RIVM National Institute for Public Health and the Environment. *PROAST* [Online]. Available: https://www.rivm.nl/en/proast [Accessed 21/11/2024].

Robertson, D. G. 2005. Metabonomics in toxicology: a review. *Toxicol Sci*, 85, 809–822.

Roy, K., et al. 2015. *Understanding the basics of QSAR for applications in pharmaceutical sciences and risk assessment*. Academic Press.

© Burleigh Dodds Science Publishing Limited, 2025. All rights reserved.

Santiago-Rodriguez, T. M. & Hollister, E. B. 2021. Multi 'omic data integration: a review of concepts, considerations, and approaches. *Semin Perinatol*, 45, 151456.

Society for the Advancement of Adverse Outcome Pathways. 2024. *Collaborative Adverse Outcome Pathway Wiki (AOP-Wiki)* [Online]. Available: https://aopwiki.org/ [Accessed 21/11 2024].

Stella, R., et al. 2016. Targeted proteomics for the indirect detection of dexamethasone treatment in bovines. *Anal Bioanal Chem*, 408, 8343–8353.

Stella, R., et al. 2021. A novel tool to screen for treatments with clenbuterol in bovine: identification of two hepatic markers by metabolomics investigation. *Food Chem*, 353, 129366.

Stella, R., et al. 2017. Metabolomics analysis of liver reveals profile disruption in bovines upon steroid treatment. *Metabolomics*, 13, 1–13.

Sun, J., et al. 2022. A review of environmental metabolism disrupting chemicals and effect biomarkers associating disease risks: Where exposomics meets metabolomics. *Environ Int*, 158, 106941.

United States Federal Government TOX21 Collaboration. *Advancing Toxicology -to Improve Environmental Health and Pharmaceutical Safety* [Online]. [Accessed 21/11 2024].

US EPA. 2024. *Toxicity Forecasting (ToxCast)*, [Online]. Available: https://www.epa.gov/comptox-tools/toxicity-forecasting-toxcast [Accessed 08/10/2024].

van Ravenzwaay, B., et al. 2007. The use of metabolomics for the discovery of new biomarkers of effect. *Toxicol Lett*, 172, 21–28.

van Ravenzwaay, B., et al. 2012. Metabolomics: a tool for early detection of toxicological effects and an opportunity for biology based grouping of chemicals-from QSAR to QBAR. *Mutat Res*, 746, 144–50.

van Ravenzwaay, B., et al. 2014. The sensitivity of metabolomics versus classical regulatory toxicology from a NOAEL perspective. *Toxicol Lett*, 227, 20–8.

Viant, M. R., et al. 2019. Use cases, best practice and reporting standards for metabolomics in regulatory toxicology. *Nat Commun*, 10, 3041.

Vinken, M., et al. 2017. Adverse outcome pathways: a concise introduction for toxicologists. *Arch Toxicol*, 91, 3697–3707.

Wei, S., et al. 2022. Metabolomics as a valid analytical technique in environmental exposure research: application and progress. *Metabolomics*, 18, 35.

WHO. 1993. *Biomarkers and risk assessment: concepts and principles*. WHO.

WHO. 2001. *Biomarkers in risk assessment: validity and validation*. WHO.

WHO. (ed.) 2009. *Principles and methods for the risk assessment of chemicals in food*. WHO.

Wild, C. 2012. The exposome: from concept to utility. *International Journal of Epidemiology*, 41, 24–32.

Zare Jeddi, M., et al. 2021. Towards a systematic use of effect biomarkers in population and occupational biomonitoring. *Environ Int*, 146, 106257.

© Burleigh Dodds Science Publishing Limited, 2025. All rights reserved.

# Chapter 7

## Understanding the role of diet/food as a source of potentially harmful chemicals within the exposome

*Bozidar Udovicki and Ilija Djekic, University of Belgrade, Serbia*



## 1 Introduction

The exposome is the measure of all the exposures of a person in a lifetime and the effect of these exposures on human health. Since the creation of the concept (Wild, 2005) three domains of the exposome have been identified: internal (unique to the individual), specific external (chemical contaminants and pollutants, occupational exposures and lifestyle factors) and general external (broader socioeconomic and psychological contexts) (Wild, 2012; DeBord et al., 2016). This encompasses exposure from diet and dietary supplements, medicinal and recreational drugs, personal care and consumer products, air and water pollution, physical activity, noise and radiation, as well as nonphysical factors such as socioeconomic status, access to green or blue spaces, or perception of one's surroundings (Wild, 2012; Balcells et al., 2024). All these specific exposures and domains overlap and interact to a various degree which leads to specific or similar individual health outcomes. A significant subset of the exposome is commonly referred to as the chemical exposome, which includes all chemical substances and their associated transformation products, whether of biological or synthetic origin, that can enter the human body through pathways such as ingestion, inhalation, or dermal absorption (Balcells et al., 2024).

The intake of chemical substances through food is a critical component of the chemical exposome as food is a considerable source of a wide range

http://dx.doi.org/10.19103/c.2025.0158.13
© Burleigh Dodds Science Publishing Limited, 2025. All rights reserved.

of natural and man-made chemicals and the perfect vehicle for daily and continuous exposure to them. In the case of the general population that is not occupationally exposed to specific chemicals, the diet represents the main and, in some cases, the almost exclusive source of exposure to most chemicals (Jeddi et al., 2022). Food is far more than just a substance that is consumed to provide nutritional support and energy to the body. Modern humans consume a wide variety of foods, each either consisting of a single food item or containing a large diversity of compounds within each food. This vast range of food choices contributes to the complexity of human diets, making food not only serve a biological function but also fulfil cultural, social, and psychological roles. Considering that one food item can potentially have several intentionally added and unintentionally present chemicals, a well-balanced diet can also be a cocktail of various harmful chemical exposures.

Generally, there is a distinction between chemicals intentionally introduced to the food and unintentionally present chemicals, but in the broader sense of the exposure assessment they can both be considered as contaminants. For intentionally added chemicals, there is much greater potential for control on conditions of use and consequently their presence in food and they are intensively assessed in the pre-regulation stage with exposure estimates intended to be protective for the consumer's health (FAO/WHO, 2020; Udovicki and Djekic, 2024). Intentionally added chemicals such are food additives are purposefully present in the foods for various reasons, predominantly to improve their quality, safety, appearance, and/or shelf life. Residues of pesticides, veterinary medicines and additives, while intentionally added, are not expected to appear in food in the concentrations above the allowed maximum residue levels (either legally binding or serving as international guidelines). Subsequently, intentionally added chemicals are not expected to cause human health concerns, provided that the established use levels are followed. Unintentionally present chemical contaminants such as natural toxins (e.g. mycotoxins, plant toxins, and aquatic biotoxins), heavy metals, processing contaminants (e.g. acrylamide, heterocyclic amines, and polycyclic aromatic hydrocarbons) and environmental pollutants (e.g. persistent organic pollutants, polychlorinated biphenyls, and microplastics) are considered potentially harmful if they are present in the significant amounts or even when they are present below the allowed levels. While, in most cases, it is not possible to estimate harmful health effects caused by exposure to chemical contaminants in the same way as for microbial food poisoning, by estimating probability of illness and number of potential cases, chronic exposure to foodborne contaminants is associated with a wide range of health issues, including cancer, neurological disorders, endocrine disruption, reproductive and developmental effects. There is also an increasing concern about the potential consequences of simultaneous exposure to multiple chemicals, even when a dose of each

© Burleigh Dodds Science Publishing Limited, 2025. All rights reserved.

chemical would not cause harm (Kortenkamp and Faust, 2018). This concern is justified considering the sheer number of various chemicals (intentionally and unintentionally present in food), the infinite number of their combinations, the wide range of potential end-point effects they can cause, and the lack of knowledge of their possible interactions and cumulative and chronic effects.

A key factor for describing the exposome is the ability to accurately measure both exposures and their effects, which is a considerable challenge. The field of dietary exposure studies has developed significantly over time with the rise of biomonitoring and omics technologies that have the potential to predict dietary intake independent of traditional dietary assessment tools (Liang et al., 2022). Until that day, traditional dietary exposure assessment methods are still essential tools for understanding and managing risk related to the presence of chemical contaminants in food, while human monitoring can be a complementary approach for dietary exposure assessment (Jeddi et al., 2022). Relying on two basic types of data, food consumption and concentration data, dietary exposure assessments provide information that spans across simple calculations of contaminant intake. They provide information on food consumption and exposure patterns of certain regions, populations or specific population groups. They are a prerequisite of regulatory frameworks providing a basis for the development of regulatory specifications, such as maximum residue limits for pesticides, additives, veterinary medicines and biocidal products, and maximum levels for contaminants. They help identify key dietary sources of chemical contaminants, enabling targeted interventions to reduce food contamination and human exposure thus protecting public health when a health hazard is demonstrated. Performing exposure assessments over extended periods provides valuable information on exposure trends, the effectiveness of control measures, and the impact of new agricultural, safety or food production practices.

## 2 Challenges in performing dietary exposure studies

### *2.1 Challenges in food consumption data collection*

Food consumption data reflect what individuals or groups consume in terms of solid foods, beverages (including drinking water) and dietary supplements (EFSA, 2011). Food consumption data can be collected at population, household and individual levels, but in modern dietary exposure assessment, individual-level data is considered the most valuable because it provides the most accurate and specific information about food consumption. Collecting quantitative information on food consumption on the individual level is performed by conventional tools and the most common are through food records (food diary), 24-hour dietary recall (24hR) and the Food Frequency

© Burleigh Dodds Science Publishing Limited, 2025. All rights reserved.

Questionnaire (FFQ). Although these methods are widely used and are essential for understanding the dietary habits, nutritional intake and health patterns of the population, they are still prone to both random and systematic errors (Table 1). These errors can arise from challenges in accurately assessing the nature and frequency of food consumption, estimating portion sizes, accounting for daily variations in intake and failures to report typical dietary habits, either due to changes in eating patterns during the study or misreporting of food choices or quantities (Bingham, 2002). Proper participant selection is essential to ensure dietary data accurately reflects the target population. A truly representative sample can only be achieved through random sampling, which minimizes bias and enhances the reliability of exposure assessments (Andrade, 2020). Additionally, the collection of detailed dietary data is often labour-intensive, time-consuming and costly.

**Table 1** Strengths and weaknesses while performing exposure studies

| | Strengths (PROs) | Weaknesses (CONs) |
|---|---|---|
| **Food consumption data collection** | | |
| Food diary | Accurate in terms of<br>• Quantity of food<br>• Type of food<br>• Preparation method | Expensive<br>Time consuming<br>Burden during survey<br>Questionable quality of data |
| 24hR | Accurate in terms of<br>• Quantity of food<br>• Type of food<br>• Preparation method | Short recall period<br>Need for repetition to account for day-to-day variability |
| Food Frequency Questionnaire | Captures long-period habits<br>Extensive list of food presented | Less accurate<br>Burden during survey |
| **Challenges in determination of food contaminants** | | |
| Analytic methods | Development of new methods<br>Improvement of accuracy and lab<br>3Rs (reliability, repeatability and reproducibility) | Specific food matrices<br>Expensive<br>Problems in detecting specific contaminants (derivates/ metabolites/nanomaterials)<br>Need for accurate calibration and maintenance<br>Strict lab work conditions |
| **Modelling challenges** | | |
| Deterministic | Ease of use<br>Estimates of central tendency | Not fully realistic scenario<br>High level of uncertainty |
| Stochastic | Distribution fitting<br>Clear understanding of uncertainties<br>Realistic scenario | Competence in statistics |

24hR, 24-hour dietary recall

© Burleigh Dodds Science Publishing Limited, 2025. All rights reserved.

The food record is a method in which participants document all foods and beverages consumed immediately before eating, along with weighing and describing any leftovers, typically for up to 7 days (EFSA, 2009). In addition to recording the quantity and type of food consumed, information about the food's source and preparation methods is also collected. While it is considered one of the most accurate methods for collecting food consumption data, it has several drawbacks that make it less commonly used. It is expensive, time-consuming and requires participants who are both literate and motivated (Shim et al., 2014; Bailey, 2021). The method also imposes a high burden on participants, which can lead to a decline in the quality of recorded information. Furthermore, it has been observed that participants may alter their dietary patterns for the sake of easier recording or due to a social desire to report foods perceived as 'healthy' (Bailey, 2021).

The 24hR is the most used retrospective method where participants are asked to recall types and amounts of all foods and beverages consumed during the immediate past, mostly a 24- or 48-hour period (Kroes et al., 2002). Information should be as precise as possible and include brand names, preparation methods, ingredient content (e.g. fat content for milk), location of consumption (homemade, restaurant, work) and time of consumption. It is an accurate method with the potential to capture the consumption of a wide range of foods. A drawback of this method is that it needs to be repeated to incorporate non-consecutive consumption days to capture day-to-day variations (i.e. typical diet), enhancing the suitability of this method for evaluating chronic exposure (EFSA, 2009). The low response rate for the repeated survey may result in biased results and decreased data quality.

The FFQ is a retrospective method that consists of a questionnaire containing a list of foods, for which subjects are asked to estimate the habitual frequency of consumption over a relatively long period (e.g. from 3 months to 1 year) to capture habitual food intake, especially for foods that vary greatly from day to day (EFSA, 2009; Shim et al., 2014). While, FFQs are useful for assessing long-term dietary intake they can be prone to certain personal biases. For instance, respondents may overestimate the frequency of consumption of infrequently consumed foods, while underestimating the frequency and consumption quantities of foods they perceive as unhealthy (Haftenberger et al., 2010; FAO/WHO, 2020). FFQs typically include between 50 and 150 food items, which can result in a high participant burden during the survey.

A specific challenge arises from the need to establish parallels in food consumption and exposure patterns across different regions or even between different studies from the same regions. Comparing dietary exposure studies across regions can be difficult due to various factors. Different regions have distinct food cultures, cuisines and eating habits with differences in local agricultural outputs and food imports. Different foods and food ingredients are

© Burleigh Dodds Science Publishing Limited, 2025. All rights reserved.

considered typical in each region, with different methods of preparation and cooking even for the same type of dish. Different studies may use various dietary assessment methods, which can lead to misinterpretation during comparisons between regions. Different regions have different socioeconomic and religious factors and population heterogeneity. Additionally, varying environmental conditions across regions can influence the likelihood of population exposure to contaminants from multiple sources, potentially leading to biased interpretations when comparing the presence of biomarkers.

## 2.2 Challenges in contaminants determination in food

Analysing food contaminants is a complex task that presents ongoing challenges, as the field continually evolves (Table 1). A key requirement for conducting a thorough risk assessment is the availability of data produced by dependable and appropriate analytical methods to estimate consumer exposure and intake of contaminants and residues. Traditional detection techniques include spectroscopy, chromatography, mass spectrometry, liquid chromatography-mass spectrometry and gas chromatography-mass spectrometry, which are considered accurate and sensitive. Their limitations are expensive maintenance of the instruments, strict working conditions, the need for specialized personnel and no portability (Ling et al., 2024). On the other hand, immunochemical-based methods, such as enzyme-linked immunosorbent assays and lateral flow immunoassays, can analyse many samples under nonlaboratory conditions but have lower precision and sensitivity compared to instrumental analysis (Ling et al., 2024).

The growing number of contaminants requires constant adaptation of existing analytical methods or the development of new, reliable techniques. This rise in contaminants is driven by factors such as global population growth, industrialization, technological advancements, changes in food production and packaging practices, shifting consumer behaviours, environmental influences and the impacts of climate change.

Often called emerging contaminants, these contaminants may include pharmaceuticals and personal care products, per- and poly-fluoroalkyl substances, cyanotoxins and other natural toxins, pesticides and antibiotics, industrial chemicals, micro/nano plastics, nanomaterials and other external substances that are present in the environment but not yet fully understood in terms of their effects on human health and ecosystems (Wang et al., 2024). These substances can enter the environment through various channels, such as industrial discharges, agricultural runoff and improper waste disposal, resulting in contamination of air, water, soil and food. Additionally, these emerging contaminants can transform and be transported over long distances, potentially creating new, uncharacterized chemicals and contributing to

© Burleigh Dodds Science Publishing Limited, 2025. All rights reserved.

pollution far from their source (Machado et al., 2016). Examples of such transformed contaminants are pesticides, as they are being converted into their metabolites in plants, insects or the environment. While a comprehensive toxicological dossier is developed for parent compounds, before approval of substances for use within the EU, only limited information about the occurrence of different metabolites, and the toxicological properties of metabolites is available, but the residue definition for dietary risk assessment should include the active substance and all metabolites of toxicological relevance (EFSA, 2012b). Similar challenges arise with mycotoxin derivatives, as their structures can change due to plant metabolism or food processing (Berthiller et al., 2005). These modifications often result in altered toxicity profiles, which are less understood compared to the parent compounds (Ekwomadu et al., 2021). Additionally, mycotoxin derivatives are typically difficult to detect using conventional analytical techniques, thus designated as masked mycotoxins (Berthiller et al., 2005).

Foods are overly complex matrices in which some natural components can negatively influence the analysis of the targeted compounds, especially in the detection of contaminants (Lin and Liou, 2020). This so-called 'matrix effects' could lead to under- or over-estimation of the contaminant content, requiring sophisticated sample preparation and extraction procedures. Food analysis is in general labour-intensive and costly process as it demands several stages of preparation, and it is, in most cases, performed by skilled workers on expensive equipment. Next to the variety of natural components, food can be contaminated by a wide range of contaminants, which can in some cases each require a specific analytic technique. This can all lead to prolonged time and additionally, increased costs needed for the analysis. The use of different methods and equipment by different laboratories could lead to inconsistency in results, which is crucial in terms of comparability. Ensuring the reliability, repeatability and reproducibility of the analytical methods is a major challenge and includes establishing the accuracy, precision, specificity and sensitivity of the used method (Shabir, 2005; Chavan and Desai, 2022). The significance of the lack of standardized protocols is evident in the analysis of micro and nano plastics, which are among the most pressing global environmental challenges today (Udovicki et al., 2022). While recommendations and principles for analysing microplastics in the environment have been published, the absence of standardized methods hinders a comprehensive interpretation and comparison of results from different research groups (Vojnović et al., 2024). Furthermore, methods to prevent procedural contamination have yet to be fully standardized and developed, leading to inevitable procedure contamination of samples during analysis (Gwinnett and Miller, 2021).

© Burleigh Dodds Science Publishing Limited, 2025. All rights reserved.

### 2.3 Modelling challenges

Dietary exposure models are essential tools for predicting and estimating exposure based on food consumption and contaminant concentration data. Considering that both food consumption and contamination data are often associated with a certain degree of assumptions, selecting the most suitable model is crucial to reducing the level of uncertainty in the final exposure assessment results.

Regardless of the exposure assessment purpose (assessing acute or chronic exposure, assessing exposure to contaminants from a single food item or aggregate diet), two main modelling approaches are used (Table 1), particularly deterministic and stochastic (probabilistic), along with their combinations. Commonly used, the deterministic approach uses single values, or point estimates, as inputs, resulting in a point estimate of exposure that falls somewhere within the full distribution of possible exposures (Udovicki and Djekic, 2024). It does not reflect the full range of possible outcomes, but it is a good tool for creating the worst-case, best-case, high-consumer and average exposure scenario of dietary exposure, providing assessors with meaningful estimates of central tendency or high-end exposures within a defined population (EPA, 2023). Still, it is not considered a representation of the realistic scenario and might have a high level of uncertainty. These models can be enhanced by integrating the point estimate components of the deterministic model with the elements of the probabilistic model (FAO/WHO, 2020). Additionally, the observed individual mean model, which generates a distribution of average exposures by multiplying individual consumption amounts by the mean concentrations of contaminants, can serve as a compromise between deterministic and probabilistic approaches (Boon and van der Voet, 2016). This model generates a distribution of individual exposures, capturing variability by accounting for differences in individual consumption patterns. By moving away from population-level averaging, it minimizes bias and enhances the precision of exposure estimates. This approach will tend to overestimate upper tail exposures in chronic assessments because a short-term survey is likely to over-represent the frequency of individuals consuming a food every day although underestimation of upper tail exposures may also occur (EFSA, 2012).

Unlike the deterministic approach, probabilistic techniques enable the estimation of intake distributions across a specified population. These methods account for variability in both individual and group-level food consumption, as well as the variability of chemicals in food commodities. The result is a distribution of potential exposure values, offering a more detailed characterization of variability in exposure and providing a clearer understanding of uncertainty in risk estimates (EFSA, 2012a; EPA, 2023; Udovicki et al., 2023).

© Burleigh Dodds Science Publishing Limited, 2025. All rights reserved.

While probabilistic techniques offer advantages in addressing uncertainty and variability, their implementation remains challenging. These methods require the use of advanced statistical and computational tools, and the availability of the consumption data of the individual consumers of the population of interest, making them demanding and consequently less commonly applied in practice.

When modelling concentration data, special attention must be given to values below a limit of detection (LOD) and a limit of quantification (LOQ) of an analytical method (i.e. non-detects or left censored data). Analytical data are typically a mixture of positive values and left-censored data. Left-censored values are a challenge for any statistical analysis involving occurrence data. Ignoring the undetected values and analysing only the quantified ones will inevitably result in extensively biased and therefore unrepresentative estimates, while a large number of such data may lead to computational difficulties in calculating percentile or mean values (Feraldi et al., 2023). Traditionally, left-censored data are dealt by a substitution method, which is extensively used in food risk assessment (Feraldi et al., 2023; Udovicki and Djekic, 2024). In the substitution method, non-detect values are replaced with '0' or LOD/2 (LOQ/2), and LOD (LOQ) values, which will create lower, middle and upper bounds of the exposure (EFSA, 2010). This substitution approach is simple to implement but has some disadvantages. These assumptions might have a great impact on the mean of the exposure, as well as on the high percentiles when the proportion of censored data is high and it does not consider the intrinsic variability of non-detected values that, through the process of imputation, are treated as observed, thus deflating the variability of any estimator based on such completed data (Helsel, 2006; Tressou, 2006; Feraldi et al., 2023)

When raw commodities are analysed, processing factors must be integrated into the model to enhance the accuracy of dietary exposure assessments and align the results more closely with actual exposures (Udovicki and Djekic, 2024). Processing factors for contaminants in food represent the changes (decrease, increase or structural change) in contaminant levels that occur during various food processing steps, such as washing, peeling, cooking, drying, fermentation or canning. These factors are critical for accurately assessing dietary exposure to contaminants.

Finally, cumulative exposure to multiple chemicals presents a complex challenge for researchers. Traditionally, toxicology and risk assessment have focused on single chemicals and their effects on human health. However, in real life, people are simultaneously exposed to hundreds of chemical substances and their mixtures, highlighting the need for cumulative risk assessments of multiple chemicals. Exposure to chemical mixtures may pose a greater health risk than exposure to single contaminants, as chemical substances can have additive toxicological effects, and interactions between them may result in synergistic or antagonistic effects (Goodson III, 2015; Hernández and Tsatsakis,

© Burleigh Dodds Science Publishing Limited, 2025. All rights reserved.

2017; Dairkee et al., 2018). Evaluation of chemical mixture effects is considered a multifactorial task that needs an integrated and systematic approach not only for long-term scenarios but often for acute or sub-chronic exposures (Hernández and Tsatsakis, 2017). Such evaluation requires an integrated and systematic approach and close collaboration across different scientific fields, particularly toxicology, epidemiology, exposure science, risk assessment and statistics for a proper integration of data from all these disciplines. While considerable progress has been made in recent years for some chemicals, such as pesticides (Nielsen et al., 2012; EFSA, 2019a, 2019b, 2022b), there are still countless relevant chemical mixtures that need to be evaluated.

## 3 Advancement in performing dietary exposure studies

### *3.1 Advancement in food consumption data collection*

Advancements in the capabilities of technologies, such as smartphones, wearable devices and sensors, software applications and web-based tools, have enhanced the proficiencies of collection, analysis and interpretation of dietary intake data in both contexts across the spectrum of users, including consumers, clinicians and researchers (Burrows and Rollo, 2019; Das et al., 2022).

Significant advancements in modern technologies for measuring dietary intake emerged in the 2000s, marking a shift towards more accurate and user-friendly dietary assessment tools. With tens of widely used and validated methods being developed annually since then, their total number could now exceed one hundred, reflecting a growing emphasis on innovative solutions for diet monitoring. Based on the type of technology used, they can be categorised in the following manner (Burrows and Rollo, 2019; Eldridge et al., 2019):

- **Web-based or online tools:** These tools are the online equivalent of commonly used paper-based dietary assessment tools. They are usually (but not exclusively) developed as self-administered dietary assessment tools. Their use can ease collecting repeated data, reduce costs, enhance portion size estimation, simplify and maximize food lists, increase convenience, decrease the burden of dietary assessment and possibly increase the quality of dietary assessment (Brassard et al., 2019; Das et al., 2022). Examples of these tools are The Automated Self-Administered 24-Hour Dietary Recall (ASA24) (Subar et al., 2012) and Oxford WebQ Online 24-Hour Dietary Questionnaire (Greenwood et al., 2019).
- **Mobile application technologies:** Today mobile applications play a significant role in simplifying the tracking of health-related behaviours and weight management (Chen et al., 2015). While mostly used by the

© Burleigh Dodds Science Publishing Limited, 2025. All rights reserved.

general population by being focused on health and fitness (with 3 25 000 mobile health applications available in 2017 (Ferrara et al., 2019)) mobile applications can have a similar approach to web-based dietary assessment technologies. With the significant advantage of being portable and readily available, mobile applications can collect images of food, allowing users to document their meals easily. The rapid development of artificial intelligence (AI) technologies has further enhanced these tools, enabling the identification of food items directly from images. AI algorithms analyse visual data to recognize food types, estimate portion sizes and calculate nutritional content, making dietary assessment more accurate (Samad et al., 2022; Shonkoff et al., 2023; Konstantakopoulos et al., 2024).

- **Image-based, image-assisted, or camera-based tools:** These tools use photos or videos of meals to estimate portion sizes and nutrient content and are often integrated with smartphone technology and recently aided by AI. Final image analysis can be performed by trained analysts or by automated methods (Fang et al., 2019). Some of the developed image-based tools are Technology Assisted Dietary Assessment (TADA™) system (Ahmad et al., 2016), FoodCam (Raju and Sazonov, 2022) and DietCam (Kong and Tan, 2012).
- **Developing technologies:** Wearable sensor technology is the most promising area of new tools. Wearable sensors include devices with sensors on the head or neck to detect chewing or swallowing or wrist-based inertial sensors to detect hand-to-mouth gestures as a proxy for bites (Das et al., 2022).

Not all the presented emerging methods for food consumption data collection have been fully utilized in exposure assessment studies. However, they continue to be developed, validated and refined for nutritional research and are likely to see increased application in the coming years.

Differences in nomenclatures, levels of detail, food categorization and food consumption practices across various sources and regions can pose significant challenges to the standardization and harmonization of food-related data, complicating efforts to perform accurate comparisons and analyses (Durazzo et al., 2018; Micha et al., 2018). The use of a common system for classifying and describing foods is an important prerequisite for analysing data from different sources (D'Addezio et al., 2024). The collection of accurate and harmonised food consumption data at a European level is considered a primary long-term objective for the European Food Safety Authority (EFSA) (EFSA, 2014). For that reason, EFSA has initiated a project for a more harmonised pan-European food consumption survey, known as the 'What's on the Menu in Europe? (EU Menu)' which provided recommendations for the collection of more harmonised food consumption

© Burleigh Dodds Science Publishing Limited, 2025. All rights reserved.

data among the EU Member States for use in dietary exposure assessments of food-borne hazards (EFSA, 2014). The reported foods should be described following the EFSA FoodEx2 food classification system. FoodEx2 is a comprehensive food classification and description system aimed at covering the need to describe food in data collection across different food safety domains (EFSA, 2015). FoodEx2 consists of descriptions of a large number of individual food items aggregated into food groups and broader food categories in a hierarchical parent-child relationship providing a standardized approach, By this consistent and structured approach, FoodEx2 reduces inconsistencies in food coding across different datasets making international comparisons of food consumption data more straightforward and reducing misclassification errors (EFSA, 2015). It is used by EFSA, various national Food Safety Authorities from European Union Member States as well as individual researchers, universities and public health institutions. It is also used at global level with the support of the Food and Agriculture Organization of the United Nations and the World Health Organization to harmonize datasets in the FAO/WHO Global Individual Food consumption data Tool–FAO/WHO GIFT (Ferreira de Sousa et al., 2020). These prerequisites led to the development of the standardized and harmonized EFSA Comprehensive European Food Consumption Database which is today a source of information on food consumption across the European Union (EFSA, 2022a). The EFSA Comprehensive European Food Consumption Database allows for comparisons of food consumption patterns and exposure assessments between regions enhancing the accuracy and reliability of conclusions drawn from the data, which is essential for public health, regulatory decisions, and international collaborations. However, it have to be noted that public versions of the Comprehensive Database are only reporting summarised data, while the actual data on individual consumers are only used by EFSA (or the specific country that provided them), which is debilitating the wide use of this database by individual researchers.

## 3.2 Advancements in contaminants determination in food

Apart from perfecting traditional chemical contaminants detection techniques considerable effort is made to develop new, fast, simple, and convenient analytical methods capable of detecting the presence of contaminants, even at minute levels. Emerging trends include the use of rapid testing methods, real-time monitoring and the application of AI in data analysis (Sage, 2024). These new methods should always be validated in line with existing official methods to avoid discussions about the quality of the occurrence data. Additionally, the use of so-called green solvents for sample preparation and extraction has

© Burleigh Dodds Science Publishing Limited, 2025. All rights reserved.

emerged as an essential component of sustainable practices as these solvents minimize environmental impact, enhance safety and reduce the health risks associated with traditional petrochemical solvents. The simultaneous detection of several contaminants is also an important aspect of the analytical method. Some of the novel methods for the determination of contaminants in food are as follows:

- **Biosensors** – Biosensors (electrochemical, optical and piezoelectric) enable rapid, accurate and on-site detection of contaminants, revolutionizing the management of food safety and providing unique advantages in sensitivity, specificity and detection capabilities (Nath, 2024). Biosensors offer valuable tools for detecting a wide range of contaminants, allergens and adulterants in food samples. However, biosensor implementation faces challenges such as sensitivity and specificity issues, matrix interference and shelf-life concerns (Nath, 2024). Recent advances in biosensor technology, including nanotechnology integration, multiplexed detection and smartphone-based biosensors, offer opportunities to improve and enhance food monitoring (Nath, 2024).
- **Infrared spectroscopy (IR) and Raman spectroscopy** – IR technique does not require complex treatments in sample preparation, it is the fast, non-destructive, low-cost method which is used for the rapid detection and identification of the many contaminants, including mycotoxins, pesticides, microplastics, various oils, processing contaminants in a wide range of food product (Qi et al., 2022; Gruska et al., 2023). Similarly to IR, Raman spectroscopy is a promising and emerging method to detect chemicals in food as it is a non-destructive, rapid, specific, ultrasensitive, and high-throughput screening method with an option for on-site detection (Petersen et al., 2021).
- **Nanomaterials** – The use of nanomaterials offers innovative, specific, low-cost, sensitive and versatile approaches for detecting contaminants in food. A variety of nanomaterials have been developed in recent years, including quantum dots, nano-porphyrin, nano-gold, nano-silver, upconversion nanoparticles, nanozymes, metal oxide nanomaterials and organic fluorescent molecules (Chen et al., 2021). They were successfully used as active components of various optical, electrochemical, magnetic-based, spectroscopic, biosensor and enzyme-linked immunosorbent methods (Curulli, 2020; Chen et al., 2021; Zhao et al., 2021; Neng et al., 2023). The rapid development of nanotechnology has led to the invention of lab-on-a-chip (LOC) technology. LOC is a miniaturized device that integrates multiple laboratory functions onto a single chip of only millimetres to a few centimetres in size. This LOC technology is portable, easy to handle, relatively inexpensive, requires less sample volume, enables simultaneous

© Burleigh Dodds Science Publishing Limited, 2025. All rights reserved.

detection of multiple analytes and permits automatic biosensing process control (Rafiq et al., 2024).

- **Artificial intelligence:** In recent years, AI has experienced rapid growth, emerging as a transformative force across numerous industries. AI systems are now adept at performing tasks that once relied on human intelligence, including image recognition, language translation and solving complex problems. AI encompasses computational techniques like machine learning, support vector machines, random forests and neural networks, enabling computers to analyse data, identify patterns and make decisions with minimal human intervention (Aggarwal et al., 2024). AI can automatically extract intricate features from large data volumes, making it particularly suitable for image recognition, spectroscopic analysis and the interpretation of sensor data for mycotoxin detection (Choudhary et al., 2022; Aggarwal et al., 2024). AI-driven software interprets spectral data with precision, identifies contaminants with high accuracy, and reduces the need for extensive human input in mass spectrometry analysis (Kuppusamy et al., 2024). Another distinctive feature of AI-driven systems is that they can predict contamination risks, recommend preventive actions and optimize resource allocation by incorporating predictive modelling techniques (Kuppusamy et al., 2024).

### 3.3 Modelling advancements

Recent years have witnessed the rapid development of computing and information technology easing the use of probabilistic methods for dietary exposure assessment. Probabilistic exposure assessments are considered the best representation of realistic scenarios as they account for the variability and uncertainty of used data. Monte Carlo simulation is the most common probabilistic approach to estimating dietary exposure to contaminants. By using Monte Carlo simulation, probability distributions for one or more variables are repeatedly sampled, and the exposure estimate is calculated each time to generate a distribution of possible exposure values (EPA, 2023). The distributions sampled in a Monte Carlo risk assessment can be either empirical data (i.e. actual ingestion rates recorded in scientific studies) or synthesized probability/statistical distributions such as ones estimated from mathematical equations that represent possible distributions of values (EPA, 2023). A variety of Monte Carlo simulations such as Minitab, @RISK, Oracle Crystal Ball and XLSTAT, are available for commercial use, either as standalone software or as part of statistical software packages (Udovicki and Djekic, 2024). R, a free and open-source language, and environment for statistical computing (R Core Team, 2025), allows researchers to design and implement Monte Carlo simulation studies, from simple to highly complex

© Burleigh Dodds Science Publishing Limited, 2025. All rights reserved.

(Pouillot and Delignette-Muller, 2010; Robert et al., 2010). Depending on the complexity level, Monte Carlo simulation can be one- or two-dimensional. A Monte Carlo analysis that characterizes either uncertainty or variability in each input variable can be described as a one-dimensional Monte Carlo simulation, while a two-dimensional Monte Carlo simulation is a term used to describe a model that simulates both uncertainty and variability in one or more input variables (EPA, 2001).

Several methods were developed for the statistical treatment of left-censored contamination data. The most common approach to deal with left-censored values is based on maximum likelihood estimation (MLE) applied to parametric models (Helsel, 2006; Feraldi et al., 2023; Hwang et al., 2023). MLE is regarded as a robust and effective analytical approach by various authoritative bodies (FAO/WHO, 2009; EFSA, 2010). MLE assumes that values below the LOD and observation data follow a particular statistical distribution (typically lognormal, Weibull or gamma distribution) and estimates distribution parameters from observed values (Hwang et al., 2023). Another often-used parametric method is regression on order statistics or log-probit regression method. This method calculates a regression equation through log transformation and assumes that the detected data follow a lognormal distribution (Hwang et al., 2023). On the other hand, nonparametric methods such as the Kaplan–Meier method do not require any assumptions about the distribution (Hwang et al., 2023).

As mentioned before, processing factors must be integrated into the model to enhance the accuracy of dietary exposure assessments. Depending on the contaminant, there are many research and review papers covering the fate of contaminants during both industrial and household processing. In 2022 EFSA published an update of the EU database of processing factors for pesticide residues (Zincke et al., 2022). This comprehensive database contains results of 2152 processing studies resulting in 2104 acceptable/indicative median processing factors for multiple pesticide commodity combinations. This database includes processing factors for both simple processes (e.g. peeling of mandarins or oranges) and industrially processed commodities (e.g. marmalade, canned fruit, instant coffee, etc.), providing comprehensive information to evaluate how processing affects pesticide levels in food products Since most available processing factors relate to industrial processes, it is essential to include those related to household and restaurant food preparation in databases to improve the accuracy of exposure assessment calculations.

As exposure to chemical mixtures may pose a greater health risk than exposure to single contaminants EFSA gave special attention to developing cumulative exposure models, particularly to pesticides. An initial step was developing a general methodology and criteria for grouping pesticides in the cumulative assessment groups (CAGs) based on their toxicological properties

© Burleigh Dodds Science Publishing Limited, 2025. All rights reserved.

(EFSA PPR, 2013). This methodology follows a phenomenological approach based on organ or system toxicity, consisting of including in a CAG for a specific effect all pesticides causing this effect, even if the underlying mode of action is unknown and dose additivity is the default assumption within the model (EFSA PPR, 2013). As the mode of calculation and expression of cumulative risks the Total Margin of Exposure (MOET), representing the ratio of a toxicological reference dose for an index substance to the estimated cumulative exposure, was chosen. To date, CAGs for the pesticides that affect the nervous system, thyroid gland and craniofacial alteration have been established and cumulative risk assessments for these groups have been performed (EFSA, 2019a, 2019b, 2020a, 2020b, 2022b).

Over the years numerous methods, tools and databases (for both food consumption and contaminant levels data) have been developed to improve the exposure assessment process. In many cases, these databases and tools are open-access, particularly ones developed by EFSA (Ioannidou et al., 2021). Unfortunately, these databases are all structured in a different format, presented in a non-optimal format making the extraction of the appropriate data a challenge. The Monte Carlo Risk Assessment (MCRA) platform, an open web-based tool, facilitates cumulative risk assessment of chemicals. It is developed by Wageningen University & Research (WUR, Biometris) for the Dutch Institute for Public Health and the Environment (RIVM) (van Klaveren et al., 2024). In 2019, EFSA and RIVM, using the MCRA software, conducted a cumulative dietary exposure assessment of pesticides that have acute effects on the nervous system (EFSA, 2020a). The EFSA–RIVM partnership is directed at making the MCRA software accessible to all European countries and stakeholders that work on the safety assessment of pesticides.

## 3.4 Scientific challenges in employing advancement in exposure studies

When evaluating specific scientific challenges, one approach can be the application of Failure Mode and Effects Analysis (FMEA) as it's considered a risk-based analytic tool (Xiao et al., 2011). Its main benefit is ease of capturing potential failures, their causes and paves the way for analysing effects of failures (IEC, 2016). It consists of two parts: (i) classification of potential failures and (ii) risks evaluation (Papadopoulos et al., 2011). In order to evaluate failures in exposure studies, it is necessary to populate an inventory of failures associated with different stages of these studies (Table 2).

The FMEA risk priority number (RPN) involves three factors (IEC, 2016):

$$RPN = S \times O \times D \tag{1}$$

© Burleigh Dodds Science Publishing Limited, 2025. All rights reserved.

**Table 2** Inventory of phenomena associated with exposure studies

| Exposome studies stages | Phenomenon |
|---|---|
| **Food consumption data collection** | Use of web/online tools |
| | Use of mobile applications |
| | Use of on-body sensors |
| **Determination of food contaminants** | Use of biosensors |
| | Employment of spectroscopy |
| | Use of AI |
| **Modelling** | Monte Carlo simulation |
| | Treatment of left-censored data |
| | Cumulative exposure assessment |

where (S) failure's severity; (O) probability of occurrence; and (D) likelihood of detection. Risks are associated with difficulties or failures resulting in inadequate exposome studies. Table 3 depicts the weighting factors modified from Djekic et al. (2018, 2019).

As shown in Table 4, the highest risks of failures in exposure studies are associated with failures in the use of AI, Monte Carlo simulations and cumulative exposure assessment studies. Such failures could significantly impact the overall study results, and addressing these potential failures should be prioritized to enhance reliability and accuracy when conducting dietary exposure studies.

## 4 Conclusion and future trends

The intake of chemical substances through food is a critical component of the chemical exposome as food is a considerable source of a wide range of natural and man-made chemicals and the perfect vehicle for daily and continuous exposure to them. Traditional dietary exposure assessment methods are still essential tools for understanding and managing risks related to the presence of chemical contaminants in food.

Usual methods for collecting food consumption data are labour-intensive, time-consuming and costly and are still prone to both random and systematic errors. Analysing food contaminants is a complex task that presents ongoing challenges, as the field continually evolves, and new contaminants emerge. Dietary exposure models are still mostly based on a deterministic approach, debatable methods for dealing with left-censored data and focused on single contaminants.

The rapid acceleration of technological advancements has become a defining feature of the modern era, fundamentally transforming various aspects of human life, including the methodologies for dietary exposure assessments. Advancements in the capabilities of technologies, such as smartphones, wearable

© Burleigh Dodds Science Publishing Limited, 2025. All rights reserved.

**Table 3** Severity, occurrence, and detection rating scale

| | Severity | |
|---|---|---|
| Rank | Consequence | Description |
| 1 | None | No failure(s)/problem(s) in modelling |
| 2 | Minor | Failure(s) associated with results for one phenomenon |
| 3 | Low | Failure(s) associated with results within one stage of exposure studies |
| 4 | Major | Failure(s) associated with results within more than one stage of exposure studies |
| 5 | Severe | Failure(s) associated with results of the entire exposure studies |
| | **Occurrence** | |
| **Rank** | **Probability** | **Description** |
| 1 | Very unlikely | Minimal probability of occurrence of failure(s) as a result of *force majeure* |
| 2 | Unlikely | Occurrence of failure(s) only as a result of misuse/rejection of the data |
| 3 | Possible | Occurrence of failure(s) only as a result of errors in transferring experimental data into the mathematical model |
| 4 | High probability | Occurrence of failure(s) will occur only for certain types of products/regions |
| 5 | Certain | Occurrence of failure(s) will occur for all exposure studies |
| | **Detection** | |
| **Rank** | **Criteria** | **Description** |
| 1 | Very high | Exposure results fully correlate and are easily verifiable with real data |
| 2 | High | The phenomenon/phenomena can be modelled, and results are verifiable and highly correlated with real data |
| 3 | Low | The phenomenon/phenomena can be modelled, but results are hard to verify with real data |
| 4 | Remote | Model and resulting data are too far from real data |
| 5 | Never | No possibility of performing exposure studies for this type of phenomenon |

devices and sensors, software applications and web-based tools, have enhanced the proficiencies of collection, analysis and interpretation of dietary intake data. Similar progress in analytical technologies made over the last decade has led to the perfecting of traditional chemical contaminants detection techniques as well as the development of new, fast, simple and convenient analytical methods capable of detecting the presence of contaminants. The rapid development of computing and information technology eased the use of statistical and probabilistic methods for dietary exposure assessment. This is especially visible by increased use of Monte Carlo simulations and similar probabilistic methods

© Burleigh Dodds Science Publishing Limited, 2025. All rights reserved.

**Table 4** Possible Failure Mode and Effects Analysis for exposure studies

| No | Stage | Phenomenon | Failure | Potential model effect? | Severity (S) | Occurrence (O) | Detection (D) | Risk |
|---|---|---|---|---|---|---|---|---|
| 1 | Food consumption data collection | Use of web/online tools | Too long questionnaire (subject doesn't fill all the data) | Inadequate consumption pattern | 2 | 2 | 3 | 12 |
| | | | Generic questionnaire for different regions | Inadequate regional consumption pattern | 2 | 4 | 3 | 24 |
| 2 | | Use of mobile applications | Not calibrated for all types of mobile phones (hard to read) | No data collected | 3 | 2 | 3 | 18 |
| | | | Too long questionnaire (subject doesn't fill all the data) | Inadequate consumption pattern | 2 | 2 | 3 | 12 |
| 3 | | Use of on-body sensors | Human subject takes of on-body sensors | High variability between participants | 3 | 3 | 3 | 27 |
| | | | Detection of wrong patterns (chewing gum) | False positive results | 2 | 3 | 3 | 18 |
| 4 | Determination of food contaminants | Use of biosensors | Too sensitive | Questionable quality of data | 2 | 3 | 4 | 24 |
| | | | Dependent on type of food | Different accuracy depending on food matrix | 2 | 4 | 3 | 24 |
| 5 | | Employment of spectroscopy | Difference in detecting soluble and insoluble contaminants | Different accuracy depending on contaminant | 2 | 4 | 3 | 24 |
| 6 | | Use of AI | Inadequate training of AI model | False positive/negative results | 4 | 3 | 4 | 48 |
| 7 | Modelling | Monte Carlo simulation | Inadequate statistical/computer skills of user | Effect on exposure results | 4 | 3 | 4 | 48 |
| 8 | | Treatment of left-censored data | Inadequate distribution assumed | Left-censored bias | 2 | 2 | 3 | 12 |
| 9 | | Cumulative exposure assessment | Synergy of cumulative risk inadequately assumed | Effect on MOET results | 4 | 3 | 4 | 48 |

© Burleigh Dodds Science Publishing Limited, 2025. All rights reserved.

which are increasingly faster and more detailed, allowing analysis of substantial number of exposure scenarios quickly, allowing for better risk quantification.

Future trends in dietary exposure assessments will align with technological advancements, featuring enhanced analytical tools for more precise identification of dietary contaminants. Innovations in food consumption surveys and tools will enable the collection of detailed, individual-level dietary data. These data should not be kept confidential (as in the current situation) but should be made available for exposure assessment experts. Additionally, AI and machine learning will facilitate the integration of large datasets, merging information from food consumption databases, consumer behaviour studies, and food and environmental monitoring. This will allow for real-time assessments, as well as global-scale integration and comparison.

## 5 Where to look for further information

- https://www.efsa.europa.eu/en/data-report/food-consumption-data.
- https://www.epa.gov/expobox/exposure-assessment-tools-tiers-and-types-deterministic-and-probabilistic-assessments.
- .U.S. EPA. Options For Development of Parametric Probability Distributions for Exposure Factors. U.S. Environmental Protection Agency, Office of Research and Development, National Center for Environmental Assessment, Washington Office, Washington, DC, EPA/600/R-00/058, 2000.
- EFSA Panel on Plant Protection Products and their Residues (PPR); Guidance on the Use of Probabilistic Methodology for Modelling Dietary Exposure to Pesticide Residues. EFSA Journal 2012; 10(10):2839. [95 pp.]. https://doi.org/10.2903/j.efsa.2012.2839.

## 6 References

Aggarwal, A., et al. (2024) 'Detection of mycotoxin contamination in foods using artificial intelligence: a review', *Foods*. https://doi.org/10.3390/foods13203339.

Ahmad, Z., et al. (2016) 'A mobile food record for integrated dietary assessment', *Proceedings of the 2nd International Workshop on Multimedia Assisted Dietary Management*, pp. 53–62.

Andrade, C. (2020) 'The Inconvenient Truth About Convenience and Purposive Samples', *Indian Journal of Psychological Medicine*, 43(1), pp. 86–88. https://doi.org/10.1177/0253717620977000.

Bailey, R. L. (2021) 'Overview of dietary assessment methods for measuring intakes of foods, beverages, and dietary supplements in research studies', *Current Opinion in Biotechnology*, 70, pp. 91–96. https://doi.org/10.1016/j.copbio.2021.02.007.

Balcells, C., et al. (2024) 'Blurred lines: crossing the boundaries between the chemical exposome and the metabolome', *Current Opinion in Chemical Biology*, 78, p. 102407. https://doi.org/10.1016/j.cbpa.2023.102407.

© Burleigh Dodds Science Publishing Limited, 2025. All rights reserved.

Berthiller, F., et al. (2005) 'Masked mycotoxins: determination of a deoxynivalenol glucoside in artificially and naturally contaminated wheat by liquid chromatography–tandem mass spectrometry', *Journal of Agricultural and Food Chemistry*, 53(9), pp. 3421–3425. https://doi.org/10.1021/jf047798g.

Bingham, S. A. (2002) 'Biomarkers in nutritional epidemiology', *Public Health Nutrition*. 2006/12/22, 5(6a), pp. 821–827. https://doi.org/10.1079/PHN2002368.

Boon, P. E. and van der Voet, H. (2016) *Probabilistic dietary exposure models: Relevant for acute and chronic exposure assessment of adverse chemicals via food*. Rijksinstituut voor Volksgezondheid en Milieu RIVM.

Brassard, D., et al. (2019) 'Examining the advantages of using multiple web-based dietary assessment instruments to measure population dietary intake: the PREDISE study', *Current Developments in Nutrition*, 3(5), p. nzz014.

Burrows, T. L. and Rollo, M. E. (2019) 'Advancement in dietary assessment and self-monitoring using technology', *Nutrients*. https://doi.org/10.3390/nu11071648.

Chavan, S. and Desai, D. (2022) 'Analytical method validation: a brief review', *World Journal of Advanced Research and Reviews*, 16, pp. 389–402. https://doi.org/10.30574/wjarr.2022.16.2.1165.

Chen, H., et al. (2021) 'Nanomaterials as optical sensors for application in rapid detection of food contaminants, quality and authenticity', *Sensors and Actuators B: Chemical*, 329, p. 129135. https://doi.org/10.1016/j.snb.2020.129135.

Chen, J., et al. (2015) 'The most popular smartphone apps for weight loss: a quality assessment', *JMIR mHealth uHealth*, 3(4), p. e104. https://doi.org/10.2196/mhealth.4334.

Choudhary, K., et al. (2022) 'Recent advances and applications of deep learning methods in materials science', *npj Computational Materials*, 8(1), p. 59. https://doi.org/10.1038/s41524-022-00734-6.

Curulli, A. (2020) 'Nanomaterials in electrochemical sensing area: applications and challenges in food analysis', *Molecules*. https://doi.org/10.3390/molecules25235759.

D'Addezio, L., et al. (2024) 'FoodEx2 harmonization of the food consumption database from the Italian IV SCAI children's survey', *Nutrients*. https://doi.org/10.3390/nu16071065.

Dairkee, S. H., et al. (2018) 'A ternary mixture of common chemicals perturbs benign human breast epithelial cells more than the same chemicals do individually', *Toxicological Sciences*, 165(1), pp. 131–144. https://doi.org/10.1093/toxsci/kfy126.

Das, S. K., et al. (2022) 'Perspective: opportunities and challenges of technology tools in dietary and activity assessment: bridging stakeholder viewpoints', *Advances in Nutrition*, 13(1), pp. 1–15. https://doi.org/10.1093/advances/nmab103.

DeBord, D. G., et al. (2016) 'Use of the "exposome" in the practice of epidemiology: a primer on -omic technologies', *American Journal of Epidemiology*, 184(4), pp. 302–314. https://doi.org/10.1093/aje/kwv325.

Djekic, I., et al. (2018) 'Hygienic design of a unit for supercritical fluid drying – case study', *British Food Journal*, 120(9), pp. 2155–2165. https://doi.org/10.1108/BFJ-01-2018-0052.

Djekic, I., et al. (2019) 'Scientific challenges in performing life-cycle assessment in the food supply chain', *Foods*. https://doi.org/10.3390/foods8080301.

© Burleigh Dodds Science Publishing Limited, 2025. All rights reserved.

Durazzo, A., et al. (2018) *Food groups and individual foods: nutritional attributes and dietary importance*. Amsterdam: Elsevier. https://doi.org/10.1016/B978-0-08-100596-5.21337-1.

EFSA (2009) 'General principles for the collection of national food consumption data in the view of a pan-European dietary survey', *EFSA Journal*, 7(12), p. 1435.

EFSA (2010) 'Management of left-censored data in dietary exposure assessment of chemical substances', *EFSA Journal*, 8(3), p. 1557. https://doi.org/10.2903/j.efsa.2010.1557.

EFSA (2011) 'Overview of the procedures currently used at EFSA for the assessment of dietary exposure to different chemical substances', *EFSA Journal*, 9(12), p. 33.

EFSA (2012a) 'Guidance on the use of probabilistic methodology for modelling dietary exposure to pesticide residues', *EFSA Journal*, 10(10), p. 2839.

EFSA (2012b) 'Scientific opinion on evaluation of the toxicological relevance of pesticide metabolites for dietary risk assessment', *EFSA Journal*, 10(7), p. 2799. https://doi.org/10.2903/j.efsa.2012.2799.

EFSA (2014) 'Guidance on the EU menu methodology', *EFSA Journal*, 12(12), p. 3944. https://doi.org/10.2903/j.efsa.2014.3944.

EFSA (2015) 'The food classification and description system FoodEx 2 (revision 2)', *EFSA Supporting Publications*, 12(5), p. 804E. https://doi.org/10.2903/sp.efsa.2015.EN-804.

EFSA (2019a) 'Establishment of cumulative assessment groups of pesticides for their effects on the nervous system', *EFSA Journal*, 17(9), p. e05800. https://doi.org/10.2903/j.efsa.2019.5800.

EFSA (2019b) 'Establishment of cumulative assessment groups of pesticides for their effects on the thyroid', *EFSA Journal*, 17(9), p. e05801. https://doi.org/10.2903/j.efsa.2019.5801.

EFSA (2020a) 'Cumulative dietary risk characterisation of pesticides that have acute effects on the nervous system', *EFSA Journal*, 18(4), p. e06087. https://doi.org/10.2903/j.efsa.2020.6087.

EFSA (2020b) 'Cumulative dietary risk characterisation of pesticides that have chronic effects on the thyroid', *EFSA Journal*, 18(4), p. e06088. https://doi.org/10.2903/j.efsa.2020.6088.

EFSA (2022a) *Food consumption data, The EFSA comprehensive european food consumption database*. https://www.efsa.europa.eu/en/data-report/food-consumption-data.

EFSA (2022b) 'Retrospective cumulative dietary risk assessment of craniofacial alterations by residues of pesticides', *EFSA Journal*, 20(10), p. e07550.

EFSA PPR (2013) 'Scientific opinion on the identification of pesticides to be included in cumulative assessment groups on the basis of their toxicological profile', *EFSA Journal*, 11(7), p. 3293. https://doi.org/10.2903/j.efsa.2013.3293.

Ekwomadu, T. I., et al. (2021) 'Fusarium mycotoxins, their metabolites (Free, emerging, and masked), food safety concerns, and health impacts', *International Journal of Environmental Research and Public Health*. 18, 11741. https://doi.org/10.3390/ijerph182211741.

Eldridge, A. L., et al. (2019) 'Evaluation of new technology-based tools for dietary intake assessment–an ILSI Europe dietary intake and exposure task force evaluation', *Nutrients*. https://doi.org/10.3390/nu11010055.

EPA (2001) *Risk assessment guidance for superfund: volume III-part A, process for conducting probabilistic risk assessment*. Washington, DC: EPA.


© Burleigh Dodds Science Publishing Limited, 2025. All rights reserved.

EPA (2023) *Exposure assessment tools by tiers and types - Deterministic and probabilistic assessments, EPA ExpoBox*. https://www.epa.gov/expobox/exposure-assessment-tools-tiers-and-types-deterministic-and-probabilistic-assessments.

Fang, S., et al. (2019) 'An end-to-end image-based automatic food energy estimation technique based on learned energy distribution images: protocol and methodology', *Nutrients*. https://doi.org/10.3390/nu11040877.

FAO/WHO (2009) *Principles and methods for the risk assessment of chemicals in food*. Geneva, Switzerland: World Health Organization.

FAO/WHO (2020) 'Dietary exposure assessment for chemicals in food', in *Environmental health criteria 240: principles and methods for the risk assessment of chemicals in food*. Geneva, Switzerland: World Health Organization, p. 141.

Feraldi, A., et al. (2023) 'Evaluation of statistical treatment of left-censored contamination data: example involving deoxynivalenol occurrence in pasta and pasta substitute products', *Toxins*. https://doi.org/10.3390/toxins15090521.

Ferrara, G., et al. (2019) 'A focused review of smartphone diet-tracking apps: usability, functionality, coherence with behavior change theory, and comparative validity of nutrient intake and energy estimates', *JMIR Mhealth Uhealth*, 7(5), p. e9232. https://doi.org/10.2196/mhealth.9232.

Ferreira de Sousa, R. et al. (2020) 'FoodEx2 enhances the harmonization of food consumption, composition and carbon footprint databases', *European Journal of Public Health*, 30(Supplement_5), p. ckaa165.769. https://doi.org/10.1093/eurpub/ckaa165.769.

Goodson III, W. H. (2015) 'Assessing the carcinogenic potential of low-dose exposures to chemical mixtures in the environment: the challenge ahead', *Carcinogenesis*, 36(Suppl_1), pp. S254–S296. https://doi.org/10.1093/carcin/bgv039.

Greenwood, D. C., et al. (2019) 'Validation of the Oxford WebQ online 24-hour dietary questionnaire using biomarkers', *American Journal of Epidemiology*, 188(10), pp. 1858–1867. https://doi.org/10.1093/aje/kwz165.

Gruska, R. M., et al. (2023) 'Fourier Transform Mid-Infrared Spectroscopy (FT-MIR) as a method of identifying contaminants in sugar beet production process–case studies', *Molecules*. https://doi.org/10.3390/molecules28145559.

Gwinnett, C. and Miller, R. Z. (2021) 'Are we contaminating our samples? A preliminary study to investigate procedural contamination during field sampling and processing for microplastic and anthropogenic microparticles', *Marine Pollution Bulletin*, 173, p. 113095. https://doi.org/10.1016/j.marpolbul.2021.113095.

Haftenberger, M., et al. (2010) 'Relative validation of a food frequency questionnaire for national health and nutrition monitoring', *Nutrition Journal*, 9, pp. 1–9.

Helsel, D. R. (2006) 'Fabricating data: how substituting values for nondetects can ruin results, and what can be done about it', *Chemosphere*, 65(11), pp. 2434–2439. https://doi.org/10.1016/j.chemosphere.2006.04.051.

Hernández, A. F. and Tsatsakis, A. M. (2017) 'Human exposure to chemical mixtures: challenges for the integration of toxicology with epidemiology data in risk assessment', *Food and Chemical Toxicology*, 103, pp. 188–193. https://doi.org/10.1016/j.fct.2017.03.012.

Hwang, M., et al. (2023) 'Statistical methods for handling nondetected results in food chemical monitoring data to improve food risk assessments', *Food Science & Nutrition*, 11(9), pp. 5223–5235. https://doi.org/10.1002/fsn3.3481.

© Burleigh Dodds Science Publishing Limited, 2025. All rights reserved.

IEC (2016) *IEC 60812:2016 Failure modes and effects analysis (FMEA and FMECA)*. Geneva, Switzerland: Commission Electrotechnique Internationale.

Ioannidou, S., et al. (2021) 'European Food Safety Authority open access tools to estimate dietary exposure to food chemicals', *Environment International*, 149, p. 106357. https://doi.org/10.1016/j.envint.2020.106357.

Jeddi, M. Z., et al. (2022) 'A vision on the "foodture" role of dietary exposure sciences in the interplay between food safety and nutrition', *Trends in Food Science & Technology*, 120, pp. 288–300. https://doi.org/10.1016/j.tifs.2022.01.024.

Kong, F. and Tan, J. (2012) 'DietCam: automatic dietary assessment with mobile camera phones', *Pervasive and Mobile Computing*, 8(1), pp. 147–163. https://doi.org/10.1016/j.pmcj.2011.07.003.

Konstantakopoulos, F. S., et al. (2024) 'A review of image-based food recognition and volume estimation artificial intelligence systems', *IEEE Reviews in Biomedical Engineering*, 17, pp. 136–152. https://doi.org/10.1109/RBME.2023.3283149.

Kortenkamp, A. and Faust, M. (2018) 'Regulate to reduce chemical mixture risk', *Science*, 361(6399), pp. 224–226. https://doi.org/10.1126/science.aat9219.

Kroes, R., et al. (2002) 'Assessment of intake from the diet', *Food and Chemical Toxicology*, 40(2–3), pp. 327–385. https://doi.org/10.1016/S0278-6915(01)00113-2.

Kuppusamy, S., et al. (2024) 'Integrating AI in food contaminant analysis: enhancing quality and environmental protection', *Journal of Hazardous Materials Advances*, 16, p. 100509. https://doi.org/10.1016/j.hazadv.2024.100509.

Liang, S., et al. (2022) 'Biomarkers of dietary patterns: a systematic review of randomized controlled trials', *Nutrition Reviews*, 80(8), pp. 1856–1895.

Lin, C.-H. and Liou, Y.-W. (2020) *Chapter 7 - Design and construction of paper-spray ionization/mass spectrometry*, Zaitsu, K. B. T.-A. I. M. S. (ed.). Amsterdam: Elsevier, pp. 207–239. https://doi.org/10.1016/B978-0-12-817220-9.00007-2.

Ling, Z., et al. (2024) 'Detection of food contaminants: a review of established rapid analytical techniques and their applications and limitations', *Food Safety and Health*, 2(1), pp. 72–95. https://doi.org/10.1002/fsh3.12032.

Machado, K. C., et al. (2016) 'A preliminary nationwide survey of the presence of emerging contaminants in drinking and source waters in Brazil', *Science of The Total Environment*, 572, pp. 138–146. https://doi.org/10.1016/j.scitotenv.2016.07.210.

Micha, R., et al. (2018) 'Global dietary surveillance: data gaps and challenges', *Food and Nutrition Bulletin*, 39(2), pp. 175–205. https://doi.org/10.1177/0379572117752986.

Nath, S. (2024) 'Advancements in food quality monitoring: integrating biosensors for precision detection', *Sustainable Food Technology*, 2(4), pp. 976–992. https://doi.org/10.1039/d4fb00094c.

Neng, J., et al. (2023) 'Trace analysis of food by surface-enhanced Raman spectroscopy combined with molecular imprinting technology: principle, application, challenges, and prospects', *Food Chemistry*, 429, p. 136883. https://doi.org/10.1016/j.foodchem.2023.136883.

Nielsen, E., et al. (2012) 'Identification of cumulative assessment groups of pesticides', *EFSA Supporting Publications*, 9(4), p. 269E. https://doi.org/10.2903/sp.efsa.2012.EN-269.

Papadopoulos, Y., et al. (2011) 'Engineering failure analysis and design optimisation with HiP-HOPS', *Engineering Failure Analysis*, 18(2), pp. 590–608. https://doi.org/10.1016/j.engfailanal.2010.09.025.

© Burleigh Dodds Science Publishing Limited, 2025. All rights reserved.

Petersen, M., et al. (2021) 'Application of Raman spectroscopic methods in food safety: a review', *Biosensors*. https://doi.org/10.3390/bios11060187.

Pouillot, R. and Delignette-Muller, M. L. (2010) 'Evaluating variability and uncertainty separately in microbial quantitative risk assessment using two R packages', *International Journal of Food Microbiology*, 142(3), pp. 330–340. https://doi.org/10.1016/j.ijfoodmicro.2010.07.011.

Qi, W., et al. (2022) 'Research progress of applying infrared spectroscopy technology for detection of toxic and harmful substances in food', *Foods*. https://doi.org/10.3390/foods11070930.

R Core Team (2025) *R: a language and environment for statistical computing*. Vienna, Austria: R Foundation for Statistical Computing. https://www.r-project.org/.

Rafiq, S. M., et al. (2024) 'Lab-on-a-chip device for food quality control and safety', *Food Control*, 164, p. 110596. https://doi.org/10.1016/j.foodcont.2024.110596.

Raju, V. B. and Sazonov, E. (2022) 'FOODCAM: a novel structured light-stereo imaging system for food portion size estimation', *Sensors*. https://doi.org/10.3390/s22093300.

Robert, C. P., Casella, G. and Casella, G. (2010) *Introducing Monte Carlo methods with R*. New York: Springer.

Sage, H. (2024) 'The importance of food analysis: Ensuring safety, quality, and nutritional value'. *Arch Food Nutr*. 7(5), p. 235.

Samad, S., et al. (2022) 'Smartphone apps for tracking food consumption and recommendations: evaluating artificial intelligence-based functionalities, features and quality of current apps', *Intelligent Systems with Applications*, 15, p. 200103. https://doi.org/10.1016/j.iswa.2022.200103.

Shabir, G. A. (2005) 'Step-by-step analytical methods validation and protocol in the quality system compliance industry', *Journal of Validation Technology*, 10, pp. 314–325.

Shim, J.-S., et al. (2014) 'Dietary assessment methods in epidemiologic studies', *Epidemiology and Health*, 36, p. e2014009.

Shonkoff, E., et al. (2023) 'AI-based digital image dietary assessment methods compared to humans and ground truth: a systematic review', *Annals of Medicine*, 55(2), p. 2273497.

Subar, A. F., et al. (2012) 'The Automated Self-Administered 24-hour dietary recall (ASA24): a resource for researchers, clinicians, and educators from the national cancer institute', *Journal of the Academy of Nutrition and Dietetics*, 112(8), pp. 1134–1137. https://doi.org/10.1016/j.jand.2012.04.016.

Tressou, J. (2006) 'Nonparametric modeling of the left censorship of analytical data in food risk assessment', *Journal of the American Statistical Association*, 101(476), pp. 1377–1386. https://doi.org/10.1198/016214506000000573.

Udovicki, B., et al. (2022) 'Microplastics in food: scoping review on health effects, occurrence, and human exposure', *International Journal of Food Contamination*, 9(1), p. 7. https://doi.org/10.1186/s40550-022-00093-6.

Udovicki, B., et al. (2023) 'Second order probabilistic assessment of chronic dietary exposure to aflatoxin M1 in Serbia', *Food and Chemical Toxicology*, 178, p. 113906. https://doi.org/10.1016/j.fct.2023.113906.

Udovicki, B. and Djekic, I. (2024) 'Quick roadmap for exposure assessment of contaminants in food', *Standards*, 4(1), pp. 25–38. https://doi.org/10.3390/standards4010002.

© Burleigh Dodds Science Publishing Limited, 2025. All rights reserved.

van Klaveren, J. D., et al. (2024) 'Update of the MCRA platform: enhancing data connectivity, security, interoperability, and accessibility', *EFSA Supporting Publications*, 21(7), p. 8941E. https://doi.org/10.2903/sp.efsa.2024.EN-8941.

Vojnović, B., et al. (2024) 'Approaches for sampling and sample preparation for microplastic analysis in laundry effluents', *Sustainability*. https://doi.org/10.3390/su16083401.

Wang, F., et al. (2024) 'Emerging contaminants: a one health perspective', *The Innovation*, 5(4), p. 100612. https://doi.org/10.1016/j.xinn.2024.100612.

Wild, C. P. (2005) 'Complementing the genome with an "exposome": the outstanding challenge of environmental exposure measurement in molecular epidemiology', *Cancer Epidemiology, Biomarkers & Prevention*, 14(8), pp. 1847–1850. https://doi.org/10.1158/1055-9965.EPI-05-0456.

Wild, C. P. (2012) 'The exposome: from concept to utility', *International Journal of Epidemiology*, 41(1), pp. 24–32. https://doi.org/10.1093/ije/dyr236.

Xiao, N., et al. (2011) 'Multiple failure modes analysis and weighted risk priority number evaluation in FMEA', *Engineering Failure Analysis*, 18(4), pp. 1162–1170. https://doi.org/10.1016/j.engfailanal.2011.02.004.

Zhao, Q., et al. (2021) 'Recent improvements in enzyme-linked immunosorbent assays based on nanomaterials', *Talanta*, 223, p. 121722. https://doi.org/10.1016/j.talanta.2020.121722.

Zincke, F., et al. (2022) 'First update of the EU database of processing factors for pesticide residues', *EFSA Supporting Publications*, 19(9), p. 7453E. https://doi.org/10.2903/sp.efsa.2022.EN-7453.

© Burleigh Dodds Science Publishing Limited, 2025. All rights reserved.

# Chapter 8

## Advances in human biomonitoring to assess levels of dietary exposure to potentially harmful chemicals from food or other sources

*Pascal Sanders and Christophe Rousselle, Agency for Food, Environmental and Occupational Health and Safety (ANSES), France*



### 1 Introduction

In recent decades, human biomonitoring (HBM) has emerged as a pivotal tool in environmental health research, providing critical insights into the levels of potentially harmful chemicals in the human body. HBM of biomarkers of exposure and effects provide an efficient means of measuring human exposure to chemical substances. Initially applied in occupational medicine for health protection, the development of sensitive analytical methods in blood and urine provided the capacity to measure much lower concentrations in human population to assess environmental exposures (Angerer et al., 2007).

HBM represents the gold standard in distinguishing between potential and actual exposure and identifying chemicals of emerging concern through information on internal exposures. The data collected by HBM studies about

http://dx.doi.org/10.19103/AS.2025.0158.15
© Burleigh Dodds Science Publishing Limited, 2025. All rights reserved.

biomarkers of exposure are complementary to those generated by the monitoring of chemical contaminants in food and potable water or in indoor and outdoor air. Collecting biomarkers of effect also complements toxicological studies and contributes to the hazard assessment of chemicals. HBM is identified as a tool that can serve the chemicals agenda by providing 'authorities with a more comprehensive view of actual exposure of the population to pollutants, especially sensitive groups such as children, and can provide better evidence from guiding appropriate responses' (EU, 2013).

As globalization and industrialization continue to reshape food production and consumption patterns, the risk of exposure to harmful chemicals through diet is a pressing public health concern. In the context of risk analysis, identification of the contribution of different sources of exposures (diet, air, water, consumer products, and occupational activities) is also crucial for risk managers to guide their actions. HBM is a powerful tool for assessing aggregate exposures and risks but do not recognize the different sources. So, HBM data also contributes to validate environmental and exposure models. They are helpful for retrospectively verifying the efficiency of measures taken to manage exposures or to prioritize new actions (von Stackelberg & Williams, 2021). Analysis of human samples using suspect screening and non-targeted analytical methods are under development for identifying emerging contaminants (Andra et al., 2017).

The chapter will present the recent advances in HBM methodologies. With different case studies performed in Europe, we will illustrate the practical applications of biomonitoring data to study exposure in different populations. We will show this by comparing exposure models to estimate intake (cadmium, arsenic), discuss the revision of hazard assessment (bisphenol A), review the impact on risk communication and summarize the progress done in mixture risk assessment based on HBM data. Then, we will discuss the limitations and uncertainties of the different processes to identify the next steps for the future.

## 2 The Human Biomonitoring for Europe (HBM4EU) Initiative

### 2.1 Introduction

The human biomonitoring for Europe (HBM4EU) initiative represents a landmark effort to assess and manage chemical exposures in European populations (Ganzleben et al., 2017). HBM4EU ran from 2017 to 2022 as a joint effort of 30 countries and 116 partner institutions from Europe and Israel with the aim to support and improve the EU's policies for chemicals, the environment, and health (Kolossa-Gehring et al., 2023). The initiative focuses on understanding

© Burleigh Dodds Science Publishing Limited, 2025. All rights reserved.

the human body's burden of various harmful chemicals through comprehensive biomonitoring studies.

The HBM4EU-aligned studies are built on existing HBM capacity in Europe by aligning national or regional HBM studies. The HBM4EU-aligned studies focus on three age groups: children, adolescents, and adults. The aim of these studies is to collect comparable internal exposure data for priority substances in a sample of the European population that is representative of sex (males, females) and includes individuals from cities, towns/suburbs, and rural areas, with different socio-economic status (SES) and is recruited from the four geographical regions of Europe (north, east, south, and west) as defined by the UN geo-scheme (Gilles et al., 2022). The population sample was selected by utilizing a sampling scheme designed to guide the inclusion of eligible studies.

The participants were recruited between 2014 and 2021 in 11–12 primary sampling units that were geographically distributed across Europe. Urine samples were collected in all age groups, and blood samples were collected in children and teenagers. Auxiliary information on socio-demographics, lifestyle, health status, environment, and diet was collected using questionnaires.

To align results across different countries, HBM4EU has developed standardized protocols and templates for study documents. This standardization facilitates the production of harmonized data, promoting comparability and robustness of the findings (Pack et al., 2023). The initiative also involves systematic development of questionnaires, incorporating evaluation tools and previous questionnaires to ensure high-quality data collection (González-Alzaga et al., 2022).

One HBM4EU key aspect is the integration of HBM with health examination surveys, as demonstrated in feasibility studies conducted in Finland, Germany, and the UK. These studies benefit from shared infrastructures, increased participation rates, and the provision of additional health information to participants (Elonheimo et al., 2023).

## 2.2 Progress in analytical methods

The progress in analytical methods within HBM4EU is substantial, enabling the detection and quantification of low levels of chemicals in human samples with high accuracy. These studies addressed two groups of priority substances (Ougier et al., 2021a). The first group had been selected in the preparation of HBM4EU according to policy-relevant questions defined by HBM4EU partner countries and European authorities, and the second group was defined in a prioritization process developed in HBM4EU (Table 1).

New chemical analyses of the priority compounds were conducted in HBM4EU, based on the prerequisite of coordinated and harmonized

© Burleigh Dodds Science Publishing Limited, 2025. All rights reserved.

**Table 1** Human biomonitoring for Europe groups of substances prioritized

| | |
|---|---|
| First group | Phthalates and 1,2-cyclohexane dicarboxylic acid diisononyl ester (DINCH), bisphenols, per- and polyfluoroalkyl substances (PFAS), organophosphorous flame retardants (OPFRs), halogenated flame retardants (HFRs), polycyclic aromatic hydrocarbons (PAHs), cadmium |
| Second group | Acrylamide, mycotoxins, pesticides, UV filters, arsenic |

approaches, except for some existing data from DEMOCOPHES (Schwedler et al., 2017) or national and regional HBM programmes.

A quality assurance and control (QA/QC) programme was designed in HBM4EU, open to all European laboratories with an interest in performing HBM analyses of exposure biomarkers of the priority substances (Esteban López et al., 2021). It was organized and coordinated to perform interlaboratory comparison investigations and/or external quality assurance schemes. The QA/QC programme resulted in the qualification of 75 laboratories from 25 countries for HBM analyses of different biomarkers related to the priority substances.

All the HBM4EU aligned studies harmonized their data according to codebooks which defined variable names, units, formats, and coding rules (Govarts et al., 2023), and the harmonized data of each study was uploaded to a common platform where a quality control was performed. Descriptive statistics were calculated centrally for all biomarkers when they were obtained from at least four countries and can be pooled. The following descriptive statistics were calculated for the HBM4EU population: the geometric mean (GM) and 95th percentile (P95), and their 95% confidence intervals. Biomarker concentrations are expressed in volume-based (μg/L) and additionally standardized for creatinine (μg/g creatinine) and blood lipids (μg/g lipid) for urinary biomarkers and lipid-soluble markers analysed in blood, respectively. Summary statistics (P5, P10, P25, P50, P75, P90, P95) are made public via the European HBM dashboard (https://hbm.vito.be/eu-hbm-dashboard) and the HBM module of IPCHEM (https://ipchem.jrc.ec.europa.eu).

HBM4EU has also focused on developing HBM guidance values (HBM-GVs) for various chemicals, which might serve as benchmarks for assessing exposure levels and potential health risks (Apel et al., 2020; Ougier et al., 2021b), helping to answer to policy questions (Gerofke et al., 2023).

Biological samples were collected for three age groups (Table 2) with different list of biomarkers analysed. Description of the sex ratio and origin of the participants are described in Table 3. The study population approaches the characteristics of the general European population based on age-matched EUROSTAT EU-28, 2017 data; however, individuals who obtained no to lower educational level (ISCED 0–2) are underrepresented.

© Burleigh Dodds Science Publishing Limited, 2025. All rights reserved.

**Table 2** Sampling size and biomarker analyses

| Age group | Number of samples collected | Biomarkers analysed |
|---|---|---|
| Children (6–12 years) | 3137 | Phthalates, DINCH, flame retardants |
| Adolescents (12–18 years) | 2950 | Phthalates, DINCH, PFAS |
| Adults (20–39 years) | 3522 | Cadmium, bisphenols, PAH metabolites |

**Table 3** Description of the groups

| Age group | Sex ratio (M/F) | Living in cities (%) | Living in towns/ suburbs (%) | Living in rural areas (%) |
|---|---|---|---|---|
| Children (6–12 years) | 50.4%/49.5% | 44.1 | 29.0 | 26.8 |
| Teenagers (12–18 years) | 49.4%/50.6% | 37.7 | 31.2 | 30.2 |
| Adults (20–39 years) | 47.4%/52.6% | 71.9 | 14.2 | 13.4 |

DINCH, 1,2-cyclohexane dicarboxylic acid diisononyl ester; PAH, polycyclic aromatic hydrocarbons; PFAS, per- and polyfluoroalkyl substances.

### 2.3 Use of human biomonitoring data in risk assessment

A review of the uses of HBM4EU data in risk assessment or estimations of the environmental burden of disease has been recently published (Santonen et al., 2023).

The HBM-GV for the general population (HBM-$GV_{GenPop}$) indicates the concentration of a compound or its metabolite(s) in a biological matrix at and below where a health risk is not anticipated, according to current knowledge (Apel et al., 2020). A lifelong exposure is usually considered for the derivation. For bioaccumulating substances, $HBM_{GenPop}$ can be given according to age ranges. They are equivalent to the HBM-I values from the German Human Biomonitoring Commission (Agency, 2007, 2014). When external exposure guidance values (toxicity reference values or TRVs) are used to estimate the concentration in human biological matrices, they correspond to the Biomonitoring Equivalents (BEs) (Hays & Aylward, 2009). For vulnerable population groups or for certain periods of life, specific HBM-GVs can be defined (e.g. women and child-bearing age, children, elderly) and can be published with specific recommendations for action. For reprotoxic substances, the critical time window (children, adolescent and young adult) is considered as exposure period.

Similarly, HBM-GV for workers (HBM-$GV_{Worker}$) can be derived to support occupational risk assessment and occupational health surveys.

For the derivation of HBM-$GV_{GenPop}$ or of HBM-$GV_{Worker}$, three options for the HBM-GV's derivation are available. The selection of one of these options is conditioned by the availability of data, their quality, and their relevance in

© Burleigh Dodds Science Publishing Limited, 2025. All rights reserved.

deriving HBM-GVs. The three options and the process of derivation are detailed in Apel (2020) (see Fig. 1).

The first option is to derive the HBM-GV from epidemiological studies or clinical studies describing the relationship between internal exposure (internal concentrations of biomarkers) and health effects observed in humans. The methodology will involve checking the relevance and quality of the studies. Depending on the most sensitive effect observed in human studies, a key study is chosen to determine a 'point of departure' (PoD). This approach is considered the most robust as it uses direct human data linking internal biomarker concentrations to adverse health effects (IPCS, 2001). However, uncertainties can arise from variability in individual responses, limited sample sizes, and the precision of exposure assessments. Confidence levels are generally higher when well-conducted epidemiological studies are available but can be lower if the data is scarce or inconsistent.

The second option is based on defined external TRVs (such as tolerable daily intake (TDI) or occupational exposure limits (OEL)) to convert them to internal values using toxicokinetic data and models. The workflow of derivation was described and is dependent of the data availabilities (Hays & Aylward, 2012). The process is based on four steps:

- the identification of the PoD (TDI);
- applying any uncertainty factors (UF) used in the derivation of TDI;
- estimating the biomarker level in the human matrix (urine, blood); and
- applying human intraspecies UF and any database deficiency factors if necessary.

The report will include the date when the value was established, the critical effect that informed the value's determination, the PoD, and the UFs that guided the selection of the TRVs for deriving the HBM-GV. Each decision to choose one value over others must be justified on a case-by-case basis. The uncertainties involved stem from assumptions in toxicokinetic modelling, the precision of external exposure data, and the process of extrapolating external doses to internal doses. Additionally, individual differences in metabolism and excretion rates contribute to the overall uncertainty.

The third option is applied when human data and TRVs are not available. The guidance values are based on extrapolating a critical dose (PoD) identified in a key animal study into a human internal concentration of a selected biomarker. Defaults UF are applied to account for uncertainties in the extrapolation process. Different modalities can be applied for this derivation and are discussed in previous papers (Apel et al., 2020; Hays et al., 2008). This approach is described in different technical recommendations (Agency, 2014; Hays et al., 2007). Selecting a key study requires the assessment of the study

© Burleigh Dodds Science Publishing Limited, 2025. All rights reserved.

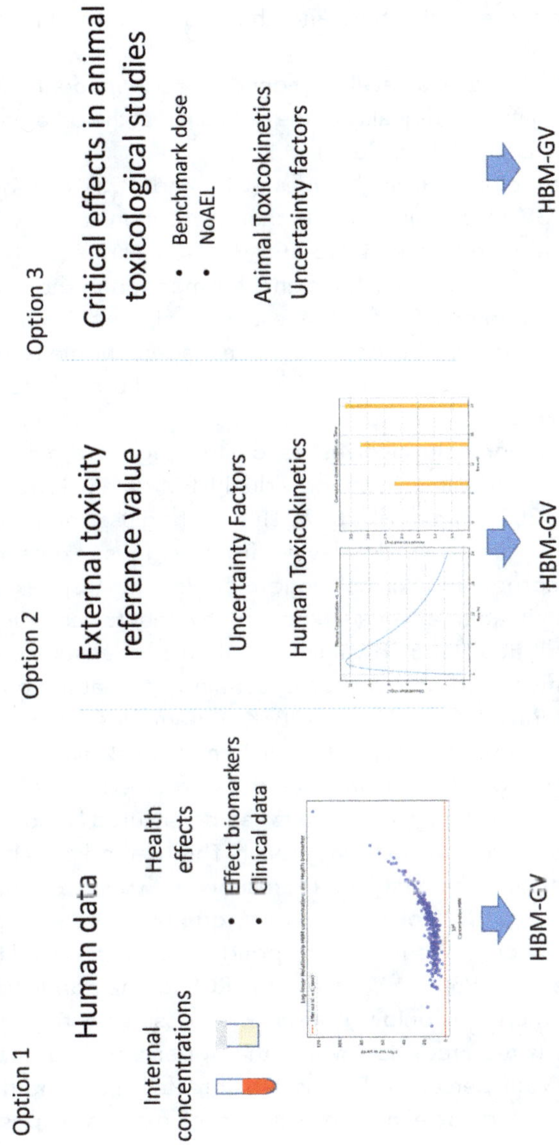


**Figure 1** Options for the derivation of human biomonitoring guidance value.

© Burleigh Dodds Science Publishing Limited, 2025. All rights reserved.

quality and reliability. For the choice of PoD, the preferred option is to use a benchmark dose approach but use of NOAEL/LOAEL can be also applied. The uncertainties are significant due to species differences in metabolism, exposure routes, and sensitivity to chemicals. Additional UFs are applied to account for these differences and the variability within animal studies themselves.

The HGM-BVs have a level of confidence dependent on the option chosen and the nature and quality of the data used at the different steps of the derivation process (Apel et al., 2020).

Aggregated biomonitoring values (GM and P95) can be compared with the HBM-GV to calculate a risk characterization ratio (RCR). The RCR calculated as the ratio between P95 and HBM-GV corresponds to the worst case. When RCR is higher than 1, a signal for monitoring the health status of the population is warned if 5% pf HBM results are above the $\text{HBM-GV}_{\text{GenPop}}$. The identification of sources of exposure and management measures to reduce the exposure of the population must be studied and brought to the attention of decision-makers.

A second approach involves reconstructing external exposure by performing reverse dosimetry on individual HBM data (Clewell et al., 2008; Quindroit et al., 2021; Tan et al., 2007). This method uses numerical inversion techniques to estimate intake levels from various exposure pathways and routes, incorporating toxicokinetic models. The RCR is calculated by dividing the estimated external exposure by the toxicological reference value. The distribution of RCR values can then be derived from individual HBM measurements. To account for uncertainty, probabilistic modelling of Physiologically Based Toxicokinetic (PBTK) parameters is employed during the reverse dosimetry process (Fig. 2), enabling the calculation of uncertainty intervals around both the reconstructed external exposure and the RCR. Consistency in methodology and results can be ensured by comparing these findings with those from the initial approach. This method can also be applied in a tiered manner. The first tier represents a 'worst-case but plausible' scenario, assuming 100% of the exposure originates from a specific route (ingestion, inhalation, or dermal) corresponding to the selected toxicological reference value. If the 95th percentile of the RCR for the population, including its uncertainty interval, is below 1, the risk is considered negligible, and no further analysis is required. However, if the upper bound of the uncertainty interval of the 95th percentile RCR is 1 or greater, a second-tier approach can be implemented to refine the estimate of external exposure and the associated RCR.

© Burleigh Dodds Science Publishing Limited, 2025. All rights reserved.

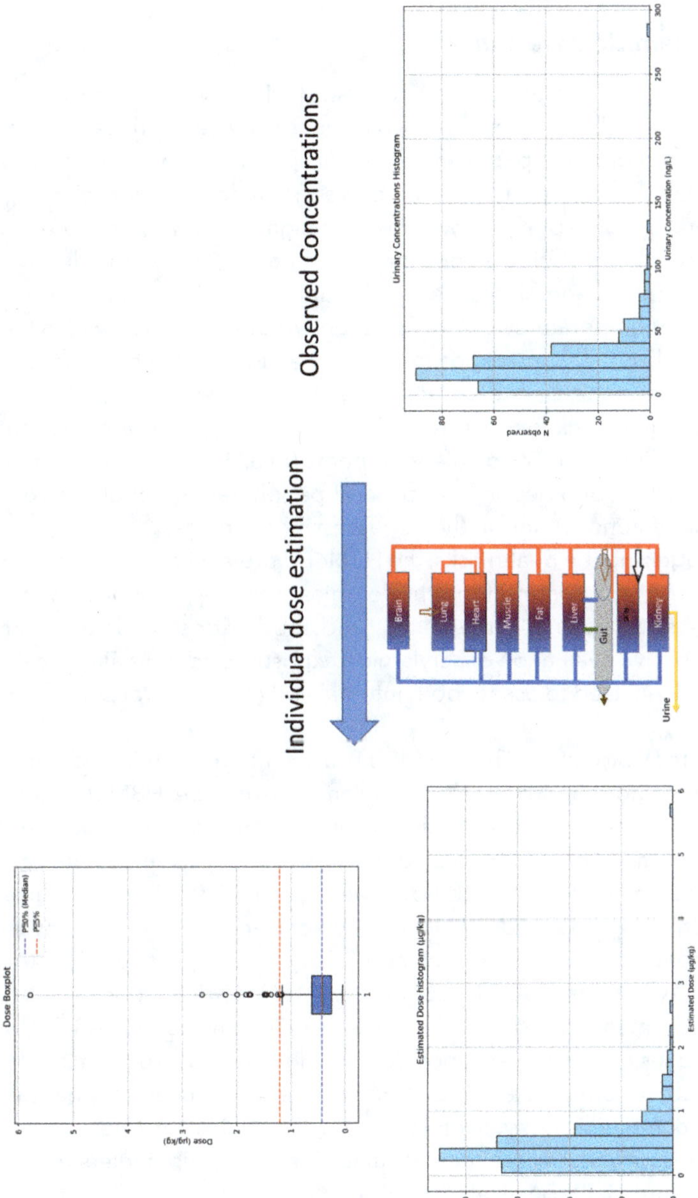


**Figure 2** Schema of a reverse dosimetry approach.

© Burleigh Dodds Science Publishing Limited, 2025. All rights reserved.

# 3 Acrylamide and cadmium case studies: methods to detect intake from dietary exposure and assess health impacts

## *3.1 Acrylamide case study*

Acrylamide is a significant industrial chemical primarily utilized in the production of polyacrylamides, which have applications in water purification, as well as in the petroleum and paper industries. In research laboratories, acrylamide is commonly used for preparing polyacrylamide gels for electrophoresis (Bergmark, 1997). Notably, it was the investigation of acrylamide exposure in laboratory settings during an occupational study that led to the discovery of its presence in the general population through food. This compound is formed during high-temperature cooking of carbohydrate-rich foods, which has raised concerns about dietary exposure to acrylamide (Albiach-Delgado, Esteve-Turrillas, Fernández, Garlito, & Pardo, 2022).

Acrylamide has known neurotoxic, carcinogenic, genotoxic, and mutagenic effects. The European Food Safety Authority (EFSA) provided benchmark dose levels ($BMDL_{10}$) for assessing cancer and peripheral neuropathy risks, serving as critical reference points in the risk assessment process.

Recognizing the health risks, the HBM4EU project has employed HBM data to enhance risk assessment methodologies. This approach leverages urinary biomarkers, toxicokinetic modelling, and food exposure data to provide a comprehensive evaluation of acrylamide exposure and its health impacts. HBM data has been used to assess both internal and external exposure:

- **Internal exposure**: The HBM4EU project integrates multiple data sources and methodologies to assess acrylamide exposure. HBM data are pivotal, with urinary metabolites such as acrylamide mercapturic acid (AAMA) and glycinamide mercapturic acid (GAMA) serving as indicators of internal exposure. LC-MS/MS methods were applied for the quantification of the mercapturic acids in urine (Sabbioni et al., 2022). The lowest limits of detection were 0.1 ng/mL and 0.3 ng/mL, respectively, for AAMA and GAMA. External QA/QC, both AAMA and GAMA in urine are included in the German External Quality Assurance Scheme (G-EQUAS, https://www.g-equas.de) and were included in the QA/QC program for HBM4EU (Esteban López et al., 2021). Reference substances for the calibration are commercially available for both the native mercapturic acids and their isotope-labelled internal standards. These biomarkers reflect recent exposure (Sabbioni et al., 2022), enabling the estimation of internal doses. AAMA and GAMA as urinary biomarkers for acrylamide were quantified in almost all 1198 children and 1180 adults gathered from five countries, with AAMA being the most abundant in both age groups.

© Burleigh Dodds Science Publishing Limited, 2025. All rights reserved.

- **External exposure**: External exposure estimates are calculated from dietary intake data, focusing on acrylamide levels in commonly consumed foods, such as fried and baked goods, and coffee. Tobacco smoking, as well as passive smoking, is also an important source of human exposure. Occupational exposure is also important for workers at the industrial site and manufacturing (Panel, 2015).

A simple toxicokinetic model was employed to convert external exposure into internal concentrations and vice versa. The urinary mass-balance method was used to estimate acrylamide intake by calculating the excreted amount of urinary AAMA. This estimation was made using default body weights (30 kg for children and 70 kg for adults) and 24-hour urine excretion volumes (0.66 L for children and 1.7 L for adults) specific to different population groups. After few days, at steady state, approximately 50% of the acrylamide intake is eliminated daily as urinary AAMA on a molar basis (Mahiout & Santonen, 2022).

The risk characterization process leverages biomonitoring data to estimate health risks associated with acrylamide exposure. For cancer risk assessment, a linear extrapolation from animal studies is applied, which, although conservative, provides a crucial estimate of potential risks. Based on the lowest $BMDL_{10}$ of 0.17 mg/kg as reference point, the margin of exposure (MOE) values was calculated. As the calculated MOE are below 10 000, the CONTAM Panel indicated a concern with respect of neoplastic effect for both children and adults, particularly at higher exposure levels (Chain, 2015).

The risk characterization was performed for the national aligned studies reported in HBM4EU (four for children and five for adults (Mahiout & Santonen, 2022)). Table 4 reports the range of value obtained in the aligned studies for the GM and P95 values of AAMA urinary concentration, the mean acrylamide daily intake and the cancer risk and risk characterization for peripheral neuropathy.

**Table 4** Range of values for geometric mean and P95 for acrylamide mercapturic acid urinary concentrations, the mean acrylamide intake and cancer risk and risk characterization ratio

| Group | GM Urin. Conc. (μg/L) | P95 Urin. Conc. (μg/L) | Mean acrylamide intake (μg/kg bw/day) | Cancer risk (1:X) | Peripheral neuropathy RCR based on GM | Peripheral neuropathy RCR based on P95 |
|---|---|---|---|---|---|---|
| Children adolescents (6–18 years) | 55–70 | 180–209 | 0.75–0.92 | 1:570–1:464 | 0.17–0.21 | 0.56–0.65 |
| Adults | 21–100 | 67-511 | 0.31–1.47 | 1:1384–1:288 | 0.07–0.34 | 0.23–1.75 |

GM, geometric mean; RCR, risk characterization ratio.

© Burleigh Dodds Science Publishing Limited, 2025. All rights reserved.

Peripheral neuropathy risk assessment uses the same urinary AAMA data, with results showing that most exposure levels are below the defined health-based limit.

The estimated mean acrylamide intake ranges derived from HBM data can be compared with the food exposure to acrylamide ranges calculated by EFSA (Panel, 2015). For infants and toddlers, the mean food exposure ranged between 0.8 µg/kg and 1.4 µg/kg bw/day, while estimated mean daily intakes are between 0.75 µg/kg and 0.92 µg/kg bw/day. For adults, the food exposure ranged between 0.4 µg/kg and 0.6 µg/kg bw/day, while estimated mean daily intakes are between 0.3 µg/kg and 1.5 µg/kg bw/day.

The two approaches lead to comparable levels of exposure on average. However, certain high-exposure population exhibit increased risks, underscoring the importance of targeted risk management strategies. For example, children and adolescent smokers had about 2.5-fold higher AAMA levels than non-smoking participants as reported in Germany (Schwedler et al., 2021).

Several uncertainties impact the risk assessment process of the cancer risk of acrylamide as described above. Linear extrapolation from animal data may overestimate human cancer risk, especially at low exposure levels. Individual differences in acrylamide metabolism and endogenous acrylamide formation introduce variability into biomarker data, affecting the accuracy of exposure estimates. Furthermore, dietary habits and smoking status significantly influence exposure levels, highlighting the need for comprehensive data collection and analysis.

### 3.2 Cadmium case study

Cadmium (Cd) exposure can come from inhalation of tobacco smoke and occupational exposure. In non-occupationally exposed populations, the primary source of cadmium is food, with leafy vegetables, cereals, and organ meats being significant contributors. Tobacco smoking significantly increases cadmium exposure as tobacco plants naturally accumulate cadmium from the soil.

Determination of Cd concentrations in urine samples were determined by ICP-MS. Limit of detection (LOD) ranged from 0.002 µg/L to 0.01 µg/L and limit of quantification (LOQ) between 0.0016 µg/L and 0.1 µg/L. Creatinine concentrations in urine were available for all data collections (n = 2500) to derive a concentration of cadmium by g of creatinine. Determination of Cd in urine samples was performed in laboratories that successfully passed the HBM4EU QA/QC program, except for the ESTEBAN study (France), for which measurements were performed before the establishment of the HBM4EU QA/QC program (Snoj Tratnik et al., 2022)

© Burleigh Dodds Science Publishing Limited, 2025. All rights reserved.

The derivation of HBM-GVs for cadmium involves several key epidemiological studies that link cadmium exposure to adverse health effects, particularly nephrotoxicity (Lamkarkach et al., 2021). The European Food Safety Authority (EFSA) conducted a comprehensive meta-analysis of 35 studies (Authority, 2009). This meta-analysis identified a Benchmark Dose Lower Confidence Limit of 4 µg/g creatinine for urinary cadmium (U-Cd), associated with elevated levels of beta-2-microglobulin (β2M) in urine, an indicator of kidney damage.

The French Agency for Food, Environmental and Occupational Health & Safety (ANSES) developed a PBPK model to predict age-specific cadmium accumulation in the body and corresponding urinary levels. The PBPK model used by Anses (ANSES, 2017) to estimate threshold biological values of U-Cd in general population was used to predict limits or so-called 'alert' values of U-Cd according to age. Based on the analysis of different risk assessment, a HBM-$GV_{GenPop}$ for cadmium of 1 µg/g creatinine at age 55–60 years (concentrations are predicted to reach a peak at age 55 years) was derived (ANSES, 2017). HBM data has been used to assess both internal and external exposure:

- **Internal exposure**: A detailed review of the HBM4EU aligned studies was published (Snoj Tratnik et al., 2022). The aggregated results for adult general population were compared with HBM-$GV_{GenPop}$ (adults over 50 years) of 1 µg/g creatinine. The studies from 9 European countries (Iceland, Denmark, Poland, Czech Republic, Croatia, Portugal, Germany, France, and Luxembourg) were included and urine of 20–39 years old adults sampled in the years 2014–2021 (n = 2510). Study participants' questionnaire data were post-harmonized.
- **External exposure**: Spatially-resolved external data were Cd concentrations in soil, agricultural areas, phosphate fertilizer application, traffic density and point source Cd release. They were collected for the respective statistical territorial unit (NUTS).

The data revealed some differences between specific study sites for Cd levels in urine but without distinct geographic patterns. Levels of exposure were similar with a previous study (Den Hond et al., 2015). The lowest mean concentrations were observed in a Portuguese study (0.09 µg/g creatinine) and the highest in Polish (0.36 µg/g creatinine) and French (0.39 µg/g creatinine) studies. However, the concentrations adjusted for basic influencing factors (sample type, sex, age, smoking, education, and sampling year) showed more unified concentrations across Europe, with smaller differences in GMs between the respective study areas. The highest proportion of study participants exceeding the HBM-$GV_{GenPop}$ value of 1 µg/g of creatinine was observed in the Polish and French studies (10.1% and 8.7%, respectively).

© Burleigh Dodds Science Publishing Limited, 2025. All rights reserved.

The age-dependent values (Lamkarkach et al., 2021) for relevant age groups of our study participants were exceeded by 16.4% participants of the total (pooled) population. The highest proportion of exceedances was again observed in Poland (33%) and France (43%), but also in Germany (36%). In one area in France, the percentage exceeded 50%. In other countries, the exceedance spanned between 1.4% (Denmark) and 8.7% (Czech Republic). The multiple regression models confirmed higher levels in women (33 %) than in men and increasing Cd urinary levels with age (by 3% each year of life) in the majority of the studies. Higher levels of urinary Cd in smokers (25 %) than in non-smokers were observed. The levels were 14% lower in the participants with medium or high educational level in comparison to the ones with low level of education. With regard to the sampling year, a slight decreasing trend, 4% each year, was observed. Despite the fact that only 3% of the subjects declared themselves as vegetarians (among those that provided data on vegetarian food), vegetarians had 35% higher levels of Cd in urine than non-vegetarians with strong statistical significance. Other types of environmental exposure were investigated (traffic density, degree of urbanization) were not found to be significantly associated with Cd levels in urine. When variables collected at the NUTS2 level (namely Cd concentration in topsoil, percent cropland, application of phosphate fertilizer, Cd release to air and water, road density, and Cd point source release) were added independently into the basic model. All, except Cd concentration in soil, were associated positively and significantly with the urinary Cd levels. The study showed an important contribution of the vegetarian diet to the overall exposure with 35% higher levels in vegetarians as compared to non-vegetarians. Smoking increases the exposure of 25% in comparison to non-smokers. The study supports the recommendation by EFSA to reduce Cd exposure by diet which is close or slightly exceeding the tolerable weekly intake.

## 4 Bisphenols and pyrethroids case studies: the development of risk assessments to characterize hazards

### 4.1 Bisphenols case study

Bisphenols, including Bisphenol A (BPA) and its analogues such as Bisphenol S (BPS), bisphenol F (BPF), are synthetic chemicals widely used in the production of plastics and epoxy resins (Meslin et al., 2022). The toxicologically relevant form for BPA is free unconjugated bisphenol as it is available for interaction with the oestrogen receptor. However, blood and urine concentrations of free BPA are much lower than of conjugated BPA. Laboratory contamination of omnipresent BPA (Buscher et al., 2015) and/or degradation of conjugated

© Burleigh Dodds Science Publishing Limited, 2025. All rights reserved.

BPA along the different steps from sampling to processing are the main challenges for analysts and requires appropriate quality assurance provisions to prevent them. The analysis of total bisphenol in urine has been shown to be a sensitive and robust approach for a quantitative of exposure to BPA and related compounds (Biomonitoring, 2012; Dekant & Völkel, 2008). A HBM4EU QA/QC programme was applied for measurements in urine of BPA, BPS, and BPF total concentrations (Vaccher et al., 2022) with four rounds of tests. Different analytical methods were applied to quantify total bisphenols in urine. Most of them used enzymatic deconjugation and analysis by GC or LC coupled to MS/MS with isotopic dilution before extraction for calibration (Vaccher et al., 2022).

Bisphenols mimic oestrogen and can interfere with the body's hormonal balance, potentially leading to reproductive, developmental, and metabolic disorders. In 2015, EFSA (E. F. EFSA Panel on Food Contact Materials & Processing, 2015) reassessed BPA exposure and toxicity and proposed a temporary TDI of 4 µg/kg bw/day, based on the increase in relative kidney weight in the F0 generation of a two-generation mouse seen as the most critical effect.

### 4.1.1 External vs internal exposure

EFSA performed an external assessment of BPA and showed that diet is the main source with canned food and non-canned meat and meat products as main contributors. Infants (6–12 months) and toddlers had the highest estimated external dietary exposure if expressed per kg of body weight. Thermal paper was the second largest source of external exposure in all population groups above three years of age. Dust was the second largest source of external exposure for children under the age of three years. Using available HBM data, the average internal exposure was estimated by forward modelling and were in good agreement with the backward modelling (E. EFSA Panel on Food Contact Materials, 2015).

The EFSA report concluded that the risk to consumers was controlled. The same year, the Risk Assessment Committee (RAC) of the European Chemicals Agency (ECHA) also issued an opinion following a request for the restriction of BPA in thermal paper. Based on the same approach as that used by EFSA, the RAC concluded that the risk of exposure to BPA from handling thermal paper was controlled for consumers but insufficiently controlled for workers. As disrupting properties of BPA at low doses was strongly suspected with concern for sensitive population groups, such as pregnant women and young children, BPA was included on the candidate list of substance of very high concern owing to its reproductive toxicity.

In 2023, EFSA reassessed the hazard based on the new scientific evidence to publish a TDI of 0.2 ng BPA/kg bw/day. Based on animal data and

© Burleigh Dodds Science Publishing Limited, 2025. All rights reserved.

human observational studies, the immune system was identified as the most susceptible. The critical effect was identified on Th17 cells in mice which are involved in cellular immune mechanisms. A reference point of 8.2 ng/kg bw/day, expressed as human equivalent dose, was identified for the critical effect (E. Efsa Panel on Food Contact Materials et al., 2023) and an overall UF of 50 was applied to derive the TDI of 0.2 ng BPA/kg bw/day.

### *4.1.2 Internal exposure and risk*

Based on the t-TDI of 4 µg/kg bw/day, HBM-GVs$_{GenPop}$ in urine of 230 µg/L for adults and 135 µg/L for children were derived (Ougier et al., 2021b) and leaded to low level of RCR according the P95 values ranged from 0.01 to 0.12 reported in HBM-aligned studies (Meslin et al., 2022). With the updated t-TDI of 0.2 ng BPA/kg bw/day the 95th percentile dietary exposure estimates of EFSA exceeded the TDI by two or three orders of magnitude (E. Efsa Panel on Food Contact Materials et al., 2023). A simple multiplication by a factor of 20 000 conducts to a range of RCR from 200 to 2400 today.

BPS is used in a variety of industrial applications as a substitute for BPA. It is used as a substitute of BPA in thermal paper (Meslin et al., 2022). BPS is considered a presumed reproductive toxicant for fertility and development. A scientific literature search and a review of its toxicity were performed (Beausoleil et al., 2022) to assess its toxicokinetics and to establish a PoD. The review concludes in discrepancies with LOAELs between 60 mg/kg bw/day and 300 mg/kg bw/day for developmental exposure in regulatory studies versus 0.5–10 µg/kg bw/day for diverse endpoints in academic studies. A LOAEL of 2 µg/kg bw/day for mammary gland and neurobehavioural toxicity endpoints, respectively, in mice and rats, was proposed as a PoD for the derivation of the HBM-GVs (Meslin et al., 2022). An UF of 3 was applied to extrapolate a NOAEL from the LOAEL. The interspecies differences between animals and humans are determined by PBPK modelling using an UF of 2.5 for interspecies differences. A factor of 10 for intra-species differences is applied to the human-adjusted NOAEL for the general population and workers.

A 100% oral exposure scenario was considered for the general population and the workers. Applying assessment factors of 75 (3 × 2.5 × 10), a constant exposure to 0.0266 µg/kg bw/day was used to estimate the maximal concentration of total PBS (free and glucuronidated) in plasma of 2.42 ng/L and the total BPS concentration in urine of 1 µg/L used as HBM-GV$_{GenPop}$. Taking into account the same assessment factors, the worker exposition scenario based on 8-hour exposure per day during 5 working days lead to a HBM-GV$_{Worker}$ of 3 µg/L.

For the general population, RCR was higher than 1 for adults from the Esteban study in France, for the children from the Euro mix study in Norway

© Burleigh Dodds Science Publishing Limited, 2025. All rights reserved.

and for the adolescents from the SLO-CRP study in Slovenia. RCR was below 1 for adults from the Euro mix study in Norway and for pregnant women in the TH pregnant women In Israel (Meslin et al., 2022).

For workers, values of P50 (median), P95, and RCR were calculated with the data provided by a French study on occupational biomonitoring for cashiers (Ndaw et al., 2018). The P50, P95, and RCR are higher in professional cashiers (2.53 µg/L, 19.9 µg/L, 6.6) than in professional control group (0.67 µg/L, 12.6 µg/L, 4.2).

### 4.1.3 Conclusion

The case studies on bisphenol A and S demonstrate how risk assessment practices are evolving, with greater account being taken of all the data from academic publications and regulatory data in identifying and characterizing the hazards. The development of this risk characterization by EFSA, with the implementation of an open and transparent assessment leading to the inclusion of a new critical study, remains a matter of debate (Prueitt & Goodman, 2024; Vom Saal et al., 2024; Zoeller et al., 2023). Communication on population exposure levels and risk characterization via comparison with guidance values and HBM data are very useful for risk managers. But changes in reference points are a challenge for risk communication and must be accompanied throughout the implementation of the process in a transparent and open manner towards stakeholders and society, in such a way that it can be understood by people without extensive knowledge of toxicology and risk assessment.

## 4.2 Pyrethrolds case study

Pyrethroids area major class of insecticides and are used worldwide not only in both agriculture- and horticulture but also for biocidal purposes, including household applications and as veterinary drugs to control ectoparasites of pets. Pyrethroids are synthetic analogues of the natural pyrethrins. Pyrethroids have similarities between them in terms of modes of action and toxicity, but differences in terms of metabolism or potency. Neurotoxicity is the health hazard of concern. Acute neurotoxicity is a common feature and developmental neurotoxicity received particular attention and is supported by epidemiological studies associating exposure biomarkers with neonatal effects, such as prematurity and decreased gestational age (Gimenez-Asensio et al., 2023). According to a recent systematic review of epidemiologic studies, pyrethroids are probably human developmental neurotoxicants. Adverse impacts of pyrethroid exposure on neurodevelopment are likely at exposure levels occurring in the general population (Andersen et al., 2022) .

© Burleigh Dodds Science Publishing Limited, 2025. All rights reserved.

The different uses within different regulatory framework (pesticides, biocides, veterinary, and human pharmaceuticals) generates the needs for assessing the aggregate and combined exposure to pyrethroids. They belong to a family of pesticides that are subject to regular regulatory monitoring to ensure food safety. Consumer exposure through food can be estimated using monitoring and consumption data.

### 4.2.1 Internal exposure

Pyrethroids are studied in HBM programme according to their widespread use and the public health concern. Well-established analytical methods are available. The difficulty in interpreting these data is that the measurement of internal exposure is based on the measurement of the main metabolites 3-phenoxybenzoic acid (3-BPA) and analogues such as the fluorinated analogue 4-fluor-3-phenoxybenzoic acid (4-FPBA, also abbreviated as F-3-PBA or F3PBA), analysis of metabolites (e.g. 3-(2,2-dichlorovinyl)2,2-dimethylcyclopropane -1-carboxylic acid (cis-DCCA + trans-DCCA)), ClF3CA (chlorotrifluorovinylcyclo-propane carboxylic acid (or cis-3-[2-chloro-3,3,3-trifluoroprop-1-enyl]-2,2-di methylcyclopropanecarboxylic acid)), and cis-(2,2-dibromovinyl)-2,2-dimethyl cyclopropanecarboxylic acid (DBCA). Most the laboratories cleave conjugates of 3-BPA, 4-FBPA *cis*-DCCA, *trans*-DCCA, and DBCA by acid hydrolysis. Other groups used enzymatic treatment. Most of the methods are based on liquid–liquid extraction and analysis are performed by GC-MS/MS methods or LC-MS/MS methods with different internal standards (Sabbioni et al., 2022). Concentrations are expressed as µg/L urine. 3PBA, 4F3PBA, *cis*-DCCA, *trans*-DCCA, and DBCA are measured regularly in the German EQUAS test program (https://www.g-equas.de). *cis*-DBCA, *cis*-DCCA, *trans*-DCCA, 3PBA, 4F3PBA, and ClF3CA were also included in the laboratory QA/QC program of HBM4EU (Esteban López et al., 2021).

In a study (Tarazona et al., 2022), a tiered approach was proposed for risk assessments, using HBM data. The first tier was a conservative screening of overall pyrethroid exposure based on phenoxy benzoic acid metabolites. The second step was a refinement of the risk estimates using probabilistic approaches and more specific metabolites.

For the first step, the acceptable daily intake (ADI) is used to calculate a provisional HBM-GV$_{GenPop}$ value for the urine concentration of metabolite according to the following equation:

$$HBM-GV_{GenPop} = \frac{ADI.\left[\dfrac{MW(Metabolite).Fue(Metabolite)}{MW(Parent\ Compound)}\right]}{Daily\,urine\,volume\,adjusted\,to\,the\,bw} \tag{1}$$

© Burleigh Dodds Science Publishing Limited, 2025. All rights reserved.

Where MW is the molecular weight and Fue, the fraction of metabolite excreted in urine. It assumes a steady-state conditions with an equilibrium between the intake of the active substance and the quantity of metabolite excreted in urine, used as a biomarker. Average daily urinary volumes adjusted to bodyweight of 0.03 L/kg bw/day and 0.02 L/kg bw/day for children and adults, respectively, have been assumed.

The data (ADI, MW, Fue) were extracted from the most recent reports from EFSA, ECHA, and scientific publications (Tarazona et al., 2022).

Aggregated HBM data (percentiles and confidence intervals for the full dataset) were used (Govarts et al., 2023). The 95th percentiles ($HBM_{95}$) were selected as conservative estimates of the population exposure levels. The RCR was estimated as the ratio $HBM_{95}$ with the provisional HBM-GV. A RCR higher than 1 represent potential concerns requiring refinement. The second step of the analysis was based on the reconstruction of the full distribution HBM concentrations to perform Monte Carlo analysis.

According to equation (1), a screening level was obtained for 3-BPA based on conservative approach for several pyrethroïds (e.g. cypermethrin, deltamethrin, etofenprox, fenpropathrin, fenvalerate, esfenvalerate, lambda-cyhalothrin, permethrin, and tau-fluvalinate). The use of the lowest ADI (0.0025 mg/kg bw) within the group of pyrethroïds and the lowest urinary fraction (9%) would be the most conservative assumptions giving the lowest possible screening value. The resulting urinary screening values of 3-PBA are 3.25 μg/L for children and 4.8 μg/L for adults.

Based on the screening values for 3-BPA for adults, the highest value of 2.87 μg/L is reported from Israel. Thus, all values were below the screening value and a low concern, with RCR < 1 for adults was identified. A similar situation was observed for the combined screening based on the sum of 3-PBA and 4-FPBA. The situation is different for children, for which, in general, higher monitoring values were reported, ranging from 3.72 μg/L in Slovenia to 7.05 μg/L in Belgium. The 95th percentiles exceeded the 3-PBA screening value for all children databases (RCRs between 1.14 and 2.17). The highest value was calculated for Belgium, followed by Cyprus (RCRs 1.95). Measured levels for 4-FPBA, in contrast, were well below the proposed screening value for this biomarker. When the screening is conducted for the sum of 3-PBA and 4-FPBA, the RCR is slightly higher, ranging from 1.33 for France and 2.19 for Belgium, with 3-PBA as the most relevant component in all cases.

A refined analysis was performed using probabilistic scenarios, considering variability in human toxicokinetics. A potential risk for about 2% of Belgian children is linked to the more hazardous pyrethroid, lambda-cyhalothrin. The risk of exceedance is reduced to 1–0.1% for assumptions considering a mixed exposure to several pyrethroids with different ADIs (Table 5).

© Burleigh Dodds Science Publishing Limited, 2025. All rights reserved.

**Table 5** Pyrethroid, biomarker of exposure, acceptable daily intake and derived human biomonitoring guidance value for children and adults

| Pyrethroid name | Biomarker of exposure | ADI Mg/ kg bw | Children HBM-GV (μg/L) | Adults HBM-GV (μg/L) | Reference |
|---|---|---|---|---|---|
| Deltamethrin | DBCA (cis-3-(2, 2-dibromovinyl) -2,2-dimethylcy clopropanoic acid) | 0.1 | 90 | 130 | Apel et al., 2023 |
| Cyfluthrin/ß-cyfluthrin | 4-FPBA (4-fluoro-3-phenoxybenzoic acid) | 0.2 | 80 | 130 | Apel et al. 2023; Leng, 1997 |
| Cypermethrin | DCCA (cis/trans isomers) | 0.005 | 30 | 45 | Apel et al., 2023 Authority et al., 2018 |
| Lambda-Cyhalothrin | ClF3CA | 0.0025 | 9 | 14 | Authority, 2014 Khemiri et al., 2017 |
| Permethrin | DCCA (trans/cis isomers) | 0.05 | 320 | 480 | Apel et al., 2023 *JMPR* |
| Bifenthrin | ClF3CA | 0.015 | 60 | 90 | Apel et al., 2023 Authority, 2010 |
| Tau-Fluvalinate; | 3-PBA (3-phenoxybenzoic acid); | 0.005 | 6.4 (conservative) 22 (realistic) | 9.6 (conservative) 33 (realistic) | Apel et al. 2023 Authority |

ADI, acceptable daily intake; HBM-GV, human biomonitoring guidance value.

A substance-specific risk assessment was conducted for each pyrethroid, comparing the HBM-$GV_{GenPop}$ with the 95th percentiles for the relevant biomarker as obtained in the various HBM4EU aligned studies. The results, expressed as RCRs show very low concerns for deltamethrin, cyfluthrin, cypermethrin, lambda-cyalothrin, permethrin, bifenthrin, and Tau-flavulinate.

The results of this study demonstrated that no exceedances of the risk levels have been observed in the HBM4EU-aligned studies. The combined risk is below but close to the acceptability threshold, particularly for children. As other studies have reported higher values, both in the EU and abroad, monitoring campaigns should be maintained and extended to other areas in more countries to ensure proper consumer protection.

### 4.2.2 External exposure

A second study was performed (de Alba-Gonzalez et al., 2024) to estimate the expected levels of urinary pyrethroid metabolites for each of the 36 diets

© Burleigh Dodds Science Publishing Limited, 2025. All rights reserved.

covered by the PRIMo model (Brancato et al., 2018). Food monitoring data covering the 2016–2020 period were extracted from the EFSA annual reports for cyfluthrin, cypermethrin, permethrin, deltamethrin, lambda-cyalothrin, and bifenthrin. The distribution measured in European adult populations are within the exposure levels derived from the low and medium scenarios from PRIMo adult diets. The study discusses the relevance of a low conservative value of Fue for 3-BPA which is the most influential parameter in the sensitive analysis of a PBPK model. It demonstrates the interest of comparison of external exposure scenario with HBM data to confirm the capacity of EFSA model PRIMo,for supporting regulatory risk assessments.

## 4.3 General discussion of the case studies

The case studies presented demonstrate the complementary nature of biomonitoring studies and models for estimating external exposure. Toxicokinetic models are essential for characteriszng the processes of absorption, distribution, metabolism, and excretion, so that internal or urine concentrations can be estimated. They are a better predictor of the internal exposure than the external intake. In the simplest case, for a substance that is rapidly eliminated in humans by the renal route in the form of a parent compound or a marker metabolite, the dose can be estimated on the basis of urinary concentrations. Even if the approach is simple, several factors are variable between individuals, groups and HBM studies. The reduction of inter-individual variability must be taken into account in the experimental design of the HBM study by the collect of individual information, alignment of sampling conditions and analytical methods.

In the case of pyrethroids, knowledge of the metabolism of the various substances has been well developed (Quindroit et al., 2021), and modelling of the different substances has made it possible to develop approaches for assessing exposure. Variability of some parameters (bioavailability, clearances, protein binding) and variance effect of different factors (sex, age, weight) determined during controlled toxicokinetic studies should be integrated in the models (Sanghavi et al., 2024) to improve the estimation of the external exposure.

When toxicokinetics in humans shows slow elimination of the substance such as cadmium, a physiologically based kinetic model is needed to take account of physiological changes and variations in exposure over a lifetime. When toxicokinetics in humans are not documented, extrapolation from animals to humans is possible, taking into account inter-species UF.

© Burleigh Dodds Science Publishing Limited, 2025. All rights reserved.

## 5 Contribution of human biomonitoring to risk assessment of chemical mixtures

HBM4EU data has significantly enhanced the documentation of human exposure to chemical mixtures and facilitated discussions on associated health risks (Luijten et al., 2023; Socianu et al., 2022). HBM is a suitable method for depicting real-life exposure to these mixtures, contributing valuable insights to current debates on mixture assessment factors (MAFs). Case studies (Sprong et al., 2020; Tavares et al., 2022) from HBM4EU indicate that exposure to chemical mixtures is widespread across various groups, involving substances with diverse uses and characteristics covered by different EU regulations. However, owing to the program's constraints, the number of individuals undergoing extensive analysis is limited, and the substances analysed vary between groups, preventing comprehensive studies on co-exposure networks and mixture identification (Rodriguez Martin et al., 2023; Socianu et al., 2022). Despite these limitations, the case studies underscore that exposure to chemical mixtures poses a significant public health issue, with identifiable risk factors, particularly legacy compounds (Rodriguez Martin et al., 2023). These studies also demonstrate that PoD indexes are more effective for assessing risks, revealing that exposure to certain substances often surpasses acceptable thresholds.

The discussions within HBM4EU have led to several recommendations for future research and regulatory strategies (Luijten et al., 2023). There is a general consensus on the necessity for further work to better characterize exposure to mixtures and their health impacts. Future biomonitoring programs should aim to document the co-occurrence of exposure markers and effect biomarkers more comprehensively (Rodríguez-Carrillo et al., 2023). Achieving this will require the development of a common European research infrastructure with robust quality assurance to ensure reproducible and comparable measurements across different populations. Additionally, there is a need to advance data analysis methods for exposure and risk characterization and to promote their application in various regulatory frameworks in line with One Substance One Assessment principles. Enhanced exposure monitoring methods should increase the capability to measure substance co-occurrence using targeted methods and to detect emerging substances via suspect screening approaches (Tkalec et al., 2024). Continuous research on methods for real-life mixture characterization, risk assessment, and toxic effect weighting, along with the identification of risk factors and exposure sources, is crucial for informed risk management decisions. These approaches are being furthered through the partnership for assessment of risks from chemicals (PARC, www.eu-parc.eu) and exposome research projects (www.humanexposome.eu) to support the implementation of the Chemical Strategy for Sustainability (Lobo Vicente et al., 2023).

© Burleigh Dodds Science Publishing Limited, 2025. All rights reserved.

## 6 Conclusion and future trends

The integration of biomonitoring into the risk analysis cycle demonstrates the complementary nature of the tool, with those in place as part of the risk assessment and monitoring process for products regulated in Europe. Biomonitoring in humans is holistic and agnostic, integrating the different sources of exposure sources of exposure in everyday life and at work. It covers all ages and critical periods from conception to adulthood. Combined with the monitoring of biomarkers of effects, it provides a link with epidemiological studies in public health and toxicology. Associated with data about food consumption, occupational activities or environmental health data, will contribute to generate complementary datasets. The data from monitoring in humans is very useful for communicating known risks, emerging risks, and hazards (Rodríguez-Carrillo et al., 2023).

The work of HBM4EU has been taken over by the PARC, which was set up as part of Horizon Europe. PARC is continuing its work to align the various studies on a common set of protocols for both sampling and data collection, by carrying out harmonized analyses under quality assurance. The new sampling campaign began in 2024. PARC implements environmental monitoring and develops methodological approaches for integrating monitoring data into exposure and risk assessment models. PARC is also continuing work on hazard characterization. Through an integrated approach between the various activities, PARC aims to develop the concepts and provide the tools needed to support the Chemical Strategy for Sustainability (Marx-Stoelting et al., 2023).

## 7 Where to look for further information

Human biomonitoring (HBM) is a rapidly evolving field with a wealth of resources available for those interested in exploring it further. Below are some key resources, including seminal articles, professional organizations, websites, journals, conferences, research projects, and top research centers.

### 7.1 Seminal Articles and Books

- Ganzleben, C., Antignac, J. P., Barouki, R., Castaño, A., Fiddicke, U., Klánová, J., and Kolossa-Gehring, M. (2017). Human biomonitoring as a tool to support chemicals regulation in the European Union. *International Journal of Hygiene and Environmental Health, 220*(2), 94–97. doi:10.1016/j.ijheh.2017.01.007
- Hays, S. M., and Aylward, L. L. (2009). Using Biomonitoring Equivalents to interpret human biomonitoring data in a public health risk context. *Journal of Applied Toxicology, 29*(4), 275–288. doi:10.1002/jat.1410

© Burleigh Dodds Science Publishing Limited, 2025. All rights reserved.

- Human Biomonitoring in Health Risk Assessment: Current Practices and Recommendations for the Future, Edited by Christophe Rousselle, https://www.mdpi.com/bookfiles/book/7111/Human_Biomonitoring_in_Health_Risk_Assessment.pdf, last accessed 3/9/2025

## *7.2 Key Societies and International Organizations*

- **International Society of Exposure Science (ISES)**: A professional society dedicated to advancing exposure science. ISES Europe | Together, we build a European Exposure Science Strategy
- **World Health Organization (WHO)**: Offers resources and guidelines on environmental health and toxicology. New WHO educational course on human biomonitoring helps countries assess human exposure to chemicals

## *7.3 Websites*

- **PARC-HBM4EU Dashboard**: Provides harmonized HBM data from across Europe (https://hbm.vito.be/eu-hbm-dashboard)
- **IPCHEM**: Provides information on chemical monitoring and assessment (https://ipchem.jrc.ec.europa.eu)
- **European Environment Agency**: Provides data and reports on environmental health (https://www.eea.europa.eu)
- **US CDC National Biomonitoring program:** National Biomonitoring Program | NBP | CDC
- **Canada National Human biomonitoring:** Human Biomonitoring of Environmental Chemicals - Canada.ca

## *7.4 Key Journals and Conferences*

### 7.4.1 Journals

- *Science of the Total Environment*
- *International Journal of Hygiene and Environmental Health*
- *Environmental Health Perspectives*
- *Environment International*
- *Environmental Resarch*

### 7.4.2 Conferences

- International Society of Exposure Science Annual Meeting
- International Conference on Environmental Epidemiology and Exposure

© Burleigh Dodds Science Publishing Limited, 2025. All rights reserved.

### 7.5 Major International Research Projects

- **HBM4EU**: A former European initiative to coordinate and advance human biomonitoring in Europe (https://www.hbm4eu.eu)
- **Partnership for the Assessment of Risks from Chemicals (PARC)**: Aims to develop next-generation chemical risk assessment (https://www.eu-parc.eu) the on-going initiative in charge o implementation of aligned HBM studies in general population and in occupational settings.

## 8 References


Agency, H. C. o. t. G. E. (2007). Ableitung von Human-Biomonitoring-(HBM-)Werten auf der Basis tolerabler Aufnahmemengen–Teil II: Grund lagen und Ableitungsweg. *Bundesgesundheitsblatt - Gesundheitsforschung - Gesundheitsschutz, 50*(2), 251–254. https://doi.org/10.1007/s00103-007-0146-5Aalpha

Agency, H. C. o. t. G. E. (2014). Position paper on the derivation of HBM values: opinion of the German Human Biomonitoring Commission. *Bundesgesundheitsblatt - Gesundheitsforschung - Gesundheitsschutz, 57*(1), 138–147. https://doi.org/10.1007/s00103-013-1867-

Albiach-Delgado, A., et al. (2022). Review of the state of the art of acrylamide human biomonitoring. *Chemosphere, 295*. https://doi.org/10.1016/j.chemosphere.2022.133880

Andersen, H. R., et al. (2022). Pyrethroids and developmental neurotoxicity: a critical review of epidemiological studies and supporting mechanistic evidence. *Environmental Research, 214*. 1-20 https://doi.org/10.1016/j.envres.2022.113935

Andra, S. S., et al. (2017). Trends in the application of high-resolution mass spectrometry for human biomonitoring: an analytical primer to studying the environmental chemical space of the human exposome. *Environment International, 100*, 32–61. https://doi.org/10.1016/j.envint.2016.11.026

Angerer, J., et al. (2007). Human biomonitoring: state of the art. *International Journal of Hygiene and Environmental Health, 210*(3–4), 201–228. https://doi.org/10.1016/j.ijheh.2007.01.024

ANSES. (2017). *Exposition au cadmium.Propositions de valeurs toxicologiques de référence par ingestion, de valeurs sanitaires repères dans les milieux biologiques (sang, urines, …)*. Retrieved from Maisons-Alfort: https://www.anses.fr/fr/system/files/VSR2015SA0140Ra-1.pdf

Apel, P., et al. (2020). Human biomonitoring initiative (HBM4EU): strategy to derive human biomonitoring guidance values (HBM-GVs) for health risk assessment. *International Journal of Hygiene and Environmental Health, 230*. 1-11 https://doi.org/10.1016/j.ijheh.2020.113622

Apel, P., et al. (2023). "Human biomonitoring guidance values (HBM-GVs) for priority substances under the HBM4EU initiative – New values derivation for deltamethrin and cyfluthrin and overall results." International Journal of Hygiene and Environmental Health 248.


© Burleigh Dodds Science Publishing Limited, 2025. All rights reserved.

Authority, E. F. S. (2009). Cadmium in food - Scientific opinion of the panel on contaminants in the food chain. *EFSA Journal*, *7*(3), 980. https://doi.org/https://doi.org/10.2903/j.efsa.2009.980

Beausoleil, C., et al. (2022). Regulatory and academic studies to derive reference values for human health: The case of bisphenol S. *Environmental Research*, *204*. 1-16 https://doi.org/10.1016/j.envres.2021.112233

Bergmark, E. (1997). Hemoglobin adducts of acrylamide and acrylonitrile in laboratory workers, smokers and nonsmokers. *Chemical Research in Toxicology*, *10*(1), 78–84. https://doi.org/10.1021/tx960113p

Biomonitoring, C. F. H. (2012). *Substance monograph on bisphenol A (BPA) - Reference and Human Biomonitoring (HBM) values for BPA in urine. Opinion of the human biomonitoring commission of the German federal environment agency (UBA).* Bundesgesundheitsbl 2012 · 55:1215 –1231. DOI 10.1007/s00103-012-1525-0 © Springer-Verlag 2012.

Brancato, A., et al. (2018). Use of EFSA pesticide residue intake model (EFSA PRIMo revision 3). *EFSA Journal*, *16*(1). 1-43 https://doi.org/10.2903/j.efsa.2018.5147

Buscher, B., et al. (2015). Quantitative analysis of unconjugated and total bisphenol A in human urine using solid-phase extraction and UPLC-MS/MS: method implementation, method qualification and troubleshooting. *Journal of Chromatography B: Analytical Technologies in the Biomedical and Life Sciences*, *1005*, 30–38. https://doi.org/10.1016/j.jchromb.2015.09.020

Chain, E. P. o. C. i. t. F. (2015). Scientific Opinion on acrylamide in food. *EFSA Journal*, *13*(6), 4104. https://doi.org/10.2903/j.efsa.2015.4104

Clewell, H. J., et al. (2008). Quantitative interpretation of human biomonitoring data. *Toxicology and Applied Pharmacology*, *231*(1), 122–133. https://doi.org/10.1016/j.taap.2008.04.021

de Alba-Gonzalez, M., et al. (2024). Applicability of food monitoring data for assessing relative exposure contributions of pyrethroids in retrospective human biomonitoring risk estimations. *Toxics*, *12*(1). 1-13 https://doi.org/10.3390/toxics12010024

Dekant, W. & Völkel, W. (2008). Human exposure to bisphenol A by biomonitoring: Methods, results and assessment of environmental exposures. *Toxicology and Applied Pharmacology*, *228*(1), 114–134. https://doi.org/10.1016/j.taap.2007.12.008

Den Hond, E., et al. (2015). First steps toward harmonized human biomonitoring in Europe: Demonstration project to perform human biomonitoring on a European scale. *Environmental Health Perspectives*, *123*(3), 255–263. https://doi.org/10.1289/ehp.1408616

EFSA Panel on Food Contact Materials, E. (2015). Scientific Opinion on the risks to public health related to the presence of bisphenol A (BPA) in foodstuffs. *EFSA Journal*, *13*(1), 3978. https://doi.org/10.2903/j.efsa.2015.3978

EFSA Panel on Food Contact Materials, E., et al. (2023). Re-evaluation of the risks to public health related to the presence of bisphenol A (BPA) in foodstuffs. *EFSA Journal*, *21*(4), e06857. https://doi.org/https://doi.org/10.2903/j.efsa.2023.6857

Elonheimo, H. M., et al. (2023). HBM4EU feasibility studies: lessons learned in combining health and human biomonitoring studies. *International Journal of Hygiene and Environmental Health*, *248*, 114100. https://doi.org/https://doi.org/10.1016/j.ijheh.2022.114100

© Burleigh Dodds Science Publishing Limited, 2025. All rights reserved.

Esteban López, M., et al. (2021). The European human biomonitoring platform: design and implementation of a laboratory quality assurance/quality control (QA/QC) programme for selected priority chemicals. *International Journal of Hygiene and Environmental Health, 234*. https://doi.org/10.1016/j.ijheh.2021.113740

EU. (2013). *Decision No 1386/2013/EU of the European Parliament and of the Council of 20 November 2013 on a General Union Environment Action Programme to 2020 'Living well, within the limits of our planet'. Official Journal of the European Union, L354/171, Brussels.*

European Food Safety, A. (2010). "Conclusion on the peer review of the pesticide riskassessment of the active substance tau-fluvalinate." *EFSA Journal* 8(7):1645. https://doi.org/10.2903/j.efsa.2010.1645

European Food Safety, A. (2011). "Conclusion on the peer review of the pesticide risk assessment of the active substance bifenthrin." *EFSA Journal* 9(5): 2159. https://doi.org/10.2903/j.efsa.2011.2159

European Food Safety, A., et al. (2018). "Peer review of the pesticide risk assessment of the active substance cypermethrin." *EFSA Journal* 16(8): e05402. https://doi.org/10.2903/j.efsa.2018.5402

Ganzleben, C., et al. (2017). Human biomonitoring as a tool to support chemicals regulation in the European Union. *International Journal of Hygiene and Environmental Health, 220*(2), 94–97. https://doi.org/10.1016/j.ijheh.2017.01.007

Gerofke, A., et al. (2023). From science to policy: how European HBM indicators help to answer policy questions related to phthalates and DINCH exposure. *International Journal of Hygiene and Environmental Health, 247*. https://doi.org/10.1016/j.ijheh.2022.114073

Gilles, L., et al. (2022). Harmonization of human biomonitoring studies in Europe: characteristics of the HBM4EU-aligned studies participants. *International Journal of Environmental Research and Public Health, 19*(11). https://doi.org/10.3390/ijerph19116787

Gimenez-Asensio, M. J., et al. (2023). Effect of prenatal exposure to organophosphates and pyrethroid pesticides on neonatal anthropometric measures and gestational age. *Environmental Research, 232*. https://doi.org/10.1016/j.envres.2023.116410

González-Alzaga, B., et al. (2022). The questionnaire design process in the European Human Biomonitoring Initiative (HBM4EU). *Environment International, 160*, 107071. https://doi.org/https://doi.org/10.1016/j.envint.2021.107071

Govarts, E., et al. (2023). Harmonized human biomonitoring in European children, teenagers and adults: EU-wide exposure data of 11 chemical substance groups from the HBM4EU Aligned Studies (2014–2021). *International Journal of Hygiene and Environmental Health, 249*. https://doi.org/10.1016/j.ijheh.2023.114119

Hays, S. M. & Aylward, L. L. (2009). Using biomonitoring equivalents to interpret human biomonitoring data in a public health risk context. *Journal of Applied Toxicology, 29*(4), 275–288. https://doi.org/10.1002/jat.1410

Hays, S. M. & Aylward, L. L. (2012). Interpreting human biomonitoring data in a public health risk context using biomonitoring equivalents. *International Journal of Hygiene and Environmental Health, 215*(2), 145–148. https://doi.org/10.1016/j.ijheh.2011.09.011

Hays, S. M., et al. (2007). Biomonitoring equivalents: a screening approach for interpreting biomonitoring results from a public health risk perspective. *Regulatory*


© Burleigh Dodds Science Publishing Limited, 2025. All rights reserved.

*Toxicology and Pharmacology*, *47*(1), 96–109. https://doi.org/10.1016/j.yrtph.2006.08.004

Hays, S. M., et al. (2008). Guidelines for the derivation of biomonitoring equivalents: report from the biomonitoring equivalents expert workshop. *Regulatory Toxicology and Pharmacology*, *51*(3 SUPPL), S4–S15. https://doi.org/10.1016/j.yrtph.2008.05.004

IPCS. (2001). Biomarkers in risk assessment: validity and validation. Retrieved from https://www.inchem.org/documents/ehc/ehc/ehc222.htm

JMPR 1987 –767. Permethrin (Pesticide residues in food: 1987 evaluations Part II Toxicology) Last accessed 3/9/2025

Khemiri, R., et al. (2017). "Documenting the kinetic time course of lambda-cyhalothrin metabolites in orally exposed volunteers for the interpretation of biomonitoring data." *Toxicology Letters,* 276: 115-121. https://doi.org/10.1016/j.toxlet.2017.05.022

Kolossa-Gehring, M., et al. (2023). Special issue editorial: key results of the european human biomonitoring initiative - HBM4EU. *International Journal of Hygiene and Environmental Health*, *253*. https://doi.org/10.1016/j.ijheh.2023.114197

Lamkarkach, F., et al. (2021). Human biomonitoring initiative (HBM4EU): human biomonitoring guidance values (HBM-GVs) derived for cadmium and its compounds. *Environment International*, *147*. https://doi.org/10.1016/j.envint.2020.106337

Leng, G., et al. (1997). "Human dose-excretion studies with the pyrethroid insecticide cyfluthrin: Urinary metabolite profile following inhalation." *Xenobiotica* 27(12): 1273-1283. https://doi.org/10.1080/004982597239859

Lobo Vicente, J., et al. (2023). HBM4EU results support the chemicals' strategy for sustainability and the zero-pollution action plan. *International Journal of Hygiene and Environmental Health*, *248*. https://doi.org/10.1016/j.ijheh.2023.114111

Luijten, M., et al. (2023). Mixture risk assessment and human biomonitoring: lessons learnt from HBM4EU. *International Journal of Hygiene and Environmental Health*, *249*. https://doi.org/10.1016/j.ijheh.2023.114135

Mahiout, S. & Santonen, T. (2022). *Annex to the Deliverable Report D5.11: Human biomonitoring in risk assessment: 4th set of examples on the use of HBM in risk assessments of HBM4EU priority chemicals*. https://doi.org/10.5281/zenodo.6980132.

Marx-Stoelting, P., et al. (2023). A walk in the PARC: developing and implementing 21st century chemical risk assessment in Europe. *Archives of Toxicology*, *97*(3), 893–908. https://doi.org/10.1007/s00204-022-03435-7

Meslin, M., et al. (2022). Human biomonitoring guidance values (HBM-GVs) for bisphenol S and assessment of the risk due to the exposure to bisphenols A and S, in Europe. *Toxics*, *10*(5). https://doi.org/10.3390/toxics10050228

Ndaw, S., et al. (2018). Occupational exposure of cashiers to bisphenol S via thermal paper. *Toxicology Letters*, *298*, 106–111. https://doi.org/10.1016/j.toxlet.2018.05.026

Ougier, E., et al. (2021a). Chemical prioritisation strategy in the European Human Biomonitoring Initiative (HBM4EU) – Development and results. *International Journal of Hygiene and Environmental Health*, *236*. https://doi.org/10.1016/j.ijheh.2021.113778

© Burleigh Dodds Science Publishing Limited, 2025. All rights reserved.

Ougier, E., et al. (2021b). Human biomonitoring initiative (HBM4EU): human biomonitoring guidance values (HBM-GVs) derived for bisphenol A. *Environment International, 154*. https://doi.org/10.1016/j.envint.2021.106563

Pack, L. K., et al. (2023). A step towards harmonising human biomonitoring study setup on European level: Materials provided and lessons learnt in HBM4EU. *International Journal of Hygiene and Environmental Health, 249*. https://doi.org/10.1016/j.ijheh.2023.114118

Panel, E. C. (2015). Scientific opinion on acrylamide in food. *EFSA Journal, 13*(6), 4104. https://doi.org/https://doi.org/10.2903/j.efsa.2015.4104

Prueitt, R. L. & Goodman, J. E. (2024). Evidence evaluated by European Food Safety Authority does not support lowering the temporary tolerable daily intake for bisphenol A. *Toxicological Sciences, 198*(2), 185–190. https://doi.org/10.1093/toxsci/kfad136

Quindroit, P., et al. (2021). Estimating human exposure to pyrethroids' mixtures from biomonitoring data using physiologically based pharmacokinetic modeling. *Environmental Research, 192*. https://doi.org/10.1016/j.envres.2020.110281

Rodríguez-Carrillo, A., et al. (2023). Implementation of effect biomarkers in human biomonitoring studies: a systematic approach synergizing toxicological and epidemiological knowledge. *International Journal of Hygiene and Environmental Health, 249*. https://doi.org/10.1016/j.ijheh.2023.114140

Rodriguez Martin, L., et al. (2023). Identification of real-life mixtures using human biomonitoring data: a proof of concept study. *Toxics, 11*(3). https://doi.org/10.3390/toxics11030204

Sabbioni, G., et al. (2022). Literature review and evaluation of biomarkers, matrices and analytical methods for chemicals selected in the research program Human Biomonitoring for the European Union (HBM4EU). *Environment International, 169*. https://doi.org/10.1016/j.envint.2022.107458

Sanghavi, K., et al. (2024). Covariate modeling in pharmacometrics: general points for consideration. *CPT: Pharmacometrics and Systems Pharmacology, 13*(5), 710–728. https://doi.org/10.1002/psp4.13115

Santonen, T., et al. (2023). How to use human biomonitoring in chemical risk assessment: methodological aspects, recommendations, and lessons learned from HBM4EU. *International Journal of Hygiene and Environmental Health, 249*. https://doi.org/10.1016/j.ijheh.2023.114139

Schwedler, G., et al. (2017). Human biomonitoring pilot study DEMOCOPHES in Germany: contribution to a harmonized European approach. *International Journal of Hygiene and Environmental Health, 220*(4), 686–696. https://doi.org/10.1016/j.ijheh.2017.01.012

Schwedler, G., et al. (2021). Benzene metabolite SPMA and acrylamide metabolites AAMA and GAMA in urine of children and adolescents in Germany – human biomonitoring results of the German Environmental Survey 2014–2017 (GerES V). *Environmental Research, 192*. https://doi.org/10.1016/j.envres.2020.110295

Snoj Tratnik, J., et al. (2022). Cadmium exposure in adults across Europe: results from the HBM4EU aligned studies survey 2014–2020. *International Journal of Hygiene and Environmental Health, 246*. https://doi.org/10.1016/j.ijheh.2022.114050

Socianu, S., et al. (2022). Chemical mixtures in the EU population: composition and potential risks. *International Journal of Environmental Research and Public Health, 19*(10). https://doi.org/10.3390/ijerph19106121

© Burleigh Dodds Science Publishing Limited, 2025. All rights reserved.

Sprong, C., et al. (2020). Cumulative dietary risk assessment overarching different regulatory silos using a margin of exposure approach: a case study with three chemical silos. *Food and Chemical Toxicology*, *142*. https://doi.org/10.1016/j.fct.2020.111416

Tan, Y. M., et al. (2007). Reverse dosimetry: interpreting trihalomethanes biomonitoring data using physiologically based pharmacokinetic modeling. *Journal of Exposure Science and Environmental Epidemiology*, *17*(7), 591–603. https://doi.org/10.1038/sj.jes.7500540

Tarazona, J. V., et al. (2022). A tiered approach for assessing individual and combined risk of pyrethroids using human biomonitoring data. *Toxics*, *10*(8). https://doi.org/10.3390/toxics10080451

Tavares, A. M., et al. (2022). Occupational exposure to Hexavalent Chromium, Nickel and PAHs: a mixtures risk assessment approach based on literature exposure data from european countries. *Toxics*, *10*(8). https://doi.org/10.3390/toxics10080431

Tkalec, Ž., et al. (2024). Innovative analytical methodologies for characterizing chemical exposure with a view to next-generation risk assessment. *Environment International*, *186*. https://doi.org/10.1016/j.envint.2024.108585

Vaccher, V., et al. (2022). European interlaboratory comparison investigations (ICI) and external quality assurance schemes (EQUAS) for the analysis of bisphenol A, S and F in human urine: results from the HBM4EU project. *Environmental Research*, *210*. https://doi.org/10.1016/j.envres.2022.112933

Vom Saal, F. S., et al. (2024). The Conflict between regulatory agencies over the 20,000-fold lowering of the tolerable daily intake (TDI) for bisphenol A (BPA) by the European Food Safety Authority (EFSA). *Environmental Health Perspectives*, *132*(4). https://doi.org/10.1289/EHP13812

von Stackelberg, K. & Williams, P. R. D. (2021). Evolving science and practice of risk assessment. *Risk Analysis*, *41*(4), 571–583. https://doi.org/10.1111/risa.13647

Zoeller, R. T., et al. (2023). European medicines agency conflicts With the European Food Safety Authority (EFSA) on bisphenol a regulation. *Journal of the Endocrine Society*, *7*(9). https://doi.org/10.1210/jendso/bvad107

© Burleigh Dodds Science Publishing Limited, 2025. All rights reserved.

# Part 3

## Developments in risk assessment: risk characterisation

# Chapter 9

# Developments in toxicokinetic modeling of chemical residues in food

*Dimosthenis Sarigiannis, National Hellenic Research Foundation/Aristotle University of Thessaloniki, Greece and University School for Advanced Study (IUSS), Italy; and Spyros Karakitsios, National Hellenic Research Foundation/Aristotle University of Thessaloniki, Greece*



## 1 Introduction

Food may hide some unwanted surprises. Pesticide residues, environmental toxins, processing leftovers, and even certain additives can show up in what we eat, and these traces sometimes pose health risks that need careful checking. Experts often compare what people eat (external exposure estimates like dietary intake levels) with toxicological doses drawn from animal studies (Reale et al., 2024) to decide if exposure to a chemical is hazardous. However, this approach misses one very important element – what happens once chemicals are in our bodies.

This is where toxicokinetics (TK) comes in. It looks at how a substance is absorbed, distributed, metabolized, and excreted – basically, it shows us how much of a chemical reaches target tissues. There is a growing view that

http://dx.doi.org/10.19103/AS.2025.0158.18
© Burleigh Dodds Science Publishing Limited, 2025. All rights reserved.

knowing the internal dose is key to a more realistic assessment of exposure risk, especially for long-term low-level exposures from chemicals in foods (Reale et al., 2024). However, shifting our focus to internal dose metrics requires a comprehensive understanding of TK.

A TK modeling approach takes things a step further by blending external exposure data with predictions about the internal dose. This lets risk assessors foresee levels in tissues or biomarkers resulting from our everyday diet. It is a crucial method considering that people are often exposed to several chemicals simultaneously – from food, water, and even the air – so adding exposures from these different sources together might make sense (Reale et al., 2024). For example, if someone ingests a pesticide via food and again through their drinking water, the TK calculations can combine these to predict the overall chemical load in the body.

Sometimes, substances with long half-lives – e.g. persistent organic pollutants or heavy metals – accumulate steadily, while chemicals that do not last long tend to spike briefly before clearing from the body. Generally speaking, internal dose modeling not only reflects the aggregate impact of multiple exposure routes, it also captures the complex differences between how various chemicals behave in our bodies (Reale et al., 2024). In many cases, this approach brings us a bit closer to tackling the real-life complexity of chemical exposures in our food. Toxicokinetic models, when you really factor in their dynamic behavior, give us a more reliable picture of how doses relate to effects – far beyond just looking at what is outside the body.

The past decade has seen a lot happen in understanding food-related chemicals. New physiologically based toxicokinetic (PBTK) models can now simulate how chemicals move through our bodies in a way that mirrors real life, covering everything from infancy to pregnancy. Often these models get paired with detailed exposure information – for instance, diet surveys or environmental monitoring – to build an overall picture that connects what we are exposed to with the internal dose we actually get. For example, a recent European study on lead put this idea into practice. The researchers crafted an "integrative modeling framework" that merged long-term data on dietary lead intake with a PBTK model to predict blood lead levels in various countries (Sy et al., 2024). In that framework, external lead – coming from past diets and the environment – was linked to blood concentrations, which let them study trends across populations and compare with biomonitoring data (Sy et al., 2024). It really shows that when you blend exposure details with TK modeling, you end up with a well-rounded view of how food contaminants might affect public health.

On another note, new methods have boosted both the predictive power and efficiency of these models. Researchers have started considering population variability – recognizing that age, gender, genetics, and overall

© Burleigh Dodds Science Publishing Limited, 2025. All rights reserved.

health can make internal doses vary widely even when everyone is exposed to the same dose of a chemical. In vitro tests are now combined with computer models to bridge the gap from cell assays to what happens in real life (in vivo to in vitro extrapolation, IVIVE), meaning data from high-throughput screens now help shape human kinetic models. And then you have a whole host of innovative computational tools – ranging from quantitative structure-activity relationship (QSAR) models to machine learning (ML) and Bayesian inference – being used to refine predictions and manage uncertainty. Such innovations tie into a bigger move toward next-generation risk assessment (NGRA), an approach aiming to meld new testing methods and human-focused data (like biomonitoring) to reduce reliance on animal tests (Reale et al., 2024).

Risk managers and risk assessors are now discussing the use of TK models in risk assessment to improve the quality of the assessment of exposure to chemical substances in food. Chemicals are looked at in several different ways – sometimes by modifying how we extrapolate doses, other times by weighing overall exposures, or even by figuring out what biomonitoring data means in a health risk setup (Sarigiannis & Hansen, 2012).

## 2 Integrating toxicokinetics with exposure assessment

For a proper risk appraisal of chemicals in food, we need to look at both what we are exposed to externally and how much actually builds up inside us. Exposure assessments usually reckon how much of a chemical is taken in (the so-called intake fraction) – like through what we eat – while TK is all about tracking how much of that chemical finds its way into systemic circulation and important tissues. When we combine these two aspects, we end up with a much more robust view of the associated health risk than if we had looked at them separately (Sarigiannis et al., 2011). In this kind of integrated setup, data on external sources – like how concentrated contaminants are in food and how much of that food we eat – get fed into a TK model that predicts what the internal concentrations look like over time.

That mix of exposure science and TK lets risk assessors tackle several key questions. For instance, how do various dietary scenarios end up translating into different tissue doses? What kind of chemical buildup might occur after long-term consumption? And, on a more detailed level, how do peaks and slow accumulations of chemicals relate to our occasional spurts of intake?

A net benefit of this integrating process is the chance to run aggregate exposure assessments that consider multiple routes of entry, not just food alone. After all, humans usually are exposed to chemicals in several ways. Take, for example, a lipophilic pesticide – you might ingest it with your meal, breathe it in if it is used indoors, or even pick it up through your skin. An integrated model adds up these varied exposures to simulate their overall contribution

© Burleigh Dodds Science Publishing Limited, 2025. All rights reserved.

to the internal dose (Sarigiannis et al., 2011). In one comprehensive framework for both aggregate and cumulative exposure, Sarigiannis et al. even brought in modules for environmental emissions, dispersion, and human contact to widen the picture. Here, we define aggregate exposure as the total exposure to a single chemical from all possible sources and through all possible routes. Cumulative exposure refers to the combined exposure to multiple chemicals that may have similar or related health effects, often considering all relevant sources, pathways, and routes of exposure. Imagine using a PBTK model that links everyday exposures – from different microenvironments – to what goes on inside the body (Sarigiannis & Karakitsios, 2011). Assessing health risks solely by measuring external environmental exposures, such as air pollutants or substances touched, may inaccurately represent actual health risk. This is because external exposure does not directly equate to internal dose – the actual chemical concentration reaching organs and tissues – which depends heavily on individual biological processes like metabolism and excretion. Moreover, individual variability in physiology, influenced by factors such as age, sex, genetics, lifestyle, and underlying health conditions, leads to significant differences in internal doses between individuals exposed to the same external conditions. Additionally, everyday exposures usually involve multiple concurrent sources and pathways, meaning focusing on a single external exposure route might overlook the cumulative risk from combined sources. Finally, temporal exposure patterns – such as short-term spikes versus long-term chronic exposures – also affect internal accumulation differently, emphasizing that measuring external exposure alone is insufficient to fully understand true health risks (Cohen Hubal et al., 2019). Modeling internal dose, which takes stock of daily concentration swings and even interactions when exposures are put together, gives us a much clearer picture. Take benzene, for instance; when we consider inhalation together with dietary exposure in a PBTK model, we see that blood concentrations are not constant during the day. In fact, exposure to other solvents like toluene or xylene can even reduce the rate of benzene metabolism, shifting the levels of its metabolites – insights only available via an integration of exposure and TK.

Now, consider the complex reality of cumulative exposures to multiple chemicals – a scenario of high importance, for example, in food safety with mixtures like pesticide residues or additive cocktails. Risk assessors increasingly refer to the "cocktail effect," where combined exposures might bring greater risk than the sum of individual parts, though they still wrestle with the complexity of it all. In most cases, approaches range from assuming a simple additivity of doses to modeling intricate TK and toxicodynamic (TD) synergistic or antagonistic interactions. Toxicokinetic modeling, by reproducing the biological journey of each compound and their interactions, adds significant evidence in this endeavor. Advanced models capture those

© Burleigh Dodds Science Publishing Limited, 2025. All rights reserved.

subtle biochemical interactions affecting both the internal exposure and the toxic punch of the mixture. For example, if two compounds end up competing for the same metabolic enzyme, a PBTK model can often predict a rise in a compound concentration. By considering these interactions – even if not in an overly perfect way – such models help move risk assessment from basic hazard quotients to a more science-rooted understanding. No one can really say there is a magic formula for gauging the risks of chemical mixtures. Policymakers have wrestled with the challenge for ages, but for decisions e.g. authorization of use of pesticides, additives, and veterinary drugs there is an agreement on how the risk assessment should look. Improvements of the methods are however welcome, if practically feasible and understandable for the people involved. Sarigiannis and Hansen (2012) argue that using an approach which mixes TK with dose-addition ideas strikes a practical balance. Instead of a neat, one-size-fits-all recipe, experts are leaning toward a more flexible approach that recognizes the many different levels at which biological systems get affected.

Another twist comes from trying to tie in TK models with everyday population data. Real-world food exposure varies wildly because of differences in diet, geography, and personal habits, while the human body processes xenobiotics in a variety of ways too. Considering how people get exposed and how the human body reacts, the range of internal doses that might build up over time can be estimated. A powerful tool in this area is two-dimensional Monte Carlo simulation: one track covers variability in facets like differing diets, concentrations, and durations of exposure, and the other handles the natural fluctuation in factors such as metabolism and body size. Sy et al. (2024) used this idea to explore long-term lead exposure in four countries in Europe. They ran a simulation that captured not only the natural differences among individuals but also the uncertainties linked with external exposure levels, giving us a window into how lead might flow through populations over time.

Researchers even went further by feeding in country-specific dietary lead data spanning decades into a PBTK model, which let them predict the spread of blood lead levels for each population and year (as detailed in the same study). Such integrated population modeling, validated against actual human biomonitoring (HBM) data – which includes measurements from blood, urine, hair, etc. – lends extra credibility to these predictions. It can show, for example, that certain groups (like children or other sensitive demographics) might end up with disproportionately higher blood lead levels if they face both elevated exposure and greater kinetic susceptibility. Biomonitoring data, despite some minor imperfections, act as a crucial bridge linking ambient exposure to internal doses, making them a key part of these integration efforts. Biomonitoring data, however, are only a snapshot of internal exposure taken at a given moment – they do not tell the full story. Toxicokinetic modeling steps in to make sense of the numbers from biomonitoring, linking them to actual exposure or risk. One

© Burleigh Dodds Science Publishing Limited, 2025. All rights reserved.

way to tackle this is to start with a forward model that estimates what biomarkers we should see based on known exposures. Then reverse dosimetry steps in, which tries to work backward from an observed biomarker level to guess the exposure behind it (Reale et al., 2024).

Huuskonen et al. (2022) showed – using data on an industrial chemical, ortho-toluidine detected in urine – that you can really use this approach to put risk assessment into practice. They opted for what they called a generic exposure reconstruction PBTK modeling method (Sarigiannis et al., 2014), basically figuring out the external intake matching the urinary measurements (Huuskonen et al., 2022). By lining these reconstructed intake figures up with regulatory risk thresholds, they ended up evaluating cancer risks for both workers and the public at large. Interestingly, the outcomes from the PBTK method were within about 30% of a simpler mass-balance calculation, indicating that even under the worst-case scenarios, population exposures stayed below regulatory limits.

This blending of biomonitoring with TK is fast becoming a cornerstone of next-generation chemical risk assessment. In practice, it creates a very useful feedback loop: models predict the inner biomarker profile, while the actual measurements then help check or fine-tune those predictions. Aligning these internal dose forecasts with clinical or epidemiological data further helps in setting biologically meaningful risk metrics – for instance, blood concentration thresholds that really matter in terms of effects.

In short, mixing TK modeling with exposure assessment gives a fuller picture of risk from foodborne chemicals, better anchored on human physiology. It allows aggregating exposures along different pathways, considering cumulative mixtures, grasping how exposures can vary across populations, and interpreting raw biomonitoring data in terms of potential adverse health effects. Transforming exposure levels into internal dose metrics lays the foundation for dose-response analysis and overall risk characterization. Up next, the discussion turns to how fresh methods in TK modeling – ranging from PBTK models and population variability techniques to IVIVE and even AI/Bayesian methods – are adding new dimensions to these integrated exposure-risk assessments, particularly as the field evolves.

## 3 Predicting toxicokinetic risk using physiologically based toxicokinetic modeling

Currently the use of PBTK modeling is an important topic in modern-day TK – it is the way we predict how much of a chemical from food might end up where inside the body. Instead of thinking about the body as one giant entity, PBTK models split it into different compartments – organs and tissues interconnected via blood flow – and use sets of differential equations to track what happens to

© Burleigh Dodds Science Publishing Limited, 2025. All rights reserved.

chemicals as they get absorbed, distributed, metabolized, and finally excreted. Rather than the old one-compartment models that are overly simple, current-day PBTK models bring in real-life details like organ sizes, blood flow rates, and enzyme activities to adapt across different species and even across a person's life (Upton et al., 2016). This extra detail comes in really handy when dealing with food contaminants; after all, it is often necessary to jump from animal tests to understanding human impacts or to bridge the gap from high-dose laboratory studies to everyday low-level exposures (Rowland Yeo et al., 2024).

Lately, there has been a significant push to make these models even more versatile for the many ways people might pick up chemicals through their diet. A critical development in this context is the emergence of generic PBTK models, where users may introduce a wide range of chemicals without starting model development from scratch each time. For instance, Sarigiannis et al. (2020) rolled out a lifelong generic PBTK model – designed pretty much for exposome studies – that covers all life stages and can be quickly adjusted for different compounds (Sarigiannis et al., 2020). This system even separates out the blood flow into arterial, venous, and portal streams, so it is easier to adapt parameters from other data. Each tissue is not just lumped together in the model; it is broken down into plasma and interstitial fluid, red blood cells, and the actual cells themselves, which is quite important when you are tracking chemicals like lead that tend to stick with red blood cells or persistent organics that slowly make their way into fat.

In addition, the model also accounts for the intrinsic dynamics of the human body itself. Body weight, the size of organs, and blood flow details all change with age, allowing researchers to simulate everything from fetal development and infancy right through to adulthood (Sarigiannis et al., 2020). This means that scenarios like how a chemical crosses the placenta during pregnancy or appears during lactation can be tackled using the same overall computational framework. In fact, Sarigiannis's team even set aside special compartments to represent the placenta and carefully tracked the timing when metabolic enzymes start showing up in fetuses and young children – vital for taking into account the details of maternal–fetal kinetics (Sarigiannis et al., 2020). Anthropometric factors keep shifting over time. They give us a kind of "lifetime dose check" that mirrors body growth and development (Sarigiannis et al., 2020). Gastellu et al. (2025) also pointed out that organs do not grow in a straight-line way – they pulled together a big set of organ growth equations for both males and females to facilitate the widespread use of PBTK modeling in chemical risk assessment (Gastellu et al., 2025). In fact, they incorporated these equations into a PBTK model spanning from newborn days to old age, then used it to figure out mercury exposure from seafood – simulating levels in hair and urine at various life stages (Gastellu et al., 2025). In most cases, this shift moves us away from unchanging, adult-only models to ones that walk hand in

© Burleigh Dodds Science Publishing Limited, 2025. All rights reserved.

hand with life stages and susceptibility windows – particularly important when foodborne pollutants might hit hardest in sensitive stages (like neurotoxicants affecting children's neurological development).

Another interesting advance in PBTK modeling is that we can now factor in metabolism and elimination routes that are unique to each chemical. Old models usually assumed a simple first-order clearance, but now it is possible to incorporate enzyme details like Michaelis–Menten parameters, several of which can be estimated using artificial intelligence (AI) algorithms (Gastellu et al., 2025) in the face of experimental data paucity or even set up a mix of metabolic routes (Alhaji Saganuwan, 2021). Take the generic model by Sarigiannis et al. for example – it can switch between using Michaelis–Menten kinetics for when metabolism gets saturated or simply assume linear clearance, based on what we know about the compound (Sarigiannis et al., 2020). Moreover, linking the original chemicals with their metabolites in the simulation is key, especially for substances that only turn toxic after being metabolized or when metabolites are what we actually measure in biomonitoring (Sarigiannis et al., 2020). In their system, xenobiotics and their metabolites join forces through different metabolizing sites – mainly in the liver, but also in additional sites like the gut, skin, or even brain when the right enzymes are available (Sarigiannis et al., 2020). This setup, for instance, makes it possible to model a pesticide that is rapidly metabolized. When a compound is turned into a toxic byproduct, the PBTK model kicks off by tracking both the original substance and its metabolite over time in tissues. Multi-compound modeling is not set in stone – it can, for example, show how two food chemicals sharing the same metabolic route can interfere with each other's clearance; in one case with benzene mixtures, the presence of one slowed down the other's removal (Sarigiannis & Gotti, 2008).

Lately, calibration and testing of these models have gotten a boost from better data and updated methods. Researchers, in most cases, now lean on HBM data to check if internal doses are predicted accurately. Take the exposome model by Sarigiannis et al.; they predicted urinary biomarker levels for bisphenol A (BPA), DEHP (the plasticizer), and cadmium, then compared those numbers to what was actually seen in human studies (Sarigiannis et al., 2020). Urine concentration predictions lined up well with the observed data, which has provided enhanced confidence in these tools – especially if they are to be used for food safety regulation. There is even growing support for open-source PBTK resources, like the R package httk that bundles general models and chemical-specific parameter libraries, paving the way for quick deployment across many chemicals, even ones with little available data (Pearce et al., 2017).

Interestingly, advanced PBTK modeling also steps in to tackle persistent, bioaccumulative toxins (Caldwell et al., 2012). Chemicals such as methylmercury, polychlorinated biphenyls, and dioxins – which often accumulate in the body from regular consumption of fish or dairy – can be hard to manage, since

© Burleigh Dodds Science Publishing Limited, 2025. All rights reserved.

they have long half-lives and tend to accumulate. The PBTK models are well-suited for simulating this slow buildup and tracking how these compounds disperse into parts like hair, brain, or fat. For example, Gastellu et al. (2025) merged sub models for methylmercury and inorganic mercury to simulate the lifelong mercury burden from fish eating. They coupled dietary exposure data with a lifelong physiology model to estimate levels in target tissues (hair for MeHg, urine for inorganic Hg) and compared those figures against health-based guidance levels based on HBM data (Gastellu et al., 2025). All in all, these advances show promise, even if the process sometimes gets a bit messy and less tidy than one might expect. A closer look at PBTK methods reveals that – even when seafood mercury is at the maximum allowed levels – internal markers hardly reach any dangerous internal concentrations. It is a good way of turning strict exposure limits into predictions of what builds up in the human body. This shows just how far modeling of food contaminants has come.

Simply put, today's PBTK models for chemicals in food are not the same old ones. They work over the whole lifespan, mix everyday science with complex biochemical insights, and can be used again and again on different chemicals, thanks to clever parameterization tools, including AI-based ones (both generative AI and artificial neural networks (ANNs). By linking what we eat with what ends up inside the human body, these models have become central to connecting exposure with harmful effects. Next up, we are going to explore how these models handle human differences and how laboratory tests and computer simulations give us an extra boost in predictive power.

## 4 Accounting for individual differences and uncertainty in toxicokinetics

No two individuals are alike – and that matters a lot for how toxins are processed. Two people eating the same contaminated food might end up absorbing and handling it in very different ways. Factors like age, body size, organ function, small genetic differences that affect enzyme action (Sarigiannis & Karakitsios, 2019; Sarigiannis et al., 2019), overall health, even the gut microbes you've got, and even any medications can all make a difference. This mix of differences means that figuring out who might end up with higher toxin levels or be more at risk is a pretty key issue. Risk assessors add safety factors to protect vulnerable subpopulations, but as we refine our models, those broad safety margins could be swapped out for ones based on real, data-driven differences. This does not necessarily mean that the safety margins will become smaller – rather they will be based on more realistic datasets.

Recent advances in TK modeling have really allowed us to measure these differences – and the uncertainties in our predictions – more carefully. Researchers often use Monte Carlo simulations (Thomas et al., 1996), treating

© Burleigh Dodds Science Publishing Limited, 2025. All rights reserved.

things like hepatic clearance rate not as a fixed value but as a range of values that capture real-world variability. By randomly sampling from such ranges and running the model thousands of times, a spread of outcomes (for example, different blood concentration levels) is obtained that genuinely reflects how varied people are. In one version of this approach, one simulation layer handles the natural inter-individual variability, while another tackles the gaps in our exact knowledge of the numbers. The result is a robust set of predictions, complete with median values, percentile bands, and confidence intervals for internal doses.

Scientists like Sarigiannis and his team have put these ideas into practice. In a study mapping lead exposure across Europe, they ran a population-based simulation that captured real differences in both exposure levels and how individuals process lead. Such work, found in several case studies, highlights not only the evolving sophistication of these methods but also their promise for better risk assessments built on human data rather than guesswork. Researchers looked at lead exposure across four European countries from the 1970s until recent years (Sy et al., 2024). They did not just settle on one predicted blood lead value – instead, they offered a range from the typical person at the 50th percentile to those at the 95th percentile for each country and time period. They went with a two-dimensional Monte Carlo (MC2D) algorithm, which essentially helped them tease apart uncertainty in inputs – say, doubtful diet intake estimates or variable kinetic rates – from real differences among people. This approach shows just how important it is to run an uncertainty analysis; if there is only limited information on how a chemical is metabolized, the model might predict a very wide range for internal doses, while well-known kinetics tighten those predictions. In this study, by channeling the uncertainty in external exposure data through their PBTK model, the team was able to gauge the confidence in the model predicted blood lead levels and even check if mismatches with biomonitoring data might just stem from data gaps (Sy et al., 2024).

Another interesting twist was their use of Bayesian TK modeling (Qing et al., 2021). Bayesian TK modeling incorporates existing knowledge (e.g. from literature or expert judgment) about model parameters (e.g. uptake/elimination rates) as probability distributions (priors). Their likelihood function quantifies how model predictions match observed experimental data (e.g. chemical concentrations in tissues over time). This connects parameters to data via statistical models (e.g. normal or log-normal distributions). Finally, its posterior inference algorithm uses Bayes' theorem to update prior beliefs with new data, yielding posterior distributions for model parameters. This method uses Bayesian statistical approaches – like Markov Chain Monte Carlo (MCMC) sampling – to estimate distributions for model parameters based on actual data (Sarigiannis et al., 2016). In practice, a Bayesian PBTK calibration might

© Burleigh Dodds Science Publishing Limited, 2025. All rights reserved.

start with some initial ranges for metabolism and then refine those numbers to better match observed blood concentrations. For example, consider the integrated benzene exposure framework put together by Sarigiannis et al.; they kind of applied Bayesian thinking by running MCMC at each step, which let the uncertainty in parameters flow through in a neat, probabilistic way. The result was not just one risk number but a whole distribution that shows the confidence bounds, a feature especially handy when dealing with sparse human data since it yields credible intervals for things like absorption fraction or enzyme $V_{max}$.

On top of that, research into population variability shows that some groups might have higher internal exposures given the same external dose. Take infants and children – they often get a higher dose relative to their body weight, plus they have immature detox systems and dietary habits (like relying on breast milk or formula, then later eating more per kilogram) that can lead to higher internal chemical levels. Models that incorporate age-dependent physiology address this issue head on. And do not forget genetic differences; for instance, variations in CYP450 enzymes that break down pesticides can be captured by assigning different metabolic clearance values to parts of the population. If there are data on how common slow metabolizers are, Monte Carlo simulations can sample from a bimodal clearance distribution, thereby reflecting that natural variation in a realistic, if somewhat unpredictable, manner. Fast and slow metabolizers mix in everyday simulations. In a group study, some folks may end up with a much higher pesticide blood AUC – that is the area under the curve – even when the overall risk feels low. This mix of kinetics is not surprising when you consider that everyone's metabolism varies; sometimes you just have a subgroup that stands out for potential risk.

These days, TK models do not stick to one rigid approach. They often run sensitivity and uncertainty checks to see which parts of the model really drive the results (McNally et al., 2011). For example, they might look at things like partition coefficients or metabolic rates and see what makes the biggest difference in internal doses. In most cases, if the liver clearance variance accounts for most differences in blood levels, it tells you that differences in metabolism – because of factors like genetics or age – are the main culprit (Marin et al., 2020). That finding might prompt targeted studies or even some focused biomonitoring of metabolic traits.

There is also a very interesting regulatory impact. One concrete use is a discussion between risk assessors on which default uncertainty factors to use in risk assessments. Typically, risk assessors apply a tenfold factor to cover human differences from an average person to a sensitive one if no data are available to indicate a different value. But if a robust and reliable PBTK model reveals that the 95th percentile dose is only three times the median level under a given exposure scenario, risk assessors might be comfortable with a smaller adjustment – or

© Burleigh Dodds Science Publishing Limited, 2025. All rights reserved.

even just use readily the model output. This model-driven approach, which has been used for years in drug safety via population pharmacokinetics (Zhang et al., 2024), is slowly making its way into environmental and dietary chemical risk work. For instance, Health Canada and the US Environmental Protection Agency (EPA) have looked at PBTK model output distributions when setting exposure guidelines for certain solvents, rather than relying only on these blanket factors.

All in all, modern TK modeling now openly accounts for human variability (Kreutz et al., 2024; Testai et al., 2021). By using probabilistic simulations – and sometimes even Bayesian calibration with real-life data – it not only predicts the average dose but also shows the spread among individuals, along with how confident we are in those numbers. Such insights help in refining risk assessments, especially for foodborne chemicals, by pointing out who might be at greater risk, like children or those with specific genetic predispositions. This marks a shift away from one-size-fits-all models toward a more personalized safety net. As more biomonitoring data and studies involving controlled dietary exposures with blood sampling come in, these variability-focused models are bound to get even sharper (Béchaux et al., 2014).

## 5 In vitro to in vivo extrapolation and high-throughput modeling

Getting hold of accurate chemical-specific parameters – like metabolic clearance rates, the fractions absorbed orally, or tissue partition ratios – for all the food-relevant chemicals (especially when data are thin) is a real challenge. Relying solely on animal studies is not only slow and expensive, but sometimes it just is not possible for ethical or logistic reasons (Lilienblum et al., 2008). The IVIVE has stepped in as an essential workaround to bridge these data gaps (Breen et al., 2021). Scientists have been using laboratory experiments – like testing with cell cultures, tissue slices, or even tiny fractions such as liver microsomes – to try to guess how a chemical behaves in the human body. In many cases, this approach lets us estimate key TK traits in vivo by starting with in vitro data (Choi et al., 2019).

Take, for instance, hepatic clearance. Researchers often watch how quickly human liver cells or enzymes break down a chemical in a laboratory assay, then scale those numbers up to cover an entire liver and, ultimately, the whole body. In doing so, we cannot ignore the fact that the size of the liver, the blood flow, and even how much the chemical clings to proteins differ between laboratory conditions and reality. With careful adjustments, we wind up with an estimated clearance or half-life that fits right into a PBTK model. In fact, several pharmaceutical and pesticide compounds have had their human clearance successfully predicted from in vitro microsomal assays when the right scaling

© Burleigh Dodds Science Publishing Limited, 2025. All rights reserved.

factors are applied (Reale et al., 2024). Moreover, IVIVE methods can also help figure out partition coefficients from solubility or binding tests and even absorption rates via cell monolayer models of the gut wall.

Looking at food safety, IVIVE now gets coupled with newer testing strategies – often called new approach methodologies – and high-throughput screening programs. Big projects like ToxCast and Tox21 have generated in vitro bioactivity data on thousands of chemicals, posing the question of what in vivo exposure would lead to similar concentrations at the target site. Essentially, this means turning an in vitro concentration–response into an in vivo dose-response. Physiologically based toxicokinetic modeling combined with IVIVE becomes the engine for that translation. Reale et al. (2024) illustrate this by explaining that PBTK models can be run in reverse – predicting the external exposure needed to yield an internal concentration that matches an in vitro effect. This "reverse dosimetry" approach (Georgopoulos et al., 2009) turns out to be really important when it comes to understanding laboratory assay results and calculating human-equivalent doses. As they put it, it is "*important for in vitro–in vivo extrapolation (IVIVE): PBTK modelling can extrapolate in vitro effect readouts … to the in vivo situation*" (Reale et al., 2024). Imagine you have a laboratory result showing liver cell toxicity measured in micromolar units. That finding is then translated into the kind of blood concentration one would expect in a person, and the oral dose needed to reach that level is reckoned. This step, simply put, helps regulators get a rough idea if what people normally eat is safely below doses that trigger cell damage even before clinical symptoms of adverse health effects are observed.

The IVIVE has really come a long way, mostly because of procedure standardization and novel computational tools. For example, the US Environmental Protection Agency (EPA) rolled out its httk system – essentially a toolkit that houses several in vitro parameter values (e.g. the fraction of a chemical that stays free in plasma and how fast the liver clears it). It even performs Monte Carlo style simulations to capture how different individuals might behave. Researchers like Breen et al. (2021) have used screening data on environmental chemicals to swiftly flag which ones could, under real-life exposure conditions, reach levels that might stir up biological responses. And so, the work often picks out chemicals with weak data coverage but whose predicted dose for effect is not all that far above current exposure estimates.

That said, in vitro tests do not always catch every aspect of what happens in the human body. Laboratory setups can miss the complex interplay of multiple organs, blood protein binding, or even active transport processes. Sometimes, in vitro clearance seems to under-predict what is going on in vivo – either because some enzyme pathways get skipped over or cellular models are not as robust as needed; while in other instances, a compound might break down in the laboratory assay yet persist when inside an organism (Yadav et al., 2021).

© Burleigh Dodds Science Publishing Limited, 2025. All rights reserved.

IVIVE estimates come with a built-in margin of uncertainty factors or ranges to account for such gaps.

A recent review even points out that differences in metabolic activity between cellular models and an actual human liver – mixed with experimental variability and imprecise scaling assumptions – add extra layers of uncertainty. In most cases, getting laboratory methods more harmonized (say, by ensuring protein binding or enzyme kinetics are measured in a consistent way) can help narrow those uncertainties. Improved physiologically based models really help the whole process. When you start modeling specifics like protein binding and how cells absorb chemicals, it becomes easier to match what we observe in laboratory tests with what happens in living systems. Still, IVIVE is not solely about kinetics (TK); it is gradually becoming more about coupling TD with those kinetics. Adverse outcome pathway development often kicks off with in vitro assays that track key events at a cellular level. To estimate risk, one really needs to know what kind of exposure drives that important concentration into the body. That is when TK modeling via IVIVE steps in – it takes an effective concentration measured in vitro (EC) and turns it into an in vivo effective dose (Chang et al., 2022). For instance, if a chemical disrupts neuronal signaling in an in vitro assay at a given concentration, an IVIVE-PBTK analysis might reveal that such a blood level only happens at very high doses in humans, suggesting present exposures are, in most cases, safe. On the flip side, if the dose required is low enough to be reached by usual dietary intake, that chemical would quickly move up the list for risk management.

Another angle is the use of in silico predictions either alongside or instead of laboratory experiments. This approach, which touches on realms like AI and QSAR (as noted in the next section), comes in handy especially when experimental in vitro data are not available. Sarigiannis et al. even pointed out that a generic PBTK model can easily borrow parameters from complex, ad-hoc models or fill in the gaps using advanced QSAR techniques. In practice, that means you can estimate partition coefficients with algorithms based on lipophilicity and tissue composition and predict metabolism through ML models that have been trained on known chemicals. These computed parameters then plug into the PBTK framework to simulate kinetics for new food contaminants – even before any lab data shows up. While a fully in silico version may be a bit less certain initially, it generally offers a starting point for risk assessment that can be refined as more data come in.

Indeed, IVIVE has become a key part of modern TK modeling. It taps into the huge pool of in vitro and in silico data to help set up human kinetic models. Especially in the context of NGRA and the collective push to reduce animal testing, IVIVE-enabled PBTK modeling serves as a handy bridge – linking high-throughput screening and computational chemistry with real-world human risk predictions, particularly when it comes to food safety. We can now project how

© Burleigh Dodds Science Publishing Limited, 2025. All rights reserved.

a new food contact material monomer, a processing-induced compound, or even a completely novel pesticide might behave in the body – and we can do it without starting off with human or animal dosing tests. IVIVE methods keep getting better, and when you pair them with robust PBTK models, our early risk assessments for foodborne chemicals only get stronger in most cases.

## 6 Application of artificial intelligence and Bayesian frameworks in toxicokinetics

Toxicokinetic modeling is tricky, especially considering the wild variety of food contaminants in the field. That is why insights from AI and other computational approaches are becoming useful aids. Currently, ML algorithms are being used to boost prediction accuracy, automatically generate models, and sift through enormous datasets coming from exposome research or biomonitoring studies. At the same time, Bayesian frameworks and other probability-based methods are being used to formally mix in data with uncertainty. Putting these pieces together, we end up with computational TK – a space where data-driven algorithms complement solid mechanistic models.

One of AI's most direct impacts shows up in QSARs feeding into PBTK models. For instance, take Sarigiannis's exposome PBTK framework; it uses QSAR to patch up gaps in chemical-specific details (Papadaki et al., 2017; Sarigiannis et al., 2017). Back in the day, QSAR models – sometimes called quantitative structure–property relationships in the context of ADME (absorption,distribution, metabolism and excretion) – were usually simple linear models or basic regressions. Now, with much larger training sets, more advanced ML methods like random forests, support vector machines, and even neural networks are thrown into the mix to predict partition coefficients, plasma protein binding, and metabolic clearance directly from molecular features. Imagine an ML model trained on hundreds of compounds with known human hepatic clearance, then used to forecast clearance for brand-new chemicals based solely on their structure. Of course, every prediction carries some uncertainty, but these estimates can be improved gradually. Moreover, AI models sometimes provide an early alert when they encounter a compound that strays outside their training dataset – a useful signal in risk assessments when extra expert judgment might be in order.

Beyond QSARs, researchers are also exploring surrogate models (Pruett & Hester, 2016) as quick approximations of heavy, full PBTK simulations. Running an entire PBTK model for thousands of dose combinations or individual profiles can be demanding on computational resources. So, neural networks are being trained on the outputs of full simulations; these ANNs then predict outcomes like peak blood concentration much faster for any new set of input parameters (doses, body weight, and similar anatomical and physiological

© Burleigh Dodds Science Publishing Limited, 2025. All rights reserved.

factors affecting TK behavior of xenobiotics). This speedy shortcut is useful, for instance, in global sensitivity analyses or when fine-tuning parameters – cases where the model must be evaluated repeatedly. Similarly, AI techniques are now supporting parameter fitting; algorithms such as genetic algorithms or Bayesian optimization can comb through the parameter space much more efficiently than endless manual trial-and-error.

Another promising frontier is tapping into Bayesian networks and even other ML classifiers to fuse exposure and kinetic data for better decision-making. Picture constructing a Bayesian network with nodes that represent sources of exposure, internal dose markers, and eventual health outcomes, all derived from epidemiological and biomonitoring studies (Orak et al., 2019; Sarigiannis et al., 2018). Although this kind of approach branches a bit into the broader realm of risk modeling, it still manages to weave in the essential principles of TK in a natural way. Linking external and internal nodes can sometimes look a bit strange. Imagine asking a network, "*If someone is eating in a particular way and their biomarker levels hit a certain mark, what are the odds they surpass a toxicity line?*" It is less about replacing traditional PBTK models and more about showing how AI can address complex, multilayer challenges – even if exploratory in nature.

AI in regulatory science – especially in TK – is still finding its footing. A good example is the GARDskin assay for skin sensitization. Essentially a ML tool, it reads a set of genomic biomarker clues to make a prediction (Wittwehr et al., 2020). When it comes to chemical residues, one might even imagine similar AI approaches that process high-dimensional data – say, metabolomic responses to exposure – and make predictions on kinetics or toxicity. Even so, these ideas are largely on the horizon; recent applications, like QSAR tools merged into current platforms, are more about supporting the traditional methods rather than overhauling them outright.

Then there is the Bayesian side of things, which has had a direct impact on TK modeling practice. Bayesian inference is being embraced for its ability to consider all kinds of data and lay out uncertainty in predictions. Picture a Bayesian PBTK model where every parameter is treated as a random variable – its prior based on earlier studies or similar chemicals – and then updated when fresh data (from human volunteer studies or biomonitoring, for instance) comes along. The outcome is not a single point but rather a whole range (a credible interval) that gives a clear picture of inherent uncertainty. For example, with ortho-toluidine, one might start with metabolic rate priors drawn from animal data and then refine them using human urinary biomonitoring results through Bayesian calibration. In effect, this yields a spread of plausible metabolic rates that agrees with the observed urine levels (Huuskonen et al., 2022).

Another interesting angle involves biomonitoring equivalents and HBM guidance values. One way to set a "safe" level of a chemical in blood or urine

© Burleigh Dodds Science Publishing Limited, 2025. All rights reserved.

is to invert the TK model and find the concentration that matches an accepted safe dose (say, an acceptable daily intake – (ADI)). Naturally, this method faces significant uncertainty due to data limits and variability; still, it can be used to better estimate the applied Safety Factors. Bayesian approaches tend to handle these issues well, propagating the uncertainty from both the toxicological reference dose and the kinetic extrapolation. In most cases, as Reale et al. (2024) note, proper handling of uncertainties from in vitro tests and PBTK predictions allows the generation of HBM-GV – that is, human biomonitoring guidance values. Bayesian computation fits nicely when we need to set reliable confidence intervals for guidance values. In one effort – just look at the European Union's (EU) HBM4EU project (Reale et al., 2024) – researchers mixed what we already know about TK with population biomonitoring data. There was only about a 5% chance of the value being exceeded, given normal variability, reducing thus the overall uncertainty in the guidance value.

Big data are now being drawn into TK with the help of advanced AI and modern computing. Exposome initiatives, for instance, gather expansive records on everything from environmental chemicals and diets to biomarkers and even health outcomes. Multivariate methods and ML sort through these sprawling datasets – often clustering individuals by their exposure profiles and corresponding biomarker levels. Sometimes we end up asking things like, "Why do some people seem to accumulate more of a chemical?" Perhaps an unknown source of exposure or even a metabolic quirk, such as a slower elimination rate, is at play. Research efforts like those spearheaded by the HEALS project (Sarigiannis, 2019; Sarigiannis et al., 2018), coupled informatics and data analytics with modeling to flag population subgroups with unusually high biomarker readings. Then, a PBTK model (Sarigiannis et al., 2020) might be set up to see if a plausible physiological reason exists.

All in all, mixing AI with Bayesian approaches is mostly about enhancing both development and interpretation of TK modeling. Artificial intelligence brings in predictive modeling, surrogate models for efficiency, and data mining on the complex interplay between exposures and outcomes. Meanwhile, Bayesian frameworks – often considered a systematic way to handle uncertainty – keep welcoming new data as they come along, which is valuable in the ever-evolving world of risk assessment. These techniques do not aim to replace classic mechanistic TK; instead, they work alongside it, making scientific conclusions with risk relevance more robust. With computational power surging and datasets exploding worldwide, we might soon see automated pipelines that go straight from chemical structure to risk characterization, even if toxicology experts keep a close eye on the algorithms. For now, though, the regulatory community stays cautious. Transparency and explainability remain non-negotiable, as regulatory bodies like the US EPA, the Food and Drug Administration (FDA), and various national health agencies insist on a clear

© Burleigh Dodds Science Publishing Limited, 2025. All rights reserved.

science-backed process. Recent research efforts – such as those discussed in frameworks on aggregate modeling (Sarigiannis et al., 2011) are steadily laying the groundwork for these advanced methods to become more widely accepted as the basis of setting health-based guidance values and improving risk assessment.

From a regulatory perspective, TK modeling has gradually moved on from purely academic research into practical use for improving the quality of the risk assessment of chemical substances in food and feed. Agencies now recognize that bringing TK data into the picture really strengthens risk assessments. Even if some minor inconsistencies arise here and there in the way data are handled, the overall trend points to a more integrated, transparent, and effective method of safeguarding public health. Risk estimates are our starting point – this part discusses how TK modeling gradually moves into regulatory practice these days, and it might even shape the next wave of risk assessments and policies for handling cumulative risks. A few examples that show just how useful these modeling approaches can be from a regulatory standpoint (Sarigiannis et al., 2016) are given below.

# 7 Current practices in regulatory toxicokinetics

## *7.1 Physiologically based toxicokinetic modeling in the chemical risk assessment arena*

To date, TK modeling in food chemical risk assessment tends to be used quite incidentally – usually on a case-by-case basis when the necessary data are available and really suit the problem at hand. One common use of TK models is for bridging interspecies comparisons or extrapolating high-dose results to lower doses. For instance, if a toxicity study in rats determines a NOAEL for a contaminant, a human PBTK model can be used to estimate the level of human exposure that would deliver a similar internal dose (WHO, 1990). This, in most cases, refines the ADI or tolerable daily intake (TDI) by easing the need to lean solely on default uncertainty factors.

Human PBTK models for different life stages were employed by the US EPA to derive reference doses for perchlorate, essentially replacing the usual default 10× intraspecies factor with estimates generated by the model (Cohen Hubal et al., 2019). Similarly, both the FDA and European Food Safety Authority (EFSA) have delved into using PBTK models for certain food contact substances to simulate worst-case internal exposures in sensitive subgroups (like infants), thereby guiding their safety checks.

In Europe, coupling HBM with TK is increasingly seen as a vital part of future risk assessment strategies (Cohen Hubal et al., 2019). Although this outlook is forward-looking, some current practices already hint at it. An example is the

© Burleigh Dodds Science Publishing Limited, 2025. All rights reserved.

derivation of HBM-GVs for priority chemicals under the HBM4EU initiative, which clearly relies on TK modeling. Ougier et al. (2021) managed to derive an HBM-GV for BPA by first finding the plasma concentration that corresponds to the TDI and then using a PBTK model to estimate the urinary BPA level that would result. This turns an external limit into an internal biomarker threshold. These guidance values, in turn, help risk assessors and public health officials to interpret biomonitoring data – if the typical urinary levels stay below the HBM-GV, then risk is assumed to be under control. It should be noted here that such guidance values are only "predictive" of nature, in assessing existing situations of exposure (e.g. regional contamination). Health-based guidance values such as ADIs and TDIs are used in a "preventive" manner, to avoid future exposure, and therefore fundamentally conservative of nature. Ganzleben et al. (2017) have even argued that pairing HBM with a solid understanding of TK results in a powerful tool for chemicals regulation in the EU; and it looks like regulators in Europe are slowly but surely moving in that direction based on current experience. When it comes to setting limits – whether for worker exposures or even contaminants in our food – the differences in how a chemical enters and is processed by the body really matter. Take a scenario where a solvent ends up in food by accident. If most of the available toxicity data come from breathing studies, then a PBTK model is brought in to close the gap between inhaling a substance and swallowing it, factoring in first-pass metabolism. Take ortho-toluidine, for instance. Its occupational limits were established based on inhalation data, yet risk assessors used a PBTK model to check if environmental contamination through oral exposure might carry similar health risks (Huuskonen et al., 2022). Interestingly, the model revealed that exposures for the general population – mostly via ingestion – were considerably lower relative to internal dose limits compared to those experienced occupationally through inhalation, which generally backs current worker-protection regulations (Huuskonen et al., 2022).

A major advantage of using TK in the risk assessment process is that additional insights regarding population vulnerabilities are accounted for. This is critical because exposure scenarios that eventually present an adequate margin of safety when accounting only for bottom-up intake could eventually be closer to the safety borderline when incorporating internal exposure assessment metrics. This is clearly illustrated below in the case of BPA. The evaluation of risk of the most important scenarios considered by the EFSA based on external exposure estimates is given in Fig. 1. Our analysis indicates that exposure for all consumer exposure scenarios (including premature neonates and bottle-fed infants) is below the EFSA temporary TDI (t-TDI) of 4 μg/kg_bw/d, with the exception of the upper part of the exposure distribution of premature neonates hosted in intensive care units (which exceeds the t-TDI value). It has also to be noted that the values of the upper part of the intake distributions (95%, max)

© Burleigh Dodds Science Publishing Limited, 2025. All rights reserved.

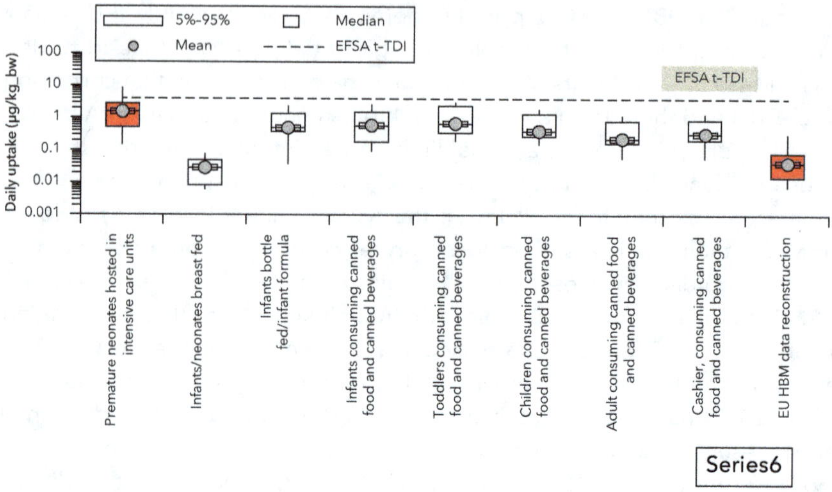


**Figure 1** Daily uptake for the most important exposure scenario combinations. The reference dose is 4 μg/kg_bw/d (EFSA t-TDI).

are the result of the highest food residues identified in some studies from the literature review. These values do not significantly affect the means of the food residue distribution; yet they result in significant deviation from the mean in the overall distribution, thereby reducing the margin of safety for a small part of the population. Nevertheless, we need to take also into account that these estimates are quite conservative. In practice, it is very unlikely that someone consumes only canned food and beverages.

To associate the risk of the several exposure scenarios based on the Biologically Effective Dose (BED) derived by the PBTK model, two different exposure metrics were used:

1 The EFSA t-TDI of 4 μg/kg_bw/d (EFSA 2015) was translated into internal concentration levels, found to correspond to a concentration of 0.013 μg/L of free plasma BPA.
2 The use of internal dosimetry metrics allows the use of in vitro toxicological data for risk characterization. In this case, instead of translating an external reference dose such as the EFSA t-TDI (obtained from in vivo animal NOAEL extrapolation), an in vitro reference dose was used. The dose used is the biological pathway altering dose (BPAD), which is analogous to current risk assessment metrics in that it combines dose-response data with analysis of uncertainty and population variability to derive exposure limits (Judson et al., 2010; Judson et al., 2011). The analogy is closest when perturbation of a pathway is a key event in the mode of action leading to a specified adverse outcome. The ToxCast BPA in vitro assays provided six ER agonist or binding AC50 values for BPA,

© Burleigh Dodds Science Publishing Limited, 2025. All rights reserved.

ranging from 0.6 to 1.7 μM. To calculate a conservative BPAD, the lowest ToxCast AC50 was selected, which is 0.64 μM for Attagene Factorial cis ERE assay (Judson et al., 2010, 2011). Incorporating uncertainty factors related to population response to xenobiotics, two different values are produced, namely the $BPAD_{99}$, which is the permissible exposure level that accounts for population variability, and $BPADL_{99}$, which is the permissible exposure level additionally accounting for uncertainty. By using the reverse TK approach that accounts for the concentration at steady state divided by the dose rate, the respective estimated population parameters give a $BPAD_{99}$ of 0.44 μg/kg_bw/d, with lower one-sided confidence limit, $BPADL_{99}$, of 0.16 mg/kg/day (Judson et al., 2011). Using these external exposure values in our PBTK model, we derive equivalent internal doses of 1.44 and 0.52 μg/L respectively. These concentrations are almost two orders of magnitude higher than the BED derived from the EFSA t-TDI (0.013 μg/L).

Evaluating risk based on internal exposure estimates resulted in reduced margins of safety for the scenarios involving neonates and infants; in this case, the upper distribution parts of internal exposure are closer to the respective BED derived from the EFSA t-TDI, but significantly lower compared to the BED that corresponded to the former TDI of 50 μg/kg_bw/d. Moreover, the median of the concentration of free BPA in the plasma as reckoned from the HBM data in Europe is far below the internal dose reference value (0.013 μg/L) derived by the EFSA t-TDI. This outcome corresponds to the lower estimates of the already considered exposure scenarios (Fig. 2).

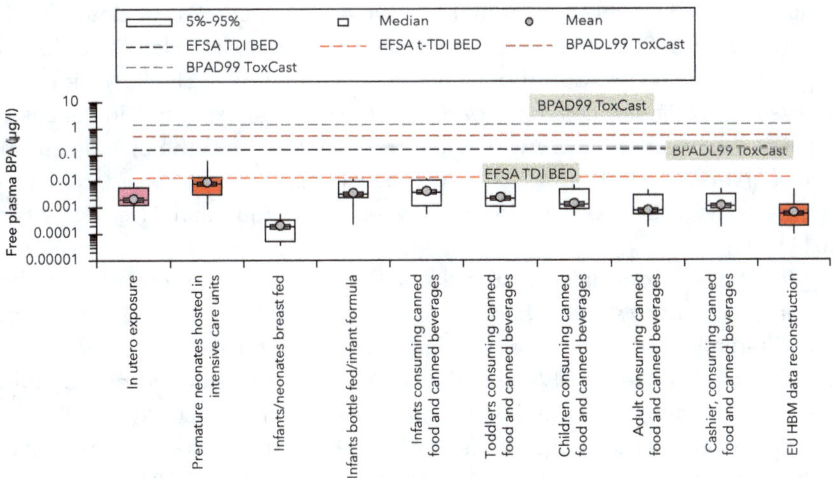


**Figure 2** Free plasma BPA under most important plausible exposure scenario combinations (refined internal exposure analysis).

© Burleigh Dodds Science Publishing Limited, 2025. All rights reserved.

Incorporation of internal dosimetry alters the overall exposure assessment outcome when age- and route-dependent variations are considered in the actual estimation of internal response. Thus, when TK is taken into account, the outcome of specific exposure scenarios such as premature neonates hosted in intensive care units is differentiated with respect to the other plausible exposure scenarios considered by EFSA; accounting for the immaturity of the detoxification pathway results in a higher probability of neonates to be exposed to an internal dose higher than the BED derived from EFSA t-TDI. Other exposure scenarios, where exposure outcome changes when age- and route-dependent differences in internal dose are taken into account, include neonates/infants consuming milk formula and canned food and beverages. However, for these scenarios, the internal dose remains significantly lower than the BED derived from the EFSA t-TDI.

Using the BPAD derived from in vitro BPA toxicity assessment as the internal exposure reference value, the maximum derived internal exposure values of the worst-case exposure scenarios (premature neonates) are ten times lower than the $BPADL_{99}$. This shows that there is no reason for concern based on either individual or aggregate scenarios of BPA exposure.

Another major advantage of incorporating TK in the risk assessment process is the capability to re-run in a forward manner the PBTK model, accounting for all potential parameters that induce bioavailability differences. Briefly, these include age, route of exposure, and genetic variability. The effect of age is related to differences in (a) physiological parameters such as tissue composition, blood flows, cardiac and respiratory rate and (b) the maturity of the conjugation activity. Taking into account the physiologically-based approach for scaling to children (Edginton et al., 2006) and the most recent findings regarding the ontogeny of enzymes involved in BPA detoxification (Court et al., 2012; Leeder, 2009) an age-dependent bioavailability difference factor of 2–3.5 is considered between infants and adults. Similarly, significant bioavailability differences occur based on the route of exposure. In the case of inhalation, BPA enters directly into the bloodstream from the alveoli, and the lack of first-pass metabolism results in bioavailability differences up to six times compared to a similar dose administrated orally (Sarigiannis & Karakitsios, 2011).

In the case of dermal exposure, the lack of first-pass metabolism is somehow decompensated by slower absorption (Zalko et al., 2011). Finally, a major factor that differentiates BED within a population is the presence of genetic variants of key enzymes involved in BPA conjugation; *in vitro* kinetic studies have identified that D85Y substitution in UGT2B15 decreases enzymatic function (Hanioka et al., 2011) and that the polymorphic alleles of UGT2B15 are translated into variations in the metabolism of BPA (Partosch et al., 2013). The effect of all the parameters described above is summarized in Fig. 3, where for the same body weight

© Burleigh Dodds Science Publishing Limited, 2025. All rights reserved.

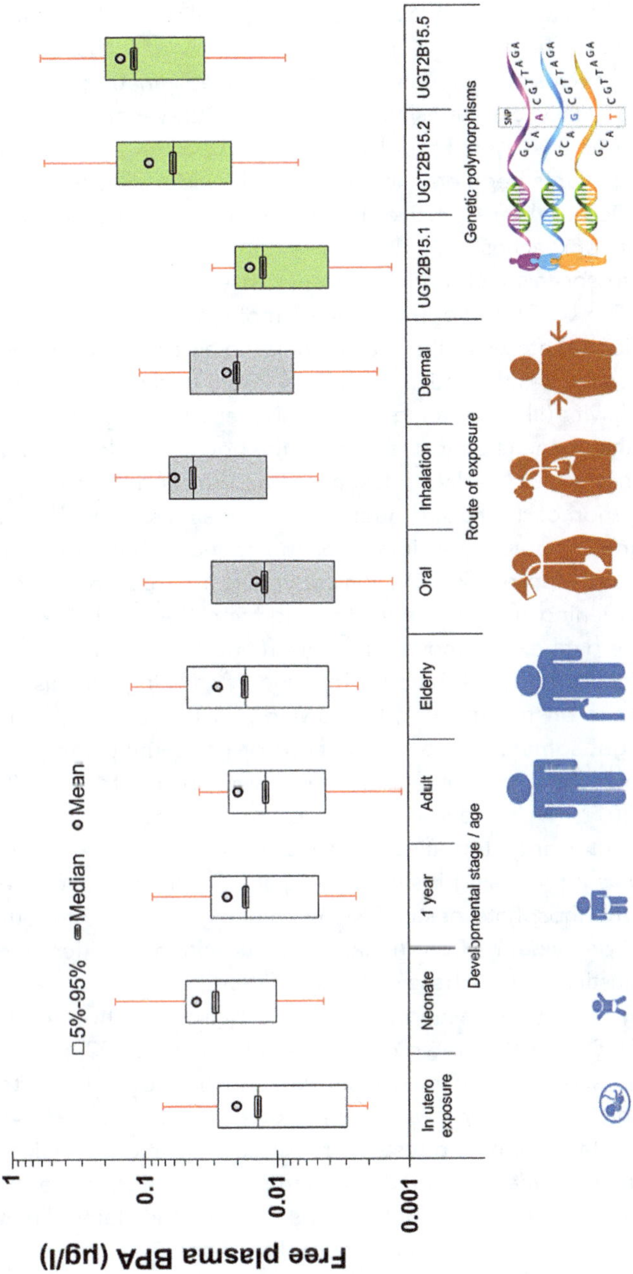


**Figure 3** Understanding the major parameters inducing bioavailability differences.

© Burleigh Dodds Science Publishing Limited, 2025. All rights reserved.

normalized dose, free plasma BPA is estimated accounting for differences in age, route of exposure and genetics. The combination of early infancy and the presence of alleles associated with slower metabolism could result in a BED in a neonate or an infant up to 15 times higher compared to that of an adult. This leads to the conclusion that a phenomenally limited variability in biomonitored data could be realistically translated into a wider BED variability. Had additional information (such as genetic polymorphisms) become available or younger children and infants been enrolled in the study, wider differences would have been identified in the respective risk characterization ratio outcomes among the different approaches.

Risk assessors also rely on TK data when assessing the combined risks from pesticides. The US EPA and EFSA, for example, have developed methods to group pesticides into what they call cumulative assessment groups – namely, clusters of pesticides that behave similarly in the body. Usually, these groups rely on additivity (like summing hazard quotients), though there is growing awareness that if one chemical remains in the body much longer than another, that difference should really be taken into account if not already considered in the estimation of the hazard quotients. In certain cases, PBTK models even step in to modify potency estimates or run sequential exposure simulations. Sarigiannis and Hansen (2012) even proposed a tiered approach: start with an initial screening using a basic dose addition (hazard index), and then – where ample data exist – move into more refined mechanistic modeling that captures kinetics and interactions. This line of thinking has also influenced EU discussions on mixture risk, acknowledging that a one-size-fits-all dose addition might sometimes be too cautious or not cautious enough, whereas PBTK models offer a more nuanced, mechanistic look at the risks (Sarigiannis & Hansen, 2012). We're now beginning to see early moves – like the tentative steps of EFSA toward integrating PBTK models in cumulative risk assessments – which marks a promising if still evolving shift in the approach (Testai et al., 2021). Organophosphate pesticides sometimes get lumped together or split apart based on evidence from metabolic competition – indeed, the data can sometimes justify treating these chemicals differently.

Phasing out animal testing is pushing a big shift in how we look at TK. The new EU Chemicals Strategy for Sustainability from 2020, for instance, is inducing the scientific and regulatory communities toward smarter ways to assess chemicals with way fewer animal studies (Reale et al., 2024). This shift means that data from in vitro tests and in silico simulations – which we talked about before when mentioning IVIVE – start playing a bigger role in regulatory dossiers. In many cases, TK modeling ends up being the element that connects laboratory assay data with human risk considerations.

Take, for example, a scenario where a company argues that a new food additive is safe because it has a liver toxicity test done in vitro plus a PBPK

© Burleigh Dodds Science Publishing Limited, 2025. All rights reserved.

model backing it up. Regulators will then have the trustworthiness of the model being examined (most probably by risk assessors from national authorities) and check the steps used in IVIVE. The Organization for Economic Co-operation and Development (OECD) is already delivering guidance in this area (OECD, 2021). Models that tick all the boxes – with a clear structure, believable biological parameters, and validation against real data – are more likely to be welcome as solid evidence.

Overall, risk assessors are cautious but gradually accepting TK modeling in food chemical evaluations. High-confidence models are used here and there to fine-tune risk assessments – whether it is about considering interspecies differences, setting biomonitoring-based standards, or dealing with cumulative effects. Increasingly, advisory panels of risk assessors/EFSA/national risk assessment bodies are inviting TK experts to the table so that model outputs are not misinterpreted.

## 7.2 The role of generic physiologically-based toxicokinetic models in accelerating their regulatory acceptance

The emergence of generic PBTK models greatly advances their utilization in regulatory risk assessment of situations where chemical substances are found in food, e.g. not complying with the existing legal EU standards or Codex Alimentarius standards, due to their standardized, reusable nature, offering consistent, transparent, and scientifically sound methods. These generic models minimize the complexity and variability associated with developing specific models for each compound, facilitating the quick application to various chemicals, thereby streamlining the risk assessment process and improving regulatory decision-making. Generic PBTK models use common physiological and biochemical parameters, such as organ volumes, blood flows, and metabolic enzyme activities, applicable to a wide range of substances. This universal approach aids regulatory reviews by providing a familiar, validated framework for assessing new data. Agencies encompassing risk assessors and/or risk managers like the US EPA, EFSA, and FDA increasingly support the use of generic PBTK models for chemical risk assessments, especially when specific data is scarce or lacking for certain compounds (Grimstein et al., 2018; Tan et al., 2018). Some of the generic PBTK models used so far are presented below:

- The INTEGRA generic PBTK model (Sarigiannis et al., 2014, 2016) covers diverse chemicals including BPA, benzene, phthalates, pesticides, and perfluorinated compounds. INTEGRA has been implemented as an online resource available to the scientific, regulatory, and industrial community. It includes already preexisting data, but users are also able to upload and use their own confidential datasets. The software encompasses major ADME

© Burleigh Dodds Science Publishing Limited, 2025. All rights reserved.

processes across all life stages, emphasizing pregnancy and mother-fetus interactions. Organs and blood compartments (arterial, venous, portal) are explicitly represented. Each organ compartment includes distinct equations for red blood cells, plasma/interstitial tissue, and cells. Liver and brain compartments are subdivided to reflect enzyme distributions and regional permeability. Metabolism follows saturable kinetics primarily in the liver but includes other tissues such as skin, brain, and placenta. Routes of exposure (inhalation, oral, dermal) are comprehensively modeled, including particle deposition, gastrointestinal absorption, and dermal permeability. Parameterization combines experimental data and QSAR models, covering 136 chemicals and continues to expand.

- The IndusChemFate generic model developed by Jongeneelen and Berge (2011) focuses on neutral and ionic organic compounds. It incorporates eleven body compartments: lungs, heart, liver, kidneys, skin, adipose tissue, muscles, bones, stomach/intestines, brain, and bone marrow. Exposure routes modeled include inhalation, dermal absorption, and oral intake, featuring detailed algorithms for dermal diffusion and gastrointestinal absorption. Metabolism occurs mainly in the liver following Michaelis–Menten saturable kinetics but can be considered in all compartments. The model incorporates enterohepatic circulation and explicitly describes renal clearance, urinary excretion, and exhalation based on compound volatility.
- The generalized PBTK model MENTOR-3P, adapted by Georgopoulos and Lioy (2006), simulates arsenic dynamics, tracking arsenates, arsenites, and their metabolites (MMA, DMA). This flow-limited model features defined body compartments and accounts explicitly for dietary and water exposure. Metabolism sites and physiological distributions are carefully modeled, capturing realistic physiological dynamics for accurate arsenic risk assessment.
- MERLIN-Expo's PBTK model (Ciffroy et al., 2016) includes 22 compartments, suitable for inorganic compounds and persistent organic pollutants. It considers detailed physiological and biochemical variations during individual development. The model describes inhalation and ingestion absorption processes explicitly, including alveolar gas exchange and gastrointestinal partitioning. Metabolism is represented by saturable or linear kinetics across multiple compartments, excluding gut lumen and alveolar space. Excretion occurs via urine, feces, bile, and milk. Validation is performed through comparison with experimental data.

However, determining chemical-specific parameters, like metabolism rates, partition coefficients, and absorption rates in generic PBTK models, presents challenges. Traditionally, parameter estimation relies on experimental methods

© Burleigh Dodds Science Publishing Limited, 2025. All rights reserved.

that are costly and time-consuming. Recent advancements in AI and ML offer promising solutions to expedite this critical step (Hartung, 2016; Madden et al., 2019).

AI methods, particularly supervised learning algorithms and deep neural networks, use vast public or private databases of existing physicochemical, biochemical, and TK data to accurately predict essential PBTK parameters. Machine learning models trained on organized chemical datasets can uncover intricate connections between molecular descriptors and TK behaviors, providing reliable parameter estimates for diverse chemicals without exhaustive empirical testing.

Furthermore, AI-powered QSPR models speed up parameter estimation by predicting properties like partition coefficients and intrinsic clearance rates directly from chemical structures (Singh et al., 2024). Advanced neural network architectures, such as convolutional (Xu et al., 2017) and graph neural networks, capture meaningful representations of chemical structures, identifying complex molecular interactions vital for accurate parameter prediction.

Artificial intelligence also enables the continuous updating and refinement of generic PBTK models (Chou & Lin, 2022). With AI systems' continuous learning capabilities, models remain precise and responsive to new data, enhancing their predictive performance over time. This flexibility boosts regulatory confidence by ensuring models are maintained at cutting-edge levels, supporting timely and accurate risk assessments.

Integrating AI into generic PBTK modeling frameworks enhances transparency and reproducibility. AI-based parameter estimation provides detailed documentation and metadata, enabling regulators to thoroughly evaluate model assumptions and uncertainties. Improved transparency fosters regulatory review and public trust, essential for widespread acceptance.

In conclusion, the development of generic PBTK models speeds their adoption in regulatory risk assessment by providing standardized, validated tools, assuming that the scientific community of risk assessors agrees on a certain degree of standardization. Simultaneously, advancements in AI greatly accelerate model parameterization, reducing reliance on resource-intensive experimental methods, improving predictive accuracy, and increasing transparency and regulatory confidence.

## 8 Future trends: next-generation risk assessment and other developments

Looking ahead, TK modeling is likely to jump right to the center of next-generation risk assessment (also known as NGRA or next generation safety assessment). The NGRA is about piecing together risks in a more holistic, human-centered way – using a mix of mechanistic insights, in vitro work, computer

© Burleigh Dodds Science Publishing Limited, 2025. All rights reserved.

models, and human data like biomonitoring and epidemiology (Reale et al., 2024). The aim here is to complete risk assessments faster, depend less on animal tests, and better capture complex, real-world exposures. Toxicokinetics – and PBTK models, in particular – turn out to be a key piece of this puzzle, linking experimental findings with the regulatory decision-making process. Imagine a system that knits together various streams of information, linking what enters the body with its actual effects on physiological systems. Next-generation risk assessment is not just about measuring external doses – it relies heavily on what happens once chemicals are inside the human body. Instead of merely measuring what humans are exposed to, this approach puts the spotlight on internal doses merging data from multiple routes and substances. Thus, when risk assessors decide on safety, they are increasingly checking if specific biomarkers or tissue concentrations exceed a flexible threshold – one that is informed by experimental models and laboratory tests. Reale et al. (2024) even proposed a concept diagram that shows how risk assessment is drifting away from the old animal test and external dose routines to a model where PBTK modeling ties exposure seamlessly to effect. In practice, this could well lead to more routine biomonitoring to keep an eye on internal levels as is currently under consideration under the One Substance One Assessment legislative package of the European Commission (EC, 2023). Indeed, some agencies are already exploring the idea of setting biomonitoring equivalents not just for individual chemicals but for a cumulative index of exposures.

Cumulative and aggregate risk assessments are bound to lean on TK models as chemical safety and consumer protection policies evolve. When regulators start demanding that we account for combined exposures – as it happens under the US watch on "cocktail effects" from pesticides and other contaminants – we really need tools that can sum up or compare internal doses in a straightforward manner. A forward-thinking idea is to harness PBTK models to hammer out a hazard index that is based on actual internal concentrations rather than just what is applied externally. So, if several chemicals induce hepatotoxicity, the contribution of each substance could be simulated and then all summed together to see if they cross a safety threshold. Sarigiannis and Hansen (2012) have described that as the "connectivity approach," which blends mechanistic insights from various scales into one picture. Even though such methods are not yet the everyday norm, early pilot studies and evolving risk assessments seem to suggest they will eventually push future guidelines in that direction.

Then there is the whole idea of the exposome – a concept that covers every exposure from conception, including everyday lifestyle choices and the body's own internal processes. Toxicokinetic modeling serves as a kind of quantitative backbone here, turning external exposome data into a clearer picture of what is really happening within the human body in terms of distribution and

© Burleigh Dodds Science Publishing Limited, 2025. All rights reserved.

biotransformation of xenobiotics. Large projects in Europe, like HEALS (Health and Environment-wide Associations based on Large Population Surveys), have already shown how lifelong PBTK models can be coupled with environmental exposure data to unpack health outcomes (Sarigiannis et al., 2020). Likewise, initiatives such as PARC (Partnership for the Assessment of Risks from Chemicals) are experimenting with similar integrative strategies (Marx-Stoelting et al., 2023). In PARC, researchers are exploring these models further – even if the connections between external exposures and internal burdens sometimes come across as immature. All of this, imperfect as it may be at times, is gradually laying the groundwork for a future where our risk assessments are as much about what is happening inside the human body as about what happens out in the field. In a 2024 report, Gastellu et al (2025) compiled a range of physiologically-based kinetic models spanning various life stages (Gastellu et al., 2025). They picture a time when authorities will not just claim "this food level is safe" but will explain that it produces a blood concentration of X – typically a value well below levels that might perturb pathways like A, B, or even C. Getting there calls for a lot of model building, careful double-checking, and mixing data from different sources; still, the first steps are already under way.

A very interesting application showcasing how TK models support chemical risk assessment, especially assimilating high-throughput data such as the ones from ToxCast21, is the study on BPA by Sarigiannis et al. (2016). The assessment discussed highlights the critical role of incorporating TK into modern chemical risk assessments, using BPA exposure as a case study. Traditional risk evaluations, primarily based on external exposure levels, generally suggested acceptable risk; however, adding TK data refined these assessments by accounting for internal dosimetry, offering crucial insights, particularly for sensitive groups like premature babies and infants. This method draws attention to potential risks linked to age-specific metabolic processes, as these younger populations have less developed detoxification pathways, which result in lower safety margins compared to assessments based solely on external exposures. By using internal dose metrics, the risk evaluations became more precise, converting external TDIs into biomonitoring equivalents, which enabled direct comparisons with HBM data. The results indicated that even in worst-case internal exposure scenarios, levels remained below concerning thresholds as identified by the EFSA t-TDI and BPAD from in vitro studies. Thus, TK is crucial for accurately evaluating chemical risks, especially for vulnerable populations, providing a clearer understanding of the real-world implications of chemical exposures.

On another front, using TK models can really support emergency reactions. Imagine a case where a new contaminant suddenly appears in the food supply – a case of adulteration or even an accident. A robust PBTK model, put together carefully, can rapidly give an idea of how much of the substance ends up in

© Burleigh Dodds Science Publishing Limited, 2025. All rights reserved.

the human population, and whether those amounts verge on dangerous. Regulators might then issue advice on consumption and take other swift risk management measures. Moreover, thanks to widely available generic PBTK tools and extensive parameter databases, even chemicals that have not been studied in much depth can get a first-cut model built in just hours or a few days. In most cases, this kind of proactive modeling looks set to become a regular part of the process as confidence in the tools grows steadily.

From the policy perspective, there is a clear move toward accepting TK models in regulatory circles based on my personal experience. A number of guidance documents and case studies now show real-world success with these approaches. For roughly twenty years, bodies like the International Programme on Chemical Safety (IPCS) and OECD have promoted the use of these models, but only recently have we seen more solid frameworks emerging. For instance, the WHO/IPCS guidance from 2014 even talks about handling model uncertainty (WHO/IPCS, 2014), while the OECD guidelines from 2021 offer straightforward reporting templates (OECD, 2021). When regulators see a model that sticks to these best practices – and that is backed by human data – they tend to be much more willing to accept its findings. At the same time, there is an understanding that a badly built model can mislead decisions. So, transparency, some peer review, and even early evaluation steps (akin to what happened with QSAR models validated under OECD principles) might become essential, perhaps leading to a kind of certification process down the line.

Focusing on food safety, there is a forward-thinking idea to use these models for optimizing dietary exposures and assessing cumulative impacts. Future dietary guidelines would incorporate knowledge about how contaminants behave in the body – for example, suggesting that people space out their consumption of certain fish to avoid sudden mercury spikes. If models show that eating all high-mercury fish in one short period creates a temporary spike in blood mercury that eventually drops, whereas spreading consumption keeps the levels smoother, then that kind of advice could really cut down risk while still offering nutritional benefits (Katsonouri et al., 2023). In a way, this strategy moves beyond simple regulatory limits on foods and enters the realm of public health advice, blurring the lines between assessing risks and managing them. Ultimately, the aim is to use our growing kinetic understanding to protect human health as effectively as possible when exposure to these substances through consumption of food is unavoidable.

To sum up, the use of TK modeling by risk assessors is steadily rising (Andersen, 2003; Tan et al., 2020). We are moving away from occasional, case-by-case assessments toward a future where internal dose modeling becomes an everyday part of risk assessments for all key foodborne chemicals. Sarigiannis and his colleagues have really helped shake up the field. Their work brings together exposure data with PBTK frameworks aimed at addressing cumulative

© Burleigh Dodds Science Publishing Limited, 2025. All rights reserved.

risks (Sarigiannis et al., 2011). Then, they went on to build general-purpose TK models including the respective operating procedures and put them against real-world biomonitoring data (Sarigiannis et al., 2020). They even mapped out smart, strategic plans for handling the risks from chemical mixtures (Sarigiannis & Hansen, 2012). All these efforts are pushing risk assessors toward more self-driven assessments instead of the old checklist routine. In many cases, the idea is that TK modeling is not just an extra tool; it is becoming the central pillar in chemical risk assessment – a shift already hinted at in NGRA initiatives that will probably end up backed by future guidance and part of the risk assessment process.

## 9 Conclusion

Toxicokinetic modeling of chemical residues in food has come to be considered essential for unraveling health risks. It works by linking external exposures, like what we eat, directly to the doses our tissues take in – basically serving as that "missing link" in our risk-thinking. This discussion has shown that mixing TK with exposure assessments opens the door to a more comprehensive evaluation. It captures how exposures add up across different routes, how mixtures can interact cumulatively, and how individual differences among populations really count – all of which makes food safety assessments better predict actual (adverse) health effects. Modern PBTK models now wander across life stages and process multiple chemicals. They allow scientists to measure differences between populations instead of just assuming them. In combination with both laboratory (*in vitro*) and computer-simulated (*in silico*) data to fill the gaps via IVIVE, plus emerging AI and Bayesian algorithms that sharpen precision and boost model efficiency, they are geared to bring about a new era of risk analysis. These advances should be included in the current methodology of authorization of chemical substances in food and animal feed. These methods should also be included in the safety evaluations of food safety issues (local or regional) that are usually done by national risk assessment authorities and EFSA. There, TK insights help shape safety guidelines, form the backbone of cumulative risk frameworks, and support how we interpret biomonitoring data. Research on chemical management has shown us, for example, that looking at cumulative chemical risks is no small feat. Risk assessment is changing rapidly – toward adoption of methods that are more predictive, mechanistic, and protective. Toxicokinetic modeling has rapidly become a key player in this paradigm shift. In the near future, evaluating exposure to foodborne chemicals will probably rely more on internal dose metrics and biologically rooted models, alongside the classic exposure estimates and toxicity endpoints. This revised approach lets us pose deeper, more subtle questions: are chemical mixtures we consume likely to interact in unexpected ways? Do fast metabolizers get the same level

© Burleigh Dodds Science Publishing Limited, 2025. All rights reserved.

of protection as slow ones? And what exactly do falling levels of a persistent pollutant in blood really tell us about the success of our interventions? Models that merge exposure data with kinetic behavior offer a solid, quantitative framework to answer these puzzles.

When we consider the course of TK modeling, the field is moving toward greater integration, sophistication, and overall impact. Recent breakthroughs – whether they are advancements in PBTK modeling, population studies the data of which are analyzed with Bayesian algorithms, new IVIVE methods, or even early AI integration – have jointly helped to establish a more evidence-driven and human-centric approach to analyzing chemical residues in our food. In most cases, future work appears set to better capture human variability (e.g. by weaving in genetic and microbiome factors), broaden high-throughput modeling to span the vast chemical universe, and tighten the bond between TK and adverse outcome pathways for improved risk predictions from internal doses.

Overall, these developments mean that risk assessors, regulators, food producers, and food producing industries will soon be better equipped to secure food safety in an era of complex exposures and rapid innovations across the food and chemical industries. By embracing TK modeling together with exposure science, we are moving toward ensuring that those regulatory limits genuinely protect every member of our population with robust, evidence-based measures. The merging of exposure assessments with insights from human TK marks a significant, if not a game-changing, evolution in how we guard public health against the countless chemical residues found in our everyday diet.

## 10 References

Alhaji Saganuwan, S. (2021). Application of modified Michaelis – Menten equations for determination of enzyme inducing and inhibiting drugs. *BMC Pharmacology and Toxicology*, *22*, 29. https://doi.org/10.1186/s40360-021-00521-x

Andersen, M. E. (2003). Toxicokinetic modeling and its applications in chemical risk assessment. *Toxicology Letters*, *138*(1–2), 9–27. https://doi.org/10.1016/s0378-4274(02)00375-2

Béchaux, C., et al. (2014). Improving dietary exposure models by imputing biomonitoring data through ABC methods. *The International Journal of Biostatistics*, *10*(2), 277–287. https://doi.org/10.1515/ijb-2013-0062

Breen, M., et al. (2021). High-throughput PBTK models for in vitro to in vivo extrapolation. *Expert Opinion on Drug Metabolism & Toxicology*, *17*(8), 903–921. https://doi.org/10.1080/17425255.2021.1935867

Caldwell, J. C., et al. (2012). Evaluation of evidence for association of human bladder cancer with drinking-water chlorination disinfection by-products. Environmental Health Perspectives, 120(10), 1551–1561. https://doi.org/10.1155/2012/852384

Chang, X., et al. (2022). IVIVE: facilitating the use of in vitro toxicity data in risk assessment and decision making. *Toxics*, *10*(5), 232. https://doi.org/10.3390/toxics10050232

© Burleigh Dodds Science Publishing Limited, 2025. All rights reserved.

Choi, G. W., et al. (2019). Interpretation of non-clinical data for prediction of human: in vitro-in vivo extrapolation and allometric scaling. *Pharmaceutics*, *11*(4), 168. https://doi.org/10.3390/pharmaceutics11040168

Chou, W.-C. & Lin, Z. (2022). Machine learning and artificial intelligence in physiologically based toxicological sciences. *Toxicological Sciences*, *191*, 1–14. https://doi.org/10.1093/toxsci/kfac101

Ciffroy, P., et al. (2016). Modelling the exposure to chemicals for risk assessment: a comprehensive library of multimedia and models for integration, prediction, uncertainty and sensitivity analysis – the MERLIN-Expo tool. *Science of The Total Environment*, *568*, 770–784. https://doi.org/10.1016/j.scitotenv.2016.03.191

Cohen Hubal, E. A., et al. (2019). Advancing internal exposure and physiologically-based toxicokinetic modeling for 21st-century risk assessments. *Journal of Exposure Science & Environmental Epidemiology*, *29*(1), 11–20. https://doi.org/10.1038/s41370-018-0046-9

Court, M. H., et al. (2012). Quantitative distribution of mRNAs encoding the 19 human UDP-glucuronosyltransferase enzymes in 26 adult and 3 fetal tissues. *Xenobiotica*, *42*(3), 266–277. http://www.scopus.com/inward/record.url?eid=2-s2.0-84863042741&partnerID=40&md5=b90a1c78da52024ebbaf42b6971bbfe1

EC. (2023). *Regulation of the European Parliament and of the Council establishing a common data platform on chemicals, laying down rules to ensure that the data contained in it are findable, accessible, interoperable and reusable and establishing a monitoring and outlook framework for chemicals*, Commission E., ed. EC.

Edginton, A. N., et al. (2006). A mechanistic approach for the scaling of clearance in children. *Clinical Pharmacokinetics*, *45*(7), 683–704. http://www.scopus.com/inward/record.url?eid=2-s2.0-33745599327&partnerID=40&md5=325be3478d1e4b34036a2e60034d0e5a

EFSA. (2015). Scientific Opinion on the risks to public health related to the presence of bisphenol A (BPA) in foodstuffs. *EFSA Journal*, *13*, 3978.

Fisher, J. W., et al. (2011). http://www.scopus.com/inward/record.url?eid=2-s2.0-81255149595&partnerID=40&md5=0b43eb359475e86e38616496a2a49dd8

Ganzleben, C., et al. (2017). Human biomonitoring as a tool to support chemicals regulation in the European Union. *International Journal of Hygiene and Environmental Health*, *220*(2 Pt A), 94–97. https://doi.org/10.1016/j.ijheh.2017.01.007

Gastellu, T., et al. (2025). A comprehensive library of lifetime physiological equations for models: enhancing dietary exposure modeling with mercury as a case study. *Environmental Research*, *265*, 120393. https://doi.org/10.1016/j.envres.2024.120393

Georgopoulos, P. G. & Lioy, P. (2006). From a theoretical framework of human exposure and dose assessment to computational system implementation: the modeling ENvironment for TOtal Risk Studies (MENTOR). *Journal of Toxicology and Environmental Health, Part B*, *9*, 457–483. https://doi.org/10.1080/10937400600755929

Georgopoulos, P. G., et al. (2009). Reconstructing population exposures to environmental chemicals from biomarkers: Challenges and opportunities. *Journal of Exposure Science and Environmental Epidemiology*, *19*(2), 149–171. http://www.scopus.com/inward/record.url?eid=2-s2.0-58449099575&partnerID=40&md5=ff34389e15004292a8116d3deb94baef

Grimstein, M., et al. (2018). Physiologically based pharmacokinetic modeling in regulatory science: An update from the U.S. Food and Drug Administration's Office of Clinical


© Burleigh Dodds Science Publishing Limited, 2025. All rights reserved.

Pharmacology. Journal of Pharmaceutical Sciences, 108(1), 21–25. https://doi.org/10.1016/j.xphs.2018.10.033

Hanioka, N., et al. (2011). Effect of UDP-glucuronosyltransferase 2B15 polymorphism on bisphenol A glucuronidation. *Archives of Toxicology*, *85*(11), 1373–1381. https://doi.org/10.1007/s00204-011-0690-5

Hartung, T. (2016). Making big sense from big data in toxicology by read-across. *Altex*, *33*(2), 83–93. https://doi.org/10.14573/altex.1603091

Huuskonen, P., et al. (2022). Health risk assessment of ortho-toluidine utilising human biomonitoring data of workers and the general population. *Toxics*, *10*(5), 217. https://www.mdpi.com/2305-6304/10/5/217

Jongeneelen, F. J. & Berge, W. F. (2011). A generic, cross-chemical predictive PBTK model with multiple entry routes running as application in MS Excel; design of the model and comparison of predictions with experimental results. *The Annals of Occupational Hygiene*, *55*(8), 841–864. https://doi.org/10.1093/annhyg/mer075

Judson, R. S., et al. (2010). In vitro screening of environmental chemicals for targeted testing prioritization: the ToxCast project. Environ. *Health Perspec*. 118: 484–492.

Judson, R. S., et al. (2011). Estimating toxicity-related biological pathway altering doses for chemical risk assessment. *Chem Res. Toxicol*. 24: 451–462.

Katsonouri, A., et al. (2023). HBM4EU-MOM: Prenatal methylmercury-exposure control in five countries through suitable dietary advice for pregnancy–Study design and characteristics of participants. *International Journal of Hygiene and Environmental Health*, *252*, 114213. https://doi.org/10.1016/j.ijheh.2023.114213

Kreutz, A., et al. (2024). Advancing understanding of human variability through toxicokinetic modeling, in vitro-in vivo extrapolation, and new approach methodologies. *Hum Genomics*, *18*(1), 129. https://doi.org/10.1186/s40246-024-00691-9

Leeder, J. S. (2009). Developmental pharmacogenetics: a general paradigm for application to neonatal pharmacology and toxicology. *Clinical Pharmacology and Therapeutics*, *86*(6), 678–682. http://www.scopus.com/inward/record.url?eid=2-s2.0-70450270796&partnerID=40&md5=e4f08026895f371cac29fe5e9f64d071

Lilienblum, W., et al. (2008). Alternative methods to safety studies in experimental animals: role in the risk assessment of chemicals under the new European Chemicals Legislation (REACH). *Archives of Toxicology*, *82*(4), 211–236. https://doi.org/10.1007/s00204-008-0279-9

Madden, J. C., et al. (2019). In silico resources to assist in the development and evaluation of physiologically-based kinetic models. *Computational Toxicology*, *11*, 33–49. https://doi.org/10.1016/j.comtox.2019.03.001

Marin, J. J. G., et al. (2020). Role of genetic variations in the hepatic handling of drugs. *International Journal of Molecular Sciences,International Journal of Molecular Sciences 21*(8), 2775. https://doi.org/10.3390/ijms21082884

Marx-Stoelting, P., et al. (2023). A walk in the PARC: developing and implementing 21st century chemical risk assessment in Europe. *Archives of Toxicology*, 97(7), 2069–2084. https://doi.org/10.1007/s00204-022-03435-7

McNally, K., et al. (2011). A workflow for global sensitivity analysis of physiologically based pharmacokinetic models. *Frontiers in Pharmacology*, *2*, 31. https://doi.org/10.3389/fphar.2011.00031

OCED. (2021). *Guidance document No. 331, Environment, health and safety, environment directorate*. OECD.

© Burleigh Dodds Science Publishing Limited, 2025. All rights reserved.

Orak, N. H., et al. (2019). Bayesian network-based framework for exposure-response study design and interpretation. *Environmental Health*, *18*(1), 23. https://doi.org/10.1186/s12940-019-0461-y

Ougier, E., et al. (2021). Human biomonitoring initiative (HBM4EU): Human biomonitoring guidance values (HBM-GVs) derived for bisphenol A. *Environment International*, *154*, 106563. https://doi.org/10.1016/j.envint.2021.106563

Papadaki, K., et al. (2017). Modeling of adipose/blood partition coefficient for environmental chemicals. *Food and Chemical Toxicology*, *110*, 274–285. https://doi.org/10.1016/j.fct.2017.10.044

Partosch, F., et al. (2013). Functional UDP-glucuronyltransferase 2B15 polymorphism and bisphenol A concentrations in blood: results from physiologically based kinetic modelling. *Archives of Toxicology*, 87(2), 295–302. http://www.scopus.com/inward/record.url?eid=2-s2.0-84873517244&partnerID=40&md5=20e3d660513927f7640f61436600a0ac

Pearce, R. G., et al. (2017). httk: R package for high-throughput toxicokinetics. *Journal of Statistical Software*, *79*(4), 1–26. https://doi.org/10.18637/jss.v079.i04

Pruett, W. & Hester, R. (2016). The creation of surrogate models for fast estimation of complex model outcomes. *PLoS One*, *11*, e0156574. https://doi.org/10.1371/journal.pone.0156574

Qing, Y., et al. (2021). Bayesian toxicokinetic modeling of cadmium exposure in Chinese population. *Journal of Hazardous Materials*, *413*, 125465. https://doi.org/10.1016/j.jhazmat.2021.125465

Reale, E., et al. (2024). Human biomonitoring and toxicokinetics as key building blocks for next generation risk assessment. *Environment International*, *184*, 108474. https://doi.org/10.1016/j.envint.2024.108474

Rowland Yeo, K., et al. (2024). Modeling: state-of-the art and future. *Clinical Pharmacology & Therapeutics*, *116*(3), 563–576. https://doi.org/10.1002/cpt.3289

Sarigiannis, D. (2019). The HEALS project. In S. Dagnino & A. Macherone (Eds.), *Unraveling the exposome: a practical view* (pp. 405–422). Springer. https://doi.org/10.1007/978-3-319-89321-1_16

Sarigiannis, D. & Gotti, A. (2008). Biology-based dose-response models for health risk assessment of chemical mixtures. *Fresenius Environmental Bulletin*, *17*(9 B), 1439–1451.

Sarigiannis, D. & Hansen, U. (2012). Considering the cumulative risk of mixtures of chemicals: a challenge for policy makers. *Environmental Health: A Global Access Science Source*, *11*(SUPPL.1), S18. http://www.scopus.com/inward/record.url?eid=2-s2.0-84863309869&partnerID=40&md5=6bc972da6ae1e3a068c01251a892ae0f

Sarigiannis, D. & Karakitsios, S. (2011). Perinatal exposure to bisphenol a: the route of administration makes the dose. *Epidemiology*, *22*(1), S172. http://journals.lww.com/epidem/Fulltext/2011/01001/Perinatal_Exposure_to_Bisphenol_A__The_Route_of.510.aspx

Sarigiannis, D. & Karakitsios, S. (2019). Advancing chemical risk assessment through human physiology-based biochemical process modeling. *Fluids*, *4*(1), 4. https://doi.org/10.3390/fluids4010004

Sarigiannis, D., et al. (2011). A computational framework for aggregate and cumulative exposure assessment. *Epidemiology*, *22*(1), S96–S97. http://journals.lww.com/epidem/Fulltext/2011/01001/A_Computational_Framework_for_Aggregate_and.270.aspx


© Burleigh Dodds Science Publishing Limited, 2025. All rights reserved.

Sarigiannis, D., et al. (2014). *Integra: from global scale contamination to tissue dose*. 7th International Congress on Environmental Modelling and Software International Environmental Modelling and Software Society.

Sarigiannis, D. A., et al. (2016). Integrated exposure and risk characterization of bisphenol-A in Europe. *Food and Chemical Toxicology*, *98*(Pt B), 134–147. https://doi.org/10.1016/j.fct.2016.10.017

Sarigiannis, D., et al. (2017). Development of QSARs for parameterizing physiology based toxicokinetic models. *Food and Chemical Toxicology*, *106*, 114–124. https://doi.org/10.1016/j.fct.2017.05.029

Sarigiannis, D., et al. (2018). Informatics and data analytics to support exposome-based discovery: part 2-computational exposure biology. In M. D. Lytras & P. Papadopoulou (Eds.), *Applying big data analytics in bioinformatics and medicine* (pp. 145–187). IGI Global. https://doi.org/10.4018/978-1-5225-2607-0.ch007

Sarigiannis, D., et al. (2019). Risk characterization of bisphenol-A in the Slovenian population starting from human biomonitoring data. *Environmental Research*, *170*, 293–300. https://doi.org/10.1016/j.envres.2018.12.056

Sarigiannis, D., et al. (2020). Development of a generic lifelong physiologically based biokinetic model for exposome studies. *Environmental Research*, *185*, 109307. https://doi.org/10.1016/j.envres.2020.109307

Singh, S., et al. (2024). Advances in Artificial Intelligence (AI)-assisted approaches in drug screening. *Artificial Intelligence Chemistry*, *2*(1), 100039. https://doi.org/10.1016/j.aichem.2023.100039

Sy, M., et al. (2024). Assessment of the long-term exposure to lead in four European countries. *Exposure and Health*, *16*(1), 21–39. https://doi.org/10.1007/s12403-023-00535-2

Tan, Y. M., et al. (2018). Challenges associated with applying physiologically based pharmacokinetic modeling for public health decision-making. Toxicological Sciences, *162*(2), 341–348. https://doi.org/10.1093/toxsci/kfy010

Tan, Y. M., et al. (2020). model reporting template for chemical risk assessment applications. *Regulatory Toxicology and Pharmacology*, *115*, 104691. https://doi.org/10.1016/j.yrtph.2020.104691

Testai, E., et al. (2021). Modelling human variability in toxicokinetic and toxicodynamic processes using Bayesian meta-analysis, physiologically-based modelling and in vitro systems. *EFSA Supporting Publications*, *18*(4), 6504E. https://doi.org/10.2903/sp.efsa.2021.EN-6504

Thomas, R., et al. (1996). Using advanced continuous simulation language (ACSL): a computational method. *Toxicological Sciences*, *31*, 19–28. https://doi.org/10.1093/toxsci/31.1.19

Upton, R. N., et al. (2016). An introduction to physiologically-based pharmacokinetic models. *Pediatric Anesthesia*, *26*(11), 1036–1046. https://doi.org/10.1111/pan.12995

WHO. (1990). *Report of the WHO Consultationon dioxins*. Bilthoven.

WHO/IPCS. (2014). *Guidance document on evaluating and expressing uncertainty in hazard characterization*. World Health Organization International Program on Chemical Safety. http://www.who.int/ipcs/methods/harmonization/areas/hazard_assessment/en.

Wittwehr, C., et al. (2020). Artificial Intelligence for chemical risk assessment. *Computational Toxicology*, *13*, 100114. https://doi.org/10.1016/j.comtox.2019.100114

© Burleigh Dodds Science Publishing Limited, 2025. All rights reserved.

Xu, Y., et al. (2017). Deep learning based regression and multiclass models for acute oral toxicity prediction with automatic chemical feature extraction. *Journal of Chemical Information and Modeling*, *57*(11), 2672–2685. https://doi.org/10.1021/acs.jcim.7b00244

Yadav, J., et al. (2021). Recent developments in in vitro and in vivo models for improved translation of preclinical pharmacokinetics and pharmacodynamics data. *Drug Metabolism Reviews*, *53*(2), 207–233. https://doi.org/10.1080/03602532.2021.1922435

Zalko, D., et al. (2011). Viable skin efficiently absorbs and metabolizes bisphenol A. *Chemosphere*, *82*(3), 424–430. http://www.scopus.com/inward/record.url?eid=2-s2.0-78650054660&partnerID=40&md5=0cee1c07b8cb915c859fd82d3cd5821e

Zhang, L., et al. (2024). Integrated models of population pharmacokinetics and exposure response to optimize dosage regimen for anaprazole sodium in duodenal ulcer. *European Journal of Pharmaceutical Sciences*, *198*, 106781. https://doi.org/10.1016/j.ejps.2024.106781

© Burleigh Dodds Science Publishing Limited, 2025. All rights reserved.

# Chapter 10

# The integration of adverse outcome pathways and aggregate exposure pathways for food safety testing: pesticides as a case study

*Alexandra-Costina Avîrvarei*, Ellen Callewaert*, Kyriaki Xiftou, Mathieu Vinken† and Julen Sanz-Serrano†, Vrije Universiteit Brussel, Belgium*



## 1 Introduction

Chemical safety assessment still predominantly relies on animal studies that focus on apical endpoints of toxicity. However, these traditional animal-centred approaches are being challenged owing to ethical and scientific concerns, including poor human predictivity and lack of mechanistic understanding (Vinken, 2024). To address these limitations, the early twenty-first century marked a paradigm shift in toxicology and moved away from animal-dependent assessment towards next-generation risk assessment. This exposure-led and

* These authors share equal first authorship.
† These authors share equal seniorship.

http://dx.doi.org/10.19103/AS.2025.0158.19
© Burleigh Dodds Science Publishing Limited, 2025. All rights reserved.

hypothesis-driven approach strives to replace and/or reduce animal testing by applying new approach methodologies (NAMs) (Cronin et al., 2022; EFSA, 2018; Judson, 2013). NAMs refer to *in vitro*, *in silico*, and *in chemico* methods, as well as to approaches that refine *in vivo* methodologies. This supports the implementation of the 3R principle (Replacement, Reduction, and Refinement of animal experimentation) and facilitates risk assessment through human-relevant techniques (Luijten et al., 2022; Schmeisser et al., 2023). NAMs move away from the 'black box' approach inherent to animal-based testing towards one where understanding the molecular mechanisms of toxicity becomes central (Kramer et al., 2019; Becker et al., 2017).

The adverse outcome pathway (AOP) framework is a promising tool to provide a solid mechanistic understanding of toxicological processes and can serve as the conceptual basis for setting up NAMs (Ankley and Edwards, 2018; Stucki et al., 2022; Arnesdotter et al., 2022; Vinken, 2024). Complementary to the chemical-agnostic AOP framework, the chemical-specific aggregate exposure pathways (AEPs) framework has been proposed to organize relevant exposure data. Thereby, AEPs trace the pathway of a chemical from the source to the site of action, through environmental and biological pathways, and AOPs provide mechanistic information from the specific event happening at the site of action that derives into the adverse effect. The AEP–AOP framework therefore provides a versatile structure to predict chemical-induced toxicity and enables risk-based decision-making (Tan et al., 2018a; Clewell et al., 2020).

One of the major sources of chemical contamination to the general population is derived from plant protection products, commonly known as pesticides (Stucki et al., 2022). Humans are directly exposed to pesticides during agricultural practices; public health initiatives for vector control; and pest management in various settings, including commercial facilities and transport. Residues of active pesticides, metabolites, and degradation products can also be detected on crops and in animal-derived foods (Kalyabina et al., 2021). As the liver is involved in all metabolic processes immediately after the absorption of any chemical from the gastrointestinal tract, the exposure to pesticides poses significant risks of adverse hepatotoxic effects (Manfo et al., 2020). Therefore, each active substance in plant protection products must be proven safe in terms of human health, therefore making risk assessment a crucial process. The present chapter focuses on the set-up, development, assessment, and applications of the AOP and AEP frameworks, with an emphasis on food toxicology. Through a case study centred on the metabolites and environmental residues of pesticides, we explore how these frameworks can jointly enhance our understanding of hepatotoxicity risks associated with the various exposure pathways of these contaminants. The combined AEP–AOP approach is illustrated using three widely studied organophosphorus pesticides–chlorpyrifos, diazinon, and malathion–to demonstrate the

© Burleigh Dodds Science Publishing Limited, 2025. All rights reserved.

translational value of these frameworks in assessing and mitigating potential human health risks in the context of food safety.

## 2 The structure, development, assessment, and quantification of adverse outcome pathways

### *2.1 Structure of adverse outcome pathways*

An AOP represents a sequence of key events (KEs) that are causally linked upon exposure to a stressor. KEs are measurable biological changes that capture relevant perturbations occurring at various levels of biological organization, ranging from the cellular to the population level. Specialized KEs, namely the molecular initiating events (MIEs) and the adverse outcome (AO), structurally anchor AOPs from one end to the other (Ankley and Edwards, 2018).

The MIE is defined as the initial interaction of a chemical stressor with a biological target at the molecular level that creates a perturbation. The AO refers to the apical toxicological endpoint (Ankley and Edwards, 2018; Callewaert et al., 2024). The connection between upstream and downstream KEs is described through key event relationships (KERs). In this way, KERs represent causal associations and therefore imply that changes in the biological state of an upstream KE influence the downstream KE (Fig. 1) (Svingen et al., 2021). Moreover, modulating factors like genetic polymorphisms, disease states, and nutritional or environmental factors can act on these KERs and thereby mediate the response of the KEs by changes in sensitivity, duration, and magnitude of the response, without directly changing the causal response–response relationship between KEs (Knapen et al., 2018; Sleiman et al., 2024). Any type of data describing linkages between pathological events can be used for AOP generation, such as *in vivo*, *in vitro*, *in silico*, or epidemiological data. Each of these studies give insights at different biological levels, e.g. *in vitro* studies give molecular and cellular mechanistic information to be integrated as MIE, while epidemiological studies give more insights at the organ or human level and, therefore, the AO (Ankley et al., 2009).

AOPs linking a single MIE to an AO provide a solid mechanistic understanding of the specific toxicological process (Paini et al., 2022). However,

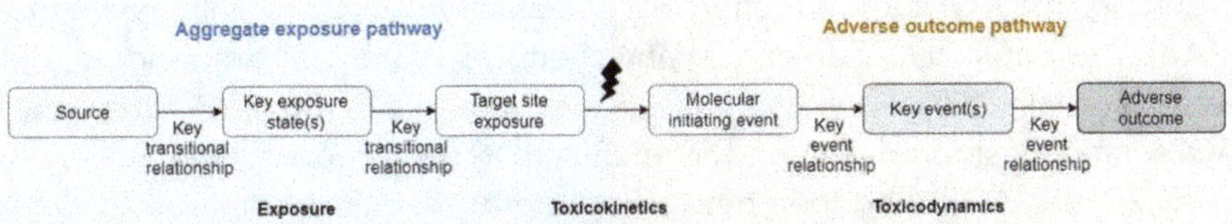


**Figure 1** Generic structure integrating exposure, toxicokinetic and toxicodynamic information. Source: Adapted from: Clippinger et al. (2018).

© Burleigh Dodds Science Publishing Limited, 2025. All rights reserved.

in view of more appropriately reflecting the actual complexity of real-life toxicity, single AOPs sharing at least 1 KE can be merged into an AOP network. AOP networks provide a more realistic representation of complex biological interactions and, as such are a versatile tool in current predictive toxicology and risk assessment of chemicals and chemical mixtures (Knapen et al., 2018; Lambert, 2022).

## 2.2 Development and assessment of adverse outcome pathways

Currently, over 400 AOPs are established for a plethora of human and ecotoxicological endpoints (https://aopwiki.org/). Owing to the increasing use of AOPs, the Organization for Economic Co-operation and Development (OECD) launched the 'OECD AOP Development Program' on a web-based platform, namely the OECD AOP Knowledge Base (AOP–KB). The AOP–KB includes the AOP-Wiki, an open-access AOP repository that allows collaborative input and review of fellow experts (https://aopkb.oecd.org/index.html; https://aopwiki.org/). The first phase of AOP development begins when developers submit the AOP to the AOP-Wiki. The Extended Advisory Group for Molecular Screening and Toxicogenomics (EAGMST) then provide feedback and coaching. The second phase involves the review of the AOP, which includes a compliance check and a scientific review. In the third phase, also named the OECD-specific endorsement phase, the AOP requires approval from the EAGMST and other responsible OECD committees. In case confidence in the scientific review process is expressed, the AOP is labelled as 'endorsed' and is published in the 'OECD dedicated series on AOPs' (OECD, 2021). The publication provides a stable version over time (i.e. reviewed and revised version), as even endorsed AOPs are continuously evolving after their publication (Fig. 2). On this note, endorsement of an AOP does not indicate that the AOP is ready or useful for direct regulatory applications (Jaylet et al., 2024; Patlewicz et al., 2015).

The regulatory acceptance of AOPs and their networks is hampered by a lack of confidence in the quality of the provided information (Bajard et al., 2023). Ideally, AOP developers comply with the *'Guidance Document on Developing and Assessing Adverse Outcome Pathways'* issued by the OECD and a set of fundamental principles published in scientific literature, including that AOPs are not chemical specific (OECD, 2013; Villeneuve et al., 2018). This implicitly assumes that any chemical capable of triggering the MIE can potentially initiate the chain of downstream KEs in the AOP. Furthermore, AOPs are considered living documents that evolve over time (Villeneuve et al., 2014a). AOPs and their networks therefore require continuous updating by incorporating new available data, but manually retrieving data from the plethora of textual evidence is a tedious and time-consuming process (Bajard

© Burleigh Dodds Science Publishing Limited, 2025. All rights reserved.

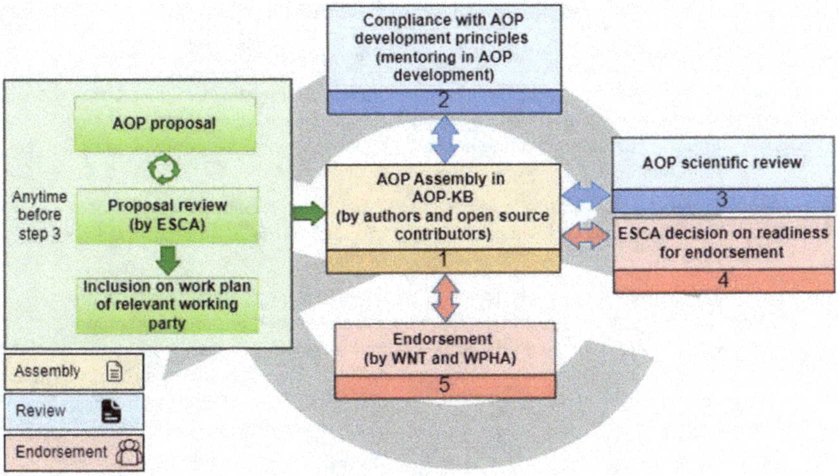


**Figure 2.** Adverse outcome pathway development and coordination process at the Organization for Economic Cooperation and Development. ESCA, The Advisory Group on Emerging Science in Chemicals Assessment; WNT, Test Guidelines Programme; WPHA, Working Party on Hazard Assessment. Source: Adapted from: https://www.oecd.org/en/topics/sub-issues/testing-of-chemicals/adverse-outcome-pathways.html.

et al., 2023). In this regard, artificial intelligence–assisted article selection and data collection, specifically machine learning approaches, such as natural language processing, have emerged as a useful and rapidly evolving tools for AOP development (Corradi et al., 2022, Hartung, 2023). Furthermore, AOPs are consistently critically evaluated through a weight-of-evidence assessment. The assessment of the confidence in the AOP relies on the tailored Bradford-Hill criteria (Table 1) (https://aopwiki.org/handbooks/5). Each of these criteria consists of several driving questions, that are answered with direct or indirect evidence for each of the KEs and/or KERs (https://aopwiki.org/handbooks/5). Furthermore, the extent of quantitative understanding for the KERs is assessed by defining how much change in the upstream KE, and/or for how long, is needed to trigger a detectable and defined change in the downstream KE (https://aopwiki.org/handbooks/5). The output of these criteria is classified as strong, moderate, or weak. The biological domain of applicability (i.e. taxa, life-stage, and sex) for the whole AOP is also defined during the assessment and is based on the groups of organisms for which the measurements represented by the KEs and KERs are relevant and operative. Overall, the aim of the weight-of-evidence assessment is not to reiterate all descriptions and data on KE and/or KERs but rather to provide a detailed overview of the relative confidence in the AOP. Determination of the confidence in the overall AOP indicates the suitability for different applications, and therefore, aids the regulatory acceptance (https://aopwiki.org/handbooks/5). Additionally, during

© Burleigh Dodds Science Publishing Limited, 2025. All rights reserved.

**Table 1** Description of tailored Bradford-Hill criteria with driving questions

| Criteria | Description | Driving questions |
|---|---|---|
| Biological plausibility of KERs | The understanding of the fundamental biological processes and the consistency of the causal relationships. | Is the mechanistic relationship between the upstream KE and downstream KE consistent with the established biological knowledge? |
| Essentiality of KEs | The impact of upstream MIE/KE perturbations on downstream KE/AOs. | What is the impact on downstream KEs and/or the AO if an upstream KE is modified or prevented? |
| Empirical support for KERs | The amount of toxicological data derived by 1 or more prototypical stressors related to the dose, temporal, and incidence concordance of KERs. | Does the upstream KE occur at lower doses and earlier time points than the downstream KE?<br>Is the incidence of the upstream KE greater than that for the downstream KE for the same dose of the tested stressor?<br>Are there inconsistencies in empirical support across taxa, species and prototypical stressor that don't align with expected pattern for hypothesized AOP? |

AO, adverse outcome; AOP, adverse outcome pathway; KE, key events; KER, key event relationships; MIE, molecular initiating events.

the development and assessment of AOPs, uncertainties and gaps in the AOPs are revealed by incomplete KERs and KEs. These gaps point to unexplored endpoints or overlooked toxic effects that can guide further targeted research (Bajard et al., 2023).

## 2.3 Quantification of adverse outcome pathways

Quantitative AOPs (qAOPs) are AOPs supported with quantitative understanding by using mathematical descriptors for the KEs and KERs. In a qAOP, the magnitude and/or duration of change in the upstream KE required to induce a corresponding change in the downstream KE is quantified (Coady et al., 2019). These quantitative relationships can be quantified in various ways, including correlations, response–response relationships, dose-dependent transitions, or points of departure. Additionally, the timescale can provide information regarding the approximate time-course of these changes, and the knowledge regarding modulating factors and feedback mechanisms can provide additional information on the KERs (https://aopwiki.org/handbooks/5).

The development of qAOPs utilizes different approaches ranging from simple mathematical equations to sophisticated biologically based computational models that consider other modulating factors, such as compensatory responses or interactions with other biological or environmental variables (https://aopwiki.org/handbooks/5).

© Burleigh Dodds Science Publishing Limited, 2025. All rights reserved.

A framework to guide development and assessment of qAOPs is currently lacking. Scientific literature describes two development stages for full qAOPs, namely probabilistic and mechanistic qAOPs (Spinu et al., 2020). Probabilistic qAOPs use statistical approaches to quantify KERs and to predict the relationship between KEs, whereas mechanistic qAOPs use mathematical functions to determine the extent of modulation of a KER, based on specified initial conditions of the KEs (Spinu et al., 2020). As such, the development of qAOPs can be achieved using various modelling strategies, ranging from Bayesian network and regression modelling to ordinary differential equations (Perkins et al., 2019).

The Bayesian network model is a frequently adopted modelling strategy for directional and functional KERs, used to predict the probability that a downstream KE will occur based on the change of an upstream KE. These models are of special interest in qAOP development and have already been applied to quantify an AOP network on liver steatosis (Villeneuve et al., 2014b; Perkins et al., 2019). Regression models predict how different levels of perturbation in an upstream KE affect a downstream KE, including saturable response. Other types of mathematical models, such as ordinary differential equations or individual-based modelling, are commonly applied for predicting temporal and time-resolved responses (Perkins et al., 2019). Any type of information on the causality of KEs can be fed into an qAOP, including KE expression knock-down or activity inhibition, and/or response–response dependence, such as dose–response measurement of upstream and downstream KE activities.

## 3 The structure and development of aggregate exposure pathways and their integration with adverse outcome pathways

### 3.1 *Structure of adverse exposure pathways*

The AEP framework has been proposed to organize exposure data, from the source of a chemical to its concentration at a site of action in a human real-life scenario (Clewell et al., 2020; Tan et al., 2018a). This data-driven framework shares common key features to the AOP framework, such as the structure and terminology, and complements it (Tan et al., 2018a). Mimicking a KE, a key exposure state (KES) defines the state of a stressor in space and time. There are two specialized KESs, namely the source and target site exposure (TSE). Key transitional relationships (KTRs) describe how a stressor either moves between different environmental media within the same KES or transforms into a different form as it progresses through two distinct KES (Di Nicola et al., 2023). Therefore, AEPs structure exposure data by linking the introduction of a stressor from its sources, transport, and transformation through environmental

© Burleigh Dodds Science Publishing Limited, 2025. All rights reserved.

media, patterns of external exposure, and the biokinetic processes that lead to the TSE (Fig. 1).

The AEP provides a comprehensive view of how chemicals behave in the environment, including their transformation into metabolites. This interconnectedness enhances our understanding of the broader ecological impacts of human activities by illustrating how chemical transformations can affect human health, highlighting interactions that may not be apparent when assessing individual chemicals (Clippinger et al., 2018; Leist et al., 2017; Peng et al., 2022).

## 3.2 Development of adverse exposure pathways

AEPs are developed by integrating exposure data from multiple pathways, including food, water, and residential environments. This development process begins with conservative estimates of exposure, which can later be refined using more precise data and probabilistic techniques. A critical aspect of this development is ensuring that exposure data is pathway-specific, allowing for accurate tracing of contaminants from their environmental sources to target sites. Importantly, exposures should only be aggregated when they are linked to a common toxicological endpoint (PMRA, 2003).

The exposure data used in AEP development encompasses a wide range of information regarding how individuals and populations interact with environmental stressors, including chemicals found in air, water, soil, food, and consumer products. This data is collected across various biological levels and incorporates both temporal and spatial dimensions, enabling the tracking of changes over time and across regions. Sources of this data include research studies, governmental databases, monitoring reports, and industry data. However, the integration of this information is often complicated by fragmentation across institutions, inconsistent formats, and confidentiality of the data. To address these challenges, AEP development focuses on enhancing data accessibility and integration by adhering to the FAIR guiding principles: making data Findable, Accessible, Interoperable, and Reproducible (Wilkinson et al., 2019; Tan et al., 2018b).

Toxicokinetic models provide a mechanistic understanding of the absorption, distribution, metabolism, and excretion processes that determine the internal dose of a chemical at the target site, making it essential to integrate exposure data within the AOP. Therefore, it can detail how factors such as solubility, volatility, and chemical reactivity influence the bioavailability of the chemical once it enters an organism. This integration allows for the identification of KESs in the exposure pathway that correspond to toxicokinetic processes (Hines et al., 2018; Tan et al., 2018a).

© Burleigh Dodds Science Publishing Limited, 2025. All rights reserved.

### 3.3 Adverse exposure pathway-adverse outcome pathway integration

The combination of the joint AEP–AOP frameworks offers a mechanism-based perspective that supports the understanding of chemical interactions across the entire source-to-outcome continuum. These joint frameworks leverage the systematic approach of AEPs to address each stage of chemical release, environmental transport, transformation processes, human exposure pathways, and dosimetry and links these to the AOP network relevant for each chemical (Fig. 1). This AEP–AOP integration supports the characterization of complex chemical interactions that AOP alone cannot capture, supporting the development of workflows for experimental design, as well as systematic reviews, switching the context from hazard identification to risk assessment (Tan et al., 2018b; Di Nicola et al., 2023).

AEPs and AOPs are complementary frameworks that share key structural elements, such as KE and KES, which capture the causal relationships in their respective framework. KEs and KES describe the ordered progression from the initial chemical exposure, as captured by AEPs, to the eventual AO described by AOPs. The integration of AEP and AOP occurs at the TSE, where the concentration of the chemical at the biological location of the MIE, marks the transition between exposure pathways and biological response pathways. Toxicokinetics are an important aspect that underpins both AEPs and AOPs. While AOP are chemical agnostic, AEPs are not (Clippinger et al., 2018; Price et al., 2020). Therefore, understanding how chemicals are absorbed, distributed, metabolized, and excreted within the organism is important for determining the internal dose of a chemical that reaches the TSE and contributes to the initiation of AOPs (Robb et al., 2024; Bouchard et al., 2005).

## 4 Applications of adverse exposure pathway-adverse outcome pathway integration

### 4.1 Generation of new approach methodologies and integrated approaches to testing and assessments

The use of NAMs in toxicology represents a transformative advancement toward more predictive and human-relevant testing strategies. Their effective use in risk assessment aims to improve not only the accuracy of hazard identification but also significantly reduce reliance on traditional animal testing (Schmeisser et al., 2023; Vinken, 2024). The generation of NAMs is greatly supported by the joint AEP–AOP framework by identifying the toxicity pathways based on chemical exposure routes. In this regard, NAMs are defined by two fundamental components in relation to AOPs and AEPs: a mechanistically anchored battery of assays, with each KE monitored by at

© Burleigh Dodds Science Publishing Limited, 2025. All rights reserved.

least one assay, and an *in vitro* system or complementary *in silico* approaches and pharmacokinetic/pharmacodynamic (PK/PD) modelling (Vinken et al., 2014; Vinken, 2024).

The AEP framework integrates various data sources and models to provide an understanding of exposure levels, routes, and biologically relevant doses. Firstly, simple, less detailed approaches are used to estimate exposure using relatively high model performance uncertainty (Teeguarden et al., 2016; Cherrie et al., 2020; Hristozov et al., 2024). As the assessment progresses, more advanced, higher-tier tools are employed. In fact, AEP can support the generation of exposure NAMs by feeding the exposure data into the models. Human biomonitoring data, PK/PD modelling, and data from reviews of relevant EU projects (e.g. Partnership for the Assessment of Risks from Chemicals) are integrated into exposure assessments to further refine predictions. Higher-tier models are essential for estimating aggregate exposure from multiple sources and routes (Teeguarden et al., 2016; Karrer et al., 2020; EFSA, 2024a). These models consider not only direct measurements but also variations in exposure owing to factors like consumer behaviour, environmental conditions, and pre-existing health conditions (Cherrie et al., 2020). This step is important as it determines the biologically relevant dose of the chemical resulting from the aggregated exposure, specifically the internal dose at the TSE of the chemical or its metabolites (EFSA, 2024b; Lim et al., 2019).

The AOP framework is structured around identifying KEs and their KERs, which describe a causal pathway leading to an AO. The first level of AOP focuses on identifying KEs that represent measurable biological changes resulting from chemical exposure. A mechanistically anchored battery of assays designed to monitor each KE is then used, typically applied to *in vitro* systems, but it can also be augmented by *in silico* and modelling approaches, such as machine learning algorithms (Vinken et al., 2014; Jiang et al., 2023). The second level involves validation of the identified KEs within a biological context, using transcriptional, translational, and functional profiling to confirm the relevance of the KE. This approach focuses on understanding how biological systems regulate these events across various biological processes, ensuring that the pathway is relevant (Vinken, 2024; Gijbels et al., 2021). AEP and AOP frameworks are dynamic and modular, designed to evolve as new data becomes available. Moreover, they can incorporate advanced modelling techniques and weight-of-evidence approaches to integrate data across different levels of biological and chemical understanding. These approaches enhance the predictive power of both frameworks and related NAMs, making them increasingly reliable tools for evaluating the potential risks of chemicals and their adverse effects (van Ertvelde et al., 2023). For example, in hepatotoxicity research,

© Burleigh Dodds Science Publishing Limited, 2025. All rights reserved.

a quantitative weight-of-evidence approach using the Bradford-Hill criteria has recently been applied to evaluate confidence in KERs within AOP networks (van Ertvelde et al., 2023; Verhoeven et al., 2024).

While the generation of NAMs has significantly advanced predictive toxicology by providing a human-relevant data, these methods alone often fall short in addressing the complexities of regulatory decision-making. To bridge this gap, integrated approaches to testing and assessment (IATAs) can be developed. These combine data from NAMs with other sources to create a more comprehensive and evidence-based framework for chemical safety assessments (Caloni et al., 2022). This pragmatic approach then integrates the analysis of pre-existing data through a weight-of-evidence framework supplemented by the generation of new empirical data via testing strategies (Clippinger et al., 2018). By following an iterative approach to answer specific regulatory questions, the acceptable level of uncertainty within the decision-making context is considered. IATAs are therefore evidence-based methodologies designed for chemical hazard or risk characterization for regulatory decision-making. The AOP framework has been adopted as the preferred method for integrating toxicity data required for hazard characterization within IATAs, while AEPs could play a similar role in structuring exposure information (Bajard et al., 2023; Tan et al., 2018a; Rooney et al., 2018). Linking assays to KES and respectively KEs anchored in the AOP and AEP would help to design targeted NAMs that are integrated within IATAs. Moreover, data from AEPs also contributes to a more human-relevant exposure assessment within the framework (OECD, 2016; Clippinger et al., 2018). All in all, the level of confidence in the AEP–AOP can be used to ensure weight of evidence in the application of the developed testing strategy (https://www.oecd.org/en/topics/sub-issues/assessment-of-chemicals/integrated-approaches-to-testing-and-assessment.html). In 2015, the OECD launched the 'OECD IATA Case Study Project' as part of its ongoing effort to develop IATA methods applicable to the regulatory framework for chemical safety assessment (OECD, 2016). Successful development of IATAs has been demonstrated in evaluating the toxic endpoints of skin irritation, skin sensitization, and eye irritation (OECD, 2016). However, AOP–AEPs could provide a systematic framework that is currently lacking (OECD, 2016). European Food Safety Authority (EFSA) has published case studies on two pesticides, flufenacet and deltamethrin, that demonstrate the value of integrating IATA and AOP frameworks in developmental neurotoxicity assessments. The assessment combined two approaches: one integrating *in vitro* testing with a literature review, and the other incorporating mechanistic data from NAMs into a postulated AOP network (EFSA, 2021; Hernández-Jerez et al., 2021).

© Burleigh Dodds Science Publishing Limited, 2025. All rights reserved.

## 4.2 Grouping of chemicals for prioritization and read-across of chemicals and mixture assessment

This approach assumes that chemicals with common characteristics, such as physicochemical and structural properties, exposure considerations, and mechanistic concepts, like mode of action or AOP, also exhibit similar toxicological effects (OECD, 2014; EFSA, 2019; Wohlleben et al., 2023). The use of *in silico* tools, such as quantitative structure–activity relationship approaches, offer great potential to group chemicals (Hemmerich and Ecker, 2020). In this context, the OECD provides the free computer software 'Quantitative Structure-Activity Relationship Toolbox' (https://qsartoolbox.org/). This software helps to predict the various KEs triggered by chemicals based on their structural properties (Schultz et al., 2018). For example, QSAR studies on LXR ligands identified the presence of phenyl rings, chloro-groups and methyl-substituents, as determinants of LXR binding and activation, an important MIE linked to liver steatosis (Honório et al., 2008).

Prioritization of chemicals has become a highly relevant strategy owing to the high number of chemicals humans are exposed to, making full assessment of each chemical a virtually unfeasible task (Judson et al., 2010; Hubbard et al., 2022). Chemicals are first screened using less complex, cheaper and faster assays or methodologies that can be guided by relevant AOPs. Consequently, the chemicals with a higher likelihood in posing specific health risks are identified and ranked according to their potency. The most potent chemicals receive highest priority for thorough assessment (Callewaert et al., 2024; Oki et al., 2019). The use of a risk-driven grouping approach (i.e. hazard and exposure) provides a robust basis for determination of high-priority assessment groups (Wohlleben et al, 2023). The EFSA has published a refined methodology for prioritization, implying a two-step probabilistic approach that focuses on substances and target organ systems (Di Piazza et al, 2024). The first step involves individual exposure assessment. The risk metrics is calculated based on pesticide residue occurrence and food consumption data, and consequently compared with health-based guidance values, such as the acceptable daily intake and acute reference dose. Pesticides, whose ratio of exposure to the health-based guidance values exceed the safety thresholds, are considered high-priority chemicals. In the second step, high-priority target organ systems associated with the prioritized pesticides are identified based on toxicological data. Each substance is assessed to determine which organ systems may be adversely affected. Toxicological data on KEs aid to identify plausible underlying mechanisms of toxicity, and an AEP–AOP-driven strategy can refine subsequent ranking of high-priority chemicals, especially for chemicals with relatively low data availability (Bajard et al., 2019).

© Burleigh Dodds Science Publishing Limited, 2025. All rights reserved.

Read-across strategies extrapolate the toxic endpoint of tested chemicals to similar untested chemicals within the same group (Lizarraga et al., 2023; Callewaert et al., 2024). The process of developing a read-across is typically done in a case-by-case manner, requiring a fit-for-purpose hypothesis and justification for the similarity rationale (OECD, 2017). This is commonly performed by applying the analogue or the category grouping approach. The former refers to using a very limited number of chemicals, whereas the latter typically contains more chemicals, thus making it a more robust and reliable approach (OECD, 2017; Banerjee et al., 2022). Hereby, it is of utmost importance to justify the grouping strategy, as sufficient evidence must be provided that the shared properties will result either in similar toxicodynamic properties or a consistent, predictable trend within the grouped compounds (Schultz and Cronin, 2017). AEP–AOPs can be incorporated in the read-across design as an additional layer of information. Using the set-up of a read-across study on liver steatosis as example, the hypothesis assumed that non-tested chemicals within the group of structurally similar aliphatic carboxylic acids could potentially induce hepatic steatosis. However, prior to the actual read-across, an AOP-based tiered testing strategy was developed to test this hypothesis and showed that within the biological similarity analysis, long-chain analogues are more homogenous in their activity than short-chain analogues (Escher et al., 2022). This highlights the major challenge to provide convincing evidence on the robustness of the read-across hypothesis and to quantify the remaining uncertainty of the estimated toxicity for the non-tested chemicals (Lizarraga et al., 2023; Usmani et al., 2024).

Traditionally, risk assessment still relies on testing the effects of individual chemicals. However, this may not adequately address the potential combined toxic effects of chemical mixtures. When two chemicals affect the same MIE, their combined effect is generally considered dose additive, as they act on a shared toxicological pathway. Whereas, when chemicals share common KEs but affect different MIEs, they can result in response addition, where each chemical influences a distinct MIE, but both contribute to the same AO (Price et al., 2020; Lambert, 2022). The integration of AEP and AOP therefore determines the nature of the interaction, whether additive, synergistic, or antagonistic, ultimately influencing the likelihood of an AO. Prior or concurrent exposures to different chemicals or mixtures can modulate these toxicokinetic processes, thereby altering the internal dose and bioavailability of the chemicals at the TSE (Hines et al., 2018; Price et al., 2020; Di Piazza et al., 2024). The first step in the cumulative risk assessment is to identify groups that induce a common toxic effect by a common mechanism of toxicity. This refers to chemicals causing the same critical effect, acting on the same molecular target tissue, acting by the same biochemical mechanism of action, or sharing a common toxic intermediate (Nielsen et al., 2012). The EFSA has established cumulative assessment groups

© Burleigh Dodds Science Publishing Limited, 2025. All rights reserved.

for chemicals that act on the same target organ, among others, for liver-specific endpoints like hypertrophy or fatty acid changes (i.e. steatosis) (Nielsen et al., 2012). The grouping guidelines issued by EFSA describe hazard-driven criteria (More et al., 2021). Hereby, chemicals that share a common AO and their AOPs are known are grouped together (More et al., 2021). These assessment groups include chemicals with the same AO but distinct MIEs, as combined exposure might still result in potential combined toxic effects (Cattaneo et al., 2023). In case of chemicals with a similar mode of action, the principle of dose-additivity is generally assumed and as such has been set as the default approach for mixture risk assessment (Borgert et al., 2004; Ankley et al., 2009). Recently, an AOP-wise testing strategy was developed to assess steatotic mixture effects for mixtures of pesticides. Based on the relative potency of the individual pesticides, equipotent mixtures were tested using a series of liver-based *in vitro* assays that monitor MIEs and KEs anchored in the AOPs on liver steatosis. The observed mixture effects were congruent with the assumption of dose additivity, and the pesticide-specific potency remained similar over the different KEs studied downstream the AOP (Lichtenstein et al., 2020).

## 5 Case-study: hepatotoxicity of pesticides

### *5.1 Regulatory framework*

The shift towards next-generation risk assessment has found acceptance not only in the pharmaceutical and cosmetic industry but also in the food industry. The publication of the 'Roadmap for Action on NAMs' by EFSA aims to refine risk assessment processes by transitioning towards human-centred models, reducing reliance on animal testing, promoting methodological harmonization and standardization, and advocating for sustainable development initiatives (Escher et al., 2022; EFSA, 2024a).

EFSA is responsible for the peer review of pesticide active substances, which are chemicals intended to disrupt the physiological functions and affect the nervous and endocrine systems of pests such as insects, rodents, microbes, fungi, and weeds (Hassaan and El Nemr, 2020). Additionally, EFSA proposes maximum residue levels (MRLs) of these pesticides in food and feed. In 2021, MRLs of organophosphates were exceeded in 3.9% of the total 87 863 European food samples (Bellisai et al., 2023). Exposure to such pesticide residue represents a significant health risk to various organs, including the liver, hence continuous risk assessment is pivotal (Manfo et al., 2020).

Pesticide assessment is governed by a regulatory framework primarily under European Commission Regulation No 1107/2009 and European Commission Regulation No 396/2005 (EC, 2005, 2020). The evaluation of pesticide active substances begins with the manufacturer's submission of safety data. These

© Burleigh Dodds Science Publishing Limited, 2025. All rights reserved.

primarily include toxicological studies and residue data focusing on potential risks to human health and the environment, such as acute and chronic toxicity, endocrine disruption, and ecological impact. A 'Draft Assessment Report' is then carried out by a *rapporteur* EU *member state*, followed by peer review by the EFSA, member states, and public consultation. If the pesticide is considered safe when used according to the recommendations, EFSA issues a conclusion, after which the European Commission consults the member states and decides on approval (EFSA, 2018). The process, however, involves distinct workflows for either new pesticide active substances or amendment of approval conditions such as renewal of pesticide active substances, setting MRLs, and approval of basic substances.

## 5.2 Exposure to organophosphate pesticides

Humans are directly exposed to several classes of pesticides. Organophosphate pesticides (OPs), including chlorpyrifos, diazinon, and malathion, are commonly used insecticides that pose potential health risks owing to their widespread application in agricultural settings and residential pest control practices. Humans can therefore be exposed to pesticide-derived contaminants, such as their metabolites and environmental residues. Moreover, exposure to these chemicals can occur via environmental and dietary pathways, aside from consumer products, which might lead to health hazards (Fig. 3) (Hung et al., 2018; Ubaid ur Rahman et al., 2021).

Exposure to OPs via the air mainly results from pesticide drift during agricultural spraying and pest control (van Wendel de Joode et al., 2012; Hung et al., 2018). The movement of pesticides from agricultural fields into the atmosphere occurs primarily through volatilization. Once airborne, these pesticides undergo chemical transformations, such as photo-chemical oxidation by atmospheric compounds, leading to the formation of degradation compounds (Raina et al., 2010). Their residues accumulate in both indoor and outdoor dust, becoming an important exposure source to the environment and consequently to humans. While ingestion of contaminated foods is the primary route of exposure, children and pregnant women living near application sites or during application periods are particularly vulnerable, as ingestion of dust may account of up to 5% of overall exposure (Raina et al., 2010; Quirós-Alcalá et al., 2012; Gibbs et al., 2017).

Detectable amounts of active pesticides, metabolites, and degradation products are also found on harvested crops or in animal-derived foods. These pesticide residues indirectly reach humans by ingestion of contaminated food and water, as well as inhalation during drift (Bueno and Cunha, 2020). OPs can settle on the surface during crop irrigation and leach into ground water from agricultural fields during rainfall, leading to potential contamination of drinking

© Burleigh Dodds Science Publishing Limited, 2025. All rights reserved.

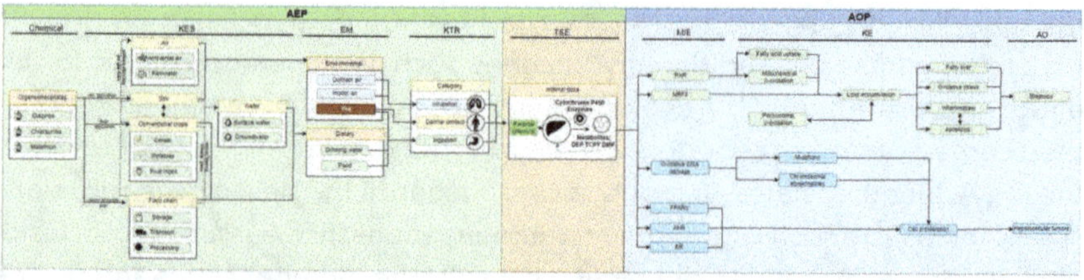

**Figure 3.** Aggregate exposure pathway and adverse outcome pathway construct applied to the hepatotoxicity of organophosphate pesticides. Key exposure states include air, soil, crops, and food, with secondary contamination in water from drift, runoff, and leaching. Organophosphate pesticides move through environmental media via the key transitional relationships inhalation, dermal absorption, and ingestion, primarily through contaminated food and water. Once absorbed, OPs reach the target site exposure in hepatocytes and activate nuclear receptors and induce oxidative DNA damage and lead to the adverse outcomes liver steatosis and hepatocellular tumors through diverse key events. AHR, aryl hydrocarbon receptor; DEP, diethyl phosphate; DMP, dimethyl phosphate; ER, estrogen receptor; NRF2, nuclear factor erythroid 2–related factor 2; TCPY, 3,5,6-trichloro-2-pyridinol; PPARα, peroxisome proliferator activated receptor alpha; PXR, pregnane X receptor.

water sources. Detectable concentrations of these pesticides have been found in groundwater for several weeks (Dolatabadi et al., 2022). However, OPs were undetectable in treated effluents owing to efficient water treatment processes. This indicates that environmental persistence in soil is mainly caused by runoff from agricultural areas (Rousis et al., 2017).

Repeated OP applications can lead to soil accumulation, posing risks of dermal exposure and inadvertent ingestion to the population (Bose et al., 2021; Zhong et al., 2022). Soil-bound pesticides can persist for extended periods, particularly in environments with low microbial activity and unfavourable pH levels. These conditions can alter the degradation pathways of pesticides, sometimes leading to the formation of more toxic metabolites, such as diethyl phosphate and 3,5,6-trichloro-2-pyridinol (Baskaran et al., 2003; Sasikala et al., 2012; Wołejko et al., 2022). The transfer of pesticides from soil to plants is an important pathway for human exposure through the food chain, affecting crops, fruits, and vegetables (Fu et al., 2015; Li et al., 2015; Noore et al., 2021). The general population is primarily exposed to OPs through contaminated food and water, while occupational groups face higher risks from inhalation and dermal contact (Garfitt et al., 2002; Bouvier et al., 2006; Bradman et al., 2009; Nematy et al., 2016; Wang et al., 2016). Importantly, temporal and spatial considerations, such as application intervals, environmental conditions, proximity to agricultural fields, and pesticide movement, can greatly influence the exposure levels. Seasonal variations are particularly relevant, as they affect pesticide concentrations and exposure durations in both occupational and non-occupational setting (Lin et al., 1996;

© Burleigh Dodds Science Publishing Limited, 2025. All rights reserved.

Lentz et al., 2015). Owing to concerns about pesticide residues in food, several OPs in food have been heavily regulated, with MRLs set as low as the limit of detection (LOD) (Mojsak et al., 2018; Kermani et al., 2021). However, the 'European Union Annual Report' on pesticide residues found that several OPs and their metabolites exceeded MRLs in European products, and imported commodities such as cumin seeds, oranges, grapefruit, cabbage, and rice (Carrasco Cabrera et al., 2023).

Depending on the source of exposure, OPs are absorbed through the lungs, gastrointestinal tract, and skin. Their lipophilicity allows to penetrate biological membranes contributing to systemic absorption (Robb et al., 2024) and accumulation in lipid-rich tissues, such as the nervous system and adipose tissue. OPs are primarily metabolized in the liver by cytochrome P450 enzymes into active metabolites that can be more toxic than the parent compounds, increasing the potential for harm and highlighting the importance of an integral evaluation (ATSDR, 2008; Bicker et al., 2005; Bouchard et al., 2005). These metabolites and their parent compounds are excreted primarily through urine and are detectable for days after exposure (Poet et al., 2004; Sharbidre et al., 2011; Wang et al., 2016).

Biomonitoring for OP exposure primarily focuses on detecting dialkyl phosphate metabolites in urine, which reflect internal exposure levels of these pesticides. 3,5,6-trichloro-2-pyridinol is a key biomarker for chlorpyrifos, while diethyl phosphate and dimethyl phosphate, are used for assessing exposure to diazinon and malathion (Karanth, 2005; Choi et al., 2006; Rathod and Garg, 2017).

## 5.3 Adverse outcome pathways and organophosphate pesticides

Exposure to OPs can pose significant health risks to several organs. Besides the well-characterized neurotoxic adverse effects, increased hepatic toxicity biomarkers, such as serum aminotransferase, (in)direct bilirubin, and alkaline phosphatase, have been observed in several traditional toxicological studies on animals (Mikhail et al., 1979; Venkateswara Rao, 2006; Tripathi and Srivastav, 2010; Karami-Mohajeri et al., 2017). As the liver is the primary target for chemical-induced toxicity, this case study will focus on OP-induced hepatotoxicity (Timbrell, 2020).

Exposure to OPs is frequently associated with changes in liver lipid profiles, including the onset of steatosis. Histopathological observations in animal studies report macro vesicular steatosis after repeated exposure to malathion and diazinon, and liver fatty deposition at the periphery after repeated exposure to chlorpyrifos (Baconi et al., 2013; Saraei et al., 2016). However, at the molecular level, the mechanisms underlying OP-induced steatosis are

© Burleigh Dodds Science Publishing Limited, 2025. All rights reserved.

not yet fully understood. *In vitro* studies report chlorpyrifos-induced gene expression of pregnane X receptor (PXR) and protein expression of nuclear factor erythroid 2 (De Anna et al., 2021; Montanarí et al., 2024). Both events provide a plausible mechanism for liver steatosis and are considered frequent MIEs in the recently updated AOP network on liver steatosis (Verhoeven et al., 2024). Moreover, modulation of PXR is described as the most frequent MIE targeted by pesticides (Verhoeven et al., 2024).

Increased intracellular triglyceride accumulation, the most frequent KE in the AOP network on chemical-induced liver steatosis, is observed after direct chlorpyrifos exposure, along with other relevant KEs. These include increased intracellular lipid accumulation through increased extracellular free fatty acid accumulation, increased hepatic *de novo* lipogenesis, and decreased triglyceride efflux (Howell III et al., 2016). Moreover, OPs induce oxidative stress, a central KE in the liver steatosis AOP, evidenced by altered levels of superoxide dismutase, catalase, glutathione peroxidase, and glutathione S-transferase (Kovacic, 2003; Karami-Mohajeri et al., 2017). Oxidative stress can further lead to disruption of membrane lipids by degradation of membrane phospholipids and proteins through peroxidation of cellular macromolecules (Akhgari et al., 2003; Moore et al., 2010; Topal et al., 2014; Toledo-Ibarra et al., 2016). Several studies have shown that antioxidants, such as selenium and N-acetylcysteine, can counteract oxidative stress induced by OPs (Shokrzadeh et al., 2013; Tehrani et al., 2013). Other relevant KEs to OPs in the liver steatosis AOP network include mitochondrial disruption, inflammation, and apoptosis. OPs increase the release of cytochrome C from mitochondria to the cytosol, and activate caspase-3 by disrupting cellular antioxidant defences (Kaur et al., 2007; Karami-Mohajeri et al., 2017). OPs can directly trigger apoptotic pathways through caspase 3 and 9, increasing the B-cell lymphoma 2-associated X protein/B-cell lymphoma 2 ratio and protein disulphide isomerase (Kashyap et al., 2010; Karami-Mohajeri et al., 2017). Mid-zonal necrosis, apoptosis and infiltration have also been observed in rodents (Karami-Mohajeri et al., 2017).

Diazinon and malathion are classified as probably carcinogenic to humans by the International Agency for Research on Cancer (IARC, 2024). Chlorpyrifos is currently regarded as an endocrine disrupting chemical, a common mechanism of non-genotoxic carcinogens (IARC, 2024, Mu et al., 2021). In fact, OPs have been associated with increased risk of liver cancer through various mechanisms (VoPham et al., 2017). Rodent experiments suggest the DNA-damaging potential of diazinon and malathion by aneugenic effects, and double-strand breaks (Moore et al., 2010; Ezzi et al., 2016). Similar results for chlorpyrifos, diazinon, and malathion have been observed in hepatocytes *in vivo* and *in vitro* (Ojha et al., 2013; Yahia and Ali, 2018). It is suggested that OP-induced oxidative stress might be a relevant underlying mechanism for indirect DNA damage (Ojha et al., 2013). The aromatic hydrocarbon receptor

© Burleigh Dodds Science Publishing Limited, 2025. All rights reserved.

(AHR), estrogen receptor (ER), and peroxisome proliferator-activated receptor pathways have been described as MIEs in the induction of liver cancer, although the latter is unlikely to be relevant to humans (Yager and Leihr, 1996; Corton et al., 2014; Becker et al., 2015; Sanz-Serrano et al., 2024). These pathways have been shown to be altered *in vitro* by Ops, such as chlorpyrifos, malathion, and diazinon (Sammi et al., 2023). Chlorpyrifos and its metabolites alter AHR and subsequently cytochrome P450 1A1 expression, including in the liver, leading to cell proliferation and ultimately cancer (Moyano et al., 2021a; Miret et al., 2023; Ünlü Endirlik et al., 2023). Moreover, cell proliferation is also partially induced by chlorpyrifos through ER activation (Moyano et al., 2020, 2021b Doan et al., 2020; Lasagna et al., 2022). Likewise, malathion has also shown to modulate ER expression and increase cell proliferation, while also agonized the AHR function (Ghisari et al., 2015; Calaf et al., 2021) (Fig. 3).

## 6 Application of an adverse exposure pathway-adverse outcome pathway construct for pesticide assessment

The AEP–AOP construct on OP hepatotoxicity organizes the data linking sources of contamination to the AOs relevant to humans (Fig. 3). This facilitates the systematic examination of the OP exposure and toxicological pathways and enables subsequent development of effective strategies for their safety assessment. As the interest of public health to incorporate *in vitro* assays, test systems, and computational models into chemical safety assessment grows. The presented AEP–AOP construct outlines the specific MIEs and other KEs that occur during OP exposure and serve as a basis to develop NAMs. PXR, nuclear factor erythroid 2–related factor 2, peroxisome proliferator activated receptor alpha, AHR, ER, and oxidative DNA damage were identified as major MIEs in hepatotoxicity related to OP pesticides (Fig. 3). As AOPs are comprised of causal events and not just effect-related KEs, and especially MIEs, could be tested as biological endpoints that causally relate to AOs. In fact, an AEP–AOP driven *in vitro* strategy, based on the most common hepatotoxicity-related MIEs in humans, can be proposed for pesticides. An index integrating the ability of OPs to activate these MIEs (and/or other KEs) allows to rank them for their carcinogenicity or steatogenic potential. Moreover, OPs with similar mechanisms of action, metabolic pathways, and molecular targets can be grouped into categories for hazard assessment. Candidates can be subsequently made for read-across, ultimately reducing the need for testing. In a similar way, OPs that interact with the highlighted MIEs or show highly conserved and potent mechanisms of toxicity can be flagged as high priority for testing.

The approach hereby presented could also set the basis for the development of an AEP–AOP-informed IATA to support the regulatory

© Burleigh Dodds Science Publishing Limited, 2025. All rights reserved.

decisions regarding the hepatotoxicity of diazinon, malathion, and chlorpyrifos. As proof of concept, a case study on an AOP-based IATA, to address the developmental neurotoxicity of deltamethrin and flufenacet, was previously proposed by the EFSA (EFSA, 2021). In this regard, the *in vitro*, *in silico*, and *in chemico* assays can potentially estimate points of departure for OPs by establishing a dose–response relationship in the abovementioned MIEs. Using AEPs as a basis, the *in vitro*-to-*in vivo* extrapolation of these concentrations can subsequently be estimated and compared to internal concentrations related to human-relevant exposures. Estimated dietary and environmental intake rates can be then integrated in physiologically based pharmacokinetic models that predict the TSE of both parent chemicals and their metabolites in hepatocytes (Fig. 3). Alternatively, some studies can use concentration of OPs measured in blood as it is less invasive than the liver. Liver concentrations are then estimated from *in vivo* assays. However, focus should be put on estimating accumulated concentrations rather than exposure concentrations. By joining the TSE and MIEs, and therefore combining AEP–AOP data, the estimated liver concentrations would then act as the dose metric in AOPs.

The application of AEP–AOPs in risk assessment requires quantitative time and dose–response information (Villeneuve et al., 2014b). While qualitative AEP–AOPs help to assess approaches such as hazard identification, prioritization, or screening of chemicals, a quantitative understanding of the KEs and KERs is essential to match the MIE with the consequent AO. The different approaches to develop qAOPs range from Bayesian network and regression modelling to ordinary differential equation and individual-based models. To feed the models, information between KEs can be derived from experimental data indicating causality, such as KE expression knock-down, KE activity inhibition, or inter-KE dependence. In this regard, Bayesian models have already been applied to quantify an AOP network on liver steatosis, although a framework to guide such development and assessment is currently lacking (Perkins et al., 2019).

## 7 Conclusion

An AEP–AOP framework on OP pesticides, based on liver exposure and toxicity data from diazinon, malathion, and chlorpyrifos, is displayed as a pragmatic visualization of the intricate pathways related to OP hepatotoxicity. The joint AEP–AOP provides a basis for a robust, evidence-based framework for risk assessment, and management, with a primary focus on human health. This can modernize the process by providing a clearer link between effector and effect and can be applicable to a broader evaluation of chemical contaminants requiring risk assessment. Current policies could be updated to require mechanistic evidence, supported by guidelines for validating such AOP-based data. This would align data collection with MIEs and other KE that led to the AO,

© Burleigh Dodds Science Publishing Limited, 2025. All rights reserved.

prioritizing endpoint-directed *in vitro*, *in silico*, and *in chemico* testing methods. Moreover, this approach promotes *in silico* modelling by integrating AEPs with dietary databases and ultimately allows cumulative exposure across all food sources and different exposure scenarios. By supporting the development of informed policy decisions, the AOP–AEP frameworks facilitate progress towards a safer and animal-free future.

## 8 Where to look for further information

Further resources on AOPs, AEPs, their synergies, and their application can be found here:

- OECD AOP Knowledge Base provides access to the AOP-Wiki (open repository of AOPs) and the Effectopedia is a collaborative open knowledge tool foraggregation and visualization of AOPs. Available at: https://aopkb.oecd.org/index.html and https://sourceforge.net/projects/effectopedia/

European Food Safety Authority publishes opinions and guidance documents relevant to cumulative pesticide risk assessment, chemical grouping, NAMs, and IATAs. Key examples include:

- Guidance on grouping chemicals for human risk assessment (EFSA Journal, 2021) https://doi.org/10.2903/j.efsa.2021.7033.
- Roadmap for action on NAMs (EFSA Supporting Publications, 2022) https://doi.org/10.2903/sp.efsa.2022.EN-7341.
- Roadmap on advancing aggregate exposure assessment to chemicals(EFSA Supporting Publications, 2024) https://doi.org/10.2903/sp.efsa.2024.EN-8971.
- Case studies on IATA for developmental neurotoxicity (EFSA Journal, 2021) https://doi.org/10.2903/j.efsa.2021.6599.
- Prioritisation of pesticides and target organs for cumulative risk assessment (EFSA Journal, 2024) https://doi.org/10.2903/j.efsa.2024.8554.

Specialized reviews and case studies of AEP–AOP framework applications:

- Clewell, R.A. et al. (2020). Doi: 10.1016/j.tiv.2020.104855.
- Hines, D.E. et al. (2018). Doi: 10.1021/acs.est.7b04940.

The EU-funded research projects ONTOX (Grant ID: 963845), which develops AI- and ontology-driven animal-free approaches for repeated dose toxicity testing, and PARC (Grant ID: 101057014), which builds a European partnership

© Burleigh Dodds Science Publishing Limited, 2025. All rights reserved.

to advance next-generation chemical risk assessment and support the EU Chemicals Strategy for Sustainability.

# 9 References

Adverse Outcome Pathway Knowledge Base (AOP-KB). Available at: https://aopkb.oecd.org/index.html (Accessed 15 August, 2024).

Adverse Outcome Pathway Wiki (AOP-Wiki). Available at: https://aopwiki.org (Accessed 6 June, 2024)

Akhgari, M., et al. (2003) 'Biochemical evidence for free radicalinduced lipid peroxidation as a mechanism for subchronic toxicity of malathion in blood and liver of rats', *Human & Experimental Toxicology*, 22(4), pp. 205–211. Available at: https://doi.org/10.1191/0960327103ht346oa.

Ankley, G.T. and Edwards, S.W. (2018) 'The adverse outcome pathway: a multifaceted framework supporting 21st century toxicology', *Current Opinion in Toxicology*, 9, pp. 1–7. Available at: https://doi.org/10.1016/j.cotox.2018.03.004.

Ankley, G.T., et al. (2009) 'Adverse outcome pathways: a conceptual framework to support ecotoxicology research and risk assessment', *Environmental Toxicology and Chemistry*, 29(3), pp. 730–741. Available at: https://doi.org/10.1002/etc.34.

Arnesdotter, E., et al. (2022) 'Adverse outcome pathways as versatile tools in liver toxicity testing', *Methods in Molecular Biology*, 2425, pp. 521–535. Available at: https://doi.org/10.1007/978-1-0716-1960-5_20.

ATSDR (2008) Toxicological profile for diazinon. Available at: https://www.atsdr.cdc.gov/ToxProfiles/tp86.pdf.

Baconi, D.L., et al. (2013) 'Investigation of the toxicity of some organophosphorus pesticides in a repeated dose study in rats', *Romanian Journal of Morphology and Embryology*, 54(2), pp. 349–356.

Bajard, L., et al. (2019) 'Prioritization of hazards of novel flame retardants using the mechanistic toxicology information from ToxCast and adverse outcome pathways', *Environmental Sciences Europe*, 31(1), p. 14. Available at: https://doi.org/10.1186/s12302-019-0195-z.

Bajard, L., et al. (2023) 'Application of AOPs to assist regulatory assessment of chemical risks – Case studies, needs and recommendations', *Environmental Research*, 217, p. 114650. Available at: https://doi.org/10.1016/j.envres.2022.114650.

Banerjee, A., et al. (2022) 'Quantitative predictions from chemical read-across and their confidence measures', *Chemometrics and Intelligent Laboratory Systems*, 227, p. 104613. Available at: https://doi.org/10.1016/j.chemolab.2022.104613.

Bradman, A., et al. (2009) 'Community-based intervention to reduce pesticide exposure to farmworkers and potential take-home exposure to their families', *Journal of Exposure Science & Environmental Epidemiology*, 19(1), pp. 79–89. Available at: https://doi.org/10.1038/jes.2008.1.

Baskaran, S., et al. (2003) 'Contrasting behaviour of chlorpyrifos and its primary metabolite, TCP (3,5,6-trichloro-2-pyridinol), with depth in soil profiles', *Soil Research*, 41(4), p. 749. Available at: https://doi.org/10.1071/SR02062.

Becker, R.A., et al. (2015) 'The adverse outcome pathway for rodent liver tumor promotion by sustained activation of the aryl hydrocarbon receptor', *Regulatory Toxicology and*

© Burleigh Dodds Science Publishing Limited, 2025. All rights reserved.

*Pharmacology*, 73(1), pp. 172–190. Available at: https://doi.org/10.1016/j.yrtph.2015.06.015.

Becker, R.A., et al. (2017) 'Quantitative weight of evidence to assess confidence in potential modes of action', *Regulatory Toxicology and Pharmacology*, 86, pp. 205–220. Available at: https://doi.org/10.1016/j.yrtph.2017.02.017.

Bellisai, G., et al. (2023) 'Targeted review of maximum residue levels (MRLs) for diazinon', *EFSA Journal*, 21(11), p. e08426. Available at: https://doi.org/10.2903/j.efsa.2023.8426.

Bicker, W., et al. (2005) 'A case study of acute human chlorpyrifos poisoning: novel aspects on metabolism and toxicokinetics derived from liquid chromatography–tandem mass spectrometry analysis of urine samples', *Toxicology Letters*, 159(3), pp. 235–251. Available at: https://doi.org/10.1016/j.toxlet.2005.05.013.

Borgert, C.J., et al. (2004) 'Can mode of action predict mixture toxicity for risk assessment?', *Toxicology and Applied Pharmacology*, 201(2), pp. 85–96. Available at: https://doi.org/10.1016/j.taap.2004.05.005.

Bose, S., et al. (2021) 'A review on the microbial degradation of chlorpyrifos and its metabolite TCP', *Chemosphere*, 283, p. 131447. Available at: https://doi.org/10.1016/j.chemosphere.2021.131447.

Bouchard, M., et al. (2005) 'Determination of biological reference values for chlorpyrifos metabolites in human urine using a toxicokinetic approach', *Journal of Occupational and Environmental Hygiene*, 2(3), pp. 155–168. Available at: https://doi.org/10.1080/15459620590922407.

Bouvier, G., et al. (2006) 'Environmental and biological monitoring of exposure to organophosphorus pesticides: Application to occupationally and non-occupationally exposed adult populations', *Journal of Exposure Science & Environmental Epidemiology*, 16(5), pp. 417–426. Available at: https://doi.org/10.1038/sj.jes.7500473.

Bueno, M.R. and Cunha, J.P.A.R. (2020) 'Environmental risk for aquatic and terrestrial organisms associated with drift from pesticides used in soybean crops', *Anais da Academia Brasileira de Ciências*, 92(suppl 1). Available at: https://doi.org/10.1590/0001-3765202020181245.

Calaf, G., et al. (2021) 'Signs of carcinogenicity induced by parathion, malathion, and estrogen in human breast epithelial cells (Review)', *Oncology Reports*, 45(4), p. 24. Available at: https://doi.org/10.3892/or.2021.7975.

Callewaert, E., et al. (2025) 'Adverse outcome pathways mechanistically describing hepatotoxicity', *Methods in Molecular Biology*, 2834 pp. 249–273. Available at: https://doi.org/10.1007/978-1-0716-4003-6_12.

Caloni, F., et al. (2022) 'Replacement of animal testing by integrated approaches to testing and assessment (IATA): a call for in vivitrosi', *Archives of Toxicology*, 96(7), pp. 1935–1950. Available at: https://doi.org/10.1007/s00204-022-03299-x.

Carrasco Cabrera, L., et al. (2023) 'The 2021 European Union report on pesticide residues in food', *EFSA Journal*, 21(4). Available at: https://doi.org/10.2903/j.efsa.2023.7939.

Cattaneo, I., et al. (2023) 'Risk assessment of combined exposure to multiple chemicals at the European food safety authority: principles, guidance documents, applications and future challenges', *Toxins*, 15(1), p. 40. Available at: https://doi.org/10.3390/toxins15010040.

© Burleigh Dodds Science Publishing Limited, 2025. All rights reserved.

Cherrie, J.W., et al. (2020) 'Exposure models for REACH and occupational safety and health regulations', International Journal of Environmental Research and Public Health, 17(2), p. 383. Available at: https://doi.org/10.3390/ijerph17020383.

Choi, K., et al. (2006) 'Metabolism of chlorpyrifos and chlorpyrifos oxon by human hepatocytes*', *Journal of Biochemical and Molecular Toxicology*, 20(6), pp. 279–291. Available at: https://doi.org/10.1002/jbt.20145.

Clewell, R.A., et al. (2020) 'Application of a combined aggregate exposure pathway and adverse outcome pathway (AEP-AOP) approach to inform a cumulative risk assessment: a case study with phthalates', *Toxicology in Vitro*, 66, p. 104855. Available at: https://doi.org/10.1016/j.tiv.2020.104855.

Clippinger, A.J., et al. (2018) 'Pathway-based predictive approaches for non-animal assessment of acute inhalation toxicity', *Toxicology in Vitro*, 52, pp. 131–145. Available at: https://doi.org/10.1016/j.tiv.2018.06.009.

Coady, K., et al. (2019) 'When are adverse outcome pathways and associated assays "fit for purpose" for regulatory decision-making and management of chemicals?', *Integrated Environmental Assessment and Management*, 15(4), pp. 633–647. Available at: https://doi.org/10.1002/ieam.4153.

Corradi, M.P.F., et al. (2022) 'Natural language processing in toxicology: delineating adverse outcome pathways and guiding the application of new approach methodologies', *Biomaterials and Biosystems*, 7, p. 100061. Available at: https://doi.org/10.1016/j.bbiosy.2022.100061.

Corton, J.C., et al. (2014) 'Mode of action framework analysis for receptor-mediated toxicity: the peroxisome proliferator-activated receptor alpha (PPAR α) as a case study', *Critical Reviews in Toxicology*, 44(1), pp. 1–49. Available at: https://doi.org/10.3109/10408444.2013.835784.

Cronin, M.T.D., et al. (2022) 'A review of in silico toxicology approaches to support the safety assessment of cosmetics-related materials', *Computational Toxicology*, 21, p. 100213. Available at: https://doi.org/10.1016/j.comtox.2022.100213.

De Anna, J.S., et al. (2021) 'The insecticide chlorpyrifos modifies the expression of genes involved in the PXR and AhR pathways in the rainbow trout, Oncorhynchus mykiss', *Pesticide Biochemistry and Physiology*, 178, p. 104920. Available at: https://doi.org/10.1016/j.pestbp.2021.104920.

Di Nicola, M.R., et al. (2023) 'The use of new approach methodologies for the environmental risk assessment of food and feed chemicals', *Current Opinion in Environmental Science & Health*, 31, p. 100416. Available at: https://doi.org/10.1016/j.coesh.2022.100416.

Di Piazza, G., et al. (2024) 'Prioritisation of pesticides and target organ systems for dietary cumulative risk assessment based on the 2019–2021 monitoring cycle', *EFSA Journal*, 22(2), p. e8554. Available at: https://doi.org/10.2903/j.efsa.2024.8554.

Doan, T.Q., et al. (2020) 'In vitro profiling of the potential endocrine disrupting activities affecting steroid and aryl hydrocarbon receptors of compounds and mixtures prevalent in human drinking water resources', *Chemosphere*, 258, p. 127332. Available at: https://doi.org/10.1016/j.chemosphere.2020.127332.

Dolatabadi, M., et al. (2022) 'Adsorption characteristics in the removal of chlorpyrifos from groundwater using magnetic graphene oxide and carboxy methyl cellulose composite', *Separation and Purification Technology*, 300, p. 121919. Available at: https://doi.org/10.1016/j.seppur.2022.121919.


© Burleigh Dodds Science Publishing Limited, 2025. All rights reserved.

EC (2005) 'Regulation (EC) No 396/2005 of the European Parliament and of the Council of 23 February 2005 on maximum residue levels of pesticides in or on food and feed of plant and animal origin and amending Council Directive 91/414/EEC', Official Journal of the European Union [Preprint].

EC (2020) Report from the Commission to the European Parliament and the Council: Evaluation of Regulation (EC) No 1107/2009 on the placing of plant protection products on the market and of Regulation (EC) No 396/2005 on maximum residue levels of pesticides. Available at: eur-lex.europa.eu/legal-content/EN/TXT/PDF/?uri=CELEX:52020DC0208.

EFSA (2018) How pesticides are regulated in the EU - EFSA and the assessment of active substances. Available at: https://www.efsa.europa.eu/en/corporate/pub/pesticides-ebook-180424.

EFSA (2019) 'Technical report of the public consultation on the draft "Guidance on harmonised risk assessment methodologies for human health, animal health and ecological risk assessment of combined exposure to multiple chemicals"', *EFSA Supporting Publications*, 16(3), pp. 22–59. Available at: https://doi.org/10.2903/sp.efsa.2019.EN-1589.

EFSA (2021) 'Development of integrated approaches to testing and assessment (IATA) case studies on developmental neurotoxicity (DNT) risk assessment', *EFSA Journal*, 19(6), pp. 10–23. Available at: https://doi.org/10.2903/j.efsa.2021.6599.

EFSA (2024a) 'Roadmap for action for advancing aggregate exposure to chemicals in the EU', *EFSA Supporting Publications*, 21(7), pp. 24–42. Available at: https://doi.org/10.2903/sp.efsa.2024.EN-8971.

EFSA (2024b) Pesticides. Available at: https://www.efsa.europa.eu/en/topics/topic/pesticides.

Escher, S.E., et al. (2022) 'Integrate mechanistic evidence from new approach methodologies (NAMs) into a read-across assessment to characterise trends in shared mode of action', *Toxicology in Vitro*, 79, p. 105269. Available at: https://doi.org/10.1016/j.tiv.2021.105269.

Escher, S.E., et al. (2022) 'Development of a roadmap for action on new approach methodologies in risk assessment', *EFSA Supporting Publications*, 19(6), p. 7341. Available at: https://doi.org/10.2903/sp.efsa.2022.EN-7341.

Ezzi, L., et al. (2016) 'Toxicopathic changes and genotoxic effects in liver of rat following exposure to diazinon', *Environmental Science and Pollution Research*, 23(11), pp. 11163–11170. Available at: https://doi.org/10.1007/s11356-016-6314-0.

Fu, Y., et al. (2015) 'Distribution of chlorpyrifos in rice paddy environment and its potential dietary risk', *Journal of Environmental Sciences*, 35, pp. 101–107. Available at: https://doi.org/10.1016/j.jes.2015.02.015.

Garfitt, S. J., et al. (2002) 'Exposure to the organophosphate diazinon: data from a human volunteer study with oral and dermal doses', *Toxicology Letters*, 134(1–3), pp. 105–113. Available at: https://doi.org/10.1016/S0378-4274(02)00178-9.

Ghisari, M., et al. (2015) 'Effects of currently used pesticides and their mixtures on the function of thyroid hormone and aryl hydrocarbon receptor in cell culture', *Toxicology and Applied Pharmacology*, 284(3), pp. 292–303. Available at: https://doi.org/10.1016/j.taap.2015.02.004.

Gibbs, J.L., et al. (2017) 'Passive sampling for indoor and outdoor exposures to chlorpyrifos, azinphos-methyl, and oxygen analogs in a rural agricultural community',

© Burleigh Dodds Science Publishing Limited, 2025. All rights reserved.

*Environmental Health Perspectives*, 125(3), pp. 333–341. Available at: https://doi.org/10.1289/EHP425.

Gijbels, E., et al. (2021) 'Dataset on transcriptomic profiling of cholestatic liver injury induced by food additives and a cosmetic ingredient', *Data in Brief*, 38, p. 107373. Available at: https://doi.org/10.1016/j.dib.2021.107373.

Hartung, T. (2023) 'Artificial intelligence as the new frontier in chemical risk assessment', *Frontiers in Artificial Intelligence*, 6, p. 1269932. Available at: https://doi.org/10.3389/frai.2023.1269932.

Hassaan, M.A. and El Nemr, A. (2020) 'Pesticides pollution: classifications, human health impact, extraction and treatment techniques', *Egyptian Journal of Aquatic Research*, 46(3), pp. 207–220. Available at: https://doi.org/10.1016/j.ejar.2020.08.007.

Hemmerich, J. and Ecker, G.F. (2020) 'In silico toxicology: from structure–activity relationships towards deep learning and adverse outcome pathways', *WIREs Computational Molecular Science*, 10(4), p. e1475. Available at: https://doi.org/10.1002/wcms.1475.

Hernández, A.F., et al. (2017) 'Toxicological interactions of pesticide mixtures: an update', *Archives of Toxicology*, 91(10), pp. 3211–3223. Available at: https://doi.org/10.1007/s00204-017-2043-5.

Hines, D.E., et al. (2018) 'A case study application of the aggregate exposure pathway (AEP) and adverse outcome pathway (AOP) frameworks to facilitate the integration of human health and ecological end points for cumulative risk assessment (CRA)', *Environmental Science & Technology*, 52(2), pp. 839–849. Available at: https://doi.org/10.1021/acs.est.7b04940.

Honório, K.M., et al. (2008) 'Two- and three-dimensional quantitative structure-activity relationships studies on a series of liver X receptor ligands', *The Open Medicinal Chemistry Journal*, 2(1), pp. 87–96. Available at: https://doi.org/10.2174/1874104500802010087.

Howell III, G.E., et al. (2016) 'Exposure to chlorpyrifos increases neutral lipid accumulation with accompanying increased de novo lipogenesis and decreased triglyceride secretion in McArdle-RH7777 hepatoma cells', *Toxicology in Vitro*, 32, pp. 181–189. Available at: https://doi.org/10.1016/j.tiv.2016.01.002.

Hristozov, D., et al. (2024) 'Next generation risk assessment approaches for advanced nanomaterials: current status and future perspectives', *NanoImpact*, 35, p. 100523. Available at: https://doi.org/10.1016/j.impact.2024.100523.

Hubbard, H.F., et al. (2022) 'Exposure prioritization (Ex Priori): a screening-level high-throughput chemical prioritization tool', *Toxics*, 10(10), p. 569. Available at: https://doi.org/10.3390/toxics10100569.

Hung, C.-C., et al. (2018) 'Pesticides in indoor and outdoor residential dust: a pilot study in a rural county of Taiwan', *Environmental Science and Pollution Research*, 25(23), pp. 23349–23356. Available at: https://doi.org/10.1007/s11356-018-2413-4.

IARC (2024) Agents classified by the IARC monographs. Available at: https://monographs.iarc.who.int/agents-classified-by-the-iarc/.

Jaylet, T., et al. (2024) 'Comprehensive mapping of the AOP-Wiki database: identifying biological and disease gaps', *Frontiers in Toxicology*, 6, 1285768. Available at: https://doi.org/10.3389/ftox.2024.1285768.

Jiang, J., et al. (2023) 'Unraveling the mechanisms underlying drug-induced cholestatic liver injury: identifying key genes using machine learning techniques on human in

© Burleigh Dodds Science Publishing Limited, 2025. All rights reserved.

vitro data sets', *Archives of Toxicology*, 97(11), pp. 2969–2981. Available at: https://doi.org/10.1007/s00204-023-03583-4.

Judson, R. (2013) 'Perspectives on validation of high-throughput assays supporting 21st century toxicity testing', *ALTEX*, 30(1), pp. 51–66. Available at: https://doi.org/10.14573/altex.2013.1.051.

Judson, R.S., et al. (2010) 'Screening of environmental chemicals for targeted testing prioritization: the ToxCast project', *Environmental Health Perspectives*, 118(4), pp. 485–492. Available at: https://doi.org/10.1289/ehp.0901392.

Kalyabina, V.P., et al. (2021) 'Pesticides: formulants, distribution pathways and effects on human health – a review', *Toxicology Reports*, 8, pp. 1179–1192. Available at: https://doi.org/10.1016/j.toxrep.2021.06.004.

Karami-Mohajeri, S., et al. (2017) 'Adverse effects of organophosphorus pesticides on the liver: a brief summary of four decades of research', *Archives of Industrial Hygiene and Toxicology*, 68(4), pp. 261–275. Available at: https://doi.org/10.1515/aiht-2017-68-2989.

Karanth, S. (2005) 'Diazinon*', In Anderson, B., et al. (Eds.), *Encyclopedia of Toxicology*. New York, NY: Elsevier, pp. 785–786. Available at: https://doi.org/10.1016/B0-12-369400-0/00304-5.

Karrer, C., et al. (2020) 'The EuroMix human biomonitoring study: source-to-dose modeling of cumulative and aggregate exposure for the bisphenols BPA, BPS, and BPF and comparison with measured urinary levels', *Environment International*, 136, p. 105397. Available at: https://doi.org/10.1016/j.envint.2019.105397.

Kashyap, M.P., et al. (2010) 'Caspase cascade regulated mitochondria mediated apoptosis in monocrotophos exposed PC12 cells', *Chemical Research in Toxicology*, 23(11), pp. 1663–1672. Available at: https://doi.org/10.1021/tx100234m.

Kaur, P., et al. (2007) 'Impaired mitochondrial energy metabolism and neuronal apoptotic cell death after chronic dichlorvos (OP) exposure in rat brain', *NeuroToxicology*, 28(6), pp. 1208–1219. Available at: https://doi.org/10.1016/j.neuro.2007.08.001.

Kermani, M., et al. (2021) 'A global systematic review, meta-analysis and health risk assessment on the quantity of malathion, diazinon and chlorpyrifos in vegetables', *Chemosphere*, 270, p. 129382. Available at: https://doi.org/10.1016/j.chemosphere.2020.129382.

Knapen, D., et al. (2018) 'Adverse outcome pathway networks I: development and applications', *Environmental Toxicology and Chemistry*, 37(6), pp. 1723 1733. Available at: https://doi.org/10.1002/etc.4125.

Kovacic, P. (2003) 'Mechanism of organophosphates (Nerve gases and pesticides) and antidotes: electron transfer and oxidative stress', *Current Medicinal Chemistry*, 10(24), pp. 2705–2709. Available at: https://doi.org/10.2174/0929867033456314.

Kramer, N.I., et al. (2019) 'Characterizing the coverage of critical effects relevant in the safety evaluation of food additives by AOPs', *Archives of Toxicology*, 93(8), pp. 2115–2125. Available at: https://doi.org/10.1007/s00204-019-02501-x.

Lambert, J. (2022) 'Adverse outcome pathway "footprinting": a novel approach to the integration of 21st century toxicology information into chemical mixtures risk assessment', *Toxics*, 11(1), p. 37. Available at: https://doi.org/10.3390/toxics11010037.

© Burleigh Dodds Science Publishing Limited, 2025. All rights reserved.

Lasagna, M., et al. (2022) 'Endocrine disruptor chlorpyrifos promotes migration, invasion, and stemness phenotype in 3D cultures of breast cancer cells and induces a wide range of pathways involved in cancer progression', *Environmental Research*, 204, p. 111989. Available at: https://doi.org/10.1016/j.envres.2021.111989.

Leist, M., et al. (2017) 'Adverse outcome pathways: opportunities, limitations and open questions', *Archives of Toxicology*, 91(11), pp. 3477–3505. Available at: https://doi.org/10.1007/s00204-017-2045-3.

Lentz, T.J., et al. (2015) 'Aggregate exposure and cumulative risk assessment–integrating occupational and non-occupational risk factors', *Journal of Occupational and Environmental Hygiene*, 12(sup1), pp. S112–S126. Available at: https://doi.org/10.1080/15459624.2015.1060326.

Li, R., et al. (2015) 'Chlorpyrifos residue levels on field crops (rice, maize and soybean) in China and their dietary risks to consumers', *Food Control*, 51, pp. 212–217. Available at: https://doi.org/10.1016/j.foodcont.2014.11.023.

Lichtenstein, D., et al. (2020) 'An adverse outcome pathway-based approach to assess steatotic mixture effects of hepatotoxic pesticides in vitro', *Food and Chemical Toxicology*, 139, p. 111283. Available at: https://doi.org/10.1016/j.fct.2020.111283.

Lim, M., et al. (2019) 'Receptor-based aggregate exposure assessment of phthalates based on individual's simultaneous use of multiple cosmetic products', *Food and Chemical Toxicology*, 127, pp. 163–172. Available at: https://doi.org/10.1016/j.fct.2019.03.031.

Lin, S.-Y., et al. (1996) 'Adsorption kinetics of diazinon at the air–water interface', *The Journal of Physical Chemistry*, 100(41), pp. 16678–16684. Available at: https://doi.org/10.1021/jp9614299.

Lizarraga, L.E., et al. (2023) 'Advancing the science of a read-across framework for evaluation of data-poor chemicals incorporating systematic and new approach methods', *Regulatory Toxicology and Pharmacology*, 137, p. 105293. Available at: https://doi.org/10.1016/j.yrtph.2022.105293.

Luijten, M., et al. (2022) 'Prioritization of chemicals in food for risk assessment by integrating exposure estimates and new approach methodologies: a next generation risk assessment case study', *Frontiers in Toxicology*, 4, p. 933197. Available at: https://doi.org/10.3389/ftox.2022.933197.

Manfo, F.P.T., et al. (2020) 'Evaluation of the effects of agro pesticides use on liver and kidney function in farmers from Buea, Cameroon', *Journal of Toxicology*, 2020, pp. 1–10. Available at: https://doi.org/10.1155/2020/2305764.

Mikhail, T.H., et al. (1979) 'Acute toxicity of organophosphorus and organochlorine insecticides in laboratory animals', *Zeitschrift für Ernährungswissenschaft*, 18(4), pp. 258–268. Available at: https://doi.org/10.1007/BF02020516.

Miret, N.V., et al. (2023) 'Mechanisms of breast cancer progression induced by environment-polluting aryl hydrocarbon receptor agonists', *Biochemical Pharmacology*, 216, p. 115773. Available at: https://doi.org/10.1016/j.bcp.2023.115773.

Mojsak, P., et al. (2018) 'Estimating acute and chronic exposure of children and adults to chlorpyrifos in fruit and vegetables based on the new, lower toxicology data', *Ecotoxicology and Environmental Safety*, 159, pp. 182–189. Available at: https://doi.org/10.1016/j.ecoenv.2018.05.006.

Montanarí, C., et al. (2024) 'Chlorpyrifos induces cytotoxicity via oxidative stress and mitochondrial dysfunction in HepG2 cells', *Food and Chemical Toxicology*, 192, p. 114933. Available at: https://doi.org/10.1016/j.fct.2024.114933.

© Burleigh Dodds Science Publishing Limited, 2025. All rights reserved.

Moore, P.D., et al. (2010) 'Malathion-induced oxidative stress, cytotoxicity, and genotoxicity in human liver carcinoma', *Environmental Toxicology*, 25(3), pp. 221–226. Available at: https://doi.org/10.1002/tox.20492.

More, S.J., et al. (2021) 'Guidance document on scientific criteria for grouping chemicals into assessment groups for human risk assessment of combined exposure to multiple chemicals', *EFSA Journal*, 19(12), pp. 8–20. Available at: https://doi.org/10.2903/j.efsa.2021.7033.

Moyano, P., et al. (2020) 'Chlorpyrifos-induced cell proliferation in human breast cancer cell lines differentially mediated by estrogen and aryl hydrocarbon receptors and KIAA1363 enzyme after 24 h and 14 days exposure', *Chemosphere*, 251, p. 126426. Available at: https://doi.org/10.1016/j.chemosphere.2020.126426.

Moyano, P., et al. (2021a) 'Aryl hydrocarbon receptor activation produces heat shock protein 90 and 70 overexpression, prostaglandin E2/Wnt/β-catenin signaling disruption, and cell proliferation in MCF-7 and MDA-MB-231 cells after 24 h and 14 days of chlorpyrifos treatment', *Chemical Research in Toxicology*, 34(9), pp. 2019–2023. Available at: https://doi.org/10.1021/acs.chemrestox.1c00258.

Moyano, P., et al. (2021b) 'Chlorpyrifos induces cell proliferation in MCF-7 and MDA-MB-231 cells, through cholinergic and Wnt/β-catenin signaling disruption, AChE-R upregulation and oxidative stress generation after single and repeated treatment', *Food and Chemical Toxicology*, 152, p. 112241. Available at: https://doi.org/10.1016/j.fct.2021.112241.

Mu, Y., et al. (2021) 'Prognostic value of the neutrophil-to-lymphocyte ratio in acute organophosphorus pesticide poisoning', *Open Life Sciences*, 16(1), pp. 703–710. Available at: https://doi.org/10.1515/biol-2021-0069.

Nematy, M., et al. (2016) 'Does diazinon-sprayed market melon alter cholinesterase activity in healthy consumers? A randomized control trial', *European Review for Medical and Pharmacological Sciences*, 20(12), pp. 2607–2612.

Nielsen, E., et al. (2012) 'Identification of cumulative assessment groups of pesticides', *EFSA Supporting Publications*, 9(4), pp. 116–152. Available at: https://doi.org/10.2903/sp.efsa.2012.EN-269.

Noore, S., et al. (2021) 'Persistence and diffusion behaviour of chlorpyrifos in five different species of vegetables: a comparative analysis', *Ecotoxicology and Environmental Safety*, 217, p. 112208. Available at: https://doi.org/10.1016/j.ecoenv.2021.112208.

OECD (2013) *Revised Guidance Document on Developing and Assessing Adverse Outcome Pathways: Series on Testing & Assessment No. 184*. ENV/JM/MONO(2013)6. Paris: OECD Publishing. Available: https://one.oecd.org/document/env/jm/mono(2013)6/en/pdf.

OECD (2014) *Guidance on Grouping of Chemicals*. Paris: OECD. Available at: https://doi.org/10.1787/9789264085831-en.

OECD (2016) 'OECD guidance on the reporting of defined approaches and individual information sources to be used within integrated approaches to testing and assessment (IATA) for skin sensitization', *Toxicology Letters*, 258, p. S2. Available at: https://doi.org/10.1016/j.toxlet.2016.06.027.

OECD (2017) *Guidance on Grouping of Chemicals* (2nd ed.). Paris: OECD. Available at: https://doi.org/10.1787/9789 264274679-en.

OECD (2021) *Guidance Document for the scientific review of Adverse Outcome Pathways, OECD Series on Testing and Assessment, No. 344*. Paris: OECD Publishing. Available at: https://doi.org/10.1787/a6bec14b-en.

© Burleigh Dodds Science Publishing Limited, 2025. All rights reserved.

OECD, Integrated Approaches to Testing and Assessment (IATA). Available at: https://www.oecd.org/en/topics/sub-issues/assessment-of-chemicals/integrated-approaches-to-testing-and-assessment.html. (Accessed 28 September, 2024)

Ojha, A., et al. (2013) 'Evaluation of DNA damage and cytotoxicity induced by three commonly used organophosphate pesticides individually and in mixture, in rat tissues', *Environmental Toxicology*, 28(10), pp. 543–552. Available at: https://doi.org/10.1002/tox.20748.

Oki, N.O., et al. (2019) 'Integrated analysis of in vitro data and the adverse outcome pathway framework for prioritization and regulatory applications: an exploratory case study using publicly available data on piperonyl butoxide and liver models', *Toxicology in Vitro*, 54, pp. 23–32. Available at: https://doi.org/10.1016/j.tiv.2018.09.002.

Paini, A., et al. (2022) 'Towards a qAOP framework for predictive toxicology: linking data to decisions', *Computational Toxicology*, 21, p. 100195. Available at: https://doi.org/10.1016/j.comtox.2021.100195.

Patlewicz, G., et al. (2015) 'Proposing a scientific confidence framework to help support the application of adverse outcome pathways for regulatory purposes', *Regulatory Toxicology and Pharmacology*, 71(3), pp. 463–477. Available at: https://doi.org/10.1016/j.yrtph.2015.02.011.

Peng, G., et al. (2022) 'Aggregate exposure pathways for microplastics (mpAEP): an evidence-based framework to identify research and regulatory needs', *Water Research*, 209, p. 117873. Available at: https://doi.org/10.1016/j.watres.2021.117873.

Perkins, E.J., et al. (2019) 'Building and applying quantitative adverse outcome pathway models for chemical hazard and risk assessment', *Environmental Toxicology and Chemistry*, 38(9), pp. 1850–1865. Available at: https://doi.org/10.1002/etc.4505.

PMRA (2003) 'Pest Management Regulatory Agency'. Available at: https://www.canada.ca/en/health-canada/services/consumer-product-safety/reports-publications/pesticides-pest-management/policies-guidelines/risk-management-pest-control-products.html.

Poet, T.S., et al. (2004) 'Physiologically based pharmacokinetic/pharmacodynamic model for the organophosphorus pesticide diazinon', *NeuroToxicology*, 25(6), pp. 1013–1030. Available at: https://doi.org/10.1016/j.neuro.2004.03.002.

Price, P.S., et al. (2020) 'Organizing mechanism-related information on chemical interactions using a framework based on the aggregate exposure and adverse outcome pathways', *Environment International*, 138, p. 105673. Available at: https://doi.org/10.1016/j.envint.2020.105673.

Quirós-Alcalá, L., et al. (2012) 'Organophosphorous pesticide breakdown products in house dust and children's urine', *Journal of Exposure Science & Environmental Epidemiology*, 22(6), pp. 559–568. Available at: https://doi.org/10.1038/jes.2012.46.

Raina, R., et al. (2010) 'Occurrence and relationship of organophosphorus insecticides and their degradation products in the atmosphere in Western Canada agricultural regions', *Environmental Science & Technology*, 44(22), pp. 8541–8546. Available at: https://doi.org/10.1021/es102450k.

Rathod, A.L. and Garg, R.K. (2017) 'Chlorpyrifos poisoning and its implications in human fatal cases: a forensic perspective with reference to Indian scenario', *Journal of*

© Burleigh Dodds Science Publishing Limited, 2025. All rights reserved.

*Forensic and Legal Medicine*, 47, pp. 29–34. Available at: https://doi.org/10.1016/j.jflm.2017.02.003.

Robb, E.L., et al. (2024) Organophosphate Toxicity. Available at: https://pubmed.ncbi.nlm.nih.gov/29261901/.

Rooney, J., et al. (2018) 'Adverse outcome pathway-driven identification of rat liver tumorigens in short-term assays', *Toxicology and Applied Pharmacology*, 356, pp. 99–113. Available at: https://doi.org/10.1016/j.taap.2018.07.023.

Rousis, N.I., et al. (2017) 'Monitoring a large number of pesticides and transformation products in water samples from Spain and Italy', *Environmental Research*, 156, pp. 31–38. Available at: https://doi.org/10.1016/j.envres.2017.03.013.

Sammi, S.R., et al. (2023) 'Complementary biological and computational approaches identify distinct mechanisms of chlorpyrifos versus chlorpyrifos-oxon-induced dopaminergic neurotoxicity', *Toxicological Sciences*, 191(1), pp. 163–178. Available at: https://doi.org/10.1093/toxsci/kfac114.

Sanz-Serrano, J., et al. (2024) 'Chemical-induced liver cancer: an adverse outcome pathway perspective', *Expert Opinion on Drug Safety*, 23(4), pp. 425–438. Available at: https://doi.org/10.1080/14740338.2024.2326479.

Saraei, F., et al. (2016) 'Study of the effects of diazinon on fetal liver in BALB/c Mice', *Iranian Red Crescent Medical Journal*, 18(4), pp. 2–6. Available at: https://doi.org/10.5812/ircmj.28076.

Sasikala, C., et al. (2012) 'Biodegradation of chlorpyrifos by bacterial consortium isolated from agriculture soil', *World Journal of Microbiology and Biotechnology*, 28(3), pp. 1301–1308. Available at: https://doi.org/10.1007/s11274-011-0879-z.

Schmeisser, S., et al. (2023) 'New approach methodologies in human regulatory toxicology – Not if, but how and when!', *Environment International*, 178, p. 108082. Available at: https://doi.org/10.1016/j.envint.2023.108082.

Schultz, T.W. and Cronin, M.T.D. (2017) 'Lessons learned from read-across case studies for repeated-dose toxicity', *Regulatory Toxicology and Pharmacology*, 88, pp. 185–191. Available at: https://doi.org/10.1016/j.yrtph.2017.06.011.

Schultz, T.W., et al. (2018) 'The OECD QSAR toolbox starts its second decade', *Methods in Molecular Biology*, 1800, pp. 55–77. Available at: https://doi.org/10.1007/978-1-4939-7899-1_2.

Sharbidre, A.A., et al. (2011) 'Effect of diazinon on acetylcholinesterase activity and lipid peroxidation of poecilia reticulata', *Research Journal of Environmental Toxicology*, 5(2), pp. 152–161. Available at: https://doi.org/10.3923/rjet.2011.152.161

Shokrzadeh, M., et al. (2013) 'Potential chemoprotective effects of selenium on diazinon-induced DNA damage in rat peripheral blood lymphocyte', *Human & Experimental Toxicology*, 32(7), pp. 759–765. Available at: https://doi.org/10.1177/0960327112468179.

Sleiman, A., et al. (2024) 'AOP report: development of an adverse outcome pathway for deposition of energy leading to learning and memory impairment', *Environmental and Molecular Mutagenesis*, 65(S3), pp. 57–84. Available at: https://doi.org/10.1002/em.22622.

Spinu, N., et al. (2020) 'Quantitative adverse outcome pathway (qAOP) models for toxicity prediction', *Archives of Toxicology*, 94(5), pp. 1497–1510. Available at: https://doi.org/10.1007/s00204-020-02774-7.

Stucki, A.O., et al. (2022) 'Use of new approach methodologies (NAMs) to meet regulatory requirements for the assessment of industrial chemicals and pesticides for effects on

© Burleigh Dodds Science Publishing Limited, 2025. All rights reserved.

human health', *Frontiers in Toxicology*, 4, p. 964553. Available at: https://doi.org/10.3389/ftox.2022.964553.

Svingen, T., et al. (2021) 'A pragmatic approach to adverse outcome pathway development and evaluation', *Toxicological Sciences*, 184(2), pp. 183–190. Available at: https://doi.org/10.1093/toxsci/kfab113.

Tan, Y.-M., et al. (2018a) 'Refining the aggregate exposure pathway', *Environmental Science: Processes & Impacts*, 20(3), pp. 428–436. Available at: https://doi.org/10.1039/C8EM00018B.

Tan, Y.-M., et al. (2018b) 'Aggregate exposure pathways in support of risk assessment', *Current Opinion in Toxicology*, 9, pp. 8–13. Available at: https://doi.org/10.1016/j.cotox.2018.03.006.

Teeguarden, J.G., et al. (2016) 'Completing the link between exposure science and toxicology for improved environmental health decision making: the aggregate exposure pathway framework', *Environmental Science & Technology*, 50(9), pp. 4579–4586. Available at: https://doi.org/10.1021/acs.est.5b05311.

Tehrani, H., et al. (2013) 'Protective effects of N-acetylcysteine on aluminum phosphide-induced oxidative stress in acute human poisoning', *Clinical Toxicology*, 51(1), pp. 23–28. Available at: https://doi.org/10.3109/15563650.2012.743029.

Timbrell, J.A. (2020) 'The liver as a target organ for toxicity', In Cohen, G.M. (Ed.), *Target Organ Toxicity*. Boca Raton, FL: CRC Press, pp. 145–173. Available at: https://doi.org/10.1201/9780367811068-8

Toledo-Ibarra, G.A., et al. (2016) 'Oxidative damage in gills and liver in Nile tilapia (Oreochromis niloticus) exposed to diazinon', *Comparative Biochemistry and Physiology Part A: Molecular & Integrative Physiology*, 200, pp. 3–8. Available at: https://doi.org/10.1016/j.cbpa.2016.05.007.

Topal, A., et al. (2014) 'In vivo changes in carbonic anhydrase activity and histopathology of gill and liver tissues after acute exposure to chlorpyrifos in rainbow trout', *Archives of Industrial Hygiene and Toxicology*, 65(4), pp. 377–385. Available at: https://doi.org/10.2478/10004-1254-65-2014-2547.

Tripathi, S. and Srivastav, A.K. (2010) 'Liver profile of rats after long-term ingestion of different doses of chlorpyrifos', *Pesticide Biochemistry and Physiology*, 97(1), pp. 60–65. Available at: https://doi.org/10.1016/j.pestbp.2009.12.005.

Ubaid ur Rahman, H., et al. (2021) 'A comprehensive review on chlorpyrifos toxicity with special reference to endocrine disruption: evidence of mechanisms, exposures and mitigation strategies', *Science of The Total Environment*, 755, p. 142649. Available at: https://doi.org/10.1016/j.scitotenv.2020.142649.

Ünlü Endirlik, B., et al. (2023) 'Non-additive mixture effects of benzo[a]pyrene and pesticides in vitro and in vivo: role of AhR signaling', *Environmental Pollution*, 316, p. 120510. Available at: https://doi.org/10.1016/j.envpol.2022.120510.

Usmani, S.M., et al. (2024) 'Review of new approach methodologies for application in risk assessment of nanoparticles in the food and feed sector: status and challenges', *EFSA Supporting Publications*, 21(9). Available at: https://doi.org/10.2903/sp.efsa.2024.EN-8826.

van Ertvelde, J., et al. (2023) 'Optimization of an adverse outcome pathway network on chemical-induced cholestasis using an artificial intelligence-assisted data collection and confidence level quantification approach', *Journal of Biomedical Informatics*, 145, p. 104465. Available at: https://doi.org/10.1016/j.jbi.2023.104465.

© Burleigh Dodds Science Publishing Limited, 2025. All rights reserved.

van Wendel de Joode, B., et al. (2012) 'Indigenous children living nearby plantations with chlorpyrifos-treated bags have elevated 3,5,6-trichloro-2-pyridinol (TCPy) urinary concentrations', *Environmental Research*, 117, pp. 17–26. Available at: https://doi.org/10.1016/j.envres.2012.04.006.

Venkateswara Rao, J. (2006) 'Sublethal effects of an organophosphorus insecticide (RPR-II) on biochemical parameters of tilapia, Oreochromis mossambicus', *Comparative Biochemistry and Physiology Part C: Toxicology & Pharmacology*, 143(4), pp. 492–498. Available at: https://doi.org/10.1016/j.cbpc.2006.05.001.

Verhoeven, A., et al. (2024) 'A quantitative weight-of-evidence method for confidence assessment of adverse outcome pathway networks: a case study on chemical-induced liver steatosis', *Toxicology*, 505, p. 153814. Available at: https://doi.org/10.1016/j.tox.2024.153814.

Villeneuve, D.L., et al. (2014a) 'Adverse outcome pathway (AOP) development I: strategies and principles', *Toxicological Sciences*, 142(2), pp. 312–320. Available at: https://doi.org/10.1093/toxsci/kfu199.

Villeneuve, D.L., et al. (2014b) 'Adverse outcome pathway development II: best practices', *Toxicological Sciences*, 142(2), pp. 321–330. Available at: https://doi.org/10.1093/toxsci/kfu200.

Villeneuve, D.L., et al. (2018) 'Adverse outcome pathway networks II: network analytics', *Environmental Toxicology and Chemistry*, 37(6), pp. 1734–1748. Available at: https://doi.org/10.1002/etc.4124.

Vinken, M. (2024) 'Adverse outcome pathway networks as the basis for the development of new approach methodologies: liver toxicity as a case study', *Current Opinion in Toxicology*, p. 100504. Available at: https://doi.org/10.1016/j.cotox.2024.100504.

Vinken, M., et al. (2014) 'Primary hepatocytes and their cultures in liver apoptosis research', *Archives of Toxicology*, 88(2), pp. 199–212. Available at: https://doi.org/10.1007/s00204-013-1123-4.

VoPham, T., et al. (2017) 'Pesticide exposure and liver cancer: a review', *Cancer Causes & Control*, 28(3), pp. 177–190. Available at: https://doi.org/10.1007/s10552-017-0854-6.

Wang, L., et al. (2016) 'Chlorpyrifos exposure in farmers and urban adults: metabolic characteristic, exposure estimation, and potential effect of oxidative damage', *Environmental Research*, 149, pp. 164–170. Available at: https://doi.org/10.1016/j.envres.2016.05.011.

Wilkinson, M.D., et al. (2019) 'Addendum: the FAIR guiding principles for scientific data management and stewardship', *Scientific Data*, 6(1), p. 6. Available at: https://doi.org/10.1038/s41597-019-0009-6.

Wohlleben, W., et al. (2023) 'Lessons learned from the grouping of chemicals to assess risks to human health', *Angewandte Chemie International Edition*, 62(22), e202210651. Available at: https://doi.org/10.1002/anie.202210651.

Wołejko, E., et al. (2022) 'Chlorpyrifos occurrence and toxicological risk assessment: a review', *International Journal of Environmental Research and Public Health*, 19(19), p. 12209. Available at: https://doi.org/10.3390/ijerph191912209.

Yager, J.D. and Leihr, J.G. (1996) 'Molecular mechanisms of estrogen carcinogenesis', *Annual Review of Pharmacology and Toxicology*, 36(1), pp. 203–232. Available at: https://doi.org/10.1146/annurev.pa.36.040196.001223.

© Burleigh Dodds Science Publishing Limited, 2025. All rights reserved.

Yahia, D. and Ali, M.F. (2018) 'Assessment of neurohepatic DNA damage in male Sprague–Dawley rats exposed to organophosphates and pyrethroid insecticides', *Environmental Science and Pollution Research*, 25(16), pp. 15616–15629. Available at: https://doi.org/10.1007/s11356-018-1776-x.

Zhong, J., et al. (2022) 'Fate of chlorpyrifos bound residues in paddy soils: release, transformation, and phytoavailability', *Environment International*, 166, p. 107338. Available at: https://doi.org/10.1016/j.envint.2022.107338.

© Burleigh Dodds Science Publishing Limited, 2025. All rights reserved.

# Chapter 11

# Developments in in silico tools for safety and hazard assessment of chemicals in food

*Szabina Stice, U.S. Food and Drug Administration, USA*[1]



## 1 Introduction

Chemicals may be present in food as naturally occurring substances, contaminants, or by intentional addition. Compounds intentionally added to food are substances deliberately incorporated into food products for various purposes, such as enhancing flavor, texture, appearance, shelf life, or nutritional value. Chemical contaminants in food refer to substances that may be present in food and have the potential to cause harm. Some common types of chemical contaminants in food include heavy metals, environmental pollutants (e.g. polychlorinated biphenyls and dioxins), industrial contaminants (e.g. perfluorooctanoic acid), and natural toxins (such as mycotoxins and alkaloids, e.g. solanine). The presence of chemical contaminants in food can pose various

1 Disclaimer: This chapter reflects the views of the author and should not be construed to represent the U.S. Food and Drug Administration's (FDA) views and policies. The mention of commercial products, their sources, or their uses in connection with material reported herein is not to be construed as an actual or implied endorsement of such products by the FDA.

http://dx.doi.org/10.19103/AS.2025.0158.31
© Burleigh Dodds Science Publishing Limited, 2025. All rights reserved.

health risks depending on the type of contaminant, its concentration, and the duration of exposure. The intentional use of veterinary drugs in animals, as well as pesticides and herbicides on crops, can lead to residues of these substances in food. These residues are not regarded as contaminants. Regulatory authorities establish limits for contaminants and set tolerances for pesticides, herbicides, and veterinary drugs to ensure food safety and protect public health (e.g. U.S. Code of Federal Regulations, Title 21, Chapter I, Subchapter B, Part 109: Unavoidable contaminants in food for human consumption and food-packaging material) (FDA, 2024a).

Employing computational toxicology tools is a promising approach that leverages technology to predict potential health risks associated with chemicals regardless of whether they are present naturally, as contaminants, or added intentionally to food products. Assessing the safety of these chemicals is critical for protecting public health and ensuring food safety standards are met.

Computational toxicology tools offer several advantages in this context:

(1) **Efficiency**: Traditional methods for assessing the safety of chemicals in food can be time-consuming and expensive, often requiring animal studies to meet regulatory authorities' criteria. Computational tools can rapidly screen large numbers of chemicals and predict potential toxic effects based on their chemical structures, predicted metabolism, and/or physical-chemical properties, thereby reducing the need for extensive, time-consuming, and costly experimental testing.

(2) **Data Integration**: Computational toxicology platforms can integrate diverse sources of data, including chemical structures, toxicological endpoints, exposure levels, and biological pathways. By aggregating and analyzing this information, they provide a comprehensive understanding of the potential risks associated with chemicals in food.

(3) **Predictive Modeling**: These tools utilize various computational models, such as quantitative structure-activity relationships (QSAR), machine learning, and structural alerts[2] to predict the toxicity of chemicals based on their structural properties and known toxicity data. This predictive capability enables the assessment of chemicals with limited or no experimental toxicity data.

(4) **Risk Prioritization**: Computational models can aid in prioritizing chemicals for further testing based on their predicted toxicity and potential exposure levels in the food supply, enabling regulatory agencies to focus resources on chemicals of highest concern.

2 Structural alerts in toxicology refer to specific chemical substructures or functional groups that are known or suspected to be associated with discrete toxicological effects.

© Burleigh Dodds Science Publishing Limited, 2025. All rights reserved.

(5) **Transparency and Reproducibility**: Many computational toxicology tools are based on established algorithms and methodologies, which enhance transparency and reproducibility in safety assessments. Some platforms are transparent and allow stakeholders–including researchers, regulators, and industry professionals–to access and validate the underlying data and models.

(6) **Regulatory Compliance**: Computational toxicology tools are often designed to align with regulatory requirements and guidelines. For example, the OECD QSAR Toolbox supports regulatory assessment under various international chemical control regimes, and the VEGA QSAR tool supports regulatory frameworks such as REACH (Registration, Evaluation, Authorization and Restriction of Chemicals) in Europe. By incorporating recognized methodologies and endpoints, these tools may also facilitate compliance with food safety regulations and standards.

(7) **Reduction of Animal Testing and Inform Testing Strategies**: By offering alternative methods for toxicity prediction, computational tools contribute to the reduction of animal testing, aligning with ethical considerations and regulatory trends favoring non-animal testing approaches. Additionally, when used to form categories (compounds of related substances), they help identify data gaps for the category or for a category member of interest. Carefully designed testing strategies for the category or a member of the category can then help minimize the number of animals needed for testing.

(8) **Accessibility**: Many computational toxicology platforms are available free to the public (open access), increasing their accessibility and utility.

Despite these advantages, acknowledging the limitations of computational toxicology tools is crucial. These tools rely on available data and models, which can introduce uncertainties and limitations, such as uncertainties in model predictions, the reliability and relevance of the underlying data, and the need for validation with experimental data. Expert judgment remains essential in risk assessment. Additionally, these tools often cannot fully replace traditional toxicity testing, especially for complex endpoints or novel chemicals with limited or no experimental data, and as such, may be outside of the applicability domain of most, if not all, tools.

Regulatory agencies worldwide are increasingly incorporating computational toxicology tools into their risk assessment frameworks to enhance food safety and protect public health. In practice, a combined approach that integrates computational predictions with experimental data and expert judgment is often employed for safety assessments of chemicals

© Burleigh Dodds Science Publishing Limited, 2025. All rights reserved.

in food. This approach optimizes the efficiency of toxicity evaluations while ensuring robust and scientifically sound risk assessments.

A synergistic approach that combines computational predictions with experimental data not only strengthens safety, risk, and hazard assessments but also fosters innovation and sustainability in chemical development and regulation. By leveraging the strengths of both predictive modeling and empirical research, stakeholders can advance toward safer chemicals, reduce risks to human health and the environment, and promote responsible stewardship of chemical resources. This integrated approach is essential for effectively and sustainably addressing complex chemical safety challenges.

This chapter focuses on computational toxicology tools freely available to anyone wishing to employ them for safety or hazard assessment, but also describes some commercially available tools. Also mentioned are promising in silico (i.e. computational) tools currently under development. These software programs vary in their capabilities, regulatory acceptance, and the endpoints they cover. When selecting software for predicting toxicity, considering such factors as the specific endpoints of interest, the reliability and relevance of the underlying model's data (domain of applicability), the predictive performance of the model, and the regulatory guidance governing the use of predicted toxicity data in your field of application is essential.

## 2 Selected publicly available in silico tools

Several computational toxicology tools are publicly available (i.e. accessible for use by the public without requiring a purchase), offering various functionalities for toxicity prediction, data identification and retrieval, chemical screening, risk assessment, and other related tasks. These tools and platforms vary in scope, methodologies, and applicability, but collectively, they contribute to advancing computational toxicology and supporting chemical safety assessments. Some of these free tools and platforms along with their source/availability, are listed in Table 1.

### *2.1 CompTox Chemicals Dashboard*

The aim of this dashboard is to provide comprehensive access to chemistry, toxicity, and exposure information on chemicals. GenRA (Generalized Read-Across) module within the dashboard focuses on identifying potential analogs for the target (data-poor) chemical. Once analogs are identified, those with

© Burleigh Dodds Science Publishing Limited, 2025. All rights reserved.

**Table 1** Selected publicly available in silico tools and their availability

| Tool (hyperlinked to more info) | Available from | Where to get it from? Online/Web-based | Download |
|---|---|---|---|
| CompTox Chemicals Dashboard – GenRA | U.S. EPA[1] | https://comptox.epa.gov/dashboard/ | |
| Cheminformatics analysis modules | U.S. EPA[1] | https://www.epa.gov/comptox-tools/cheminformatics | |
| Analog Identification Methodology (AIM) | U.S. EPA[1] | | https://www.epa.gov/tsca-screening-tools/analog-identification-methodology-aim-tool |
| OECD QSAR Toolbox | OECD[2] | | https://qsartoolbox.org/download/ |
| NTP ICE (Integrated Chemical Environment) | NTP[3] | https://ice.ntp.niehs.nih.gov/ | |
| VEGA HUB | VEGA HUB | | https://www.vegahub.eu/portfolio-types/vega-software/ |
| T.E.S.T. (Toxicity Estimation Software Tool) | U.S. EPA[1] | | https://www.epa.gov/comptox-tools/toxicity-estimation-software-tool-test |
| Cefic-LRI (Long-range Research Initiative) Tools | Cefic-LRI | https://cefic-lri.org/toolbox/[4] | |
| TISBE (TIRESIA Improved on Structure-Based Explainability) | UNIBA[5] | https://prometheus.farmacia.uniba.it/tisbe | |
| PubChem | NIH[6] | https://pubchem.ncbi.nlm.nih.gov/ | |
| ADMETLab 3.0 | CBDD[7] | https://admetlab3.scbdd.com/ | |
| ToxAlerts | OCHEM[8] | https://ochem.eu//alerts/home.do | |
| ProTox 3.0 | Charite[9] | https://tox.charite.de/protox3/index.php?site=compound_input | |
| OncoLogic 9.0 | U.S. EPA[1] | | https://www.epa.gov/tsca-screening-tools/oncologictm-expert-system-evaluate-carcinogenic-potential-chemicals |

[1]U.S. Environmental Protection Agency (USA)
[2]Organisation for Economic Co-operation and Development
[3]National Toxicology Program (USA)
[4]Tools listed at this site are downloadable and/or web-based
[5]Dipartimento di Farmacia-Scienze del Farmaco, Università degli Studi di Bari Aldo Moro (Italy)
[6]National Institutes of Health (USA)
[7]CBDD team of Central South University, the HIT team of National University of Defense Technology, and the Innovation Institute for Artificial Intelligence in Medicine of Zhejiang University (China)
[8]OCHEM (Online CHEmical database with Modeling environment) platform, eADMET GmbH
[9]Charite University of Medicine, Institute for Physiology, Structural Bioinformatics Group (Germany)

© Burleigh Dodds Science Publishing Limited, 2025. All rights reserved.

data can aid in the evaluation of the data-poor substance, a process referred to as read-across.[3]

When using the CompTox Chemicals Dashboard, entering a target chemical (using systemic name, synonym, CAS number, DTXSID, or InChlKey) and selecting the GenRA tab will display the top 10 potential analogs.[4] For making in vivo toxicity studies to toxicity effect predictions, the data should be filtered by ToxRef data (ToxRef is a comprehensive database developed by the U.S. EPA that provides information on the toxicity of chemicals). To make predictions using ToxCast data, the data should be filtered by ToxCast (ToxCast contains publicly available in vitro medium- and high-throughput screening assay data for almost 10 000 chemicals). In summary, GenRA helps in identifying structurally similar chemicals with available data, which can then be used to infer potential toxicological effects for the target chemical.

## 2.2 Cheminformatics analysis modules

Cheminformatics analysis modules offer comprehensive data on chemicals, including precise chemical structures, both experimental and predicted physicochemical properties, and toxicity information; thereby supporting chemical safety decision-making. Although these modules are accessible online through the EPA's website (https://www.epa.gov/comptox-tools/cheminformatics), the tool is still under development and is expected to be integrated into the CompTox Chemicals Dashboard in the future.

As of August 2024, the tool has several tabs: Hazard, Alerts, Predict 1.0, Predict 2.0, Search, Standardize, and Toxprints.

- Hazard: Once the chemical is inputted (using DTXSID, SMILES, name, CAS number, InChl, or InChlKey, or by drawing it), this tab provides data on various endpoints relevant to chemicals in food, including acute toxicity, carcinogenicity, genotoxicity, endocrine disruption, reproductive and developmental toxicity, neurotoxicity, and systemic toxicity. Clicking on the chemical's CAS number will direct users to the CompTox Chemicals Dashboard page for the query chemical.
- Predict 1.0: This tab offers only limited predictions (oral rat LD50[5] value, mutagenicity, developmental toxicity, and estrogen receptor binding).

3 Read-across is a commonly used technique to fill data gaps by using information on an endpoint for a data-rich chemical to predict the same endpoint for a data-poor chemical based on their similarity, such as structural similarity, the same mode of toxic action, and metabolic similarity. The 'endpoint that is read-across may include target organ toxicity, physicochemical property, and others, and may be quantitative or qualitative.

4 Top 100 candidate analogs may be downloaded.

5 The amount of a substance required to cause death in 50% of a test population.

© Burleigh Dodds Science Publishing Limited, 2025. All rights reserved.

- Predict 2.0: This tab features three models–DevTox (Developmental Toxicity), Mutagenicity, and LD50–that may be useful for assessing chemicals in food.
- Search: This function helps identify potential read-across substances.
- The Standardize and Toxprints tabs are used for data standardization and generating chemical fingerprints, respectively, which can aid in chemical analysis and comparison.

## 2.3 AIM (Analog Identification Methodology) Tool

This tool is intended to facilitate the identification of analogs and relevant data sources for a given source chemical. After entering a chemical (using its name, CAS number, or SMILES) and selecting the advanced options and report settings, by clicking the Find Analogs bar, users can generate output consisting of a list of potential analogs and data sources for both the target chemical and its identified analogs.

It is important to note that while AIM remains available for download, it is no longer updated and supported by the EPA. Users are encouraged to use GenRA as an alternative. Despite of this, AIM is recognized as a very user-friendly tool that can quickly identify structural analogs for a query substance. Once analogs are identified using AIM, users can manually retrieve[6] safety data for these analogs from publicly available toxicology databases and collections of safety evaluations such as ECHA (https://echa.europa.eu/), OECD Existing Chemicals Database (https://hpvchemicals.oecd.org/ui/Search.aspx), EPA's CompTox Chemicals Dashboard (https://comptox.epa.gov/dashboard/), EFSA (European Food Safety Authority) (https://www.efsa.europa.eu/en/publications), JMPR (Joint FAO/WHO Meeting on Pesticide Residues) (https://apps.who.int/pesticide-residues-jmpr-database), JECFA (Joint FAO/WHO Expert Committee on Food Additives) (https://apps.who.int/food-additives-contaminants-jecfa-database/), EPA IRIS (https://cfpub.epa.gov/ncea/iris/search/index.cfm), and EPA RED (Registration Eligibility Decisions) (https://archive.epa.gov/pesticides/reregistration/web/html/status.html), among other sources.

## 2.4 OECD QSAR Toolbox

The OECD QSAR Toolbox is designed to support chemical hazard assessment. Upon entering a substance into the Toolbox, it identifies relevant structural features and potential mechanisms of toxicity. The tool then identifies other chemicals with the same mode of toxic action and/or structural features (i.e.

6 Links to data sources are no longer updated, and as such, many data links do not work.

© Burleigh Dodds Science Publishing Limited, 2025. All rights reserved.

category building). When building categories, the tool factors in absorption, distribution, metabolism, and excretion (ADME) and physicochemical similarity among category members. Existing experimental data from category members is then used to fill data gaps for the query substance. In addition to read-across for quantitative endpoints, e.g. LD50, and qualitative endpoints, e.g. mutagenicity (negative, equivocal, or positive), trend analysis for certain quantitative endpoints is possible. Major prediction endpoints relevant to the safety assessment of chemicals in food include genetic, acute, repeated dose, reproductive, and developmental toxicities, along with carcinogenicity, teratogenicity, immunotoxicity, neurotoxicity, and ADME.

The Open (Quantitative) Structure-activity/property Relationship App (OPERA), a free suite of QSAR models developed by the U.S. National Institute of Environmental Health Sciences (NIEHS), enabling the prediction of physicochemical and ADME endpoints along with toxicity endpoints, can be integrated as an add-in within the OECD QSAR toolbox. OPERA and various other add-ins and plugins for the OECD QSAR Toolbox are available at https://repository.qsartoolbox.org/.

Descriptions of the complex OECD Toolbox workflow and its main functionalities are available from Dimitrov et al. 2016) and at https://qsartoolbox.org/wp-content/uploads/2023/05/Toolbox-4.6-Getting-Started.pdf. The latter also serves as a quick getting-started guide. To aid in initial use, a series of instructional videos is available on YouTube under the search term "OECD QSAR Toolbox 4.6", which are recommended for users to review prior to utilizing the software.

## 2.5 NTP ICE (National Toxicology Program Integrated Chemical Environment)

NTP ICE can facilitate the safety assessment of chemicals by filling data gaps and providing predictions. The Search tool within NTP ICE allows users to find key prediction endpoints relevant to the safety assessment of chemicals in food, including acute toxicity, reproductive and developmental toxicity, endocrine disruption, cardiotoxicity, cancer, ADME parameters, and physicochemical properties. The In Vitro to In Vivo Extrapolation (IVIVE) tool enables estimation of the in vivo plasma concentration of a query compound from a given in vitro assay. Its Physiologically Based Pharmacokinetic (PBPK) tool can generate predictions of tissue-specific concentration profiles of query chemicals following a dosing event. The Chemical Quest tool can help with identifying chemicals that are structurally similar to the query chemical, supporting analog-based safety assessments. NTP ICE integrates these tools to facilitate a comprehensive evaluation of chemical safety, providing valuable insights for risk assessment and regulatory decision-making.

© Burleigh Dodds Science Publishing Limited, 2025. All rights reserved.

## 2.6 VEGA (Virtual models for property Evaluation of chemicals within a Global Architecture) HUB

VEGA HUB is a collection of in silico tools aimed at facilitating the safety evaluation of chemicals for regulatory purposes and providing a reproducible procedure for read-across. VEGA HUB offers models taken from CAESAR (Computer Assisted Evaluation of industrial chemical Substances According to Regulations), T.E.S.T. (see Section 2.1.7), and contributions from VEGA collaborators. The user may choose what they want to download based on their needs. Selected downloadable (https://www.vegahub.eu/download/) tools, models, and software that may be employed for the safety assessment of chemicals in food include, but are not limited to:

- VEGA QSAR: Offers QSAR models to predict toxicity and physicochemical properties of chemicals that facilitate read-across assessment. Predicted endpoints of toxicity include mutagenicity, carcinogenicity, endocrine disruption, hepatotoxicity, and acute and developmental toxicities. After entering the chemical(s) using SMILES, users select the desired prediction endpoints via the "Select" tab and generate predictions using the "Predict" tab.
- SILIFOOD: Facilitates risk assessment of food contact materials (FCM) by combining (Q)SAR predictions with existing toxicological data. The QSAR predictions are VEGA predictions for toxicological endpoints (indication on accumulation potential in man, genotoxicity, carcinogenicity, endocrine toxicity, and reproductive and developmental toxicity) relevant to FCMs. Users input the chemical (using CAS number, SMILES, or FCM number) and specify the output directory to generate the PDF report.
- ToxEraser FCM: Identifies potentially "risky" FCMs and suggests safer alternatives. According to its description on the site where the tool can be downloaded (https://www.vegahub.eu/portfolio-item/toxeraser-fcm/), "The approach applied for the substitution takes into account three main parameters: the level of safety, the similarity algorithm and the identification of hazardous structural groups/alerts." After entering the chemical's SMILES code, users can specify the desired maximum number of structurally related substances (in the "Max Results" tab), and run the calculation by clicking the "Calculate" button. Once the tool indicates that the results are ready, they can be downloaded using the Download tab (users must specify a directory for the output).
- VERA (Virtual Extensive Read-Across): Facilitates chemical read-across. According to the description on the site where the tool can be downloaded (https://www.vegahub.eu/portfolio-item/vera/), VERA "finds the most similar compounds with a certain feature (e.g. structural

© Burleigh Dodds Science Publishing Limited, 2025. All rights reserved.

alerts and molecular groups) and provides clusters of similar substances while comparing these similar substances within different clusters." After uploading the SMILES file, specifying the directory for the output, and choosing the endpoint of interest (e.g. carcinogenicity), users can initiate the software prediction by clicking on "Start calculation".

- RAXpy: Facilitates read-across based on structural, metabolic, and biological similarities. Users input the substance of interest using SMILES, specify the number of neighbors for each similarity type (structural, biological, and metabolic), and select a dataset (either "Human hepatotoxicity (ToxRef)" or "Drug-induced liver injury (DILI)"). RAXpy generates results based on the chosen dataset.
- ToxRead: Identifies structural alerts associated with the query compound and finds similar compounds with those alerts; that is, ToxRead is a tool to assist with read-across. After inputting the SMILES code of the compound of interest and modifying the number (3 by default) of similar compounds returned from the analysis, the user must select the endpoint of interest. Clicking on "Run read-across" will generate an interactive chart on the right and a graph on the left. In the graph, the target chemical is in the center, and similar chemicals are shown as circles, the sizes of which are proportional to their similarity with the query compound. The circle's color depends on the activity of the read-across substance, e.g. when mutagenicity is chosen as an endpoint, the intensity of the red color displayed is related to the value of mutagenicity and the intensity of green is related to the value of non-mutagenicity.
- ToxDelta: Complements ToxRead by identifying the differences between two substances (dissimilarity features) and evaluating whether these differences are related to decreased or increased toxicity. As of July 2024, only a beta version of the tool is available for download.

## *2.7 T.E.S.T. (Toxicity Estimation Software Tool)*

T.E.S.T. was developed to facilitate the estimation of the toxicity of chemicals using QSAR methodologies. It provides predictions for several endpoints relevant to the safety assessment of chemicals in food, including mutagenicity, acute toxicity, and developmental toxicity along with physicochemical properties. Additionally, T.E.S.T. can identify structural analogs for a given query chemical.

After entering one or more chemical(s) (using CAS, SMILES, Name, InChI, InChIKey, or DTXSID or by drawing the structure of the chemical), users can select the toxicity endpoint from a list. Next, they select the QSAR toxicity estimation method (hierarchical clustering by default), followed by the output folder. To generate reports in single chemical mode, users check the "Create

© Burleigh Dodds Science Publishing Limited, 2025. All rights reserved.

detailed reports" checkbox; for batch chemical mode, they check the "Create reports" checkbox. Clicking "Calculate!" initiates the toxicity estimation process.

## 2.8 Cefic-LRI (Long-range Research Initiative) Tools

This initiative of the European Chemical Industry Council aims to aid in the preparation of regulatory dossiers such as REACH and with research and development. Among its tools, Ambit, a cheminformatics data management system, facilitates chemical safety assessment using existing information and predictions about a chemical and supports read-across and category formation. It also contains the VEGA predictive tool; hence, it can predict various endpoints, such as mutagenicity, developmental toxicity, and carcinogenicity, that are relevant for chemicals in food. The Ambit tool can be downloaded at https://ambit.sourceforge.net/or the LRI Ambit Read-across tool is available for online use at https://ambitlri.ideaconsult.net/tool2. After entering a query compound (its name, identifiers, SMILES, or InChl) into the search function, users can activate the vertical sidebar to the left by hovering over it, which allows them to select and apply the desired predictive models.

## 2.9 TISBE (TIRESIA (Toxicology Intelligence and Regulatory Evaluations for Scientific and Industry Applications) Improved on Structure-Based Explainability)

TISBE builds upon the TIRESIA (Toxicology Intelligence and Regulatory Evaluations for Scientific and Industry Applications) framework, which was an earlier tool for predicting toxicological endpoints but faced challenges with prediction accuracy and interpretability. TISBE represents a significant advancement over TIRESIA by integrating more sophisticated machine learning techniques and employing a consensus approach with five independent models. This enhances the robustness and reliability of developmental toxicity predictions while providing clearer insights into the basis of these predictions. Additionally, TISBE features enhanced interpretability by explaining predictions based on chemical structure. For further details on this new tool, see Mastrolorito et al. (2024).

## 2.10 PubChem

PubChem is an open chemistry database from the U.S. National Institutes of Health. While not intrinsically a predictive toxicology tool, it can aid in the identification of read-across substances that may be used for the safety assessment of a data-poor substance. Entering the name, SMILES/SMARTS,

© Burleigh Dodds Science Publishing Limited, 2025. All rights reserved.

InChI, CID, or molecular formula of a chemical on the homepage (https://pubchem.ncbi.nlm.nih.gov/) in the search tab or manually drawing the structure in PubChem Sketcher (https://pubchem.ncbi.nlm.nih.gov/edit3/), the tool retrieves results. Clicking on the "Similar Structures Search" tab for the "Best Match" substance displays substances that are structurally closely related. Clicking on "Settings" enables the adjustment of the Tanimoto Threshold (the most popular similarity measure for comparing chemical structures). After identifying compounds with highly similar structures to the query compound, users can manually retrieve safety data for these compounds from public toxicology databases and collections of safety evaluations and from the PubChem database.

## *2.11 ADMETLab 3.0*

ADMETLab was designed primarily to aid medicinal chemists in accelerating the drug development process by providing ADMET predictions. However, it can also be applied to predict ADMET properties of food chemicals. The software has two modes: "ADMET Screening" and "ADMET Evaluation". The ADMET Screening Mode is a batch mode that allows for the evaluation of multiple chemicals simultaneously. ADMET Evaluation enables the calculation and prediction of various physicochemical properties, ADME and toxicity endpoints, and toxicophore[7] rules for a single chemical. Toxicity endpoints include Ames toxicity, genotoxicity, rat oral acute toxicity, carcinogenicity, respiratory toxicity, nephrotoxicity, neurotoxicity, hematotoxicity, immunotoxicity, and human hepatotoxicity. The software includes two toxicophore rules, the genotoxic carcinogenicity rule and the nongenotoxic carcinogenicity rule, that could be useful for evaluating the potential carcinogenicity of food chemicals. To obtain predictions, users should enter the SMILES string of the chemical into the software and press "Submit". The tool will then generate predictions for the specified properties and endpoints.

## *2.12 ToxAlerts*

The web-based public platform OCHEM (Online CHEmical database with Modeling environment) can run predictions via the ToxAlert tool, designed to screen chemicals against structural alerts. It also allows the user to upload their own alerts by using the "Upload new alerts" tab and screen chemicals. A compound can be screened for alerts using the "Screen your molecules" tab. Alerts relevant for the hazard assessment of chemicals in food may include those for non-genotoxic and genotoxic carcinogenicity, idiosyncratic toxicity,

7 A toxicophore is a chemical structure or substructure that is responsible for the toxic properties of the chemical.

© Burleigh Dodds Science Publishing Limited, 2025. All rights reserved.

developmental and mitochondrial toxicity, endocrine disruption, and "reactive, unstable, toxic" endpoint. All available alerts may be viewed by clicking on the "View available alerts" tab.

### 2.13 ProTox

This web-based tool can predict various endpoints including acute toxicity, carcinogenicity, mutagenicity, nutritional toxicity, immunotoxicity, and organ toxicity (e.g. hepatotoxicity, neurotoxicity, and nephrotoxicity), along with Tox21 nuclear receptor signaling and stress response pathways and molecular initiating events. After inputting the chemical (using its name or SMILES) and selecting prediction models, users can generate predictions for various toxicological endpoints, including prediction accuracy percentages or probabilities, as well as a list of similar compounds by clicking on "Start Tox-Prediction".

### 2.14 OncoLogic

This downloadable tool that mimics EPA's decision-making process can predict both cancer and non-cancer human health hazards. Predicted cancer hazard levels are low, marginal, low-moderate, moderate, moderate-high, and high; predicted non-cancer hazard levels are low, moderate, and high. These concern levels are defined in the tool's Interpretive Guidance Document (EPA, 2006). For example, high non-cancer hazard is defined as "Evidence of adverse effects in humans or conclusive evidence of severe effects in animal studies. Death, organ pathology (microscopic) at LOAEL[8] ≤ 100 mg/kg/day; multiple organ toxicity; NOAEL[9] ≤ 10 mg/kg/day." To obtain predictions, users identify the query chemical by its name or CAS number or draw its structure, and click on "Evaluate". The tool will then generate results that include cancer and non-cancer concern levels, the line of reasoning (an explanation of the prediction process), and an interpretation of the results.

## 3 Selected commercially available in silico tools

Commercially available in silico tools (i.e. those that are offered for purchase), like publicly available ones, utilize various approaches such as expert rule-based systems and machine learning algorithms to predict toxicity endpoints and provide insights into potential hazards associated with chemical compounds. Several commercially available predictive toxicology software tools are used in the industry and in government. Following are some examples with their major

8 LOAEL: lowest observed adverse effect level.
9 NOAEL: no observed adverse effect level.

© Burleigh Dodds Science Publishing Limited, 2025. All rights reserved.

prediction endpoints relevant for the safety or hazard assessment of chemicals in food:

### 3.1 MultiCASE

MultiCASE is a software tool designed for predicting chemical toxicity and assessing potential risks through a range of computational methods. It features QSAR models, structural alerts, and machine learning techniques to evaluate the safety of chemical compounds based on their molecular structures. MultiCASE integrates with extensive chemical and toxicological databases to enhance prediction accuracy and support regulatory compliance. The software has three major tools that are useful for safety assessments of chemicals in food:

- CASE Ultra: a software product that predicts genotoxicity, carcinogenicity, acute toxicity, endocrine disruption, reproductive and developmental toxicity, cardiotoxicity, hepatotoxicity, and renal toxicity. After inputting a chemical, users can select prediction endpoints from the "Change Folder" tab in the upper left corner. Clicking "Run Test" will generate the predictions. The "Neighborhood Analysis" feature provides the closest neighbors for the prediction endpoint and enables users to see how similar the compounds used for the prediction are to the query compound.
- META Ultra: a software product that predicts human metabolites and identifies reactive metabolites. For each metabolite, it provides the enzymes involved, reaction type, species, and probability of the metabolite formation, along with other relevant data.
- QSAR Flex: a software product that assesses the carcinogenic potency of *N*-nitrosamines.

For more detailed information on the MultiCASE software, refer to Chakravarti and Saiakhov (2022).

### 3.2 ChemTunes Tox GPS

ChemTunes ToxGPS is an advanced software tool for predicting chemical toxicity and assessing potential hazards. According to the software's webpage (https://mn-am.com/products/chemtunestoxgps/), "The predictions are based on mechanistically-informed, probabilistic QSAR models and endpoint-specific structural knowledge (expert rules) which are combined to provide a weight-of-evidence prediction together with an estimate of the associated uncertainty. Nearest neighbors of the query compounds in the training sets are identified and linked to the experimental study data in the ChemTunes databases." This software product's prediction endpoints relevant to the safety assessment

© Burleigh Dodds Science Publishing Limited, 2025. All rights reserved.

of chemicals in food are acute toxicity, carcinogenicity, genetic toxicity, reproductive and developmental toxicity, target organs (liver and mitochondria), DNA and protein binding, absorption, and liver metabolism. It also provides nearest neighbors in a training set for the various endpoints. "Chemical search" enables users to identify exact, partial, or similar compounds. Regulatory information, safety evaluations, and toxicity study records can be retrieved. "Actions" enables users to calculate molecular properties and Chemotype profiles[10] and to predict toxicity. "Find Structural Analog Compounds" helps users identify potential read-across compounds, along with their predicted toxicities and chemical similarity scores. For more detailed information on the ChemTunes ToxGPS platform, refer to Yang et al. (2023a). For information on incorporating the tool into a workflow for high-throughput read-across, see Yang et al. (2023b).

### 3.3 Leadscope

Leadscope is a versatile software suite designed for predictive toxicology, chemical safety, risk assessment, and hazard evaluation. It utilizes advanced QSAR models and machine learning algorithms to predict chemical toxicity based on structural features and biological data. Leadscope integrates with extensive chemical and toxicological databases to provide comprehensive data for risk evaluation and regulatory compliance. The software includes tools for model evaluation, validation, and visualization, facilitating detailed analysis and decision-making. Leadscope supports a wide range of endpoints, including acute toxicity, carcinogenicity, genetic toxicity, reproductive and developmental toxicity, neurotoxicity, target organs (e.g. liver, kidney, and heart), endocrine activity, and metabolism (bioactivation and CYP inhibition). This extensive coverage makes it particularly relevant for assessing chemicals in various contexts, including food safety.

After entering the query compound into "Search", users can select the database to search for various endpoints. The "Read-across" tab enables the identification and justification of analogs. Predictions may be generated by clicking on "Apply Models/Alerts". After identifying the query compound, users must select models and/or alerts before predictions can be generated. For more information on the Leadscope platform, refer to Myatt et al. (2022) and Leadscope (2022).

10 A chemotype profile provides a detailed characterization of the chemical features and structures that define a particular type of compound or class of chemicals.

© Burleigh Dodds Science Publishing Limited, 2025. All rights reserved.

### 3.4 Lhasa Nexus Suite

The Lhasa Nexus suite is a comprehensive software platform for predicting chemical toxicity and managing chemical safety data. It includes tools for QSAR modeling, structural alerts, and expert rule-based systems to assess potential hazards based on chemical structures. Nexus integrates with extensive databases and offers advanced analytics for risk assessment and regulatory compliance. The Lhasa Nexus Suite has three tools useful for filling data gaps during the safety evaluations of chemicals in food:

- Derek Nexus: predicts various endpoints of toxicity, including carcinogenicity, teratogenicity, hepatotoxicity, and acute, reproductive, and developmental toxicity. After inputting the query compound, users set up "Derek Prediction" by clicking on the triangle next to the blue "D" symbol. Users can specify species and prediction endpoints here. Clicking "D Start Prediction" will generate predictions along with "training set examples" for comparison.
- Sarah Nexus: predicts mutagenicity. After inputting the query compound, users set up the "Sarah Prediction" by clicking on the triangle next to the red "S" symbol. Clicking "S Start Prediction" will generate the mutagenicity prediction along with "training set examples" for reference.
- Meteor Nexus: predicts metabolism. After inputting the query compound, users set up the "Meteor Prediction" by clicking on the triangle next to the orange "M" symbol. Users can specify biotransformation reactions, species, and other constraints before clicking on "M Start Prediction" to generate a metabolic prediction.

For more detailed information on the Lhasa Nexus Suite, refer to Ponting et al. (2022).

### 3.5 TopKat (Toxicity Prediction by Komputer Assisted Technology)

TOPKAT is a sophisticated tool that leverages QSAR models and statistical methods to forecast a range of toxicological endpoints based on chemical structures. It integrates with extensive databases and offers a user-friendly interface for predicting and analyzing chemical safety. TopKat predictions (acute toxicity, mutagenicity, carcinogenicity, rat chronic LOAEL, hepatotoxicity, and developmental toxicity) may help with assessing the safety profiles of chemicals in food and provide insights into potential hazards associated with food chemical substances.[11]

11 The author of this chapter has no access to this tool, and no user manual could be located to describe its basic workflow.

© Burleigh Dodds Science Publishing Limited, 2025. All rights reserved.

## 3.6 ACD/Tox Suite (Advanced Chemistry Development/Tox Suite)

ACD/Tox Suite is a comprehensive software package designed to predict the toxicological properties of chemical compounds utilizing various techniques and models, including QSAR models. The suite leverages computational models and algorithms to estimate various toxicity endpoints, e.g. acute toxicity, mutagenicity, endocrine disruption, and adverse effects on organs, aiding in the safety assessment of chemicals. After inputting the query chemical and selecting the property/endpoint of interest, users can generate the results. Users can evaluate the reliability of predictions with the help of a reliability index and a display of similar structures.[10]

## 3.7 ADMET Predictor (Absorption, Distribution, Metabolism, Excretion, and Toxicity Predictor)

This tool utilizes a range of computational models and methods, such as QSAR and machine learning, to predict ADME properties and toxicity endpoints such as mutagenicity, chromosomal aberration, acute toxicity, carcinogenicity, (i.e. TD50[12], which may have only very limited utility in the safety evaluation of chemicals in food), developmental toxicity, endocrine toxicity, and some liver effects for chemicals. After inputting the chemical, users can generate ADMET predictions by clicking the Data tab, then on "Calculate ADMET Properties", and specifying the desired prediction endpoints. Clicking "Calculate" will generate results.

Please note both the commercial software tools and publicly available tools discussed may have features, prediction endpoints, and functions, (e.g. clustering, providing Cramer classifications (Cramer et al., 1976) and Threshold of Toxicological Concern values (Munro et al., 1996), Globally Harmonized System of Classification and Labelling of Chemicals (GHS) acute oral consensus (EPA, 2004) prediction, and consensus call for the ICH M7 regulatory guideline (ICH, 2017), not mentioned above that may be useful for the safety and hazard assessment of chemicals in food and for regulatory submissions. Please consult the developers' websites, user manuals, and published articles (e.g. Sushko et al., 2011; Gatnik and Worth, 2010; Patlewicz et al., 2017), for more information on the full capabilities and all functions of these tools. Detailed description of these is beyond the scope of this chapter, which is to highlight tools that might be useful for the safety and hazard assessment of chemicals in food.

12 The lifetime dose that would cause cancer in 50% of the animals that would have stayed cancer-free without dosing.

© Burleigh Dodds Science Publishing Limited, 2025. All rights reserved.

## 4 Promising in silico tools currently under development

There are several predictive toxicology software tools, platforms, and projects under various stages of development. These projects aim to enhance the efficiency and accuracy of toxicological predictions for chemical substances and expand their functionalities. Following are two notable examples:

### *4.1 RISK-HUNT3R (RISK assessment of chemicals integrating Human centric Next generation Testing strategies promoting the 3Rs)*

RISK-HUNT3R is a European Research & Innovation project funded under the European Commission's Horizon 2020 program. According to the project's website, "The project will develop, validate and implement integrated approaches to lead the way towards next-generation risk assessment (NGRA). Innovative mechanism-based new approach methods (NAMs) will exclusively be in vitro and in silico and relevant for human health. Through systematic and iterative evaluation of its NAM toolbox, the project will optimize a strategy to assess chemical exposure, toxicokinetics, and toxicodynamics. RISK-HUNT3R will provide a full sustainable framework for NGRA that is human-relevant, fully based on non-animal approaches, and fit for implementation through engagement with chemical safety regulators" (https://www.risk-hunt3r.eu/). For more information about this project, visit the official website.

### *4.2 EDT (Expanded Decision Tree)*

To support increased use of alternative methods, the Human Foods Program of the U.S. Food and Drug Administration (FDA) has developed an in silico tool, the Expanded Decision Tree (EDT) (FDA, 2024b). The EDT is a series of chemical structure-based questions leading to the classification of chemicals into one of six classes of chronic oral toxic potential (from Class I, the class of lowest toxic potential, to Class VI, aiming to capture the most toxic compounds that exist). Each of the six EDT classes has an associated threshold of toxicological concern (TTC) level. According to the TTC concept, if exposure to a compound does not exceed its class TTC, it is reasonably expected to be safe.

The EDT software currently under development is designed to enable fast, accurate, and reproducible evaluation of a large number of substances simultaneously. It will assist with the screening and prioritization of substances for (re)evaluation, read-across to help fill data gaps, and prediction of safe levels of exposure (TTC).

© Burleigh Dodds Science Publishing Limited, 2025. All rights reserved.

## 5 Assessing reliability of predictions generated by computational toxicology tools

Understanding the reliability of predictions generated by computational toxicology tools is crucial for several reasons:

- **Decision-Making in Safety Assessment**: Predictions from computational toxicology tools often inform decisions regarding the safety of chemicals and products. Reliable predictions help ensure that decisions made based on these predictions are accurate and appropriate.
- **Regulatory Compliance**: Regulatory agencies may use predictions from computational tools to classify chemicals, set exposure limits, and determine safety thresholds. Reliable predictions help ensure public safety.
- **Resource Allocation**: Resources such as time, funding, and personnel may be allocated based on toxicological predictions. Reliable predictions help prioritize resources effectively, focusing efforts on chemicals of higher concern.
- **Reducing Animal Testing**: Reliable computational predictions, when laws, rules, or regulations allow, can reduce the need for animal testing by providing alternative methods of assessing chemical safety. This aligns with ethical considerations and regulatory trends favoring the reduction, refinement, and replacement (3Rs) of animal testing (e.g. FDA, 2024c, EFSA, 2024).
- **Public Confidence**: Reliable predictions enhance public and stakeholder confidence in chemical safety assessments. Transparent communication of prediction reliability builds trust in regulatory decisions and industry practices.
- **Scientific Advancement**: Assessing the reliability of computational tools fosters ongoing improvement and validation of models, which contributes to advancing the field of computational toxicology and enhancing the accuracy of future predictions.

Computational software tools can communicate the reliability of predictions in several ways. Some tools assign probability scores or confidence levels to predictions. For example, a prediction might be accompanied by a percentage score indicating the likelihood of it being accurate; a higher score indicates greater confidence in the prediction. Tools may provide confidence intervals around numerical predictions. These intervals indicate the range within which the true value is likely to fall with a certain level of confidence (e.g. 95% confidence interval).

© Burleigh Dodds Science Publishing Limited, 2025. All rights reserved.

To assess and communicate the reliability of predictions, software tools may use internal quality metrics or validation statistics. These metrics could include measures of model performance (e.g. accuracy, sensitivity, specificity) or validation against external datasets, primarily consisting of toxicological data from in vivo and in vitro studies. Such datasets are crucial for determining how well the computational model predicts known toxicological outcomes, serving as a benchmark to confirm that the model's calculations align with experimental findings.

Some tools visualize prediction reliability through graphical representations. For example, a graph might show predicted values along with confidence bands or error bars, indicating the uncertainty associated with each prediction. They may provide explanations or transparency regarding how predictions are made and the factors influencing prediction reliability. This can include information about the data used, the model algorithms, and any assumptions or limitations.

Additionally, software tools often provide guidance to users on interpreting prediction reliability. This may include recommendations on when to trust predictions based on their reliability scores or metrics. Note that some tools will not provide predictions of certain endpoints for compounds outside their applicability domain, while others will provide predictions while noting that the query compounds are outside their applicability domain. Generally, it is not advisable to trust predictions in the latter scenario.

While in silico tools are valuable for predicting chemical properties and biological activities, including toxicity, it is essential to use them within their established applicability domains to ensure reliability and accuracy. Exercise caution when a query substance falls outside the applicability domain, and consider additional assessments or data sources for making informed decisions regarding chemical safety assessment and regulatory compliance.

In summary, ensuring and understanding the reliability of predictions from computational toxicology tools is fundamental to making informed decisions, optimizing resource use, and promoting public health and safety.

## 6 Consensus Modeling for Endpoint Prediction (CoMPARA)

When predicting various endpoints of toxicity, such as carcinogenicity or reproductive toxicity, selecting the appropriate software tool to use and whether to use predictions from more than one tool can be challenging. Different computational models may have strengths and weaknesses based on their algorithms, training datasets, and underlying assumptions. Aggregating predictions from multiple models can reduce the impact of model-specific biases and variability and provide a more robust estimation of toxicological endpoints.

© Burleigh Dodds Science Publishing Limited, 2025. All rights reserved.

Consensus Modeling for Endpoint Prediction (CoMPARA) is an approach used in computational toxicology and chemical safety, hazard, and risk assessment. It involves combining multiple computational models to predict toxicological endpoints or properties of chemical compounds. The core idea behind CoMPARA is to improve the reliability and accuracy of predictions by leveraging the strengths of various modeling techniques while minimizing their individual weaknesses. It is particularly useful in situations where individual models may have limitations due to their underlying assumptions, data availability, or algorithmic biases. By combining multiple models, CoMPARA aims to produce more robust, reliable, and broadly applicable predictions across different chemical compounds and endpoints. This approach contributes to enhancing the efficiency and accuracy of computational methods in toxicology and chemical risk assessment fields.

By following the steps below, CoMPARA enables researchers and practitioners to leverage the strengths of multiple computational models, thereby enhancing the reliability and applicability of predictions in various domains of chemical and toxicological sciences. In practical terms, CoMPARA typically involves the following steps:

- **Selection of Models**: Several computational models, each based on different algorithms or methodologies, should be selected.
- **Prediction Generation**: Each selected model independently predicts the endpoint or property of interest (such as toxicity, bioactivity, or environmental impact) for a given chemical compound.
- **Consensus Building**: Predictions from all selected models are then aggregated or combined to generate a single consensus prediction. Various methods can be used for this aggregation, such as averaging the predictions, weighted averaging based on model performance, or employing advanced ensemble techniques.
- **Evaluation and Validation**: The performance of the consensus model is evaluated to assess its reliability and accuracy. This often involves comparing it against experimental data or established benchmarks. Validation ensures that the combined prediction is more reliable than any single model prediction alone.
- **Application**: The final consensus prediction can be used in such applications as chemical safety, hazard, or risk assessment, regulatory decisions, drug discovery, or environmental impact studies.

While CoMPARA itself is not a standalone software tool, its principles are often applied in various software platforms that integrate multiple computational models for toxicological prediction. By leveraging the strengths of these models, CoMPARA enhances the reliability and accuracy of predictions in chemical

© Burleigh Dodds Science Publishing Limited, 2025. All rights reserved.

safety, hazard assessment, and risk evaluation. Users are encouraged to carefully evaluate the individual models involved, understand their performance metrics, and ensure that predictions are aggregated appropriately. Ultimately, this approach not only aids in making informed decisions but also contributes to more robust and comprehensive assessments in the fields of toxicology and chemical risk assessment.

Additional information on CoMPARA may be found in a number of publications such as Valsecchi et al. (2020), Mansouri et al. (2020), and Collins et al. (2024).

## *6.1 Case study: Using selected tools to aid in safety assessments of chemicals in food*

### *6.1.1 Selection of chemical for the case study*

Flame retardants are chemicals or materials designed to reduce the flammability of combustible materials and delay or inhibit the spread of fires. To do this, they work through several mechanisms. Fire retardant chemicals can be environmental pollutants under certain conditions (Xiong et al., 2019). While they are beneficial for fire safety in various applications, their use and disposal can lead to environmental contamination and potential risks. Some fire-retardant chemicals are designed to be persistent in order to effectively protect materials from fire. However, this persistence means they can remain in the environment for long periods after their initial use. They may accumulate and persist in soil, water bodies, and sediments, and potentially bioaccumulate in plants and animals, thus ending up in the food supply as chemical contaminants.

4-(1,2-Dibromoethyl)-1,2-dibromocyclohexane (DBE-DBCH, formerly abbreviated as TBECH, CAS 3322-93-8) is considered an emerging brominated flame retardant (BFR). Emerging flame retardants have either been recently introduced or are being used as replacements for older BFRs that have been phased out or restricted due to environmental and health concerns. Like many emerging flame retardants, only limited safety data are available for DBE-DBCH, but understanding its potential human health effects is important since it was shown to be present in fish, beef, lamb, pork, chicken, cheese, eggs, and vegetable oil (Marteinson et al., 2021; Zuiderveen et al., 2020), foods commonly consumed.

Based on in silico data, in vitro assays, and in vivo studies conducted on frogs, fish, birds, and rats, DBE-DBCH is an endocrine disruptor of sex and thyroid hormone pathways, with in vivo toxic effects on reproductive, metabolic, and some other endpoints (Marteinson et al., 2021). No chronic/carcinogenicity studies, nor dog or primate reproductive and/or developmental or subchronic

© Burleigh Dodds Science Publishing Limited, 2025. All rights reserved.

toxicity studies are available. The only available study in rodents is a 28-day dietary study in rats (Curran et al., 2017). Exposure to TBECH was 0, 0.9, 4.2, 21.3, 98.0, or 328.9 mg/kg bw/day in males and 0, 0.8, 3.9, 19.4, 91.7, or 321.4 mg/kg bw/day in females. At the high dose, rats had to be euthanized after 15–18 days due to the substance's toxic effects. Investigators observed some evidence of hormone disruption, including changes in serum testosterone levels in males and changes in serum T3 and T4 levels in females. Dose-dependent clinical and histopathological changes consistent with mild hepatic and renal inflammation were reported. The NOAEL was established as 4.2 mg/kg bw/day in males and 3.9 mg/kg bw/day in females. Sister chromatid exchange tests in Chinese hamster ovary cells in vitro and mammalian cell mutagenicity assay yielded positive results, while chromosome aberration tests and bacterial mutagenicity assays produced negative results (NTP, 2024).

There are data gaps (such as DBE-DBCH's carcinogenicity potential) when it comes to assessing the substance's potential human health risk. To demonstrate the utility of in silico tools to fill the carcinogenicity data gap, DBE-DBCH was run through various in silico tools to examine whether the carcinogenic potential of the substance can be predicted in a consistent manner.

### *6.1.2 Carcinogenicity predictions employing selected tools: OECD QSAR Toolbox*

Although this tool returned a message stating that no experimental carcinogenicity data on the chemical of interest are available (Figure 1), when the substance was profiled, it retrieved a structural alert for genotoxic carcinogenicity (see Figure 2) for the parent and its predicted metabolites. For QSAR prediction of carcinogenicity, several models are available (e.g. "FDA RCA Cancer Female Mouse – Danish QSAR DB CASE Ultra model"). Eight of the 18 models available returned a no value (i.e. no prediction), out of domain prediction, while 10 of the 18 models returned a positive, in domain prediction (see Table 2).

### 6.1.3 Carcinogenicity predictions employing selected tools: cheminformatics analysis modules (Hazard Comparison Dashboard)

The carcinogenicity potential was predicted to be "high" (corresponding to IARC group 2B (possibly carcinogenic to humans)), and according to the hazard statement, the substance is "suspected of causing cancer". The possible outputs are very high, high, medium, low, inconclusive, and not applicable.

© Burleigh Dodds Science Publishing Limited, 2025. All rights reserved.

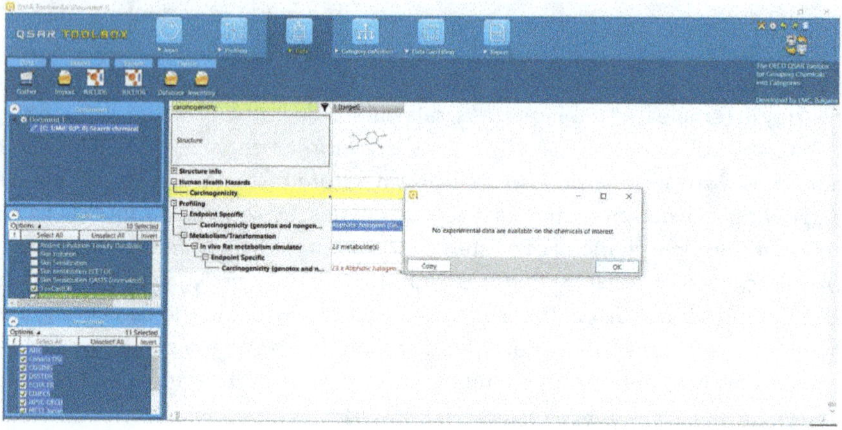

**Figure 1** OECD QSAR Toolbox output: no carcinogenicity data are available for the chemical of interest.

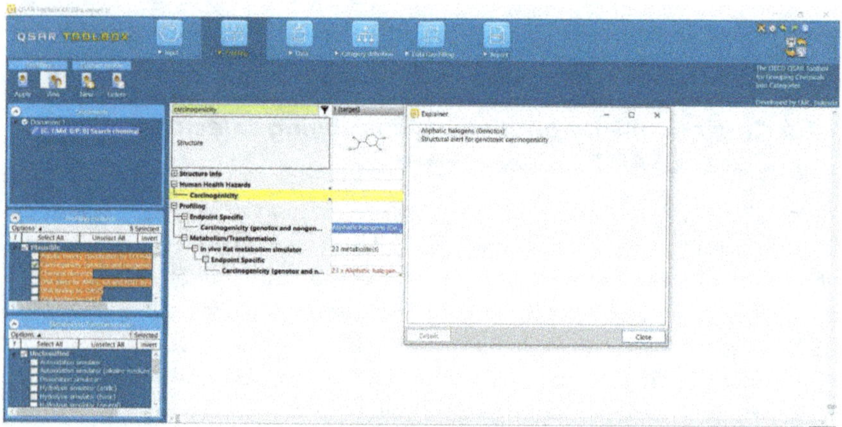

**Figure 2** OECD QSAR Toolbox output: structural alert for genotoxic carcinogenicity for the chemical of interest.

### *6.1.4 Carcinogenicity predictions employing selected tools: VEGA QSAR*

Predicted carcinogenicity using various models:

- Carcinogenicity model (CAESAR) 2.1.10: "Prediction is non-carcinogen, but the result may be not reliable...The predicted compound is outside the applicability domain of the model." Reliability: 1 out of 3 stars.
- Carcinogenicity model (ISS) 1.0.3: "Prediction is carcinogen, but the results show some critical aspects, which require to be checked. The predicted

© Burleigh Dodds Science Publishing Limited, 2025. All rights reserved.

**Table 2** OECD QSAR Toolbox partial Excel report: carcinogenicity predictions for the target chemical (Danish (Q)SAR Database)

| *Endpoint* | *Value* | *Species, duration, test type, type of method, assay, strain, test guideline, year, reference, database* |
|---|---|---|
| Carcinogenicity | | Test organisms (species): mouse<br>Endpoint: Carcinogenicity<br>Prediction approach: External SAR/QSAR<br>SAR/QSAR: FDA RCA Cancer Male Mouse – Danish QSAR DB Leadscope model |
| Carcinogenicity | Positive | Test organisms (species): mouse<br>Endpoint: Carcinogenicity<br>Prediction approach: QSAR<br>SAR/QSAR: FDA RCA Cancer Female Mouse – Danish QSAR DB Leadscope model |
| Carcinogenicity | | Test organisms (species): rat<br>Endpoint: Carcinogenicity<br>Prediction approach: External SAR/QSAR<br>SAR/QSAR: FDA RCA Cancer Male Rat – Danish QSAR DB Leadscope model |
| Carcinogenicity | Positive | Test organisms (species): rat<br>Endpoint: Carcinogenicity<br>Prediction approach: QSAR<br>SAR/QSAR: FDA RCA Cancer Rat – Danish QSAR DB CASE Ultra model |
| Carcinogenicity | | Test organisms (species): rat/mouse liver specific cancer<br>Endpoint: Carcinogenicity<br>Prediction approach: External SAR/QSAR<br>SAR/QSAR: Liver Specific Cancer in Rat or Mouse – Danish QSAR DB battery model |
| Carcinogenicity | Positive | Test organisms (species): rodent<br>Endpoint: Carcinogenicity<br>Prediction approach: QSAR<br>SAR/QSAR: FDA RCA Cancer Rodent – Danish QSAR DB Leadscope model |
| Carcinogenicity | Positive | Test organisms (species): rat/mouse liver specific cancer<br>Endpoint: Carcinogenicity<br>Prediction approach: QSAR<br>SAR/QSAR: Liver Specific Cancer in Rat or Mouse – Danish QSAR DB SciQSAR model |
| Carcinogenicity | Positive | Test organisms (species): mouse<br>Endpoint: Carcinogenicity<br>Prediction approach: QSAR<br>SAR/QSAR: FDA RCA Cancer Mouse – Danish QSAR DB CASE Ultra model |
| Carcinogenicity | Positive | Test organisms (species): rat<br>Endpoint: Carcinogenicity<br>Prediction approach: QSAR<br>SAR/QSAR: FDA RCA Cancer Female Rat – Danish QSAR DB CASE Ultra model |

(*Continued*)

© Burleigh Dodds Science Publishing Limited, 2025. All rights reserved.

**Table 2** (*Continued*)

| *Endpoint* | *Value* | *Species, duration, test type, type of method, assay, strain, test guideline, year, reference, database* |
|---|---|---|
| Carcinogenicity | Positive | Test organisms (species): rat<br>Endpoint: Carcinogenicity<br>Prediction approach: QSAR<br>SAR/QSAR: FDA RCA Cancer Male Rat – Danish QSAR DB CASE Ultra model |
| Carcinogenicity | | Test organisms (species): rat/mouse liver specific cancer<br>Endpoint: Carcinogenicity<br>Prediction approach: External SAR/QSAR<br>SAR/QSAR: Liver Specific Cancer in Rat or Mouse – Danish QSAR DB CASE Ultra model |
| Carcinogenicity | | Test organisms (species): mouse<br>Endpoint: Carcinogenicity<br>Prediction approach: External SAR/QSAR<br>SAR/QSAR: FDA RCA Cancer Female Mouse – Danish QSAR DB CASE Ultra model |
| Carcinogenicity | Positive | Test organisms (species): rodent<br>Endpoint: Carcinogenicity<br>Prediction approach: QSAR<br>SAR/QSAR: FDA RCA Cancer Rodent – Danish QSAR DB CASE Ultra model |
| Carcinogenicity | Positive | Test organisms (species): mouse<br>Endpoint: Carcinogenicity<br>Prediction approach: QSAR<br>SAR/QSAR: FDA RCA Cancer Male Mouse – Danish QSAR DB CASE Ultra model |
| Carcinogenicity | | Test organisms (species): rat/mouse liver specific cancer<br>Endpoint: Carcinogenicity<br>Prediction approach: External SAR/QSAR<br>SAR/QSAR: Liver Specific Cancer in Rat or Mouse – Danish QSAR DB Leadscope model |
| Carcinogenicity | | Test organisms (species): rat<br>Endpoint: Carcinogenicity<br>Prediction approach: External SAR/QSAR<br>SAR/QSAR: FDA RCA Cancer Female Rat – Danish QSAR DB Leadscope model |
| Carcinogenicity | | Test organisms (species): rat<br>Endpoint: Carcinogenicity<br>Prediction approach: External SAR/QSAR<br>SAR/QSAR: FDA RCA Cancer Rat – Danish QSAR DB Leadscope model |
| Carcinogenicity | Positive | Test organisms (species): mouse<br>Endpoint: Carcinogenicity<br>Prediction approach: QSAR<br>SAR/QSAR: FDA RCA Cancer Mouse – Danish QSAR DB Leadscope model |

© Burleigh Dodds Science Publishing Limited, 2025. All rights reserved.

compound could be out of the applicability domain of the model." Reliability: 2 out of 3 stars.

- Carcinogenicity model (IRFMN-ISSCAN-CGX) 1.0.2: "Prediction is possible non-carcinogen, but the results may not be reliable...The predicted compound is outside the applicability domain of the model." Reliability: 1 out of 3 stars.
- Carcinogenicity model (IRFMN-Antares) 1.0.2: "Prediction is carcinogen, but the result shows some critical aspects, which require to be checked. The predicted compound could be out of the applicability domain of the model." Reliability: 2 out of 3 stars.
- Carcinogenicity oral classification model (IRFM) 1.0.1: "Prediction is carcinogen, the result appears reliable...The predicted compound is within the applicability domain of the model." Reliability: 3 out of 3 stars.

### 6.1.5 Carcinogenicity predictions employing selected tools: ADMETLab 3.0

The genotoxic carcinogenicity mutagenicity rule and the nongenotoxic carcinogenicity rule both returned the presence of structural alerts. The probability of the compound being carcinogenic was 0.993 (this value is 0 for a non-carcinogen and 1 for a carcinogen).

### 6.1.6 Carcinogenicity predictions employing selected tools: ProTox 3.0

The prediction for the carcinogenicity endpoint was "Active" with a probability of 0.78 (this value is zero for a non-carcinogen and 1 for a carcinogen).

### 6.1.7 Carcinogenicity predictions employing selected tools: OncoLogic 9.0

The prediction for carcinogenicity was "Moderate"; defined as "Likely to be moderately active carcinogen".

### 6.1.8 Carcinogenicity predictions employing selected tools: Leadscope

Results:

- Female mouse carcinogenicity: positive with a positive probability or precision of 0.727.

© Burleigh Dodds Science Publishing Limited, 2025. All rights reserved.

- Female rat carcinogenicity: positive with a positive probability or precision of 0.581.
- Male mouse carcinogenicity: positive with a positive probability or precision of 0.523.
- Male rat carcinogenicity: positive with a positive probability or precision of 0.586.

According to the Leadscope manual, "For most models, a test chemical is evaluated as active for a set of models if the average probability is ≥0.5 and inactive if the average probability is <0.5." (Leadscope, 2020). The same manual states that an FDA report from 2015 recommended "using a positive probability cutoff of 0.4 for the maximum negative prediction and 0.6 for the minimum positive prediction, with probabilities of 0.4–0.6 representing equivocal or indeterminate predictions."

### *6.1.9 Carcinogenicity predictions employing selected tools: CASE Ultra (MultiCASE)*

Results:

- Female mouse carcinogenicity: Two positive alerts were found in the query chemical. 72.7% of compounds containing one of the alerts and 88.9% of compounds containing the second alert were positive for carcinogenicity. Overall, the compound was predicted to be positive (i.e. carcinogen), with a QSAR calculated probability of 80.7%. The classification threshold of the model is 45.0%, with a grey zone between 35.0% and 55.0%. Predictions falling within the grey zone are deemed inconclusive.
- Male mouse carcinogenicity: One positive alert was found in the query chemical; 77.8% of compounds containing this alert were positive. The result of carcinogenicity prediction was inconclusive as the QSAR calculated probability was 52.3%, a value that falls inside the gray zone (35.0% to 55.0%) around the model's current classification threshold of 45.0%.
- Female rat carcinogenicity: One positive alert was identified for the query compound; 85.7% of compounds containing this alert were positive. The query compound was predicted to be positive with a QSAR calculated probability of 60.3%.
- Male rat carcinogenicity: Two positive alerts were found in the test chemical; 100.0% of compounds containing one of the alerts and 72.7% of compounds containing the second alert were positive for carcinogenicity. Overall, the compound was predicted to be positive (i.e. carcinogen), with a QSAR calculated probability of 83.3%.

© Burleigh Dodds Science Publishing Limited, 2025. All rights reserved.

### 6.1.10 Carcinogenicity predictions employing selected tools: Derek Nexus

The software returned a carcinogenicity alert for "Alkylating agent". Mammalian carcinogenicity was predicted to be plausible. In Derek Nexus, plausible is defined as "The weight of evidence supports the proposition." Possible predictions in Nexus are impossible, improbable, doubted, equivocal, plausible, probable, and certain. Mammalian carcinogenicity predictions in Derek associated with a reasoning level of equivocal or above have been considered positive.

### 6.1.11 Carcinogenicity predictions employing selected tools: ChemTunes ToxGPS

Results:

- Mouse tumorigenicity: positive prediction with a probability (positive) range of [0.987, 0.999].
- Rat tumorigenicity: positive prediction with a probability (positive) range of [0.650, 0.800].

In ChemTunes ToxGPS, the "Probability (positive) range" refers to the range of probabilities assigned to chemicals that indicate the likelihood of a positive outcome or classification based on the toxicity prediction model used in ToxGPS; that is, the range provided in ToxGPS indicates the spectrum of probabilities within which the predictions fall.

### 6.1.12 Carcinogenicity predictions employing selected tools: overall carcinogenicity prediction

Table 3 summarizes the predictions for carcinogenicity for DBE-DBCH. Both non-carcinogenicity predictions (highlighted in red) and the "No value" prediction (highlighted in orange) should be disregarded as the query compound is out of the applicability domain of those tools or models. While one prediction is inconclusive, all others are positive with a probability at or above the probability threshold for a positive response for the applicable model. Considering all outcomes (i.e. carcinogen or noncarcinogen), and the confidence in/probability of individual predictions and whether the query compound is in the applicability domain of the different models used or not, the overall carcinogenicity prediction for the compound is positive. This prediction was supported by high-throughput screening (HTS) data available from the NTP I.C.E. tool. DBE-DBCH was tested in numerous HTS assays aimed

© Burleigh Dodds Science Publishing Limited, 2025. All rights reserved.

**Table 3** Summary of carcinogenicity predictions for DBE-DBCH

| Software tool | Carcinogenicity Prediction | Reliability of prediction |
|---|---|---|
| OECD QSAR Toolbox | Carcinogen (10/18 models) | In Domain |
| | No value (8/18 models) | Out of Domain |
| VEGA QSAR | Non-carcinogen | Outside of the applicability domain (reliability 1/3 stars) |
| | Carcinogen | Could be out of the applicability domain (reliability: 2/3 stars) |
| | Possible non-carcinogen | Outside of the applicability domain (reliability 1/3 starts) |
| | Carcinogen | Could be out of the applicability domain (reliability 2/3 stars) |
| | Carcinogen | In applicability domain (reliability: 3/3 stars) |
| Cheminformatics analysis modules | High | Suspected of causing cancer; IARC group 2B (possibly carcinogenic to humans) |
| ADMETLab 3.0 | Carcinogen | Probability of being carcinogen: 0.993[1] |
| ProTox 3.0 | Active | Probability of being carcinogen: 0.78 |
| OncoLogic 9.0 | Moderate | Likely to be moderately active carcinogen |
| Leadscope | Positive (female mouse) | Predicted value: 0.727 |
| | Positive (female rat) | Predicted value: 0.581 |
| | Positive (male mouse) | Predicted value: 0.523 |
| | Positive (male rat) | Predicted value: 0.568 |
| CASE Ultra (MultiCASE) | Positive (female mouse) | Probability of being carcinogen: 80.7%[2] |
| | Inconclusive (male mouse) | Probability of carcinogenicity: 52.3%[2] |
| | Positive (female rat) | Probability of carcinogenicity: 60.3%[2] |
| | Positive (male rat) | Probability of being carcinogen: 83.3%[2] |
| Derek Nexus | Carcinogen | Plausible[3] |
| ChemTunes ToxGPS | Positive (mouse) | Probability (positive) range: [0.987, 0.999] |
| | Positive (rat) | Probability (positive) range: [0.650, 0.800] |
| **Overall conclusion:** | **Carcinogen** | |

[1]For carcinogens this value is 1 and for non-carcinogens the value is 0.[2]The model's current classification threshold of 45.0% with a grey zone between 35.0 and 55.0%.[3]The weight of evidence supports the proposition.

at evaluating whether it had one or more of the ten key characteristics of human carcinogens (KCC) (Smith et al., 2016). In 18.4% (7 of 38) KCC8 assays aimed at capturing carcinogenicity due to modulation of receptor-mediated effects and 11.3% (9 of 79) KCC10 assays aimed at capturing carcinogenicity due to altered cell proliferation, cell death, or energetics/nutrient supply came back as "Active" providing some experimental support for the carcinogenicity prediction. As a reminder, a synergistic approach that combines computational

© Burleigh Dodds Science Publishing Limited, 2025. All rights reserved.

predictions with experimental data to bolster safety and hazard assessments is recommended.

## 7 Summary and future trends

Computational toxicology tools offer notable advantages compared to traditional animal safety testing methods. These include considerations for ethical standards, enhanced efficiency, cost-effectiveness, and the ability to offer mechanistic insights, alongside their growing acceptance within regulatory frameworks (EPA, 2024; FDA, 2023; EFSA, n.d.; ECHA, n.d.). As integral components of modern toxicological assessments, these tools are poised to advance further with evolving technologies and methodologies.

However, along with their benefits, computational toxicology tools exhibit drawbacks when contrasted with traditional animal testing approaches. For instance, they often struggle to accurately replicate intricate biological interactions such as metabolism, bioavailability, and systemic effects. Unlike animal studies, which provide a comprehensive view of toxicity encompassing these complexities, predictive models heavily rely on the quality and scope of available data. Issues arise particularly with novel substances or complex endpoints due to limitations in data diversity and biases, leading to potential prediction inaccuracies. For example, models trained on limited or biased datasets may fail to generalize accurately to novel substances outside those biases, non-representative validation datasets can distort assessments of model performance, affecting confidence in predictions for novel substances; and sparse or low-quality data for certain chemical types or endpoints can undermine the reliability of predictions for novel substances in those categories.

Moreover, computational models may fail to anticipate unexpected or idiosyncratic toxicities that manifest only in vivo, such as immune responses, behavioral changes, or cumulative long-term effects, which pose challenges for computational simulation. While these models can predict overall toxicity outcomes, they frequently fall short in elucidating the detailed mechanistic reasons behind specific effects at the molecular or cellular levels. Understanding these mechanisms is crucial to developing safer chemicals[13] and enhancing predictive toxicology models.

To gain acceptance in regulatory frameworks, computational models must meet stringent validation and reliability criteria, navigating issues related to performance variability, data integrity, and model complexity (OECD, 2014; Patlewicz et al., 2016). Despite these challenges, the potential of computational toxicology tools remains promising. Future advancements hinge on refining

13 Safer chemicals reduce the risks of adverse health effects in humans, such as toxicity, carcinogenicity, mutagenicity, and reproductive harm; as well as exhibit reduced persistence in the environment and show decreased potential for bioaccumulation compared to more hazardous chemicals.

© Burleigh Dodds Science Publishing Limited, 2025. All rights reserved.

methodologies, enhancing data quality, and fostering interdisciplinary collaboration to effectively integrate these tools into comprehensive toxicological assessments. This evolution is driven by technological innovations, regulatory imperatives, ethical considerations, and the demand for more efficient and sustainable approaches to evaluating chemical safety.

## 8 Where to look for further information

Various resources for learning more about computational toxicology tools are available. Journals, such as *Toxicological Sciences*, *Chemical Research in Toxicology*, *Environmental Health Perspectives*, and the *Journal of Chemical Information and Modeling* are good sources of computational methods and tools. Many books and book chapters also discuss this topic and serve as great references, including Fowler (2013), Cronin and Yoon (2019), Hong (2019), and Valerio (2024). Government agencies, such as the EPA, and organizations, such as the OECD, provide resources and guidance on computational toxicology methods and tools. By leveraging these resources, one can gain a comprehensive understanding of computational toxicology tools, their applications, methodologies, and current advancements in the field. Additionally, staying updated with the latest research through conferences and courses will further enhance knowledge and skills in this rapidly evolving discipline.

## 9 Acknowledgments

I would like to express my sincere gratitude to Dr. Kirk Arvidson (US FDA) for his invaluable contributions to this chapter. His insightful feedback and meticulous edits greatly enhanced the clarity and quality of the content. His expertise and thoughtful suggestions were instrumental in refining the final manuscript. I would also like to thank Joanne Berger, FDA Library, for manuscript editing assistance.

## 10 References

Chakravarti, S.K. and Saiakhov, R.D., 2022. MultiCASE platform for in silico toxicology. In *In silico methods for predicting drug toxicity* (pp. 497–518). Springer US.

Collins, S.P., et al., 2024. Development and application of consensus in silico models for advancing high-throughput toxicological predictions. *Frontiers in Pharmacology*, *15*, p.1307905.

Cramer, G.M., et al., 1976. Estimation of toxic hazard–a decision tree approach. *Food and Cosmetics Toxicology*, *16*(3), pp.255–276.

Cronin, M.T. and Yoon, M., 2019. Computational methods to predict toxicity. In *The history of alternative test methods in toxicology* (pp. 287–300). Academic Press.

© Burleigh Dodds Science Publishing Limited, 2025. All rights reserved.

Curran, I.H., et al., 2017. Toxicologic effects of 28-day dietary exposure to the flame retardant 1, 2-dibromo-4-(1, 2-dibromoethyl)-cyclohexane (TBECH) in F344 rats. *Toxicology*, *377*, pp.1–13.

Danish (Q)SAR Database, Division of Diet, Disease Prevention and Toxicology, National Food Institute, Technical University of Denmark, http://qsar.food.dtu.dk.

Dimitrov, S.D., et al., 2016. QSAR Toolbox–workflow and major functionalities. *SAR and QSAR in Environmental Research*, *27*(3), pp.203–219.

ECHA, n.d. Alternatives to animal testing under REACH. Accessed on 2024-10-21. Available at: https://echa.europa.eu/animal-testing-under-reach

EFSA, 2024. Alternatives to animal testing. Accessed on 2024-08-25. Available at: https://www.efsa.europa.eu/en/topics/topic/alternatives-animal-testing

EFSA, n.d. Alternatives to animal testing. Accessed on 2024-10-21. Available at: https://www.efsa.europa.eu/en/topics/topic/alternatives-animal-testing

EFSA Panel on Contaminants in the Food Chain, 2012. Scientific opinion on emerging and novel brominated flame retardants (BFRs) in food. *Efsa journal*, *10*(10), p.2908.

EPA, 2004. Chemical Hazard Classification and Labeling: Comparison of OPP Requirements and the GHS. Accessed on 2024-07-24. Available at: https://www.epa.gov/sites/default/files/2015-09/documents/ghscriteria-summary.pdf

EPA, 2006. Interpretive Guidance Document for Sustainable Futures Summary Assessment. Accessed on 2024-07-26. Available at: https://www.epa.gov/sites/default/files/2015-09/documents/onco-igc1.pdf

EPA, 2024. Strategic Vision for Adopting New Approach Methodologies. Accessed on 2024-10-21. Available at: https://www.epa.gov/pesticide-science-and-assessing-pesticide-risks/strategic-vision-adopting-new-approach

FDA, 2023. Utility of Computational Toxicology at FDA Center for Tobacco Products in Evaluation of Potential Hazards Due to Ingredients in Tobacco Products. Accessed 2024-10-21. Available at: https://www.fda.gov/science-research/fda-science-forum/utility-computational-toxicology-fda-center-tobacco-products-evaluation-potential-hazards-due

FDA, 2024a. Title 21: Food and Drugs. Chapter I - Food and Drug Administration, Department of Health and Human Services. Subchapter B - Food for Human Consumption. Part 109 - Unavoidable contaminants in food for human consumption and food-packaging material. Accessed 2024-07-15. Available at: https://www.ecfr.gov/cgi-bin/text-idx?SID=[SessionID]&mc=true&node=pt21.2.109

FDA, 2024b. External Peer Review of the Expanded Decision Tree: Ranking Toxic Potential. Accessed 2024-07-15. Available at: https://www.fda.gov/science-research/peer-review-scientific-information-and-assessments/external-peer-review-expanded-decision-tree-ranking-toxic-potential

FDA, 2024c. Facts about FDA and Animal Welfare, Testing & Research. Accessed on 2024-08-2024. Available at: https://www.fda.gov/news-events/rumor-control/facts-about-fda-and-animal-welfare-testing-research

Fowler, B.A. ed., 2013. *Computational toxicology: methods and applications for risk assessment*. Academic Press.

Gatnik, M.F. and Worth, A., 2010. Review of Software Tools for Toxicity Predictions. JRC Scientific and Technical Reports. Accessed 2024-07-24. Available at: https://publications.jrc.ec.europa.eu/repository/bitstream/JRC59685/reqno_jrc59685_software_tools_for_toxicity_prediction%5B1%5D.pdf

© Burleigh Dodds Science Publishing Limited, 2025. All rights reserved.

Hong, H., 2019. *Advances in computational toxicology*. Springer.

ICH, 2017. Assessment and Control of DNA Reactive (Mutagenic) Impurities in Pharmaceuticals to Limit Potential Carcinogenic Risk. Guideline. Accessed 2024-07-24. Available at: https://database.ich.org/sites/default/files/M7_R1_Guideline.pdf

Leadscope, 2020. Leadscope Model Applier Documentation. Accessed 2024-07-24. Available at: https://www.leadscope.com/ls-manuals/pdfs/ModelApplierManual-3.0.pdf

Leadscope, 2022. Implementation of the *in silico* Toxicology Protocols Within Leadscope. Accessed on 08/23/2024. Available at: https://www.leadscope.com/ls-manuals/pdfs/Implementation_of_the_in_silico_toxicology_protocols-2022_0.pdf

Mansouri, K., et al., 2020. CoMPARA: collaborative modeling project for androgen receptor activity. *Environmental Health Perspectives*, *128*(2), p.027002.

Marteinson, S.C., et al., 2021. A review of 1, 2-dibromo-4-(1, 2-dibromoethyl) cyclohexane in the environment and assessment of its persistence, bioaccumulation and toxicity. *Environmental Research*, *195*, p.110497.

Mastrolorito, F., et al., 2024. TISBE: a public web platform for the consensus-based explainable prediction of developmental toxicity. *Chemical Research in Toxicology*, *37*(2), pp. 323–339.

Munro, I.C., et al., 1996. Correlation of structural class with no-observed-effect levels: a proposal for establishing a threshold of concern. *Food and Chemical Toxicology*, *34*(9), pp. 829–867.

Myatt, G.J., et al., 2022. Implementation of in silico toxicology protocols within a visual and interactive hazard assessment platform. *Computational Toxicology*, *21*, p.100201.

NTP. 1,2-Dibromo-4-(1,2-dibromoethyl)cyclohexane (3322-93-8). Chemical Effects in Biological Systems (CEBS). Research Triangle Park, NC (USA): National Toxicology Program (NTP). Accessed 2024-07-08. https://doi.org/10.22427/NTP-DATA-DTXSID8024947

OECD, 2014. *Guidance on the validation of (Quantitative) Structure-Activity Relationship [(Q)SAR] models*. OECD Publishing. https://doi.org/10.1787/9789264085442-en

Patlewicz, G., et al., 2016. Validation of computational methods. *Validation of Alternative Methods for Toxicity Testing*, pp.165–187. https://doi.org/10.1007/978-3-319-33826-2_6

Patlewicz, G., et al., 2017. Navigating through the minefield of read-across tools: a review of in silico tools for grouping. *Computational Toxicology*, *3*, pp.1–18.

Ponting, D.J., et al., 2022. Use of Lhasa Limited products for the in silico prediction of drug toxicity. In *In silico methods for predicting drug toxicity* (pp. 435–478). Springer US.

Smith, M.T., et al., 2016. Key characteristics of carcinogens as a basis for organizing data on mechanisms of carcinogenesis. *Environmental Health Perspectives*, 124(6), pp.713–721.

Sushko, I., et al., 2011. Online chemical modeling environment (OCHEM): web platform for data storage, model development and publishing of chemical information. *Journal of Computer-Aided Molecular Design*, *25*, pp.533–554.

Valerio Jr., L.G., 2024. *Predictive analytics for toxicology: applications in discovery science*. CRC Press.

Valsecchi, C., et al., 2020. Consensus versus individual QSARs in classification: comparison on a large-scale case study. *Journal of Chemical Information and Modeling*, *60*(3), pp.1215–1223.

© Burleigh Dodds Science Publishing Limited, 2025. All rights reserved.

Xiong, P., et al., 2019. A review of environmental occurrence, fate, and toxicity of novel brominated flame retardants. *Environmental Science & Technology*, *53*(23), pp.13551–13569.

Yang, C., et al., 2023a. The role of a molecular informatics platform to support next generation risk assessment. *Computational Toxicology*, *26*, p.100272.

Yang, C., et al., 2023b. High throughput read-across for screening a large inventory of related structures by balancing artificial intelligence/machine learning and human knowledge. *Chemical Research in Toxicology*, *36*(7), pp.1081–1106.

Zuiderveen, E.A., et al., 2020. Novel brominated flame retardants: a review of their occurrence in indoor air, dust, consumer goods and food. *Chemosphere*, *255*, p.126816.

© Burleigh Dodds Science Publishing Limited, 2025. All rights reserved.

# Part 4

## Environmental and processing contaminants and natural toxins

# Chapter 12

# Advances in assessing heavy metals in food

*Paul Williams, Simon Haughey and Yicong Li, The Queen's University of Belfast, UK; Ramón Fernández-Ruiz, Universidad Autónoma de Madrid, Spain; Alfredo de la Escosura-Muñiz, University of Oviedo, Spain; Alessandro Migliori, Nuclear Science and Instrumentation Laboratory, Division of Physical and Chemical Sciences, Department of Nuclear Sciences and Applications, International Atomic Energy Agency (IAEA), Vienna, Austria; Christina Vlachou and Britt Maestroni, Food Safety and Control Laboratory, Joint FAO/IAEA Centre of Nuclear Techniques in Food and Agriculture, Department of Nuclear Sciences and Applications, International Atomic Energy Agency (IAEA), Vienna, Austria*



http://dx.doi.org/10.19103/AS.2025.0158.21
© Burleigh Dodds Science Publishing Limited, 2025. All rights reserved.

## 1 Introduction

Elements are the fundamental materials of which all matter is composed; they can be classified as metals, metalloids, and non-metals. Approximately, three-quarters of all known chemical elements are metals. They are characterized by high electrical and thermal conductivity as well as by malleability, ductility, and high reflectivity of light [1]. Examples of metals and metalloids are lead (Pb), mercury (Hg), cadmium (Cd), and arsenic (As): they have relatively high densities, atomic weights, or atomic numbers, and they are toxic even at low concentrations. They can be the cause of environmental degradation and generate health effects in humans. Generally, they are referred to as heavy metals (HMs), although the use of this terminology is regularly questioned by the scientific community [2]. In 2002, an IUPAC report argued that the term 'heavy metal' is meaningless and misleading, suggesting that a more precise classification based on the periodic table and the chemical basis of toxicity would be more useful [3]. Nevertheless, the term HM continues to be used worldwide. Other metals, such as zinc (Zn), selenium (Se), copper (Cu), and nickel (Ni), are considered essential, at low concentrations, for metabolic/biological functions in humans and are therefore required in the diet in a certain defined amount, above which they may become toxic [4].

HM contamination in food has emerged as a pressing issue of significant global concern, primarily due to the alarming presence of hazardous elements, such as As, Pb, Cd, and Hg, which have been identified in various food items across different regions. In this chapter, we will focus mainly on very toxic elements present in food, which are non-essential for human metabolic activities. They can mimic the biological function of essential metals and perturb the processes and are classified as contaminants in foods. It is important to note that food is just one of the many routes of exposure of HM to humans [5].

This chapter explores the mechanisms of HM uptake and bioaccumulation in plants and animals and highlights global contamination trends, particularly in fish and fresh produce. Risk assessment frameworks, regulatory standards, and toxicological evaluations by international bodies like FAO, WHO, and EFSA are discussed, with emphasis on the shift toward new approach methodologies (NAMs), such as *in vitro*, *in silico*, and artificial intelligence (AI)-assisted tools. Advanced analytical techniques, especially X-ray fluorescence (XRF), are detailed for their role in detecting and quantifying HMs in food. A case study on turmeric adulteration with lead chromate illustrates the public health implications of economically motivated fraud. The chapter concludes by emphasizing the need for rapid, cost-effective screening tools, such as electrochemical sensors and lateral flow assays (LFAs), and the integration of AI for enhanced detection, data analysis, and regulatory compliance. These

© Burleigh Dodds Science Publishing Limited, 2025. All rights reserved.

innovations are essential for ensuring food safety, particularly in resource-limited settings.

## 2 Origins of heavy metals in food

Metals occur as contaminants in food due to their presence in the environment, and increasing concentrations of HM in the environment will increase concentrations in foods that are produced in a degraded environment. Some being of geogenic origin, heavy elements can be naturally omnipresent in air, soil, sediments, and water resources. Their presence in food is exacerbated by anthropogenic activities such as agronomic practices, i.e. through application of metal-containing pesticides and fertilizers, irrigation with wastewaters, emissions from industrial activities, emissions from vehicles, deposition on crops grown in contaminated air and dust, plant uptake from polluted sites close to mining operations, contaminated run-off, among others. Natural phenomena such as leaching from enriched rocks and sediments, volcanic eruptions, weathering of rocks, and forest fires can take place among other processes. Weathering and volcanic eruptions have been reported to significantly contribute to HM pollution in the environment and consequently in foods [5]. For most of the toxic metals, natural variations are small compared to emissions from industrial activities, implying that mankind has become the key agent today in the global atmospheric cycle of trace metals and metalloids [6].

Plants absorb metals from soil for their normal development [7]. However, because of the lack of specificity of the uptake mechanisms, when growing in very contaminated substrates, plants may incorporate concentrations of essential and toxic HM [8]. In agricultural systems, the origin of HM is multiple: from phosphate-based and other combined fertilizers, wastewater irrigation and agrochemical treatments using metal-containing pesticides. In certain countries, like the UK and China, As-, Hg-, and Pb-based pesticides are banned, and in other countries like France and Italy as is used largely as a fungicide for viticulture. Increased concentrations of HM may also derive from livestock manure, as oligoelements are added to animal feed to promote health and growth and excreted through urine or feces, which will then be present in manure applied to land, continuing the cycle of pollution [9]. Another source of HM in food is through packaging materials, such as aluminium cans, leak-proof bags, cardboard, tetra packs, cellophane, tissues, sachets, aluminium bags and boxes, plastic bags, and containers [10]. HM may leach out of packing materials directly into the food item, with the rate of leaching changing with temperature and food pH (low pH and high temperature) [11]. Unsuitable post-harvest handling, food processing, packaging, storage, transportation, handling, and cooking tools can also result in HM contamination of food. For instance, some traditional cooking and processing practices, such as washing,

© Burleigh Dodds Science Publishing Limited, 2025. All rights reserved.

parboiling, boiling, steaming, frying, and roasting, can extract the bound toxic metals from vegetables and cereals [12], facilitating transfer of heavy metals from the environment to foods to humans.

In plants, apart from atmospheric deposition on leaves and other plant parts, HMs are absorbed via foliar and root uptake and translocated mainly through two pathways: the apoplastic pathway (passive diffusion) and the symplastic pathway (active, energy-dependent transport) [13, 14]. Plant roots may absorb various metals from the soil. While some HMs, such as manganese (Mn), Zn, molybdenum (Mo), Co, iron (Fe), Ni, and Cu, have a positive role in plant growth development and metabolism, others, like As, Hg, Cd, and Pb, are highly toxic for the plant, leading to detrimental effects on plant morphology, physiology, and genetics.

In plants, dedicated 'transporters' belonging to different families are involved in the uptake of these metal ions from the soil and their transport to different tissues. These transporters have low ion selectivity and cannot distinguish between the metal ions. As a result, toxic and HM ions can be absorbed and accumulated in plants. This not only inhibits the growth and development of plants but also poses a risk to food security through loss of food and/or through unsafe food for the consumers [15]. Certain plant species have demonstrated remarkable tolerance to toxic metals and can naturally thrive in contaminated environments. Hyperaccumulating plants like the *Poaceae* family (grass), capable of storing metals at concentrations 100–1000 times higher than non-accumulators, are well-suited for phytoremediation efforts in HM-contaminated soils [16, 17]. Figure 1 provides examples illustrating the reasons for the presence of HMs in foods.

To understand how HM transfers from the environment to foods, it is crucial to discuss their mobility and availability. In soil, some metals are more

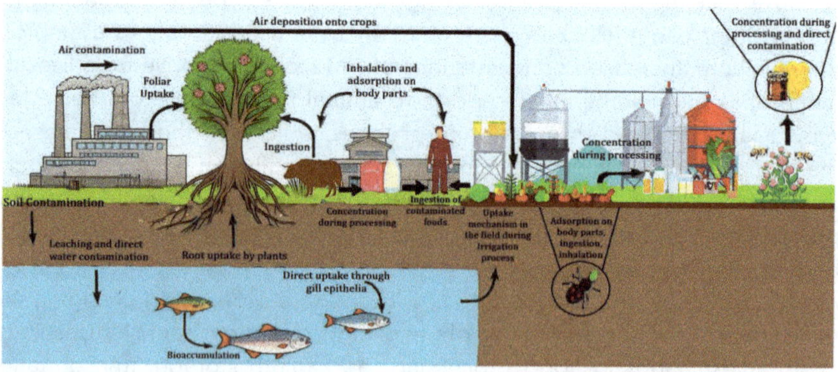


**Figure 1** HM transferred to foods, considering adsorption, absorption, uptake, translocation, direct contamination, biomagnification, bioaccumulation, among other mechanisms (Credit: Ms Pearl Bergan, IAEA).

© Burleigh Dodds Science Publishing Limited, 2025. All rights reserved.

mobile than others, e.g. Cd and Zn, while, e.g., Cu and Pb are immobile and easily form organic complexes with fulvic acids. Analytical data for total As in soil only provides an indication of the soil contamination but does not indicate how much As is available to the plant or how mobile it is. The latter depends on its chemical status and the interactions with the soil mineral component (i.e. clay soils), which affect its movement and availability for processes like adsorption, potential precipitation, and incorporation into mineral structures of the soil [18]. Table 1 provides a summary of the chemical species in which HM can be present in the environment and in foods.

These interactions can be studied, e.g. using a combination of an isotopic exchange technique and synchrotron X-ray absorption spectroscopy (XAS) [1]. Over time, the leaching of As bound to soil decreases significantly. This detailed knowledge is essential for restoring soil to a healthy and usable state by managing techniques such as additions of amendments, phytoremediation (using plants to absorb contaminants), and soil washing [20].

The uptake and bioaccumulation of HMs in plants is a very complicated subject with many reviews covering different types of plants, soil physico-chemical properties (e.g. pH, cation exchange capacity), rhizosphere microbial activities, metal transporters, among others. Uptake, absorption, translocation, and other processes shall be evaluated for individual metals [8, 21–27].

Other sources of HM in foods are from bioaccumulation and biomagnification processes along different trophic chains, whether terrestrial or aquatic [28–30]. Invertebrates can ingest metals from soils, water, and

**Table 1** Inorganic and organic forms of HMs in the environment and food

| Element | Inorganic forms | Organic forms |
|---|---|---|
| Arsenic (As) | Arsenite ($As^{3+}$ / As(III))<br>Arsenate ($As^{5+}$ / As(V)) | Dimethylarsinic acid (DMA)<br>Monomethylarsonic acid (MMA)<br>Trimethylarsine oxide (TMAO) |
| Cadmium (Cd) | $Cd^{2+}$ (Cadmium ion) | Cadmium acetate (Cd $(CH_3COO)_2$)<br>Cadmium carbonate ($CdCO_3$)<br>Cadmium stearate ($Cd(C_{36}H_{72}O_4)$) |
| Chromium (Cr) | $Cr^{2+}$ (Chromous)<br>$Cr^{3+}$ (Chromic)<br>$Cr^{6+}$ (Hexavalent chromium) | Organic Cr chelates, known as<br>Cr methionine (CrMet),<br>Cr picolinate (CrPic),<br>Cr nicotinate (CrNic),<br>Cr-yeast,<br>Cr propionate,<br>Cr histidinate |
| Mercury (Hg) | $Hg^0$ (Elemental mercury, vapor)<br>$Hg^+$ (Mercurous)<br>$Hg^{2+}$ (Mercuric) | Methylmercury ($CH_3Hg^+$) |
| Lead (Pb) | $Pb^{2+}$ (Lead ion)<br>Other oxidation states (e.g. $Pb^{4+}$) in industrial compounds | Tetramethyl lead (TML)-$Pb(C_2H_5)_4$<br>Tetraethyl lead (TEL)-$Pb(CH_3)_4$ |

© Burleigh Dodds Science Publishing Limited, 2025. All rights reserved.

plants or absorb them through their exoskeletons or other body coverings when in direct contact with metals. Different invertebrate orders and families may accumulate specific metals based on their habitat, diet, and physiological responses. Aquatic invertebrates are important feed sources for fish, birds, and mammals and may represent the cause of presence of HM in a variety of food products [9]. Large mammals important to the human diet such as grazers (i.e. cattle, deer, sheep) and omnivores (i.e. pigs and wild boar), often ingest polluted vegetation, including polluted mushrooms, small amounts of soil, contaminated drinking water, or through inhalation. The extent of exposure depends on the metal and its concentration in the vegetation and soil, the age of the animal, and how quickly the metal is metabolized by the animal [31]. In general, the sources of pollution of HM in foods can be multiple, from a mixture of different potential origins. Table 2 provides a

**Table 2** Potential sources of contamination in a generic food basket

| Type of food | Potential source of contamination | Mechanism involved |
|---|---|---|
| Animal and derived products (cheese, milk, etc.) | Feed, water, grazing pasture, soil, sediments, conservation processes, industrial processing, storage, transportation, packaging materials. | Ingestion and adsorption on body parts, air band dust deposition. Concentration during processing. |
| Cereals and grains | Soil, irrigation water, air, dust, storage silos. | Uptake mechanism in the field. |
| Drinking water | HM leached through soils into groundwater, water pipelines, storage reservoirs. | Leaching, direct contamination, deposition, geogenic nature. |
| Edible insects | Natural environment and natural water sources, air, sediments, soils. | Adsorption on body parts, ingestion, inhalation. |
| Fish and fishery products, seafoods | Water, feed, sediments, smaller prays. | Passive uptake, filtration processes, direct uptake through gill epithelia. |
| Fresh fruits and vegetables | Soil, irrigation water, air, dust. | Uptake mechanism, air deposition, irrigation process. |
| Honey and derived products | Bees and other pollinators, agro-industrial processing, packaging materials. | Concentration during processing, direct contamination. |
| Processed food products | Processing lines, water, storage containers, packaging materials. | Concentration during processing, direct contamination. |
| Pulses | Soil, irrigation water, air, dust. | Uptake mechanism in the field. |
| Table salt | Excess of Iodine, co-accumulation of other metals. | Concentration during processing, direct contamination. |
| Vegetable oils | Original seeds and fruits, packaging materials, and storage containers. | Concentration during processing, direct contamination. |

*Source*: Modified from [37].

© Burleigh Dodds Science Publishing Limited, 2025. All rights reserved.

summary of potential sources of contamination of different food products by selected HM (As, Cd, Hg, Pb).

To reduce the presence of toxic elements in food, it is necessary to promote good practices to apply modern remediation technologies as well as improved agronomic practices [32–34]. Additionally, more research is needed to fully understand uptake and bioavailability and bioaccumulation mechanisms for different HM in different plants and other foods in order to apply improved agricultural practices and mitigation and/or remediation strategies [35].

## 3 Levels of heavy metals in different food products

The growing environmental and public health concern related to the presence of HM necessitates further investigation into the sources and levels of contamination. Such information on levels of HMs in food products is difficult to find from public and published sources; however, this information is vital and international bodies such as the Food and Agricultural Organization of the United Nations (FAO) and the World Health Organization (WHO) are encouraging member states to submit analytical data, through global database such as the GEMS Food Contaminants Database [36] for contaminants to be able to carry out dietary risk assessments and implement proper risk management and communication.

A recent review of HMs in foods and beverages [37] provides an overview of the most recent studies providing information on HM presence in different foods worldwide. According to the authors, the levels of HMs in foods are heterogeneous due to the factors that contribute to their accumulation. The review identifies some values that exceed the maximum levels established by international regulatory frameworks, especially in countries with high environmental contamination and elevated industrial activity. Data from FAO/WHO evaluations of priority contaminants in food are available from WHO for Cd [38], Hg and As [39], and Pb [39].

A comprehensive mapping analysis for HM in food was carried out utilizing the FoodAkai platform, which is available from Agroknow [40]. This analysis focused on thoroughly evaluating alerts pertaining to food products worldwide between the years 2010 and 2024. During this extensive period of time, a total of 4526 alerts for HM were recorded. The majority of these alerts were announced by food safety authorities, including the European Union's Rapid Alert System for Food and Feed (EU-RASFF), the United States Food and Drug Administration (US-FDA), and various food safety authorities from individual countries around the globe. It is noteworthy that the primary focus of these alerts centred around the hazardous substances, Hg and Pb, which accounted for a substantial 65% of the total announcements made.

© Burleigh Dodds Science Publishing Limited, 2025. All rights reserved.

The key objective of this rigorous analysis was to pinpoint the main food categories that are consistently linked to these alerts. Additionally, it aimed to identify the specific countries of origin associated with the contaminated food items, thereby offering crucial insights into the extent and distribution of HM contamination in the global food supply chain. This information is essential for understanding the broader implications for public health and developing strategies aimed at mitigating the risks associated with these harmful elements in food products. To achieve this, a detailed examination and mapping of extensive data sets that span over a decade were undertaken. Through this rigorous analysis, significant revelations emerged regarding the nature of HM contamination across a wide range of food categories.

Particularly noteworthy was that the study revealed that fish and fish products are the most heavily contaminated food group among those analysed. Within this category, Hg was identified as the most concerning HM with specific fish species – all wild as opposed to commercially farmed – such as swordfish, shark, and tuna, accounting for 1038 out of 1328 alerts, which constitutes approximately 78% of the total alerts recorded in this category (Fig. 2). This statistic emphasizes the urgent need for further exposure-based assessments regarding the consumption of these particular seafood items, as well as the broader implications for food safety and public health.

The second most abundant alerts, which were identified during this mapping analysis, were predominantly found within the fresh fruits and vegetables (FFVs) category. Within this category, Pb and Cd emerged as

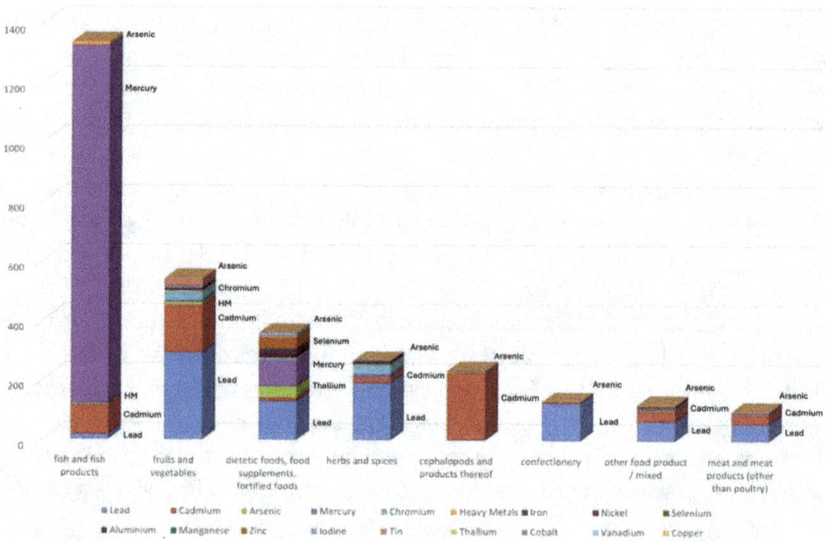


**Figure 2** General food categories and their associated heavy metal contamination issues.

© Burleigh Dodds Science Publishing Limited, 2025. All rights reserved.

significant contaminants, posing potential risks to consumer health. A total of 520 alerts were gathered in this specific category, with key contamination issues identified in several fruits and vegetables. Notably, plums accounted for 120 alerts, followed by apples with 70 alerts, prunes with 36 alerts, mushrooms with 33 alerts, and spinach with 32 alerts, indicating a concerning trend in HM presence across these commonly consumed food items. With regard to the country of origin of the foodstuffs, specific findings illustrated in Fig. 3 highlight the top countries that are connected to HM contamination. Notable among these are nations such as China and Spain, which, when combined, account for 1458 alerts related to the country of origin, representing a substantial 32% of the total alerts recorded for foodstuffs.

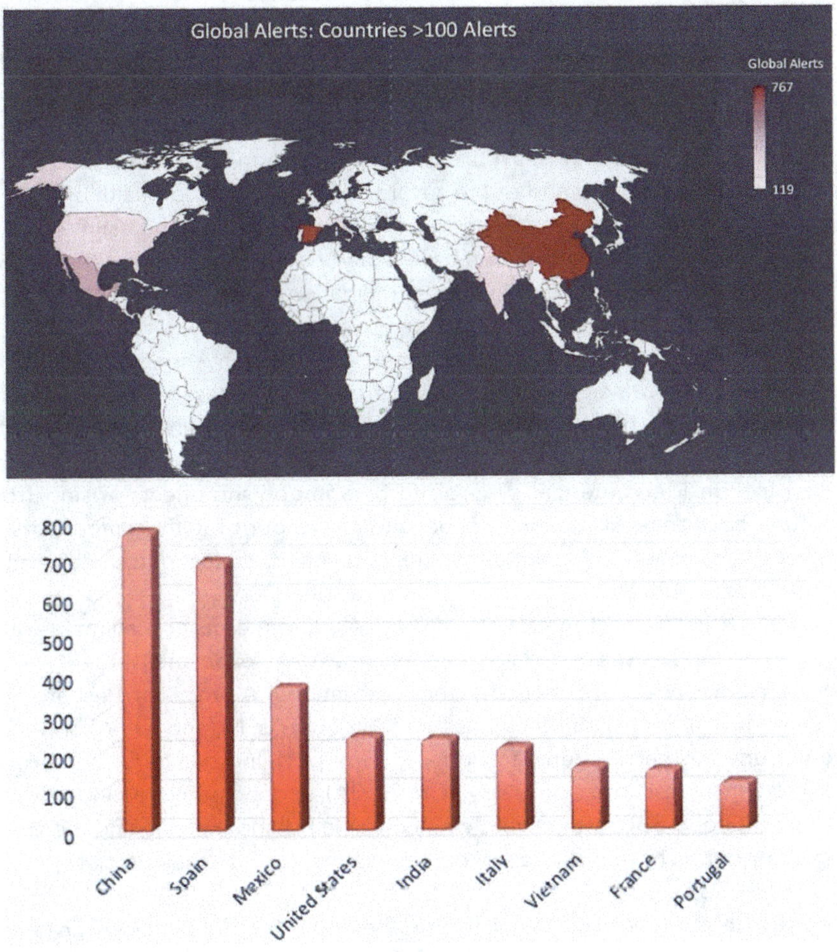


**Figure 3** Country of origin with >100 alerts from 2010 to 2024.

© Burleigh Dodds Science Publishing Limited, 2025. All rights reserved.

These insights further reinforce the necessity for collaborative efforts on a global scale to address and mitigate the challenges posed by HM contamination. Within a growing environmental pollution, continuous monitoring of HMs is essential for public health and trade purposes, and it is imperative for policymakers, industry stakeholders, and researchers to work together to develop effective strategies and solutions that can help combat this pressing issue, ensuring a safer and healthier food environment for consumers everywhere.

## 4 Current risk assessment and regulation of heavy metals in food

As mentioned previously, the ubiquitous presence of toxic metals in all environmental compartments and food webs and their bioaccumulating and deleterious potential have been historically acknowledged, necessitating the assessment of any associated human health risks. Monitoring of the occurrence and exposure assessment of toxic metals in the diet is integral part of any food control system and comprises a thorough investigation of the likelihood of the human dietary exposure and of the severity of the associated health effects (as identified in the hazard identification and characterization). The estimates and conclusions of risk characterization have a major impact on decision-making for hazard prioritization and adoption of risk mitigation measures [41].

Due to the special attributes and behaviours of metals and metal compounds, the United States Environmental Protection Agency (US-EPA) has established key principles to be considered when assessing human health risks, underlining the dependence of metal bioaccessibility, bioavailability, fate, and effects on its speciation and specific compound, along with the need to consider potential synergistic or antagonistic effects arising from the co-occurrence of metals in food and to account for the differences in vulnerability, which varies with age, sex, pregnancy status, nutritional status, and genetics [42].

Risk assessment of metals has largely relied in the past on animal testing and, in several cases, on data from occupational exposures [43–51]. Due to sufficient evidence in humans, the International Agency for Research on Cancer (IARC) has classified Cd, As and inorganic As, Ni, and Cr (VI) and their compounds as carcinogenic to humans (Group 1). Inorganic Pb and $CH_3Hg$ have been assessed as probably (Group 2A) and possibly carcinogenic to humans (Group 2B), respectively, while organic Pb and metallic and inorganic Hg compounds have not been deemed classifiable as to their carcinogenicity to humans (Group 3) [43–45].

The Joint FAO/WHO Expert Committee on Food Additives (JECFA) and the European Food Safety Authority (EFSA) have conducted risk assessments

© Burleigh Dodds Science Publishing Limited, 2025. All rights reserved.

of several metals, summarizing their toxic effects. Hazard identification and characterization was based on the evaluation of toxicological studies for the estimation, if possible, of a 'safe' level of human exposure to the metal, so that a toxicological guidance value [52] or a health-based guidance value (EFSA) could be identified [46–51]. When the human dietary exposure to the metal is higher than these values, adverse health effects cannot be excluded. In case, a 'safe' level of human exposure cannot be identified because of non-threshold critical toxic effects, risk characterization is based on the Margin of Exposure (MOE) approach [53]. However, additional toxicological evidence acquired over the years allowed for updated risk assessment and withdrawal of previously established health-based guidance values for Pb and As, as they could no longer be considered health protective and the possibility of a risk to some consumers could not be excluded [48, 49, 52].

Due to the existing scientific evidence, safety standards and monitoring systems for toxic metals are of utmost importance to safeguard public health, mitigate risks, and regulate food production and trade. Table 3 summarizes toxicological information on carcinogenic and non-carcinogenic effects of some metals, along with toxicological guidance or health-based guidance values established by JECFA and EFSA, as well as information on relevant safety standards establishing maximum levels for these metals in various food commodities [52, 54]. The MOE approach has increasingly been applied not only for substances that are both genotoxic and carcinogenic but also for various types of non-genotoxic substances. Therefore, in case of metals for which a JECFA toxicological guidance value or an EFSA health-based guidance value have not been established, Table 3 also includes information about Benchmark Dose Lower Confidence Limit (BMDL) values, which have been calculated through statistical modelling of toxicological data to estimate the dose of a substance that is associated with a specific level of adverse effect [53].

## 5 New approach methodologies for assessing risks from heavy metals in food

Traditional risk assessment approaches are being complemented by modern laboratory, computational, epidemiological, and biomonitoring methods, which have elucidated toxic effects on a wide variety of organ systems and investigated differences in susceptibility [55]. Under the *3R* concept (replacement, reduction, and refinement of animal testing), NAMs, such as *in vitro*, *in chemico*, and *in silico* methodologies, are currently being developed worldwide. Although NAMs are currently lacking standardization and regulatory approval for wider utilization in regulatory decision-making, research in this field is vibrant and is expected to support and accelerate the pace of the new generation food safety risk assessment [56–58].

© Burleigh Dodds Science Publishing Limited, 2025. All rights reserved.

**Table 3** Summary of toxicological information, along with toxicological guidance or health-based guidance values established by JECFA and EFSA, respectively, as well as information on relevant safety standards through which maximum levels have been established in various food commodities for Cd, Pb, As, Hg, Ni, Cr, and Al

| Metal | IARC evaluation of carcinogenicity in humans | Non-carcinogenic effects (IARC 1993, 2006, 2012; EFSA, 2008, 2009a, 2009b, 2010, 2012, 2020) | | | | | | JECFA Toxicological guidance value [52] | EFSA Health-based guidance value [46]-[51] | EFSA Benchmark Dose Lower Confidence Limit (BMDL) value [48, 49] | Codex Alimentarius Standard [52] | EU standard [54] | Food commodities for which maximum levels have been established [52], [54] |
|---|---|---|---|---|---|---|---|---|---|---|---|---|---|
| | | (Developmental) neurotoxicity | Oxidative stress | Kidney toxicity | Cardiovascular toxicity | Bone toxicity | Reproductive toxicity | | | | | | |
| **Cd** | Group 1[a] (Cd and Cd compounds: cancer of the lung, kidney and prostate) [45] | √ | √ | √ | √ | √ | √ | PTMI = 25 μg/kg bw per month | PTWI = 7 μg/kg bw per week | | √ | √ | Fruit; treenuts; vegetables; fungi; pulses; oilseeds; cerials; meat & offals; fish and other seafood; cocoa; chocolate; salt; infant formulae; baby & children food & drinks |
| **Pb** | Inorganic Pb compounds: Group 2A[b] Organic Pb compounds: Group 3[c] [44] | √ | √ | √ | √ | √ | √ | – | – | BMDL01 = 0.50 μg/kg b.w. per day for developmental neurotoxicity BMDL01 = 1.50 μg/kg b.w. per day for effects on blood pressure BMDL10 = 0.63 μg/kg b.w. per day for chronic kidney disease | √ | √ | Fruit; treenuts; vegetables; fungi; pulses; dried spices; cerials; meat & offals; fish and other seafood; milk; honey; fats & oils; fruit juices; wine, cider, perry and fruit wine; salt; infant formulae; baby & children food & drinks; food supplements |

© Burleigh Dodds Science Publishing Limited, 2025. All rights reserved.

| | | | | | | | | | | | | |
|---|---|---|---|---|---|---|---|---|---|---|---|---|
| **As** | Group 1[a] (As and inorganic As compounds: cancer of the lung, urinary bladder, skin, kidney, liver and prostate) [45] | √ | √ | √ | √ | √ | – | – | BMDL 01 = 0.3-8 µg/kg b.w. per day for cancers of the lung, skin and bladder, as well as for skin lesions | √ | √ | Total As: Salt; edible fats & oils; fat spreads |
| | | | | | | | – | – | BMDL05 = 0.06 µg iAs/kg bw per day for skin cancer (squamous cell carcinoma) | √ | √ | Inorganic As: Rice; fish oils; infant formulae; baby food; fruit juices |
| **Hg** | Methylmercury compounds: Group 2B[d] Metallic Hg and inorganic Hg compounds: Group 3[c] [43] | √ | √ | √ | √ | √ | Inorganic Hg: PTWI = 4 µg/kg bw per week | Inorganic Hg: PTWI = 4 µg/kg bw per week | | √ | √ | Total Hg: Fish and other seafood; salt; food supplements |
| | | | | | | | Methyl Hg: PTWI = 16 µg/kg bw per week (expressed as Hg) | Methyl Hg: PTWI = 13 µg/kg bw per week (expressed as Hg) | | √ | | Methyl Hg: Fish (Alfonsino; Marlin; Orange roughy; Pink cusk-eel; Shark; Tuna) |
| **Ni** | Group 1[a] (Mixtures of Ni compounds and Ni metal: cancer of the lung, nasal cavity and paranasal sinuses) [45] | √ | √ | √ | √ | √ | – | TDI = 13 µg/kg bw per day | | | | |

(*Continued*)

© Burleigh Dodds Science Publishing Limited, 2025. All rights reserved.

**Table 3** (*Continued*)

| Metal | IARC evaluation of carcinogenicity in humans | Non-carcinogenic effects (IARC 1993, 2006, 2012; EFSA, 2008, 2009a, 2009b, 2010, 2012, 2020) | | | | | | JECFA Toxicological guidance value [52] | EFSA Health-based guidance value [46]-[51] | EFSA Benchmark Dose Lower Confidence Limit (BMDL) value [48, 49] | Codex Alimentarius Standard [52] | EU standard [54] | Food commodities for which maximum levels have been established [52], [54] |
|---|---|---|---|---|---|---|---|---|---|---|---|---|---|
| | | (Developmental) neurotoxicity | Oxidative stress | Kidney toxicity | Cardiovascular toxicity | Bone toxicity | Reproductive toxicity | | | | | | |
| **Cr** | Group 1[a] (Cr (VI) compounds: cancer of the lung, nose and nasal sinuses) [45] | | √ Cr(VI) | √ Cr(VI) | | | √ Cr(VI) | – | TDI of 0.3 mg/kg b.w. per day for Cr(III) | | | | |
| **Al** | - | √ | √ | √ | √ | √ | √ | – | TWI = 1 mg/kg bw per week | | | | |

JECFA: Joint FAO/WHO Expert Committee on Food Additives.
PTWI: provisional tolerable weekly intake, PTMI: provisional tolerable monthly intake, TDI: tolerable daily intake, TWI: tolerable weekly intake.
[a]Carcinogenic to humans.
[b]Probably carcinogenic to humans.
[c]Not classifiable as to their carcinogenicity to humans.
[d]Possibly carcinogenic to humans.

© Burleigh Dodds Science Publishing Limited, 2025. All rights reserved.

NAMs comprise alternative testing for toxicity assessment, such as bacteria-, cell line-, and plant tissue-based studies, lab-on-chips or organ-on-chips systems [56–62], and modelling approaches, such mode of action (MoA), adverse outcome pathways (AOPs), toxicokinetic and toxicodynamic (TKTD) considerations and quantitative structure-activity relationship (QSAR) [63, 64], as alternatives to *in vivo* methods and to confirm and evaluate epidemiological data [65, 66].

The term MoA refers to a functional or anatomical change following exposure to a chemical, while a mechanism of action refers to changes at the molecular level [67]. Studies on the MoA of metals report the interference with antioxidant defence mechanisms through interaction with key antioxidant enzymes, the disruption of signalling pathways and of a variety of cellular processes (cell growth, proliferation, survival, metabolism, and apoptosis/cell death), the substitution of essential metals such as Ca, Cu, and Fe with structurally similar HMs such as Cd and Pb in the metal-binding sites of proteins, the generation of damaging reactive oxygen species (ROS) and oxidative stress as the main mechanisms of HM toxicity [65, 68–70]. AOPs are being built to better understand the mechanism of toxicity.

AOPs describe the sequential chains of causally linked events that conduce to an adverse health effect and comprise the molecular initiating event (MIE), the key events (KEs), and the adverse outcome(s) (AO(s)) as main components [71]. Each KE or KE Relationship is supported by evidence from different assays and tests, which the AOP integrates, showing up where there is evidence from specific tests, and where there may be gaps in knowledge [72]. AOPs have also been used as quantitative tools to understand toxicity mechanisms induced by metals [73]. (Developmental) neurotoxicity, generation of ROS, and kidney, cardiovascular, and reproductive toxicity are common AOs recognized for Cd, Pb, As, Hg, Ni, and Al [74–85], while bone toxicity stands out for Cd, Pb, and Al [69, 70, 83]. Taking neurotoxicity as an AO arising from exposure to different metals, one can observe that the KEs, such as competition with or disruption of essential metals, direct binding to proteins, impairment of neurotransmission, neuronal death, and blood–brain barrier damage leading in increased exposure of the brain to other toxicants, are very similar [70, 79, 81, 83].

Co-exposure to multiple hazards is one of the challenges in health risk assessment. AOPs are particularly important for addressing exposures to multiple stressors that lead to the same AO or otherwise to individual stressors that perturb multiple MIEs [86]. An AOP exploring the association between multiple metal co-exposures and chronic kidney disease was constructed based on data mining and expert knowledge, integrating and modelling occupational epidemiological evidence from various databases. The study considered the common MIEs (generation of ROS) and KEs (MAPK signalling pathway, oxidative

© Burleigh Dodds Science Publishing Limited, 2025. All rights reserved.

stress, mitochondrial dysfunction, DNA damage, inflammation, hypertension, cell death, and kidney toxicity) and revealed that Pb, Cd, Hg, and As may be critical metals in chronic kidney disease [87].

*In silico* tools, such as quantitative structure–activity relationship models (QSAR), employed usually in combination with other non-testing methods, are gaining increased attention in risk assessment studies for modelling the interaction of chemicals with biomolecules and predicting toxicity, especially as screening tools for prioritization [88]. Mathematical models are used to establish a relation between molecular structure and activity, employing various descriptors, e.g. physicochemical properties, to predict effects, including absorption, distribution, metabolism, excretion, and toxicity (ADMET) profiling [89].

Recent applications of QSARs in food risk assessment refer to the evaluation of the toxicity of metal/metal oxide nanoparticles (NPs) such as Ag NPs, $TiO_2$ NPs, and ZnO NPs used as antimicrobials and antioxidants in food processing, as well as for improving thermal, mechanical, and barrier capabilities of food packaging materials, thus extending the shelf life of food products [90]. QSAR models have evolved, incorporating machine learning algorithms to process complex data sets. Sang et al. [91] evaluated the cytotoxicity of mixtures of Nano-$TiO_2$ and HMs applying machine by learning algorithms to a QSAR model. The inflammatory potential of 30 metal oxide NPs, including $TiO_2$ and ZnO, was evaluated and ranked through a QSAR using the proinflammatory cytokine IL-1β in THP-1 cells as an index [92]. The electron transfer between the metal oxide NPs and their biological pairs, leading to the increased oxidative stress of cells, and the release and transport of metal ions were used to build classification QSAR models for a multicomponent mixture of metal nanomaterials, an approach that can lead to a more knowledge-driven hazard assessment, assisting the synthesis of safe-by-design metal NPs for various applications [93]. Generation of realistic estimates of the human exposure to toxicants is one of the pillars of health risk assessment.

From a human intake perspective, there are different sources of intake of HM that lead to human exposure [68]. Food is only one of the many exposure routes. Human biomonitoring has been developed to assess internal exposures to metals from all sources and via all routes of exposure also considering metal accumulation potential, complementing the external exposure assessment approaches that are widely applied and approved by the regulation. Metal levels in blood, urine, faeces, liver, kidney, hair, and other tissues have been used as biomarkers of exposure to estimate the biologically effective dose in the target organ or tissue and investigate inter-individual differences in absorption [94, 95].

© Burleigh Dodds Science Publishing Limited, 2025. All rights reserved.

A few biomarkers of effect, such as the urinary beta-2-microglobulin for Cd [47], related to molecular or cellular effects in mechanistic pathways have been validated to consider inter-individual differences in sensitivity toward metal exposure, whereas biomarkers of susceptibility are still lacking. Biomonitoring from less-invasive biological matrices, such as exhaled breath condensate, is currently being explored [95, 96]. The major challenges of biomonitoring are related to the derivation of biomonitoring limit and guidance values and the interpretation of the results from a risk assessment perspective. Human biomonitoring guidance values have been derived for the general population and/or workers for exposures to metals such as Cd, Hg, and Cr(VI) within the HBM4EU [97] and other projects [98, 99].

The fact that humans are exposed to an extremely wide variety of anthropogenic sources of exposure to hazardous metals has long raised concerns about their health effects. Despite recent scientific achievements in elucidating their mode of action and advances in analytical chemistry for estimating their levels in food and the environment, gaps still exist mainly regarding the mechanisms of systemic and targeted toxicity, especially for metal species, and regarding the characterization of the toxicity of mixtures of metals, which impedes accurate risk assessment [100]. Broad monitoring programmes in the food chain, based on proper sampling strategies and fit-for-purpose analytical methods, are of utmost importance to estimate baseline metal and metal species occurrence levels, support the generation of actual exposure estimates and the setting of food safety standards that will be protective for both the general population and vulnerable population groups, and provide the scientific justification for risk mitigation.

## 6 Advanced analytical techniques to detect and quantify heavy metals in food: x-ray florescence

Historically, trace elemental analysis was based on chemical methods, mainly gravimetry and titrimetric methods. Nowadays, physical methods, based on physical properties of elements and products of their reactions, play the main role in elemental analysis [101, 102]. Physical methods for elemental analysis are based on electrochemical, spectroscopic, mass spectrometric methods, and XRF.

XRF is an analytical technique that utilizes the emission of characteristic X-rays from the atoms of a material that has been ionized through irradiation with primary X-rays. After ionization, the atom is excited and unstable, and an outer electron can fill the vacancy created. This mechanism can produce the emission of an X-ray photon with an energy characteristic of that specific atomic species. The ionizations of interest for the technique are those that occur in the shells closest to the atomic nucleus, primarily the K, L, and M shells. The K

© Burleigh Dodds Science Publishing Limited, 2025. All rights reserved.

shell is the closest to the nucleus; therefore, the energy of X-rays emitted as a consequence of this ionization process is the highest, followed by the L and M shells. Within the same shell, the energy of the characteristic X-rays increases with the atomic number; therefore, heavy elements have higher energies compared to light ones. In Fig. 4, the energy of Kα and Lα characteristic X-rays is shown as a function of the atomic number Z.

The instrumentation needed to perform this technique is, in principle, quite simple: a source of primary X-rays and an X-ray detector. Whereas in the past, an X-ray-emitting radioactive source was used, nowadays, the source of primary X-rays is mostly an X-ray tube. The detection system is crucial, especially in terms of energy resolution, namely the capability to select between different characteristic lines. Following the interaction of an X-ray in the crystal of the detectors, a signal is generated that has an amplitude proportional to the initial energy of the X-ray. Due to statistical and noise effects, the signals related to the X-rays, which initially have the same energy, are represented by a peak with a certain width. Since in the past such a width was too big to allow a proper selectivity, the most used approach was the wavelength-dispersive XRF (WDXRF), where the selectivity is achieved by using the diffraction of the characteristic radiation on a specific crystal. This approach provides excellent energy resolution, and it is still employed in many applications.

In the last 30 years, thanks to the improvements in semiconductor detectors, a different approach has been being more and more employed, the EDXRF, where the width obtained is still bigger than that one of WDXRF, but small enough to allow a proper selectivity. Furthermore, the EDXRF approach

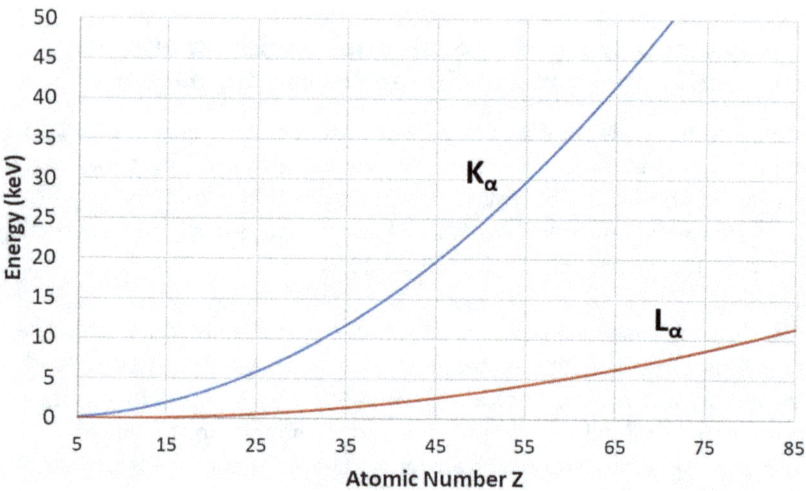


**Figure 4** Energy of the Kα and Lα characteristic X-rays is shown as a function of the atomic number Z.

© Burleigh Dodds Science Publishing Limited, 2025. All rights reserved.

allows for the measurement in a shorter time, as there is no need to scan every single line by the diffracting crystal. The result of an EDXRF measurement is a spectrum with every peak corresponding to different elements, where the horizontal axis is the energy of the characteristic X-ray, and the vertical axis is the number of counts. The technique can, therefore, first provide information about which elements are present in the sample and, second, at which level of concentration. The quantification of element concentration is mainly done through the calibration of the instrument with reference materials that have a matrix similar to the one investigated. An example of a spectrum obtained with a TXRF spectrometer is reported in Fig. 5.

Among other features, XRF offers the possibility of obtaining information about a wide range of elements (Na-U) in a short time with good selectivity and sensitivity. The technique is also relatively affordable when compared to other elemental techniques and can be portable, thanks to the technological achievements of the last 20 years that allow the production of miniaturized X-ray tubes and detectors powered by batteries. The detection limits (DLs) achievable by XRF depend on many different factors, such as the element of interest, the matrix of the sample, and others. The background in the spectra represents a strong handicap to the DLs. Therefore, many spectrometers are designed to minimise this contribution as much as possible while maintaining high sensitivity. The most commonly used strategies for background reduction are filters (the only feasible solution for portable instrumentation)

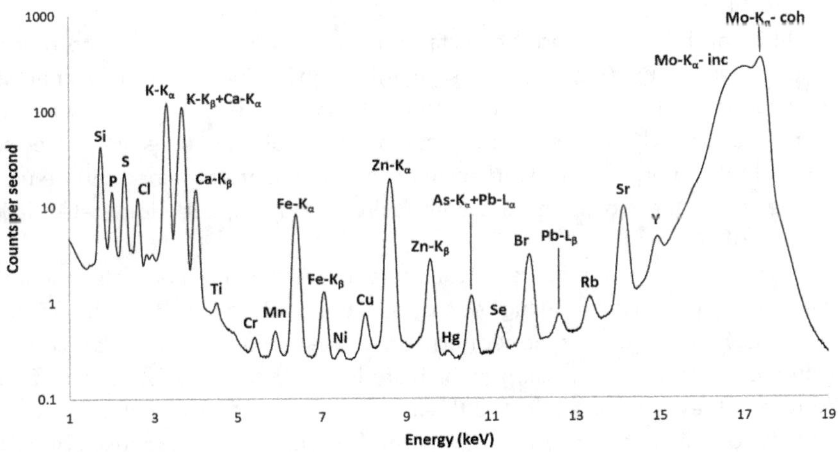


**Figure 5** Spectrum obtained with a TXRF spectrometer on a fish tissue sample. The lines of the different elements detected are indicated, together with the scattering lines (coherent and incoherent) of the monochromatic primary radiation (molybdenum K$\alpha$). Please note that the vertical scale is logarithmic in order to allow the visibility of the small peaks in the spectrum.

© Burleigh Dodds Science Publishing Limited, 2025. All rights reserved.

and secondary targets. An extended review of the XRF technique and instrumentation can be found in [103–105].

A peculiar variant of XRF is the total-reflection XRF (TXRF), where a low amount of sample, in the order of micrograms, is deposited on an adequate sample carrier and irradiated with a monochromatic X-ray beam at a very low incidence angle below the critical angle for total reflection. For a conventional Mo X-ray source monochromatized to Kα (17.44 keV) impinging over a flat quartz reflector, the critical angle is 0.1°, and the TXRF measurement angle will be 0.07°. Working in this peculiar geometry has an essential consequence that matrix effects can be negligible for most of the elements of interest [106]. This setup enables the achievement of high sensitivity and low background, resulting in significantly lower DLs for various materials, including biological ones [107, 108] and food samples [109]. Furthermore, the negligibility of matrix effects for most of the elements makes the quantification for this technique relatively easy, and it is unnecessary to use reference materials for the spectrometer calibration.

On the other hand, since the technique is very sensitive and being the best way of depositing the sample on the carrier to have it in liquid form, the sample preparation is much more time consuming and aggressive; in most cases, solid samples need to be digested by open vessel method or, more frequently, by using high pressure and temperature microwave oven for acid digestion [110]. An alternative to this approach is the suspension of the sample, which can also provide satisfactory results [111, 112]. An extended overview of the application of TXRF in food research can be found in [109, 113].

XRF can be performed, both standard and TXRF, with synchrotron light, benefiting all features of this kind of source, as high brilliance, monochromaticity, polarization, and others [114]. Due to the limited accessibility to synchrotron beamtime, limited and specific representative samples are usually selected for XRF or TXRF compositional studies. In addition, synchrotron lines can be used to investigate the stoichiometric composition and the oxidation state of the element of interest through XAS.

XRF is typically more suitable for characterizing heavy elements than light ones, mainly because of the higher cross-section for X-ray production of high-Z elements. Furthermore, as mentioned above, heavy elements are characterized by higher-energy XRF emissions, which are less affected by absorption effects. Therefore, they are less subject to the so-called matrix effects that can result in a much-complicated procedure for quantification. This book focuses on food samples characterized by a light organic-based matrix that makes these effects not so prominent, especially for high-energy X-rays. This fact allows therefore a more straightforward calibration of the spectrometers. In the case of light element quantification, which imply low-energy X-rays, some precautions

© Burleigh Dodds Science Publishing Limited, 2025. All rights reserved.

should be taken into consideration, as the study of coke beverages by TXRF carried out by Von Bohlen and Fernández-Ruiz shows [115].

This chapter focuses mainly on detecting HMs like Cd, As, Hg, and Pb. For Pb and Hg, and in most cases also for Cd, the K-lines have energies that are too high to be excited and detected by standard spectrometers. Therefore, they are quantified using the L-lines. Quite often, these L-lines overlap with the K-lines of lighter elements that can have higher concentrations, thus hindering their quantification and raising the DLs.

Concerning sample preparation, if the material is solid, it can be analysed as it is, in the form of loose powder or as a pressed pellet [116, 117]. Otherwise, it can be chemically treated, involving acid digestion, preconcentration of the elements of interest, and filtering. In the case of liquid samples, they can be analysed as they are in special cups (with helium flow for instruments operating in a vacuum) or filtered. The liquid form is most suitable for TXRF, even if, in some cases, the sample is too dense and needs to be diluted or digested.

DLs of XRF for food samples depend a lot, as will be explained below, on the spectrometer and XRF variant used and also on the composition of the sample itself. In some cases, the detection and quantification limits for a specific element are below its maximum levels. Nevertheless, even when the DLs are above this amount, XRF can be used for a qualitative approach or a first fast and affordable screening of samples. For instance, as mentioned in the previous paragraph, XRF can also be performed with portable instrumentation that has higher (worse) DLs when compared to laboratory equipment but can provide crucial support for what concerns a first in-situ screening of the samples of interest or, alternatively, of the soil where plants are growing, and animals are grazing [118]. Finally, XRF performed with synchrotron light can reach DLs below 1 ppm (μg/g) for the elements of interest even without sample treatment, thanks to the peculiar features of this source of primary radiation already mentioned above, as high brilliance, monochromaticity and polarization and others [119].

## 7 Applying x-ray florescence: detecting cadmium, lead, arsenic, and mercury

We can now review in more detail the features of XRF when dealing with the elements of interest in this chapter, considering different kinds of instrumentation and methods.

### *7.1 Cadmium*

The first distinction to report when dealing with the characterization of Cd is whether the spectrometer used can properly excite and detect its K-lines. In

© Burleigh Dodds Science Publishing Limited, 2025. All rights reserved.

such case, the lines would be well isolated and free of overlapping, allowing relatively low DLs at the order of 0.5 ppm. However, very few spectrometers are designed to excite these lines efficiently; also, the detector should be thick enough to have a good efficiency at the energy of Cd-Kα (23.2 keV). Otherwise, Cd must be characterized using the L-lines that can be hidden by much more abundant elements like potassium that have K-lines partially overlapping and can suffer much more, being at relatively low energy, from matrix effects.

The TXRF spectrometry, in routine laboratory work, can detect Cd and other contaminant elements, such as Cr, As, Pb, or Hg, at limits of detection (LOD) at the level of 1 ppm (μg/g) or slightly lower in food after a complete microwave acid digestion and internal standardization [120]. These DLs are possible if the digestion process of the organic matter in the food is decomposed at the molecular level, that is, breaking the organic macromolecules and transforming them into mainly $CO_2$ and $H_2O$. High-pressure and high-temperature microwave oven systems are necessary for a sufficient degree of molecular breaking. As mentioned above, to improve the DLs, it is necessary to use extraction and preconcentration procedures with the advantage of discriminately concentrating Cd and reducing the presence of the disturbing element, potassium in this case. One of the most specific techniques for TXRF analyses, taking advantage of its microanalytical character, is liquid–liquid microextraction (LLME), which uses microliters of organic solvent to extract metals from aqueous samples, commonly using chelating agents (e.g. APDC, DDC) to form hydrophobic complexes with Cd and also Pb, Hg, and As. Then, the metal complexes are extracted and quantified using various organic solvents, such as chloroform, toluene, or MIBK. In this way, the achievable DLs (LOD) can be very small, in the order of 0.1–10 ppb (μg/kg) [121].

This analytical strategy enables us to achieve the lowest DLs. However, it is necessary to properly evaluate the procedure efficiency and the possible recovery factors for each matrix in advance. For TXRF, the extraction methods also have the drawback of not allowing, in most cases, the quantification to be performed through the standard routine of exploiting an internal standard.

## *7.2 Lead and arsenic*

These two elements are treated together since they show one of the strongest overlaps in XRF spectra. The main line of As (As-Kα) is extremely close to the main line of Pb (Pb-Lα), and the separation is only about 11 eV. In this case, however, the two elements generally have similar concentration levels, unlike Cd and potassium, where the latter typically has a much higher concentration than the former. Nevertheless, the overlap is so strong that it is practically impossible to separate the contributions of the two elements. Fortunately, the second line of Pb (Pb-Lβ) is well separated from the Lα and it is still quite

© Burleigh Dodds Science Publishing Limited, 2025. All rights reserved.

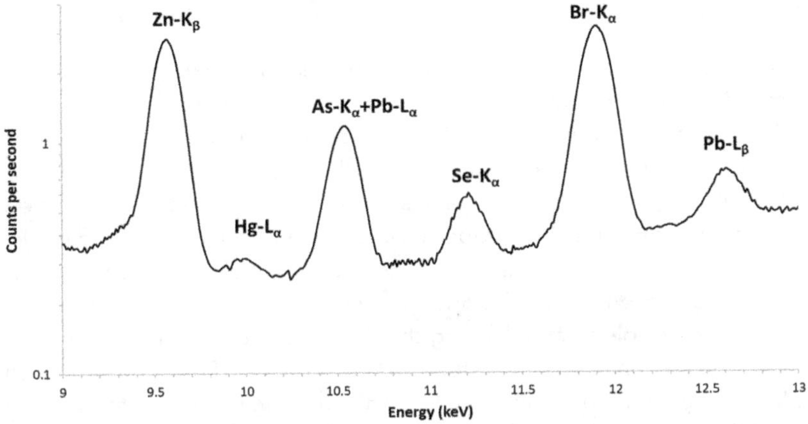


**Figure 6** Area of the spectrum of Fig. 5 where the characteristic lines of As and Pb are present. The strong overlapping of As-Kα and Pb-Lα peaks is clearly visible. Also to note, even if not shown, that the signal of As-Kβ at about 11.7 keV is hidden by the presence of the peak of bromine Kα, which partially overlaps with As-Kβ.

intense, having about half its intensity [122]. The area of the spectrum of Fig. 2 where the As-Kα and Pb-Lα peaks are present is reported in Fig. 6.

Nowadays, fitting codes can efficiently solve the issue by exploiting updated databases for the relative intensity of the lines and quantifying Pb from the Lβ and deducting As contribution as a difference; the quantification of As through its second line (As-Kβ) is more difficult being the intensity of this line about 1/8 of the main one [123]. Another even more reliable approach would be the one exploiting an exciting energy that is above the As K absorption edge but below all the L ones of Pb; this is the case of some secondary targets XRF exploiting a KBr target (Br-Kα have such energy between the edges) or even better, at synchrotrons where the monochromatic energy can be set at the proper level. This way, the As can be quantified without the overlapping of Pb. Then Pb can be determined at higher energy using its Lβ or as a difference since As was already determined. In the most common case where this approach cannot be used, the DLs for these elements are affected not only by their overlapping but also by the partial overlapping of other elements. Indeed, in this spectrum region, the lines of many quite common elements are present; for instance, if the sample contains a high amount of selenium, its Kβ line can partially cover the Lβ of Pb, making the quantification even more challenging. Therefore, the achievable DLs depend on the sample under analysis. Modern TXRF spectrometers can achieve DLs in the ppb range for liquid samples [124]; this means that, for digested and diluted samples, the limits are, as stated above, in the range of 1 ppm (μg/g) or slightly lower when considering the concentrations in the original material.

© Burleigh Dodds Science Publishing Limited, 2025. All rights reserved.

### 7.3 Mercury

Among the four elements considered in this chapter, Hg can be considered the most difficult to quantify by XRF. In this case, no common elements strongly overlap with the main line of Hg (Hg-Lα). Still, this element is often present at very low concentrations, which are frequently below the DLs of the technique. Furthermore, most reference materials, except those based on fish, have very low amounts of Hg, making proper calibration particularly challenging. In the case of TXRF, where lower DLs can be achieved, if the sample is being digested in a microwave oven or with the open vessel method, care shall be taken not to overheat the sample to avoid losing this element because of its high volatility. In most cases, even low-temperature cycles performed for a longer time cause the loss of Hg, and the recovery factor can be quite difficult to determine.

Different sensitive methodologies have recently been developed for determining Hg in various kinds of samples, including water, beverages, seafood, plants, and biological materials, using TXRF and EDXRF spectrometry, overcoming previously reported analytical challenges. One of the more recent and outstanding methodologies involves the selective preconcentration and separation of Hg(II) ions through dispersive micro-solid-phase extraction (DMSPE) using graphene oxide functionalized with thiosemicarbazone. The combined DMSPE/EDXRF and DMSPE/TXRF approaches are notable for their high enrichment factors and low DLs. Specifically, DMSPE/TXRF achieves DLs as low as 2.1 pg $mL^{-1}$ (ppt) for liquid samples and 1.8 ng $g^{-1}$ (ppb) for solids. In contrast, the DMSPE/EDXRF method yields higher DLs – 60 pg $mL^{-1}$ for liquids and 73 ng $g^{-1}$ for solids – owing to the lower sensitivity of EDXRF relative to TXRF [125].

## 8 Case study: detecting adulteration of turmeric with heavy metals

The herb and spice industry is under constant threat of economically motivated fraud due to the high economic value of the commodities it produces and the sector's complex supply chains [126]. Colour, taste, antioxidant content, and aroma of spices, e.g. turmeric powder, are key characteristics that determine their quality/grade and influence the market price and consumers' perception. Furthermore, aside from flavour, colour is responsible for the wide use of turmeric powder as a food ingredient in many food preparations [127] and influences customers' decisions about brand quality when making purchases [128]. Thus, fraudsters illegally add similarly coloured chemical substances that may be toxic to humans to make the turmeric appear more attractive to consumers, along with masking additions of bulking agents; these processes are collectively termed economically motivated adulteration (EMA) [129].

© Burleigh Dodds Science Publishing Limited, 2025. All rights reserved.

This results in the turmeric being adulterated and potentially unfit for human consumption. The safety of turmeric is an important aspect to be considered alongside economically motivated fraud.

Turmeric (*Curcuma longa* L.), a botanical species from the *Zingiberaceae* family, has gained popularity in many parts of the world due to its extensive use in culinary practices, medicinal uses, and as a dietary supplement [130]. Historically, cultivation and use of turmeric have been centred in the Middle East and Asia [131]; lately, the African continent has also become an important exporter of turmeric. Of substantial concern is the increasing evidence that turmeric from leading producing areas is frequently adulterated with coloured dyes (11% of imports to the EU) [132]. This is one of the factors that is creating emerging markets for export, especially when regulatory frameworks are less restrictive. Some of the adulterants identified, used, e.g., as bulking agents, e.g. starch, are relatively non-toxic [133], whereas others are acutely toxic [134, 135] such as lead chromate ($PbCrO_4$) [136]. Figure 7 provides an overview of the turmeric paradigm: from 'turmeric-for-food-and-health' to 'turmeric-fraud -harm-to-consumers-health'.

The use of $PbCrO_4$ to enhance the yellow colour appearance of turmeric is increasingly being reported. However, Pb can be present in the soil where turmeric rhizomes are growing. The uptake of Pb from the soil into the turmeric rihizome is a possibility, but it is unlikely source of contamination, as maximum uptake of Pb into the root of the plant is only ~10% [138]. According to Forsyth et al. [139], Pb to Cr molar ratio of approximately 1:1 in turmeric may suggest $PbCrO_4$ adulteration. This scenario was found in turmeric adulterated with

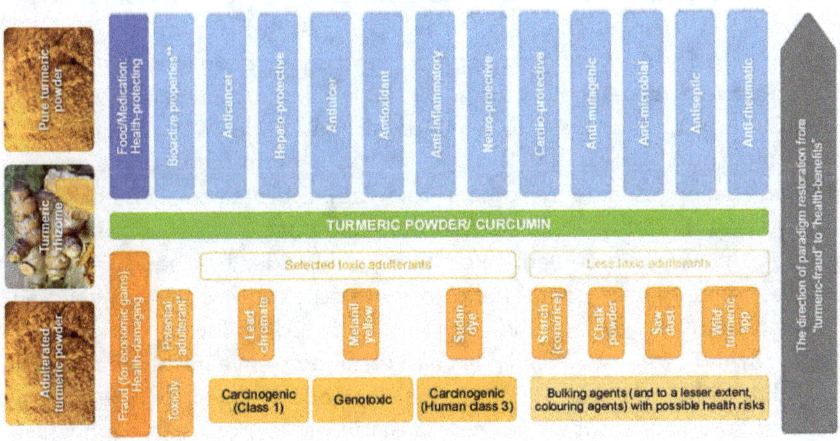


**Figure 7** The good and bad of turmeric powder: An overview of the turmeric (Curcumin longa) shift (from authentic turmeric powder, i.e., 'turmeric-for-food-and-health' to not-authentic turmeric powder, i.e. 'turmeric-fraud-harm-to-consumers-health') paradigm. Source: From [137].

© Burleigh Dodds Science Publishing Limited, 2025. All rights reserved.

**Figure 8** Typical turmeric processing after harvest.

$PbCrO_4$ across South Asia [135]. The consumption of Pb in spices, e.g. turmeric, is a public health concern worldwide when considering the associated adverse health effects [140]. In general, such adulterations only take place when turmeric rhizomes are transformed into powders (refer to Fig. 8, for a typical processing scheme of turmeric). It is, well known that the addition of bulking agents and colourants is only applicable to homogenized turmeric. $PbCrO_4$ is the primary adulterant in turmeric powder, partly due to its low price and non-specific regulatory/maximum limit (ML) in turmeric powder [134]. In the EU, updated MLs for Pb in root and tuber vegetables (0.1 mg/kg in fresh turmeric rhizomes) and dried root and rhizome spices (1.5 mg/kg) were recently established. These limits are under the purview of Commission Regulation (EU) 915 of 2023, superseding Regulations 2021/1317 and Regulation (EC) No 1881/2006 [141].

It is worth noting that Cr is also particularly toxic in its hexavalent state (Cr(VI)), e.g. it may serve as a risk factor for lung cancer and was reported in turmeric in one study in India [142–145]. However, the primary toxicity concern centres on Pb.

Exposure to Pb from turmeric powder adulterated with $PbCrO_4$ has been shown to cause neurological abnormalities, developmental delays, and cognitive impairments among exposed individuals [146]. In South Asian countries, Pb has been recognized as a risk factor for preterm birth due to increased consumption of turmeric powder by pregnant women, aiming at improving their nutrition and health [139, 147]. Furthermore, prolonged exposure to Pb may exert damage to the bones, brain development, and reproductive and respiratory organs [148–150]. Exposures to low levels of Pb (<3.5 μg/dL) have been associated with deterioration in cognitive ability and behavioural change [151] as well as poor performances in school and reading challenges, amongst others [152–155]. There is no safe blood Pb level in children [156, 157] and unfortunately, anthropogenic Pb exposure, partly through intake of Pb-contaminated turmeric powder has been indicated to be the primary factor contributing to elevated Pb levels among children in Bangladesh (N = 309; aged 20–40 months; blood Pb levels >5 μg/dL) [139, 147, 158], India (an est. 275 million children, aged 0–9 years, blood Pb levels ≥ 5 μg/dL) [159], and the US [160]. For the US, blood Pb levels of 2.9–12.7 μg/dL in five children in Kansas City were associated with exposure through intake of Pb (6.86 μg/g) from adulterated turmeric powder [161]. Likewise, in Las Vegas in 2019, it was speculated that childhood Pb poisoning might be associated with the consumption of turmeric contaminated with Pb [162]. When considering

© Burleigh Dodds Science Publishing Limited, 2025. All rights reserved.

that blood Pb in a gestational exposure may cross through the placenta to the foetus [163–165], pregnant women directly place their unborn babies in danger of Pb exposure, although disclaimed by a recent review [166] that calls for more evidence in this direction.

The adulteration of turmeric with $PbCrO_4$ raises concerns not just for regions involved in its production and consumption but also for those linked to its importation [167]. For example, turmeric powder bought from India (median: 0.71, max: 6504 mg/kg) revealed more than three times higher median Pb levels when compared with those bought from the US (median: 0.19 mg/kg) [168]. Additionally, the consequences of toxic metals, particularly $PbCrO_4$ adulteration, extend beyond endangering public health as they also impact the integrity of global food supplies and companies that unwittingly become involved in supply chains that sell such products [169]. This underlines the necessity of improving food safety and preventing fraud crimes, including strategies to reward farmers and implementing stringent monitoring initiatives along with more robust regulation enforcement. Going forward, there is a need for more studies to investigate the source of Pb contamination in spices [170] in general and turmeric powder in particular. Although this requires highly sophisticated techniques to perform such analysis, simple 'direct-scan' methods, linked to chemometrics models established with real authentic samples, may be speculated as the technique of choice for rapid detection of adulterants, especially for Pb, in the field, thus, eliminating the chances of adulterated turmeric spreading in local markets, and/or crossing trade borders.

Generally, turmeric powder adulterants reported in the literature include starch (from diversified tubers such as cassava and cereal-based foods, e.g., corn), diversified types and classes of Curcuma powder (including spent turmeric, synthetic turmeric, and *Curcuma* spp., e.g. *C. zedoria* and *C. mangga*) that closely mimic *Curcuma longa* L. powder, yellow-mimicking substances (such as Yellowstone powder, and sawdust), and the azo dyes (including metanil yellow, Sudan dye, aniline dye), and $PbCrO_4$. The methods (either single or in combination) used to detect these adulterants in turmeric powder include physical and chemical methods of the Food Safety and Standards Authority of India (FSSAI), microscopy, molecular, chromatographic (gel permeation chromatography (GPC), and thin-layer chromatography (TLC)), spectrometry (e.g. ICP-MS), and spectroscopy (such as Raman, laser-induced breakdown (LIB), near-infrared (NIR), Fourier-transform infrared (FT-IR), XRF, etc.). In some cases, the method used was in combination with multivariate analysis tools (specifically the principal component analysis (PCA) and the partial least squares discriminant analysis (PLS-DA)). Some of these techniques employed in the detection of adulterants and authentication of turmeric have been reviewed by Sasikumar et al. [171]. Spectroscopy is amongst the most frequently used approaches for detecting fraud and authentication of turmeric powder. It has

© Burleigh Dodds Science Publishing Limited, 2025. All rights reserved.

the advantage of being a non-destructive, field-friendly analytical technique and thus provides an option for rapid field decisions.

General applications, performance criteria, and the challenges of various physical, chemical, and molecular techniques used for adulterant authentication in food, spices, and herbs have been described in detail [172–175]. Some of the methods include but are not limited to LIB spectroscopy (LIBS) [176], Raman spectroscopy [177], X-ray powder diffraction (XRD) [178], FT-IR spectroscopy coupled to chemometric analysis [133], and flame atomic absorption spectrometry (FAAS) [179]. In addition, some of these techniques, such as UV-visible spectroscopy, developed as a simple and rapid detection tool, have inherent limitations, such as measurement errors due to high matrix interferences and poor limits of detection [180, 181]. Likewise, FT-IR and NIR exhibit similarly high limits of detection to UV-vis spectroscopy, which exceeds the regulatory threshold with poor sensitivity for detecting Pb adulteration in turmeric and a lack of validation capabilities for quantitative analysis of toxic metals [167, 182]. LIBS [183] and XRD [178] have been trialled recently for Pb detection in spices. However, thus far, they have only been validated for screening purposes.

The reliability and usefulness of the information generated by these analytical techniques increase when exploited in combination with other non-destructive techniques used to detect illegal adulteration of spices with various harmful adulterants. Therefore, the application, characteristics, advantages, limitations, and development of multi-source and non-destructive analytical techniques to assess the quality and authenticity of spices have been reviewed extensively. These techniques include vibrational spectroscopy (including FT-IR, FT-MIR, NIR, Raman, and hyperspectral imaging HSI), and electronic sensor technology (e.g. electronic nose and electronic tongue), as well as techniques such as multivariate analysis [126, 175, 184]. Furthermore, the purpose of each step in the process of authentication, including the use of non-destructive analytical techniques (to measure chemical information in samples), data processing (to remove artifacts and improve the accuracy of results), data fusion (for multi-sourced data to further improve the accuracy of results), and building of models (for exploration, prediction or classification of samples as authentic or suspicious) have been highlighted [175, 184]. The application of these methods, which are mostly laboratory-based, is to provide reliable measurements for rapid decisions on whether the tested turmeric is adulterated or not.

However, rapid detection systems for field testing by food industries and food safety agencies are ideal for reducing exposure to toxic adulterants. Considering the complexity of developing such techniques, the emphasis for method development should be on the most frequently reported adulterants. Turmeric powder adulteration with $PbCrO_4$ is increasingly being assessed

© Burleigh Dodds Science Publishing Limited, 2025. All rights reserved.

**Table 4** Reported Pb adulterants in turmeric

| Country (sample origin) | Method of detection | Adulterant | Frequency (%) | Mean level mg/kg | Range or max, mg/kg |
|---|---|---|---|---|---|
| Bangladesh | XRF | Pb‡ | / | 80 | <LOD-483 |
| US (purchased from Bangladesh, India, Nepal, Pakistan, Morocco, US) | ICP-MS | Pb | 76/105 (72) | 160 | 2700 |
| Bangladesh (nine districts) | ICP-MS, XRF | Pb‡ | 16/140 (11) | 1152 | / |
| Bangladesh | ICP-MS, XRF | Pb‡ | 52/200 (26) | 690 | / |
| India | ICP-MS, XRF | Pb‡ | | 0.2 | 0.7 |
| Republic of Korea | ICP-MS | Pb | 13 (100) | 0.09 | 0.04–0.97 |
| Republic of Georgia | ICP-MS, pXRF | Pb | 2** | 1897.06* | 466–3328 |
| India – Penta, Bihar | pXRF | Pb | 128** | 114.6 | <44–>363 |
| Bangladesh, Pakistan | ICP-MS, pXRF | Pb‡ | 11** | 857 | 1.04–4221 |
| India | ICP-MS, pXRF | Pb‡ | 11** | 1118 | 0–5279 |
| US (Kansas) | ICP-MS | Pb | 1 (100) | 6.86 | 6.86 |
| US (also purchased from India) | ICP-OES | Pb | 71** | 205.33Ł | 0.06Ł–6504 |
| | ICP-OES/-MS | Pb | 13** | 0.12 | 0.07–0.28 |
| South Asia (India, Pakistan, Sri Lanka, and Nepal) | ICP-OES/-MS pXRF | Pb | 51/356 (14) | 3* (>LOD: 2) | 2936 |

*Median; **number of observations; Łmultiyear sampling 2011-2020; ‡Pb: Pb value estimated from $PbCrO_4$ adulteration based on a 1:1 molar ratio of Pb to Cr in the samples; ICP-MS: inductively coupled plasma-mass spectrometry; ICP-OES: inductively coupled plasma-optical emission spectrometry; pXRF: portable X-ray fluorescence. Note: ICP-MS is a laboratory-based instrument, and pXRF is both a laboratory-based and portable (or handheld) or field-based instrument. Source: Adapted from [137].

© Burleigh Dodds Science Publishing Limited, 2025. All rights reserved.

using costly, time-consuming, and non-field-friendly techniques, which may not significantly contribute to rapid field decisions. Therefore, potentially non-destructive analytical tools for rapid field decision-making are of high interest in reducing potential exposures to adulterants. Table 4 summarizes the various reports on Pb adulteration in turmeric powder examined using ICP-MS and/ or XRF.

Generally, within the spices industry and amongst regulatory bodies, validated and accredited mass spectrometry, particularly ICP-MS, is widely used for multi-elemental analysis due to its specificity and sensitivity with high sample throughput [185]. ICP-MS, which has been used to measure $PbCrO_4$ levels in turmeric (Table 4), is not only time-consuming, but it involves elaborate sample preparation steps and requires skilled laboratory staff [186, 187]. In addition, its substantial purchase/installation costs, coupled with the high resource requirements (i.e. Ar gas, electricity, high purity reagents) of ICP-MS, may pose a challenge for use as a routine analysis approach for many low-income countries, which include many of the primary producers of turmeric [170]. Considering its relative effectiveness, it is suitable as a reference instrument for the confirmation of the presence and level of toxic Pb in any suspicious turmeric samples.

XRF is gaining wide recognition as the most practical solution for detecting toxic metal adulteration in spices [188]. Among XRF techniques, energy dispersive pXRF has gained interest due to its portability, allowing for *in-situ* analysis in the field, and improved DLs between 1 mg/kg Pb and 8 mg/kg Pb, which are close to the regulatory limits (between 1 mg/kg Pb and 10 mg/kg Pb) set by various international bodies including India and the EU [145, 189, 190]. Furthermore, pXRF is relatively cost and time efficient for screening and does not necessarily require skilled laboratory staff to operate. Forsyth et al. [134] and Lopez et al. [145] have demonstrated that handheld XRF analysis provides information that can enhance field detection of Pb in turmeric powder and thus may contribute to a reduction in Pb exposure amongst various consumer populations. However, there is a need to explore the functionality of the pXRF to improve calibration to detect Pb levels below the ML of 1.5 mg/kg [141], thereby rendering it suitable for rapid screening analysis and, thus, improving the decision-making approach in the spices industry.

From the perspective of the 'Food Fortress' system [191], a framework designed to enhance food safety and traceability through advanced monitoring and testing protocols, and one that is practically applicable in large-scale food production and export operations where consistent quality control in critical technology diffusion is feasible through government-backed policies, industry incentives, and partnerships with testing laboratories, especially in regions with emerging food industries aiming to meet international standards (e.g. Africa). In this context, portable technologies such as pXRF have shown potential as Tier

© Burleigh Dodds Science Publishing Limited, 2025. All rights reserved.

1 screening tools within a 'Food Fortress' systems approach, complemented by validation through mass spectrometry-based Tier 2 testing.

## 9 Rapid screening techniques to detect heavy metals in food: electrochemical approaches

Advanced analytical techniques, like the ones described above, require time, trained analysts and sophisticated and expensive laboratory equipment, and facilities. These limitations can become a key problem, especially when large farming areas must be controlled and particularly when such areas are located in developing countries with low resources and no access to the above-mentioned analytical techniques. In this context, rapid and cheap analytical techniques are strongly required by producers, regulators, and retailers for the screening of samples, reducing the number of them sent to laboratories. Both electrochemical approaches using screen-printed electrodes (SPEs) and optical approaches based on LFAs represent the hotspot screening techniques for HM analysis in food. Such strategies have been pushed in the last decade thanks to the advances in nanotechnology, which offer novel materials that improve analytical performance. In the next sections, an overview of the use of both analytical strategies for food analysis using nanomaterials is given. Finally, emerging trends consisting in the use of smartphones coupled with colorimetric and fluorescent approaches, together with the perspectives of applying AI in this field are also commented.

Electrochemical methods provide numerous advantages for detecting trace amounts of HMs, including rapid analysis, high selectivity, and outstanding sensitivity [192]. Anodic stripping voltammetry (ASV) is the preferred highly sensitive electroanalytical technique used for detecting trace amounts of HMs (Fig. 9a). It is a two-step process involving: (1) an accumulation/preconcentration step, where the target HM ions are first reduced and deposited onto the working electrode surface by applying a negative potential. This step effectively concentrates the metal ions from the solution onto the electrode; and (2) a stripping step, where the potential is swept in the positive direction, causing the deposited metals to re-oxidize and dissolve back into the solution. This oxidation process generates a current that is measured. The current produced during the stripping step is proportional to the concentration of the HMs in the sample (Fig. 9b). By analysing the current peaks, the type and amount of HMs present can be determined. Moreover, each metal is re-oxidized at a different potential, allowing the simultaneous detection of many metals in a single scan (Fig. 9c). ASV is valued for its ability to detect HMs at very low concentrations, making it ideal for food safety applications [193].

The use of SPEs for such purposes has strongly emerged in the last decade due to their unique properties and numerous advantages [194]. SPEs are

© Burleigh Dodds Science Publishing Limited, 2025. All rights reserved.

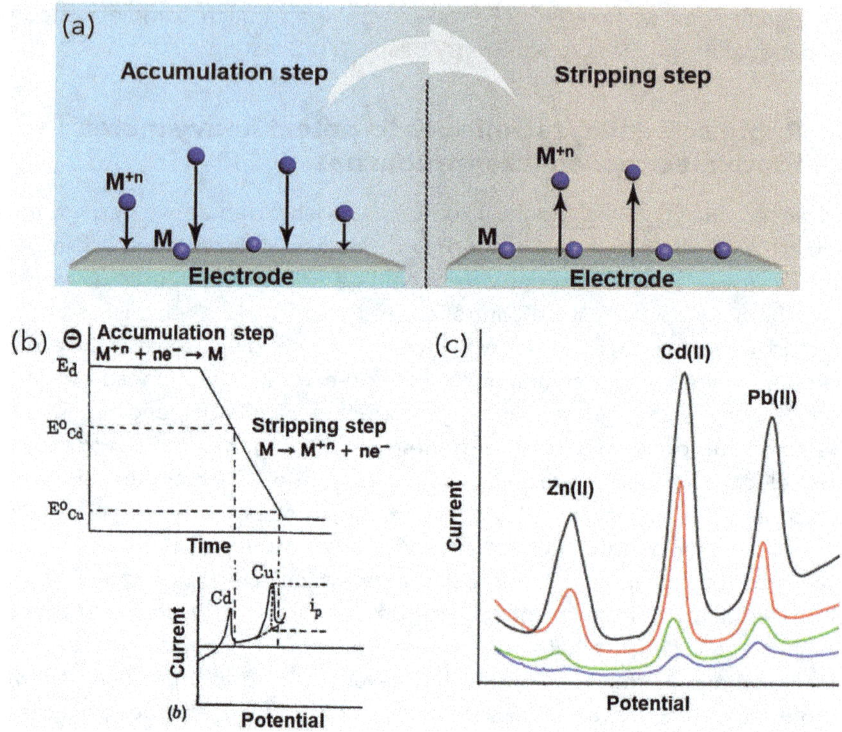


**Figure 9** (a) Schematic illustration of the deposition and stripping process in anodic stripping voltammetry for the detection of HMs; (b) graphs illustrating the potential applied during the accumulation and stripping steps (up) and the signals obtained during the stripping step (down) for Cd and Cu; (c) Voltammogram showing the simultaneous detection of Zn(II), Cd(II), and Pb(II) in a single scan. Source: Adapted from: [196] with permission.

typically made by printing various inks (such as carbon, silver, gold, or platinum) onto plastic or ceramic substrates. They usually consist of three electrodes: a working electrode (typically made of carbon), a reference electrode (typically made of silver/silver chloride), and a counter electrode (typically made of carbon), all of them in a single substrate of approximately 3.4 × 1.0 × 0.05 cm (length, width, height) (Fig. 10a).

Some of the advantages of the use of SPEs are the following: (1) versatility: the composition of the inks can be modified by adding metals, enzymes, polymers, or nanoparticles to tailor the electrodes for specific applications; (2) cost-effectivity: SPEs are relatively inexpensive to produce, making them suitable for disposable applications; (3) small sample volume: they require only a small volume of the sample for analysis, which is beneficial for applications where sample availability is limited; (4) high sensitivity and selectivity: the

© Burleigh Dodds Science Publishing Limited, 2025. All rights reserved.

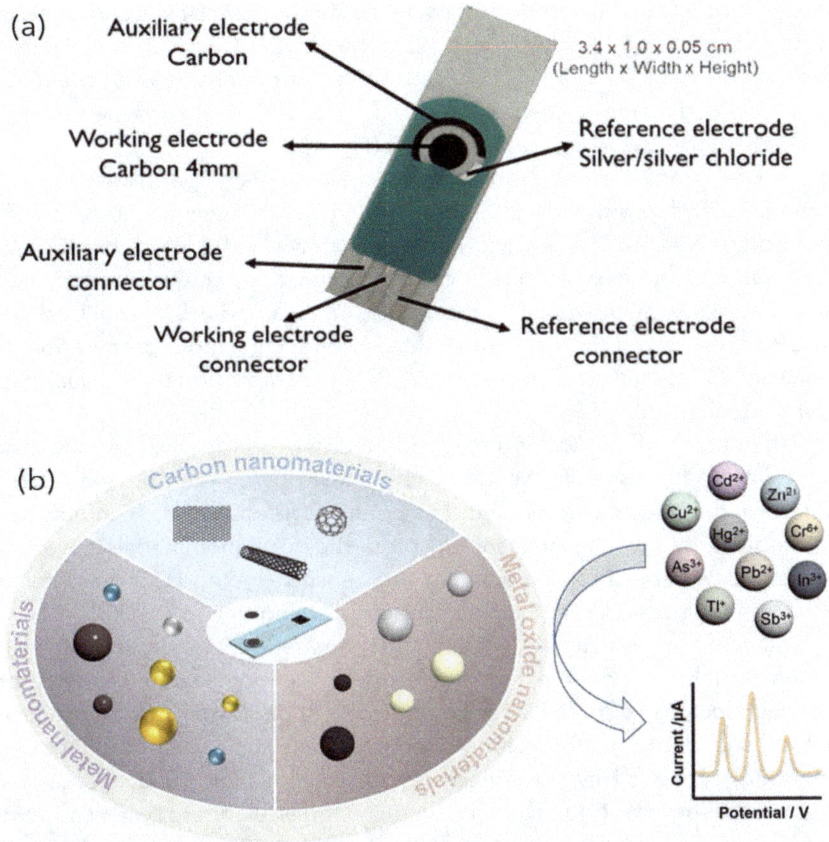


**Figure 10** (a) Picture of a screen-printed electrode, detailing the main components and the full dimensions; (b) the state-of-the-art of nanomaterial-enabled screen-printed electrochemical systems. Source: Adapted from [196] with permission.

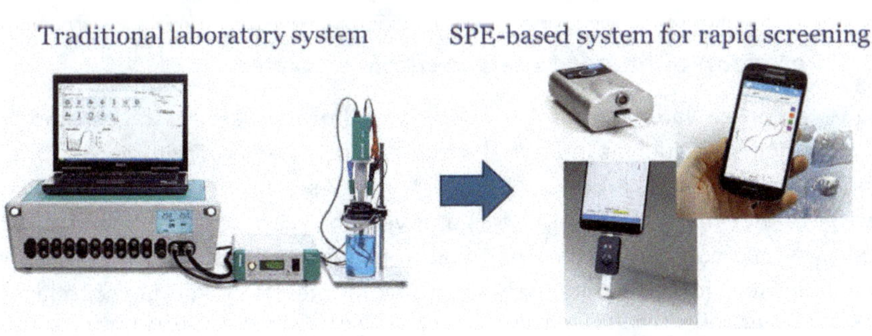


**Figure 11** Pictures comparing a traditional electrochemical analysis system based on the standard three-electrode system and bench potentiostats (left) and a system based on SPEs and portable potentiostats (right).

© Burleigh Dodds Science Publishing Limited, 2025. All rights reserved.

ability to modify the electrode surface allows for high sensitivity and selectivity in detecting specific analytes; and (5) portability and ease of use: SPEs are compact, portable, and easily coupled with smartphones, making them ideal for on-site and real-time analysis, as advantageous alternative to the traditional three-electrode based laboratory set-ups (Fig. 11).

These properties and advantages make SPEs a popular choice for screening and rapid analytes detection in fields such as environmental monitoring, medical diagnostics, and food safety [195]. It is worth noting that solid samples typically require a previous digestion, either in acidic media or assisted by microwaves, while liquid ones are directly analysed after adjusting the pH to the optimum value. After such sample preparation, the electrochemical analysis only requires 50 μL of sample and typically lasts less than 2–3 minutes.

The choice of the working electrode material is crucial for sensitivity and selectivity. In the case of HMs detection, a key driver in the development of detection methods based on SPEs has been provided by the recent advances in nanotechnology. The outstanding properties of nanomaterials have been combined with the SPEs for obtaining very sensitive methods for the detection of HMs in food samples, at the levels required by legislation [196].

Carbon nanomaterials, metal nanomaterials, and metal oxide nanomaterials have been extensively used for the modification of the carbon working electrode of the SPEs, showing excellent performance for HM analysis in food samples (Fig. 10b). Carbon nanomaterials such as carbon nanotubes offer high conductivity, low electrical resistance, high surface area, high mechanical strength, good chemical stability, and an affinity to bind with various nanomaterials. Graphene is also an outstanding carbon nanomaterial that shows good conductivity, fast electron transfer rate, large specific surface area, strong mechanical strength, high chemical and thermal stability, and excellent transparency. Finally, metal nanoparticles exhibit good electrical conductivity, large surface area, and favourable catalytic activity.

## *9.1 Detection of heavy metals in drinking water*

HMs can enter water supplies through natural processes like soil erosion and human activities such as industrial waste disposal and agricultural runoff. This means that metals like Pb, Cd, Hg, Zn, Cu, or As may be present in drinking water (either mineral or tap water) at levels that can pose significant health risks. In this context, electrochemical approaches based on SPEs offer an excellent tool for the rapid analysis of several metals in a single scan lasting a few seconds.

SPEs modification with different nanomaterials allows to reach DLs at μg $L^{-1}$–ng $L^{-1}$ levels (Table 5). In particular, metal nanomaterials (bismuth, gold,

© Burleigh Dodds Science Publishing Limited, 2025. All rights reserved.

**Table 5** Detection of HMs in drinking water using SPEs modified with nanomaterials

| Sample | Analyte | Linear range | LOD | Nanomaterial | Ref. |
|---|---|---|---|---|---|
| Tap water | Zn(II) | 0.1–100 μg $L^{-1}$ | 90 ng $L^{-1}$ | GR | [197] |
| | Cd(II) | | 60 ng $L^{-1}$ | | |
| | Pb(II) | | 80 ng $L^{-1}$ | | |
| Tap water | Pb(II) | 4–10 μg $L^{-1}$ | 280 μg $L^{-1}$ | BiNPs | [198] |
| | Cd(II) | 300–900 μg $L^{-1}$ | 40 μg $L^{-1}$ | | |
| Tap water | Pb(II) | 1–50 μg $L^{-1}$ | 2.3 μg $L^{-1}$ | BiNPs-porous carbon | [199] |
| | Cd(II) | | 1.5 μg $L^{-1}$ | | |
| Tap water | Pb(II) | 10–100 μg $L^{-1}$ | 2.1 μg $L^{-1}$ | AuNPs | [200] |
| | Cu(II) | 10–150 μg $L^{-1}$ | 1.4 μg $L^{-1}$ | | |
| Tap water | Cd(II) | 40–150 μg $L^{-1}$ | 12.1 μg $L^{-1}$ | AuNPs and AgNPs | [201] |
| | Pb(II) | 1–90 μg $L^{-1}$ | 0.12 μg $L^{-1}$ | | |
| Tap water | Zn(II) | 1–180 μg $L^{-1}$ | 39 ng $L^{-1}$ | MWCNTs-AuNPs | [202] |
| | Pb(II) | | 37 ng $L^{-1}$ | | |
| | Cu(II) | | 17 ng $L^{-1}$ | | |
| Tap water | Hg(II) | 2–20 μM | 0.191 μM | $Fe_2O_3$ NPs | [203] |
| | Cd(II) | | 0.167 μM | | |
| | Pb(II) | | 0.177 μM | | |
| Tap water | Pb(II) | 20–120 μg $L^{-1}$ | 3.2 μg $L^{-1}$ | $Co_3O_4$ NPs | [204] |
| | Cd(II) | | 3.5 μg $L^{-1}$ | | |
| Mineral water | As(III) | Not specified | 5.68 μg $L^{-1}$ | PtNPs | [205] |

*GR: graphene; BiNPs: bismuth nanoparticles; AuNPs: gold nanoparticles; AgNPs: silver nanoparticles; MWCNTs: multi-walled carbon nanotubes; $Fe_2O_3$ NPs: iron oxide nanoparticles; $Co_3O_4$ NPs: cobalt oxide nanoparticles; PtNPs: platinum nanoparticles.

silver, and platinum nanoparticles) and metal oxide nanomaterials (iron oxide, cobalt oxide nanoparticles) modification allows to reach the μg $L^{-1}$ levels, while the modification with carbon nanomaterials like graphene and multi-walled carbon nanotubes has shown to lower the DLs to the ng $L^{-1}$ level.

## 9.2 Detection of heavy metals in cereals, fruits, and vegetables

Growing cereals, fruits, and vegetables in contaminated soil can result in the transfer of HMs into edible parts, potentially posing health risks to humans. Electrochemical approaches based on SPEs modified with different nanomaterials have been recently proposed for the detection of HMs in such samples, at the levels required for the legislation (Table 6). Cd(II), Pb(II), Cu(II), and Zn(II) have been detected in cereals like rice, sorghum, wheat, and corn using SPEs modified with nanomaterials. Fruits and vegetables have also

© Burleigh Dodds Science Publishing Limited, 2025. All rights reserved.

**Table 6** Detection of HMs in cereals, fruits, and vegetables using SPEs modified with nanomaterials

| Food sample | Analyte | Linear range | LOD | Nanomaterial | Ref. |
|---|---|---|---|---|---|
| Rice | Cd(II) | Not specified | $10^{-7}$ M | GR | [206] |
| Rice | Cd(II) | 1–80 μg $L^{-1}$ | 80 ng $L^{-1}$ | GR | [207] |
| | Pb(II) | | 100 ng $L^{-1}$ | | |
| | Cu(II) | | 64 ng $L^{-1}$ | | |
| Rice, sorghum, wheat, and corn | Zn(II) | 3–900 μg $L^{-1}$ | 0.48 μg $L^{-1}$ | GR | [208] |
| | Cd(II) | | 0.45 μg $L^{-1}$ | | |
| | Pb(II) | | 0.29 μg $L^{-1}$ | | |
| Herb items | Pb(II) | 2–100 μg $L^{-1}$ | 0.2 μg $L^{-1}$ | MWCNTs | [209] |
| | Cd(II) | 2–100 μg $L^{-1}$ | 0.8 μg $L^{-1}$ | | |
| | Zn(II) | 12–100 μg $L^{-1}$ | 11.0 μg $L^{-1}$ | | |
| Tea and vegetables | Cd(II) | 0.1–200 μg $L^{-1}$ | 90 ng $L^{-1}$ | MWCNTs (aptasensor) | [210] |
| | Hg(II) | 0.05–20 μg $L^{-1}$ | 100 ng $L^{-1}$ | | |
| Fruit juice, preserved eggs, black tea, green tea | Pb(II) | 5–200 μg $L^{-1}$ | 1.0 μg $L^{-1}$ | GR | [211] |
| Salt, fruit juice | As(III) | 25–500 μg $L^{-1}$ | 3.0 μg $L^{-1}$ | GR-AuNPs (smartphone-assisted) | [212] |
| | Cr(VI) | 250–25 000 μg $L^{-1}$ | 40 μg $L^{-1}$ | | |
| | Hg(II) | 100–1500 μg $L^{-1}$ | 16 μg $L^{-1}$ | | |
| | Cd(II) | 25–750 μg $L^{-1}$ | 2.0 μg $L^{-1}$ | | |
| | Pb(II) | 25–750 μg $L^{-1}$ | 0.95 μg $L^{-1}$ | | |
| Lettuces and potatoes | Cd(II) | 0.2–500 μg $L^{-1}$ | 0.07 μg $L^{-1}$ | Bi-MOF-MWCNTs (smartpone-assisted) | [213] |

*GR: graphene; MWCNTs: multi-walled carbon nanotubes; AuNPs: gold nanoparticles; MOF: metal-organic frameworks.

been object of the detection of the same metals, been the method extended in some cases to the detection of Hg(II), As(II), and Cr(VI). Graphene, multi-walled carbon nanotubes, and composites of both nanomaterials with metal nanoparticles have been proposed in such cases for the modification of SPEs to reach typically μg $L^{-1}$ detection levels, although some works achieve DLs at the ng $L^{-1}$. In most cases, the simultaneous detection of two or three metals is performed in the same scan, leading to a very rapid analysis of the samples. ASV is the detection method of choice in almost all approaches, although an aptasensor has also been found in the literature. The combination with smartphones proposed in some works evidences the suitability of this methodology for *in situ* and decentralized screening of samples.

© Burleigh Dodds Science Publishing Limited, 2025. All rights reserved.

## 9.3 Detection of heavy metals in milk, honey, beverages, fish, seafood, and canned food

HMs can enter the milk through contaminated feed, water, and soil consumed by lactating cows. They can also be found in honey since bees forage across large areas and interact with polluted air, water, soil, nectar, and pollen. Environmental contamination and agricultural practices can also lead to their presence in tea and coffee. Fish living in contaminated water are also important carriers of HMs. Beverages, like beer, and canned food, can contain high levels of HMs which originate from the primary materials used in the product Cd(II), Pb(II), Hg(II), Cr(III), Cr(VI), and Sn(II), in most cases in multiplexing mode, have been detected in this kind of samples. As in the case of water analysis, SPEs modified with different nanomaterials have been proposed, reaching DLs at legislation-relevant levels (Table 7). Again, carbon nanotubes and graphene, alone or combined with metals, are the preferred choice for SPEs modification for previous HMs detection by anodic stripping voltammetry. The detection levels achieved are at the μg $L^{-1}$, being in some cases lowered to ng $L^{-1}$.

## 9.4 Detection of heavy metals in urine used for food monitoring

The analysis of HMs in urine has consistently attracted attention in biomedical and environmental sciences. Urine is a valuable medium for metal analysis because it is easy to collect, store, and prepare samples. Consequently, it is widely used to estimate exposure levels. Urine sample analysis provides a more accurate reflection of the total body intake of certain elements compared to other biological fluids, so is an outstanding tool for human monitoring in relation to consumption of contaminated food. It is worth mentioning that HMs are resistant to degradation by microorganisms, leading to their accumulation in the human kidneys, lungs, and liver.

Few works have recently reported the detection of HMs in urine using SPEs modified with nanomaterials. Cd(II), Pb(II), Cu(II), and Hg(II) have been detected in such samples at μg $L^{-1}$ levels using graphene, gold nanoparticles, and MOFs as electrode modifiers (Table 8).

Cartoons and graphs illustrating the electrochemical approaches are shown at Fig. 12.

# 10 Rapid screening techniques for detecting heavy metals in food: lateral flow assay approaches

LFA test strips are effective and convenient tools for detecting food contaminants. They offer several advantages, including ease of use portability

© Burleigh Dodds Science Publishing Limited, 2025. All rights reserved.

**Table 7** Detection of HMs in milk, honey, beverages, fish, seafood, and canned food using SPEs modified with nanomaterials

| Food sample | Analyte | Linear range | LOD | Nanomaterial | Ref. |
|---|---|---|---|---|---|
| Milk and honey | Cd(II) | 1.0–60.0 μg L$^{-1}$ | 0.2 μg L$^{-1}$ | SWCNTs | [214] |
| | Pb(II) | | 0.4 μg L$^{-1}$ | | |
| Milk | Pb(II) | 1.0–60.0 μg L$^{-1}$ | 0.8 μg L$^{-1}$ | GR | [215] |
| | Cd(II) | | 0.5 μg L$^{-1}$ | | |
| | Cd(II) | | 0.06 μg L$^{-1}$ | | |
| | Pb(II) | | 0.08 μg L$^{-1}$ | | |
| Tea and coffee | Cr(VI) | 0.5–10 μM | 0.01 μM | AuNPs | [216] |
| | Cr(III) | | | | |
| Beer | Pb(II) | 31.3–2000.0 μg L$^{-1}$ | 6.7 μg L$^{-1}$ | Bi-BDD | [217] |
| | Hg(II) | | 7.5 μg L$^{-1}$ | | |
| Fish | Cd(II) | 10.0–80.0 μg L$^{-1}$ | 0.16 μg L$^{-1}$ | GR-ZnO | [218] |
| | Pb(II) | | 0.17 μg L$^{-1}$ | | |
| Seafood | Pb(II) | 1.0–100.0 μg L$^{-1}$ | 0.1 μg L$^{-1}$ | BiNPs | [219] |
| | Cd(II) | | 0.2 μg L$^{-1}$ | | |
| Seafood | Pb(II) | 0.1–150 μg L$^{-1}$ | 0.02 μg L$^{-1}$ | BiNPs/porous carbon | [220] |
| | Cd(II) | | 0.03 μg L$^{-1}$ | | |
| | Cd(II) | | 1.5 μg L$^{-1}$ | | |
| Canned food | Sn(II) | 10.0–250.0 μg L$^{-1}$ | 0.26 μg L$^{-1}$ | BiNPs/GR | [221] |
| | Pb(II) | | 0.44 μg L$^{-1}$ | | |

*GR: graphene; SWCNTs: single-walled carbon nanotubes; AuNPs: gold nanoparticles; BiNPs: bismuth nanoparticles.

and affordability. These tests have become widely used for identifying harmful substances in food and water due to their numerous benefits [225, 226]. They require minimal sample preparation, which reduces labor and the risk

**Table 8** Detection of HMs in urine using SPEs modified with nanomaterials

| Sample | Analyte | Linear range | LOD | Nanomaterial | Ref. |
|---|---|---|---|---|---|
| Urine | Cd(II) | 1–2500 μg L$^{-1}$ | 7.1 μg L$^{-1}$ | GR-DTT | [222] |
| | Pb(II) | | 1.9 μg L$^{-1}$ | | |
| | Cu(II) | | 0.4 μg L$^{-1}$ | | |
| | Hg(II) | | 0.7 μg L$^{-1}$ | | |
| Urine | Pb(II) | 5–25 μg L$^{-1}$ | 1 μg L$^{-1}$ | AuNP | [223] |
| Urine | Cd(II) | 0.5–100 μg L$^{-1}$ | 0.17 μg L$^{-1}$ | MOF | [224] |

* GR: graphene; DTT: diaminoterthiophene; AuNPs: gold nanoparticles; MOF: metal-organic frameworks.

© Burleigh Dodds Science Publishing Limited, 2025. All rights reserved.

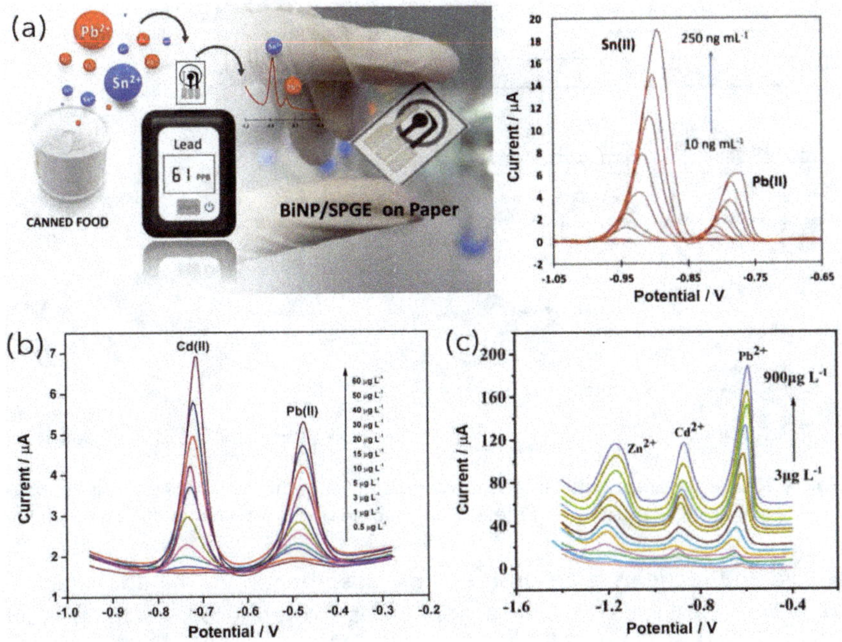


**Figure 12** Screen-printed electrodes (SPEs) for rapid screening of HMs in food. (a) Simultaneous detection of Sn(II) and Pb(II) in canned food using a portable potentiostat. Source: Adapted from [221] with permission; (b) simultaneous detection of Cd(II) and Pb(II) in milk and honey. Source: Adapted from [214] with permission; (c) Simultaneous detection of Zn(II), Cd(II), and Pb(II) in rice, sorghum, wheat, and corn. Source: Adapted from [208] with permission.

of errors. Additionally, LFAs are cost-effective and do not require specialized equipment, making them accessible to a broader user. These assays can be designed for either qualitative or quantitative analysis, making them versatile for various applications, being ideal for the rapid screening of samples out of the laboratory.

The LFA consists of several key components: sample pad, conjugate pad, nitrocellulose (NC) membrane, and absorbent pad. The NC membrane is equipped with multiplexed capture probes for control (C) lines and test (T) lines [227, 228] (Fig. 13).

The fluid sample placed at the sample pad migrates through the conjugate pad, where typically nanoparticle/bioreceptor conjugates are immobilized. During this step, the analyte interacts with the bioreceptor, forming a complex. The sample continues to migrate through the detection pad due to capillary forces, where a complementary bioreceptor was previously deposited. Test and control lines are typically constituted by antibodies or other receptors, and their role is to capture the analyte and any excess of conjugated particles. The excess

© Burleigh Dodds Science Publishing Limited, 2025. All rights reserved.

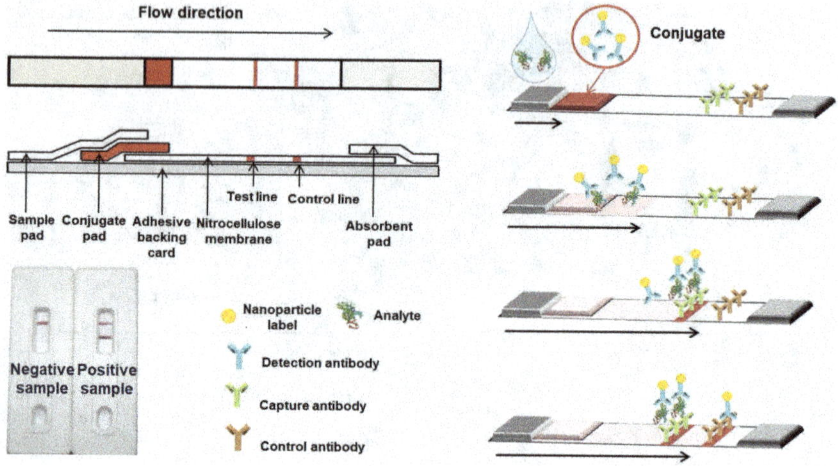


**Figure 13** (Left) Components of a lateral flow strip, together with a picture of the results for a negative and a positive sample; (right) scheme of a typical lateral flow assay (LFA).

of fluid and reagents is absorbed by the absorbent pad. The appearance of colour in the test line due to the nanoparticles allows one to know the presence or absence of the analyte by the naked eye. The control line is always visible, ensuring that the fluid migration is successful. Strip readers can also be used for quantitative analysis. Interestingly, the use of multiple receptors immobilized in different lines allows multiplexing in a single assay. These systems can be classified as either sandwich-type or competitive-type, depending on the detection strategy employed (Fig. 14).

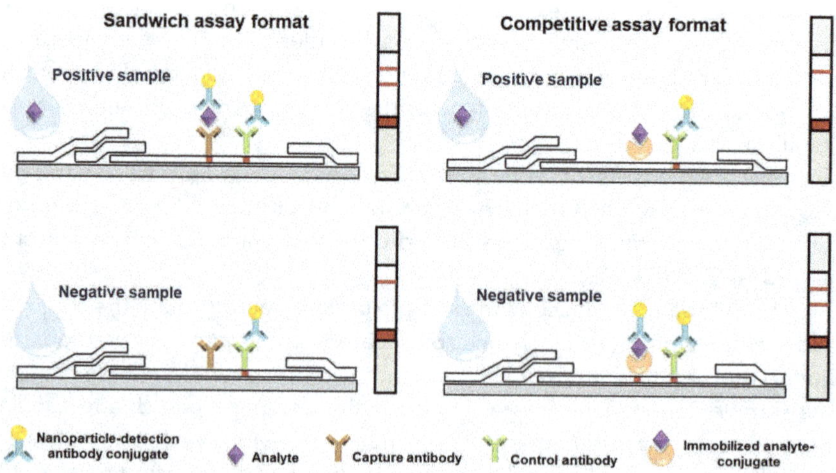


**Figure 14** Typical formats of lateral flow assays: (Left) direct and (Right) competitive assays.

© Burleigh Dodds Science Publishing Limited, 2025. All rights reserved.

The above-mentioned characteristics make LFA ideal tools for rapid screening of food samples. The sandwich-type is the most popular choice in LFAs, including those for the detection of food contaminants like food-borne pathogens. However, this format is not suitable for the detection of low-molecular-weight molecules. This is the case of HMs (also of pesticides, allergens, antibiotics, and toxins), for which detection competitive assays, are typically required. Liquid samples are directly analysed by placing in the strip a volume of around 100 μL. Solid samples require previous digestion, as in the analysis by electrochemical approaches. After 5–10 minutes, the results are clearly visible. The most relevant approaches for the detection of HMs in food samples using LFA based on nanomaterials are summarized in Table 9. Pb(II), Hg(II), and Cd (II) are the HMs that have been detected in food using competitive LFAs. Drinking water, milk, tea, and grain are the samples analysd with these systems.

Gold nanoparticles are the preferred labels for colorimetric detection (Fig. 15a). In some cases, gold is combined with silver in core–shell structures that have excellent characteristics for their detection using surface-enhanced Raman spectroscopy (SERS) that is performed on the strip (Fig. 15b).

**Table 9** Detection of HMs in food samples using LFA based on nanomaterials

| Sample | Analyte | Linear range | LOD | Approach | Ref. |
|---|---|---|---|---|---|
| Drinking water | Pb(II) | 0.1–5 μg $L^{-1}$ | 50 ng $L^{-1}$ | AuNPs<br>Aptamers | [229] |
| Milk | Hg(II) | 1 nM–100 μM | 1 nM | AuNPs<br>Aptamers | [230] |
| Tea | Hg(II) | 0.01–1 nM | 0.36 pM | Au@Ag NPs<br>T–Hg–T strategy;<br>SERS enhanced | [231] |
| Green tea | Pb(II) | 2.5–100 mg $kg^{-1}$ | 1.375 mg $k^{-1}$ | AuNPs<br>MIP | [232] |
| Grain | Cd(II) (and toxin in multiplex) | 0.01–60 μg $L^{-1}$ | 24 ng $L^{-1}$ | Fluorescent NPs<br>Antibodies<br>anti-Cd/EDTA | [233] |
| Milk | Hg(II) | 0.1–200 μg $L^{-1}$ | 30 ng $L^{-1}$ | AuNPs<br>Antibodies<br>Electrochemistry | [234] |
| Drinking water | Cd(II) | 0.4–2000 ppb | 0.1 μg $L^{-1}$ | AuNPs<br>Antibodies | [235] |
| Drinking water | Pb(II) (and toxin, antibiotic, hormone and pesticides in multiplex) | Not specified | 4 μg $L^{-1}$ | AuNPs<br>Antibodies | [236] |

*AuNPs: gold nanoparticles; Au@Ag NPs: core-shell gold–silver nanoparticles; SERS: surface-enhanced Raman spectroscopy; MIP: molecularly imprinted polymer; EDTA: ethylenediaminetetraacetic acid (EDTA).

© Burleigh Dodds Science Publishing Limited, 2025. All rights reserved.

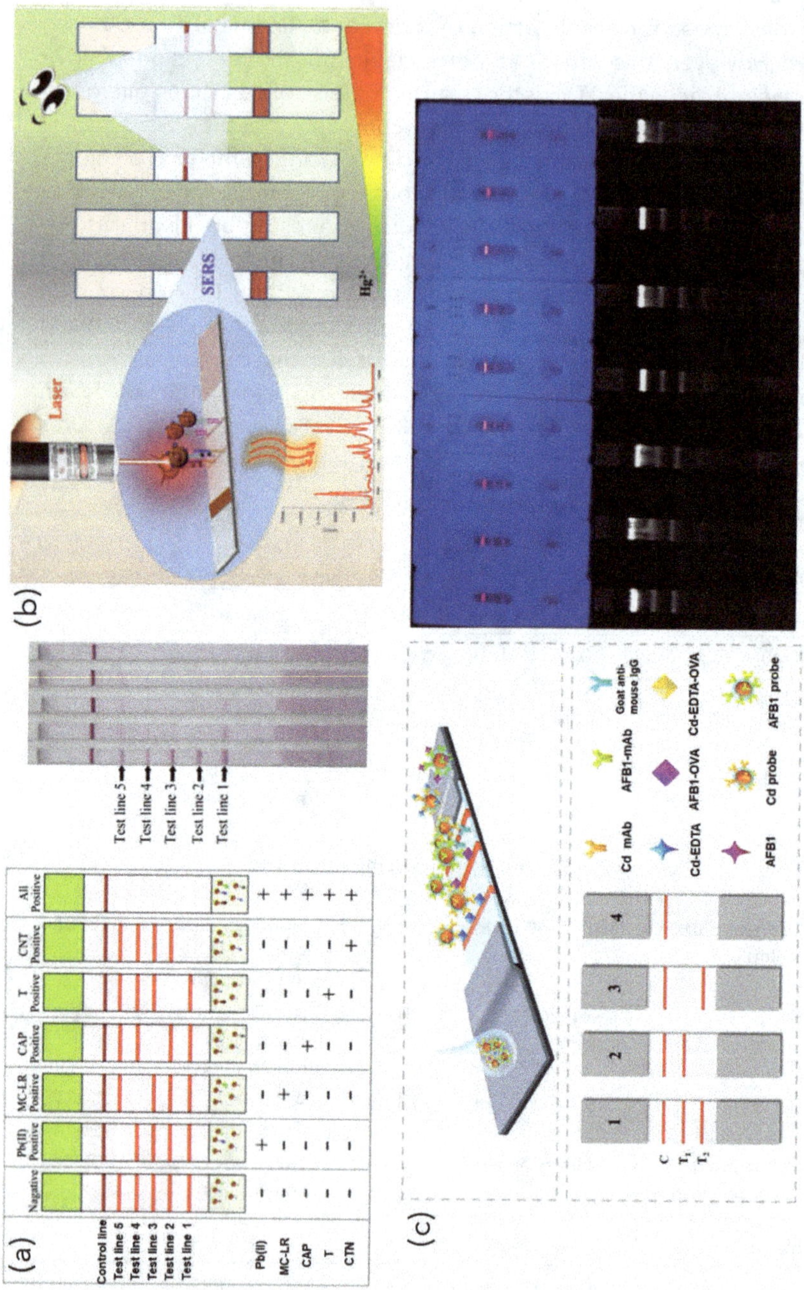


**Figure 15** Lateral flow assays (LFA) for HMs detection. (a) LFA based on gold nanoparticles and colorimetric detection for the simultaneous analysis of Pb(II) and other four contaminants (microcystin–leucine–arginine (MC–LR), chloramphenicol (CAP), testosterone (T) and chlorothalonil (CTN)) in drinking water. Source: Adapted from [236] with permission. (b) LFA based on gold@silver core@shell nanoparticles and surface-enhanced Raman spectroscopy (SERS) detection for HG(II) analysis in tea samples. Source: Adapted from [231] with permission. (c) LFA based on fluorescent nanoparticles for the simultaneous detection of Cd(II) and aflatoxin B1 in grain. Source: Adapted from [233] with permission.

© Burleigh Dodds Science Publishing Limited, 2025. All rights reserved.

Fluorescent nanoparticles have also been proposed for lowering the DLs (Fig. 15c). However, it is worthy noting that in both SERS and fluorescent-based approaches, the naked-eye detection is not possible.

Regarding the receptors, specific antibodies and aptamers are the preferred choice, requiring in some cases the previous conjugation of the HM with complexing agents like EDTA. Molecularly imprinted polymers (MIPs) have also been proposed as alternative specific recognition elements. Interestingly, the thymine–Hg(II)–thymine (T–Hg–T) recognition system has also been proposed for the detection of Hg(II) in LFA. In this case, nucleic acids are immobilized in both the strip and the nanoparticle surface, and the sensing of the Hg(II) is performed taking advantage of the target-induced formation of T–Hg(II)–T mismatch structures on the interface that modify the test line composition after the assay. It also deserves to be highlighted that the integration of electrodes in the test line has been reported, allowing the sensitive electrochemical detection of the nanoparticle tags.

Finally, it deserves to be highlighted that the multiplexing abilities of the LFA have been applied for the detection of toxins, antibiotics, hormones, and pesticides together with HMs in a single assay, in samples like grain and drinking water.

Regarding the detection levels, colorimetric LFA approaches typically allow to detect HMs in food at $\mu g\ L^{-1}$ levels. The combination with fluorescent detection or electrochemistry has shown the higher sensitivity, reaching the detection of these contaminants at $ng\ L^{-1}$ levels.

## 11 Rapid screening using smartphone-based systems

Smartphone-based point-of-care tests (POCTs) are emerging as cost-effective, accurate, and easy-to-use solutions, making them ideal for resource-limited areas. As a result, they are seen as a viable alternative to conventional diagnostic tests. The smartphone connection to the internet, as well as wireless NFC technology, also allows rapid communication for sending results from the field to central laboratories [237] (Fig. 16a).

Smartphones can be used to control portable potentiostats in electrochemical approaches, and as an alternative to lateral flow strip readers by using the camera and adequate apps for quantitative analysis. Colorimetric and fluorescent strategies have also been coupled with smartphone-assisted visual platforms for food safety applications [238–240], including metal ions detection [241, 242]. In the case of HMs, in addition to enzyme-linked immunosorbent assays (ELISAs) [243], colorimetric assays based on gold nanoparticle aggregation [244] (Fig. 16b) and fluorescent assays based on quenching of quantum dots nanoparticle fluorescence [245, 246] (Fig. 16c), both induced by metal ions, are emerging approaches suitable for rapid food screening.

© Burleigh Dodds Science Publishing Limited, 2025. All rights reserved.

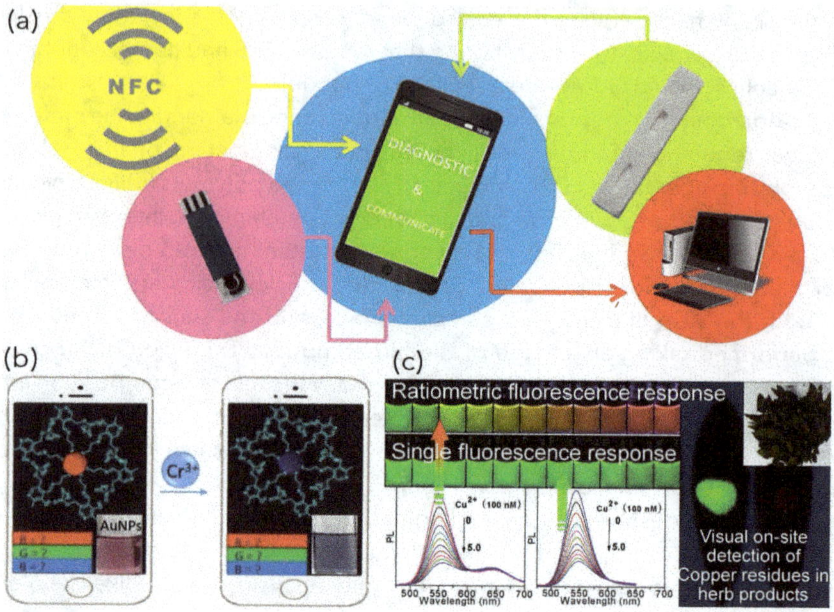


**Figure 16** Smartphone-based systems coupled to colorimetric and fluorescent strategies. (a) Cartoon illustrating the use of smartphones for point-of-care (POC) testing. Source: Adapted from [237] with permission. (b) Colorimetric assays based on gold nanoparticles aggregation for Cr(III) detection in water. Source: Adapted from [244] with permission. (C) Fluorescent assays based on quenching of quantum dots nanoparticles' fluorescence for Cu(II) detection in herb products. Source: Adapted from [246] with permission.

## 12 The use of artificial intelligence in detecting heavy metals in food

While rapid screening techniques discussed in this chapter are simple, portable, and cost-effective, it is crucial to enhance their reproducibility and stability to meet the stringent demands of the food bioprocess industry and/or official control regulations. Research in this area is focusing on improving the tracing, detection, and extraction of features from complex matrices. Acquiring and analysing food signals helps to identify contaminants and optimize food processing.

In this context, AI approaches such as machine learning (ML) allow for correcting errors in the data analysis due to interferences from other compounds in food samples [247]. The use of certain algorithms reduces the interferences, helping in the processing of large amounts of data in different samples, which is of key relevance in the case of rapid screening techniques for HMs detection in food [248].

© Burleigh Dodds Science Publishing Limited, 2025. All rights reserved.

## 12.1 Artificial intelligence for feature extraction

AI techniques have been extensively employed for the automated selection of representative variables and extraction of key features from high-dimensional spectral data, thereby enhancing the stability and generalizability of analytical models. For example, feature extraction methods such as the ReliefF algorithm and PCA have been successfully applied to hyperspectral data (400–1000 nm) from kale and basil [249], enabling accurate detection of Cd contamination when combined with predictive models like artificial neural networks (ANN) and support vector machines (SVM). Feature extraction can be achieved not only through traditional variable selection algorithms but also via end-to-end feature learning using deep neural networks. Specifically, wavelet transforms combined with a stacked convolutional autoencoder have been utilized to extract spectral features from lettuce leaves [250]. When integrated with support vector regression (SVR), this approach has facilitated the simultaneous prediction of Cd and Pb concentrations with high accuracy, yielding coefficients of determination ($R^2$) in the range of approximately 0.93–0.94.

## 12.2 Artificial intelligence for integrating multiple data

An emerging frontier in AI-assisted HM detection involves enhancing robustness and sensitivity to complex sample matrices through integrating multimodal data. By fusing outputs from multiple sources, AI models effectively capture complementary information and overcome the limitations inherent in single-source data, particularly in spectroscopic applications where the feasibility for in-field use is notably high. For instance, the fusion of laser-induced breakdown spectroscopy (LIBS) and NIR reflectance spectroscopy has been successfully employed to quantify HMs in lily tissues. Employing partial least squares regression (PLSR) algorithms to integrate these modalities substantially improved detection accuracy compared to single-spectrum-based models [251]. Additionally, SERS, combined with micro-Raman imaging and a one-dimensional stacked autoencoder convolutional neural network (1D-SAE-CNN), has been effectively utilized to measure copper concentrations in apple rootstocks [252]. Moreover, the integration of SERS and fluorescence spectroscopy using chemometric fusion modelling coupled with PLSR has successfully detected trace levels of Pb in complex food matrices, effectively correcting baseline drifts and minimizing spectral interferences. The performance of the fused model outperforms each technique individually [253]. Consequently, AI-driven fusion technology can be used to integrate all types of data, such as atomic, molecular, electrochemical, and fluorescence data, thereby enhancing model stability, adaptability, and detection capability.

© Burleigh Dodds Science Publishing Limited, 2025. All rights reserved.

This approach provides practical and reliable monitoring solutions for HM contamination in agricultural products.

## 12.3 Artificial intelligence for heavy metal detection

AI techniques have emerged as powerful tools for identifying and quantifying HM contamination in food products, enhancing analytical precision, efficiency, and practical applicability. For example, laser-induced breakdown spectroscopy (LIBS), combined with a pre-trained XGBoost transfer learning model, significantly improved Cd detection in rice husks [254]. Similarly, the extreme learning machine (ELM) neural network, when applied to LIBS data from rice straw, facilitated rapid and efficient Cd quantification, demonstrating suitability for real-time, in-field monitoring [255]. Beyond plant matrices, AI methods have also been effectively adopted for quantifying HMs in animal feed. A study employing Vis–NIR hyperspectral imaging data from distillers' dried grains with solubles (DDGS), a widely-used animal feed by-product, successfully quantified Cu and Zn concentrations using support vector regression (SVR), simultaneously suppressing spectral noise and redundancy, thereby enhancing model robustness and generalizability [256]. AI-driven portable sensing systems represent another critical advancement. For instance, a fluorescence biosensor integrated with an Hg ion-responsive protein chimera yielded measurable signal quenching upon $Hg^{2+}$ binding. Machine learning regression directly translated these signals into precise concentration measurements, eliminating the need for conventional calibration curves and improving specificity by distinguishing between Hg-induced signals and interferences from other metals. Similarly, carbon quantum dot-based nanosensors, coupled with AI models accounting for environmental variables such as pH and co-contaminants, enabled real-time, low-level detection of Cr(VI), significantly reducing false positives in water quality assessments [257]. Additionally, CNNs were effectively used to extract critical electrochemical features from gold nanoparticle-modified electrodes using differential pulse voltammetry (DPV), enabling simultaneous detection of $Cd^{2+}$, $Pb^{2+}$, $Cu^{2+}$, and $Hg^{2+}$. Integration with an Internet-of-Things (IoT) platform facilitated real-time, remote monitoring, achieving low DLs (0.6–1.4 μM) and illustrating the strong potential for multiplexed HM analysis in complex aqueous environments [258]. Portable Raman spectroscopy coupled with PLS-DA demonstrated high classification accuracy (~89%) in identifying As-stressed rice leaves, highlighting its efficacy for early agricultural pollution detection [259]. In seafood monitoring, machine vision techniques have also shown promise. Systems employing image-derived colour and texture features from fish gills and muscle tissues, combined with transfer learning through pre-trained CNN models, have facilitated rapid, non-invasive, and

© Burleigh Dodds Science Publishing Limited, 2025. All rights reserved.

cost-effective screening for Hg contamination [260]. Collectively, these advancements highlight AI's role in HM detection across the food supply chain. Employing classical machine learning algorithms or sophisticated techniques such as transfer learning and deep neural networks, AI effectively translates complex, multimodal data into actionable quantitative insights, significantly enhancing detection speed, sensitivity, and field applicability.

### 12.4 Artificial intelligence for analytical instrument calibration and optimization

AI has been applied in instrument calibration and parameter optimization in analytical methodologies [261]. Within spectroscopic calibration processes, AI-driven techniques have significantly enhanced the mitigation of matrix interference effects inherent to complex sample compositions. Specifically, both supervized and unsupervized machine learning algorithms have proven effective within inductively coupled plasma optical emission spectrometry (ICP-OES) systems, enabling real-time adjustments to calibration strategies to counteract interferences caused by intricate sample matrices. Supervized learning methods, including generalized linear models (GLMs), regularized GLM, and partial least squares (PLS) regression, have effectively correlated emission signals from non-target plasma lines with recovery efficiencies of the analyte elements [262]. In energy-dispersive X-ray fluorescence (EDXRF), an innovative autoencoder neural network integrating fundamental parameter models – such as mass absorption coefficients and excitation efficiencies – has been developed. Training the model using limited experimental data supplemented by simulated spectral data resulted in accurate multi-element quantitative analyses across diverse sample matrices [263]. Similarly, confocal micro-X-ray fluorescence (μXRF) analysis has benefited from the development of a combined 'forward physics simulation plus neural network' workflow. In this approach, extensive simulated spectra reflecting numerous elemental concentration profiles and density conditions were utilized to train a deep neural network capable of directly predicting concentrations of up to 53 elements without traditional calibration processes [264].

## 13 Conclusion

This chapter outlines the key aspects of assessing HMs in food, emphasizing the importance of detection, risk assessment, technological advancements, and future prospects. HM in foods may pose significant risks to consumer health and can lead to trade restrictions and financial losses for producers when the levels do not comply with the regulatory maximum levels.

© Burleigh Dodds Science Publishing Limited, 2025. All rights reserved.

The classification and characteristics of chemical elements, focusing on metals and metalloids are described, highlighting their environmental and health impacts, particularly HMs like Pb, Hg, Cd, and As. These elements can be toxic even at low concentrations and can cause significant environmental degradation and health issues. The origins of HMs in food can be both natural and from anthropogenic sources. Natural sources include geogenic processes like volcanic eruptions and weathering of rocks, while human activities such as industrial emissions, agricultural practices, and the use of metal-containing pesticides and fertilizers contribute significantly to HM contamination. HMs enter the food chain through soil, water, and air, via metal uptake mechanisms, and bioaccumulation in plants and animals. Mitigating HM levels in foods necessitates reducing their presence in the environment through biotechnological processes such as bioremediation and phytoremediation.

This chapter discusses the levels of HM in various food products, highlighting the growing concern for public health and the environment. It emphasizes the importance of data collection and exposure assessment by international bodies like FAO/WHO. A comprehensive review and mapping analysis from 2010 to 2024 revealed significant contamination alerts, primarily for Hg in fish products and Pb and Cd in FFVs. Fish species such as swordfish, shark, and tuna were notably affected, while plums, apples, prunes, mushrooms, and spinach showed high contamination levels. Countries like China and Spain were frequently linked to these alerts, underscoring the need for global collaboration to address sources of HM contamination in the food supply chain.

The bioaccumulation and harmful potential of certain metals, which are widely present in food webs, have long been recognized. This necessitates toxicological and exposure assessment studies to evaluate health risks, alongside the establishment of safety standards and monitoring programmes to protect public health and regulate food production and trade. Historically, risk assessment of chemical substances has relied heavily on animal testing and data from occupational exposures. However, recent advances in *in vitro*, *in silico*, epidemiological, and biomonitoring methods offer promising alternatives to traditional animal models. Despite progress in understanding metal toxicity pathways and advancements in analytical chemistry for detecting their presence in food, gaps remain in risk assessment, particularly concerning the mechanisms of systemic and targeted toxicity for metal species. There is also a need to consider potential synergistic or antagonistic effects in metal mixtures. Comprehensive monitoring programmes throughout the food chain are crucial to- establish food safety standards and provide scientific justification for risk mitigation measures.

HMs are just one aspect of food safety; official monitoring programmes must adopt a risk-based approach to address all potential contaminants and

© Burleigh Dodds Science Publishing Limited, 2025. All rights reserved.

residues in foods to ensure public health and trade. Efforts should also focus on comprehensive legislation to prevent HM buildup in agricultural soils and groundwater, thereby preventing their transfer to the food web.

Various techniques are employed to detect HM, with physical methods preferred for their rapid results and lower costs. XRF is a powerful technique for detecting and quantifying elements in food, offering rapid, non-destructive analysis with high sensitivity, and selectivity. Portable XRF instruments are valuable for *in-situ* screening of food and environmental samples, although detecting HMs at very low concentrations remains challenging due to overlapping spectral lines and matrix effects. Recent advancements, such as the use of synchrotron light and preconcentration techniques, have improved DLs and accuracy. Continued technological advancements are expected to enhance XRF capabilities, making it an indispensable tool in ensuring food safety. A case study was discussed for turmeric powder, a popular culinary and medicinal spice, that is prone to adulteration, like oregano and other herbs and spices.

Various laboratory-based techniques, such as physicochemical analysis, microscopy, chromatography, spectrophotometry, and spectroscopy, are used to detect fraud, but these methods may only identify issues after the product is on the market. Therefore the 'Food Fortress' system [137], designed to enhance food safety and traceability, suggests that technology diffusion be through government policies, industry incentives, and partnerships with testing labs that can help maintain quality control and minimize food safety hazards. Portable technologies like pXRF can serve as initial screening tools, with mass spectrometry providing further validation and implementation in official control regulations and guidance documents.

As in the case of herbs and spices, rapid and cost-effective analytical techniques are essential for screening food samples, reducing the number sent to laboratories. Electrochemical approaches using SPEs and optical approaches based on LFAs are prominent screening techniques for HMs analysis in food. Advances in nanotechnology have driven these strategies forward, offering novel materials that enhance analytical performance. Electrochemical methods can analyze HMs at μg $L^{-1}$ levels in about 2 minutes, requiring only 50 μL of sample. Advanced nanomaterials like graphene and multi-walled carbon nanotubes have lowered DLs to the ng $L^{-1}$ level, enabling simultaneous detection of various HMs in a single scan.

LFAs based on nanomaterials are ideal for rapid screening, requiring around 100 μL of sample and 10–15 minutes of analysis. They allow for qualitative naked-eye detection and quantitative strip reader-based analysis, though LFAs for HMs are limited by the need for specific receptors for such low-weight analytes. Smartphones are emerging as outstanding tools for controlling potentiostats, reading LFA strips, and conducting colorimetric

© Burleigh Dodds Science Publishing Limited, 2025. All rights reserved.

assays. Smartphone connectivity to the internet and wireless NFC technology facilitates rapid communication of results from the field to central laboratories.

Finally, AI approaches are emerging in this field, correcting errors in data analysis due to interferences from other compounds, which is crucial for rapid screening techniques for HMs detection. AI and the Internet of Things can enhance risk prediction and monitoring of metal exposure. Integrated approaches, under the One Health concept, are ultimately necessary to address HM toxicity, considering human health, animal health, ecological safety, and food security.

## 14 Acknowledgments

The European Union's Horizon Europe Research and Innovation Programme (FOODSAFER Project), grant number 101060698 and Research Ireland, Northern Ireland's Department of Agriculture, Environment and Rural Affairs (DAERA), UK Research and Innovation (UKRI) via the International Science Partnerships Fund (ISPF) under Grant number 22/CC/11147 at the Co-Centre for Sustainable Food Systems.

## 15 References

1. Britannica, Chemical element. https://www.britannica.com/science/chemical-element (accessed 19 March 2025)
2. Pourret, O., et al. 2021. Heavy Metal: a misused term? *Acta Geochimica*, 40, 466–471.
3. IUPAC. 2002. IUPAC technical report. *Pure and Applied Chemistry*, 74(5), 793–807.
4. Rietjens, I.M.C.M., et al. 2025. The definition of chemical contaminants in food: ambiguity and consequences. *Regulatory Toxicology and Pharmacology*, 155, 105739. https://doi.org/10.1016/j.yrtph.2024.105739
5. Tchounwou, P.B., et al. 2012. Heavy Metal Toxicity and the Environment. In: Luch, A. (eds) *Molecular, Clinical and Environmental Toxicology. Experientia Supplementum*, vol. 101. Springer, Basel. https://doi.org/10.1007/978-3-7643-8340-4_6
6. Nriagu, J., 1989. A global assessment of natural sources of atmospheric trace metals. *Nature*, 338, 47–49. https://doi.org/10.1038/338047a0
7. Sánchez-Castro, I., et al. 2023. Past, present and future trends in the remediation of heavy-metal contaminated soil – Remediation techniques applied in real soil-contamination events. *Heliyon*, 9(6), e16692. https://doi.org/10.1016/j.heliyon.2023.e16692
8. Clemens, S. and Ma, J.F. 2016. Toxic heavy metal and metalloid accumulation in crop plants and foods. *Annual Review of Plant Biology*, 67, 489–512. https://doi.org/10.1146/annurev-arplant-043015-112301
9. Gall, J., et al. 2015. *Environmental Monitoring and Assessment*, 187, 201. https://doi.org/10.1146/10.1007/s10661-015-4436-3
10. Mukhi, S., et al. 2023. Assessing the impact of heavy metal residues in food and drug packaging on the standard bacterial strains: a biofilm perspective. *CyTA - Journal of Food*, 21(1), 711–717. https://doi.org/10.1080/19476337.2023.2279187.

© Burleigh Dodds Science Publishing Limited, 2025. All rights reserved.

11. Dong, Z., et al. 2014. Migration of toxic metals from ceramic food packaging materials into acid food simulants. *Mathematical Problems in Engineering*, 2014, Article 759018. https://doi.org/10.1155/2014/759018
12. Mwove, J., et al. 2023. Environmental exposure assessment of lead and cadmium in street vended foods sold in selected locations in Kenya. *Food Science & Nutrition*, 11, 2610–2619. https://doi.org/10.1002/fsn3.3344
13. Zheng, T., et al. 2023. Magnetic iron-based nanoparticles biogeochemical behavior in soil-plant system: a critical review. *Science of The Total Environment*, 904, 166643. https://doi.org/10.1016/j.scitotenv.2023.166643
14. Li, Y., et al. 2023. Toxic effects of cadmium on the physiological and biochemical attributes of plants, and phytoremediation strategies: a review. *Environmental Pollution*., 15, 325. https://doi.org/10.1016/j.envpol.2023.121433
15. Panda, A., et al. 2025. Uptake, impact, adaptive mechanisms, and phytoremediation of heavy metals by plants: role of transporters in heavy metal sequestration. *Plant Physiology and Biochemistry*, 221, 109578. https://doi.org/10.1016/j.plaphy.2025.109578
16. Oosten, M. and Maggio, A. 2015. Functional biology of halophytes in the phytoremediation of heavy metal contaminated soils. *Environmental and Experimental Botany*, 111, 135–146. https://doi.org/10.1016/j.envexpbot.2014.11.010
17. Aziz, I. and Mujeeb, A. 2022. Halophytes for phytoremediation of hazardous metal(Loid)s: a terse review on metal tolerance, bio-indication and hyperaccumulation. *Journal of Hazardous Materials*, 424(Pt A), 127309–127309.
18. Angon, P.B., et al. 2024. Sources, effects and present perspectives of heavy metals contamination: soil, plants and human food chain. *Heliyon*, 10, 7. https://doi.org/10.1016/j.heliyon.2024.e28357
19. Aucour, A.M., et al. 2023. Coupling metal stable isotope compositions and X-ray absorption spectroscopy to study metal pathways in soil–plant systems: a mini review. *Metallomics*, 15, 4. https://doi.org/10.1093/mtomcs/mfad016
20. Rahman, M.S., et al. 2017. Arsenic solid-phase speciation and reversible binding in long-term contaminated soils. *Chemosphere*, 168, 1324–1336. https://doi.org/10.1016/j.chemosphere.2016.11.130
21. Pasricha, S., et al. 2012. Molecular mechanisms underlying heavy metal uptake, translocation and tolerance in hyperaccumulators-an analysis: heavy metal tolerance in hyperaccumulators. *Environmental Challenges*, 4, 100197. https://doi.org/10.1016/j.envc.2021.100197
22. Angulo-Bejarano, P.I., et al. 2021. Metal and metalloid toxicity in plants: an overview on molecular aspects. *Plants (Basel)*, 10(4), 635. https://doi.org/10.3390/plants10040635
23. Pachura, P., et al. 2015. Assessment of the availability of heavy metals to plants based on the translocation index and the bioaccumulation factor. *Desalination and Water Treatment*, 57, 1–6. https://doi.org/10.1080/19443994.2015.1017330
24. Shahid, M., et al. 2017. Foliar heavy metal uptake, toxicity and detoxification in plants: a comparison of foliar and root metal uptake. *Journal of Hazardous Materials*, 325, 36–58. https://doi.org/10.1016/j.jhazmat.2016.11.063
25. Zakaria, Z., et al. 2021. Understanding potential heavy metal contamination, absorption, translocation and accumulation in rice and human health risks. *Plants*, 10(6), 1070. https://doi.org/10.3390/plants10061070

© Burleigh Dodds Science Publishing Limited, 2025. All rights reserved.

26. Greger, M. 2004. Metal availability, uptake, transport and accumulation in plants. https://doi.org/10.1007/978-3-662-07743-6_1
27. Zhang, R., et al. 2025. Recent advances in understanding the mechanisms of plant cadmium accumulation as affected by grafting in vegetable production. *Frontiers in Plant Science*, 16, 1526041. https://doi.org/10.3389/fpls.2025.1526041
28. Szynkowska-Jóźwik, M., et al. 2018. Bioaccumulation and biomagnification of trace elements in the environment. https://doi.org/10.1002/9781119133780.ch13
29. Godebo, T.R., et al. 2025. Metals in honey from bees as a proxy for environmental contamination in the United States. *Environmental Pollution*, 364(2), 125221. https://doi.org/10.1016/j.envpol.2024.125221
30. Ebong, G.A., et al. 2024. Risk potentials of disease-causing metals in poultry-related foods and products in Uyo, southern Nigeria. *Journal of Materials and Environmental Science*, 15(6), 850–865.
31. Briffa, J., et al. 2020. Emmanuel Sinagra, Renald Blundell, Heavy metal pollution in the environment and their toxicological effects on humans. *Heliyon*, 6(9), 2405–8440. https://doi.org/10.1016/j.heliyon.2020.e04691
32. Etesami, H. and Maheshwari, D.K. 2018. Use of plant growth promoting rhizobacteria (PGPRs) with multiple plant growth promoting traits in stress agriculture: action mechanisms and future prospects. *Ecotoxicology and Environmental Safety*, 156, 225–246. https://doi.org/10.1016/j.ecoenv.2018.03.013
33. Wang, X., et al. 2024. Effect of biochars on the immobilization and form of Cadmium (Cd) in simulated Cd deposition of iron rich soils. *Ecotoxicology and Environmental Safety*, 272, 116045. https://doi.org/10.1016/j.ecoenv.2024.116045
34. Rizwan, M., et al. 2016. Cadmium minimization in wheat: a critical review. *Ecotoxicology and Environmental Safety*, 130, 43–53. https://doi.org/10.1016/j.ecoenv.2016.04.001
35. Afzal, M.R., et al. 2023. Insights into the mechanisms involved in lead (Pb) tolerance in invasive plants-the current status of understanding. *Plants (Basel)*, 12(11), 2084. https://doi.org/10.3390/plants12112084
36. World health organization (WHO), Global Environment Monitoring System (GEMS) / Food Contamination Monitoring and Assessment Programme. https://www.who.int/teams/nutrition-and-food-safety/databases/global-environment-monitoring-system-food-contamination (accessed January 2025)
37. Scutaras, U., et al. 2023. Heavy metals in foods and beverages: global situation, health risks and reduction methods. *Foods*, 12, 3340. https://doi.org/10.3390/foods12183340
38. WHO (World Health Organization). 2012. Safety evaluation of certain food additives and contaminants: prepared by the seventy-third meeting of the Joint FAO/WHO Expert Committee on Food Additives (JECFA) (WHO Food Additives Series No. 64). World Health Organization. https://www.who.int/publications/i/item/9789241660648
39. WHO (World Health Organization). 2011. Safety evaluation of certain contaminants in food: prepared by the seventy-second meeting of the Joint FAO/WHO Expert Committee on Food Additives (JECFA) (WHO Food Additives Series No. 63). World Health Organization. https://www.who.int/publications/i/item/9789241660631
40. Agroknow. Use AI to prevent food recalls. https://agroknow.com/ (accessed April 2025)

© Burleigh Dodds Science Publishing Limited, 2025. All rights reserved.

41. World Health Organization. 2010. WHO human health risk assessment toolkit: chemical hazards (IPCS harmonization project document; no. 8 https://apps.who.int/iris/bitstream/handle/10665/44458/9789241548076_eng.pdf;jsessionid=06F72D6403A31B82B28A6AF5AEB03907?sequence=1 (accessed 9 April 2025)
42. US EPA. United States Environmental Protection Agency, Office of the Science Advisor, Risk Assessment Forum. 2007. Framework for metals risk assessment, EPA 120/R-07/001. https://www.epa.gov/sites/default/files/2013-09/documents/metals-risk-assessment-final.pdf (accessed 19 April 2025)
43. IARC. International Agency for Research on Cancer. 1993. IARC Monographs on the Evaluation of Carcinogenic Risks to Humans. Beryllium, Cadmium, Mercury and Exposures in the Glass Manufacturing Industry. Volume 58. https://publications.iarc.fr/Book-And-Report-Series/Iarc-Monographs-On-The-Identification-Of-Carcinogenic-Hazards-To-Humans/Beryllium-Cadmium-Mercury-And-Exposures-In-The-Glass-Manufacturing-Industry-1993 (accessed 8 March 2025)
44. IARC. International Agency for Research on Cancer. 2006. IARC Monographs on the Evaluation of Carcinogenic Risks to Humans. Inorganic and Organic Lead Compounds. Volume 87. https://www.inchem.org/documents/iarc/vol87/volume87.pdf (accessed 8 March 2025)
45. IARC. International Agency for Research on Cancer. 2012. IARC Monographs on the Evaluation of Carcinogenic Risks to Humans. Arsenic, metals, fibres, and dusts, Volume 100 C. A Review of Human Carcinogens. https://publications.iarc.fr/Book-And-Report-Series/Iarc-Monographs-On-The-Identification-Of-Carcinogenic-Hazards-To-Humans/Arsenic-Metals-Fibres-And-Dusts-2012 (accessed 8 March 2025)
46. EFSA. European Food Safety Authority. 2008. Scientific Opinion of the Panel on Food Additives, Flavourings, Processing Aids and Food Contact Materials on a request from European Commission on Safety of aluminium from dietary intake. *The EFSA Journal*, 754, 1–34.
47. EFSA. European Food Safety Authority. 2009a. Scientific opinion of the panel on contaminants in the food chain on a request from the European Commission on cadmium in food. *The EFSA Journal*, 980, 1–139. https://efsa.onlinelibrary.wiley.com/doi/epdf/10.2903/j.efsa.2009.980 (accessed 8 March 2025)
48. EFSA. European Food Safety Authority. 2009b. EFSA Panel on Contaminants in the Food Chain (CONTAM); scientific Opinion on Arsenic in Food. *EFSA Journal*, 7(10), 1351. https://doi.org/10.2903/j.efsa.2009.1351
49. EFSA. European Food Safety Authority. 2024. EFSA Panel on Contaminants in the Food Chain (CONTAM); update of the risk assessment of inorganic arsenic in food. *EFSA Journal*, 22, e8488. https://doi.org/10.2903/j.efsa.2024.8488
50. EFSA. European Food Safety Authority. 2012. EFSA Panel on Contaminants in the Food Chain (CONTAM); scientific opinion on the risk for public health related to the presence of mercury and methylmercury in food. *EFSA Journal*, 10(12), 2985. https://doi.org/10.2903/j.efsa.2012.2985
51. EFSA. European Food Safety Authority, et al. 2020. Scientific Opinion on the update of the risk assessment of nickel in food and drinking water. *EFSA Journal*, 18(11), 6268. https://doi.org/10.2903/j.efsa.2020.6268
52. FAO and WHO. Food and Agriculture Organization of the United Nations (FAO), World Health Organization (WHO). 2023. Codex Alimentarius International Food Standards. General Standard for Contaminants and Toxins in Food and Feed CXS

© Burleigh Dodds Science Publishing Limited, 2025. All rights reserved.

193-1995. Adopted in 1995. Revised in 1997, 2006, 2008, 2009. Amended in 2010, 2012, 2013, 2014, 2015, 2016, 2017, 2018, 2019, 2021, 2022, 2023. https://www.fao.org/fao-who-codexalimentarius/sh-proxy/fr/?lnk=1&url=https%253A%252F%252Fworkspace.fao.org%252Fsites%252Fcodex%252FStandards%252FCXS%2B193-1995%252FCXS_193e.pdf (accessed 8 March 2025)

53. EFSA. European Food Safety Authority. 2005. Opinion of the scientific committee on a request from EFSA related to a harmonised approach for risk assessment of substances which are both genotoxic and carcinogenic. *EFSA Journal*, 3(10), 282. https://doi.org/doi.org/10.2903/j.efsa.2005.282
54. European Commission. 2023. Commission Regulation (EU) 2023/915 of 25 April 2023 on maximum levels for certain contaminants in food and repealing Regulation (EC) No 1881/2006 (OJ L 119, 5.5.2023, pp. 103–157).
55. Nordberg, G.F., et al. 2014. *Handbook on the Toxicology of Metals*, Academic Press, Cambridge.
56. Cattaneo, I., et al. 2023. Implementing New Approach Methodologies (NAMs) in food safety assessments: strategic objectives and actions taken by the European Food Safety Authority. *Trends in Food Science & Technology*, 133, 277–290. https://doi.org/10.1016/j.tifs.2023.02.006
57. US EPA, United States Environmental Protection Agency. 2021. New approach methods work plan (v2). Washington, DC. EPA/600/X-21/209. New Approach Methods Work Plan.
58. US FDA, United States Food and Drug Administration. 2021. Advancing new alternatives methodologies at FDA. Advancing New Alternative Methodologies at FDA January 2021.
59. Bessa, M.J., et al. 2024. Toxicity Assessment of Food and By-Products: In Vitro Methods, Springer Protocols, Series: Methods and Protocols in Food Science. Extraction, Characterization, and Functional Assessment of Bioactive Compounds. https://doi.org/10.1007/978-1-0716-3942-9_17
60. Costa, L.G., et al. 2007. An in vitro approach to assess the toxicity of certain food contaminants: methylmercury and polychlorinated biphenyls. *Toxicology*, 237(1–3), 65–76. https://doi.org/10.1016/j.tox.2007.05.003
61. Manful, M.E., et al. 2023. New Approach Methodologies (NAMs) for safety testing of complex food matrices: a review of status, considerations, and regulatory adoption. *Trends in Food Science & Technology*, 142, 104191. https://doi.org/10.1016/j.tifs.2023.104191
62. US EPA. United States Environmental Protection Agency, Accelerating the pace of chemical risk assessment (APCRA). www.epa.gov/chemical-research/accelerating-pace-chemical-risk-assessment-apcra (accessed 4 May 2025)
63. von Hellfeld, R., et al. 2023. Considerations for future quantitative structure-activity relationship (QSAR) modelling for heavy metals: a case study of mercury. *Toxicology*, 499, 153661. https://doi.org/10.1016/j.tox.2023.153661
64. Zhang, Y., et al. 2023. Health risk assessment of cadmium exposure by integration of an in silico physiologically based toxicokinetic model and in vitro tests. *Journal of Hazardous Materials*, 443(Part A), 130191. https://doi.org/10.1016/j.jhazmat.2022.130191
65. Nordberg, G.F. and Fowler, B.A. 2018. *Risk Assessment for Human Metal Exposures, Mode of Action and Kinetic Approaches*, Academic Press, Cambridge.

© Burleigh Dodds Science Publishing Limited, 2025. All rights reserved.

66. Yang, D., et al. 2023. Advancing food safety risk assessment in China: development of new approach methodologies (NAMs). *Frontiers in Toxicology*, 5(Regulatory Toxicology). https://doi.org/10.3389/ftox.2023.1292373
67. Grant, R.L., et al. 2010. Experimental Models for the Investigation of Toxicological Mechanisms. In McQueen, C.A. (ed.). *Comprehensive Toxicology* (2nd ed.). Oxford, Elsevier, p. 204.
68. Balali-Mood, M., et al. 2021. Toxic mechanisms of five heavy metals: mercury, lead, chromium, cadmium, and arsenic. *Frontiers in Toxicology*, 12(Predictive Toxicology). https://doi.org/10.3389/fphar.2021.643972
69. Jomova, K., et al. 2024. Heavy metals: toxicity and human health effects. *Archives of Toxicology*, 99, 153–209. https://doi.org/10.1007/s00204-024-03903-2
70. Zhang, S., et al. 2020. Adverse impact of heavy metals on bone cells and bone metabolism dependently and independently through anemia. *Advanced Science*, 7, 2000383. https://doi.org/10.1002/advs.202000383
71. OECD, Organisation for Economic Co-operation and Development. 2017. Revised guidance document on developing and assessing adverse outcome pathways, series on testing & assessment No. 184. https://one.oecd.org/document/env/jm/mono(2013)6/en/pdf (accessed 3 May 2025)
72. Carusi, A., et al. 2022. Adverse Outcome Pathway – Study Report. In Wittwehr C. and Whelan, M. (eds.). *EUR 30925 EN*, Publications Office of the European Union, Luxembourg. https://doi.org/10.2760/476003
73. Yang, L., et al. 2022. Elucidating the differences in metal toxicity by quantitative adverse outcome pathways. *Environmental Science & Technology*, 56, 13233–13244. https://doi.org/10.1021/acs.est.2c03828
74. Angrish, M., et al. 2018. Epigenetic applications in adverse outcome pathways and environmental risk evaluation. *Environmental Health Perspectives*, 126(4), 045001. https://doi.org/10.1289/EHP2322
75. Brix, K.V., et al. 2017. The mechanisms of nickel toxicity in aquatic environments: an adverse outcome pathway analysis. *Reviews of Environmental Contamination and Toxicology*, 36(5), 1128–1137. https://doi.org/10.1002/etc.3706
76. Chai, Z., et al. 2021. Generating adverse outcome pathway (AOP) of inorganic arsenic-induced adult male reproductive impairment via integration of phenotypic analysis in comparative toxicogenomics database (CTD) and AOP wiki. *Toxicology and Applied Pharmacology*, 15, 115370. https://doi.org/10.1016/j.taap.2020.115370
77. Clewell, H.J., et al. 2018. Application of the adverse outcome pathway (AOP) approach to inform mode of action (MOA): a case study with inorganic arsenic. *Journal of Toxicology and Environmental Health, Part A*, 81(18), 893–912. https://doi.org/10.1080/15287394.2018.1500326
78. Hadjadji, C., et al. 2025. Evidence linking cadmium and/or lead exposure to immunomodulatory effects in mammals based upon an adverse outcome pathways approach, and research perspectives. *Chemosphere*, 371, 144056. https://doi.org/10.1016/j.chemosphere.2024.144056
79. Hooper, M.J., et al. 2013. Interactions between chemical and climate stressors: a role for mechanistic toxicology in assessing climate change risks. *Environmental Toxicology and Chemistry*, 32(1), 32–48. https://doi.org/10.1002/etc.2043


© Burleigh Dodds Science Publishing Limited, 2025. All rights reserved.

80. Ito, S., et al. 2024. Proof of concept for quantitative adverse outcome pathway modeling of chronic toxicity in repeated exposure. *Scientifc Reports*, 14, 4741. https://doi.org/10.1038/s41598-024-55220-4
81. Kang, B., et al. 2024. Mercury-induced toxicity: mechanisms, molecular pathways, and gene regulation. *Science of the Total Environment*, 943, 173577. https://doi.org/10.1016/j.scitotenv.2024.173577
82. Klose, J., et al. 2023. Application of the adverse outcome pathway concept for investigating developmental neurotoxicity potential of Chinese herbal medicines by using human neural progenitor cells in vitro. *Cell Biology and Toxicology*, 39, 319–343. https://doi.org/10.1007/s10565-022-09730-4
83. Sahoo, A.K., et al. 2024. An integrative data-centric approach to derivation and characterization of an adverse outcome pathway network for cadmium-induced toxicity. *Science of the Total Environment*, 920, 170968. https://doi.org/10.1016/j.scitotenv.2024.170968
84. Yokel, R.A. 2020. Aluminum reproductive toxicity: a summary and interpretation of scientific reports. *Critical Reviews in Toxicology*, 50(7), 551–593. https://doi.org/10.1080/10408444.2020.1801575
85. Zilliacus, J., et al. 2024. Building an adverse outcome pathway network for estrogen-, androgen- and steroidogenesis-mediated reproductive toxicity. *Frontiers in Toxicology*, 6(Regulatory Toxicology). https://doi.org/10.3389/ftox.2024.1357717
86. Barnes, D.A., et al. 2024. Development of an adverse outcome pathway network for nephrotoxicity. *Archives of Toxicology*, 98, 929–942. https://doi.org/10.1007/s00204-023-03637-7
87. Wang, Y., et al. 2024. Investigating the relationship of co-exposure to multiple metals with chronic kidney disease: an integrated perspective from epidemiology and adverse outcome pathways. *Journal of Hazardous Materials*, 480, 135844. https://doi.org/10.1016/j.jhazmat.2024.135844
88. BfR, German Federal Institute for Risk Assessment, Department Food Safety, Berlin, Germany, et al. 2018. Identification and evaluation of potentially mutagenic and carcinogenic food contaminants. *EFSA Journal*, 16(8), e16085. https://doi.org/10.2903/j.efsa.2018.e16085
89. Soares, T.A., et al. 2022. The (Re)-Evolution of Quantitative Structure–Activity Relationship (QSAR) studies propelled by the surge of machine learning methods. *Journal of Chemical Information and Modeling*, 62(22), 5317–5320. https://doi.org/10.1021/acs.jcim.2c01422
90. Joshi, N.C., et al. 2024. A review on metal/metal oxide nanoparticles in food processing and packaging. *Food Science and Biotechnology*, 33, 1307–1322. https://doi.org/10.1007/s10068-023-01500-0
91. Sang, L., et al. 2022. Machine learning for evaluating the cytotoxicity of mixtures of Nano-TiO2 and heavy metals: QSAR model apply random forest algorithm after clustering analysis. *Molecules*, 27, 6125. https://doi.org/10.3390/molecules27186125
92. Huang, Y., et al. 2020. Quantitative structure–activity relationship models for predicting inflammatory potential of metal oxide nanoparticles. *Environmental Health Perspectives*, 128(6), 067010. https://doi.org/10.1289/EHP6508
93. Gakis, G.P., et al. 2023. A structure–activity approach towards the toxicity assessment of multicomponent metal oxide nanomaterials. *Nanoscale*, 15, 16432. https://doi.org/10.1039/d3nr03174h


© Burleigh Dodds Science Publishing Limited, 2025. All rights reserved.

94. Martinez-Morata, I., et al. 2023. A state-of-the-science review on metal biomarkers. *Current Environmental Health Reports*, 10(3), 215–249. https://doi.org/10.1007/s40572-023-00402-x
95. Santonen, T., et al. 2022. Biological Monitoring of Metals and Biomarkers. In *Handbook on the Toxicology of Metals* (5th ed.), Volume I: General Considerations, pp. 217–235. https://doi.org/10.1016/B978-0-12-823292-7.00007-3
96. Rodríguez-Carrillo, A., et al. 2023. Implementation of effect biomarkers in human biomonitoring studies: a systematic approach synergizing toxicological and epidemiological knowledge. *International Journal of Hygiene and Environmental Health*, 249, 114140. https://doi.org/10.1016/j.ijheh.2023.114140
97. Apel, P., et al. 2020. Human biomonitoring initiative (HBM4EU) – Strategy to derive human biomonitoring guidance values (HBM-GVs) for health risk assessment. https://doi.org/ 10.1016/j.ijheh.2020.113622
98. ACGIH, American Conference of Governmental Industrial Hygienists, Threshold Limit Values (TLVs®) and Biological Exposure Indices (BEIs®). https://www.acgih.org/science/tlv-bei-guidelines/ (accessed 10 May 2025)
99. HSE, Health and Safety Executive Science and Research Centre, Biological Monitoring Guidance Values, BM Guidance Values.
100. Khan-Mayberry, N. 2013. Opinion: challenges and safeguards - filling the gaps in toxicological research of heavy metals. *Journal of Clinical Toxicology*, 3(5), 1000e121. https://doi.org/10.4172/2161-0495.1000e121
101. Laitinen, H.A. 1977. History of trace analysis. *Journal of research of the National Bureau of Standards*, 93(3), 175–85. https://doi.org/10.6028/jres.093.013
102. Colorado State University. A brief history of inorganic classical analysis. https://diverdi.colostate.edu/all_courses/brief%20history%20of%20inorganic%20classical%20analysis.pdf (accessed April 2025)
103. Van Grieken, R.E. and Markowicz, A.A. 2002. *Handbook of X-Ray Spectrometry*, Marcel Dekker, Inc, New York.
104. Beckhoff, B., et al. 2006. *Handbook of Practical X-Ray Fluorescence Analysis*, Springer, Berlin.
105. Tsuji, K., et al. 2004. *X-Ray Spectrometry: Recent Technological Advances*, Wiley, Chichester.
106. Klockenkämper, R. and von Bohlen, A. 2015. *Total-Reflection X-Ray Fluorescence Analysis and Related Methods*, John Wiley & Sons, Inc., Hoboken, 2nd edn. https://doi.org/10.1002/9781118985953
107. Szoboszlai, N., et al. 2009. Recent trends in total reflection X-ray fluorescence spectrometry for biological applications, *Analytica Chimica Acta*, 633, 1–18, https://doi.org/10.1016/j.aca.2008.11.009
108. Fernández-Ruiz, R. 2025. Bioanalytical application of the total-reflection X-ray fluorescence spectrometry. *International Journal of Molecular Sciences*, 26(3), 1049. https://doi.org/10.3390/ijms26031049
109. Borgese, L., et al. 2015. Total reflection X-ray fluorescence as a tool for food screening. Spectrochimica Acta Part B, 113, 1–15. https://doi.org/10.1016/j.sab.2015.08.001
110. De La Calle, I., et al. 2013. Sample pretreatment strategies for total reflection X-ray fluorescence analysis: a tutorial review. *Spectrochimica Acta Part B*, 90, 23–54. https://doi.org/10.1016/ j.sab.2013.10.001


© Burleigh Dodds Science Publishing Limited, 2025. All rights reserved.

111. Marguí, E., et al. 2022. Potential of total-reflection X-ray spectrometry for multielement analysis of biological samples using dilution or suspension sample preparation techniques. *X-Ray Spectrometry*, 51, 230–240. https://doi.org/10.1002/xrs.3230
112. Fernández-Ruiz, R. 2025. A brief revision of the suspension-assisted direct solid analysis by TXRF spectrometry. *Spectrochimica Acta Part B: Atomic Spectroscopy*, 229, 107214. https://doi.org/10.1016/j.sab.2025.107214
113. Marguí, E., et al. 2024. An overview of the applications of total reflection X-ray fluorescence spectrometry in food, cosmetics, and pharmaceutical research. *Journal of Analytical Atomic Spectrometry*, 39, 1700. https://doi.org/10.1039/d4ja00096j
114. Vives, A., et al. 2006. Synchrotron radiation total reflection X-ray fluorescence (SR-TXRF) for evaluation of food contamination. *Journal of Radioanalytical and Nuclear Chemistry*, 270, 147–153. https://doi.org/10.1007/s10967-006-0322-3
115. von Bohlen, A. and Fernández-Ruiz, R. 2020. Experimental evidence of matrix effects in Total-reflection X-ray Fluorescence analysis: coke case. *Talanta*, 209, 120562. https://doi.org/10.1016/j.talanta.2019.120562
116. Perring, L. and Blanc, J. 2008. Validation of quick measurement of mineral nutrients in milk powders: comparison of energy dispersive X-ray fluorescence with inductively coupled plasma-optical emission spectroscopy and potentiometry reference methods. *Sensing and Instrumentation for Food Quality and Safety*, 2, 254–261.
117. Perring, L. and Blanc, J. 2008. Faster measurement of minerals in milk powders: comparison of a high power wavelength dispersive XRF system with ICP-AES and potentiometry reference methods. *Food Analytical Methods*, 1, 205–213. https://doi.org/10.1007/s12161-008-9030-7
118. Gutiérrez-Ginés, M.J., et al. 2013. Assessment of field portable X-ray fluorescence spectrometry for the in situ determination of heavy metals in soils and plants. *Environmental Science: Processes & Impacts*, 15(8), 1545. https://doi.org/10.1039/c3em00078h
119. Jackson, B. and Punshon, T. 2015. Recent advances in the measurement of arsenic, cadmium, and mercury in rice and other foods. *Current Environmental Health Reports*, 2(1), 15–24. https://doi.org/10.1007/s40572-014-0035-7
120. Beltrán, B.G., et al. 2020. Total Reflection X-Ray Fluorescence Spectroscopy (TXRF) method validation: determination of heavy metals in dietary supplements. *Journal of Chemistry*, Article ID 8817393. https://doi.org/10.1155/2020/8817393
121. Marguí, E., et al. 2013. Determination of cadmium at ultra-trace levels in environmental water samples by means of total reflection X-ray spectrometry after dispersive liquid-liquid microextraction. *Journal of Analytical Atomic Spectrometry*, 28, 266. https://pubs.rsc.org/en/content/articlelanding/2013/ja/c2ja30252g (accessed 10 May 2025)
122. Byers, H.L., et al. 2018. XRF techniques to quantify heavy metals in vegetables at low detection limits. *Food Chemistry X*, 1, 100001. https://doi.org/10.1016/j.fochx.2018.100001
123. Moya-Riffo, L.B., et al. 2013. A procedure for overlapping deconvolution and the determination of its confidence interval for arsenic and lead signals in TXRF spectral analysis. *X-Ray Spectrometry*, 42, 93–99. https://doi.org/10.1002/xrs.2439


© Burleigh Dodds Science Publishing Limited, 2025. All rights reserved.

124. Contaminants in Water. www.bruker.com/es/applications/detection-and-environmental/environmental/water-analysis-environmental/contaminants-in-water.html (accessed 10 May 2025)
125. Musielak, M., et al. 2022. Ultrasensitive and selective determination of mercury in water, beverages and food samples by EDXRF and TXRF using graphene oxide modified with thiosemicarbazone, *Food Chemistry*, 390, 133136. https://doi.org/10.1016/j.foodchem.2022.133136
126. Castillejos-Mijangos, L.A., et al. 2022. Uses of FT-MIR spectroscopy and multivariate analysis in quality control of coffee, cocoa, and commercially important spices. *Foods*, 11, 579.
127. Yadav, R.P. and Tarun, G. 2017. Versatility of turmeric: a review the golden spice of life. *Journal of Pharmacognosy and Phytochemistry*, 6, 41–46.
128. Rajain, P. and Rathee, R. 2019. Role colour plays in influencing consumer behaviour. *International Research Journal of Business Studies*, 12, 209–222.
129. Johnson, R. Food fraud and "economically motivated adulteration" of food and food ingredients. www.sgp.fas.org/crs/misc/R43358.pdf (accessed 24 April 2025)
130. Hay, T., et al. 2022. Review of edible australian flora for colour and flavour additives: appraisal of suitability and ethicality for Bushfoods as natural additives to facilitate new industry growth. *Trends in Food Science and Technology*, 129, 74–87.
131. Tayyem, R.F., et al. 2006. Curcumin content of turmeric and curry powders. *Nutrition and Cancer*, 55, 126–131. https://doi.org/10.1207/s15327914nc5502_2
132. European Commission. 2021. Joint Research Centre. Results of an EU Wide Coordinated Control Plan to Establish the Prevalence of Fraudulent Practices in the Marketing of Herbs and Spices; Publications Office: LU.
133. Shannon, M., et al. 2022. The detection and determination of adulterants in turmeric using Fourier-Transform Infrared (FTIR) spectroscopy coupled to chemometric analysis and micro-FTIR imaging. *Food Control*, 139, 109093. https://doi.org/10.1016/j.foodcont.2022.109093
134. Forsyth, J.E., et al. 2019. Turmeric means "yellow" in Bengali: lead chromate pigments added to turmeric threaten public health across Bangladesh. *Environmental Research*, 179, 108722. https://doi.org/10.1016/j.envres.2019.108722
135. Forsyth, J.E. 2024. Evidence of turmeric adulteration with lead chromate across South Asia. *Science of the Total Environment*, 949, 175003. https://doi.org/10.1016/j.scitotenv.2024.175003
136. IARC Working Group on the Evaluation of Carcinogenic Risks to Humans. 1990. *Chromium, Nickel and Welding*, International Agency for Research on Cancer, Lyon, France.
137. Abia, W.A. 2025. Africa, an emerging exporter of turmeric: combating fraud with rapid detection systems. *Foods*, 14(9), 1590. https://doi.org/10.3390/foods14091590
138. Craig, A.W. 1976. IARC monographs on the evaluation of carcinogenic risk of chemicals to man Vol. 8. Some aromatic azo compounds (1975). *British Journal of Cancer*, 33, 572.
139. Forsyth, J.E., et al. 2019. Sources of blood lead exposure in Rural Bangladesh. *Environmental Science & Technology*, 53, 11429–11436. https://doi.org/10.1021/acs.est.9b00744
140. Ishida, M.L., et al. 2022. Regulatory policies for heavy metals in spices–a New York approach. *Journal of Regulatory Science*, 10, 1–12.


© Burleigh Dodds Science Publishing Limited, 2025. All rights reserved.

141. Commission Regulation (EU) 2023/915 of 25 April 2023 on maximum levels for certain contaminants in food and repealing Regulation (EC) No 1881/2006.
142. Wilbur, S., et al. 2012. *Toxicological Profile for Chromium; Agency for Toxic Substances and Disease Registry (ATSDR) Toxicological Profiles*, Agency for Toxic Substances and Disease Registry (US), Atlanta.
143. Chromium: Toxicological Overview. GOV.UK. https://www.gov.uk/government/publications/chromium-general-information-incident-management-and-toxicology/chromium-toxicological-overview (accessed 27 February 2025).
144. International Programme on Chemical Safety & Inter-Organization Programme for the Sound Management of Chemicals. 2013. *Inorganic Chromium(VI) Compounds*, World Health Organization, Geneva.
145. Lopez, A.M., et al. 2022. Assessing analytical methods for the rapid detection of lead adulteration in the global spice market. *Environmental Science & Technology*, 56, 16996–17006. https://doi.org/10.1021/acs.est.2c03241
146. Di Anibal, C.V., et al. 2011. High-resolution 1H nuclear magnetic resonance spectrometry combined with chemometric treatment to identify adulteration of culinary spices with Sudan Dyes. *Food Chemistry*, 124, 1139–1145. https://doi.org/10.1016/j.foodchem.2010.07.025
147. Forsyth, J.E., et al. 2023. Food safety policy enforcement and associated actions reduce lead chromate adulteration in turmeric across Bangladesh. *Environmental Research*, 232, 116328. https://doi.org/10.1016/j.envres.2023.116328
148. Collin, M.S., et al. 2022. Bioaccumulation of lead (Pb) and its effects on human: a review. Journal of Hazardous Materials Advances*Journal of Hazardous Materials Advances*, 7, 100094.
149. Hossini, H., et al. 2022. A comprehensive review on human health effects of chromium: insights on induced toxicity. *Environmental Science and Pollution Research*, 29, 70686–70705. https://doi.org/10.1007/s11356-022-22705-6
150. Witkowska, D., et al. 2021. Heavy metals and human health: possible exposure pathways and the competition for protein binding sites. *Molecules*, 26, 6060.
151. CDC, CDC Updates Blood Lead Reference Value. Childhood Lead Poisoning Prevention. https://www.cdc.gov/lead-prevention/php/news-features/updates-blood-lead-reference-value.html (accessed 27 February 2025)
152. Lanphear, B.P., et al. 2016. Prevention of childhood lead toxicity. *Pediatrics*, 138, e20161493.
153. Lanphear, B.P. 2007. The conquest of lead poisoning: a pyrrhic victory. *Environmental Health Perspectives*, 115, A484–A485. https://doi.org/10.1289/ehp.10871
154. Marshall, A.T., et al. 2020. Association of lead-exposure risk and family income with childhood brain outcomes. *Nature Medicne*, 26, 91–97.
155. Ruckart, P.Z. 2021. Update of the blood lead reference value–United States, 2021. *MMWR. Morbidity and Mortality Weekly Report*, 70, 1509–1512.
156. Abbasi, A., et al. 2021. Highly selective and sensitive benzimidazole based bifunctional sensor for targeting inedible azo dyes in red chilli, red food color, turmeric powder, and Cu(Ii) in coconut water. *Journal of Fluorescence*, 31, 1353–1361. https://doi.org/10.1007/s10895-021-02766-5
157. Bellinger, D.C. 2008. Neurological and behavioral consequences of childhood lead exposure. *PLoS Medicne*, 5, e115.

© Burleigh Dodds Science Publishing Limited, 2025. All rights reserved.

158. Gleason, K., et al. 2014. Contaminated turmeric is a potential source of lead exposure for children in Rural Bangladesh. *Journal of Environmental and Public Health*, 2014, 1–5. https://doi.org/10.1155/2014/730636
159. Brown, M.J., et al. 2022. Prevalence of elevated blood lead levels and risk factors among children living in Patna, Bihar, India 2020. *PLoS Global Public Health*, 2, e0000743. https://doi.org/10.1371/journal.pgph.0000743
160. Cowell, W., et al. 2017. Ground turmeric as a source of lead exposure in the United States. *Public Health Reports*, 132, 289–293. https://doi.org/10.1177/0033354917700109
161. Becker, F., et al. 2022. Tracking the source of contaminant lead in children's blood. *Environmental Research*, 212, 113307. https://doi.org/10.1016/j.envres.2022.113307
162. Kappel, M. 2021. Notes from the field: childhood lead poisoning associated with turmeric spices–Las Vegas, 2019. *MMWR. Morbidity and Mortality Weekly Report*, 70, 1584–1585.
163. Cheng, L., et al. 2017. Fetal exposure to lead during pregnancy and the risk of preterm and early-term deliveries. *International Journal of Hygiene and Environmental Health*, 220, 984–989.
164. Gundacker, C. and Hengstschläger, M. 2012. The role of the placenta in fetal exposure to heavy metals. *Wiener Medizinische Wochenschrift*, 162, 201–206.
165. Rísová, V. 2019. The pathway of lead through the mother's body to the child. *Interdisciplinary Toxicology*, 12, 1–6. https://doi.org/10.2478/intox-2019-0001
166. Santana, A.B., et al. 2023. Gestational lead exposure and its effects on fetal/infant development–a systematic review. *Reproductive Toxicology*, 117, 108342. https://doi.org/10.1016/j.reprotox.2023.108342
167. Erasmus, S.W., et al. 2021. Real or fake yellow in the vibrant colour craze: rapid detection of lead chromate in turmeric. *Food Control*, 121, 107714. https://doi.org/10.1016/j.foodcont.2020.107714
168. Angelon-Gaetz, K.A., et al. 2023. Lead levels in spices from market basket and home lead investigation samples in North Carolina. *Public Health Reports*, 138, 91–96. https://doi.org/10.1177/00333549211066152
169. Esteki, M., et al. 2019. Tackling fraudsters with global strategies to expose fraud in the food chain. *Comprehensive Reviews in Food Science and Food Safety*, 18, 425–440. https://doi.org/10.1111/1541-4337.12419
170. Ericson, B., et al. 2020. Elevated levels of lead (Pb) identified in Georgian spices. *Annals of Global Health*, 86, 124. https://doi.org/10.5334/aogh.3044
171. Sasikumar, B. 2019. Advances in adulteration and authenticity testing of turmeric (Curcuma longa L.). *Journal of Spices and Aromatic Crops*, 28, 96–105.
172. Bansal, S., et al. 2017. Food adulteration: sources, health risks, and detection methods. *Critical Reviews in Food Science and Nutrition*, 57, 1174–1189. https://doi.org/10.1080/10408398.2014.967834
173. Das, C., et al. 2020. Comparative Study for the Impedimetric Detection and Quantification of Adulterants in Different Bio-Consumables. In Proceedings of the 2020 International Symposium on Devices, Circuits and Systems (ISDCS), Howrah, India, 4–6 March 2020, IEEE, New York, pp. 1–6.
174. Haji, A., et al. 2023. Selected food items adulteration, their impacts on public health, and detection methods: a review. *Food Science & Nutrition*, 11, 7534–7545. https://doi.org/10.1002/fsn3.3732


© Burleigh Dodds Science Publishing Limited, 2025. All rights reserved.

175. Negi, A., et al. 2021. Emerging techniques for adulterant authentication in spices and spice products. *Food Control*, 127, 108113. https://doi.org/10.1016/j.foodcont.2021.108113
176. Tiwari, M., et al. 2013. Laser-induced breakdown spectroscopy: an approach to detect adulteration in turmeric. *Spectroscopy Letters | JournalSpectroscopy Letters*, 46, 155–159. https://doi.org/10.1080/00387010.2012.702707
177. Khongkaew, P., et al. 2020. Quantitative models for detecting the presence of lead in turmeric using raman spectroscopy. *Chemometrics and Intelligent Laboratory Systems*, 200, 103994. https://doi.org/10.1016/j.chemolab.2020.103994
178. Paranthaman, R., et al. 2021. Development of a method for qualitative detection of lead chromate adulteration in turmeric powder using X-ray powder diffraction. *Food Control*, 126, 107992. https://doi.org/10.1016/j.foodcont.2021.107992
179. Alam, M.N.E., et al. 2023. Lead exposure of four biologically important common branded and non-branded spices: relative analysis and health implication. *Biological Trace Element Research*, 201, 4972–4984. https://doi.org/10.1007/s12011-022-03553-4
180. Cloutis, E., et al. 2016. Spectral reflectance (350–2500 Nm) properties of historic artists' pigments. II. Red–Orange–Yellow Chromates, Jarosites, Organics, Lead(–Tin) Oxides, Sulphides, Nitrites and Antimonates. *Journal of Near Infrared Spectroscopy*, 24, 119–140. https://doi.org/10.1255/jnirs.1207
181. Soliman, S.S., et al. 2022. Experimentally designed chemometric models for the assay of toxic adulterants in turmeric powder. *RSC Advances*, 12, 9087–9094. https://doi.org/10.1039/D2RA00697A
182. Neng, J., et al. 2020. Application of surface-enhanced raman spectroscopy in fast detection of toxic and harmful substances in food. *Biosensors and Bioelectronics*, 167, 112480.
183. Kumar, S., et al. 2022. Lead (Pb) contamination in agricultural products and human health risk assessment in Bangladesh. *Water, Air, & Soil Pollution*, 233, 257. https://doi.org/10.1007/s11270-022-05711-9
184. Xu, Y., et al. 2023. Recent trends of multi-source and non-destructive information for quality authentication of herbs and spices. *Food Chemistry*, 398, 133939. https://doi.org/10.1016/j.foodchem.2022.133939
185. Wilschefski, S. and Baxter, M. 2019. Inductively coupled plasma mass spectrometry: introduction to analytical aspects. *CBR*, 40, 115–133. https://doi.org/10.33176/AACB-19-00024
186. McIntosh, K.G., et al. 2016. Evaluation of portable XRF instrumentation for assessing potential environmental exposure to toxic elements. *International Journal of Environmental Analytical Chemistry*, 96, 15–37. https://doi.org/10.1080/03067319.2015.1114104
187. Nordin, N. and Selamat, J. 2013. Heavy metals in spices and herbs from wholesale markets in Malaysia. *Food Additives & Contaminants: Part B*, 6, 36–41. https://doi.org/10.1080/19393210.2012.721140
188. Lai, L., et al. 2023. Recent progress on fluorescent probes in heavy metal determinations for food safety: a review. *Molecules*, 28, 5689. https://doi.org/10.3390/molecules28155689
189. Shim, J., et al. 2019. Heavy metals in spices commonly consumed in Republic of Korea. *Food Additives & Contaminants: Part B*, 12, 52–58. https://doi.org/10.1080/19393210.2018.1546772


© Burleigh Dodds Science Publishing Limited, 2025. All rights reserved.

190. Zeiner, M., et al. 2022. Multielement determination in turmeric (Curcuma longa L.) using different digestion methods. *Molecules*, 27, 8392. https://doi.org/10.3390/molecules27238392
191. The Food Fortress–From Crisis To An Innovative Food Quality Assurance Scheme|Case Studies|Research|Queen's University Belfast. www.qub.ac.uk/Research/case-studies/leading-feed-quality-assurance-scheme.html (accessed 14 April 2025)
192. Bansod, B., et al. 2017. A review on various electrochemical techniques for heavy metal ions detection with different sensing platforms. *Biosensors and Bioelectronics*, 94, 443–4455. https://doi.org/10.1016/j.bios.2017.03.031
193. Meng, R., et al. 2023. The innovative and accurate detection of heavy metals in foods: a critical review on electrochemical sensors. *Food Control*, 150, 109743. https://doi.org/10.1016/j.foodcont.2023.109743
194. Crapnell, R.D. and Banks, C.E. 2024. Electroanalytical overview: screen-printed electrochemical sensing platforms. *ChemElectroChem*, 11, e202400370. https://doi.org/10.1002/celc.202400370
195. Rubino, A. and Queirós, R. 2023. Electrochemical determination of heavy metal ions applying screen-printed electrodes based sensors: a review on water and environmental samples analysis. *Talanta Open*, 7, 100203. https://doi.org/10.1016/j.talo.2023.100203
196. Liu, X., et al. 2019. Recent advances in nanomaterial-enabled screen-printed electrochemical sensors for heavy metal detection. *Trends in Analytical Chemistry*, 115, 187e202. https://doi.org/10.1016/j.trac.2019.03.021
197. Mehmeti, E., et al. 2016. Electrochemical sensors for the simultaneous determination of zinc, cadmium and lead using a Nafion/ionic liquid/graphene composite modified screen-printed carbon electrode. *Analytica Chimica Acta*, 918, 26–34. https://doi.org/10.1016/j.aca.2016.03.026
198. Palisoc, S., et al. 2019. Fabrication of a bismuth nanoparticle/Nafion modified screen-printed graphene electrode for in situ environmental monitoring. *Analytical Methods*, 11, 1591. https://doi.org/10.1039/c9ay00221a
199. Niu, P., et al. 2016. Screen-printed electrodes made of a bismuth nanoparticle porous carbon nanocomposite applied to the determination of heavy metal ions. *Microchimica Acta*, 183, 617–623. https://doi.org/10.1007/s00604-015-1684-4
200. Kanyong, P., et al. 2016. Gold nanoparticle modified screen-printed carbon arrays for the simultaneous electrochemical analysis of lead and copper in tap water. *Microchimica Acta*, 183, 2361–2368. https://doi.org/10.1007/s00604-016-1879-3
201. Torres-Rivero, K., et al. 2023. Functionalization of screen-printed electrodes with grape stalk waste extract-assisted synthesized silver and gold nanoparticles: perspectives of electrocatalytically enhanced determination of uranyl ion and other heavy metals ions. *Nanomaterials*, 13(6), 1055. https://doi.org/10.3390/nano13061055
202. Dong, Y., et al. 2021. Application of gold nanoparticles/polyaniline-multi-walled carbon nanotubes modified screen-printed carbon electrode for electrochemical sensing of zinc, lead, and copper. *Microchemical Journal*, 170, 106726.
203. Singh, D., et al. 2024. Chitosan-assisted self-assembly of flower-shaped ε-Fe2O3 nanoparticles on screen-printed carbon electrode for Impedimetric detection of Cd2+, Pb2+, and Hg2+ heavy metal ions in various water samples. *International*


© Burleigh Dodds Science Publishing Limited, 2025. All rights reserved.

*Journal of Biological Macromolecules*, 265(2), 130867. https://doi.org/10.1016/j.ijbiomac.2024.130867

204. Yogeeshwari, R.T., et al. 2022. Ultra-trace detection of toxic heavy metal ions using graphitic carbon functionalized Co3O4 modified screen-printed electrode. *Carbon Letters*, 32, 181–191. https://doi.org/10.1007/s42823-021-00265-y
205. Sanllorente-Méndez, S., et al. 2009. Determination of arsenic(III) using platinum nanoparticle-modified screen-printed carbon-based electrodes. *Electroanalysis*, 221(3–5), 635–639. https://doi.org/10.1002/elan.200804389
206. Teng, Y., et al. 2018. Screen graphene-printed electrode for trace cadmium detection in rice samples combing with portable potentiostat. *International Journal of Electrochemical Science*, 13(7), 6347–6357. https://doi.org/10.20964/2018.07.61
207. Wang, Z., et al. 2014. Electrochemical determination of lead and cadmium in rice by a disposable bismuth/electrochemically reduced graphene/ionic liquid composite modified screen-printed electrode. *Sensors and Actuators B: Chemical*, 199, 7–14. https://doi.org/10.1016/j.snb.2014.03.092
208. Chen, Y., et al. 2023. A sensitive electrochemical sensor based on 3D porous melamine-doped rGO/MXene composite aerogel for the detection of heavy metal ions in the environment. *Talanta*, 256, 124294. https://doi.org/10.1016/j.talanta.2023.124294
209. Noyrod, P., et al. 2010. Determination of trace heavy metals in herbs by sequential injection analysis-anodic stripping voltammetry using screen-printed carbon nanotubes electrodes. *Analytica Chimica Acta*, 668(1), 54–60. https://doi.org/10.1016/j.aca.2010.01.018
210. Pan, Y., et al. 2024. A target-triggered ultra-sensitive aptasensor for simultaneous detection of Cd2+ and Hg2+ using MWCNTs-Au NPs modified electrode. *Food Chemistry*, 440, 138185. https://doi.org/10.1016/j.foodchem.2023.138185
211. Jian, J.M., et al. 2013. Fast and sensitive detection of Pb2+ in foods using disposable screen-printed electrode modified by reduced graphene oxide. *Sensors*, 13(10), 13063–13075. https://doi.org/10.3390/s131013063
212. Pungjunun, K., et al. 2022. Smartphone-based electrochemical analysis integrated with NFC system for the voltammetric detection of heavy metals using a screen-printed graphene electrode. *Microchimica Acta*, 189, 191. https://doi.org/10.1007/s00604-022-05281-x
213. Rao, L., et al. 2024. Smartphone-based portable sensor with Bi-MOF nanocomposite for Cd (II) in vegetable samples. *Journal of Electroanalytical Chemistry*, 973, 118681. https://doi.org/10.1016/j.jelechem.2024.118681
214. Yao, Y., et al. 2019. Simultaneous determination of Cd(II) and Pb(II) ions in honey and milk samples using a single-walled carbon nanohorns modified screen-printed electrochemical sensor. *Food Chemistry*, 274, 8–15. https://doi.org/10.1016/j.foodchem.2018.08.110
215. Ping, J., et al. 2014. Development of an electrochemically reduced graphene oxide modified disposable bismuth film electrode and its application for stripping analysis of heavy metals in milk. *Food Chemistry*, 151, 65–71. https://doi.org/10.1016/j.foodchem.2013.11.026
216. Filik, H. and Avan, A.A. 2019. Dextran modified magnetic nanoparticles based solid phase extraction coupled with linear sweep voltammetry for the speciation of Cr(VI)

© Burleigh Dodds Science Publishing Limited, 2025. All rights reserved.

and Cr(III) in tea, coffee, and mineral water samples. *Food Chemistry*, 292, 151–159. https://doi.org/10.1016/j.foodchem.2019.04.058

217. Silva, L.R.G., et al. 2022. A simple and reliable electroanalytical method employing a disposable commercial electrode for simultaneous determination of lead(II) and mercury(II) in beer. *Journal of Food Composition and Analysis*, 110, 104564. https://doi.org/10.1016/j.jfca.2022.104564
218. Zhang, W., et al. 2017. A ZnO–RGO-modified electrode coupled to microwave digestion for the determination of trace cadmium and lead in six species fish. *Analytical Methods*, 9, 4418–4424. https://doi.org/10.1039/C7AY01574G
219. Huang, H., et al. 2024. A screen-printed carbon electrode modified with bismuth nanoparticles and conjugated mesoporous polymer for simultaneous determination of Pb(II) and Cd(II) in seafood samples. *Journal of Food Composition and Analysis*, 125, 105837. https://doi.org/10.1016/j.jfca.2023.105837
220. Huang, H., et al. 2024. Synthesis and application of bismuth nanoparticle-loaded longan porous carbon for simultaneous electrochemical determination of Pb(II) and Cd(II) in seafoods. *Food Chemistry*, 452, 139572. https://doi.org/10.1016/j.foodchem.2024.139572
221. Pungjunun, K., et al. 2020. Enhanced sensitivity and separation for simultaneous determination of tin and lead using paper-based sensors combined with a portable potentiostat. *Sensors and Actuators B: Chemical*, 318, 128241. https://doi.org/10.1016/j.snb.2020.128241
222. Choi, S.M., et al. 2015. A disposable chronocoulometric sensor for heavy metal ions using a diaminoterthiophene-modified electrode doped with graphene oxide. *Analytica Chimica Acta*, 892, 77–84. https://doi.org/10.1016/j.aca.2015.08.037
223. Fernández, E., et al. 2020. Magnetic dispersive solid-phase extraction using a zeolite-based composite for direct electrochemical determination of lead(II) in urine using screen-printed electrodes. *Microchimica Acta*, 187, 87. https://doi.org/10.1007/s00604-019-4062-9
224. Zhao, X., et al. 2019. Core-shell Self-doped polyaniline coated metal-organic-framework (SPAN@UIO-66-NH2) screen printed electrochemical sensor for Cd2+ ions. *Journal of the Electrochemical Society*, 166(12), B873–B880. https://doi.org/10.1149/2.0251912jes
225. Ince, B., et al. 2023. Lateral flow assays for food analyses: food contaminants, allergens, toxins, and beyond. *Trends in Analytical Chemistry*, 169, 117418. https://doi.org/10.1016/j.trac.2023.117418
226. Wu, Y., et al. 2021. Ensuring food safety using fluorescent nanoparticles-based immunochromatographic test strips. *Trends in Food Science and Technology*, 118, 658–678. https://doi.org/10.1016/j.tifs.2021.10.025
227. Rivas, L., et al. 2014. Lateral Flow Biosensors Based on Gold Nanoparticles. In Valcárcel, M. and López-Lorente, A.I. (eds.) *Gold Nanoparticles in Analytical Chemistry* (Comprehensive Analytical Chemistry Series). Chapter 14, pp. 569–605. Elsevier, The Netherlands.
228. Parolo, C., et al. 2020. Tutorial: design and fabrication of nanoparticle-based lateral-flow immunoassays. *Nature ProtocolsNature Protocols*, 15, 3788–3816. https://doi.org/10.1038/s41596-020-0357-x

© Burleigh Dodds Science Publishing Limited, 2025. All rights reserved.

229. Berlina, A.N., et al. 2021. Combination of phenylboronic acid and oligocytosine for selective and specific detection of lead(ii) by lateral flow test strip. *Analytica Chimica Acta*, 1155, 338318. https://doi.org/10.1016/j.aca.2021.338318
230. Zhong, Z.T., et al. 2022. Terminal deoxynucleotidyl transferase associated with split G-quadruplex/hemin deoxyribozyme amplification detection for various contaminants in milk based on pregnancy test strip platform. *Biosensors and Bioelectronics*, 216, 114644. https://doi.org/10.1016/j.bios.2022.114644
231. Yao, L., et al. 2022. Rapid and sensitive detection of Hg2+ with a SERS-enhanced lateral flow strip. *Analyst*, 147, 4337. https://doi.org/10.1039/d2an01100j
232. Lou, Y., et al. 2022. Lateral flow analysis of Pb (II) in green tea integrated with ionic imprinted paper-based chip. *Microchemical Journal*, 176, 107235. https://doi.org/10.1016/j.microc.2022.107235
233. Wang, L., et al. 2022. One-step extraction and simultaneous quantitative fluorescence immunochromatography strip for AFB1 and Cd detection in grain. *Food Chemistry*, 374, 131684. https://doi.org/10.1016/j.foodchem.2021.131684
234. Wang, Y., et al. 2016. 'Green' immunochromatographic electrochemical biosensor for mercury(II). *Microchimica Acta*, 183, 2509–2516. https://doi.org/10.1007/s00604-016-1866-8
235. López-Marzo, A., et al. 2013. High sensitive gold-nanoparticle based lateral flow Immunodevice for Cd2+ detection in drinking waters. *Biosensors and Bioelectronics*, 47, 190–198. https://doi.org/10.1016/j.bios.2013.02.031
236. Xing, C., et al. 2015. Ultrasensitive immunochromatographic assay for the simultaneous detection of five chemicals in drinking water. *Biosensors and Bioelectronics*, 66, 445–453. https://doi.org/10.1016/j.bios.2014.12.004
237. Quesada-González, D. and Merkoçi, A. 2017. Mobile phone-based biosensing: an emerging "diagnostic and communication" technology. *Biosensors and Bioelectronics*, 92, 549–562. https://doi.org/10.1016/j.bios.2016.10.062
238. Nelis, J.L.D., et al. 2020. Smartphone-based optical assays in the food safety field. *Trends in Analytical Chemistry*, 129, 115934. https://doi.org/10.1016/j.trac.2020.115934
239. Lu, Y., et al. 2019. Smartphone-based biosensors for portable food evaluation. *Current Opinion in Food Science*, 28, 74e81. https://doi.org/10.1016/j.cofs.2019.09.003
240. Rateni, G., et al. 2017. Smartphone-based food diagnostic technologies: a review. *Sensors*, 17(6), 1453. https://doi.org/10.3390/s17061453
241. Jain, R., et al. 2020. Advances in imaging-assisted sensing techniques for heavy metals in water: trends, challenges, and opportunities. *Trends in Analytical Chemistry*, 123, 115758. https://doi.org/10.1016/j.trac.2019.115758
242. Sivakumar, R. and Lee, N.Y. 2021. Recent progress in smartphone-based techniques for food safety and the detection of heavy metal ions in environmental water. *Chemosphere*, 275, 130096. https://doi.org/10.1016/j.chemosphere.2021.130096
243. Yu, S., et al. 2016. A portable chromium ion detection system based on a smartphone readout device. *Analytical Methods*, 8(38), 6877e6882. https://doi.org/10.1039/C6AY01994C
244. Upadhyay, Y., et al. 2018. Smartphone-assisted colorimetric detection of Cr3þ using vitamin B 6 cofactor functionalized gold nanoparticles and its applications in real sample analyses 2018. *ChemistrySelect*, 3(24), 6892e6896. https://doi.org/10.1002/slct.201801289


© Burleigh Dodds Science Publishing Limited, 2025. All rights reserved.

245. He, K., et al. 2020. Ratiometric fluorescent paper-based sensor based on CdTe quantum dots and graphite carbon nitride hybrid for visual and rapid determination of Cu2+ in drinks. *Journal of Photochemistry and Photobiology A: ChemistryJournal of Photochemistry and Photobiology A: Chemistry*, 96(5), 1154–1160. https://doi.org/10.1111/php.13271
246. Yao, J., et al. 2013. Efficient ratiometric fluorescence probe based on dual-emission quantum dots hybrid for on-site determination of copper ions. *Analytical Chemistry*, 85(13), 6461–6468. https://doi.org/10.1021/ac401011r
247. Hassan, M.M., et al. 2025. Progress of machine learning-based biosensors for the monitoring of food safety: a review. *Biosensors and Bioelectronics*, 267, 116782. https://doi.org/10.1016/j.bios.2024.116782
248. Yadav, N., et al. 2023. Artificial intelligence in heavy metals detection: Methodological and ethical challenges. *Hygiene and Environmental Health Advances*, 7, 100071. https://doi.org/10.1016/j.heha.2023.100071
249. Souza, A., et al. 2022. Classifying cadmium contaminated leafy vegetables using hyperspectral imaging and machine learning. *Heliyon*, 8, e12256.
250. Zhou, X., et al. 2020. Hyperspectral technique combined with deep learning algorithm for detection of compound heavy metals in lettuce. *Food Chemistry*, 321, 12650.
251. Zhao, Q., et al., 2023. Data fusion of Laser-induced breakdown spectroscopy and Near-infrared spectroscopy to quantitatively detect heavy metals in lily. *Microchemical Journal*, 190, 108623.
252. Zhou, X., et al. 2022. Detection of heavy metal lead in lettuce leaves based on fluorescence hyperspectral technology combined with deep learning algorithm. *Spectrochimica Acta. Part A, Molecular and Biomolecular Spectroscopy*, 266, 120460.
253. Jin, Y., et al. 2023. Simultaneous quantitative determination of low-concentration preservatives and heavy metals in tricholoma matsutakes based on SERS and FLU spectral data fusion. *Foods*, 12, 1234.
254. Xie, W., et al., 2024. Improved Cd detection in rice grain using LIBS with Husk-based XGBoost transfer learning. *Agriculture*, 14, 567.
255. Wang, W., et al., 2020. Application of laser-induced breakdown spectroscopy in detection of cadmium content in rice stems. *Frontiers in Plant Science*, 11, 599616.
256. Feng, L., et al., 2023. Machine-learning-facilitated prediction of heavy metal contamination in distiller's dried grains with solubles. *Environmental Pollution*, 333, 122043.
257. Zhang, M., et al., 2023. Machine learning integrated high quantum yield blue light carbon dots for real-time and on-site detection of Cr(VI) in groundwater and drinking water. *Science of the Total Environment*, 904, 166822.
258. Lahari, S.A., et al. 2025. IoT integrated and deep learning assisted electrochemical sensor for multiplexed heavy metal sensing in water samples. NPJ Clean Water, 8, 12.
259. Juarez, I.D., et al. 2024. Diagnosing arsenic-mediated biochemical responses in rice cultivars using Raman spectroscopy. *Frontiers in Plant Science*, 15, 1371748.
260. Maurya, R., et al. 2023. Computer aided detection of mercury heavy metal intoxicated fish: an application of machine vision and artificial intelligence technique. *Multimedia Tools and Applications*, 82, 20517–20536.


© Burleigh Dodds Science Publishing Limited, 2025. All rights reserved.

261. Pan, H., et al., 2022. Calibration strategies for laser ablation ICP-MS in biological studies: a review. *TrAC Trends in Analytical Chemistry*, 156, 116751.
262. Carter, J.A., et al. 2020. Non-analyte signals and supervised learning to evaluate matrix effects and predict analyte recoveries in inductively coupled plasma optical emission spectrometry. *Journal of Analytical Atomic Spectrometry*, 35, 679–692.
263. Andric, V., et al. 2024. Deep learning assisted XRF spectra classification. *Scientific Reports*, 14, 3666.
264. Forste, F., et al., 2025. Neural networks for quantifying laboratory confocal micro X-ray fluorescence measurements. *Analytical Chemistry*, 97, 7177–7185.
265. EFSA. 2014. European Food Safety Authority. 2014. EFSA Panel on Contaminants in the Food Chain (CONTAM); Scientific Opinion on the risks to public health related to the presence of chromium in food and drinking water. *EFSA Journal*, 12(3), 3595. https://doi.org/10.2903/j.efsa.2014.3595

© Burleigh Dodds Science Publishing Limited, 2025. All rights reserved.

# Chapter 13

# Advances in chemical analysis of flame-retardant residues in food

*Gary Codling, University of Exeter, UK*



## 1 Introduction

Since humanity first harnessed fire, various materials have been used to control or extinguish it. In the UK after the Great Fire of 1666, regulations on building materials used in London limited the use of wood or other combustibles. However, it was not until the early twentieth century that the industrial-scale production of chemical flame retardants (FRs) began. FRs are a diverse group of chemical substances incorporated into flammable products to inhibit the risk of combustion.[1]

It was during the 1960s, however, that a rise in fire-related deaths attributed to combustible upholstery and toxic gases emitted from synthetic materials prompted the introduction of fire safety legislation and the large-scale

http://dx.doi.org/10.19103/AS.2025.0158.22
© Burleigh Dodds Science Publishing Limited, 2025. All rights reserved.

introduction of FR chemicals into the home. The increase in electronic equipment, such as personal computers and televisions that operated at relatively high temperatures and contained toxic materials, also saw an additional need for combustion prevention.

Despite their value in saving lives, and in prevention of damage and loss, some of the most widely-used FRs have been found to be persistent, bioaccumulate, and toxic (PBT), prompting concerns regarding human and environmental safety.[2–7] In addition, a major cause of domestic fires was linked to smoking and open fires in the home, both of which have drastically reduced. Around 50% of the global population was estimated to smoke in the 1960s, but estimates are as low as 10% today with smoking now seen as an outdoor activity even at home in many countries. The clean air acts in the UK of the 1950s and 1960s also saw a massive reduction in open fires in the UK and other countries around the world have applied similar restrictions. Not only have these restrictions improved air quality, particularly in cities, they have also seen a reduction in fire risk. These changes in domestic fire risk and the PBT nature of many FRs have led to the revaluations in the use of these compounds. The changing balance between safety from fire and safety from chemical contaminants has made these compounds less desirable in the home. This is not to say that there is not still a need for FRs today with many people owing their lives to fire prevention resulting from FRs. However, there is an acknowledged need to also understand the environmental and human health risks of these compounds.

## 2 Types of flame retardants

A FR is any compound that is added to a material for the purpose of reducing its combustibility. The exact number of compounds that are classed as FRs varies depending on the definition and the use. Some compounds have FR properties but multiple uses, including plasticizers, e.g. tris(2-chloroethyl) phosphate (TCEP), tris(2-chloroisopropyl) phosphate, and tris(1,3-dichloro-2-propyl) phosphate (TDCPP). Since their primary use is not as FRs, they are not discussed in detail in this chapter.

Today, it is estimated that there are in excess of 175 compounds specifically used for their flame-retarding properties,[9] along with a suite of legacy FRs that still impact our environment and diet. As this suggests, FRs are chemically diverse, with different properties, environmental behaviours, and toxicological profiles. The primary classes of FRs found in food fall into the following main categories:

- brominated FRs (BFRs);
- chlorinated FRs (CFRs) (also known as the halogenated FRs);

© Burleigh Dodds Science Publishing Limited, 2025. All rights reserved.

- organophosphorus FRs (OPFRs);
- nitrogen FRs (NFRs); and
- inorganic FRs (dominated by aluminium hydroxide, magnesium hydroxide, and antimony oxides).

Each presents unique challenges for detection and risk assessment. In 2022, BFRs made up 5% and 11% of the EU and US market compared to 28% across Asia.[26] OPFRs were introduced after 2009 as having good fire-retarding properties but with lower toxicity. The current market also includes metallic, nitrogenous, nanomaterials, and other FRs (Table 1).

Modes of action for BFRs, CFRs, and OPFRs classes are similar. Under high temperatures, they shed their respective element: bromine, chlorine, or phosphorus. This creates an elemental radical that can bind to the hydroxyl and hydrogen radicals within a fire. A fire is a chain reaction of radical formation. By starving the fire of radicals, an FR can slow the spread or prevent the fire entirely. The FR may also create a layer of char on a material surface that is resistant to heat and prevents volatile gas escape, thus starving a fire of potential fuel.

However, all three chemical classes of FR (BFR, CFR, and OPFR) have individual compounds that have also been associated with numerous adverse health outcomes including endocrine and thyroid disruption; immunotoxicity; reproductive toxicity, diabetes cancer, neurodevelopment effects, metabolic disruption, behavioural, and developmental effects.[31,32] The weight of evidence for most health outcomes places the CFRs as the greatest risk as a class, followed by BFRs and OPFRs. Another factor is that, in extreme situations where fires reach higher temperatures, CFRs and BFRs can create their own toxic compounds thus making the fire more of a health risk than had no FR been applied. Investigations into firefighter training to extinguish these fires found significant levels of dioxin compounds on clothing attributed to chemicals within the source materials, including FRs.

One of the main reasons for OPFRs being considered significantly less risky is lower biological persistence compared with BFRs and CFRs. This is also reflected in the restrictions already in place and the monitoring regulations, with many CFRs and BFRs being restricted while most OPFRs are still under investigation for potential risks. However, while CFRs and BFRs tend to bind to lipids (which can keep them away from vulnerable organs), OPFRs are associated with blood and proteins, which means they are potentially capable of reaching a greater number of tissues across the body.

NFRs, e.g., release a cloud of nitrogen gas under extreme heat, diluting the fires' ability to find oxygen. Some are considered as safer alternatives due to lower toxicity and environmental persistence. However, each is not without risks in use. Melamine and its derivatives, melamine cyanurate, e.g., are popular

© Burleigh Dodds Science Publishing Limited, 2025. All rights reserved.

**Table 1** Common types of flame retardant.

| FR group | Example compounds | Structure type | Uses/sources | MRLs regulated | Banned/restricted | Persistence/ bioaccumulation | Found in food | Analytical methods |
|---|---|---|---|---|---|---|---|---|
| Organic flame retardants | | | | | | | | |
| PBDEs (BFRs) | BDE-47, BDE-99, BDE-209 | Polybrominated diphenyl ethers | Electronics, furniture, textiles | Yes (e.g. EU, EFSA) | Yes (penta, octa, deca under Stockholm Convention) | High persistence, bioaccumulative | Yes (fish, dairy, meat, eggs) | GC-MS, HRMS, LC-MS/MS |
| Other BFRs | TBBPA, DBDPE, BTBPE | Brominated phenols, ethers | Plastics, electronics | Partially | Some restricted (e.g. EU 2023 proposal) | Persistent, some bioaccumulative | Yes | GC-MS, LC-MS |
| CFRs | Dechlorane Plus, HBB | Chlorinated aromatics | Electronics, insulation | Limited | Some under POPs | Persistent, bioaccumulative | Yes | GC-MS, HRMS |
| OPFRs (PFRs) | TCEP, TDCPP, TDBPP | Organophosphates | Plastics, foams, packaging | Some (e.g. REACH SVHC list) | TDBPP banned (EU, US) | Low persistence, mobile in body | Yes (ubiquitous) | GC-MS LC-MS/MS, |
| Inorganic flame retardants | | | | | | | | |
| Inorganic FRs | $Al(OH)_3$, $Mg(OH)_2$, $Sb_2O_3$ | Metal hydroxides, oxides | Plastics, coatings, textiles | Rarely | Antimony trioxide flagged for inhalation risk | Low persistence, low bioaccumulation | Rarely | ICP-MS, XRF |
| NFRs | Melamine, melamine cyanurate | Nitrogen-based | Textiles, foams | No | No | Low persistence, low bioaccumulation | Rarely (packaging contact) | LC-MS, FTIR |
| Intumescent FRs | APP, pentaerythritol | Phosphorus/ nitrogen-based | Coatings, construction | No | No | Stable in materials | Rarely | LC-MS, thermal analysis |
| SFRs | Sulphur-based FRs | Sulphur compounds | Specialized uses | No | Limited data | Toxic at high exposure | Rarely | Limited data |
| Nano-FRs | Nanoclays, CNTs, graphene oxide | Nanomaterials | Packaging, electronics | No | No | Unknown, under study | Potentially (packaging) | Raman, XRD, DLS, ICP-MS |

© Burleigh Dodds Science Publishing Limited, 2025. All rights reserved.

NFRs. They tend not to bioaccumulate. However, melamine has been linked with kidney problems.

Currently, inorganic materials such as aluminium-based FRs dominate the global market. These FRs rely on mineral compounds like aluminium hydroxide (ATH), magnesium hydroxide, or antimony trioxide that release water vapour when heated. This water release cools combustion area, dilutes flammable gasses, and reduces their intensity. Like BFR, CFR, and OPFR, they can also form a protective char. ATH and magnesium hydroxide are common inorganic FRs, typically used in plastics, textiles, and coatings. Most inorganic FRs are considered low risk for toxicity and this may go some way to explain why they are now dominating the FR market. One inorganic FR, antimony trioxide, is known to be carcinogenic if inhaled. The majority of these FRs are considered to have low potential to be found in food. A recent study suggested that hexa-substituted cyclotriphosphazene is an emerging and yet unstudied group of FRs.[30]

Other classes of FR include intumescent FRs, which expand upon exposure to heat, forming an insulating foamy char. These are often OPFRs or NFRs but can also be carbon based. Ammonium polyphosphate (APP) and pentaerythritol are commonly used in intumescent formulations for coatings, foams, and construction materials. There is some concern that excessive use and improper disposal of building materials could be a likely route of transfer to the environment followed by transfer through food webs.

Sulphur FRs (SFRs) are typically specialized in use, given that exposure to sulphur causes gastrointestinal irritation in high concentrations and has toxic endpoints. It is likely that day-to-day exposure to these is low when compared to other FRs.

Nanomaterials such as carbon nanotubes, clay nanoparticles, and graphene oxide are growing in use as FRs since they exhibit high thermal stability. Nanocomposite films containing nanoclay, e.g., are coming to the market due to their thermal stability and FR behaviour in applications such as food packaging.[8] Some of the latest materials include smart colorimetry, able to identify pH changes within the material for monitoring spoilage. Currently, the cost of material is prohibitive for everyday application though montmorillonite clay and graphene oxide have been used for plastics and textiles in a limited capacity. There is some concern regarding the use of nanomaterials since, once damaged, they may shed particles capable of entering individual cells if ingested.

It is important not to overlook legacy FRs, though now restricted or less used. Their widespread previous use, particularly in long-lasting products, such as furniture, as well as their persistence in the environment, makes them a continuing problem. Legacy compounds also our greatest repository of knowledge on FRs and their behaviour whilst newer FR compounds are less well studied.

© Burleigh Dodds Science Publishing Limited, 2025. All rights reserved.

The legacy FR market is dominated by the polybrominated diphenyl ethers (PBDEs). Though there are 209 potential congeners (related chemicals) in this group, there were three main commercial formulations:

- penta-brominated diphenyl ethers (BDEs);
- octa-BDEs; and
- deca-BDEs.

In the EU, in 2000, it is estimated that up to 100 000 tonnes of deca-BDE was being used, reducing to less than 1000 tonnes by 2004. In China, during 2000 deca-BDE made up 45% of all BFRs produced and octa-BDE 10%, whilst today these make up a small fraction (under 5%) of BFRs.[5]

The deca-BDE mix consists largely of a single fully brominated biphenyl molecule ($C_{12}Br_{10}O$). Exposure to ultraviolet light and other environmental processes causes the loss of bromines from the molecule. This may create a range of more toxic breakdown products with longer biological half-lives than the parent molecule. The detection of a breakdown product such as BDE-209 is also more more difficult. Penta- and octa-mixes were restricted under the Stockholm Convention in 2004, but it took until 2009 for deca-BDE to be restricted.

This complexity of current and legacy compounds, coupled with metabolites from parent molecules, presents particular challenges in identification of compounds in the study of dietary exposure and whether the sources are from the food itself, the material it is packed in or the environment.

## 3 Pathways for contamination by flame retardants

FRs enter the food chain through a number of pathways, including contact and environmental contamination.

### 3.1 Contact contamination

Contact contamination includes migration from packaging materials and contamination during food processing. In many instances, FRs are impregnated into the material. This allows contaminants to become mobile. In food packaging, e.g. for ultra-processed products, the contact period can be extended allowing greater risk of transfer. Packaging expected to be exposed to heat such as microwave and oven-proof material may be impregnated with 'food safer' FRs to provide thermal stability but which may still constitute a potential risk. Contact contamination can also occur during processing. Many food processing plant buildings and equipment use FRs for fire safety. Ships, planes, airports, and docks are known sources during transportation.

© Burleigh Dodds Science Publishing Limited, 2025. All rights reserved.

### 3.2 Environmental contamination

When FRs are produced at an industrial scale, the production environment can result in compounds entering the atmosphere, soil, and the aquatic environment. There is often a halo associated with high-volume production sites measurable in the soil and air. Regulations are in place for emissions and monitoring in many countries. The US, e.g., has 1340 Superfund Sites listed on the EPAs National Priority List (NPL) representing locations prioritized for monitoring and clean-up due to hazardous contamination at levels that pose a health risk. A significant proportion of these sites are linked to industrial contamination, and some are linked to the production and use of FRs, particularly PBDEs, TDCPP, and Firemaster 550.

Electronics products, furnishing, clothing, building materials, and packaging materials will all leach FRs with time. Studies of dust in homes and work locations have all identified FRs at relatively high concentrations that can be traced to these products. Dust and air contamination can then contaminate foodstuffs. In one case, chickens were found to have eaten the polystyrene insulation containing FRs that formed part of their building. However, most contamination is indirect. Given the semi-volatile nature of many FRs, the air and soil around waste sites (particularly for the disposal of electronic products) can be major sites for heavier FRs. Plants growing close to waste sites shown greater concentrations of FRs, and humans living in the vicinity of some electronic waste sites with poor air protection measures also carry a notable body burden.

Semi-volatile FRs can enter the atmosphere or aquatic systems. FRs can be subject to long-range atmospheric transport. The Log Air-Water Partition Coefficient (Log $K_{aw}$) is a thermodynamic parameter that quantifies the relative affinity of a chemical compound for air versus water. It is a critical factor in environmental science, particularly for understanding the behaviour of volatile organic compounds (VOCs), semi-volatile organic compounds (SVOCs), and other pollutants. A high $K_{aw}$ implies a greater tendency to be present in the air. This value is temperature dependent because temperature changes the vapour pressure. At lower temperatures, compounds tend to remain dissolved in water. In colder temperatures, a greater volume of FRs will be present in aquatic systems.

### 3.3 Biological half-lives

The wide diversity of molecular structures in FRs leads to highly variable biological half-lives. The biological half-life, the time required for 50% of an ingested compound to be excreted or transformed, can range from hours to several years (Table 2).

© Burleigh Dodds Science Publishing Limited, 2025. All rights reserved.

**Table 2** Estimated biological half-lives and primary tissue accumulation sites for selected FRs in humans

| Flame retardant | Chemical class | Estimated biological half-life | Primary tissue accumulation site |
|---|---|---|---|
| Polybrominated diphenyl ethers (PBDEs) | Brominated | 1–10 years | Fat |
| BDE-47 | Brominated | 1.8 years | Fat |
| BDE-99 | Brominated | 5.4 years | Fat |
| BDE-209 | Brominated | 2 weeks | Blood/liver |
| Hexabromocyclododecane (HBCDD) | Brominated | 1–2 months | Blood/fat |
| Tetrabromobisphenol A (TBBPA) | Brominated | 2–4 days | Blood/liver |
| Decabromodiphenyl ethane (DBDPE) | Brominated | ~1 year | Fat/blood |
| Dechlorane plus | Chlorinated | ~1 year | Fat |
| Tris(1,3-dichloro-2-propyl) phosphate (TDCPP) | Organophosphate | 16 hours | Blood/liver |
| Tris(2-chloroethyl) phosphate (TCEP) | Organophosphate | 10 hours | Blood/liver |
| Tris(2,3-dibromopropyl) phosphate (TDBPP) | Organophosphate | 12 hours | Blood/liver |

These values reflect current understanding based on available toxicokinetic data and may vary depending on exposure conditions and individual physiology.

A biological half-life does not imply that a compound disappears completely. For example, if an organism has a body burden of 100 ng/g of a compound with a 1-year half-life, this decreases to 50 ng/g after 1 year, 25 ng/g after 2 years, and continues halving each year ad infinitum. The compound persists in diminishing amounts over time. Estimating the half-life of FRs in humans, however, is challenging and requires complex modelling supported by laboratory experiments. In an ideal scenario, an organism exposed to a compound would have its excreta and body burden measured over time to directly observe the compound's loss. For compounds with short half-lives, this process is feasible. However, for compounds with multi-year half-lives, estimation often assumes a linear relationship over extended timeframes.

For human studies, exposure to high, localized concentrations of an FR provides an opportunity to estimate its half-life. A notable example occurred in 1973, when a packaging error resulted in polybrominated biphenyls (PBBs), sold under the trade name FireMaster, being mixed with NutriMaster, a livestock feed supplement. This error led to widespread contamination of livestock and human diets, resulting in acute PBB exposure. Studies of affected individuals estimated the biological half-life of PBBs in adults to be approximately 10–15 years.

© Burleigh Dodds Science Publishing Limited, 2025. All rights reserved.

In contrast, assessing chronic low-level exposure to FRs is more complex. For instance, if an individual takes up 12 ng of a compound annually with a 1-year half-life, at year-end, half of the first January's 1 ng will remain, while February's 1 ng will retain more than half, and so forth. Additionally, at the start of the next year, the person takes up another 1 ng. Measuring such a compound's presence in the body requires detailed knowledge of exposure rates and numerous physiological factors to accurately determine its half-life in real-world scenarios.

### 3.4 Tracking contamination routes

The bio-accumulative nature of some FRs, coupled with biological half-lives running into many years, means that even trace levels in the environment can build up to significant concentrations. In the 1970s, brominated compounds (particularly PBDEs due to their excellent flame-retarding properties) dominated the market.[5,10] BDE contamination of the marine organisms was first reported in the 1970s.[11,12] In the early 2000s, a study looking at breastmilk in Sweden collected over a 30-year period identified an exponential increase in the human body burden of BDS[13] with predictions of toxic concentrations being reached within decades. BDEs were subsequently measured in breast milk in Australia,[14] the US,[15] and the UK.[16]

In terms of contamination routes, hand-to-mouth activity identified dust as a significant source for toddlers,[17] whilst breastmilk can act as a primary source of FRs for newborns.[13,16,18] In adults, the majority of our FR body burden is linked to foodstuffs, FRs have identified in food from the US,[15] Europe[19,20] (including Spain, Italy, The Netherlands, Denmark, Finland, Iceland, Norway, and Sweden), France,[21] Belgium[22] and the UK.[23–25] Many of these studies identified marine fish as a major focus of contaminants,[15,20–22] FRs were also found in eggs, meat, freshwater fish, dairy products (milk and cheese), and baby formula.

Studies in the UK found that all of the eight organophosphate esters (OPEs) in the study were detected in the food samples investigated.[29] These studies along with human exposure data, and emerging toxicity studies have placed seven OPEs on the REACH list (see below) as substances of very high concern (SVOCs) and one tris(2,3-dibromopropyl) phosphate (TDBPP) is banned in the US and EU due to concerns about carcinogenicity.

Comparing the biological half-life of PBDEs to later use BFRs and then to OPFRs, we see a significant shift in persistence within an organism (Table 1). However, when it comes to health risk. BDEs bind to lipids once in the body, whilst some OFFRs (e.g. TDCPP, TCEP, TDBPP) are more mobile so are capable of reaching all parts of the organism. If an FR is mobile in the blood during pregnancy, it may pass the placental barrier and thus impact a developing

© Burleigh Dodds Science Publishing Limited, 2025. All rights reserved.

foetus, BDEs may also do this if fat is mobilized for the high energy needed for development. After birth, lipids are mobilized for breast feeding, and babies may acquire BDEs through this route.[15,16]

## 4 The regulatory framework for flame retardants

The regulatory framework for managing FRs in food spans various international, regional, and national organizations. These frameworks aim to balance fire safety with the protection of public health and the environment. There are ranges of approaches including outright bans, restrictions on use along with guidelines pertaining to specific compounds and usage in areas such as food packaging.

### *4.1 International and regional bodies*

The Codex Alimentarius Commission (CAC), a joint body of the World Health Organization (WHO) and the Food and Agriculture Organization (FAO) of the United Nations, is responsible for establishing international food safety standards. The General Standard for Contaminants and Toxins in Food and Feed (CODEX STAN 193-1995), developed by the CAC, provides guiding principles for establishing national limits for FRs and other contaminants in food. However, the CAC does not set specific MRLs for FRs in food. These substances are classified as chemical contaminants and are addressed under frameworks such as the Stockholm Convention on Persistent Organic Pollutants (UN-POPs).

The UN-POPs is a global treaty aimed at eliminating or restricting the production and use of persistent organic pollutants, including certain FRs. The treaty targets compounds that are persistent in the environment, toxic, bioaccumulative (or pseudopersistent), and capable of long-range transport (LRT). The LRT criterion is particularly important, as it ensures that substances with transboundary environmental impacts are regulated through international cooperation. The convention regularly evaluates new chemicals for potential risks to human health and the environment. When substances are identified as harmful, global strategies are implemented to reduce or eliminate their use. However, since many FRs are not yet subject to comprehensive international regulation, their governance primarily falls under regional or national authorities.

In the EU, the use of chemicals, including FRs, is regulated under the Registration, Evaluation, Authorisation and Restriction of Chemicals (REACH) framework. The European Chemicals Agency (ECHA) plays a central role in assessing FRs and recommending restrictions or bans. For example, in 2023, the EU proposed a ban on all aromatic BFRs due to their toxicity and environmental persistence.

© Burleigh Dodds Science Publishing Limited, 2025. All rights reserved.

At the national level, individual countries set MRLs for specific chemicals in food products. In the UK, e.g., the Health and Safety Executive (HSE) oversees chemical regulations. In the United States, the Toxic Substances Control Act (TSCA) regulates chemical substances, including FRs, with the Environmental Protection Agency (EPA) evaluating and restricting high-risk FRs. Additionally, state-level regulations, such as California's Proposition 65, provide further protections by identifying and limiting exposure to harmful chemicals.

It is important to note that most MRLs for food focus on chemicals that come into direct contact with food, such as veterinary drugs and pesticides. For FRs, the emphasis is on source control and regulating how these substances transfer to food, as exposure typically occurs through environmental contamination rather than direct application.

Under the UN-POPs, penta- and octa-BDE formulations were the first to be regulated. 2009 saw regulation of BDE-209, hexabromocyclododecane (HBCDD), and hexabromobiphenyl (HBB).[27] The ECHA proposed a a ban of all aromatic BFRs in 2023.[28] Although BDEs are regulated, other BFRs such as BEH-TBPH (bis(2-ethylhexyl)tetrabromophthalate), BTBPE (1,2-bis(2,4,6-tribromophenoxy)ethane), and DBDPE (decabromodiphenylethane) are still in use. The legacy of BDEs persists with a much slower rate of loss from the environment than the PCBs with whom they share similar structural characteristics.

There are also international guidelines provided by the International Standards Organisation (ISO) including ISO/IEC17025 for laboratory detection of FRs, ISO22000 for food safety management and ISO21469 Safety of Machinery and Food Grade Lubricants.

### 4.2 Food safety authorities

The European Food Safety Authority (EFSA) conducts scientific evaluations of chemical contaminants, including FRs, sets MRLs, and provides guidance on dietary exposure. The UK Committee on Toxicology of Chemicals in Food, Consumer Products, and the Environment (COT) offers advice on FRs and other chemical risks in food and consumer products. The United States Consumer Product Safety Improvement Act regulates chemicals in consumer products, including provisions related to FRs.

## 5 Sample preparation: overview

The objective of any laboratory method is to analyse a matrix (e.g. food sample) to detect contaminants of interest such as FRs. When using gas or liquid chromatography, the sample must be prepared in an injection-ready form. Given that the average concentration of FRs in food is approximately 10.9

© Burleigh Dodds Science Publishing Limited, 2025. All rights reserved.

ng/kg[15,19–22,25] for organic FRs, a concentration and clean-up step is typically required. The majority of the food matrix consists of non-target materials, which act as noise in the detection of FRs, which typically constitute only about 0.001% of the total mass.

The challenge in contaminant assessment is to measure all FRs of interest while removing the remaining 99.999% of irrelevant material. There are two approaches:

- targeted and
- non-targeted.

In targeted approaches, specific chemicals or groups of chemicals are identified within the matrix, focusing on their chemical properties to exclude irrelevant material. Non-targeted approaches aim to remove as many interfering molecules as possible while retaining trace molecules for comprehensive analysis.

Some techniques to measure FRs are non-destructive or require limited sample preparation. However, most methods for determining FRs that can detect trace levels require a series of steps:

- preparation (pre-extraction);
- extraction;
- clean-up;
- concentration;
- filtration; and
- detection.

These steps can overlap, occur out of sequence, or serve dual functions. For example, sample extraction can also act as a bulk clean-up process.

When a food sample is collected for analysis, the timing and source of collection are vital considerations. For example, a carrot collected directly from a field will present the FRs for that environment. However, does this reflect the same burden if the carrot is collected from a supermarket after transport, washing, and packing? A significant question in analysis is whether the item should be tested as purchased or cooked to reflect usual exposure. Some food items, such as kidney beans, are not safe to eat raw, so it is logical to assess contamination levels of FRs in their cooked state. Carrots, on the other hand, can be eaten raw, peeled, unpeeled, or cooked in multiple ways, which affects the best option for presenting the FR risk. Many food regulations focus on the concentrations in raw samples. In a study of fish comparing raw and cooked samples for BDEs, there was little difference between fresh and cooked concentrations. However, fish may not be the ideal foodstuff to identify variance, as cooking times for many fish are

© Burleigh Dodds Science Publishing Limited, 2025. All rights reserved.

short (minutes) compared to other foodstuffs where a variety of processes can alter the FR load. Processed food items may be at greater risk of contamination than raw ingredients.[19] However, since some FRs are SVOCs, heating during food preparation, either at home or during industrial processes, may reduce FRs.

A primary consideration is quality control and assurance (QA/QC) for samples. The aim of QA/QC is to demonstrate that concentrations are accurate to a given level of precision. There will always be some degree of uncertainty. During interlaboratory trials, the relative standard deviation (RSD) can be 10–30% or greater if the matrix is highly complex. In a trial of FRs in biofluids (serum and urine) across Europe, just 70% of laboratories provided satisfactory results for BDE 47 and 153 whilst, for other halogenated FRs, it was as low as 44%.[33] The main advancement in this area is not in technology but more in the collaboration and development of standardized methodologies and testing to ensure compliance between laboratories. Most food analysis for regulatory monitoring is completed under ISO accreditation, but these studies highlight the complexities of FR analysis and the risk of errors.

Advancements in detection technology and reproducibility have significantly influenced QA/QC requirements for analysing FRs in food. Improved sensitivity and accuracy of instruments like mass spectrometers demand greater precision. Standardization and reproducibility ensure cross-country parity, aided by automation and the use of machine learning (ML) and artificial intelligence (AI) for data-driven analysis.

Strict adherence to the use of blanks in food sample analysis is essential. There is a growing library of certified reference materials (CRMs) from groups like NIST, Sigma-Aldrich, and IRMM, which meet ISO-30-35 standards. These materials are continually refined with additional compound classes and matrices.

The EFSA has established criteria for the analysis of FRs in food using mass spectrometry (MS) to ensure accuracy, reliability, and consistency. These criteria specify that the method must be targeted only for FRs, CRMS must be used, blanks must ensure proof of quality of methods, and sensitivity must meet the levels determined for safety. As methods for FRs become more robust and sensitive, the criteria for QA/QC need to become tighter to reflect the precision and detection quality available.

## 6 Sample preparation: pre-extraction

Some advances in FR detection are unique for organic, inorganic or nano-FRs, whilst others are more universal. Even within subgroups, there are methods and processes more suitable for specific classes.

Traditional methods still in use include lyophilization (freeze-drying) and mechanical homogenization, such as using a blender or hand grinding of wet

© Burleigh Dodds Science Publishing Limited, 2025. All rights reserved.

or dry material. Recent advancements in homogenization techniques include cryogenic grinding, precision milling, and nano-level methods.

Cryogenic grinding involves freezing samples in liquid nitrogen and grinding them at ultra-low temperatures. This prevents the loss of FRs due to thermal degradation or volatilization. It is particularly useful for materials like plastics, which become brittle when frozen and are easily broken down, increasing the surface area available for extraction.

Precision milling systems create uniform particle sizes, leading to more robust and reproducible methods. Many of these systems are designed and calibrated to prevent cross-contamination, a risk in manual homogenization processes.

Nano-level homogenization employs ultrasonics and high shear mixers to ensure that nanomaterials are uniformly distributed within the sample, resulting in more consistent measurements. This is especially beneficial for non-destructive or point measurements, such as matrix-assisted laser desorption/ionization time-of-flight mass spectrometry (MALDI-TOF), where a small area is analysed. Non-uniform dispersion can lead to data that does not accurately represent the true concentration.

Considerations for pre-extraction preparation include the sample matrix, target compounds, cross-contamination, particle size uniformity, and environmental conditions. The physical and chemical properties of the sample (e.g. moisture content, hardness, and matrix composition) determine the most suitable preparation method. Sensitive compounds may require gentle methods like lyophilization to avoid degradation, while robust compounds may tolerate more intensive methods like ball grinding. Care should be taken to clean equipment thoroughly between samples to prevent cross-contamination. Smaller and more uniform particle sizes lead to more efficient and reproducible extraction samples. For heat-sensitive analytes, cryogenic grinding or other low-temperature techniques are preferable to prevent loss or degradation.

By selecting the appropriate preparation method – ranging from simple manual cutting to advanced techniques like cryogenic grinding – analysts can ensure better accessibility of solvents to target compounds, enhance extraction yields, and maintain the integrity of the analytes. The greater improvement we are seeing comes from techniques more able to adapt to specific food types.

## 7 Extraction for organic analysis

Extraction is a critical step in analytical food chemistry. It involves separating target compounds from the food matrix, ideally leaving most of the matrix behind. The majority of routine methods use either Soxhlet, accelerated solvent extraction (ASE), solid-phase extraction (SPE), or liquid–liquid extraction (LLE). Each has established protocols, adapted to specific sample types and laboratory

© Burleigh Dodds Science Publishing Limited, 2025. All rights reserved.

conditions and are not discussed in detail here. This chapter explores some key emerging extraction methods, highlighting their advantages, limitations, and applications (Table 3).

### 7.1 Solid-to-liquid extraction

This is essential for converting solid samples into liquid extracts suitable for downstream analysis, such as chromatography. Common techniques include Soxhlet extraction, ASE, and others enhanced by microwave digestion, vortexing, or sonication. For accurate mass determination of compounds, samples should be dried (e.g. lyophilized) and homogenized prior to extraction.

### 7.2 Soxhlet and accelerated solvent extraction

Invented in 1879, Soxhlet remains a cornerstone of food extraction methods for FRs and other contaminants. However, it has limitations such high solvent

**Table 3** Summary of the key extraction techniques used in FR extraction

| Technique | Key features | FR-specific notes |
|---|---|---|
| Soxhlet extraction | Classical method; high solvent use; long duration | Still widely used for lipid-rich samples; automated versions reduce time and solvent |
| Accelerated solvent extraction (ASE) | High temp/pressure; fast; low solvent use | Effective for high-fat matrices; inline clean-up possible |
| QuEChERS (Quick, Easy, Cheap, Effective, Rugged, and Safe) | Quick, low solvent, modular clean-up | Modified for FRs with cryo-baths, PSA sorbents, and automation |
| | ACN-water systems; phase separation via salts/sugars | Eco-friendly variants like Sugaring-Out used for aqueous FR matrices |
| Microwave-assisted extraction (MAE) | Rapid heating; efficient for multiple samples | Suitable for robust FRs; not ideal for thermally sensitive compounds |
| Ultrasound assisted extraction (UAE) | Cavitation-enhanced; energy-efficient | Used in both extraction and clean-up phases |
| Supercritical fluid extraction (SFE) | Uses $CO_2$; minimal solvent; selective for non-polar compounds | Green method; effective for non-polar FRs |
| Solid-phase extraction (SPE) | Clean-up/enrichment; customizable sorbents | Inline systems automate FR extraction; cartridges tailored for viscous matrices |
| Solid-phase microextraction (SPME) | Solvent-free; direct desorption into detectors | Ideal for trace-level FRs; minimal sample prep |
| Stir-bar sorptive extraction (SBSE) | High efficiency for liquids; sorbent-coated stir bar | Effective for aqueous FR samples |
| Liquid–liquid extraction (LLE) | Simple; uses immiscible solvents | Greener solvents (e.g. ACN) replacing chlorinated ones for FR safety |

© Burleigh Dodds Science Publishing Limited, 2025. All rights reserved.

consumption, long extraction times (12+ hours), and potential thermal degradation of sensitive compounds. Advances such as Automated Hot Soxhlet Extraction (AHSE) reduce extraction times to 2–3 hours, minimize solvent usage, and incorporate solvent recovery. Though this is a legacy method, optimization steps are still being developed to make approaches more robust or greener, e.g. less use of methylene chloride (DCM), there is always adjustments. Another relatively old approach is ASE that uses high pressure and heat to speed up solvent extraction. This has been used for FR analysis in food with many modifications to methods and much like Soxhlet is a mainstay of routine lab methods. Though as with Soxhlet volatiles and less robust chemicals may be degraded or lost.

## 7.3 Modern solvent-based techniques

QuEChERS (Quick, Easy, Cheap, Effective, Rugged, and Safe) techniques were initially developed for pesticide residue analysis but have evolved to handle a broad range of analytes,[34,35] including FRs. The process involves acetonitrile (ACN) extraction with salts (e.g. $MgSO_4$, NaCl) and a clean-up step using dispersive SPE (d-SPE). Advantages include high throughput, minimal solvent use, and adaptability for automation. Challenges include limited recovery for specific polarities and small sample sizes that may compromise detection limits. Modifications to QuEChERS are continuous with adjustments to extraction and clean-up materials and actions such as placing samples in cryo-baths to minimizse the thermal buildup, and automated systems.[35]

Homogeneous liquid–liquid extraction (HLLE), a variation of QuEChERS for aqueous samples, uses ACN–water mixtures. Separation is achieved by additives (e.g. salts, sugars) or cooling, forming distinct phases. Variants include salt-assisted liquid–liquid extraction (SALLE), commonly used with aqueous matrices, and Sugaring-Out, an eco-friendly method utilizing sugar for phase separation. Microextraction variants are suitable for trace-level analytes, offering minimal solvent use.[36–38]

## 7.4 Advanced and specialized techniques

Microwave-assisted extraction (MAE) uses microwave energy to heat solvents rapidly, improving extraction efficiency and reducing time. It is suitable for multiple samples but may not be ideal for molecules sensitive to microwave energy. Ultrasound-assisted extraction (UAE) uses ultrasonic waves to create cavitation bubbles in the solvent, enhancing compound release. UAE is energy-efficient and reduces extraction time but may not match the precision of other methods.

© Burleigh Dodds Science Publishing Limited, 2025. All rights reserved.

Supercritical fluid extraction (SFE) uses supercritical $CO_2$ to dissolve non-polar compounds effectively.[39] It is environmentally friendly, requiring minimal solvents, but energy intensive.

SPE is widely used for sample clean-up and enrichment. A sorbent binds target compounds, which are then eluted with a solvent of higher affinity. Advances in automated systems streamline extraction, reducing operator error and enhancing reproducibility. SPE also offers options for tailored elution in food chemistry the industrially packed SPE provide uniform reliable extractions, and emerging SPE cartridges are adjusted to meet issues such as high matrix or viscous samples. For SPE the latest options include fully automated inline systems capable of loading samples and eluting to a detection systems (mass spectrometer) removing the operator entirely.

Solid-phase microextraction (SPME), a micro-scale variant, uses a fibre coated with an extraction phase, which absorbs analytes. The fibre is then desorbed directly into the detection system, minimizing clean-up steps. Stir-bar sorptive extraction (SBSE) utilizes a sorbent-coated stir bar that adsorbs compounds as it stirs the sample. It is particularly effective for liquid samples and provides high extraction efficiency.

LLE separates compounds based on their affinity for two immiscible liquids, typically an aqueous matrix and an organic solvent (e.g. hexane or dichloromethane). Advantages include simplicity and effectiveness for aqueous samples like juices. Greener alternatives, such as acetonitrile, are replacing chlorinated solvents due to safety concerns. Automation in modern systems detects phase boundaries and streamlines extraction.

Emerging techniques include sugaring-out LLE, an eco-friendly HLLE alternative using sugars for phase separation, and automated SPE and LLE, which integrate with analytical systems to reduce manual steps and enhance reproducibility. Green solvents and novel processes minimize environmental impact while maintaining efficacy. However, for many laboratories, the tried and tested approaches of Soxhlet and ASE, e.g., remain the mainstay of extraction methods as they have decades of reproducible results from these applications and though a method is older does not always imply that it is less robust than an emerging technique.

In conclusion, the choice of extraction method depends on the target compounds, sample matrix, and analytical requirements. While traditional methods like Soxhlet remain robust, advances in automation, eco-friendliness, and efficiency are shaping the future of food chemical analysis, with multiple labs opting for combined systems that merge the extraction directly with the analysis removing operator error.

© Burleigh Dodds Science Publishing Limited, 2025. All rights reserved.

## 7.5 Clean-up

The degree of clean-up required for a sample depends on the matrix, detection instrumentation, extraction, and target compounds. Chromatography instruments like GC and LC-MS, commonly used for detecting FRs, have saturation limits where detectors can become overloaded. This issue is exacerbated as samples are concentrated to reach detection limits for target FRs (Table 4).

Most of the clean-up methods are not new techniques but used and adapted, depending if the analysis is focused on specific FR classes, the approach can vary, e.g. sulphuric acid treatment when assessing chlorinated compounds is viable but will be destructive for some BFRs. However, if the approach is focused, destructive clean-up or exclusion methods will remove interferences. In current detection systems, the need for ultra-clean samples

**Table 4** Summary of the most common clean-up methods used for FR organic samples

| Technique | Key features | FR-specific notes |
|---|---|---|
| Solid phase extraction (SPE) | Uses cartridges with sorbents; customizable; can trap target or interference compounds | Silica, Florisil, and black carbon cartridges used; solvent choice critical for loading/elution.[40] |
| Liquid–liquid extraction (LLE) | Uses immiscible solvents to separate analytes | Adapted for clean-up with greener solvents; used post-extraction for matrix simplification |
| Gel permeation chromatography (GPC) | Size-based separation using porous beads | Removes lipids; reproducible with commercial columns; target elution window can be defined |
| Freezing-lipid filtration | Lipid precipitation at subzero temperatures | Effective for fatty samples like liver and dairy; allows subsampling of lipid-free extract |
| Silica gel | Binds polar compounds; elution with non-polar solvents | Used to retain interferences; solvent polarity tailored to FRs |
| Multilayer silica gel | Layered acidic, basic, and neutral silica | Removes broad range of interferences; used in EPA methods |
| Florisil | Magnesium silicate; traps interferences | Solvent polarity used to retain non-targets and elute FRs |
| Activated carbon (Black Carbon) | Amorphous carbon; binds pigments and contaminants | Used in aqueous processes; removes color and broad contaminants |
| Primary secondary amine (PSA) | Removes lipids, organic acids, sugars | Used with QuEChERS; PSA mass affects FR recovery[41] |
| Sulphuric acid treatment | Oxidative clean-up; removes organic matrix | Effective for lipid-rich samples; aggressive method |
| Ultrasound-assisted extraction (UAE) | Cavitation-enhanced; energy-efficient | Used in both extraction and clean-up phases |

© Burleigh Dodds Science Publishing Limited, 2025. All rights reserved.

can be offset by the higher mass resolution and accuracy available to avoid excessive methods.

## 8 Extraction for inorganic analysis

Metallic FRs (MFRs) are now the dominant FR used with the most common being aluminium hydroxide, magnesium hydroxide, and antimony trioxide. As with organic FRs, there is the same concerns over the leaching of these compounds to food from contact materials, though to-date the toxicology concerns are less. Many MFRs are used in concert with organic FRs and so for a complete understanding of exposure both MFRs and OFRs may be assessed.

Similar challenges exist as in organic FRs that of the complexity of food matrices, the regulatory compliances both in the methods used and in the health risk of exposure. There are also challenges in migration analysis to understand the routes of transfer and what the risk of the use around food stuff is. A challenge that is also observed in some organic FRs is around the stability of the molecules. Compounds like ammonium phosphate (APP) and red phosphate absorb moisture leading to analytical variance.

Traditional sample preparation techniques, such as LLE and SPE, have been widely used for the extraction of MFRs from food matrices. However, these methods are often time-consuming, labour-intensive, and require large volumes of organic solvents. Extraction techniques such as SPME, QuEChERS, and MAE are the most common emerging techniques.

## 9 Advances in separation technologies: chromatography

Once a sample has been extracted and cleaned up, the detection method is where much of the emerging and cutting-edge technology is focused. The traditional separation is column chromatography (liquid or gas) separating compounds by their interactions with the stationary phase. For GC, the typical column is a DB-5 MS, 30 m or 15 m long and, in most instances, for FR analysis, this is as much as is needed. For LC amenable FRs, the go to is typically a column with a silica bond phase such as C18, though manufacturers may use differing carbon chains or other materials. It is worth noting that the majority of accredited methods (ISO/IEC 17025) for FRs in food use either GC or LC coupled to a mass spectrometer.

For many years, FR analysis was the province of gas chromatography, PBDEs, hexabromocyclododecane (HBCDD), tetrabromobisphenol A (TBBPA), DBDPE, BTBPE, BEH-TBPH, hexabromobenzene (HBB), pentabromotoluene (PBT), 2,4,6-tribromophenol (TBP) are all classes that are suited to gas

© Burleigh Dodds Science Publishing Limited, 2025. All rights reserved.

chromatography at high sensitivity and selectivity. For some methods, a short column (<15 m) was used to prevent degradation within the oven. But it is also shown that HBCD, TBBPA, hydroxylated-BDEs (OH-BDEs), brominated phenols, DBDPE, TBPH, and BTBPE are also detectable by LC.[41] LC is also a versatile and sensitive technique for many of the MFR compounds that are now starting to dominate the FR market.

There are two main components of analytical determination of FRs:

- separation (chromatography) and
- detection (typically mass spectrometry).

Chromatography can be divided into three phases:

- a stationary phase;
- a mobile phase; and
- a sample.

How the sample interacts with the stationary phase and how the mobile phase reacts with the sample are the key factors. Over the last 100 years, chromatography has developed to include a range of technically sophisticated methods to enhance the separation of small molecules. In sample clean-up GPC and silica gel and other processes are all chromatographic techniques. But for separation science, there are three main routes: gas (GC), liquid (LC), and supercritical fluid (SCF). Each has a value in detection of FRs.

## 9.1 Gas chromatography

### 9.1.1 Gas choices

For a GC-MS technology, improvements have occurred at all areas of the mass spectrometer. There are five main areas of the MS (Fig. 1). The gas used is typically helium, though, as this is a finite resource, other inert gasses have been used for pesticide in food analysis mainly hydrogen[42] and nitrogen.[43] For the most part, the gas choice only focuses on meeting the same levels as seen in the use of helium, though hydrogen can allow faster spectroscopy.

### 9.1.2 Injector

The injector on a GC-MS consists of two parts: the robot that delivers the sample and the injector itself. A traditional robot uses a syringe to take up a fixed volume of sample. Accurate volume uptake ensures reproducible results. Recently, many groups have adopted a modular rail system with

© Burleigh Dodds Science Publishing Limited, 2025. All rights reserved.

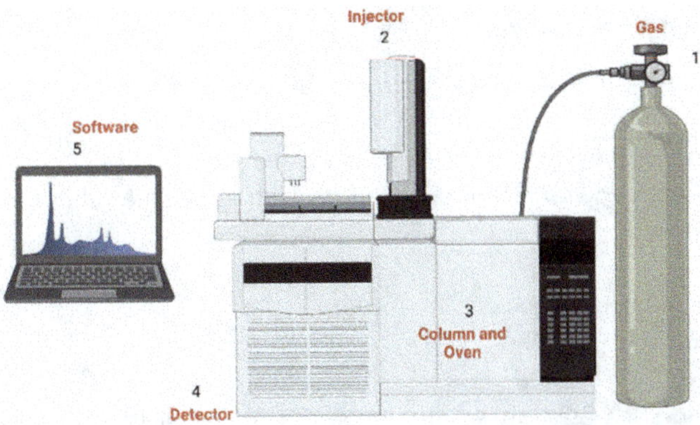


**Figure 1** A simplified diagram of a GC-MS, showing (1) Carrier gas, (2) injector, (3) column and oven, (4) detector, and (5) software (image created via Biorender.com).

attachments like a chilled sample rack, vortex, incubator, and automated derivatization. Chilled sample racks are particularly useful for FRs, which are volatile or semi-volatile.

Some injector types can bypass extensive sample prep. Headspace (HS) analysis heats sealed vials to release volatiles into the gas phase, which are then injected directly or collected on a sorbent. SPME, including SPME Arrow, uses a coated fibre-to-bind analytes, which are thermally desorbed in the injector. Though SPME has higher detection limits than traditional methods, it is rapid, solvent-free, and effective for phenolic BFRs in aqueous samples.[44 45]

The injector's role is to volatilize the sample for GC analysis. Common injection systems include the following:

- Split/Splitless (SSL): Default mode. Splitless is preferred for trace FRs.
- On-Column (OC) & Cold On-Column (COC): Direct injection to avoid thermal degradation (e.g. BDE-209).
- Programmed Temperature Vaporizing (PTV): Gradual heating allows solvent venting and selective FR volatilization.
- Headspace (HS) and SPME: For volatiles and semi-volatiles.
- Pyrolysis: Decomposes solids to identify FRs in materials like plastics and foams.
- Thermal Desorption (TD): Releases analytes from sorbents for injection.

Innovations focus on liner design to improve vaporization and reduce contamination. While SSL is most common, OC and PTV are valuable for thermally sensitive FRs.

© Burleigh Dodds Science Publishing Limited, 2025. All rights reserved.

### 9.1.3 Oven and column

The stationary phase for GC-MS is the GC-Column for the most part, there have been few revolution in the stationary phase though multiple options for columns focused upon specific classes are available.

There are two significant adaptations to the GC oven that have emerged in FR analysis:

- two-dimensional gas chromatography and
- cryo-freezing.

Cryo-freezing of a GC column, also known as cryogenic focusing, is a technique used to enhance the analysis of FRs by improving the separation and detection of volatile and semi-volatile compounds.[46] There are different options as to how it is applied, in one instance, a cryogenic cooling system using liquid nitrogen or carbon dioxide rapidly cools the whole GC oven to below −50°C. This system is integrated into the GC instrument and can be controlled to achieve precise temperature settings. The compounds injected volatilize in the injector but bunch up again in the column, focusing them. Then heating as fast as 5°C $s^{-1}$ is used to remobilize the sample. Cryo-freezing has been used in FR analysis especially for the determination of OPFRs.

Two-dimensional gas chromatography (GC × GC-MS) has been a revolution in the identification of FRs by applying a second column with a different stationary phases, molecules can be split by volatility and polarity (Fig. 2). A modulator between the columns splits samples sending a fraction to the second column, with detection by MS. Why 2-D GC-revolutionized FR analysis relates to the number isomers and molecules with similar fragmentation spectra that can co-elute. Visualization of 2-D GC output can be as a heat map with the elution time of the two columns making the X and Y axes and the colour depth the signal intensity or as a three-dimensional graphic.[47,48]

## 9.2 Liquid chromatography

In comparison to the changes that have occurred in GC for injector, in liquid chromatography, the injector has changed little. For the majority of methods for FRs, where LC is used, the column choice is C18 (18 carbon chains) or C8 (8 carbon chain length) attached to silica,[49] the shorter chain is less hydrophobic and can create a shorter run time and improve peak height. An additional option is HILIC (Hydrophilic Interaction Liquid Chromatography) column.[50]

Non-volatile and polar FRS, such as some OPFRs and BFRs, are predominantly separated by LC, including TCEP and TDCPP. Semi-volatiles like HBCD and TBBPA

© Burleigh Dodds Science Publishing Limited, 2025. All rights reserved.

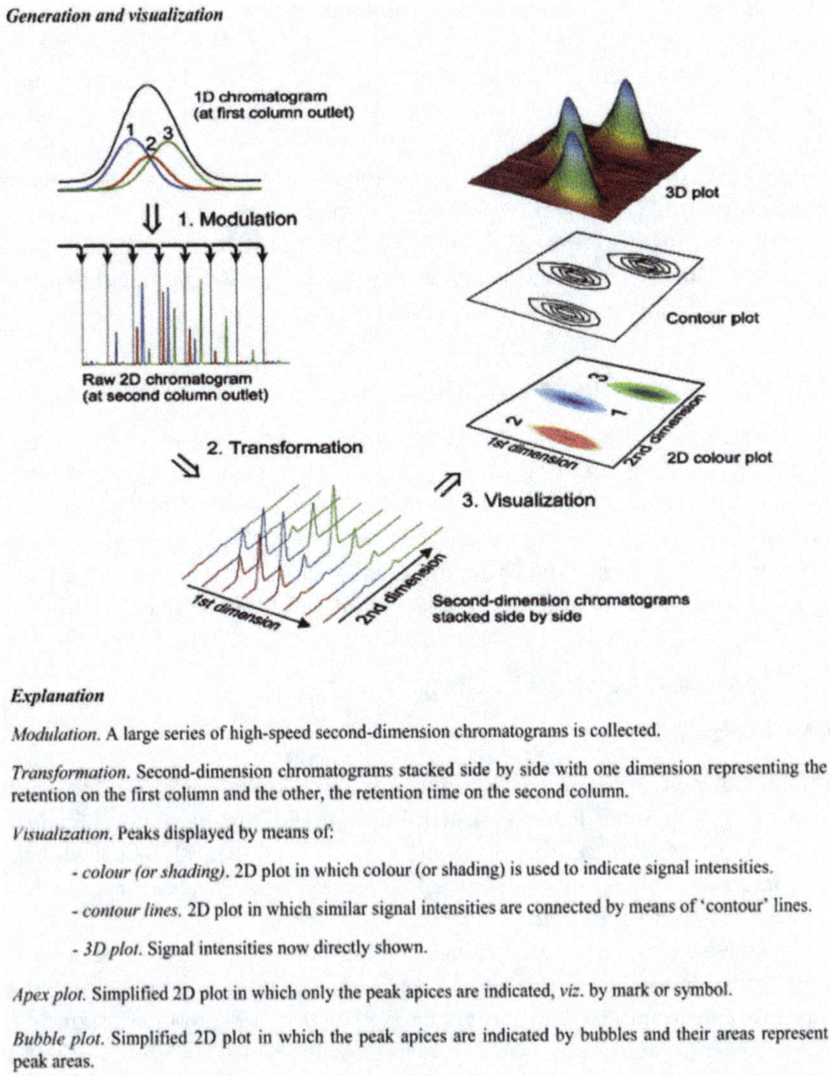


**Figure 2** The principles of 2-D chromatography and the visualization options for identifications.[48] What is presented comes from a focused work on gas chromatography but LC-2D systems are also available.

are capable of separation by either LC or GC. In most cases, for the separation on a standard column, the LC side has shifted little. LC is also suitable for the analysis of metal-organic complexes, such as organoantimony compounds, in food.

Two-dimensional LC, much like described for the GC, this configuration has two columns with differing stationary phase that can separate a sample

© Burleigh Dodds Science Publishing Limited, 2025. All rights reserved.

into a second dimension this extra separation provides further information particularly in complex samples. There are different 2D-LC techniques, such as comprehensive 2D-LC (LC x LC) and heart-cutting 2D-LC. Comprehensive 2D-LC involves continuous sampling from the first dimension to the second, while heart-cutting 2D-LC selectively transfers specific fractions from the first dimension to the second. For the two columns, there is often the use of standard reversed-phase LC followed by a HILIC column.

For the mobile phase in LC, most of the current published methods use water and either methanol (MeOH) or acetonitrile (ACN) with acid and buffer so in terms of innovation, there is limited changes.

### 9.3 Supercritical fluid chromatography

SCF sits between gas and liquid chromatography offering an interesting alternative to either method. Using liquid carbon dioxide on a biphenyl column, the SCF approach is able to capture highly volatile and structurally delicate FRs that may be lost by GC or not separated well by LC.[51] Though the main use of SCF is in the extraction of FRs from materials such as printed circuit boards and GC or LC separation is used in the determination of the extract.[39]

## 10 Ionization and detection

### 10.1 Ionization

After a molecule is separated chromatographically, there is a need to apply a charge to the molecule for mass spectral analysis. For the majority of FR analysis, the go-to options is electron ionization (EI) or negative chemical ionization for gas chromatography, or electrospray ionization (ESI) either in positive or negative for liquid chromatography.

However, there are some novel options for GC, including atmospheric pressure chemical ionization (APCI). This creates gas phase ions at atmospheric pressure. Compared to EI, it can enhance sensitivity to some compounds with high boiling point compounds. It is also possible to ionize some neutral and highly polar compounds. APCI methods have in most cases been shown to meet the quantification limits recommended in the 2014/118/EU methods for FRs in food.[52]

For LC-MS options such as APCI and atmospheric pressure photoionization (APPI) have been used for non-polar and low polar FRs.[49,53]

### 10.2 Detection

Each part of the mass spectrometer from the injector to the ionization is useful, but the biggest changes come from the mass spectrometer detection. The

© Burleigh Dodds Science Publishing Limited, 2025. All rights reserved.

future is in high-resolution mass spectrometry (HRMS) for both GC and LC, and irrespective of how the sample is ionized.

MS initially relied upon a single quadrupole, this consisted of four rods, two with direct current and two with AC. The AC is also referred to as the radio frequency (RF). By adjusting the FR, ionized molecules can collide or be ejected from the quadruple and hit a detector that converts the impact of ions to electrical signal. There was a limit on mass accuracy by this method, with the latest systems accurate to ±0.1 Da (Daltons).

As EI on a GC-MS is considered a hard ionization method, fragmenting molecules at the source a single quadrupole was the mainstay for use for a long time. For LC-MS, where the ionization was softer, the parent molecule was preserved and thus a secondary fragmentation and quadruple could be used (LC-MS/MS) providing greater information on the structure of the molecule.

A relatively new approach for GC is to also use GC-MS/MS as a cheaper alternative to high-resolution MS. This can provide additional chemical information, but as it uses already fragmented molecules, the sensitivity is typically less.

HRMS in the past was used as FRs in food on a system such as a magnetic sector system that have a mass accuracy of 0.0002 Da and a mass resolution as high as 60 k allowing separation of relatively close ion masses (Fig. 3). Where these systems were limited is in how they switched between masses. The scan rate was 1–2 Hz, compared to a low-resolution GC or LC, where it was up to 20 Hz. This meant that more compounds could be scanned for at any time at low mass accuracy or high accuracy on a magnetic sector GC but less compounds. There was also a huge cost difference in a magnetic sector compared to a single quadrupole system.

Over the last 20 years, the time-of-flight (TOF) and the orbitrap, both with a mass accuracy as low as 0.0001 Da, have revolutionized FR analysis in food and other sectors. The biggest addition is also the mass resolution, a TOF can have mass resolution of 10–50 000, while an orbitrap can be 60 to over 500 000. What that means is how well two co-eluting ions can be separated, e.g., on a TOF masses between 400.00 and 400.01 can be distinguished if they co-elute while an orbitrap at 60 k can separate between 400.000 and 400.007 and at 500 K this is between 400.000 and 400.001 (Fig. 3). Where these systems differ to magnetic sector is the scan rate an orbitrap can be up to 20 Hz and a TOF as high as 50 000 Hz. This meant that it was possible to run in full scan at high mass resolution and mass accuracy.

For FR in food analysis, this firstly changed the landscape of detection, Commission Recommendation 2014/118/EU recommended trace levels of BFRs in food and feed to be monitored, and it included the need for congener-specific determination of PBDEs and diastereoisomer-specific determination of HBCDDs

© Burleigh Dodds Science Publishing Limited, 2025. All rights reserved.

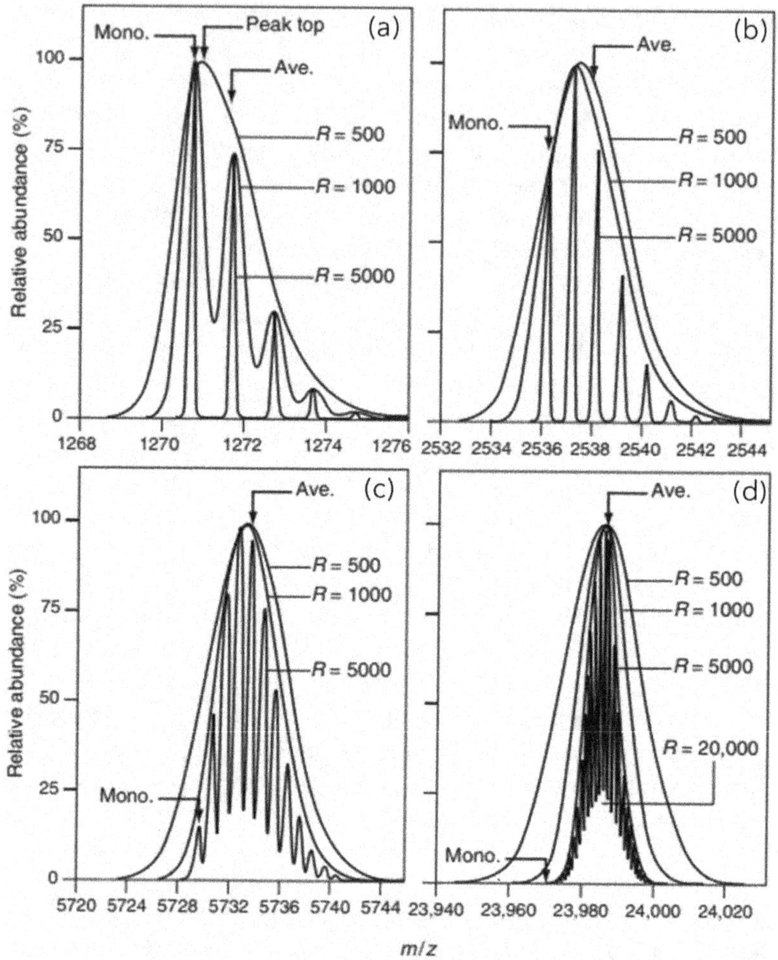


**Figure 3** The effects of resolution on the peak intensity and appearance of molecules at differing mass resolutions, these presented are all peptides that due to the isotopes of oxygen, carbon, nitrogen, and sulphur at low resolution appear as single molecules but as resolution increase multiple isotopes are evident.[54]

in various matrices. It included the need for high selectivity and sensitivity that, though no quantification limits were imposed, the levels must meet those determined for food safety. It also recommended that only high-resolution or tandem mass spectrometry (MS/MS) data is valid. There are no EU mandatory legal MRLs in the EU; however, the EFSA has started to develop guidelines.

What the EU programme and similar means for food safety and FRs in food is that HRMS especially if it can include MS/MS is the go-to option in food safety. However, the HRMS also has an additional card it can play. The high

© Burleigh Dodds Science Publishing Limited, 2025. All rights reserved.

Hertz rates also allow non-targeted screening approaches, and these can be done in parallel with the targeted approach. Non-target approaches can use a range of chromatographic separation options and ionization steps that will create a range of compounds, but the mass spectrometer can be programmed to detect every ion between a given mass range. For example, you can screen for every feature between 50 Da and 1000 Da. The wider the screening range, the less sensitive, however, in a single injection, it is possible to get a fingerprint of the FRs within a sample.

In both GC and LC FR analysis full scan non-target is opening up the landscape of FRs. By assessing the chemical databases, looking at metabolites and structural factors, a database of 7922 organophosphorus compounds was created some of which may be FRs and testing samples against this found over 500 OPs in the environmet[53] These approaches of non-target using high resolution allow for the potential detection of all FRs in food samples with greater mass accuracy than used previously. Target applications with HRMS are capable of being accredited to meet guidelines in analysis, while the non-target approaches can provide the emerging and unscreened for compounds. It also means that FRs do not have to be the only compounds screened for, a food sample ran in non-target by HRMS can provide information on the plant protection products, dioxins, and PCBs in the same sample.

## 11 Alternative methods

A large proportion of this chapter has focused on the chromatographic mass spectral analysis, and to date these remain the baseline for regulatory investigations of FRs in food. However, there are some additional FR tools.

### 11.1 Non-destructive testing methods

Portable and On-Site Detection Methods: systems such as X-ray fluorescence (XRF) are becoming useful tools for rapid screening. The system that can be handheld uses high-energy x-rays to excite atoms in a sample. The secondary fluorescence unique to each atom is then measured and used to provide information. Elements such as bromine, chlorine, and phosphorus used in FRs are capable of being detected by XRF. This would not provide individual component-level detail but has been used, e.g., to identify FR residues in food and feed.[55] Where XRF is particularly useful is in the detection of MFRs.

Infrared spectroscopy (IR) measures the absorbance of infrared light by a sample, providing molecular structure information of the compounds in the sample. While less sensitive[56] than MS, IR systems can be handheld and rapidly screen multiple samples.

© Burleigh Dodds Science Publishing Limited, 2025. All rights reserved.

Fourier transform infrared spectroscopy (FTIR) is a powerful analytical technique used to identify FR chemicals in materials by comparing the sample's spectrum to reference spectra of known FRs. This method allows for precise identification of compounds such as TBBPA and other BFRs. FTIR is particularly useful for its ability to provide detailed information about the chemical bonds and molecular structure of the sample.

In comparison, MS measures the distribution of mass in a sample, providing an elemental analysis but not the chemical composition. While FTIR excels in identifying specific chemical compounds and their structures, MS is more suited for determining the elemental composition and molecular weight of the sample. Both techniques are complementary and often used together to provide a comprehensive analysis of FR materials.[56]

Nuclear magnetic resonance (NMR) uses the magnetic properties of atomic nuclei in a given sample. It can provide very detailed information on the structure, and dynamics of environmental compounds. While non-destructive NMR provides only structural dynamic information, while in general, the MS is more sensitive and able to identify a wider range of compounds. In food analysis, the combination of NMR with MS for a powerful data set is considered, but NMR alone is limited in what it can provide. In terms of cost, commercial NMR systems can be bought for as low as 30 000 pounds (GBP), but depending on the sensitivity, the cost can be into the millions.

Raman spectroscopy (RS) is an analytical technique used to identify and characterize FR chemicals by measuring the inelastic scattering of light (Raman scattering) from the sample. Providing detailed information about the vibrational modes of the molecules, which can be used to identify specific chemical compounds and their molecular structures. Raman spectroscopy is particularly useful for analyzing solid and liquid samples without the need for extensive sample preparation.

In the context of FR analysis, Raman spectroscopy can detect and quantify various FR additives, such as phosphorus-based and brominated compounds. It is also effective in studying the distribution and interaction of FRs within polymer matrices. The non-destructive nature of Raman spectroscopy makes it an excellent tool for both qualitative and quantitative analyses of FRs in various materials.

## 11.2 Alternative ionization for mass spectrometry

An emerging option in MS and food analysis is the Rapid Evaporative Ionization Mass Spectrometry (REIMS) system also known as the iKnife. Initially developed for distinguishing between cancerous and non-cancerous tissues in hospitals, REIMS is now used in food analysis for rapid screening of food product

© Burleigh Dodds Science Publishing Limited, 2025. All rights reserved.

adulteration. It can also identify specific ions, including FRs. However, since REIMS cannot separate molecules, it is not ideal for isomers but can be set up for FR screening.

MALDI-TOF MS is another powerful technique used in FR analysis. MALDI-TOF is particularly effective for identifying and characterizing large, complex molecules, such as polymers and additives in FR formulations. The technique involves embedding the sample in a light-absorbing matrix, which is then ionized by a laser. The ions are accelerated into a flight tube, and their TOF is measured to determine their mass-to-charge ratio. MALDI-TOF offers high sensitivity and resolution, allowing for the detection of specific FR compounds even in complex mixtures. Additionally, it can analyse large molecules without significant fragmentation, preserving the integrity of the sample. This makes MALDI-TOF a valuable tool for both qualitative and quantitative analyses of FRs in various materials.

Inductively coupled plasma mass spectrometry (ICP-MS) is a highly sensitive analytical technique used to detect trace metals and identify metallic nanoparticles in various samples, including water and soil. ICP-MS works by ionizing the sample with an inductively coupled plasma, creating ions that are then detected based on their mass-to-charge ratio. This method allows for the precise quantification of trace elements at very low concentrations, making it invaluable for environmental analysis. In the context of FR analysis, ICP-MS can detect and quantify metallic FR additives, such as antimony and aluminium compounds. Its high sensitivity and ability to handle complex matrices make it an excellent tool for ensuring the safety and effectiveness of FR materials.

## 11.3 Immunoassays and biosensors

Immunoassays and biosensors are alternative methods for detecting FR residues. These techniques are typically faster and more cost-effective than chromatography and MS but lack the sensitivity and often selectivity for trace detection.

### 11.3.1 Immunoassays

Immunoassays rely on the specific binding of antibodies to FR molecules. These assays are easy to use and provide rapid results, making them suitable for screening large numbers of food samples. However, immunoassays suffer from cross-reactivity and are generally less sensitive and selective than chromatography-based methods.

© Burleigh Dodds Science Publishing Limited, 2025. All rights reserved.

### 11.3.2 Biosensors

Biosensors use biological elements, such as enzymes or antibodies, to detect chemical changes in a sample. Biosensors can be highly specific and provide real-time data, but their application to FR detection is still in the early stages of development.

### 11.4 Labelling techniques for detection

Labelling methods are effective for identifying less easily detected FRs, such as nano-FRs, and for tracing their presence. Fluorescent labelling is particularly useful for tagging nanoparticles with fluorescent dyes, facilitating their detection in environmental samples using fluorescence microscopy or flow cytometry. Magnetic labelling, involving iron oxide nanoparticles, allows molecules to be separated and detected with magnet-based tools. Isotopic labelling is employed in MS for quality assurance and quality control (QA/QC) during the development of extraction methods and validation of recovery. Additionally, labelling can assist in source attribution of FRs from contact materials in closed systems. Radioactive isotope labelling, such as with C14, has been used more for observing the degradation and utilization of molecules within a system rather than for direct detection.

## 12 Nanomaterial flame retartants

This group of FRs has been separated from the organic FRs in this chapter as they represent an advancement in FR technology and also a challenge in detection and identification. Nano-FRs include carbon-nanotubes, nanoclays, and layered double hydroxides, these all improve the thermal stability and flame retardancy of the material as with traditional FRs.

Nano clays can create a barrier that slows down the spread of fire and reduces dripping, which helps prevent the fire from spreading to other objects. Carbon nanotubes are effective in reducing heat release and can work together with other FRs to enhance performance. Layered double hydroxides and other 2D nanomaterials contribute to improved flame retardancy by promoting char formation and acting as thermal insulators.

The use of nano-FR is growing because they can achieve significant improvements in fire resistance with relatively low loadings, minimizing any negative impact on the mechanical properties of the host material. Additionally, many nano-FRs are considered environmentally friendly, offering a safer alternative to traditional FRs.

However, though used in low volumes per application and taking up a low percentage of the global market, their potential use is set to grow. When

© Burleigh Dodds Science Publishing Limited, 2025. All rights reserved.

compared to traditional FRs, methods outlined for extraction and identification are less well suited for nanomaterials such that the extent of exposure is largely unknown.

Though there are limited methods in use, we can extrapolate from environmental methods for liquid foodstuffs, filtration, or solid–phase–extraction are potential concentration steps. While in solid samples after lyophilizing, homogenizing, and filtration through gradients of sieves, centrifuge and resuspension in fluids or solvent extraction methods may be applicable. Nano-level homogenization using high shear mixers and ultrasonic homogenizers have been proven to effectively disperse nano-FRs in materials for even distribution when performing extractions. These methods have been trialled for soil and sediment studies, but there is little published specifically for food.

For detection, nanomaterials may be best detected by electron or atomic force microscopy, x-ray diffraction, and dynamic light scattering can also be used.

For carbon nanotubes and other carbon nanomaterials, Raman spectroscopy can provide molecular information from vibrational modes, while x-ray photoelectron spectroscopy can identify the elemental makeup of nanomaterials. Other spectroscopic methods include FTIR capable of providing functional group information and identifying chemical bonds in organic nano-molecules, while ICP-MS can detect elements.

Methodologies such as labelling techniques are also potential factors in identification of nanomaterial FRs. There is concern that there is imbalance between the current use of nanomaterials and our understanding of the toxicological risk and ability to measure that risk.[57] There have been developments of micro-toxicity systems to observe the risk of nanomaterials on cells and organs to understand the impact of their use.

## 13 Data analysis and modelling

The analysis of FRs in food requires robust data processing and validation to ensure accuracy, precision, and reproducibility. With the increasing complexity of food matrices and the diversity of FRs – ranging from BFRs to OPFRs and metal-based FRs – advanced data analysis tools are essential. Emerging techniques such as chemometrics and ML are now integral to FR research, enabling deeper insights into compound behaviour, matrix interactions, and exposure risks.

### *13.1 Chemometrics*

Chemometrics applies statistical and mathematical tools to extract meaningful patterns from chemical data. In FR analysis, it has been used to

© Burleigh Dodds Science Publishing Limited, 2025. All rights reserved.

- differentiate FR types in complex food matrices;
- quantify trace levels of BFRs and OPFRs using multivariate calibration; and
- identify contamination sources through pattern recognition.

Techniques like Principal Component Analysis (PCA) and Partial Least Squares Regression (PLSR) have been applied to distinguish between FRs in lipid-rich versus aqueous matrices and to optimize extraction and clean-up protocols for maximum recovery and minimal interference.

### *13.2 Machine learning and artificial intelligence*

ML and AI are increasingly used to model and predict FR behaviour in food systems. These tools can

- predict FR concentrations based on spectral or chromatographic data;
- classify food types based on FR contamination profiles; and
- optimize extraction and detection parameters using training data from previous experiments.

Recent studies have used support vector machines (SVMs) and artificial neural networks (ANNs) to predict the presence and concentration of FRs in food, achieving high accuracy even with small datasets.[58] More advanced models, such as self-enforcing Deep Neural Networks (SDNN), have been developed to predict fireproof performance and FR behaviour in polymers, with potential applications in food-contact materials.[59]

AI also enables non-targeted screening of FRs by processing large, complex datasets from high-resolution MS. These models can detect unknown or emerging FRs by identifying patterns across multiple matrices and analytical techniques, even when traditional peak-based analysis fails.

## 14 Conclusion

FRs are a diverse group of chemicals with varying properties, environmental behaviours, and toxicological profiles. They can be categorized into brominated, chlorinated, organophosphorus, nitrogen-based, inorganic, and nanomaterial-based FRs.

Legacy FRs, such as PBDEs, continue to persist in the environment despite regulations, while newer FRs like OPFRs are emerging as alternatives, though they also pose health risks.

The bioaccumulative nature of FRs, particularly in marine organisms, leads to higher concentrations in the food chain, especially in fatty tissues. This bioaccumulation is a significant concern for human exposure, as FRs have

© Burleigh Dodds Science Publishing Limited, 2025. All rights reserved.

been linked to endocrine disruption, neurotoxicity, reproductive toxicity, and other adverse health effects. FRs enter the food chain through multiple routes, including direct contact (e.g. food packaging), environmental contamination (e.g. from industrial sites or e-waste), and bioaccumulation in marine and terrestrial organisms. The semi-volatile nature of many FRs allows them to travel long distances through the atmosphere, leading to contamination in remote regions, including polar areas.

The regulation of FRs in food involves international, regional, and national bodies, including the SUN-POPs and the EU's REACH framework. These regulations aim to balance fire safety with public health protection by restricting or banning harmful FRs.

MRLs for FRs in food are set by various authorities, though the focus is often on source control rather than direct application in food. Until now, no MLs for FRs in foods are set by the Codex Alimentarius or by the EU.

Detecting FRs in food is challenging due to their low concentrations (often in the ng/kg range) and the complexity of food matrices. Advanced analytical techniques are required to separate and identify these compounds accurately.

Sample preparation is critical, involving steps, such as homogenization, extraction, clean-up, and concentration. Techniques like QuEChERS, SPE, and cryogenic grinding are commonly used to prepare samples for analysis.

Chromatography (gas and liquid) coupled with MS is the gold standard for FR detection. Advances in HRMS and two-dimensional chromatography have significantly improved the sensitivity and specificity of FR analysis.

Non-destructive methods like XRF, IR, and Raman spectroscopy are being developed for rapid screening of FRs in food, though they lack the sensitivity of MS-based methods.

Immunoassays and biosensors offer faster and more cost-effective alternatives for FR detection but are less sensitive and selective compared to chromatography-based methods.

Labelling techniques, such as fluorescent and isotopic labelling, are being explored to trace the presence of FRs, particularly in nanomaterials.

The analysis of FRs in food generates complex data sets, requiring advanced chemometric and ML techniques for accurate interpretation. These methods help identify patterns and relationships in the data, improving the quantification of FRs in complex food matrices.

## 15 References

1. Watanabe, I. & Sakai, S. Environmental release and behavior of brominated flame retardants. *Environment International* 29, 665–682 (2003).

© Burleigh Dodds Science Publishing Limited, 2025. All rights reserved.

2. Bellés, M. et al. Behavioral effects and oxidative status in brain regions of adult rats exposed to BDE-99. *Toxicology Letters* 194, 1–7 (2010).
3. Linares, V., et al. Human exposure to PBDE and critical evaluation of health hazards. *Archives of Toxicology* 89, 335–356 (2015).
4. Reverte, I., et al. Neurodevelopmental effects of decabromodiphenyl ether (BDE-209) in APOE transgenic mice. *Neurotoxicology and Teratology* 46, 10–17 (2014).
5. Yu, G. et al. Brominated flame retardants (BFRs): a review on environmental contamination in China. *Chemosphere* 150, 479–490 (2016).
6. Yogui, G. & Sericano, J. Polybrominated diphenyl ether flame retardants in the US marine environment: a review. *Environment International* 35, 655–666 (2009).
7. Cressey, D. Cancer-causing flame retardants linger on in California. *Nature News* (2012).
8. Abu-Thabit, N. et al. Preparation of pH-indicative and flame-retardant nanocomposite films for smart packaging applications. *Sensors* 20, 7064 (2020).
9. Birnbaum Linda S & Staskal Daniele F. Brominated flame retardants: cause for concern? *Environmental Health Perspectives* 112, 9–17 (2004).
10. Zhang, Y., et al. Release, transformation, and risk factors of polybrominated diphenyl ethers from landfills to the surrounding environments: a review. *Environment International* 157, 106780 (2021).
11. de Boer, J. & Wells, D. E. Pitfalls in the analysis of brominated flame retardants in environmental, human and food samples – including results of three international interlaboratory studies. *TrAC Trends in Analytical Chemistry* 25, 364–372 (2006).
12. Jansson, B., et al. Brominated flame retardants – Ubiquitous environmental pollutants? *Chemosphere* 16, 2343–2349 (1987).
13. Norén, K. & Meironyté, D. Certain organochlorine and organobromine contaminants in Swedish human milk in perspective of past 20–30 years. *Chemosphere* 40, 1111–1123 (2000).
14. Toms, L.-M. L. et al. Polybrominated diphenyl ethers (PBDEs) in human milk from Australia. *Chemosphere* 68, 797–803 (2007).
15. Schecter Arnold et al. Polybrominated diphenyl ethers (PBDEs) in U.S. mothers' milk. *Environmental Health Perspectives* 111, 1723–1729 (2003).
16. Kalantzi, O. I. et al. Different levels of polybrominated diphenyl ethers (PBDEs) and chlorinated compounds in breast milk from two UK regions. *Environmental Health Perspectives* 112, 1085–1091 (2004).
17. Sugeng, E. J., et al. Brominated and organophosphorus flame retardants in body wipes and house dust, and an estimation of house dust hand-loadings in Dutch toddlers. *Environmental Research* 158, 789–797 (2017).
18. Torres-Moreno, A. C. et al. Polybrominated diphenyl ethers (PBDEs), polychlorinated biphenyls (PCBs), organochlorine pesticides (OCPs) in human breast milk from Colombia: a probabilistic risk assessment approach. *Chemosphere* 339, 139597 (2023).
19. Alves, R. N. et al. Preliminary assessment on the bioaccessibility of contaminants of emerging concern in raw and cooked seafood. *Food and Chemical Toxicology* 104, 69–78 (2017).
20. EFSA Panel on Contaminants in the Food Chain (CONTAM). Scientific opinion on brominated flame retardants (BFRs) in food: brominated phenols and their derivatives. *EFSA Journal* 10, 2634 (2012).


© Burleigh Dodds Science Publishing Limited, 2025. All rights reserved.

21. Vénisseau, A. et al. Occurence of legacy and novel brominated flame retardants in food and feed in France for the period 2014 to 2016. *Chemosphere* 207, 497–506 (2018).
22. Xu, F. et al. Multi-contaminant analysis of organophosphate and halogenated flame retardants in food matrices using ultrasonication and vacuum assisted extraction, multi-stage cleanup and gas chromatography–mass spectrometry. *Journal of Chromatography A* 1401, 33–41 (2015).
23. Fernandes, A. R. et al. Bromine content and brominated flame retardants in food and animal feed from the UK. *Chemosphere* 150, 472–478 (2016).
24. Tao, F. et al. Emerging and legacy flame retardants in UK human milk and food suggest slow response to restrictions on use of PBDEs and HBCDD. *Environment International* 105, 95–104 (2017).
25. Rose, M., et al. Contamination of fish in UK fresh water systems: risk assessment for human consumption. *Chemosphere* 122, 183–189 (2015).
26. Ceresana. *Flame Retardants Market Report*. https://ceresana.com/en/produkt/flame-retardants-market-report.
27. Wang, X. et al. A review of organophosphate flame retardants and plasticizers in the environment: analysis, occurrence and risk assessment. *Science of the Total Environment* 731, 139071 (2020).
28. ECHA. *ECHA Identifies Certain Brominated Flame Retardants as Candidates for Restriction*. (2023). https://echa.europa.eu/-/echa-identifies-certain-brominated-flame-retardants-as-candidates-for-restriction.
29. Gbadamosi, M. R., et al. Organophosphate esters in UK diet; exposure and risk assessment. *Science of the Total Environment* 849, 158368 (2022).
30. Zhang, S. et al. Vanillin-based flame retardant enables polylactic acid high-efficiency fireproof, anti-UV and oxygen barrier for food packaging. *International Journal of Biological Macromolecules* 278, 134558 (2024).
31. Kim, Y. R., et al. Health consequences of exposure to brominated flame retardants: a systematic review. *Chemosphere* 106, 1–19 (2014).
32. Feiteiro, J., et al. Health toxicity effects of brominated flame retardants: from environmental to human exposure. *Environmental Pollution* 285, 117475 (2021).
33. Dvorakova, D. et al. Interlaboratory comparison investigations (ICIs) and external quality assurance schemes (EQUASs) for flame retardant analysis in biological matrices: Results from the HBM4EU project. *Environmental Research* 202, 111705 (2021).
34. Trent, T. Veterinary drug residue analysis using the AutoMate-Q40: an automated solution to QuEChERS. *LC-GC North America* 32, S24–S24 (2014).
35. Kaewsuya, P., et al. Automated QuEChERS tips for analysis of pesticide residues in fruits and vegetables by GC-MS. *Journal of Agricultural and Food Chemistry* 61, 2299–2314 (2013).
36. Valente, I. M. & Rodrigues, J. A. Recent advances in salt-assisted LLE for analyzing biological samples. *Bioanalysis* 7, 2187–2193 (2015).
37. Farajzadeh, M. A., et al. Development of a simple and efficient pretreatment technique named pH-dependent continuous homogenous liquid–liquid extraction. *Analytical Methods* 8, 5676–5683 (2016).
38. Wang, B., et al. Sugaring-out: a novel phase separation and extraction system. *Chemical Engineering Science* 63, 2595–2600 (2008).

© Burleigh Dodds Science Publishing Limited, 2025. All rights reserved.

39. Altwaiq, A., et al. Extraction of brominated flame retardants from polymeric waste material using different solvents and supercritical carbon dioxide. *Analytica Chimica Acta* 491, 111–123 (2003).
40. Gauthier, L. T. & Letcher, R. J. Isomers of dechlorane plus flame retardant in the eggs of herring gulls (*Larus argentatus*) from the laurentian great lakes of North America: temporal changes and spatial distribution. *Chemosphere* 75, 115–120 (2009).
41. Baduel, C., et al. Development of sample extraction and clean-up strategies for target and non-target analysis of environmental contaminants in biological matrices. *Journal of Chromatography A* 1426, 33–47 (2015).
42. Use of hydrogen as a carrier gas for the analysis of steroids with anabolic activity by gas chromatography–mass spectrometry – ScienceDirect. https://www.sciencedirect.com/science/article/pii/S0021967311011745?via%3Dihub#tbl0015.
43. Saito-Shida, S., et al. Comparison of nitrogen and helium as carrier gases for determination of pesticide residues in foods via gas chromatography–tandem mass spectrometry with atmospheric pressure chemical ionization. *Journal of Food Composition and Analysis* 134, 106492 (2024).
44. Kalogiouri, N. P. et al. A novel headspace solid-phase microextraction arrow method employing comprehensive two-dimensional gas chromatography–mass spectrometry combined with chemometric tools for the investigation of wine aging. *Analytica Chimica Acta* 1304, 342555 (2024).
45. Development of a solid-phase microextraction method for the analysis of phenolic flame retardants in water samples – ScienceDirect. https://www.sciencedirect.com/science/article/pii/S0021967306006145.
46. Sadaunykas, A. *Enhancing gas chromatographic analysis: a novel cryo-enrichment module for GC analysis*. Vilniaus Universitetas (2023).
47. What Is GCxGC? Comprehensive two-dimensional gas chromatography. https://phenomenex.blog/2018/05/10/gcxgc/.
48. Adahchour, M., et al. Recent developments in comprehensive two-dimensional gas chromatography (GC × GC). *TrAC Trends in Analytical Chemistry* 25, 438–454 (2006).
49. Pantelaki, I. & Voutsa, D. Organophosphate flame retardants (OPFRs): a review on analytical methods and occurrence in wastewater and aquatic environment. *Science of The Total Environment* 649, 247–263 (2019).
50. Shi, Y. et al. Occurrence, distribution and seasonal variation of organophosphate flame retardants and plasticizers in urban surface water in Beijing, China. *Environmental Pollution* 209, 1–10 (2016).
51. A. MacKay, G. & M. Smith, R. Supercritical fluid extraction and chromatography–mass spectrometry of flame retardants from polyurethane foams. *Analyst* 118, 741–745 (1993).
52. Zacs, D., et al. Evaluation of analytical performance of gas chromatography coupled with atmospheric pressure chemical ionization Fourier transform ion cyclotron resonance mass spectrometry (GC-APCI-FT-ICR-MS) in the target and non-targeted analysis of brominated and chlorinated flame retardants in food. *Chemosphere* 225, 368–377 (2019).
53. Huang, J., et al. Non-target screening and risk assessment of organophosphate esters (OPEs) in drinking water resource water, surface water, groundwater, and seawater. *Environment International* 168, 107443 (2022).


© Burleigh Dodds Science Publishing Limited, 2025. All rights reserved.

54. Carr, S. A. & Annan, R. S. Overview of peptide and protein analysis by mass spectrometry. *Current Protocols in Protein ScienceCurrent Protocols in Protein ScienceCurrent Protocols in Protein Science* Chapter 16, Unit 16.1 (2001).
55. Feng, X., et al. X-ray fluorescence application in food, feed, and agricultural science: a critical review. *Critical Reviews in Food Science and Nutrition* 61, 2340–2350 (2021).
56. Paseiro-Cerrato, R., et al. Brominated flame retardants (BFRs) in contaminated food contact articles: identification using DART-HRMS and GC-MS. *Food Additives & Contaminants: Part A* 38, 350–359 (2021).
57. Lu, R. X. Z. & Radisic, M. Organ-on-a-chip platforms for evaluation of environmental nanoparticle toxicity. *Bioactive Materials* 6, 2801–2819 (2021).
58. Yan, C. et al. Advancing flame retardant prediction: a self-enforcing machine learning approach for small datasets. *Applied Physics Letters* 122, 081901 (2023).
59. Schenk, C., et al. Data-driven design and green preparation of bio-based flame retardant polyamide composites. Journal of Materials Chemistry A. Advance online publication (2025). https://doi.org/10.1039/D5TA02511G.

© Burleigh Dodds Science Publishing Limited, 2025. All rights reserved.

# Chapter 14

## Advances in assessing residues of perfluoroalkyl substances in foods

*Thimo Groffen, University of Antwerp, Belgium; Chaymae El Amraoui Aarab, Mélanie Di Mario, Els Van Hoeck and Laure Joly, Sciensano, Belgium; Marc Hanikenne, University of Liège, Belgium; and Els Van Pamel, Flanders Research Institute for Agriculture, Fisheries and Food (ILVO), Belgium*



## 1 Introduction

This chapter focusses on the sources of per- and polyfluoroalkyl substances (PFASs) in food, the state-of-the-art of analytical methods for food matrices, and on mitigation and remediation strategies for agriculture. Furthermore, the PFAS concentrations in food items in Europe will be addressed, incorporating differences in concentrations depending on the origin of food (i.e. supermarket, home-grown, bio-food, etc.) and the effects of food processing in blue food (i.e. anything we eat that comes from the water). Finally, key knowledge gaps and future challenges of PFAS contamination in food are discussed.

### *1.1 General introduction to per- and polyfluoroalkyl substances*

PFASs are a large family of millions of synthetic fluorinated alkyl compounds (Schymanski et al., 2023) that have been extensively used since the 1940s

http://dx.doi.org/10.19103/AS.2025.0158.23
© Burleigh Dodds Science Publishing Limited, 2025. All rights reserved.

in a wide array of products (Glüge et al., 2020). The definition of PFAS has been changed over the past years (Buck et al., 2011; Wang et al., 2021), and nowadays PFAS are defined as *fluorinated substances that contain at least one fully fluorinated methyl or methylene carbon atom (without any H/Cl/Br/I atom attached to it)* (Wang et al., 2021). PFAS can be further divided into non-polymer and polymer PFAS, which each contain different subclasses, groups, and subgroups (Buck et al., 2011), including the frequently studied group of the perfluoroalkyl acids (PFAAs). The latter comprises the subgroup of the perfluoroalkyl sulfonic acids (PFSA) like perfluorooctane sulfonic acid (PFOS) and the perfluoroalkyl carboxylic acids (PFCA) like perfluorooctanoic acid (PFOA) (Figure 1).

Their physico-chemical stability and surfactant properties give PFAS the water- and oil-repellent characteristics that make them suitable in many industrial branches and as consumer products, including amongst other firefighting foams, non-sticky cookware, textiles, and food contact materials (FCMs) (Glüge et al., 2020). Their large production volumes and global use, in combination with their intrinsic properties, have led to the global environmental PFAS-contamination issue (Ackerman Grunfeld et al., 2024; Brusseau et al., 2020; Kurwadkar et al., 2022). The perfluoroalkyl moiety of many PFAS is extremely resistant to environmental and metabolic degradation, leading to high persistence in the environment (Cousins et al., 2020). In addition, many PFAS precursors may degrade into these persistent chemicals (Buck et al., 2011).

Once released into the environment, PFASs end up in all environmental compartments (Fig. 2) (Brusseau et al., 2020; Faust, 2023; Groffen et al., 2021). However, the exact fate and transport of PFAS in the environment are influenced by the characteristics of the individual PFAS itself as well as the specificities of the contaminated site. For example, shorter-chain PFAS are overall more mobile due to their higher water solubility compared to longer-chain PFAS and thus typically end up in aqueous matrices (Brendel et al., 2018). Some PFASs are known to be taken up and to accumulate in animals and plants (Brake et al., 2023; Li et al., 2022), including crops and livestock (Death et al., 2021; Ghisi et al., 2019; Jha et al., 2021; Schiavone and Portesi, 2023), which eventually leads to human exposure.

Dietary exposure is considered to be the most important pathway of human exposure, followed by inhalation and dermal exposure (Zhang et al., 2024). Consumption of PFAS-contaminated food sources has been proven to affect serum PFAS concentrations (Roth et al., 2020; Susmann et al., 2019; Thépaut et al., 2021). PFAS are overall strongly proteinophilic, meaning that they will accumulate primarily in protein-rich tissues such as the liver (Ng and Hungerbühler, 2013). PFAS exposure in humans has been related to various toxic effects, including immunotoxic effects, altered thyroid function, kidney disease, reproductive and developmental effects, and cancer (EFSA Panel on

© Burleigh Dodds Science Publishing Limited, 2025. All rights reserved.

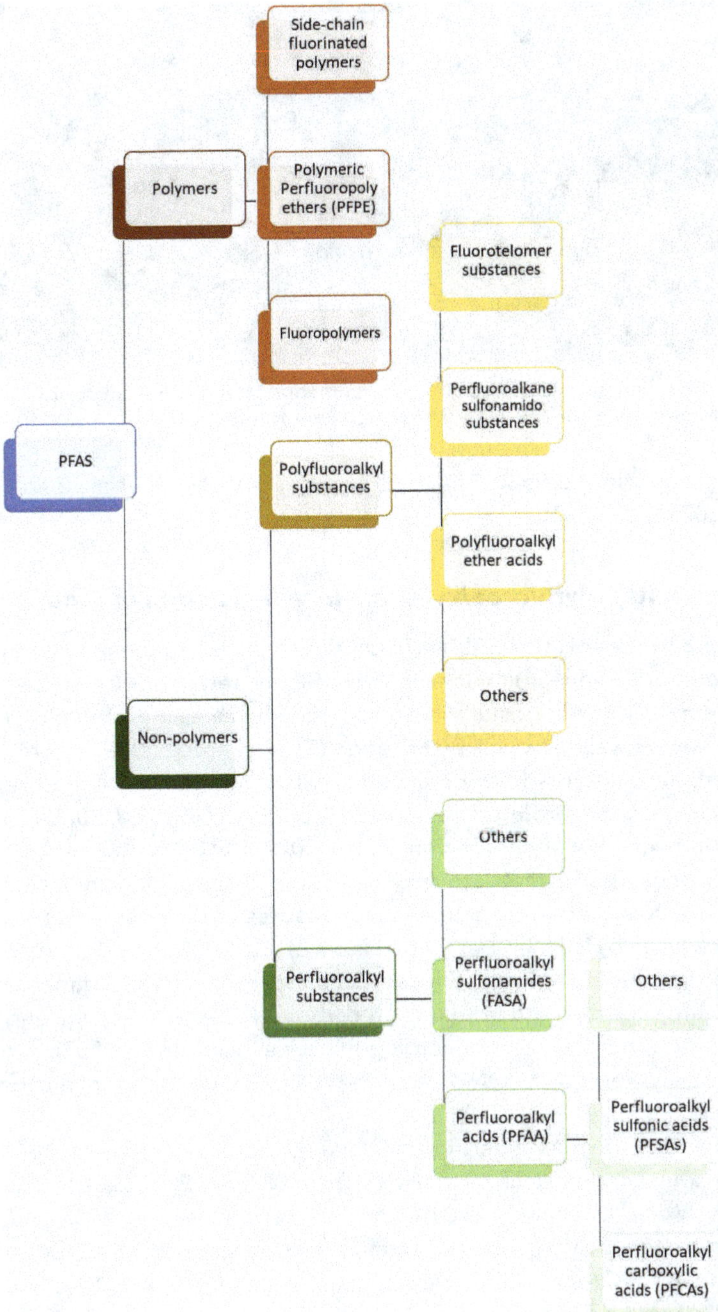


**Figure 1** A simplified family tree of PFAS. Figure made with Microsoft PowerPoint.

© Burleigh Dodds Science Publishing Limited, 2025. All rights reserved.

**Figure 2** Simplified diagram of the environmental fate and transport of PFAS. GW = groundwater; SW = surface water. Made with Photoshop v26.0.

Contaminants in the Food Chain (EFSA CONTAM Panel) et al., 2020; Fenton et al., 2021).

### 1.2 Per- and polyfluoroalkyl substance legislation in food

Since 2023, levels of four PFAS, i.e. PFOS, PFOA, perfluorononanoic acid (PFNA), and perfluorohexane sulfonic acid (PFHxS), are regulated by EU Regulation 2023/915 in certain foodstuffs of animal origin such as meat, edible offal, fish, crustaceans, bivalve mollusks and eggs (Table 1) (EU, 2023a). Furthermore, the sampling, analysis (including performance requirements), and reporting should be done according to Regulation (EU) 2022/1428 (EU, 2022a).

Moreover, Recommendation (EU) 2022/1431 established limits of quantification (LOQ) that laboratories should achieve and indicative levels above which further investigation into the causes of the contamination should be undertaken by EU Member States (Table 2) (EU, 2022b). Additionally, a list of compounds that EU Member States should monitor was established, including PFCA with a chain length from 4 to 14 carbons, PFSA with a chain length from 4 to 13 carbons, and some emerging PFAS, i.e. acid form of F53B, acid form of GenX, acid form of ADONA, capstone A and B, fluorotelomer alcohols ,and sulfonates.

In addition, other specifications are available for water intended for human consumption since Directive (EU) 2020/2184 (EU, 2020), further called Drinking Water Directive (DWD), came into force in January 2021 and should be implemented by 2026. The minimum requirements for the 'sum of PFAS' covering 20 PFCA and PFSA with a chain length from 4 to 13 carbons and 'total PFAS' were fixed at 0.10 μg/L and 0.50 μg/L, respectively. Additionally, methods of analysis for monitoring these two parameters of the DWD must comply with the technical guidelines on the methods of analysis for monitoring

© Burleigh Dodds Science Publishing Limited, 2025. All rights reserved.

**Table 1** Maximum levels (μg/kg) on a wet weight basis set for the four regulated perfluoroalkyl substances and their sum in Regulation (EU) 2023/915

| | Maximum level (μg/kg) | | | | |
|---|---|---|---|---|---|
| | PFOS* | PFOA* | PFNA* | PFHxS* | Sum of PFOS, PFOA, PFNA, PFHxS** |
| Meat (game mammals) | 0.30–1.0 (5.0) | 0.20–0.80 (3.5) | 0.20 (1.5) | 0.20 (0.60) | 1.3–1.6 (9.0) |
| Offal (game mammals) | 6.0 (50) | 0.70 (25) | 0.40 (45) | 0.50 (3.0) | 8.0 (50) |
| Fish meat | 2.0–35 | 0.2–8.0 | 0.5–8.0 | 0.2–1.5 | 2.0–45 |
| Crustaceans and bivalve mollusks | 3.0 | 0.70 | 1.0 | 1.5 | 5.0 |
| Eggs | 1.0 | 0.30 | 0.70 | 0.30 | 1.7 |

*The maximum level applies to the sum of linear and branched stereoisomers, whether or not they are chromatographically separated (PFOS: perfluorooctane sulfonic acid, PFOA: perfluorooctanoic acid, PFNA: perfluorononanoic acid, PFHxS: perfluorohexane sulfonic acid).
**For the sum, the maximum levels refer to lower bound concentrations, which are calculated assuming all the values below the limit of quantification (LOQ) are zero.
*Source*: Adapted from: EU (2023a).

**Table 2** Limits of Quantification (LOQ) and Indicative Levels Defined in Recommendation (EU) 2022/1431

| | Limit of quantification/Indicative level (μg/kg) | | | |
|---|---|---|---|---|
| | PFOS | PFOA | PFNA | PFHxS |
| Fruits | 0.002/0.010 | 0.001/0.010 | 0.001/0.005 | 0.004/0.015 |
| Vegetables | 0.002/0.010* | 0.001/0.010 | 0.001/0.005 | 0.004/0.015 |
| Starchy roots and tubers | 0.002/0.010 | 0.001/0.010 | 0.001/0.005 | 0.004/0.015 |
| Food for infants and young children | 0.002/0.050 | 0.001/0.050 | 0.001/0.050 | 0.004/0.050 |
| Milk | 0.010/0.020 | 0.010/0.010 | 0.020/0.050 | 0.040/0.060 |
| Fish meat | 0.10/-** | 0.10/-** | 0.10/-** | 0.10/-** |
| Meat of terrestrial animals | 0.10/-** | 0.10-** | 0.10/-** | 0.10/-** |
| Eggs | 0.30/-** | 0.30/-** | 0.30/-** | 0.30/-** |
| Crustaceans and mollusks | 0.30/-** | 0.30/-** | 0.30/-** | 0.30/-** |
| Edible offal of terrestrial animals | 0.50/-** | 0.50/-** | 0.50/-** | 0.50/-** |
| Fish oil | 0.50/-** | 0.50/-** | 0.50/-** | 0.50/-** |

*Except for wild fungi, i.e. 1.5 μg/kg.
**No indicative level was set.
PFOS: perfluorooctane sulfonic acid, PFOA: perfluorooctanoic acid, PFNA: perfluorononanoic acid, PFHxS: perfluorohexane sulfonic acid.

© Burleigh Dodds Science Publishing Limited, 2025. All rights reserved.

PFAS, namely EN 17892 part A or B for 'sum of PFAS' and (1) TOP assay (total oxidizable precursor assay), (2) EOF-CIC (combustion ion chromatography (CIC) after extraction of fluorine (EOF)) or (3) LC-HRMS (liquid chromatography – high-resolution mass spectrometry) suspect and non-target analysis for 'PFAS total' (EU, 2024).

The aforementioned regulations, with exception of the 2020/2184 EU Directive, focus on four PFAS since they were evaluated together by EFSA because of similar effects in animals, toxicokinetics, and observed levels in human blood (EFSA Panel on Contaminants in the Food Chain (EFSA CONTAM Panel) et al., 2020). Nonetheless, other PFAS are present in the environment. Therefore, the RIVM developed an approach based on relative potency factors (RPFs), assuming that cumulative effects of PFAS mixtures contribute to one common toxicological effect, and predictions of these risks can be investigated by dose addition (Bil et al., 2021). This method requires similar toxicity among chemicals that only differ in potency. In addition, the mixtures should not show synergism or antagonism. The relative potency of various chemicals is expressed relative to the potency of an index compounds, which is PFOA in the case of PFAS. Concentrations of individual compounds can then be multiplied with their relative potency to obtain concentrations in PFOA equivalents, which subsequently can be compared with regulations and quality standards to identify potential risks. The RPF methodology, has already been applied successfully for other chemicals such as dioxins and dioxin-like chemicals (Falandysz et al., 2014; Suzuki et al., 2020; Trnovec et al., 2013).

## 2 Sources of per- and polyfluoroalkyl substances in food

### *2.1 Uptake routes for crops*

PFAS contamination of agricultural soils typically takes place through (1) the application of sewage sludge, which itself is often contaminated with PFAS and is commonly used as a fertilizer on agricultural land, (2) irrigation with PFAS-contaminated water, e.g. near industrial sites or areas with firefighting foam applications, and (3) more marginally atmospheric deposition through air and rain (Fig. 3a) (Ghisi et al., 2019; Kosiarski et al., 2024; Tansel et al., 2024). The physico-chemical properties of the contaminated soils, including organic matter content, pH, and clay mineral content, influence the PFAS bioavailability and uptake by plants. Hence, higher organic content and clay minerals tend to display stronger PFAS-binding capacities, reducing their availability for plant uptake, in contrast to sandy soils characterized by low organic matter content where PFAS movement towards plant roots is expected to be more important

© Burleigh Dodds Science Publishing Limited, 2025. All rights reserved.

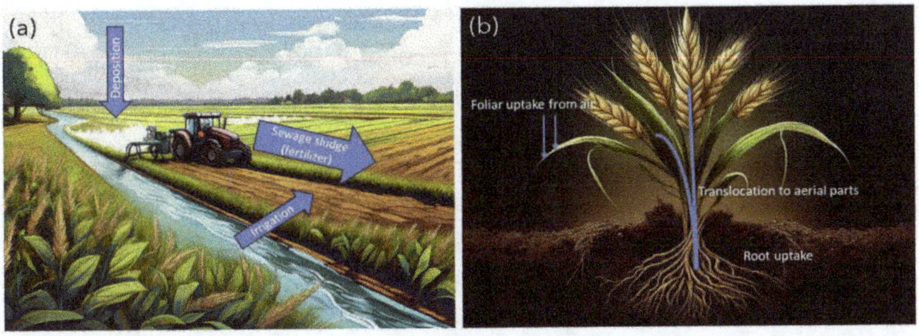


**Figure 3** Diagram of the primary sources for per- and polyfluoroalkyl substance uptake in plants (a) and the uptake and translocation routes in plants (b). Root uptake and translocation within plants depend on the type of per- and polyfluoroalkyl substance. Root uptake is also affected by soil physico-chemistry. Made with Photoshop v26.0.

(Cai et al., 2021). Moreover, PFAS mobility in the soil solution is also dependent on the type of PFAS (chain length and functional groups), with long-chain PFAS tending to bind more strongly to soil organic matter and short-chain PFAS being more mobile (Adu et al., 2023).

Fig. 3b displays a schematic, simplified, overview of the primary PFAS uptake and translocation pathways in plants. Plants primarily take up PFAS through absorption by their roots, a step which requires PFAS to cross the hydrophobic phospholipid bilayer of the plasma membrane (Lesmeister et al., 2021). Once inside root tissues, PFAS can be transported to the plant aerial parts through the xylem, a tissue of the plant's vascular system where water and minerals circulate, thanks to the evapotranspiration stream. Like for the soil, the mobility of PFAS *in planta* depends on their chemical structure, determining different translocation factors (TF, i.e. shoot/root ratios) with short-chain PFAS having higher TF and thus expected to be more mobile compared to long-chain PFAS being more restricted to the roots due to lower TF (Ghisi et al., 2019; Zhou et al., 2021). Another, more minor and less understood pathway for PFAS accumulation in plant aerial parts is PFAS foliar uptake either through gaseous/volatile forms or particle-bound PFAS (Chen et al., 2024; Li et al., 2022). The accumulation of PFAS in plant tissues ranges from ng/g to μg/g of tissue dry weight (Li et al., 2022).

In addition to the influence of soil and PFAS chemical properties, PFAS uptake and translocation to aerial parts is highly variable between plant species and even among varieties within species. In other words, the PFAS bioaccumulation factor (i.e. the ratio of PFAS concentration in plant tissues to the soil) is very variable among different plant species (Lesmeister et al., 2021). Leafy vegetables, characterized by high transpiration rates (e.g. lettuce and spinach), tend to accumulate more PFAS in aboveground tissues. Being in direct contact with the soil, root vegetables (e.g. carrots) can also accumulate

© Burleigh Dodds Science Publishing Limited, 2025. All rights reserved.

PFAS (Lesmeister et al., 2021; Li et al., 2022; Xu et al., 2024). Although PFAS contamination of crops grown on agricultural land has received a lot of attention, it is also important to consider the impact of home-grown crops on food safety (Lasters et al., 2024).

In summary, PFAS uptake and accumulation in crops depend on several factors, including the type of PFAS, the plant species, and soil properties. The full complexity and interactions of these factors are only starting to be unveiled and understanding these mechanisms is crucial for managing food safety.

## 2.2 Uptake routes for livestock

Many PFAS are known to bioaccumulate and biomagnify in food webs (Death et al., 2021; Peritore et al., 2023). There are numerous factors affecting the uptake of PFAS by livestock, including the physico-chemistry of the PFAS itself. Typically, a positive relationship can be observed between accumulated levels and the carbon chain length of the PFAS (Feng et al., 2023; Houde et al., 2011). In addition, PFSA accumulate more in organisms than PFCA (Conder et al., 2008). The types of PFAS present in products of animal origin may vary substantially among types of animal product, even within the same region (FSANZ, 2016; Van Leeuw et al., 2024b).

Livestock may encounter various sources of PFAS contamination (Fig. 4). Exposure through foraging has been reported to account for 78% of the PFAS exposure in ruminants, whereas soil exposure accounted for over 80% of the PFAS exposure in poultry and pigs (Brambilla et al., 2015). These proportions can vary depending on not only the exposure medium (i.e. type of feed, drinking water) but also housing and rotation practices and feeding habits. Overall, livestock is predominantly exposed through dietary pathways,

**Figure 4** Diagram of the primary sources for per- and polyfluoroalkyl substance uptake in animals. Made with Photoshop v26.0.

© Burleigh Dodds Science Publishing Limited, 2025. All rights reserved.

including the ingestion of contaminated soil particles, feed, and drinking water.

Commercial fishmeal for animals has been reported to contribute substantially to PFAS exposure in some farm animals, including cattle (Guruge et al., 2008). In addition, ruminants can be exposed through plant-based feed. Biosolids from wastewater treatment plants may be used as fertilizer for pastures. These biosolids could contain PFAS, which are subsequently transferred to soil and plants (Sepulvado et al., 2011), which are then consumed by grazing animals. PFAS uptake from biosolid exposure has also been reported for pigs (Fernandes et al., 2019). In addition, plant-based feed may also be already contaminated from the sites where crops were cultivated (e.g. Kowalczyk et al., 2012). Geophageous animals such as chickens may be directly exposed to PFAS from ingestion of soil particles or through feeding on earthworms living in their enclosures (Lasters et al., 2023). Drinking water of animals can also be contaminated through presence of nearby (industrial) sources or runoff (Lupton et al., 2022; Mikkonen et al., 2023). Studies from Australia have reported contamination of PFHxS in livestock and chicken eggs when exposed to a contaminated groundwater well (Bräunig et al., 2017; Drew et al., 2021).

Besides dietary exposure, PFAS-contaminated dust particles may also contribute to PFAS exposure in livestock. These particles can be either directly inhaled by the animals or first deposited onto animal feed sources such as vegetation. Animals that live in, or closer to, urbanized areas, which often contain more sources of PFAS, are typically exposed more to airborne PFAS compared to animals living in more remote areas (Cui et al., 2019). Furthermore, game animals accumulate higher concentrations of PFAS in their tissues compared to livestock due to outdoor living (Death et al., 2021). Finally, the substrate or bedding of animals could contribute to the PFAS exposure. PFAS concentrations in chicken eggs have been linked to exposure to recycled cardboard, dried paper pulp, and wood shavings (Fernandes et al., 2019).

Once taken up, most PFAS can be easily absorbed from the gastrointestinal tract. Since most PFAS are metabolically inert, they are not metabolized (Pizzurro et al., 2019). Elimination occurs primarily through urine or feces, although transplacental transfer, lactational transfer, and deposition into eggs have also been reported (Feng et al., 2023; Guruge et al., 2008; Kowalczyk et al., 2012). Being proteinophilic, PFAS typically end up in protein-rich tissues such as serum liver and kidneys (Pizzurro et al., 2019). Elimination rates and half-lives in animals differ substantially among types of PFAS and can range from several days to years depending on the animal species (as reviewed by Death et al., 2021).

© Burleigh Dodds Science Publishing Limited, 2025. All rights reserved.

## 2.3 Migration from food contact materials

FCM are materials and items designed to come into contact with food during its manufacture, transport, storage, etc. FCM should be sufficiently inert so their constituents neither adversely affect consumer health nor influence the quality of the food. Examples of FCM are containers used in the food transport industry, packaging materials, kitchenware, and tableware.

PFAS are commonly used in FCM due to their carbon-fluorine bonds, which provide high resistance to degradation, even at elevated temperatures. Additionally, their physicochemical properties offer resistance to oil and moisture. For decades, PFAS have been used to produce polytetrafluoroethylene (PTFE), commonly known as Teflon®, and to produce frying pans and pots due to its non-sticky properties. More recently, since the ban of certain single-use plastic products on the EU market in July 2021 (EU, 2019), alternatives were introduced on the market (e.g. bagasse, wheat pulp, wood, paper and board, etc.). However, these new alternatives require treatment of the material to achieve physical properties comparable to plastic. Indeed, without a coating or treatment, certain materials, such as paper and board, bagasse, etc., cannot withstand certain types of food (e.g. fatty food). Non-sticky coatings for paper and board are the most frequent application of PFAS in FCM (Bokkers et al., 2019; Peters et al., 2019). However, information about the composition of the PFAS blend in such applications is scarce (Lerch et al., 2023). When the FCM is in contact with food, migration of its constituents can occur and thus pose a risk to the consumer (EFSA Panel on Contaminants in the Food Chain (EFSA CONTAM Panel) et al., 2020). Indeed, the European Food Safety Authority (EFSA) considered that FCM can also be a source of human exposure to PFAS, although their impact is considered minor compared to other sources (EFSA Panel on Contaminants in the Food Chain (EFSA CONTAM Panel) et al., 2020). In a recent Belgian study on 78 paper and board articles and 15 pulp-based (bagasse, wheat) articles intended for takeaway, 11 PFAS (out of the 25 PFAS targeted) were detected in the samples, mainly perfluoroalkyl carboxylic acids (PFBA, PFPeA, PFHxA, PFHpA, PFOA, PFNA, PFDA, PFUnDA, PFDoDA and PFTrDA) and PFOS. While all pulp-based samples contained PFAS, 43% of the paper and board samples showed the presence of these chemicals (Van Leeuw et al., 2024a).

In order to minimize the risks of PFAS migration from FCM to food, the increased concerns about PFAS are leading regulatory authorities to act against the manufacturing and use of these chemicals on their own, as well as their placement on the market in other substances, mixtures, or articles (e.g. the PFAS restriction proposal to ECHA (ECHA, 2021)).

© Burleigh Dodds Science Publishing Limited, 2025. All rights reserved.

# 3 Analytical methods to determine per- and polyfluoroalkyl substances in food

## 3.1 *Sampling and preservation*

Guidelines for the proper sampling and preservation of PFAS-contaminated (or potentially contaminated) items have been set by multiple organizations, including the European Union (EU, 2022a). Sampling should be conducted by trained personnel who are familiar with potential contamination occurring during sampling and processing of samples. Since PFAS can be present in various consumer products, it is recommended to avoid certain PFAS-containing materials, such as PTFE, to ensure that samples are not contaminated by the sampling equipment itself (Rodowa et al., 2020). Suitable materials for sampling containers include stainless steel, polypropylene, or high-density polyethylene. These materials are recommended over other types of containers, e.g. glass, since PFAS may adsorb onto glass surfaces (ISO, 2009). In addition, clothes or gloves worn during sampling may contain fluorinated compounds (Gremmel et al., 2016) and some laboratory items may contain fluoropolymers that may cause contamination of samples during processing (Lath et al., 2019; Martin et al., 2004). Furthermore, cosmetics, such as sunscreen, have to be avoided during sampling as they may also contain PFAS (Keawmanee et al., 2015). Moreover, attention must be paid to the filters in laboratory fume hoods, especially when changing them. It is therefore important to assess the blank contamination using field, laboratory, and equipment blanks as quality control (Denly et al., 2019; Rodowa et al., 2020). In general, every material that might contain fluoropolymers has to be either avoided or cleaned with polar solvents (and tested for blank contamination), prior sampling or sample processing, but preferably PFAS-free materials have to be used (Androulakakis et al., 2022).

Food samples to be analyzed for PFAS contamination are often simply purchased or collected from supermarkets, or, in case of homegrown food, collected by hand from gardens (Bedi et al., 2023; Genualdi et al., 2022; Lasters et al., 2024, 2022; Liu et al., 2024; Piva et al., 2023; Roberts et al., 2023). To ensure PFAS stability, it is recommended to keep the samples below 0°C over time, before analysis. Analyte interconversions (increasing concentrations of two perfluorooctane sulfonamide acetic acids were correlated to decreasing concentrations of two perfluorooctane sulfonamide ethanols) were observed when water samples were stored at 4°C for 7 days, which is the typical temperature at which water samples are stored prior to analysis (Woudneh et al., 2019).

© Burleigh Dodds Science Publishing Limited, 2025. All rights reserved.

### 3.2 Extraction and clean-up

Food items are typically extracted using an organic solvent, followed by purification steps to reduce matrix-effects (Iannone et al., 2024; Ogunbiyi et al., 2023; Sadia et al., 2020; Sadia et al., 2020; Zhang et al., 2022). Commonly used methods to extract PFAS from food matrices are described below and include solid–liquid extraction (SLE) (Gao et al., 2018; Gosetti et al., 2010; Groffen et al., 2019; Powley et al., 2005; Zabaleta et al., 2015), solid-phase extraction (SPE) (Ogunbiyi et al., 2023), ion-pair extraction (IPE) (Lacina et al., 2011; So et al., 2006), and 'Quick, Easy, Cheap, Effective, Rugged and Safe' (QuEChERS) method (Genualdi et al., 2021; Hwang et al., 2024; Liu et al., 2024). Acetonitrile (ACN) and methanol (MeOH) are considered good and common solvents for the extraction of PFAS due to their medium polarity functionalities (Sadia et al., 2020). Although ACN showed better recoveries than some other solvents in the solid-liquid extraction of various food matrices (Sadia et al., 2020), degradation of PFAS homologues (primarily per- and polyfluoroalkyl ether acids) after 30 days in polar aprotic (neither accepting nor donating hydrogen atoms) solvents has been reported (Zhang et al., 2022), indicating that solvent choice is important as it can have substantial impacts on the analytical performance.

SLE aims to remove soluble components from solid matrices using large volumes of high-purity solvents such as MeOH and ACN. To overcome inefficient extraction of non-polar long-chain and hydrophilic short-chain PFAS, which has been reported for aquatic organisms (Valsecchi et al., 2013), the solvent polarities have to be adjusted based on the PFAS compounds of interest (Luque et al., 2010).

The application of SPE techniques results in the isolation and preconcentration of PFAS and is often used for direct extraction of PFAS from liquid samples due to the advantage of simultaneously extracting and cleaning-up the extracts (Ogunbiyi et al., 2023). Although these methods are often fast, they also have limitations: SPE cartridges can be expensive, whereas clogging of the SPE column with matrix residues can limit SPE efficiency. When using on-line SPE methods, carry-over effects tend to occur, requiring multiple washing steps with solvents varying in polarity (Ogunbiyi et al., 2023).

IPE uses a counter ion to facilitate the distribution of ionic PFAS into organic phases and is also commonly used to extract PFAS from solid food matrices (Iannone et al., 2024; Ogunbiyi et al., 2023; Sungur et al., 2018). Typically, the samples undergo either acid or alkaline digestion prior to ion pairing with tetra-n-butylammonium hydrogen (TBA), which helps in reducing the polarity of PFAS in the matrix (Ogunbiyi et al., 2023). On the downside, the matrix can be co-extracted during the extraction of the ion-pair, which often leads to a high lipid content in the extracts and subsequently poor PFAS recovery (Sadia et al., 2020). To address this issue, alkaline-based ion pairing extraction

© Burleigh Dodds Science Publishing Limited, 2025. All rights reserved.

coupled with SPE can been used (Sadia et al., 2020). Acid-based digestion, although already successfully used for food items before (Lacina et al., 2011), is less optimal than alkaline-based digestion as it can lead to partial oxidation of PFAS precursors resulting in impurities in the analysis of PFAS (Ogunbiyi et al., 2023; Young et al., 2012). Nonetheless, co-extraction of other compounds that interfere with short-chain PFAS also seems to occur during alkaline digestion, due to the presence of glycerol and fatty acids that are being released during digestion (Lacina et al., 2011).

The QuEChERS method, i.e. homogenized samples are agitated for a very short time (typically 1 min) with ACN and water solvents and then centrifuged with a mixture of salts to transfer PFAS in the upper phase, has the capability of extracting PFAS from a wide variety of food items, especially when coupled with an SPE cleanup step (Abafe et al., 2021; Gallocchio et al., 2022; Genualdi et al., 2021; Hlouskova et al., 2013; Meng et al., 2022b; Ogunbiyi et al., 2023; Organtini et al. 2023). The QuEChERS method can be used to extract both non-ionic and ionic PFAS, although further cleanup might hamper this (Meng et al., 2022b). To overcome the loss of some PFAS classes, a 'quick, easy, cheap, effective, rugged, safe, efficient, and robust' (QuEChERSER) method, which covers a broader range of polarities, can be used instead (Taylor and Sapozhnikova, 2022).

To eliminate matrix interference, endogenous substances must be removed after extraction and prior to analysis. Depending on the targeted analyte and matrix composition, retentive and non-retentive SPE methods have been used (Iannone et al., 2024; Ogunbiyi et al., 2023). Retentive SPE covers weak anion exchangers (WAX) and hydrophilic-lipophilic (HLB) sorbents, although the latter often shows low recoveries of short-chain PFAS due to poor retention of hydrophilic compounds on the sorbent materials (So et al., 2006). The electrostatic interactions between short-chain PFAS and WAX sorbents result in better recoveries (Pan et al., 2020). Mixed-mode WAX sorbents can be used to extract and fractionate cationic, zwitterionic and neutral PFAS and different solvents can be used to elute different PFAS homologues (Ogunbiyi et al., 2023). Non-retentive SPE methods using graphitized carbon have also been used to remove co-extracted compounds from the samples (Groffen et al., 2022; Lasters et al., 2022; Meng et al., 2022a; Wang et al., 2022a) and could eliminate matrix interferences prior to instrumental analysis. However, graphitized carbon might not be sufficient as a clean-up for high-fat and protein-based food items, for which suppression of PFAS homologues has been reported (Sadia et al., 2020). In such cases, an additional clean-up step, e.g. using WAX SPE, might be required. Graphitized carbon SPE are used, often as a first purification step, in combination with WAX SPE for dirty matrices.

© Burleigh Dodds Science Publishing Limited, 2025. All rights reserved.

### 3.3 Analytical techniques

High and ultra performance liquid-chromatography tandem mass spectrometry (HPLC-MS/MS or UPLC-MS/MS) are the most commonly used techniques to quantify PFAS concentrations in food items (Iannone et al., 2024; Ogunbiyi et al., 2023). LC-MS-based approaches often use reverse phase separation stationary phase columns or columns with polar functionalized $C_{18}$ alkyl chains, allowing for better retention of more polar chemicals at the first minutes of elution (Schiavone and Portesi, 2023). These LC-MS typically operate with an electrospray ionization (ESI) mode (Groffen et al., 2022; Lasters et al., 2022; Meng et al., 2022a; Wang et al., 2022a), being more sensitive than atmospheric pressure chemical ionization (APCI) and atmospheric pressure photoionization (APPI) (Zacs and Bartkevics, 2016). However, ESI is more susceptible to ion enhancement or suppression caused by matrix effects compared to these other ionization sources (Zacs and Bartkevics, 2016). Another targeted approach includes nano-ESI-HRMS, a high sensitivity, short run time, and low injection volume demanding technique, facilitating the detection of polar molecules, including ultrashort-chain PFAS, which are often underestimated or overlooked in LC-MS (Wu et al., 2022). However, to the best of our knowledge this technique has not yet been applied to food items. GC-MS, albeit used less often, can be applied to analyze neutral and volatile PFAS from food matrices or FCMs (Wolf et al., 2024). Furthermore, high resolution mass spectrometry (HRMS) techniques can be applied as sensitive method for trace analysis of PFAS in complex food matrices, ensuring high sensitivity and selectivity (Barola et al., 2020; Chiesa et al., 2018; Chiumiento et al., 2023; Zacs and Bartkevics, 2016).

Despite the fact that over 7 million PFAS molecules are listed in PubChem (Schymanski et al., 2023), targeted analyses of PFAS typically include panels of only 40–50 compounds detected by ESI-MS using isotopically labelled standards. Therefore, non-targeted qualitative methods are needed to close the fluorine mass balance (i.e. a tool used to determine the fraction of fluorine accounted for by known PFAS) as much as possible by measuring the PFAS concentrations that are not targeted in typical panel analyses, including PFAS precursors and polymers, as well as organo-fluorine and inorganic fluorine.

Suspect-screening and untargeted qualitative approaches, typically using LC-accurate mass MS instruments (HRMS) using time-of-flight (TOF) or Orbitrap, provide information on accurate mass and empirical formula for the identification of unknown PFAS. A generalized workflow for the identification of novel PFAS includes the selection of peaks by mass filtering (only in case of suspect-screening), followed by fitting the molecular formula to the exact mass of interest, and then the generation and ranking of structures using MS/MS databases and fragmentation prediction tools (Bletsou et al., 2015).

© Burleigh Dodds Science Publishing Limited, 2025. All rights reserved.

Both data-independent acquisition (DIA) and data-dependent acquisition (DDA) approaches are often used as acquisition mode (Guo and Huan, 2020; Ogunbiyi et al., 2023). In DDA mode, only certain precursor ions are selected, based on their intensity and abundance in full scan mode, and subsequently fragmented one at a time, whereas in DIA mode all selected ions in the mass to charge ratio (m/z) range are fragmented and ionized, allowing a more comprehensive screening and detection of all peaks without inclusion of a mass filter or threshold before injection of the samples into the MS instrument (Ogunbiyi et al., 2023). A second non-targeted approach is suspect-screening analysis, making use of databases with information on the accurate mass, molecular formula and isotopic patterns of potential PFAS candidates (Boisacq et al., 2023; Phillips et al., 2018). Although non-targeted approaches have been used to characterize PFAS contamination in various matrices, including FCM (Boisacq et al., 2023; Liguori-Bills et al., 2023; Ren et al., 2022; Stroski et al., 2024; Vera et al., 2024), they have only been used sporadically to determine unknown, novel PFAS in food items (Kaufmann et al., 2022; Wang et al., 2022b).

Other qualitative techniques to characterize unknown PFAS in samples include the TOP assay, particle-induced gamma-ray emission (PIGE) spectroscopy, fluorine nuclear magnetic resonance spectroscopy ($^{19}$F-NMR), and determination of adsorbable organic fluorine (AOF) and extractable organic fluorine (EOF) by combustion ion chromatography (CIC). Although, to the best of our knowledge, these techniques have not been applied on food samples yet, some have been used on FCM (Sapozhnikova et al., 2023; Schaider et al., 2017). Seen that they can provide complementary information to targeted, suspect-screening, and non-targeted approaches, these are promising techniques that require further method optimization and validation for food products.

The TOP assay oxidizes PFAA precursors to PFAA which can be measured using targeted analysis. By measuring the increase in PFAA concentrations before and after TOP assay, an estimate of the concentration of PFAA precursors can be obtained (Ateia et al., 2023). However, this is nothing more than just an estimate because oxidation can be incomplete (Ateia et al., 2023). The chain length of the PFAA after oxidation gives an indication of the minimum chain length of the precursor molecule. Nonetheless, precursors cannot be differentiated based on their functional group as some PFAA precursors are oxidized to the same end product. By using gamma-ray wavelength emissions of fluorine when impacted by a proton ion beam, PIGE can be used to assess total fluorine of materials isolated on a thin surface (Ritter et al., 2017; Schaider et al., 2017). Fluorine-based calibration standards can be used to perform quantification. The downside of this technique is that PIGE cannot distinguish between PFAS groups and chain lengths nor between organic and inorganic

© Burleigh Dodds Science Publishing Limited, 2025. All rights reserved.

fluorine. $^{19}$F-NMR can be used to determine total PFAS. The terminal -$CF_3$ moiety provides a unique chemical shift which can be differentiated by $^{19}$F-NMR from other organic fluorine signals and inorganic fluorine (Camdzic et al., 2023). The AOF/CIC technique is, like PIGE, not specific to PFAS, but can be used to determine the organofluorine concentration by passing samples through carbon-based sorbents on which organofluorine chemicals adsorb. After removal of inorganic fluorine (which sometimes can be difficult), the sorbent is combusted at high temperatures and the gas, after passing through deionized water, is then analyzed for fluorine by ion chromatography (Wagner et al., 2013). In the EOF method, the extract of the sample is combusted at high temperature to measure the release fluorine using ion chromatography (Miyake et al., 2007). Similar to AOF and PIGE, it does not provide details on the structure nor whether the fluorine is inorganic or organically bound, but the combination of AOF and EOF with CIC makes it possible to estimate the levels of organofluorine compounds (Aro et al., 2022).

### 3.4 Quality control/quality assurance

Thousands of PFAS are registered in databases, and the development of new methodologies is essential to monitor these compounds effectively. However, the analysis of PFAS in various matrices and the related quality assurance (QA) and quality control (QC) measures pose several challenges.

In Europe, most laboratories conducting PFAS analysis follow the 'Guidance Document on Analytical Parameters for the Determination of Per-and Polyfluoroalkyl Substances (PFAS) in Food and Feed' issued by the European Union Reference Laboratory for Halogenated POPs in Feed and Food (EURL-POPs) (EURL for halogenated POPs in feed and food, 2024). This document outlines essential guidance on analytical QA, method validation parameters, instrumentation details, and protocols for measuring and reporting results. Furthermore, for drinking water, Member States are required to adhere to the recommendations of the 'Technical guidelines on the methods of analysis for monitoring PFAS in water intended for human consumption' (EU, 2020, 2024), describing specific methodologies and analytical requirements to analyze PFAS in drinking water.

A primordial challenge is the limited availability of analytical reference standards (Wang et al., 2023) whether native or isotopically labeled, as accurate detection often depends on the existence of established reference points for comparison. Indeed, analytical reference standards exist only for a limited number of PFAS. Some examples of suppliers are Wellington Laboratories, LGC, Agilent, and Restek.

In addition, the analysis of PFAS is further complicated by the presence of isomers. Indeed, some PFAS are a mixture of linear and branched structural

© Burleigh Dodds Science Publishing Limited, 2025. All rights reserved.

isomers. This is mainly the case for sulfonate PFAS, such as PFOS, where the branched isomers are challenging to quantify. The EU guidance documents have provided approaches to ensure the comparability of the results obtained by different laboratories. To harmonize the quantification, the EURL-POPs recommend quantifying the branched isomers by one of the three following approaches: using (1) an analytical PFOS standard that includes a known ratio of branched and linear PFOS, (2) a linear PFOS standard and calculating the concentration based on the average of the two mass transitions (m/z 499 → 99 and m/z 499 → 80), or (3) a linear PFOS standard and calculating the concentration on the parent ion (m/z 498.93022) by HRMS. In the drinking water technical guidelines, it is highlighted that the quantitation of PFAS should include its linear and branched forms (EU, 2024).

Laboratories aiming to be qualified in PFAS analysis should validate their methods according to the criteria set out in Commission Implementing Regulation (EU) 2022/1428 (described in section B.3.; Methods of analysis: specific performance requirements) and the available guidelines, such as the applicability of the foodstuffs specified in Regulation (EC) No. 1881/2006 (linearity, limit of quantification, recovery, selectivity of the method, specificity, trueness, or precision of the analytical method). Method validation is a crucial process to ensure that a method is suitable for its intended use. Based on the results of the validation, the measurement uncertainty can be determined, which is a crucial parameter when reporting analytical results and evaluating compliance.

Furthermore, it is recommended to include QC parameters during each analysis sequence to evaluate the quality of the data obtained and evaluate potential errors and/or contamination (EU, 2009, 2022a; EURL for halogenated POPs in feed and food, 2024). It is recommended to routinely check the reagent blanks (i.e. procedural blanks, analyzed according to the validated analytical method) to ensure that no contamination has occurred during the procedure or sample injection. If contamination continuously occurs, it should be recorded on a QC chart to plot the occurrence of the contamination over time graphically. It includes statistical control limits that indicate the acceptability or invalidity of the results. In addition, control samples should be used as an internal QC measure and recorded on QC charts. A control sample is a sample that has been fortified with an analyte of known concentration and is used to verify the precision and accuracy of the analytical method. Certified reference materials (CRMs) are ideally recommended for these purposes. However, the availability of CRMs is limited (Pelch et al., 2023; Valsecchi et al., 2013; Zhou et al., 2023) but exists for some matrices like drinking water (e.g. IRMM-428 WATER from Institute for Reference Materials and Measurements (IRMM) and fish (e.g. IRMM-427 PIKE-PERCH from IRMM)). When CRMs are not available, materials from successful interlaboratory studies can be used. Participating

© Burleigh Dodds Science Publishing Limited, 2025. All rights reserved.

in interlaboratory proficiency tests (PT) is essential for laboratories involved in chemical analysis. These tests help to ensure that laboratories maintain high standards of accuracy and reliability in their testing methods. For PFAS, different PT schemes are, among others, available at FAPAS Proficiency Testing, ERA Waters, LGC Standards, DRRR or Phenova. Furthermore, the EURL-POPs organizes different PTs focusing on the analysis of PFAS in food, which are commercially available. If PT or reference material is not available, laboratories have alternative options. They can initiate comparisons with other laboratories independently or perform a blind sample test that has been fortified by an external party not involved in the analysis.

In summary, although some guidance documents and directives are available in view of PFAS analysis, significant progress is still required. Specifically, the lack of certain reference materials and standards poses challenges limiting the effectiveness and consistency of testing methods.

## *3.5 Analytical challenges related to legislation*

The analysis of PFAS poses significant challenges, particularly in reaching low LOQ (Brunn et al., 2023; EURL for halogenated POPs in feed and food, 2024; PerkinElmer, 2023; Stecconi et al., 2024) with adequate specificity and selectivity, avoiding contamination, and controlling potential interferences from the matrix.

Lowering the LOQ for ultra-short- and short-chain PFAS, such as trifluoroacetic acid (TFA) (Björnsdotter et al., 2020; Wang et al., 2024) and perfluorooctanoic acid (PFBA) (Stecconi et al., 2024), is mainly complicated due to contamination potentially issued from laboratory equipment or instruments, consumables, and solvents (Björnsdotter et al., 2020; Wang et al., 2024). In addition, contamination can also occur for long-chain PFAS (Nahar et al., 2023) and are visible, e.g. in extraction blank samples (Nahar et al., 2023; Organtini et al., 2023) or in the instrumental blanks when issued from the LC instrument, LC solvent or mass-labeled standards (Phomsopha et al., 2022). Although the exact source may differ and remain unknown, it is crucial to take appropriate measures to avoid or drastically reduce laboratory contamination before analysis (Björnsdotter et al., 2020; Liang et al., 2023) to meet the LOQ criteria outlined in the Commission Implementing Regulation (EU) 2022/1428 (EU, 2022a) and Commission Recommendation (EU) 2022/1431 (EU, 2022b).

Analysis of PFAS is typically conducted by LC coupled to a triple quadrupole mass spectrometer (MS/MS), which offers enhanced sensitivity and specificity due to the detection of multiple MS/MS transitions for confirmation. However, short-chain PFAS like PFBA and perfluorooctanoic acid (PFPeA) suffer from limited fragmentation, yielding only one MS/MS

© Burleigh Dodds Science Publishing Limited, 2025. All rights reserved.

transition and no secondary transitions for verification (Bangma et al., 2023, 2021; Genualdi et al., 2022). This limitation can be resolved by a second analysis on an LC column with different characteristics to confirm the detection of the short-chain PFAS (Ballesteros-Gómez et al., 2010), although this can be time-consuming.

Alternatively, HRMS (Chiumiento et al., 2023; Ogunbiyi et al., 2023) could be applied in which the precision and accuracy of the data can benefit largely from the high level of accurate mass-to-charge (m/z) ratio precision (<5 ppm). However, other issues may arise, such as the lack of sensitivity of hexafluoropropylene oxide-dimer acid (HFPO-DA or GenX) (Brase et al., 2021; Mullin et al., 2019; Vughs et al., 2019) at low concentrations in HRMS due to in-source fragmentation.

Another challenge is the analysis of ultra-short-chain PFAS (e.g. TFA) due to their polar and water-soluble nature, which leads to non-retention on reversed-phase LC columns and inadequate separation (Björnsdotter et al., 2020), something which can be solved by using ion-exchange columns, with superior retention properties for hydrophilic substances, together with optimized mobile phase pH and flow rates (Taniyasu et al., 2008).

Finally, the matrix in which PFAS are present may also interfere with the analysis. For example, the bile salt taurodeoxycholic acid can compromise the analysis of PFOS in the liver and eggs since these two molecules coelute and share the same MS/MS transitions (Ballesteros-Gómez et al., 2010; Bangma et al., 2023; Benskin et al., 2007; Chiumiento et al., 2023). This issue can be addressed by quantifying PFOS using the other MS/MS transition and identifying PFOS with a HRMS instrument or another LC column with different characteristics.

The trade-off between sensitivity and specificity is a common consideration in analytical chemistry, and researchers must carefully evaluate the analytical requirements of their study to determine the most suitable instrumentation for achieving accurate and reliable results in PFAS analysis. In addition, avoiding and reducing the risk of PFAS contamination during sampling and analysis should be aimed (see below).

Overall, monitoring various PFAS across various matrices is crucial for generating new and up-to-date data and facilitating risk assessment and the establishment of new maximum levels in European legislation, potentially extending to other PFAS or foodstuffs than those currently regulated. However, setting new and lower maximum limits requires laboratories to establish lower LOQ, which in turn requires the development of more sensitive analytical methods to meet the criteria outlined in the regulations. Therefore, it can be concluded that the analysis of PFAS following regulations remains challenging.

© Burleigh Dodds Science Publishing Limited, 2025. All rights reserved.

# 4 Mitigation strategies for agriculture

## 4.1 Phytomanagement

Phytomanagement strategies in view of PFAS prevention and/or reduction may vary from soil amendment applications, biological (microbes and/or plants) remediation, crop, and variety selection (adapted to the level of soil contamination and specific crop requirements), etc. In the paragraphs below, some examples and their limitations are described.

Limiting the application of PFAS-contaminated biosolids such as sewage sludge, monitoring, and if needed filtering, irrigation water, and reducing atmospheric deposition to prevent PFAS from entering agricultural land are important steps in phytomanagement. In addition, a key aspect to prevent PFAS accumulation in crops is to reduce PFAS mobility in contaminated soils (Bolan et al., 2021). This can be achieved by adding matrices/substances that can adsorb PFAS present in the soil, such as activated carbon (Navarro et al., 2023), biochar (Ramos and Ashworth, 2024; Sørmo et al., 2021), or organically modified clays (Dong et al., 2024; Mukhopadhyay et al., 2021). If these strategies prove to efficiently limit PFAS uptake by plants, it remains to be assessed if this positive effect is stable over time (see (Navarro et al., 2023)) and what are the long-lasting impacts of such amendments on soil health, which is a key aspect of the Mission Soil of the EU Green Deal (EU, 2023b). High-temperature treatments (e.g. pyrolysis) (Alinezhad et al., 2022; Keller et al., 2024) and electrochemical degradation techniques (Mirabediny et al., 2023) have been developed to break down PFAS, but, in addition to being soil-health unfriendly, these methods are energy-intensive and expensive and are therefore not sustainable to remediate large areas of contaminated agricultural land.

Remediation strategies based on the exploitation of biological activities suffer not only from the fact that PFAS are generally very stable and resistant to breakdown but also because of the complexity of PFAS mixtures in soils. Hence, developing remediation methods based on microbial degradation of PFAS in soil, as have been successfully applied for other organic contaminants (e.g. petroleum hydrocarbons (Chunyan et al., 2023; Wu et al., 2017)), is challenging, as most naturally occurring microbes lack enzymes capable of breaking down the stable carbon–fluorine bonds found in PFAS. However, new approaches in environmental biotechnology are being explored to leverage certain microbial (Berhanu et al., 2023; Smorada et al., 2024) or enzyme-assisted (Harris et al., 2024; Marciesky et al., 2023) processes that may hold promise for biodegradation.

Phytoremediation, using plants to remove, bind or degrade contaminants from soils, is also being explored as a natural and cost-effective alternative way to manage PFAS contamination in agricultural land (Kavusi et al., 2023).

© Burleigh Dodds Science Publishing Limited, 2025. All rights reserved.

Although several studies have reported some success (e.g. hemp (Nason et al., 2024), or grass species (He et al., 2023)), such approaches encounter similar limitations. Indeed, phytoextraction approaches, where the aim is to accumulate PFAS in harvestable parts of the plants (e.g. roots or stems), are limited due to the overall limited uptake and mobility of PFAS in most plant species. In addition, PFAS uptake by plants is chain length dependent, as was mentioned elsewhere. A more favorable outcome is expected for the more mobile short-chain PFAS (e.g. in sunflower and mustard (Nassazzi et al., 2023)). The effectiveness of specific soil amendment applications, including surfactants and desorbing agents, enhancing PFAS mobility is also being examined (Bolan et al., 2021).

A more viable option is to implement phytomanagement solutions (Burges et al., 2018) at the contaminated sites in which a plant cover is used to minimize PFAS movement, preventing wind and water erosion that could otherwise disperse PFAS to adjacent areas or to underground water (Evangelou and Robinson, 2022). If the applied plant species limits the accumulation of PFAS in above-ground tissues, the plant biomass can then be valorized for non-food applications, such as biomaterials or energy, restoring some economic values to heavily contaminated agricultural land.

Another effective strategy for minimizing PFAS contamination in cultivated crops is selecting plant species or varieties that naturally limit PFAS uptake and translocation to edible tissues. Since PFAS uptake and translocation vary widely among plant species (as described elsewhere), selecting crops with lower bioaccumulation potential (e.g. cereals grains instead of leafy vegetables) for cultivation on marginally contaminated soils can help in reducing the risk of PFAS entry into the food chain. Furthermore, natural variation in the extent of PFAS uptake and accumulation within crop species (e.g. Yu et al., 2021, 2018) could be exploited to identify varieties with lower PFAS-uptake risk, and potentially initiate selective breeding programs (Xiang et al., 2019).

In summary, mitigating PFAS contamination in agriculture is complex due to the persistence and mobility of PFAS in various environmental compartments and to the high variability of PFAS uptake and translocation observed among plant species and PFAS molecules themselves. Therefore, a comprehensive PFAS management strategy in agriculture should typically combine several approaches, i.e., soil amendment application, biological (microbes and/or plants) remediation, crop and variety selection, adapted to the level of soil contamination and specific crop requirements. Although these offer promising future directions for PFAS remediation, much work is still required to design efficient and cost-effective phytomanagement solutions. Tailoring these approaches to specific agricultural contexts will furthermore be essential to minimize the negative impact of PFAS on food safety and to support sustainable farming practices in affected areas.

© Burleigh Dodds Science Publishing Limited, 2025. All rights reserved.

### 4.2 Reduction of livestock exposure

As described elsewhere, farm practices and housing conditions may have a substantial impact on the concentrations of PFAS in livestock and may thus also support exposure reduction. To gain a better insight into the most optimal risk assessment and management strategies, both the farm information, e.g. housing conditions, and environmental monitoring data should be available. For example, provided that the quantification and detection limits are below relevant health-based criteria for PFAS, identification of PFAS in very low environmental concentrations or PFAS with short half-lives are unlikely to have a large influence in the risk assessment of a specific contaminated site (Mikkonen et al., 2023).

There are currently few mitigation strategies available to reduce PFAS contamination already present in affected livestock. To identify the food-safety risks of PFAS contamination of livestock, the concentrations of PFAS in the tissues should be compared to the maximum levels according to regulation (e.g. Table 1 for European maximum levels) or to the indicative levels above which further investigation into the causes of the contamination should be undertaken by EU Member States (Table 2). However, these levels are based on concentrations of PFAS in meat and offal, meaning that animals have already been slaughtered then and that attempts to reduce the PFAS concentrations in those animals by managing the livestock is no longer possible. Only for the remaining, living part of that livestock this can be aimed. Using tissue partitioning coefficients and half-time elimination information, acceptable serum concentrations for cattle meat (based on the maximum tolerable levels according to regulation) can be calculated. These acceptable serum levels can then be compared with reported livestock serum levels from PFAS-contaminated sites (Mikkonen et al., 2023). Then, by using clearance models, the duration of depuration needed to achieve the acceptable levels can be estimated (Mikkonen et al., 2023).

There are multiple ways to minimize exposure of livestock, of which most primarily focus on remediation of soil and water, being the basis of the animal food chain. Each technique has its own pros and cons, and all require further monitoring. The conventional and emerging remediation technologies for soil and water have been reviewed elsewhere (Concawe, 2024; Mahinroosta and Senevirathna, 2020; Ross et al., 2018; Wanninayake, 2021).

The uptake routes for livestock have been described elsewhere. Regarding risk assessment and mitigation, the determinants of PFAS exposure in livestock should be addressed as measures on the level of such factors (e.g. foraging/grazing and animal feed, dust, substrate, and bedding) can help to reduce exposure. For example, a positive association between clay fraction in the outdoor substrate of a chicken enclosure and PFAS concentrations in the

© Burleigh Dodds Science Publishing Limited, 2025. All rights reserved.

chicken eggs has been reported (Lasters et al., 2023). In addition, the presence of earthworms in the soil was considered a significant determinant of presence of PFAS in chicken eggs. To reduce exposure via these sources, it has been recommended to introduce a sandy soil which reduces PFAS adsorption to this sandy layer as well as the amount of earthworms present in the soil (Lasters et al., 2023). The number of sources and types of drinking water (e.g. tap water, rainwater, surface water, etc.) should be considered as PFAS concentrations may vary across the different types. Finally, PFAS contamination in the feed, and substrates/bedding should be monitored. As some animals may spend a substantial amount of time in feedlots prior to slaughter, a reduction in the exposure to the feedlots compared to the pastures allows to estimate the depuration timeframes needed to achieve acceptable levels for human consumption (Mikkonen et al., 2023).

# 5 Case studies

## 5.1 Per- and polyfluoroalkyl substance *levels in Belgian food: effect of origin*

In 2021, the federal research institute for public health, animal health, and food safety in Belgium (Sciensano) launched in Belgium the project FLUOREX 'Exposure assessment of perfluoroalkyl substances as a follow-up on the concerns raised in the recent opinion of EFSA', financed by the Federal Public Service Health, Food Chain Safety and Environment, Contractual Research (FLUOREX, RF 21/6350) (Van Leeuw et al., 2024a). The PFAS selection (25 molecules in total) included the four regulated PFAS (PFOS, PFOA, PFNA, PFHxS), other carboxylate-PFAS (C4-C14), sulfonate-PFAS (C4-C13), and PFAS substitutes (DONA, F53B minor and major forms and HFPO-DA) based on Recommendation (EU) 2022/1431 (EU, 2022b) and Regulation (EU) 2023/915 (EU, 2023a). First, a comprehensive sampling yielded 283 samples, reflecting all beverages and foods relevant for PFAS exposure and Belgian consumption habits. Next, the selected samples were analyzed with validated methods using an Orbitrap Q Exactive (LC-HRMS, Thermo), thereby taking into account the QC measures from the EURL guidance (EURL for halogenated POPs in feed and food, 2024). It should be noted that these methods are very sensitive, with most of the LOQs, defined as the lowest validated level, ranging from 0.002 ng/g to 0.05 ng/g, depending on the compounds and matrices considered.

With this very sensitive analytical method, widespread PFAS contamination in various foodstuffs was demonstrated. It should be noted that water samples were not included (Van Leeuw et al., 2024b). Although approximately 57% of the samples contained none of the 25 studied PFAS, about 23% contained only one PFAS, 15% between 2 and 5 PFAS and less than 5% between 6 and 11 PFAS (Fig. 5).

© Burleigh Dodds Science Publishing Limited, 2025. All rights reserved.

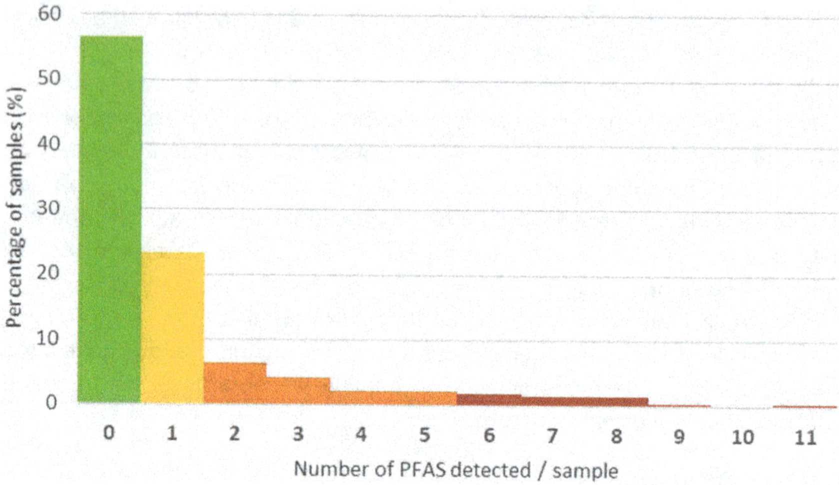


**Figure 5** Overview of the percentage of PFAS detections in the samples (out of 25 PFAS in total) (FLUOREX project) (Van Leeuw et al., 2024a, 2024b).

PFOS was the most frequently detected compound (19% of the samples), followed by PFOA (15%). Nine PFAS were below the LOQ (i.e. PFTeDA, PFPeS, PFHpS, PFDS, PFUnDS, PFDoDS, PFTrDS, Minor F53B, HFPO-DA) in the samples. For the food commodities considered, at least one PFAS was quantified in 74%, 68%, 60%, and 57% of the fish, vegetables, water, and composite dishes samples, respectively (Fig. 6). No PFAS were detected (i.e. above the LOQ) in the nine egg samples independently of their origin (caged, free-range, organic), neither in five seasonings and sauces nor in three sugar and confectionery. The average relative contribution of the four regulated PFAS to the sum of the 25 analyzed PFAS was 53%, ranging from 0% to 100%. In 35 samples (i.e. about 28% of the samples with detected PFAS), mainly plant-based products, only the most short-chain PFAS (< 7 carbons) other than those regulated were found (PFBA, PFPeA, PFHxA, PFHpA, and PFBS). For animal-based products, PFOS was mostly detected and, specifically for fish and other seafood and game animals, also carboxylic molecules with long chains (8–13 carbons) (PFOA, PFNA, PFDA, PFUnDA, PFDoDA, and PFTrDA) were detected.

As for the concentrations reported, the total concentration for the 25 PFAS analyzed in this study ranged from <LOQ to 5.4 μg/kg. Individual detections higher than 1.0 μg/kg were found for PFOA, PFUnDA, and PFTrDA (maximum of 2.9 μg/kg) in crab samples, PFOS in grey shrimp and PFBA in red pepper.

Only one sample among the regulated categories (see regulation section) exceeded the maximum levels set by the Commission Regulation (EU) 2023/915

© Burleigh Dodds Science Publishing Limited, 2025. All rights reserved.

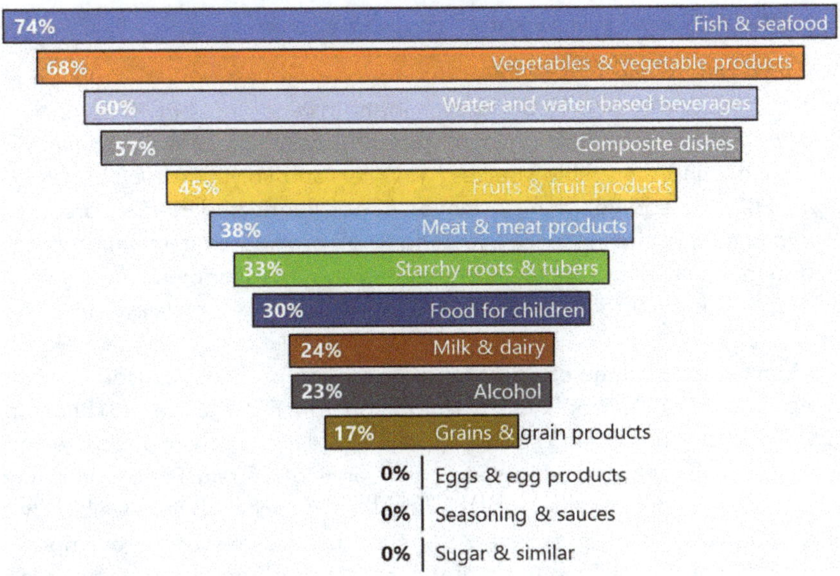


**Figure 6** Percentage of foodstuffs on the Belgian market that contains polyfluoroalkyl substance. Made with Microsoft PowerPoint. Results of the FLUOREX project (Van Leeuw et al., 2024a, 2024b).

(EU, 2023a), i.e. the crab sample from the Irish sea with a PFOA content of 1.2 µg/kg. Furthermore, indicative levels are mentioned in Recommendation (EU) 2022/1431 (EU, 2022b) for several groups, meaning that further investigation of the causes of the contamination should be carried out when the levels are exceeded. Seven samples of vegetables, mainly from Belgium (5/7), and one fruit (also from Belgium) exceeded the indicative level of 0.01 µg/kg for PFOA in fruits and vegetables, with a maximum concentration of 0.20 µg/kg in oyster mushrooms.

Despite the detection of many PFAS in food items and beverages from Belgian supermarkets, only very few exceedances of the maximum acceptable levels were reported. On the other hand, the concentrations in home-produced vegetables, fruits and chicken eggs are often much higher than those reported for items bought in supermarkets, especially close to PFAS sources and hotspot areas. This is not only the case in Belgium (Colles et al., 2022; Consortium UAntwerpen, VITO, PIH, UHasselt en VUB, 2023; Lasters et al., 2022, 2023, 2024) but also in other countries across the globe (Dimitrakopoulou et al., 2024; Scher et al., 2018). Although speculative, this is likely caused by the less controlled and regulated environment in which home-produced feed is cultivated or housed (e.g. housing of hens) compared to food destined for supermarkets.

© Burleigh Dodds Science Publishing Limited, 2025. All rights reserved.

## 5.2 Effect of food processing – case blue food

Amongst others, the fate of PFAS during food processing varies depending on the PFAS compound (subclasses, subgroups, individual PFAS compounds (e.g. long-chain versus short-chain), the processing process(es) applied, the food matrix and the edible species (ITRC Per- and Polyfluoroalkyl Substances Team, 2023). In addition, transformation of polyfluorinated PFAS or precursors occurs in certain circumstances, leading to other PFAS contamination profiles in food matrices before and after processing, as well as temporal and/or spatial differences in occurrence of PFAS (ITRC Per- and Polyfluoroalkyl Substances Team, 2023).

Most studies on the effect of food processing on PFAS focused on (1) the compounds PFOS and/or PFOA, (2) migration from FCMs, and/or (3) blue food. In addition to studies being scarce, the results are also often inconclusive, more specifically about whether losses or increases of certain PFAS compounds occur. Studies on changes on the level of bioaccessibility/-availability of PFAS due to cooking/processing are especially scarce. Some examples of studies that addressed the topic of the impact food processing may have on the behavior and fate of PFAS specifically in blue food are listed below (non-limitative). Migration aspects from contact materials are not included here but were discussed before. Most available studies on the impact of food processing on PFAS concerned blue food, an important contributor of dietary exposure to PFAS (EFSA Panel on Contaminants in the Food Chain (EFSA CONTAM Panel) et al., 2020), though with inconsistent results.

In 2014, Bhavsar et al. (2014) reported increasing PFOS concentrations during cooking of fish fillets, but largely unchanged total amounts of PFOS. The cooking method used or the fish species itself appeared to lead to minor changes. Their observations did thus not confirm the earlier findings of Del Gobbo et al. (2008) who found cooking by baking, boiling, and frying 18 fish species to reduce perfluorinated acids (PFAA) with baking having the largest effect (and even no longer detectable when baked for 15 min at 163°C). In contrast, Vassiliadou et al. (2015) reported generally increased PFAS concentrations in most cases due to water evaporation after frying or grilling of fish/shellfish. Cooking seafood by steaming appeared to have different outcomes depending on the fish species and PFAS compound (though not consistently related to the compounds' chain length) (Barbosa et al., 2018). For instance, in mussels, PFDA was found to be no longer detectable after steaming, while PFBA significantly increased. In contrast, in skipjack tuna, a significant increase of PFTrDA and PFDA was observed, in contrast to a significant decrease of PFUnDA, PFDoDA, and PFOS as a result of cooking by steaming. Taylor et al. (2019) found PFAS concentrations not to be reduced in a consistent way by cooking (boiling, frying, and baking) seafood. For instance, while PFOS,

© Burleigh Dodds Science Publishing Limited, 2025. All rights reserved.

PFOA, and PFHxS doubled in concentration after boiling of prawns, boiling of Blue Swimmer crab did not significantly change the PFOS concentration but halved those of PFHxS and PFOA. After cooking (boiling, baking, frying), PFAS appeared to be liberated to oil and water; the authors suggested mass loss during cooking to be responsible for increased concentrations. Sungur et al. (2021) studied the impact of cooking method (boiling, baking, frying), cooking oil (olive, sunflower), and cooking time (10, 15, 20 min) on the level of PFOS and PFOA in eight fish species. The decrease after cooking was as follows: frying>boiling>baking, with the highest reduction (70–86% for PFOA and 82–89% for PFOS) when frying in sunflower oil for 20 min. According to the authors, fish species did not seem to play a significant role. In their systematic review and meta-analysis, Vendl et al. (2022) highlighted the reduction of PFAS concentrations due to thermal processing (cooking) of blue food. They found PFAS reduction to be favored when cooking for longer times and using larger relative volumes of cooking media, as well as a stronger decrease of shorter chains PFAS in oil-based cooking compared to PFAS with longer chains. Also, Chen et al. (2023) stated that cooking by means of boiling or frying could efficiently decrease PFAS in seafood whereas Wei et al. (2024) recommends including water-based cooking methods to minimize PFAS health risks.

Some studies focused on one fish or seafood species. For example, Kim et al. (2020) observed a reduction of the total PFAS content when grilling (91%), steaming (75%), frying (58%), and braising (47%) mackerel fillets. However, variation among the different PFAS also appeared to be the case with a significant effect for some cooking methods for some compounds, though not for all. Braising (with seasoning) together with potato decreased PFAS content in fish even more but increased it in potato. Pretreatment by washing decreased the PFAS content by 74% on average, though not significantly further when increasing the washing frequency. Pretreatment of 15 min soaking in liquid also reduced the PFAS concentrations, i.e. by 51% when using sake and by 80% when using rice-washed solution. PFBA and PFTrDA remained present after all pretreatments assessed. Unfortunately, not all EFSA-4 compounds were present in the naturally contaminated fish sample and therefore, risks could not be evaluated in this study. Also for grass carp, the effect of boiling, steaming, grilling, and frying on the levels of PFAS was studied (Hu et al., 2020). While concentrations (ng/g wet weight) of short-chain PFAS decreased, those of long-chain PFAS increased in cooked fish fillets. Expressed as amounts (ng) of PFAS, these decreased in cooked fish fillets except for PFOS. In the cooking juice, short-chain PFAS were dominant with the highest amounts of PFBS after boiling and frying.

In addition to the results obtained for fish cooking, also the majority of processes used for preparing traditional whale foods (beluga blubber) gave inconsistent results as to their influence on the concentrations of PFAS

© Burleigh Dodds Science Publishing Limited, 2025. All rights reserved.

(Binnington et al., 2017). Roasting increased the concentration of certain PFAS (through solvent depletion), whereas the solid–liquid phase separation during aging depleted some ionogenic PFAS from the lipid-rich liquid oil phase. Only one literature source reported changed bioaccessibility of PFAS (PFOS and PFUnDA) when cooking seafood, i.e. a generally higher bioaccessibility percentage (Alves et al., 2017).

## 6 Conclusion and future trends

The chapter focuses on the occurrence, analysis, and mitigation of PFAS in food. PFAS are synthetic compounds widely used for their durability and surfactant properties in industrial and consumer products, but their persistence in the environment leads to bioaccumulation in food chains and potential health risks. Key sources of PFAS in food include environmental contamination through soil, water, and air, uptake by crops and livestock, and migration from FCMs. Factors like soil properties, plant species, and PFAS chain length influence their uptake and accumulation in crops, while livestock exposure primarily occurs through contaminated feed, water, and soil.

Analytical challenges for PFAS include obtaining low detection limits, reducing contamination risks, and addressing matrix interferences. Advanced methods like LC-MS/MS and LC-HRMS are employed, though specific issues persist with short-chain PFAS and complex food matrices. QC measures, such as participation in PTs and the availability of CRMs, are essential for reliable analysis. Despite progress, gaps remain in analytical standards for emerging PFAS and untargeted screening methods to capture their wide diversity.

Mitigation strategies in agriculture include soil amendments (e.g. biochar, activated carbon) to reduce PFAS mobility, phytoremediation using plants to extract, capture and retain PFAS, and selective breeding of crop varieties with low bioaccumulation potential. For livestock, exposure can be minimized by managing contaminated soil, water, and feed sources and employing risk assessment frameworks to estimate acceptable PFAS levels in animal products. However, current remediation methods for heavily contaminated sites remain energy-intensive and costly.

Case studies, such as PFAS levels in Belgian food, reveal widespread contamination with significant variation based on food type and source. Short-chain PFAS are more prevalent in plant-based foods, while animal-based products like fish show higher levels of long-chain PFAS. Food processing can alter PFAS levels, with some methods potentially reducing the concentrations of certain PFAS, though results are inconsistent across studies.

This chapter underscores the need for more sensitive analytical techniques, PFAS-occurrence data, comprehensive regulations, and innovative remediation approaches to address PFAS contamination. Future research should prioritize

© Burleigh Dodds Science Publishing Limited, 2025. All rights reserved.

developing standards for novel PFAS, improving risk assessments, designing sustainable mitigation strategies for agriculture and food production to enhance food safety and environmental protection, and monitoring the presence of PFAS in varying food and beverage types.

Targeted analysis approaches are typically insufficient to cover all different types of PFAS that are potentially present in food items. Therefore, there is a need for analytical methods to sensitively screen a broader range of PFAS to get a better estimation on the fluorine mass balance. Furthermore, targeted analyses now typically focus on a small number of PFAS as analytical standards for the majority of PFAS are still lacking. Therefore, there is a need to synthesize new and relevant analytical standards in order to develop and optimize targeted methods that would enable identification of these novel PFAS, including ultrashort-chain PFAS, in food items, the confirmation of non-target screening data, as well as characterization of the health risks posed by these novel PFAS by using them in (eco)toxicity tests. To comply with regulatory changes, more sensitive analysis methods, with proper QA/QC procedures, are required. Furthermore, maximum acceptable levels of many PFAS are still missing and risk assessment for these emerging PFAS is required to finetune and establish regulations. Another option would be to define RPFs for a larger number of PFAS, which can then be used to determine PFAS concentrations based on toxicity equivalents to, e.g. PFOA, which is regulated. The downside of this approach may be that RPF values can be highly variable depending on the toxicity endpoint and that concentrations of PFAS equivalents, based on one endpoint, are then compared to maximum acceptable levels that were defined based on other toxicological endpoints. Finally, more research concerning remediation and mitigation strategies is needed to reduce crop and livestock exposure, and thus contribute to food safety.

## 7 Where to look for further information

- Interstate Technology and Regulatory Council (ITRC) online document on PFAS (https://pfas-1.itrcweb.org/).
- European Union Reference Laboratory for Halogenated Persistent Organic Pollutants in Feed and Food (EURL-POPs). https://eurl-pops.eu/.

## 8 Acknowledgments

The research that yielded these results was funded by the Belgian Federal Public Service Health, Food Chain Safety and Environment through the contract PFASFORWARD 1 RT 23/07. TG is a senior postdoctoral researcher funded by a postdoctoral scholarship of the Research Foundation Flanders (1205724N). We acknowledge the University of Liège, the 'Fonds de la Recherche

© Burleigh Dodds Science Publishing Limited, 2025. All rights reserved.

Scientifique-FNRS', the Walloon Region and the Waste2Bio Initiative (https://www.waste2bio.org). The authors declare no conflict of interest.

## 9 References

Abafe, O.A., et al., 2021. Confirmatory analysis of per and polyfluoroalkyl substances in milk and infant formula using UHPLC–MS/MS. *Molecules* 26, 3664. https://doi.org/10.3390/molecules26123664

Ackerman Grunfeld, D., et al., 2024. Underestimated burden of per- and polyfluoroalkyl substances in global surface waters and groundwaters. *Nat. Geosci.* 17, 340–346. https://doi.org/10.1038/s41561-024-01402-8

Adu, O., et al., 2023. Bioavailability, phytotoxicity and plant uptake of per-and polyfluoroalkyl substances (PFAS): a review. *J. Hazard. Mater.* 447, 130805. https://doi.org/10.1016/j.jhazmat.2023.130805

Alinezhad, A., et al, 2022. An investigation of thermal air degradation and pyrolysis of per- and polyfluoroalkyl substances and aqueous film-forming foams in soil. *ACS EST Eng.* 2, 198–209. https://doi.org/10.1021/acsestengg.1c00335

Alves, R.N., et al., 2017. Preliminary assessment on the bioaccessibility of contaminants of emerging concern in raw and cooked seafood. *Food Chem. Toxicol.* 104, 69–78. https://doi.org/10.1016/j.fct.2017.01.029

Androulakakis, A., et al., 2022. Current progress in the environmental analysis of poly- and perfluoroalkyl substances (PFAS). *Environ. Sci. Adv.* 1, 705–724. https://doi.org/10.1039/D2VA00147K

Aro, R., et al., 2022. Extractable organofluorine analysis: a way to screen for elevated per- and polyfluoroalkyl substance contamination in humans? *Environ. Int.* 159, 107035. https://doi.org/10.1016/j.envint.2021.107035

Ateia, M., et al., 2023. Total oxidizable precursor (TOP) assay–best practices, capabilities and limitations for PFAS site investigation and remediation. *Environ. Sci. Technol. Lett.* 10, 292–301. https://doi.org/10.1021/acs.estlett.3c00061

Ballesteros-Gómez, A., et al., 2010. Tetrahydrofuran–water extraction, in-line clean-up and selective liquid chromatography/tandem mass spectrometry for the quantitation of perfluorinated compounds in food at the low picogram per gram level. *J. Chromatogr. A* 1217, 5913–5921. https://doi.org/10.1016/j.chroma.2010.07.032

Bangma, J., et al., 2023. Analytical method interferences for perfluoropentanoic acid (PFPeA) and perfluorobutanoic acid (PFBA) in biological and environmental samples. *Chemosphere* 315, 137722. https://doi.org/10.1016/j.chemosphere.2022.137722

Bangma, J.T., et al., 2021. Identification of an analytical method interference for perfluorobutanoic acid in biological samples. *Environ. Sci. Technol. Lett.* 8, 1085–1090. https://doi.org/10.1021/acs.estlett.1c00901

Barbosa, V., et al., 2018. Effects of steaming on contaminants of emerging concern levels in seafood. *Food Chem. Toxicol.* 118, 490–504. https://doi.org/10.1016/j.fct.2018.05.047

Barola, C., et al., 2020. A liquid chromatography-high resolution mass spectrometry method for the determination of thirty-three per- and polyfluoroalkyl substances in animal liver. *J. Chromatogr. A* 1628, 461442. https://doi.org/10.1016/j.chroma.2020.461442

© Burleigh Dodds Science Publishing Limited, 2025. All rights reserved.

Bedi, M., et al., 2023. Per- and polyfluoroalkyl substances (PFAS) measured in seafood from a cross-section of retail stores in the United States. *J. Hazard. Mater.* 459, 132062. https://doi.org/10.1016/j.jhazmat.2023.132062

Benskin, J.P., et al., 2007. Simultaneous characterization of perfluoroalkyl carboxylate, sulfonate, and sulfonamide isomers by liquid chromatography–tandem mass spectrometry. *Anal. Chem.* 79, 6455–6464. https://doi.org/10.1021/ac070802d

Berhanu, A., et al., 2023. A review of microbial degradation of per- and polyfluoroalkyl substances (PFAS): biotransformation routes and enzymes. *Sci. Total Environ.* 859, 160010. https://doi.org/10.1016/j.scitotenv.2022.160010

Bhavsar, S.P., et al., 2014. Cooking fish is not effective in reducing exposure to perfluoroalkyl and polyfluoroalkyl substances. *Environ. Int.* 66, 107–114. https://doi.org/10.1016/j.envint.2014.01.024

Bil, W., et al., 2021. Risk assessment of per- and polyfluoroalkyl substance mixtures: a relative potency factor approach. *Environ. Toxicol. Chem.* 40, 859–870. https://doi.org/10.1002/etc.4835

Binnington, M.J., et al., 2017. Effects of preparation on nutrient and environmental contaminant levels in Arctic beluga whale (Delphinapterus leucas) traditional foods. *Environ. Sci. Process. Impacts* 19, 1000–1015. https://doi.org/10.1039/C7EM00167C

Björnsdotter, M.K., et al., 2020. Challenges in the analytical determination of ultra-short-chain perfluoroalkyl acids and implications for environmental and human health. *Anal. Bioanal. Chem.* 412, 4785–4796. https://doi.org/10.1007/s00216-020-02692-8

Bletsou, A.A., et al., 2015. Targeted and non-targeted liquid chromatography-mass spectrometric workflows for identification of transformation products of emerging pollutants in the aquatic environment. *TrAC Trends Anal. Chem.* 66, 32–44. https://doi.org/10.1016/j.trac.2014.11.009

Boisacq, P., et al., 2023. Assessment of poly- and perfluoroalkyl substances (PFAS) in commercially available drinking straws using targeted and suspect screening approaches. *Food Addit. Contam. Part A* 40, 1230–1241. https://doi.org/10.1080/19440049.2023.2240908

Bokkers, B., et al., 2019. Per- and polyfluoroalkyl substances (PFASs) in food contact materials. Rijksinstituut voor Volksgezondheid en Milieu RIVM. https://doi.org/10.21945/RIVM-2018-0181

Bolan, N., et al., 2021. Remediation of poly- and perfluoroalkyl substances (PFAS) contaminated soils – To mobilize or to immobilize or to degrade? *J. Hazard. Mater.* 401, 123892. https://doi.org/10.1016/j.jhazmat.2020.123892

Brake, H.D., et al., 2023. Current per- and polyfluoroalkyl substance (PFAS) research points to a growing threat in animals. *J. Am. Vet. Med. Assoc.* 1–7. https://doi.org/10.2460/javma.22.12.0582

Brambilla, G., et al., 2015. Pathways and factors for food safety and food security at PFOS contaminated sites within a problem based learning approach. *Chemosphere* 129, 192–202. https://doi.org/10.1016/j.chemosphere.2014.09.050

Brase, R.A., et al., 2021. Legacy and emerging per- and polyfluoroalkyl substances: analytical techniques, environmental fate, and health effects. *Int. J. Mol. Sci.* 22, 995. https://doi.org/10.3390/ijms22030995

Bräunig, J., et al., 2017. Fate and redistribution of perfluoroalkyl acids through AFFF-impacted groundwater. *Sci. Total Environ.* 596–597, 360–368. https://doi.org/10.1016/j.scitotenv.2017.04.095


© Burleigh Dodds Science Publishing Limited, 2025. All rights reserved.

Brendel, S., et al., 2018. Short-chain perfluoroalkyl acids: environmental concerns and a regulatory strategy under REACH. *Environ. Sci. Eur.* 30, 9. https://doi.org/10.1186/s12302-018-0134-4

Brunn, H., et al., 2023. PFAS: forever chemicals–persistent, bioaccumulative and mobile. Reviewing the status and the need for their phase out and remediation of contaminated sites. *Environ. Sci. Eur.* 35, 20. https://doi.org/10.1186/s12302-023-00721-8

Brusseau, M.L., et al., 2020. PFAS concentrations in soils: background levels versus contaminated sites. *Sci. Total Environ.* 740, 140017. https://doi.org/10.1016/j.scitotenv.2020.140017

Buck, R.C., et al., 2011. Perfluoroalkyl and polyfluoroalkyl substances in the environment: Terminology, classification, and origins. *Integr. Environ. Assess. Manag.* 7, 513–541. https://doi.org/10.1002/ieam.258

Burges, A., et al., 2018. From phytoremediation of soil contaminants to phytomanagement of ecosystem services in metal contaminated sites. *Int. J. Phytoremediation* 20, 384–397. https://doi.org/10.1080/15226514.2017.1365340

Cai, Y., et al., 2021. A review of responses of terrestrial organisms to perfluorinated compounds. *Sci. Total Environ.* 793, 148565. https://doi.org/10.1016/j.scitotenv.2021.148565

Camdzic, D., et al., 2023. Quantitation of total PFAS including trifluoroacetic acid with fluorine nuclear magnetic resonance spectroscopy. *Anal. Chem.* 95, 5484–5488. https://doi.org/10.1021/acs.analchem.2c05354

Chen, X., et al., 2024. Multimedia and full-life-cycle monitoring discloses the dynamic accumulation rules of PFAS and underestimated foliar uptake in wheat near a fluorochemical industrial park. *Environ. Sci. Technol.* 58, 18088–18097. https://doi.org/10.1021/acs.est.4c03525

Chen, Z., et al., 2023. Bioaccumulation and risk mitigation of legacy and novel perfluoroalkyl substances in seafood: insights from trophic transfer and cooking method. *Environ. Int.* 177, 108023. https://doi.org/10.1016/j.envint.2023.108023

Chiesa, L.M., et al., 2018. Detection of perfluoroalkyl acids and sulphonates in Italian eel samples by HPLC-HRMS Orbitrap. *Chemosphere* 193, 358–364. https://doi.org/10.1016/j.chemosphere.2017.10.082

Chiumiento, F., et al., 2023. A new method for determining PFASs by UHPLC-HRMS (Q-Orbitrap): application to PFAS analysis of organic and conventional eggs sold in Italy. *Food Chem.* 401, 134135. https://doi.org/10.1016/j.foodchem.2022.134135

Chunyan, X., et al., 2023. The role of microorganisms in petroleum degradation: current development and prospects. *Sci. Total Environ.* 865, 161112. https://doi.org/10.1016/j.scitotenv.2022.161112

Colles, A., et al., 2022. Per- en poly-fluoralkylstoffen in en rond de woning. Verantwoordelijke uitgever Departement Omgeving, Vlaams Planbureau voor Omgeving, versie november 2022, studie in opdracht van Departement Omgeving en OVAM.

Concawe, 2024. PFAS soil treatment processes - a review of operating ranges and constraints. Report no 8/24.

Conder, J.M., et al., 2008. Are PFCAs bioaccumulative? A critical review and comparison with regulatory criteria and persistent lipophilic compounds. *Environ. Sci. Technol.* 42, 995–1003. https://doi.org/10.1021/es070895g

© Burleigh Dodds Science Publishing Limited, 2025. All rights reserved.

Consortium UAntwerpen, VITO, PIH, UHasselt en VUB, 2023. Jongerenstudie HBM - omgeving 3M – Resultatenrapport, in opdracht van het Departement Omgeving, Vlaams Planbureau voor Omgeving, 219p.

Cousins, I.T., et al., 2020. The high persistence of PFAS is sufficient for their management as a chemical class. *Environ. Sci. Process. Impacts* 22, 2307–2312. https://doi.org/10.1039/d0em00355g

Cui, Q., et al., 2019. Per- and polyfluoroalkyl substances (PFASs) in the blood of two colobine monkey species from China: occurrence and exposure pathways. *Sci. Total Environ.* 674, 524–531. https://doi.org/10.1016/j.scitotenv.2019.04.118

Death, C., et al., 2021. Per- and polyfluoroalkyl substances (PFAS) in livestock and game species: a review. *Sci. Total Environ.* 774, 144795. https://doi.org/10.1016/j.scitotenv.2020.144795

Del Gobbo, L., et al., 2008. Cooking decreases observed perfluorinated compound concentrations in fish. *J. Agric. Food Chem.* 56, 7551–7559. https://doi.org/10.1021/jf800827r

Denly, E., et al., 2019. Per- and polyfluoroalkyl substances in environmental sampling products: Fact or fiction? *Remediat. J.* 29, 65–76. https://doi.org/10.1002/rem.21614

Dimitrakopoulou, M.-E., et al., 2024. Comprehensive analysis of PFAS presence from environment to plate. *Npj Sci. Food* 8, 80. https://doi.org/10.1038/s41538-024-00319-1

Dong, Q., et al., 2024. Adsorption of per- and polyfluoroalkyl substances (PFAS) by ionic liquid-modified clays: Effect of clay composition and PFAS structure. *J. Colloid Interface Sci.* 654, 925–934. https://doi.org/10.1016/j.jcis.2023.10.112

Drew, R., et al., 2021. Accumulation of PFAS by livestock – determination of transfer factors from water to serum for cattle and sheep in Australia. *Food Addit. Contam. Part A* 38, 1897–1913. https://doi.org/10.1080/19440049.2021.1942562

ECHA, 2021. Consultation on a proposed restriction on the manufacture, placing on the market and use of per- and polyfluoroalkyl substances (PFAS).

EFSA Panel on Contaminants in the Food Chain (EFSA CONTAM Panel), et al., 2020. Risk to human health related to the presence of perfluoroalkyl substances in food. *EFSA J.* 18. https://doi.org/10.2903/j.efsa.2020.6223

EU, 2009. Document 32009R0152. Commission Regulation (EC) No 152/2009 of 27 January 2009 laying down the methods of sampling and analysis for the official control of feed. *OJ L* 54, 1–130.

EU, 2019. Document 32019L0904. Directive (EU) 2019/904 of the European Parliament and of the Council of 5 June 2019 on the reduction of the impact of certain plastic products on the environment. *OJ L* 155, 1–19.

EU, 2020. Document 32020L2184. Directive (EU) 2020/2184 of the European Parliament and of the Council of 16 December 2020 on the quality of water intended for human consumption (recast). *OJ L* 435, 1–62.

EU, 2022a. Document 32022R1428. Commission Implementing Regulation (EU) 2022/1428 of 24 August 2022 laying down methods of sampling and analysis for the control of perfluoroalkyl substances in certain foodstuffs. *OJ L* 221, 66–73.

EU, 2022b. Document 32022H1431. Commission Recommendation (EU) 2022/1431 of 24 August 2022 on the monitoring of perfluoroalkyl substances in food. *OJ L* 221, 105–109.

© Burleigh Dodds Science Publishing Limited, 2025. All rights reserved.

EU, 2023a. Document 32023R0915. Commission Regulation (EU) 2023/915 of 25 April 2023 on maximum levels for certain contaminants in food and repealing Regulation (EC) NO 1881/2006. *OJ L* 119, 103–157.

EU, 2023b. European Commission: Directorate-General for Research and Innovation, EU missions – Soil deal for Europe – What is the EU mission – A soil deal for Europe, Publications Office of the European Union, 2023, https://data.europa.eu/doi/10.2777/171313.

EU, 2024. Document 52024XC04910. Commission Notice – Technical guidelines regarding methods of analysis for monitoring of per- and polyfluoroalkyl substances (PFAS) in water intended for human consumption. C/2024/5414. OJ C, C/2024/4910, 7.8.2024.

EURL for halogenated POPs in feed and food, 2024. Guidance Document on Analytical Parameters for the Determination of Per- and Polyfluoroalkyl Substances (PFAS) in Food and Feed, version 2.0 of 10 September2024.

Evangelou, M.W.H. and Robinson, B.H., 2022. The phytomanagement of PFAS-contaminated land. *Int. J. Environ. Res. Public. Health* 19, 6817. https://doi.org/10.3390/ijerph19116817

Falandysz, J., et al., 2014. The toxicological effects of halogenated naphthalenes: a review of aryl hydrocarbon receptor-mediated (dioxin-like) relative potency factors. *J. Environ. Sci. Health Part C* 32, 239–272. https://doi.org/10.1080/10590501.2014.938945

Faust, J.A., 2023. PFAS on atmospheric aerosol particles: a review. *Environ. Sci. Process. Impacts* 25, 133–150. https://doi.org/10.1039/d2em00002d

Feng, Q.-J., et al., 2023. Bioaccumulation, tissue distributions, and maternal transfer of perfluoroalkyl carboxylates (PFCAs) in laying hens. *Sci. Total Environ.* 905, 167008. https://doi.org/10.1016/j.scitotenv.2023.167008

Fenton, S.E., et al., 2021. Per- and polyfluoroalkyl substance toxicity and human health review: current state of knowledge and strategies for informing future research. *Environ. Toxicol. Chem.* 40, 606–630. https://doi.org/10.1002/etc.4890

Fernandes, A.R., et al., 2019. The potential of recycled materials used in agriculture to contaminate food through uptake by livestock. *Sci. Total Environ.* 667, 359–370. https://doi.org/10.1016/j.scitotenv.2019.02.211

FSANZ, 2016. *Assessment of potential dietary exposure to perfluorooctane sulfonate (PFOS), perfluorooctanoic acid (PFOA) and perfluorohexane sulfonate (PFHxS) occurring in foods sampled from Contaminated sites*. Food Standards Australia New Zealand. https://www1.health.gov.au/internet/main/publishing.nsf/content/2200FE086D480353CA2580C900817CDC/$File/Dietary-Exposure-Assesment.pdf.

Gallocchio, F., et al., 2022. Fast and sensitive analysis of short- and long-chain perfluoroalkyl substances in foods of animal origin. *Molecules* 27, 7899. https://doi.org/10.3390/molecules27227899

Gao, Y., et al., 2018. Simultaneous determination of 21 trace perfluoroalkyl substances in fish by isotope dilution ultrahigh performance liquid chromatography tandem mass spectrometry. *J. Chromatogr. B* 1084, 45–52. https://doi.org/10.1016/j.jchromb.2018.03.008

Genualdi, S., et al., 2021. Method development and validation of per- and polyfluoroalkyl substances in foods from FDA's total diet study program. *J. Agric. Food Chem.* 69, 5599–5606. https://doi.org/10.1021/acs.jafc.1c01777

© Burleigh Dodds Science Publishing Limited, 2025. All rights reserved.

Genualdi, S., et al., 2022. Analysis of per- and poly-fluoroalkyl substances (PFAS) in processed foods from FDA's total diet study. *Anal. Bioanal. Chem.* 414, 1189–1199. https://doi.org/10.1007/s00216-021-03610-2

Ghisi, R., et al., 2019. Accumulation of perfluorinated alkyl substances (PFAS) in agricultural plants: a review. *Environ. Res.* 169, 326–341. https://doi.org/10.1016/j.envres.2018.10.023

Glüge, J., et al., 2020. An overview of the uses of per- and polyfluoroalkyl substances (PFAS). *Environ. Sci. Process. Impacts* 22, 2345–2373. https://doi.org/10.1039/d0em00291g

Gosetti, F., et al., 2010. Determination of perfluorochemicals in biological, environmental and food samples by an automated on-line solid phase extraction ultra high performance liquid chromatography tandem mass spectrometry method. *J. Chromatogr. A* 1217, 7864–7872. https://doi.org/10.1016/j.chroma.2010.10.049

Gremmel, C., et al., 2016. Systematic determination of perfluoroalkyl and polyfluoroalkyl substances (PFASs) in outdoor jackets. *Chemosphere* 160, 173–180. https://doi.org/10.1016/j.chemosphere.2016.06.043

Groffen, T., et al., 2019. Development and validation of an extraction method for the analysis of perfluoroalkyl substances (PFASs) in environmental and biotic matrices. *J. Chromatogr. B* 1116, 30–37. https://doi.org/10.1016/j.jchromb.2019.03.034

Groffen, T., et al., 2021. Risks posed by per- and polyfluoroalkyl substances (PFAS) on the African continent, emphasizing aquatic ecosystems. *Integr. Environ. Assess. Manag.* 17, 726–732. https://doi.org/10.1002/ieam.4404

Groffen, T., et al., 2022. PFAS accumulation in several terrestrial plant and invertebrate species reveals species-specific differences. Environ. Sci. Pollut. Res. 30, 23820–23835. https://doi.org/10.1007/s11356-022-23799-8

Guo, J. and Huan, T., 2020. Comparison of full-scan, data-dependent, and data-independent acquisition modes in liquid chromatography–mass spectrometry based untargeted metabolomics. *Anal. Chem.* 92, 8072–8080. https://doi.org/10.1021/acs.analchem.9b05135

Guruge, K.S., et al., 2008. Species-specific concentrations of perfluoroalkyl contaminants in farm and pet animals in Japan. *Chemosphere* 73, S210–S215. https://doi.org/10.1016/j.chemosphere.2006.12.105

Harris, B.A., et al., 2024. Enzymatic degradation of PFAS: current status and ongoing challenges. *ChemSusChem* e202401122. https://doi.org/10.1002/cssc.202401122

He, Q., et al., 2023. Phytoextraction of per- and polyfluoroalkyl substances (PFAS) by weeds: effect of PFAS physicochemical properties and plant physiological traits. *J. Hazard. Mater.* 454, 131492. https://doi.org/10.1016/j.jhazmat.2023.131492

Hlouskova, V., et al., 2013. Occurrence of perfluoroalkyl substances (PFASs) in various food items of animal origin collected in four European countries. *Food Addit. Contam. Part A* 30, 1918–1932. https://doi.org/10.1080/19440049.2013.837585

Houde, M., et al., 2011. Monitoring of perfluorinated compounds in aquatic biota: an updated review: PFCs in aquatic biota. *Environ. Sci. Technol.* 45, 7962–7973. https://doi.org/10.1021/es104326w

Hu, Y., et al., 2020. Cooking methods affect the intake of per- and polyfluoroalkyl substances (PFASs) from grass carp. *Ecotoxicol. Environ. Saf.* 203, 111003. https://doi.org/10.1016/j.ecoenv.2020.111003

Hwang, S.H., et al., 2024. Development of a method using QuEChERS and LC–MS/MS for analysis of per- and polyfluoroalkyl substances in rice matrix. *Food Chem.* 445, 138687. https://doi.org/10.1016/j.foodchem.2024.138687

© Burleigh Dodds Science Publishing Limited, 2025. All rights reserved.

Iannone, A., et al., 2024. There's something in what we eat: an overview on the extraction techniques and chromatographic analysis for PFAS identification in agri-food products. *Foods* 13, 1085. https://doi.org/10.3390/foods13071085

ISO, 2009. Water quality – Determination of perfluorooctanesulfonate (PFOS) and perfluorooctanate (PFOA) – Method for unfiltered samples using solid phase extraction and liquid chromatography/mass spectrometry, International Standards Organisation, ISO 25101: 2009.

ITRC Per- and Polyfluoroalkyl Substances Team, 2023. Technical/Regulatory guidance. Per- and Polyfluoroalkyl Substances (PFAS).

Jha, G., et al., 2021. Per- and polyfluoroalkyl substances (PFAS) in integrated crop–livestock systems: environmental exposure and human health risks. *Int. J. Environ. Res. Public. Health* 18, 12550. https://doi.org/10.3390/ijerph182312550

Kaufmann, A., et al., 2022. Simplifying nontargeted analysis of PFAS in complex food matrixes. *J. AOAC Int.* 105, 1280–1287. https://doi.org/10.1093/jaoacint/qsac071

Kavusi, E., et al., 2023. Critical review on phytoremediation of polyfluoroalkyl substances from environmental matrices: need for global concern. *Environ. Res.* 217, 114844. https://doi.org/10.1016/j.envres.2022.114844

Keawmanee, S., et al., 2015. Method development and initial results of testing for perfluorooctane sulfonate (PFOS) and perfluorooctanoic acid (PFOA) in waterproof sunscreens. *Environ. Eng. Res.* 20, 127–132. https://doi.org/10.4491/eer.2014.S1.006

Keller, A.A., et al., 2024. Elimination of microplastics, PFAS, and PPCPs from biosolids via pyrolysis to produce biochar: feasibility and techno-economic analysis. *Sci. Total Environ.* 947, 174773. https://doi.org/10.1016/j.scitotenv.2024.174773

Kim, M., et al., 2020. Effect of washing, soaking, and cooking methods on perfluorinated compounds in mackerel ( *Scomber japonicus* ). *Food Sci. Nutr.* 8, 4399–4408. https://doi.org/10.1002/fsn3.1737

Kosiarski, K., et al., 2024. From wastewater to feed: understanding per- and polyfluoroalkyl substances occurrence in wastewater-irrigated crops. *J. Environ. Qual.* Advance online publication. https://doi.org/10.1002/jeq2.20630

Kowalczyk, J., et al., 2012. Transfer of perfluorooctanoic acid (PFOA) and perfluorooctane sulfonate (PFOS) from contaminated feed into milk and meat of sheep: pilot study. *Arch. Environ. Contam. Toxicol.* 63, 288–298. https://doi.org/10.1007/s00244-012-9759-2

Kurwadkar, S., et al., 2022. Per- and polyfluoroalkyl substances in water and wastewater: a critical review of their global occurrence and distribution. *Sci. Total Environ.* 809, 151003. https://doi.org/10.1016/j.scitotenv.2021.151003

Lacina, O., et al., 2011. Simple, high throughput ultra-high performance liquid chromatography/tandem mass spectrometry trace analysis of perfluorinated alkylated substances in food of animal origin: Milk and fish. *J. Chromatogr. A* 1218, 4312–4321. https://doi.org/10.1016/j.chroma.2011.04.061

Lasters, R., et al., 2022. Home-produced eggs: an important human exposure pathway of perfluoroalkylated substances (PFAS). *Chemosphere* 308, 136283. https://doi.org/10.1016/j.chemosphere.2022.136283

Lasters, R., et al., 2023. Prediction of perfluoroalkyl acids (PFAAs) in homegrown eggs: Insights into abiotic and biotic factors affecting bioavailability and derivation of potential remediation measures. *Environ. Int.* 181, 108300. https://doi.org/10.1016/j.envint.2023.108300

© Burleigh Dodds Science Publishing Limited, 2025. All rights reserved.

Lasters, R., et al., 2024. Per- and polyfluoroalkyl substances (PFAS) in homegrown crops: accumulation and human risk assessment. *Chemosphere* 364, 143208. https://doi.org/10.1016/j.chemosphere.2024.143208

Lath, S., et al., 2019. Sorption of PFOA onto different laboratory materials: filter membranes and centrifuge tubes. *Chemosphere* 222, 671–678. https://doi.org/10.1016/j.chemosphere.2019.01.096

Lerch, M., et al., 2023. Food simulants and real food – What do we know about the migration of PFAS from paper based food contact materials? *Food Packag. Shelf Life* 35, 100992. https://doi.org/10.1016/j.fpsl.2022.100992

Lesmeister, L., et al., 2021. Extending the knowledge about PFAS bioaccumulation factors for agricultural plants – A review. *Sci. Total Environ.* 766, 142640. https://doi.org/10.1016/j.scitotenv.2020.142640

Li, J., et al., 2022. Exposure routes, bioaccumulation and toxic effects of per- and polyfluoroalkyl substances (PFASs) on plants: a critical review. *Environ. Int.* 158, 106891. https://doi.org/10.1016/j.envint.2021.106891

Liang, S.-H., et al., 2023. Analysis of ultrashort-chain and short-chain (C1 to C4) per- and polyfluorinated substances in potable and non-potable waters. *J. Chromatogr. Open* 4, 100098. https://doi.org/10.1016/j.jcoa.2023.100098

Liguori-Bills, N.B., et al., 2023. Analysis of PFAS in locally acquired food containers. *LCGC N. Am.* 41, 22–31. https://doi.org/10.56530/lcgc.na.sd7179a7

Liu, S., et al., 2024. Concentrations of per- and polyfluoroalkyl substances in vegetables from Sydney and Newcastle, Australia. *J. Sci. Food Agric.* 104, 6667–6675. https://doi.org/10.1002/jsfa.13491

Lupton, S.J., et al., 2022. Plasma and skin per- and polyfluoroalkyl substance (PFAS) levels in dairy cattle with lifetime exposures to PFAS-contaminated drinking water and feed. *J. Agric. Food Chem.* 70, 15945–15954. https://doi.org/10.1021/acs.jafc.2c06620

Luque, N., et al., 2010. Analysis of perfluorinated compounds in biota by microextraction with tetrahydrofuran and liquid chromatography/Ion Isolation-based ion-trap mass spectrometry. *J. Chromatogr. A* 1217, 3774–3782. https://doi.org/10.1016/j.chroma.2010.04.014

Mahinroosta, R. and Senevirathna, L., 2020. A review of the emerging treatment technologies for PFAS contaminated soils. *J. Environ. Manage.* 255, 109896. https://doi.org/10.1016/j.jenvman.2019.109896

Marciesky, M., et al., 2023. Mechanisms and opportunities for rational in silico design of enzymes to degrade per- and polyfluoroalkyl substances (PFAS). *J. Chem. Inf. Model.* 63, 7299–7319. https://doi.org/10.1021/acs.jcim.3c01303

Martin, J.W., et al., 2004. Peer reviewed: analytical challenges hamper perfluoroalkyl research. *Environ. Sci. Technol.* 38, 248A–255A. https://doi.org/10.1021/es0405528

Meng, L., et al., 2022a. Occurrence, temporal variation (2010–2018), distribution, and source appointment of per- and polyfluoroalkyl substances (PFAS) in Mollusks from the Bohai Sea, China. *ACS EST Water* 2, 195–205. https://doi.org/10.1021/acsestwater.1c00346

Meng, P., et al., 2022b. Extraction and matrix cleanup method for analyzing novel per- and polyfluoroalkyl ether acids and other per- and polyfluoroalkyl substances in fruits and vegetables. *J. Agric. Food Chem.* 70, 4792–4804. https://doi.org/10.1021/acs.jafc.1c07665

© Burleigh Dodds Science Publishing Limited, 2025. All rights reserved.

Mikkonen, A.T., et al., 2023. Spatio-temporal trends in livestock exposure to per- and polyfluoroalkyl substances (PFAS) inform risk assessment and management measures. *Environ. Res.* 225, 115518. https://doi.org/10.1016/j.envres.2023.115518

Mirabediny, M., et al., 2023. Effective PFAS degradation by electrochemical oxidation methods-recent progress and requirement. *Chemosphere* 321, 138109. https://doi.org/10.1016/j.chemosphere.2023.138109

Miyake, Y., et al., 2007. Trace analysis of total fluorine in human blood using combustion ion chromatography for fluorine: a mass balance approach for the determination of known and unknown organofluorine compounds. *J. Chromatogr. A* 1154, 214–221. https://doi.org/10.1016/j.chroma.2007.03.084

Mukhopadhyay, R., et al., 2021. Natural and engineered clays and clay minerals for the removal of poly- and perfluoroalkyl substances from water: State-of-the-art and future perspectives. Adv. Colloid Interface Sci. 297, 102537. https://doi.org/10.1016/j.cis.2021.102537

Mullin, L., et al., 2019. Analysis of hexafluoropropylene oxide-dimer acid (HFPO-DA) by liquid chromatography-mass spectrometry (LC-MS): review of current approaches and environmental levels. *TrAC Trends Anal. Chem.* 118, 828–839. https://doi.org/10.1016/j.trac.2019.05.015

Nahar, K., et al., 2023. A review of analytical methods and technologies for monitoring per- and polyfluoroalkyl substances (PFAS) in Water. Water 15, 3577. https://doi.org/10.3390/w15203577

Nason, S.L., et al., 2024. A comprehensive trial on PFAS remediation: hemp phytoextraction and PFAS degradation in harvested plants. *Environ. Sci. Adv.* 3, 304–313. https://doi.org/10.1039/D3VA00340J

Nassazzi, W., et al., 2023. Phytoextraction of per- and polyfluoroalkyl substances (PFAS) and the influence of supplements on the performance of short–rotation crops. *Environ. Pollut.* 333, 122038. https://doi.org/10.1016/j.envpol.2023.122038

Navarro, D.A., et al., 2023. Stabilisation of PFAS in soils: long-term effectiveness of carbon-based soil amendments. *Environ. Pollut.* 323, 121249. https://doi.org/10.1016/j.envpol.2023.121249

Ng, C.A. and Hungerbühler, K., 2013. Bioconcentration of perfluorinated alkyl acids: how important is specific binding? *Environ. Sci. Technol.* 47, 7214–7223. https://doi.org/10.1021/es400981a

Ogunbiyi, O.D., et al., 2023. Analytical approaches for screening of per- and poly fluoroalkyl substances in food items: a review of recent advances and improvements. *Environ. Pollut.* 329, 121705. https://doi.org/10.1016/j.envpol.2023.121705

Organtini, K.L., et al., 2023. Ultra-trace detection of per- and polyfluoroalkyl substances (PFAS) in drinking water to meet new US EPA interim health advisory levels. https://www.waters.com/content/dam/waters/en/app-notes/2023/720007855/720007855-en.pdf

Pan, Y., et al., 2020. Analysis of emerging per- and polyfluoroalkyl substances: progress and current issues. *TrAC Trends Anal. Chem.* 124, 115481. https://doi.org/10.1016/j.trac.2019.04.013

Pelch, K.E., et al., 2023. 70 analyte PFAS test method highlights need for expanded testing of PFAS in drinking water. *Sci. Total Environ.* 876, 162978. https://doi.org/10.1016/j.scitotenv.2023.162978


© Burleigh Dodds Science Publishing Limited, 2025. All rights reserved.

Peritore, A.F., et al., 2023. Current review of increasing animal health threat of per- and polyfluoroalkyl substances (PFAS): harms, limitations, and alternatives to manage their toxicity. *Int. J. Mol. Sci.* 24, 11707. https://doi.org/10.3390/ijms241411707

PerkinElmer, 2023. Overcoming typical challenges for PFAS analysis to meet evolving regulatory requirements. https://resources.perkinelmer.com/lab-solutions/resources/docs/atr-overcoming-typical-challenges-for-PFAS-Analysis.pdf.

Peters, R.J.B., et al., 2019. Review of analytical approaches for the identification of non-intentionally added substances in paper and board food contact materials. *Trends Food Sci. Technol.* 85, 44–54. https://doi.org/10.1016/j.tifs.2018.12.010

Phillips, K.A., et al., 2018. Suspect screening analysis of chemicals in consumer products. *Environ. Sci. Technol.* 52, 3125–3135. https://doi.org/10.1021/acs.est.7b04781

Phomsopha, T., et al., 2022. Achieving low parts-per-quadrillion detection limits for PFAS analysis in drinking water. A collaborative exploration of instrument sensitivity and sample preparation for ultra-trace analysis of PFAS on the SCIEX 7500 system. https://sciex.com/content/dam/SCIEX/pdf/tech-notes/environmental-industrial/water-and-soil/PFAS_EPA_drinking-water_2022-HALs_7500_RUO-MKT-02-15208-A_final.pdf.

Piva, E., et al., 2023. Per- and polyfluoroalkyl substances (PFAS) presence in food: Comparison among fresh, frozen and ready-to-eat vegetables. *Food Chem.* 410, 135415. https://doi.org/10.1016/j.foodchem.2023.135415

Pizzurro, D.M., et al., 2019. Interspecies differences in perfluoroalkyl substances (PFAS) toxicokinetics and application to health-based criteria. *Regul. Toxicol. Pharmacol.* 106, 239–250. https://doi.org/10.1016/j.yrtph.2019.05.008

Powley, C.R., et al., 2005. Matrix effect-free analytical methods for determination of perfluorinated carboxylic acids in environmental matrixes. *Anal. Chem.* 77, 6353–6358. https://doi.org/10.1021/ac0508090

Ramos, P. and Ashworth, D.J., 2024. Per- and poly-fluoroalkyl substances in agricultural contexts and mitigation of their impacts using biochar: a review. *Sci. Total Environ.* 927, 172275. https://doi.org/10.1016/j.scitotenv.2024.172275

Ren, J., et al., 2022. Suspect screening and nontargeted analysis of per- and polyfluoroalkyl substances in a Lake Ontario Food Web. *Environ. Sci. Technol.* 56, 17626–17634. https://doi.org/10.1021/acs.est.2c04321

Ritter, E.E., et al., 2017. PIGE as a screening tool for Per- and polyfluorinated substances in papers and textiles. *Nucl. Instrum. Methods Phys. Res. Sect. B Beam Interact. Mater. At.* 407, 47–54. https://doi.org/10.1016/j.nimb.2017.05.052

Roberts, J., et al., 2023. Unwanted ingredients–highly specific and sensitive method for the extraction and quantification of PFAS in everyday foods. *Food Anal. Methods* 16, 857–866. https://doi.org/10.1007/s12161-023-02451-2

Rodowa, A.E., et al., 2020. Field sampling materials unlikely source of contamination for perfluoroalkyl and polyfluoroalkyl substances in field samples. *Environ. Sci. Technol. Lett.* 7, 156–163. https://doi.org/10.1021/acs.estlett.0c00036

Ross, I., et al., 2018. A review of emerging technologies for remediation of PFASs. *Remediat. J.* 28, 101–126. https://doi.org/10.1002/rem.21553

Roth, K., et al., 2020. Diet as an exposure source and mediator of per- and polyfluoroalkyl substance (PFAS) toxicity. *Front. Toxicol.* 2, Article 601149. https://doi.org/10.3389/ftox.2020.601149

Sadia, M., et al., 2020. Trace level analyses of selected perfluoroalkyl acids in food: method development and data generation. *Environ. Pollut.* 263, 113721. https://doi.org/10.1016/j.envpol.2019.113721

© Burleigh Dodds Science Publishing Limited, 2025. All rights reserved.

Sapozhnikova, Y., et al., 2023. Assessing per- and polyfluoroalkyl substances in globally sourced food packaging. *Chemosphere* 337, 139381. https://doi.org/10.1016/j.chemosphere.2023.139381

Schaider, L.A., et al., 2017. Fluorinated compounds in U.S. fast food packaging. *Environ. Sci. Technol. Lett.* 4, 105–111. https://doi.org/10.1021/acs.estlett.6b00435

Scher, D.P., et al., 2018. Occurrence of perfluoroalkyl substances (PFAS) in garden produce at homes with a history of PFAS-contaminated drinking water. *Chemosphere* 196, 548–555. https://doi.org/10.1016/j.chemosphere.2017.12.179

Schiavone, C. and Portesi, C., 2023. PFAS: a review of the state of the art, from legislation to analytical approaches and toxicological aspects for assessing contamination in food and environment and related risks. *Appl. Sci.* 13, 6696. https://doi.org/10.3390/app13116696

Schymanski, E., et al., 2023. Per- and polyfluoroalkyl substances (PFAS) in PubChem: 7 million and growing. https://doi.org/10.26434/chemrxiv-2023-j823z

Sepulvado, J.G., et al., 2011. Occurrence and fate of perfluorochemicals in soil following the land application of municipal biosolids. *Environ. Sci. Technol.* 45, 8106–8112. https://doi.org/10.1021/es103903d

Smorada, C.M., et al., 2024. Bacterial degradation of perfluoroalkyl acids. *Curr. Opin. Biotechnol.* 88, 103170. https://doi.org/10.1016/j.copbio.2024.103170

So, M.K., et al., 2006. Alkaline digestion and solid phase extraction method for perfluorinated compounds in mussels and oysters from South China and Japan. *Arch. Environ. Contam. Toxicol.* 50, 240–248. https://doi.org/10.1007/s00244-005-7058-x

Sørmo, E., et al., 2021. Stabilization of PFAS-contaminated soil with activated biochar. *Sci. Total Environ.* 763, 144034. https://doi.org/10.1016/j.scitotenv.2020.144034

Stecconi, T., et al., 2024. A LC-MS/MS procedure for the analysis of 19 perfluoroalkyl substances in food fulfilling recent EU regulations requests. *Talanta* 266, 125054. https://doi.org/10.1016/j.talanta.2023.125054

Stroski, K.M., et al., 2024. Non-targeted analysis of per- and polyfluorinated substances in consumer food packaging. *Chemosphere* 360, 142436. https://doi.org/10.1016/j.chemosphere.2024.142436

Sungur, Ş., et al., 2018. Determination of perfluorooctanoic acid (PFOA) and perfluorooctane sulfonic acid (PFOS) in food and beverages. *Int. J. Environ. Anal. Chem.* 98, 360–368. https://doi.org/10.1080/03067319.2018.1468440

Sungur, Ş., et al., 2021. A comparison of levels of perfluorooctanoic acid (PFOA) and perfluorooctane sulfonic acid (PFOS) in raw and cooked fish. Toxin Rev. 40, 59–64. https://doi.org/10.1080/15569543.2018.1554589

Susmann, H.P., et al., 2019. Dietary habits related to food packaging and population exposure to PFASs. *Environ. Health Perspect.* 127, 97001. https://doi.org/10.1289/ehp4092

Suzuki, G., et al., 2020. Validity of using a relative potency factor approach for the risk management of dioxin-like polychlorinated naphthalenes. *Chemosphere* 244, 125448. https://doi.org/10.1016/j.chemosphere.2019.125448

Taniyasu, S., et al., 2008. Analysis of trifluoroacetic acid and other short-chain perfluorinated acids (C2–C4) in precipitation by liquid chromatography–tandem mass spectrometry: Comparison to patterns of long-chain perfluorinated acids (C5–C18). Anal. Chim. Acta 619, 221–230. https://doi.org/10.1016/j.aca.2008.04.064

© Burleigh Dodds Science Publishing Limited, 2025. All rights reserved.

Tansel, B., et al., 2024. PFAS in biosolids: accumulation characteristics and fate profiles after land application. *J. Environ. Manage.* 370, 122395. https://doi.org/10.1016/j.jenvman.2024.122395

Taylor, M.D., et al., 2019. Do conventional cooking methods alter concentrations of per- and polyfluoroalkyl substances (PFASs) in seafood? *Food Chem. Toxicol.* 127, 280–287. https://doi.org/10.1016/j.fct.2019.03.032

Taylor, R.B. and Sapozhnikova, Y., 2022. Comparison and validation of the QuEChERSER mega-method for determination of per- and polyfluoroalkyl substances in foods by liquid chromatography with high-resolution and triple quadrupole mass spectrometry. *Anal. Chim. Acta* 1230, 340400. https://doi.org/10.1016/j.aca.2022.340400

Thépaut, E., et al., 2021. Per- and polyfluoroalkyl substances in serum and associations with food consumption and use of personal care products in the Norwegian biomonitoring study from the EU project EuroMix. *Environ. Res.* 195, 110795. https://doi.org/10.1016/j.envres.2021.110795

Trnovec, T., et al., 2013. Relative effect potency estimates of dioxin-like activity for dioxins, furans, and dioxin-like PCBs in adults based on two thyroid outcomes. *Environ. Health Perspect.* 121, 886–892. https://doi.org/10.1289/ehp.1205739

Valsecchi, S., et al., 2013. Determination of perfluorinated compounds in aquatic organisms: a review. *Anal. Bioanal. Chem.* 405, 143–157. https://doi.org/10.1007/s00216-012-6492-7

Van Leeuw, V., et al., 2024a. Exposure assessment of perfluoroalkyl substances as follow-up on the concerns raised in the recent opinion of EFSA. Sciensano. Scientific report 2024. D/2024.14.440/10.

Van Leeuw, V., et al., 2024b. Per- and polyfluoroalkyl substances in food and beverages: determination by LC-HRMS and occurrence in products from the Belgian market. *Chemosphere* 366, 143543. https://doi.org/10.1016/j.chemosphere.2024.143543

Vassiliadou, I., et al., 2015. Levels of perfluorinated compounds in raw and cooked Mediterranean finfish and shellfish. *Chemosphere* 127, 117–126. https://doi.org/10.1016/j.chemosphere.2014.12.081

Vendl, C., et al., 2022. Thermal processing reduces PFAS concentrations in blue food: a systematic review and meta-analysis. *Environ. Pollut.* 304, 119081. https://doi.org/10.1016/j.envpol.2022.119081

Vera, P., et al., 2024. The analysis of the migration of per and poly fluoroalkyl substances (PFAS) from food contact materials using ultrahigh performance liquid chromatography coupled to ion-mobility quadrupole time-of-flight mass spectrometry (UPLC- IMS-QTOF). *Talanta* 266, 124999. https://doi.org/10.1016/j.talanta.2023.124999

Vughs, D., et al., 2019. The determination of two emerging perfluoroalkyl substances and related halogenated sulfonic acids and their significance for the drinking water supply chain. *Environ. Sci. Process. Impacts* 21, 1899–1907. https://doi.org/10.1039/C9EM00393B

Wagner, A., et al., 2013. Determination of adsorbable organic fluorine from aqueous environmental samples by adsorption to polystyrene-divinylbenzene based activated carbon and combustion ion chromatography. *J. Chromatogr. A* 1295, 82–89. https://doi.org/10.1016/j.chroma.2013.04.051

Wang, Z., et al., 2021. A new OECD definition for per- and polyfluoroalkyl substances. *Environ. Sci. Technol.* 55, 15575–15578. https://doi.org/10.1021/acs.est.1c06896

© Burleigh Dodds Science Publishing Limited, 2025. All rights reserved.

Wang, S., et al., 2022a. Perfluoroalkyl substances in water, sediment, and fish from a subtropical river of China: environmental behaviors and potential risk. *Chemosphere* 288, 132513. https://doi.org/10.1016/j.chemosphere.2021.132513

Wang, Y.-Q., et al., 2022b. Per- and polyfluoralkyl substances (PFAS) in drinking water system: target and non-target screening and removal assessment. *Environ. Int.* 163, 107219. https://doi.org/10.1016/j.envint.2022.107219

Wang, Q., et al., 2023. Tracing per- and polyfluoroalkyl substances (PFASs) in the aquatic environment: target analysis and beyond. *TrAC Trends Anal. Chem.* 169, 117351. https://doi.org/10.1016/j.trac.2023.117351

Wang, Y., et al., 2024. Fast analysis of short-chain and ultra-short-chain fluorinated organics in water by on-line extraction coupled to HPLC-HRMS. *Sci. Total Environ.* 943, 173682. https://doi.org/10.1016/j.scitotenv.2024.173682

Wanninayake, D.M., 2021. Comparison of currently available PFAS remediation technologies in water: a review. *J. Environ. Manage.* 283, 111977. https://doi.org/10.1016/j.jenvman.2021.111977

Wei, T., et al., 2024. Can PFAS threaten the health of fish consumers? A comprehensive analysis linking fish consumption behaviour and health risk. *Sci. Total Environ.* 920, 170960. https://doi.org/10.1016/j.scitotenv.2024.170960

Wolf, N., et al., 2024. Thermal desorption – gas chromatography – mass spectrometry (TD-GC-MS) analysis of PFAS used in food contact materials. *Food Addit. Contam. Part A* 41, 1099–1117. https://doi.org/10.1080/19440049.2024.2370371

Woudneh, M.B., et al., 2019. Effect of sample storage on the quantitative determination of 29 PFAS: observation of analyte interconversions during storage. *Environ. Sci. Technol.* 53, 12576–12585. https://doi.org/10.1021/acs.est.9b03859

Wu, C., et al., 2022. Rapid quantitative analysis and suspect screening of per-and polyfluorinated alkyl substances (PFASs) in aqueous film-forming foams (AFFFs) and municipal wastewater samples by Nano-ESI-HRMS. *Water Res.* 219, 118542. https://doi.org/10.1016/j.watres.2022.118542

Wu, M., et al., 2017. Bioremediation of hydrocarbon degradation in a petroleum-contaminated soil and microbial population and activity determination. *Chemosphere* 169, 124–130. https://doi.org/10.1016/j.chemosphere.2016.11.059

Xiang, L., et al., 2019. Food safety concerns: crop breeding as a potential strategy to address issues associated with the recently lowered reference doses for perfluorooctanoic acid and perfluorooctane sulfonate. *J. Agric. Food Chem.* acs.jafc.9b04625. https://doi.org/10.1021/acs.jafc.9b04625

Xu, J., et al., 2024. Differential uptake and translocation of perfluoroalkyl substances by vegetable roots and leaves: Insight into critical influencing factors. *Sci. Total Environ.* 949, 175205. https://doi.org/10.1016/j.scitotenv.2024.175205

Young, W.M., et al., 2012. Determination of perfluorochemicals in Cow's milk using liquid chromatography-tandem mass spectrometry. *J. Agric. Food Chem.* 60, 1652–1658. https://doi.org/10.1021/jf204565x

Yu, P.-F., et al., 2018. Cultivar-dependent accumulation and translocation of perfluorooctanesulfonate among lettuce ( *Lactuca sativa* L.) cultivars grown on perfluorooctanesulfonate-contaminated soil. *J. Agric. Food Chem.* 66, 13096–13106. https://doi.org/10.1021/acs.jafc.8b04548

© Burleigh Dodds Science Publishing Limited, 2025. All rights reserved.

Yu, P.-F., et al., 2021. Variety-selective rhizospheric activation, uptake, and subcellular distribution of perfluorooctanesulfonate (PFOS) in lettuce (*Lactuca sativa* L.). *Environ. Sci. Technol.* 55, 8730–8741. https://doi.org/10.1021/acs.est.1c01175

Zabaleta, I., et al., 2015. Simultaneous determination of perfluorinated compounds and their potential precursors in mussel tissue and fish muscle tissue and liver samples by liquid chromatography–electrospray-tandem mass spectrometry. *J. Chromatogr. A* 1387, 13–23. https://doi.org/10.1016/j.chroma.2015.01.089

Zacs, D. and Bartkevics, V., 2016. Trace determination of perfluorooctane sulfonate and perfluorooctanoic acid in environmental samples (surface water, wastewater, biota, sediments, and sewage sludge) using liquid chromatography – Orbitrap mass spectrometry. *J. Chromatogr. A* 1473, 109–121. https://doi.org/10.1016/j.chroma.2016.10.060

Zhang, C., et al., 2022. Stability of per- and polyfluoroalkyl substances in solvents relevant to environmental and toxicological analysis. *Environ. Sci. Technol.* 56, 6103–6112. https://doi.org/10.1021/acs.est.1c03979

Zhang, J., et al., 2024. Dietary exposure to per- and polyfluoroalkyl substances: potential health impacts on human liver. *Sci. Total Environ.* 907, 167945. https://doi.org/10.1016/j.scitotenv.2023.167945

Zhou, J., et al., 2023. Development of a new matrix-certified reference material for accurate measurement of PFOA and PFOS in oyster meat powder. *Microchem. J.* 190, 108673. https://doi.org/10.1016/j.microc.2023.108673

Zhou, Y., et al., 2021. Source, transportation, bioaccumulation, distribution and food risk assessment of perfluorinated alkyl substances in vegetables: a review. *Food Chem.* 349, 129137. https://doi.org/10.1016/j.foodchem.2021.129137

© Burleigh Dodds Science Publishing Limited, 2025. All rights reserved.

# Chapter 15

# Assessing microplastics and nanoplastics in food

*Alfonso Lampen, Federal Institute for Risk Assessment (BfR), Germany*



## 1 Introduction and definitions

When classifying particles, which can mainly occur as contaminants in food, a distinction is made between micro- and nanomaterials. Nanoparticles are chemically different particles in the size range 1–100 nm (= $10^{-9}$ m) that can occur in food. According to Recommendation 2011/696/EU (EU, 2011) of the European Commission, nanomaterial is a natural, process-derived or manufactured material that contains particles in an unbound state as an aggregate or agglomerate and in which at least 50% of the particles in the number size distribution have one or more external dimensions in the range from 1 nm to 100 nm. Known nanoparticles include titanium dioxide ($TiO_2$), which is often used as a coating to improve the appearance of confectionery or to brighten salad dressings as an approved food additive E171, or silicon dioxide ($SiO_2$), also known as silicic acid, which is used to improve the flow properties of ketchup, e.g., and as a flow aid for powdered foods. In addition to inorganic nanoparticles, there are also a large number of natural nanoparticles (e.g. fat micelles), which can be found in milk, hen's eggs, fats and fat by-products or are used as carriers for functional nutrients such as poorly soluble bioactive substances. These so-called organic nanoparticles are completely degraded in the human metabolism. For some inorganic nanoparticles, such as $TiO_2$, strict restrictions have recently been imposed in the food sector, such as the

http://dx.doi.org/10.19103/AS.2025.0158.26
© Burleigh Dodds Science Publishing Limited, 2025. All rights reserved.

withdrawal of approval for E171 in Europe, as there are indications of possible genotoxicity (EU, 2022/63).

In contrast to nanoparticles, there are no generally accepted definitions for microplastics and nanoplastics: microplastics (microplastic particles, MPP) are understood to be plastic/plastic particles with a particle size of 0.1 or 1–5000 μm ($10^{-6}$ m), which can enter the environment as solid water-insoluble particles (e.g. ECHA, 2020). The size of 100 nm reflects the boundary to the nanoparticle range. Particle sizes above 5000 μm are classified as macroplastics. Plastics chemically belong to the subgroup of polymers, which includes various materials such as polyethylene (PE), polypropylene (PP), polyvinyl chloride (PVC), polyethylene terephthalate (PET) and polystyrene (PS). The MPPs are correspondingly heterogeneous in terms of shape, form and chemical properties.

To sum it up: Plastic particles are referred to as **microplastics** if they are smaller than 5 mm. If these plastic particles are further broken down into a size between 1 nm and 100 nm, they are referred to as **nanoplastics**.

Due to further developments in food-related analysis techniques, there has recently been an increase in findings of microplastics and nanoplastics in food. MPPs have been detected qualitatively and quantitatively in mineral water, honey, seafood, table salt, even beer and milk powder (Di Fiore et al., 2023; Kadac-Czapska et al., 2023; Visentin et al., 2024). This and the reports that microplastics have been detected in the human faeces (Schwabl et al., 2019), urine (Pironti et al., 2022), blood (Leslie et al., 2022), sputum (Huang et al., 2022), human placentas, breast milk and infant faeces (Liu et al., 2023) have caused great uncertainty among consumers and raised concern among the scientific community.

The problem of the presence of microplastics in the environment is not new. Global plastic production is constantly increasing. According to the Plastics Europe association, around 390 million tons of plastic were produced worldwide in 2021 and around 57 million tons in Europe alone (Plastics Europe, 2022). These are particles, fibers or films that can end up in the environment.

A distinction is made between primary and secondary microplastics: **primary microplastics** are deliberately produced in the form of plastic-based granules or pellets and used, e.g. in the cosmetics industry, food production or in consumer goods. The materials used include PP, PE, PS, polyamide (nylon), ethylene vinyl acetate (EVA), PVC and PET as well as additives such as plasticizers, flame retardants, colorants or antioxidants, which change the technical properties of the plastics used. **Secondary microplastics** are created from larger plastic parts (macroplastics) through friction and decomposition (degradation) of plastic waste. UV light contact with salt water or temperature promote the degradation process. In plastic bags, e.g., UV light excites the binding electrons in the plastics and thus splits the chemical bonds.

© Burleigh Dodds Science Publishing Limited, 2025. All rights reserved.

Secondary microplastics account for the majority of microplastics entering the environment. Progressive degradation through various mechanisms such as oxidative or hydrolytic ageing processes, fracture-mechanical or abrasive destruction as well as (partial) microbiological degradation of the microplastic can lead to the release of additives (such as bisphenol A or nonylphenol) and promote the formation of nanoplastic particles in the environment (Mitrano et al., 2021). Here, too, a distinction can be made between primary nanoplastic particles such as plastics used in cosmetics, e.g. nanoparticle-produced glitter, and nanoplastics resulting from abrasion or degradation processes from microplastics (Mitrano et al. 2021).

# 2 Occurrence of micro-/nanoplastics

## 2.1 Analysis of micro-/nanoplastics

Various analytical methods are used to analyse MPP. A standardization and harmonization of the methods has not yet been established, so that the comparison of results leads to significant differences in the determined particle numbers/contents. Spectroscopic methods such as the micro-Fourier transform infrared spectrometer (μFTIR) and μRaman spectroscopy, in which the inelastic scattering of light on solids (RAMAN scattering) is measured, have established themselves as very good methods for some matrices such as mineral water. μFTIR is suitable for particles with a lower limit of 20 μm, while μRaman spectroscopy can detect MPPs down to approx. 1 μm due to its higher resolution (Fig. 1). In general, however, it can still be stated that the analysis of particles <5 μm with imaging methods is limited and challenging, as the maximum resolution is limited downwards.

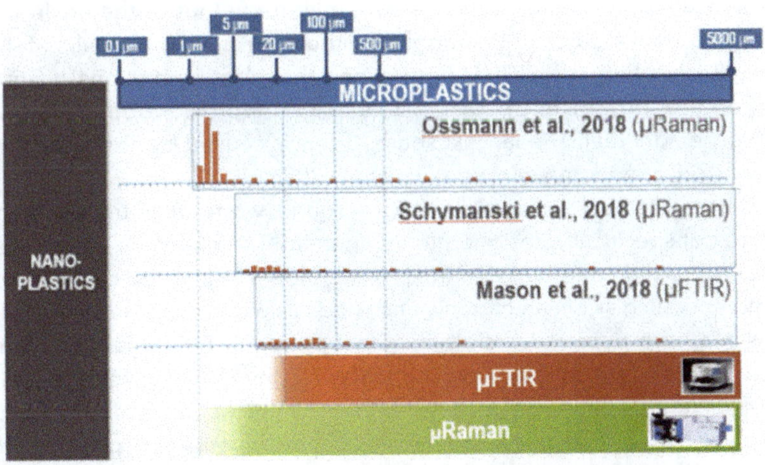


**Figure 1** Literature comparison of μFTIR and μRaman water analyses.

© Burleigh Dodds Science Publishing Limited, 2025. All rights reserved.

The exemplary comparison of the measurement curves of Oßmann et al. (2018), Schymanski et al. (2018) and Mason et al. (2018) (Fig. 1) with regard to the analyses of mineral waters for microplastics shows the clear qualitative and quantitative differences in the analyses. For example, 90% of the MPPs detected by Ossmann were in the range below 5 μm, while 90% of the MPPs detected by Schymanski and Mason were in the range below 100 μm. The number of MPPs detected also differed. Lower detection limits have a strong influence on the resulting particle size distribution. The nanoplastic range cannot yet be covered with these methods. However, newer RAMAN approaches are promising: e.g., PET nanoparticles were detected in the water of mineral water bottles using surface-enhanced RAMAN spectroscopy (SERS) (Zhang et al., 2023). Another approach is promising, i.e. induced RAMAN scattering (SRS microscopy), which has been further developed into a platform and can identify the most important plastic particles at the single-particle level with high specificity in nano sizes using an automated algorithm. This approach enabled the detection of various nanoplastic particles in mineral waters for the first time (Qian et al., 2024). Pyrolysis with gas chromatography coupled with mass spectrometry would be another example for analysing nanoplastics (Okoffo und Thomas, 2024). New and different sample concentration methods are needed to determine the number of particles per mass of nanoplastic. The methods have not yet been validated. There are still major analytical challenges for the future, which will only then make it possible to obtain real exposure data, which is essential for risk assessment.

## *2.2 Micro-/nanoplastics in the environment*

Due to the enormous global production volumes and types of use, microplastics and nanoplastics are currently ubiquitous in the environment. In a consortium study, the Fraunhofer Institute for Environmental, Safety and Energy Technology (UMSICHT) determined that plastic emissions in Germany alone in the form of microplastics amount to approx. 33 000 tons/year or 4000 g/person/year. The largest share of this is caused by abrasion from car tires at 1228 g/person/year (Bertling et al., 2021).

These plastic emissions can pass through wastewater treatment plants that are not specifically designed to retain them. Therefore, they accumulate in higher population densities and their distribution is influenced by water and air movements such as ocean currents and winds as well as particle density. Plastic emissions have become a relevant global problem. The oceans in particular accumulate plastics, as they represent a natural and definitive reservoir. Marine flora and fauna are therefore exposed to them. Physical and chemical interactions take place (Fig. 2). Examples are animals become entangled in plastic materials such as nets or strings, which can lead to wounds or strangulation. Macroplastics

© Burleigh Dodds Science Publishing Limited, 2025. All rights reserved.

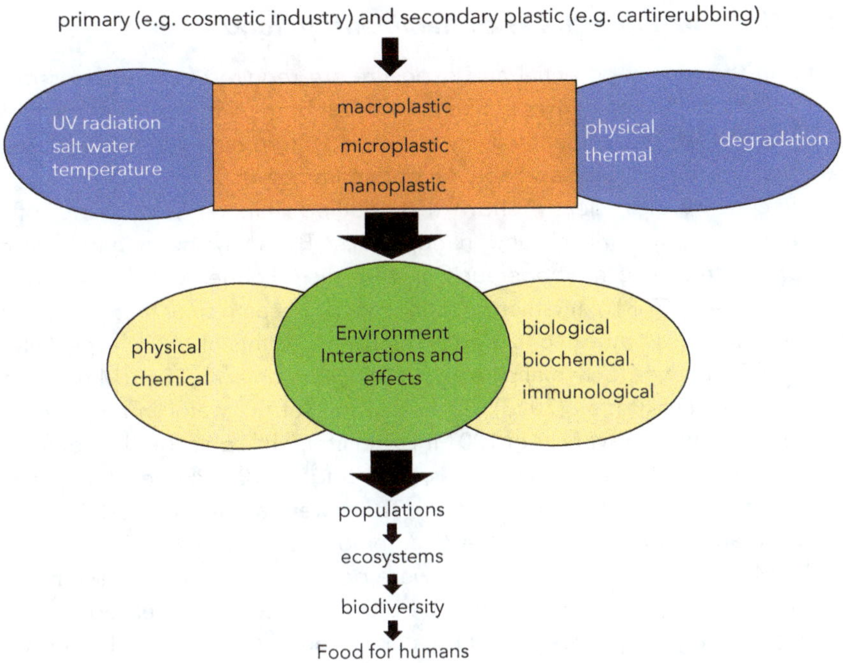


**Figure 2** Interactions of macro-, micro- and nanoplastics with the environment and their possible effects.

are swallowed by various animals (e.g. seabirds, sea turtles), which also leads to health problems for the animals (Wilcox et al., 2015). Corals, sponges and seabeds are covered by macroplastics, so that the supply of light, food and oxygen is impeded. This has a negative impact on ecosystems (Kvale et al., 2021). As the amount of plastic input has been increasing globally for many years and there is no end in sight, this has a long-term impact on biodiversity and will therefore also have a strong influence on human food sources. The top three polymer types of microplastics are PE, PP and PS. Organic and inorganic additives (e.g. metals) account for around 4% of the weight of plastic. Half of these are phthalates, acylphenols and bisphenol A, with the remainder distributed across many other substance groups such as polychlorinated biphenyls (PCBs), polycyclic aromatic hydrocarbons (PAHs) and organochlorine pesticides (EFSA, 2016).

While the ecological impact of plastic waste is obvious, there is no clear evidence to date regarding the potential effects of micro- and nanoplastics on human health. A potential risk to human health from micro- and nanoplastics can arise through ingestion via the food chain. Food can be contaminated by micro- and nanoplastics, either through direct exposure in the environment or through trophic transfer of microplastics between predators and prey in edible species.

© Burleigh Dodds Science Publishing Limited, 2025. All rights reserved.

## 2.3 Micro-/nanoplastics as contaminants in food

Data on the hazard potential and exposure are required for risk assessment with regard to the health risks of microplastics. If both parameters are available, a potential risk can be quantitatively assessed. Unfortunately, however, there is currently insufficient data available on both parameters.

Recent studies clearly show that bottled mineral waters are often contaminated with micro- and nanoplastics. By analysing mineral waters using spectroscopic methods such as the micro-Fourier transform infrared spectrometer (μFTIR) or μRaman spectroscopy, it became clear that both plastic bottles and glass bottles contained significant amounts of MPPs, especially in the 2–100 μm range (Schymanski et al., 2018; Oßmann et al., 2018; Mason et al., 2018). Schymanski et al. (2018), e.g., showed that the water in recycled PET bottles contained an average of 100 MPPs per liter, while glass bottles contained 50 MPPs per liter. In disposable bottles, an average of 10 MPPs per liter were still measured using μRaman spectroscopy. There were always more MPPs in PET bottles than in other plastic bottles (Gambino et al., 2022). Since the majority of the MPPs identified in PET bottles were polyester, the authors assume that the particles originate from the plastic bottles themselves. Increased abrasion during the washing process could be a likely cause. Interestingly, bottles with still water contained significantly less MPP than bottles with carbonated water; the latter seems to exert greater stress on the plastic material. A recent study from the University of Colombia was able to detect an average of $2.4 \times 10^5$ plastic particles/liter of water from different brands of plastic bottles using induced RAMAN scattering (SRS microscopy); of these, approx. 90% were nanoplastics and only 10% were microplastics (Qian et al., 2024). These values are several times higher than those known to date, which indicates that only the tip of the iceberg is currently known.

Current reviews summarize that MPPs are also found in tap water – but in two to three times lower concentrations. As the particle size decreases, the MPP content increases in both plastic bottles and tap water (Gambino et al., 2022). Data on microplastic concentrations in German tap water are hardly available to date. Mintenig et al. (2019) detected approx. 7 MPP/$m^3$ in Bavarian groundwater using FTIR, while only approx. 0.7 MPP/$m^3$ were measured on average in the tap water extracted from it. Similar data is available from other European countries (Gambino et al., 2022). The lower MPP frequency in tap water than in natural sources indicates a high removal rate of MPP in drinking water treatment plants.

Other sources of exposure for the oral intake of micro-/nanoplastics are orally ingested dust (e.g. in food), honey, beer, sugar, sea salt and aquatic organism (Cox et al., 2019). Cox et al. (2019) showed that the sum of possible oral intake of microplastics can be placed in the range of inhalation intake,

© Burleigh Dodds Science Publishing Limited, 2025. All rights reserved.

which was previously seen as dominant (estimates between 41 000 particles/year and 51 000 particles/year; depending on age and gender). This means that the total exposure is estimated to be in the range of 74 000–121 000 microparticles/year, with a further 90 000 particles/year being added through the consumption of water if bottles are used for this purpose. The data from Qian et al. (2024) indicate a significantly higher oral intake of nanoparticles. They measured micro- and nanoplastic concentrations of 2 40 000 particles per liter of bottled water, about 90% of which are nanoplastics (Qian et al., 2024). In addition, Visentin et al. (2024) were able to detect microplastics (mainly polypropylene, polyethylene, polystyrene and PET) in skimmed milk samples from eight EU countries.

MPPs are also frequently detected in the filter organs and gills of marine animals. So far, no relevant quantities of MPPs have been measured in the muscle tissue of fish. The EFSA has estimated that around 7 μg of MPPs are ingested orally when eating 225 g of mussels (this corresponds to around 900 particles/meal) (EFSA, 2016).

In addition to the oral intake of MPP, indirect contamination by lipophilic contaminants adhering to MPP is also possible. These can be environmental contaminants such as dioxins, PCBs, PAHs, biocides or algal toxins (EFSA, 2016). According to EFSA estimates, these additional contaminants increase exposure form food by less than 0.01%, so their toxicological relevance appears to be rather low. The study by Qian et al. (2024) suggests that exposure to nanoplastics may be significantly higher than with MPP, meaning that the additional input of environmental contaminants with nanoplastic particles is currently still open.

## 3 The oral bioavailability of micro-/nanoplastics

In order to realize an oral exposure at all, MPPs that are taken up orally would have to overcome several barriers of the intestine. Only then can a relevant oral bioavailability occur. The intestinal barrier consists of several parts: the physical barrier, which only allows certain particle sizes to pass through, the chemical barrier, which can change substances through activation or deactivation and thus strongly influence oral bioavailability, and the immunological barrier, through which particles can be transported away. Studies have shown that particles >150 μm cannot cross the intestinal barrier. They remain trapped in the mucosal layer and are sloughed off with intestinal cells within approx. 72 hours and excreted unchanged in the stool. Only particles <1.5 μm can principally overcome these intestinal barriers at all and thus enter the human body and be distributed to other organs (EFSA, 2016). This could apply to small microplastic and also to nanoplastic particles. Some *in-vitro* studies with intestinal cells show a slight uptake of MPPs up to 4 μm (Stock et al., 2019). Larger MPPs should

© Burleigh Dodds Science Publishing Limited, 2025. All rights reserved.

not be taken up as they would exceed the size of individual epithelial cells. Only particles absorbed by the intestine and distributed through the blood system can reach organs located further down, such as the liver (Barceló et al., 2023). In all tissues with which the particles come into contact, they could in principle have an undesirable effect. Microplastic and nanoplastic particles have now been detected in various human samples of the placenta and in breast milk (Liu et al., 2023), in urine (Pironti et al., 2022), in faeces (Schwabl et al., 2019), in saliva (Huang et al., 2022) and in the liver (Hovatits et al., 2022). The question of a potential health risk from microplastics in food has therefore become the focus of research and risk assessment. However, initial *in-vitro* and *in-vivo* studies (with mice) show that only very few MPPs are orally bioavailable (Stock et al., 2019), so it can currently be assumed that there is hardly any oral exposure of humans with MPPs >1.5 µm because oral bioavailability is too low. Nevertheless, with regard to nanoplastic particles, it should be noted that they could in principle be orally bioavailable due to their small size and known transport mechanisms such as vesicle transport.

In addition, it was investigated whether the digestive system has an influence on the chemical stability of MPP. For this purpose, various MPPs were incubated *in vitro* with the different digestive juices of the mouth, stomach and intestine and then analysed and characterized. It was found that the structural changes were solely due to the adhesion of the organic matrix of the digestive juices. After degradation of the organic matrix with hydrogen peroxide, hydrochloric acid or nitric acid ($H_2O_2$, HCl or $HNO_3$), the MPPs are unchanged (Stock et al., 2020). This means that the conditions of the human digestive system are not sufficient to degrade MPPs in the body. However, digestion in the body can lead to changes in the properties of the micro- and nanoparticles so that transport through the intestinal barriers is facilitated (Paul et al., 2024). In addition, it was recently reported that four plastic materials (including PET, polyethylene and styrene) with high production numbers and a size of ≥700 nm were found in the blood of healthy volunteers at concentrations of 1.6 µg/mL using pyrolysis with gas chromatography coupled with mass spectrometry (Leslie et al., 2022). This could indicate that some MPPs are obviously bioavailable to a 'small' extent. The study was later called into question because internal controls for contamination during the analysis were obviously missing (Kuhlman, 2022). This shows the difficulties of the analysis. It also makes it clear that the biological and toxicological relevance in relation of the size of the particles should be investigated. Very little is currently known about the oral bioavailability of nanoplastic particles, as these are not yet analytically controlled. However, there are indirect indications that they could be bioavailable: Marfella at al. (2024) found micro- and nanoplastic plagues of undefined size in the carotid artery of patients at greater risk of heart attacks and strokes using pyrolysis with

© Burleigh Dodds Science Publishing Limited, 2025. All rights reserved.

gas chromatography-coupled mass spectrometry (polyethylenes: approx. 21 µg/mg plague and polyvinyl chloride approx. 5 µg/mg plague, respectively). Future studies should clarify whether and to what extent nanoplastics are bioavailable due to their smaller size.

## 4 The toxicology of micro-/nanoplastics

A risk assessment for MPP is challenging because plastics are very divergent in terms of composition, material properties, particle size and shape as well as frequency (Fig. 3).

PS is best studied because chemical synthesis is relatively simple, and labelling molecules can also be easily attached. In addition, there is endogenous fluorescence and the size distribution is narrow and easy to determine. Other types of plastic present difficulties, as they are more common but very difficult to produce. The material density is also a relevant criterion for enabling measurability in light scattering methods.

With regard to toxic effects of MPP, it can be summarized that only few qualitatively acceptable studies are available and mostly no measurable reproducible effects are reported in various animal studies. Some data suggest effects on growth and the immune system. Human data regarding possible (adverse) health effects are almost completely unavailable (EFSA, 2016).

*In-vitro* influences on the redox system, the metabolism of foreign substances or the immunological state of cells have been described, which can lead to 'stress reactions', in particular, inflammatory and immune reactions. There is currently no evidence of genotoxicity, carcinogenicity or neurotoxicity.

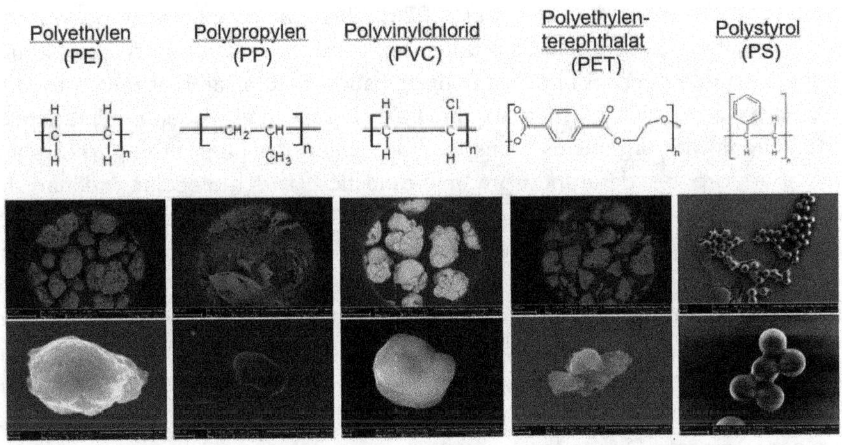


**Figure 3** Scanning electronic images (examples) of various MPPs with different magnification.

© Burleigh Dodds Science Publishing Limited, 2025. All rights reserved.

In fish, effects on the gastrointestinal tract and liver have been reported; in mice, these are inflammatory reactions (EFSA, 2016).

In a study on genetically modified mice, which are particularly sensitive to stress, the uptake and effect of orally ingested PS was investigated, and it was found that only very small amounts of MPP were detected in the intestinal mucosa but no toxic effects indicative of oxidative stress (Stock et al., 2020). With regard to the effects of nanoplastics, *in-vitro* studies have reported cytotoxic effects, membrane damage and oxidative stress in some cell types depending on size, positive charge and usually high doses, but the data situation is still too incomplete to draw conclusions for a risk assessment (Banerjee and Shelver, 2021; Barceló et al., 2023; Zhu et al., 2023).

## 5 Sustainability issues

The probable entry of microplastics and nanoplastics into the environment and thus into the food chain influences key aspects of a more sustainable diet: health, animal welfare and the environment (see Fig. 1). In addition to human health, the focus is also on animal biodiversity. While further research is needed into the negative health effects of microplastics and nanoplastics on humans, it is already clear that MPPs pose a risk to many marine animals and birds that 'accidentally' eat microplastics, e.g. For example, MPPs are mistaken for prey by aquatic organisms or ingested via prey. Therefore, MPPs are often found in the digestive tract, filter organs and gills. It has recently been shown that even low concentrations of MPPs have an impact on the marine ecosystem. The uptake of microplastics by zooplankton reduces feeding pressure and results in more algae growth, which in turn causes more organic matter to sink into deeper layers of the sea. Bacteria decompose these and thus reduce the oxygen content of the oceans (Kvale et al., 2021). This may affect the global nutrient cycle, resulting in an impact on biodiversity. Furthermore, it cannot be assumed that environmental problems with nanoplastics are the same as environmental problems with microplastics, as the particle size has a major influence on interactions with organisms, transport routes and the fate of the particles and will therefore be different from microplastics. The molecular mechanisms will differ as smaller particles can reach target cells in the organism as they can overcome biological barriers by, e.g., vesicular transport. Nanoplastic particles could therefore have a much larger negative impact on health, particularly in humans, than microplastics. In addition, the necessary analytics are fundamentally different. New analytical tools are needed to be able to assess the exposure and consequently also the potential health risks. If they are detected in food such as mineral water bottles, micro- and nanoplastic particles are contaminants. Due to the lack of exposure data for a risk assessment, they are not yet listed in the EU regulation (EU, 2023/915) intended for this purpose.

© Burleigh Dodds Science Publishing Limited, 2025. All rights reserved.

In order to tackle the major challenges posed by the ever-increasing global plastic problem, the EU Commission has initiated regulations aimed at strengthening the circular economy for plastics as part of the *European Green Deal* (EU, 2019), which aim to protect the environment and human health and thus also contribute to the sustainability goals of the United Nations (UN). These include a whole range of regulations such as marketing bans (e.g. the ban on single-use plastics), labeling requirements (e.g. plastic content), extended producer responsibility and minimum requirements (e.g. for the recycling share of PET beverage bottles in the EU; target: 90% collection of PET bottles by 2029). In addition, a comprehensive ban on intentionally added MPPs was adopted as part of the REACH chemicals regulation (EU, 2023/2055).

The environmental impact of microplastics and possibly nanoplastics is damaging resources (including the sea, water, soil and biodiversity) for the food sector (see Fig. 1). There should therefore be a social consensus to reduce this input to the bare minimum. This means that the use of plastics should be questioned much more critically, e.g. to minimize the consumption of single-use plastic products such as plastic bags and disposable bottles, to rely on alternative natural materials and to further develop recycling and to significantly reduce the abrasion of car tires. The latter could be achieved through new mobility concepts, e.g. initial approaches have recently been seen in the above-mentioned regulations in the EU. Further joint efforts by government organizations, companies and consumers are needed to take the challenges for greater sustainability further and achieve the UN Sustainable Development Goals by 2030. Interesting in this context: in 2022, Germany joined the *High Ambition Coalition to End Plastic Pollution* initiative, which has set itself the goal of ending plastic pollution by 2040.

## 6 Conclusion

A conclusive risk assessment on the impact of microplastic and nanoplastic particles on the human body is not yet possible due to a lack of data on hazard potential and exposure as well as other open questions; however, the general principles of risk assessment (hazard identification, hazard characterization, exposure estimation and risk characterization) are applicable. Microplastics accumulate in the environment and can enter the food chain. Environmental influences can further decompose and chemically alter plastic material. Ingestion via food is likely, but the oral bioavailability of MPP >1.5 μm in humans appears to be very low. The mechanisms of action of MPP after oral and inhalation uptake in the body have not yet been fully elucidated. Overall, the current data suggest a low risk for consumers with regard to orally ingested microplastics with larger size (>1.5 μm) in the diet. However, organisms living in seawater are clearly being harmed by microplastics, in some cases significantly,

© Burleigh Dodds Science Publishing Limited, 2025. All rights reserved.

and the environmental damage is evident. Over time, however, microplastics become nanoplastics. More recent data show a relevant exposure of humans. The potential risks of nanoplastics are still completely unclear, but can be classified as potentially relevant, as smaller nanoplastic particles overcome biological barriers and would therefore be bioavailable. Detailed data on this is currently lacking. New research approaches are needed here with regard to the analysis, uptake and possible molecular effects of nanoplastics in the organism as well as exposure. In addition, more intensive efforts should be made to significantly reduce the relevant environmental impact and achieve the sustainability goals.

## 7 References

Banerjee, A. and Shelver, W.L. Micro- and nanoplastic induced cellular toxicity in mammals: a review. *Sci Total Environ*. 2021 Feb 10;755(Pt 2):142518. https://doi.org/10.1016/j.scitotenv.2020.142518

Barceló, D., et al. Microplastics: detection in human samples, cell line studies, and health impacts. *Environ Toxicol Pharmacol*. 2023 Aug;101:104204. https://doi.org/10.1016/j.etap.2023.104204

Bertling, J., et al. Plastics in the environment: emissions to agricultural soils, 2021. https://publica.fraunhofer.de/entities/publication/6e3e887c-3a19-4769-b29b-9dc5e5364d9f/details (viewed on 28.11.2023)

Cox, K.D., et al. Human consumption of microplastics. *Environ Sci Technol* 2019;53:7068–7074. https://doi.org/10.1021/acs.est.9b01517

Di Fiore, C., et al. Microplastic contamination in commercial salt: an issue for their sampling and quantification. *Food Chem* 2023;404:134682. https://doi.org/10.1016/j.foodchem.2022.134682

EFSA Panel on Contaminants in the Food Chain (EFSA CONTAM Panel): statement on the presence of microplastics and nanoplastics in food, with particular focus on seafood. *EFSA J* 14;2016:4501. https://doi.org/10.2903/j.efsa.2016.4501

European Chemicals Agency (ECHA): opinion of the committee for risk assessment and opinion of the committee for socio-economic analysis on an annex XV dossier proposing restrictions of the manufacture, placing on the market or use of a substance within the EU, 2020. https://echa.europa.eu/documents/10162/a513b793-dd84-d83a-9c06-e7a11580f366 (viewed on 28.11.2023)

European Commission: Commission Recommendation of October 18, 2011 on the definition of nanomaterials. *Official Journal of the European Union L* 2011;275:38–40.

European Commission: The European Green Deal. Striving to be the first climate-neutral continent, 2019. https://commission.europa.eu/strategy-and-policy/priorities-2019-2024/european-green-deal_en (viewed on 28.11.2023)

European Commission: Commission Regulation (EU) 2023/915 of April 25, 2023 on maximum levels for certain contaminants in foodstuffs and repealing Regulation (EC) No 1881/2006 L119 (2023) 103–157.

European Commission: Commission Regulation (EU) 2023/2055; Annex XVII to Regulation (EC) No 1907/2006 of the European Parliament and of the Council

© Burleigh Dodds Science Publishing Limited, 2025. All rights reserved.

concerning the Registration, Evaluation, Authorisation and Restriction of Chemicals (REACH) as regards synthetic polymer microparticles.

European Parliament, European Council (ed.): Commission Regulation (EU) No 2022/63 of 14 January 2022 amending Annexes II and III to Regulation (EC) No 1333/2008 of the European Parliament and of the Council as regards the food additive titanium dioxide (E 171), 2022. https://eur-lex.europa.eu/legal-content/DE/TXT/PDF/?uri=CELEX:32022R0063&from=DE (consulted on 14.02.2023)

Gambino, I., et al. Occurrence of microplastics in tap and bottled water: current knowledge. *Int J Environ Res Public Health* 2022;19:5283. https://doi.org/10.3390/ijerph19095283

Horvatits, T., et al. Microplastics detected in cirrhotic liver tissue. *EBioMedicine*. 2022 Aug;82:104147. https://doi.org/10.1016/j.ebiom.2022.104147

Huang, S., et al. Detection and analysis of microplastics in human sputum. *Environ Sci Technol*. 2022 Feb 15;56(4):2476–2486. https://doi.org/10.1021/acs.est.1c03859

Kadac-Czapska, K., et al. Microplastics in food - a critical approach to definition, sample preparation, and characterization. *Food Chem*. 2023 Aug 30;418:135985.

Kvale, K., et al. Zooplankton grazing of microplastic can accelerate global loss of ocean oxygen. *Nat Commun*. 2021 Apr 21;12(1):2358. https://doi.org/10.1038/s41467-021-22554-w

Kuhlman, R.L. Letter to the editor, discovery and quantification of plastic particle pollution in human blood. *Environ Int* 2022;167:107400. https://doi.org/10.1016/j.envint.2022.107400

Leslie, H.A., et al. Discovery and quantification of plastic particle pollution in human blood. *Environ Int* 2022;163:107199. https://doi.org/10.1016/j.envint.2022.107199

Liu, S., et al. Detection of various microplastics in placentas, meconium, infant feces, breastmilk and infant formula: a pilot prospective study. *Sci Total Environ*. 2023 Jan 1;854:158699. https://doi.org/10.1016/j.scitotenv.2022.158699

Marfella, R., et al. Microplastics and nanoplastics in atheromas and cardiovascular events. *N Engl J Med* 2024 Mar 7;390(10):900–910. https://doi.org/10.1056/NEJMoa2309822

Mason, S.A., et al. Synthetic polymer contamination in bottled water. *Front Chem* 2018;6:407. https://doi.org/10.3389/fchem.2018.00407

Mintenig, S.M., et al. Low numbers of microplastics detected in drinking water from ground water sources. *Sci Total Environ* 2019;648:631–635. https://doi.org/10.1016/j.scitotenv.2018.08.178

Mitrano, D.M., et al. Placing nanoplastics in the context of global plastic pollution. *Nat Nanotechnol* 2021;16:491–500. https://doi.org/10.1038/s41565-021-00888-2

Oßmann, B.E., et al. Small-sized microplastics and pigmented particles in bottled mineral water. *Water Res* 2018;141:307–316. https://doi.org/10.1016/j.watres.2018.05.027

Okoffo, E.D. and Thomas, K.V. Mass quantification of nanoplastics at wastewater treatment plants by pyrolysis-gas chromatography-mass spectrometry. *Water Res* 2024 May 1;254:121397. https://doi.org/10.1016/j.watres.2024.121397

Paul, M.P. et al. Influence of artificial digestion on characteristics and intestinal cellular effects of micro-, submicro- and nanoplastics. *Food and Chemical Toxicology* 2024; 184: 114423.

Plastics Europe AISBL: Plastics - the Facts 2022, 2022. https://plasticseurope.org/de/wp-content/uploads/sites/3/2022/10/PE-PLASTICS-THE-FACTS_20221017.pdf

© Burleigh Dodds Science Publishing Limited, 2025. All rights reserved.

Pironti, C., et al. First evidence of microplastics in human urine, a preliminary study of intake in the human body. *Toxics* 2022 Dec 30;11(1):40. https://doi.org/10.3390/toxics11010040

Qian, N., et al. Rapid single-particle chemical imaging of nanoplastics by SRS microscopy. *Proc Natl Acad Sci U S A* 2024 Jan 16;121(3):e2300582121. https://doi.org/10.1073/pnas.2300582121

Schwabl, P., et al. Detection of various microplastics in human stool: a prospective case series. *Ann Intern Med* 2019;171:453–457. https://doi.org/10.7326/M19-0618

Schymanski, D., et al. Analysis of microplastics in water by micro-Raman spectroscopy: release of plastic particles from different packaging into mineral water. *Water Res* 2018;129:154–162. https://doi.org/10.1016/j.watres.2017.11.011

Stock, V., et al. Uptake and effects of orally ingested polystyrene microplastic particles in vitro and in vivo. *Arch Toxicol* 2019;93:1817–1833. https://doi.org/10.1007/s00204-019-02478-7

Stock, V., et al. Impact of artificial digestion on the sizes and shapes of microplastic particles. *Food Chem Toxicol* 2020;135:111010. https://doi.org/10.1016/j.fct.2019.111010

Visentin, E., et al. Characterization of microplastics in skim-milk powders. *J Dairy Sci*. 2024 Apr 10:S0022-0302(24)00731-8. https://doi.org/10.3168/jds.2023-24373

Wilcox, C., et al. Threat of plastic pollution to seabirds is global, pervasive, and increasing. *Proc Natl Acad Sci U S A* 2015 Sep 22;112(38):11899–11904. https://doi.org/10.1073/pnas.1502108112

Zhang, J., et al. Identification of Poly(ethylene terephthalate) nanoplastics in commercially bottled drinking water using surface-enhanced raman spectroscopy. *Environ Sci Technol* 2023 Jun 6;57(22):8365–8372. https://doi.org/10.1021/acs.est.3c00842

Zhu, X., et al. Micro- and nanoplastics: a new cardiovascular risk factor? *Environ Int*. 2023 Jan;171:107662. https://doi.org/10.1016/j.envint.2022.107662

© Burleigh Dodds Science Publishing Limited, 2025. All rights reserved.

# Chapter 16

# Contaminants related to high-temperature processing of food

*Youngsun Lee, University of Helsinki, Finland; and Hoonjeong Kwon, Seoul National University, Republic of Korea*



## 1 Introduction

Process contaminants are harmful by-products unintentionally generated in food during processing or cooking methods such as heating, drying, and fermentation (EFSA, 2008; FDA, 2024). Thermal process contaminants, including acrylamide (AA), heterocyclic amines (HCAs), and polycyclic aromatic hydrocarbons (PAHs), are formed in foods during high-temperature cooking processes such as frying, baking, roasting, and grilling, posing significant health risks to both human and animal consumers (Feng et al., 2025).

The formation of these toxicants often involves the Maillard reaction, a common pathway where reactions occur between carbonyl and amino groups. However, each toxicant may also be produced through additional pathways (Xiong et al., 2024). The types and quantities of these contaminants are strongly influenced by food composition – such as carbohydrates, proteins/amino acids, and lipids – as well as cooking temperature and duration (Xiong et al., 2024; Edna Hee et al., 2024;).

Acrylamide is unique in that it forms in carbohydrate-rich, plant-based foods such as potatoes and cereal grains. It is classified as a probable human carcinogen (Group 2A) by the International Agency for Research on Cancer (IARC) (Edna Hee et al., 2024). Heterocyclic amines and PAH, on the other hand, are primarily generated in high-protein and high-fat foods, respectively, such as meat, fish, and fats or oils. The HCAs are classified by IARC into Group

http://dx.doi.org/10.19103/AS.2025.0158.33
© Burleigh Dodds Science Publishing Limited, 2025. All rights reserved.

2A (probable carcinogens, e.g. imidazoquinoline (IQ)) and Group 2B (possible carcinogens, e.g. MeIQ, MeIQx, PhIP, Glu-P-1, Glu-P-2, Trp-P-1, Trp-P-2, AαC, MeAαC) (Lee et al., 2020, 2022; Edna Hee et al., 2024). The PAHs are categorized across several groups: Group 1 (carcinogenic to humans, e.g. benzo[a]pyrene), Group 2A (e.g. dibenz[a,h]anthracene), Group 2B (e.g. chrysene, benzo[a]anthracene, benzo[b]fluoranthene, benzo[k]fluoranthene, indeno[1,2,3-cd]pyrene), and Group 3 (not classifiable as to human carcinogenicity, e.g. benzo(g,h,i)perylene) (Patel et al., 2020).

The IARC (2015) has also evaluated the carcinogenic risks of red and processed meats. Red meat is classified as Group 2A, and processed meat as Group 1, based on strong evidence linking these foods to colorectal cancer. Heterocyclic amines and PAHs were identified as the primary carcinogens in red and processed meats (IARC, 2015).

While heating processes are essential for enhancing food flavor and ensuring safety, they also contribute to the formation of hazardous contaminants, posing significant health risks. This chapter explores the formation mechanisms, occurrence, and dietary exposure levels of key thermal processing contaminants, including AA, HCAs, and PAHs, across various countries, including the associated risk characterization outcomes. It also discusses management strategies aimed at mitigating these risks. Using research findings from Korea as a case study, the chapter provides insights into the occurrence, dietary exposure, and control measures for these contaminants.

## 2 Acrylamide

Acrylamide is a small, hydrophilic compound widely used in industry to produce polyacrylamides, which serve various applications (Filippini et al., 2022). In 2002, the Swedish Food Authority and a research group from the University of Stockholm first reported that AA, a potential carcinogen, was detected in various cooked foods prepared at high temperatures, typically above 120°C, and under low-moisture conditions (Swedish Food Agency; Tareke et al., 2002). The formation of AA is closely linked to the Maillard reaction, involving reducing sugars and the amino acid asparagine (Edna Hee et al., 2024).

Acrylamide has demonstrated neurotoxic properties, adversely affecting germ cells and posing risks to reproductive health. Although AA tested negative in the Ames test for mutagenicity, it has shown gene mutations in mammalian cells and chromosomal aberrations in both in vitro and in vivo studies. In long-term carcinogenicity studies on rats exposed to AA via drinking water, males developed scrotal, thyroid, and adrenal tumors, while females exhibited mammary, thyroid, and uterine tumors. Furthermore, AA has been shown to cross the placental barrier, raising concerns about potential prenatal exposure (Hogervorst et al., 2022).

© Burleigh Dodds Science Publishing Limited, 2025. All rights reserved.

## 2.1 Mechanism and occurrence

Acrylamide can form through several mechanisms, as illustrated in Fig. 1. The primary formation pathway involves the Maillard reaction, where reducing sugars react with the amino group of amino acids under high temperatures, producing a Schiff base, a key intermediate formed after dehydration (Jin, Wu & Zhang, 2013). Acrylamide formation begins around 120°C, with the optimal temperature range being between 160°C and 180°C (Mottram, Wedzicha & Dodson, 2002). Among amino group donors, free asparagine – a common amino acid in potatoes and cereals – plays a pivotal role as a substrate in AA formation (Mottram, Wedzicha & Dodson, 2002).

Acrylamide is generated via three main pathways: (1) the specific amino acid pathway (SP), (2) the generic amino acid pathway (GP), and (3) a minor pathway involving acrolein. In the SP, asparagine reacts with sugars to form a decarboxylated Schiff base, which undergoes β-elimination to produce AA. In the GP, asparagine reacts with various carbonyl intermediates, such as dicarbonyl compounds (e.g. glyoxal), hydroxycarbonyl compounds (e.g. glyceraldehyde), and 5-hydroxymethyl-2-furfural. These intermediates are produced during the Maillard reaction between sugars and amino acids, including asparagine.

The interplay between the SP and GP pathways depends on the availability of substrates. If asparagine is depleted through the SP, AA production becomes limited. Conversely, if sugars are the limiting factor and asparagine remains abundant, the GP can continue generating AA by utilizing intermediates produced during the SP. Consequently, AA formation is more likely when sugar availability is limited, highlighting the essential role of asparagine in this process (Hwang & Kwon, 2022).

Blank (2005) reported that the Amadori rearrangement compound is inefficient at converting into AA. Instead, the Schiff base undergoes transformation directly or indirectly through an oxazolidin-5-one intermediate, which is decarboxylated into azomethine ylide. This intermediate can then tautomerize into a decarboxylated Amadori compound. Acrylamide is subsequently produced either through β-elimination of the decarboxylated Amadori compound or via the deamination of 2-aminopropionamide, a critical precursor to AA (Jin, Wu & Zhang, 2013; Granvogl & Schieberle, 2006).

A minor pathway involves acrolein, which is generated through the oxidative degradation of lipids, the breakdown of amino acids, proteins, and carbohydrates, or Strecker degradation during the Maillard reaction. Acrolein can serve as a precursor to AA when it is oxidized to acrylic acid in the presence of ammonia, followed by amination (Claeys, De Vleeschouwer & Hendrickx, 2005; Jin, Wu & Zhang, 2013; Edna Hee et al., 2024).

The Joint FAO/WHO Expert Committee on Food Additives (JECFA) conducted a comprehensive evaluation of AA levels in foods using data from

© Burleigh Dodds Science Publishing Limited, 2025. All rights reserved.

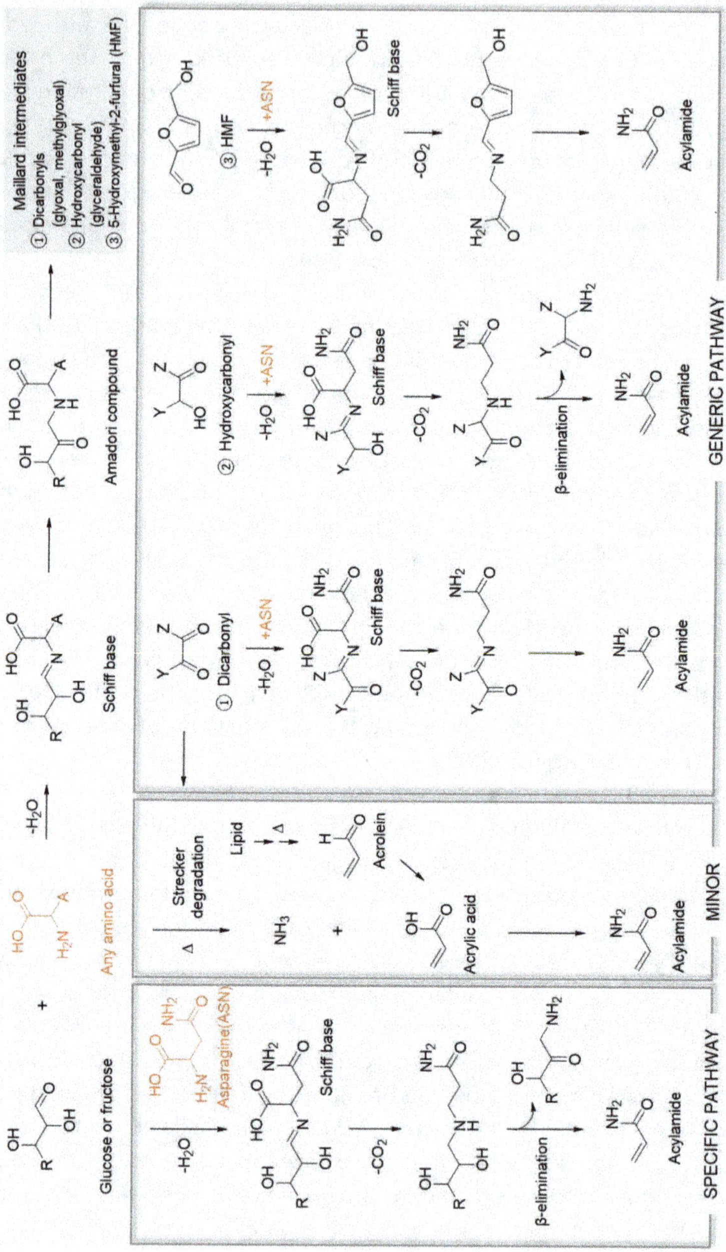


**Figure 1** Acrylamide formation mechanism adapted by Jin, Wu & Zhang (2013) and Hwang & Kwon (2022).

© Burleigh Dodds Science Publishing Limited, 2025. All rights reserved.

31 countries between 2004 and 2009. The data represented regions such as Europe (61%), Asia (28%), North America (9%), the Pacific (1%), and Latin America (1%). Major food groups contaminated with AA included potato crisps (399–1202 μg/kg), potato chips (159–963 μg/kg), biscuits (169–518 μg/kg), crispbread and crackers (87–459 μg/kg), and ready-to-drink coffee (3–68 μg/L). These findings aligned with a prior 2005 evaluation, though significant decreases in AA levels in rye products were noted (WHO, 2011).

The European Food Safety Authority (EFSA) evaluated AA levels in food samples collected from 24 European countries and six food associations. The highest concentrations were found in coffee substitutes (average medium bound level: 1499 μg/kg) and dried coffee (522 μg/kg), followed by potato crisps and snacks (389 μg/kg) and fried potato products (308 μg/kg). Lower levels were detected in processed cereal-based baby foods (73 μg/kg), soft bread (42 μg/kg), and other baby foods (24 μg/kg). Notably, AA concentrations in fresh sliced potato crisps significantly declined from 763 μg/kg in 2002 to 358 μg/kg in 2011 (EFSA, 2005).

In Korea, AA levels in foods have been analyzed through total diet studies (TDS) and processed food evaluations (MFDS, 2016a; Jeong et al. 2020). Recommended by the World Health Organization (WHO) and the Food and Agriculture Organization (FAO), TDS is widely used to estimate exposure to hazardous compounds by analyzing table-ready samples. For process-related hazards, TDS is considered the most suitable approach for exposure assessment.

In the Korean TDS, the highest detection rate of AA (11.9%) was observed in processed foods, with levels ranging from not detectable (n.d.[1]) to 12 029 ng/g. Agricultural products (8.3%, n.d.[2]–220 ng/g), livestock products (8.3%, n.d.–143 ng/g), and marine products (1.7%, n.d.[3]–305 ng/g) exhibited lower detection rates. Black pepper prepared using various cooking methods showed the highest AA levels: grilling (12 019 ng/g), air-frying (3596 ng/g), stir-frying (1313 ng/g), boiling (1169 ng/g), and pan-frying (1084 ng/g). In contrast, raw (non-cooked) black pepper contained only 324 ng/g. Other foods with notable AA concentrations included instant coffee, French fries, and potato snacks (MFDS, 2016a). Recently, over 1000 processed food samples were tested for AA in Korea, with 68% testing positive (Jeong et al. 2020).

Acrylamide levels in various food categories are summarized in Table 1. Consistent with studies from other countries, AA was detected in potato snacks (9–2835 ng/g), French fries (101.6–1175 ng/g), biscuits, wafers, cookies,

1 The method detection limit (MDL) ranged from 2.8 to 3.4 ng/g across ten representative matrices, including porridge, apple juice, corn oil, peanut butter, beef, milk, 20% ethanol, flounder, tuna, and seaweed (MFDS, 2016a). For each sample, the most appropriate representative matrix was selected based on its food composition, and the corresponding MDL was applied.

2 *Ibid.*

3 *Ibid.*

© Burleigh Dodds Science Publishing Limited, 2025. All rights reserved.

**Table 1** Acrylamide contents in Korean processed foods (adapted from Jeong, Hwang and Kwon (2020))

| Food categories | | Detection rate (%) | Range (ng/g)[1] | | |
|---|---|---|---|---|---|
| Confectioneries and bakery products | Potato snack | 100 | 9 | – | 2835 |
| | Biscuits/wafers/cookies/crackers/snacks | 100 | 2.2 | – | 1091 |
| | Soft breads | 46 | n.d. | – | 48.8 |
| | Candies | 13 | n.d. | – | 18.4 |
| | Cocoa/chocolate products | 86 | n.d. | – | 113 |
| Frozen desserts | | 50 | n.d. | – | 66 |
| Sugars and honey | Sugars | 55 | n.d. | – | 87.4 |
| Edible oils and fats | | 48 | n.d. | – | 563 |
| Beverages | Teas | 76 | n.d. | – | 7331 |
| | Coffee | 97 | n.d. | – | 766 |
| | Coffee substitutes | 100 | 121.3 | – | 2050 |
| | Other beverages | 5 | n.d. | – | 3.2 |
| Formulas and baby foods | | 51 | n.d. | – | 53.4 |
| Condiments and seasonings | Sauces | 55 | n.d. | – | 37.7 |
| | Curries | 92 | n.d. | – | 141 |
| | Stock powders | 35 | n.d. | – | 221 |
| | Spices | 92 | n.d. | – | 1235 |
| Preserved products | Kimchi | 46 | n.d. | – | 14.1 |
| | Fermented soy products | 46 | n.d. | – | 9.6 |
| | Pickled/Sugar-preserved products | 8 | n.d. | – | 8.6 |
| Fruit and vegetable products | Dried or canned fruits and vegetables | 42 | n.d. | – | 344 |
| | Fruit and vegetable juices | 54 | n.d. | – | 440 |

© Burleigh Dodds Science Publishing Limited, 2025. All rights reserved.

| | | | | | |
|---|---|---|---|---|---|
| Grain products (excluding bakery products) | Starch and wheat flour | 43 | n.d. | – | 8 |
| | Starch jellies | 14 | n.d. | – | 3.7 |
| | Noodles | 77 | n.d. | – | 40 |
| | Breakfast cereals | 100 | 7.5 | – | 237 |
| | Others | 85 | n.d. | – | 1664 |
| Legume products | Tofu | 27 | n.d. | – | 4 |
| | Roasted beans | 67 | n.d. | – | 123 |
| Processed meat | | 23 | n.d. | – | 4.1 |
| Processed seafood | Seasoned lavers (roasted) | 100 | 16.8 | – | 156 |
| | Others | 10 | n.d. | – | 6.2 |
| Egg products | | 20 | n.d. | – | 3.8 |
| Ready-to-eat/cook | French fries | 100 | 101.6 | – | 1175 |
| | Fried foods (excluding French fries) | 93 | n.d. | – | 81.1 |
| | Dumplings | 57 | n.d. | – | 44.1 |
| | Braised foods | 90 | n.d. | – | 31 |
| | Soups | 91 | n.d. | – | 29.8 |
| Other products | Peanut or nut products | 84 | n.d. | – | 268 |
| | Rice cakes | 25 | n.d. | – | 3.2 |
| | Flour from roots and tubers | 100 | 17.4 | – | 1117 |

[1]The range of method detection limit (MDL) of four matrices (apple juice, starch, corn oil and coconut) was 2.0–3.0 ng/g (Jeong, Hwang & Kwon, 2020). The most appropriate representative matrix was selected for each sample based on its food composition, and the corresponding MDL was applied to the sample.

© Burleigh Dodds Science Publishing Limited, 2025. All rights reserved.

crackers, and other snacks (2.2–1091 ng/g), coffee (n.d.–7331 ng/g), and coffee substitutes (121.3–2050 ng/g). Notably, loose teas containing Jerusalem artichoke (*Helianthus tuberosus L.*), a tuber vegetable, exhibited the highest AA levels (902–7331 ng/g) among the five samples analyzed, likely due to roasting or drying during tea preparation. Acrylamide levels in black and white pepper were 397.0 ± 144.0 ng/g and 61.8 ± 72.2 ng/g, respectively, aligning with findings from the TDS, where raw black pepper contained 324 ng/g. The higher AA content in black pepper is attributed to heating unripe berries prior to drying, while white pepper is prepared by fermenting ripe berries to remove the outer pericarp. Acrylamide was also detected in sesame, perilla, and flavored oils, while most refined cooking oils showed no detectable levels. Sesame and perilla oils, popular in Korea, are produced by roasting and pressing seeds without refinement to retain flavor, which likely contributes to AA formation. Jeong et al. (2020) reported AA in turmeric samples and curries containing turmeric powder. One turmeric sample had an AA level of 1235 ng/g, significantly higher than other samples (26.9–344 ng/g). Among grain products, AA was detected in breakfast cereals and roasted grain products, including *Misugaru*, a mixture of steamed or roasted grains. All seasoned laver (*Gim*) samples, a type of edible algae typically consumed after being roasted in salt and sesame oil seasoning, contained AA (16.8–156 ng/g). This may be attributed to the asparagine content in laver (20–90 mg/100g), which could contribute to AA formation during processing (Fig. 2).

## 2.2 Exposure and risk characterization

Acrylamide enters the human body through various sources, including food, drinking water, cigarette smoking, and occupational exposure. In industrial settings, AA is primarily used in the production of polyacrylamide, which is utilized in waterproof construction materials, dye dispersions for textiles, laboratory electrophoresis materials, and as a flocculant in drinking water treatment. According to WHO (2011), residual AA, as a monomer of polyacrylamide, has been detected in drinking water at levels up to several μg/L, exceeding the recommended guideline value of 0.5 μg/L. The WHO has recommended minimizing AA levels in drinking water due to the already high dietary exposure from cooked foods, which is challenging to control (World Health Organization, 2011). Although occupational exposure to AA is a concern, the primary exposure route for the general population is through food, particularly cereal-based foods cooked at high temperatures (Filippini et al., 2022).

Dietary exposure assessments in the EU, USA, and Korea are summarized in Table 2 by age group. Infants and children consistently show higher daily AA intakes per body weight compared to adults across all regions.

© Burleigh Dodds Science Publishing Limited, 2025. All rights reserved.

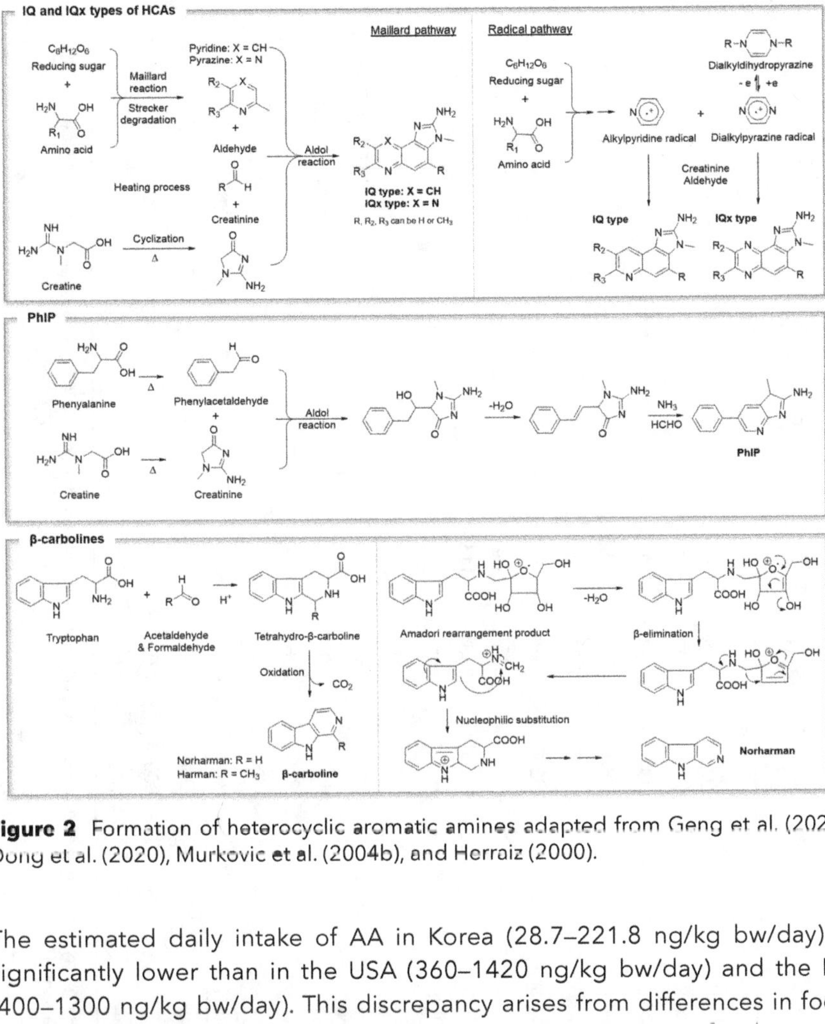


**Figure 2** Formation of heterocyclic aromatic amines adapted from Geng et al. (2024), Dong et al. (2020), Murkovic et al. (2004b), and Herraiz (2000).

The estimated daily intake of AA in Korea (28.7–221.8 ng/kg bw/day) is significantly lower than in the USA (360–1420 ng/kg bw/day) and the EU (400–1300 ng/kg bw/day). This discrepancy arises from differences in food consumption patterns, despite similar AA concentrations in foods across these regions. For example, rice, which is typically boiled, is a staple food in Korea, whereas wheat- and potato-based foods, often fried or roasted, are more common in Western diets.

In Korea, the average estimated exposure to AA is similar between the TDS (90 ng/kg bw/day) (MFDS, 2016a) and a more recent processed food study (86 ng/kg bw/day) (Jeong et al., 2020). However, the primary food contributors differ between the studies. In the TDS, black pepper accounted for 18.2% of AA exposure, followed by snacks, biscuits, and cookies (10.6%), potatoes (10.5%), beef (8.3%), and instant coffee (8.1%). Conversely, in the processed foods study, biscuits, wafers, cookies, crackers, and non-potato snacks contributed

© Burleigh Dodds Science Publishing Limited, 2025. All rights reserved.

**Table 2** Exposure to acrylamide (AA, ng/kg bw/day) and its margin of exposure (MOE) across age groups in different countries

| | | AA (ng/kg bw/day) | | | | MOE[1] | | | | |
|---|---|---|---|---|---|---|---|---|---|---|
| Country | Age | Mean | 50th | 90th | 95th | Mean | 50th | 90th | 95th | Comments |
| Korea[2] | 3–5 | 221.8 | 70.0 | – | 835.7 | 1398 | 4429 | – | 371 | Processed foods |
| | 6–11 | 169.9 | 53.5 | – | 635.5 | 1825 | 5794 | – | 488 | |
| | 12–18 | 110.7 | 33.0 | – | 461.6 | 2800 | 9394 | – | 672 | |
| | 19–64 | 83.3 | 40.8 | – | 247.0 | 3721 | 7598 | 1255 | | |
| | ≥65 | 45.6 | 24.3 | – | 114.7 | 6798 | 12757 | – | 2703 | |
| Korea[3] | 1–2 | 198.9 | 68.0 | – | 798.1 | 905* | 4559 | – | 226* | Total diet study |
| | 3–5 | 205.4 | 92.3 | – | 675.5 | 878* | 3359 | – | 267* | |
| | 6–11 | 154.2 | 64.5 | – | 595.4 | 1169* | 4806 | – | 303* | |
| | 12–18 | 119.3 | 48.3 | – | 477.3 | 1513* | 6418 | – | 377* | |
| | 19–64 | 76.0 | 26.5 | – | 313.6 | 2368* | 11698 | – | 573* | |
| | ≥65 | 28.7 | 5.0 | – | 128.3 | 6207* | 62000 | – | 1406* | |
| USA[4] | <2 | 1420 | – | 3020 | – | 218 | – | 103 | – | Based on 2-day consumption estimates |
| | >2 | 360 | – | 860 | – | 861 | – | 360 | – | |
| EU[5] | <1 | 800[1)] | – | – | 1800 | 388 | – | – | 172 | Lower bound data[6] |
| | 1–2 | 1300 | – | – | 2300 | 238 | – | – | 135 | |
| | 3–9 | 1200 | – | – | 2200 | 258 | – | – | 141 | |
| | 10–17 | 700 | – | – | 1400 | 443 | – | – | 221 | |
| | 18–64 | 500 | – | – | 1000 | 620 | – | – | 310 | |
| | 65–74 | 400 | – | – | 800 | 775 | – | – | 388 | |
| | ≥75 | 400 | – | – | 900 | 775 | – | – | 344 | |

[1]Margin of exposure (MOE) was calculated using acrylamide intake and the $BMDL_{10}$ value of 310 μg/kg bw/day (WHO, 2011). The values marked with * were obtained from the reference, while the others were calculated.
[2]Jeong et al. (2020).
[3]MFDS (2016).
[4]Abt et al. (2022).
[5]EFSA (2005).
[6]When the samples contained the acrylamide under the limit of detection levels, zero was used for calculating the estimated daily intakes.

© Burleigh Dodds Science Publishing Limited, 2025. All rights reserved.

the most (20.6%), followed by coffee (17.4%) and potato snacks (11.6%). The lower contribution of black pepper (0.7%) in the processed foods study is attributed to the analysis of uncooked samples, as AA levels in black pepper increase significantly with cooking.

The main contributors to AA exposure vary by age group. In the TDS, snacks, biscuits, and cookies were the primary contributors for infants and young children (1–5 years old). For children, adolescents, and adults (6–64 years old), black pepper was the main source, while for the elderly (65 years and older), instant coffee was the largest contributor. In the study by Jeong, Hwang, and Kwon (2020), potato snacks (53.8%) were the main contributors for children and adolescents (3–18 years old), while coffee (26.6%) and confectioneries (15.1%) were the leading sources for adults and the elderly (18–80 years old).

For individuals over 2 years old, the primary sources of AA exposure include breakfast cereals (17%), French fries (16%), potato chips (13%), cookies (10%), crackers (including pita chips) (8%), brewed coffee (7%), and corn snacks (6%). In Europe, the main contributor to AA exposure for individuals over 1 year old is fried potato products. Among infants, children, and adolescents, major contributors include confectioneries, breakfast cereals, bakery products, and other potato-based foods. For adults, the leading sources are soft bread, coffee, biscuits, crackers, crispbread, and various potato products.

The health risks of dietary exposure to AA were assessed using the margin of exposure (MOE) approach, in accordance with the guidelines outlined by the European Food Safety Authority (EFSA) for evaluating the risk of genotoxic and carcinogenic compounds (EFSA, 2005). The MOE values were calculated using the lower 95% confidence interval of the BMD corresponding to a 10% response ($BMDL_{10}$) and summarized in Table 2 (WHO, 2011). The JECFA proposed two $BMDL_{10}$ values for AA risk characterization: 180 μg/kg bw/day for tumors in the Harderian gland of male mice and 310 μg/kg bw/day for mammary tumors in female rats (WHO, 2011). In this chapter, the $BMDL_{10}$ value of 310 μg/kg bw/day for mammary tumors in female rats was used to calculate the MOE, as the Harderian gland does not exist in humans, despite the lower $BMDL_{10}$ being observed in the Harderian gland of male mice. According to JECFA, contaminants with MOE values exceeding 10000 are considered to pose a low risk to human health (WHO, 2005). The average MOE values for AA were as follows: 1398–6798 in the Korean population focusing on processed foods, 1509-–10801 in Korean TDS, 218–861 in the USA, and 238–775 in the EU. These findings suggested that the health risks associated with AA exposure are high across all populations studied. Infants and children are particularly vulnerable, exhibiting higher risks of AA exposure compared to other age groups. To reduce AA exposure, mitigating strategies such as modifying

© Burleigh Dodds Science Publishing Limited, 2025. All rights reserved.

cooking practices or food processing methods should be implemented, taking into account the mechanism of AA formation.

## 2.3 Management

The principal objective of managing AA levels in food is to reduce them to as low as reasonably achievable (ALARA), acknowledging that it is impossible to eliminate AA completely. In both the EU and Korea, AA levels are actively managed through benchmark levels and recommended specification levels, respectively, as summarized in Table 3 (EU, 2017; MFDS, 2024). The food categories with established AA levels are generally similar in both regions. However, the recommended levels in Korea are higher than those in the EU for most food categories, except for instant coffee and coffee substitutes made exclusively from chicory.

For example, the benchmark level for soft bread in the EU is 50 µg/kg for wheat-based bread and 100 µg/kg for other types, whereas Korea's recommended level is 1000 µg/kg – more than ten times higher. Although the benchmark levels are primarily based on feasible limits, Korea's local dietary culture has allowed for a more relaxed standard for AA in bread products.

The EU (2017) established mitigation measures and benchmark levels to reduce AA levels in food. Under these guidelines, manufacturers are required to implement specific procedures to meet the benchmark levels outlined in Table 3. Additionally, practical guidance is provided in the "Acrylamide Toolbox," published by FoodDrinkEurope, with the latest version released in 2019 (FoodDrinkEurope, 2019). The toolbox includes 14 parameters grouped into the following categories:

1 Agronomy (e.g. reducing sugars, asparagine content);
2 Recipe (e.g. basic formulation, ingredients, and product form, including raising/leavening agents, minor ingredients such as glycine and divalent cations, pH adjustments, rework, dilution, piece size, and fermentation);
3 Processing (e.g. thermal input and moisture control, asparaginase application, pre-treatment, finished product color, texture, and flavor); and
4 Final preparation (e.g. consumer or end-user guidance).

These measures are tailored for various food categories, including potato-based snacks, cereal/grain-based products, fine bakery wares, breakfast cereals, coffee, coffee substitutes, and baby foods, to help reduce AA levels effectively.

In Korea, AA levels have been managed through recommended levels for high-exposure food categories since January 2021. Reviews to establish

© Burleigh Dodds Science Publishing Limited, 2025. All rights reserved.

**Table 3** Guidelines for acrylamide in food from EU and Korea

| Food | EU[1] | Republic of Korea[2] |
|---|---|---|
| French fries (ready-to-eat) | 500 | ≤1000 |
| Potato crisps from fresh potatoes and from potato dough | 750 | ≤1000[3] |
| Potato-based crackers | | |
| Other potato products from potato dough | | |
| Soft bread | | ≤1000[4] |
| Wheat-based bread | 50 | |
| Soft bread other than wheat-based bread | 100 | |
| Breakfast cereals (excluding porridge) | | ≤300[5] |
| Bran products and whole grain cereals, gun puffed grain | 300 | |
| Wheat and rye-based products[6] | 300 | |
| Maize, oat, spelt, barley, and rice based products[6] | 150 | |
| Biscuits and wafers | 350 | ≤1000[3] |
| Crackers with the exception of potato-based crackers | 400 | ≤1000[3] |
| Crispbread | 350 | ≤1000[3] |
| Gingerbread | 800 | ≤1000[3] |
| Products similar to the other products in this category | 300 | ≤1000[3] |
| Roast coffee | 400 | ≤800 |
| Instant (soluble) coffee | 850 | ≤800 |
| Coffee substitutes | | ≤1000[7] |
| Coffee substitutes exclusively from cereals | 500 | |
| Coffee substitutes from a mixture of cereals and chicory | [8] | |
| Coffee substitutes exclusively from chicory | 4000 | |
| Baby foods, processed cereal-based foods for infants and young children excluding biscuits and rusks[9] | 40 | ≤300 |
| Biscuits and rusks for infants and young children[9] | 150 | ≤300 |

[1]Commission Regulation (EU) 2017/2158 and the level is expressed as benchmark level (μg/kg).
[2]MFDS (2024) and the level is expressed as recommended specification level (μg/kg).
[3]Expressed as confectionery in Korean guideline.
[4]Expressed as processed cereal product and ready-to-eat foods in Korean guideline.
[5]Expressed as cereal in Korean guideline.
[6]Non-whole grain and/or non-bran-based cereals. The cereal present in the largest quantity determines the category.
[7]Expressed as tea (solid) in Korean guideline.
[8]The benchmark level to be applied to coffee substitutes from a mixture of cereals and chicory takes into account the relative proportion of these ingredients in the final product.
[9]As defined in Regulation (EU) No 609/2013.

maximum AA levels in food are conducted every 2 years. If AA levels exceed the recommended specifications, the Ministry of Food and Drug Safety (MFDS) encourages manufacturers, processors, and importers to adhere to and improve upon the recommended guidelines. In cases where AA levels surpass the recommended thresholds, the results are made publicly available on the MFDS homepage (MFDS, 2024)

© Burleigh Dodds Science Publishing Limited, 2025. All rights reserved.

In addition to the EU and Korea, the US Food and Drug Administration (FDA) issued guidance in 2016 to help the food industry reduce AA levels in food products. Similarly, the Codex Alimentarius Commission released a Code of Practice in 2009 aimed at minimizing AA formation in potato and cereal products, which are significant contributors to AA exposure. These guidelines are intended for both regulatory authorities and food manufacturers, recommending strategies such as:

1 Management of raw materials: Selecting appropriate varieties and handling practices to reduce precursors of AA.
2 Optimization of food processing conditions: Adjusting cooking temperatures and times to minimize AA formation.
3 Addition of ingredients: Incorporating substances that inhibit the AA formation pathway.

Implementing these strategies can effectively reduce AA levels in food products, thereby mitigating potential health risks associated with its consumption.

One of the primary strategies to reduce AA formation in food, such as potatoes, involves lowering the levels of its precursors – asparagine and reducing sugars. Enzymes like asparaginase can effectively degrade asparagine, while techniques such as blanching or soaking can reduce the concentrations of both glucose and asparagine in foods. Potatoes should be stored at temperatures between 8°C and 12°C, as lower temperatures (below 8°C) can increase reducing sugar levels through a process known as low-temperature sweetening (Rifai & Saleh, 2020).

Controlling cooking temperature and time is also critical for minimizing AA formation. Cooking at temperatures below 120°C and using shorter cooking durations are recommended (MFDS, 2024). Microwave frying has been shown to effectively lower both temperature and cooking time compared to deep-frying, resulting in reduced AA levels (Ahmed & Mohammed, 2024). Additionally, low-temperature cooking methods such as boiling or steaming can further prevent AA formation (MFDS, 2024) (Fig. 3).

Incorporating specific ingredients can also help inhibit AA production. These include amino acids (e.g. glycine), vitamins, polysaccharides (e.g. pectin, alginate), lactic acid bacteria, and natural extracts (e.g. polyphenols, grapeseed extract) (Ahmed & Mohammed, 2024; Rifai & Saleh, 2020). It is also advisable to add black pepper after cooking, as AA levels in pepper increase significantly during high-temperature processes such as grilling and frying (MFDS, 2023).

© Burleigh Dodds Science Publishing Limited, 2025. All rights reserved.

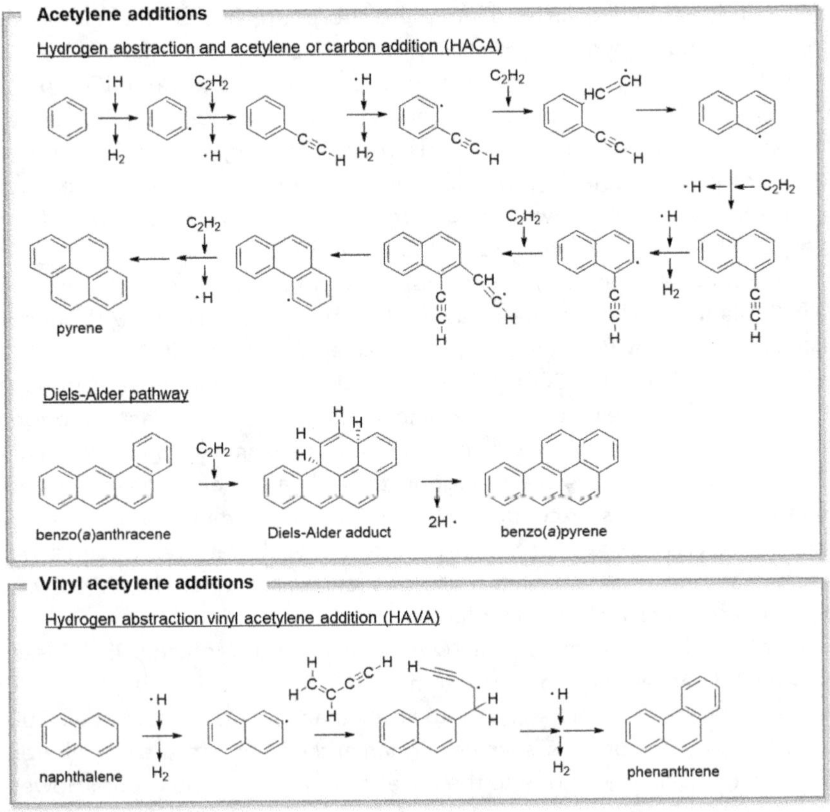


**Figure 3** Polycyclic aromatic hydrocarbon formation adapted by Reizer et al. (2022).

## 3 Heterocyclic amines

Heterocyclic amines are a group of compounds characterized by their heterocyclic structures, typically consisting of two to five aromatic rings (commonly three) containing one or more nitrogen atoms and usually an exocyclic amino group. However, certain HCAs, such as β-carbolines (harman and norharman), lack this exocyclic amino group (Dong et al., 2020; Alaejos & Afonso, 2011). The HCAs are formed under high-temperature conditions. Since their discovery in broiled fish by Dr. Sugimura and colleagues in the 1970s, over 30 HCAs have been identified (Alaejos & Afonso, 2011; Edna Hee et al., 2024; Dong et al., 2020; Sugimura et al., 2004).

Heterocyclic amines are classified into two major groups – thermic HCAs and pyrolytic HCAs – based on the heating temperature and generation process. Thermic HCAs, also known as aminoimidazoazaarenes, IQ types, or polar HCAs, are generally formed at temperatures between 150°C and 300°C. These

© Burleigh Dodds Science Publishing Limited, 2025. All rights reserved.

compounds are further divided into three subgroups based on their chemical structures: imidazopyridine (IP) types, IQ types, and imidazoquinoxaline (IQx) types. Pyrolytic HCAs, also referred to as aminocarbolines, non-IQ types, or nonpolar HCAs, are produced at temperatures exceeding 300°C through the pyrolysis of individual amino acids and proteins. Pyrolytic HCAs primarily consist of pyridoindole, pyridoimidazole, and other related structures, all of which share a common pyridoindole or dipyridoimidazole framework (Dong et al., 2020; Alaejos & Afonso, 2011; Murkovic, 2004b).

Heterocyclic amines are recognized as potent mutagens in the Ames/Salmonella test and have been classified as carcinogens by the IARC and the National Toxicology Program (NTP, 2021; Alaejos & Afonso, 2011; IARC, 1987, 1993; Sugimura et al., 2004). Table 4 provides a summary of representative HCAs, including their structures, cancer classifications, mean tumorigenic doses (TD50), and mutagenicity, enabling comparison across different HCAs. All HCAs listed in Table 4 are mutagenic toward *Salmonella typhimurium*, with mutagenic potentials exceeding $10^3$. Heterocyclic amines are considered frameshift-type mutagens, as they are more effectively detected by strain TA98, which is sensitive to frameshift mutations, compared to strain TA100, which detects base-pair substitution mutagens.

Table 4 also summarizes the TD50 values, which represent the doses of individual HCAs required to induce tumors in 50% of test animals. These values indicate that rats are more sensitive to HCAs than mice (Sugimura et al., 2004). Among HCAs, β-carbolines, such as harman and norharman, are not inherently mutagenic to *Salmonella* due to the absence of a free amino group. However, they act as comutagens when combined with non-mutagenic aromatic amines (e.g. aniline) in *S. typhimurium* TA98, enhancing mutagenic activity (Murkovic, 2004b).

### 3.1 Mechanism and occurrence

Heterocyclic amines are formed through the reaction of amino acids, reducing sugars, and/or creatine/creatinine in food exposed to high temperatures. Amino acids serve as essential precursors for HCA formation, reacting with sugars and/or creatine. Creatine, found in the muscle tissue of meat and fish, is required for the formation of thermic HCAs but is not involved in the synthesis of pyrolytic HCAs. Although HCAs begin to form at room temperature, their concentrations become more significant at elevated temperatures. Thermic HCAs are typically produced at cooking temperatures ranging from 100°C to 300°C, while pyrolytic HCAs are generated through the pyrolysis of amino acids and proteins at temperatures exceeding 300°C.

Certain compounds, such as β-carbolines (e.g. harman and norharman) and Trp-P-1, are exceptions, as they can form at lower temperatures. The

© Burleigh Dodds Science Publishing Limited, 2025. All rights reserved.

**Table 4** Classification of heterocyclic amines (Lee et al., 2020; NTP, 2021; Alaejos & Afonso, 2011; IARC, 1987, 1993; Sugimura et al., 2004)

| Abbreviated name | Chemical name | Structure | Cancer classification | TD$_{50}$ (mg/kg/day)[1] | | Mutagenicity in *S. typhimurium* with S9 mix (Revertants/μg)[1] | |
|---|---|---|---|---|---|---|---|
| | | | | Rats | Mice | TA98 | TA100 |
| Thermic HCAs/IQ type/Aminoimidazoazarenes | | | | | | | |
| IQ | 2-Amino-3-methy l-3H-imidazo[4,5-f]q uinoline | 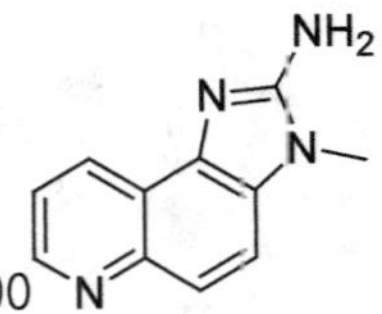 0000 | 2A[2] Reasonably anticipated to be a human carcinogen[3] | 0.7 | 14.7 | 433 000 | 7000 |
| MeIQ | 2-Amino-3,4-dimethyl -3H-imidazo[4,5-f]qu inoline | 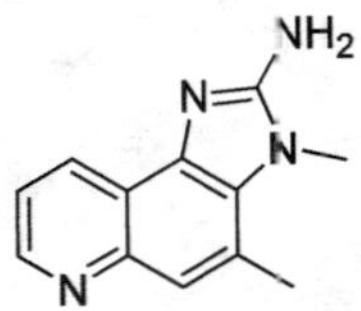 | 2B[2] Reasonably anticipated to be a human carcinogen[3] | 0.1 | 8.4 | 661 000 | 30 000 |
| MeIQx | 2-Amino-3,8-dimethyl imidazo[4,5-f]quinox aline | 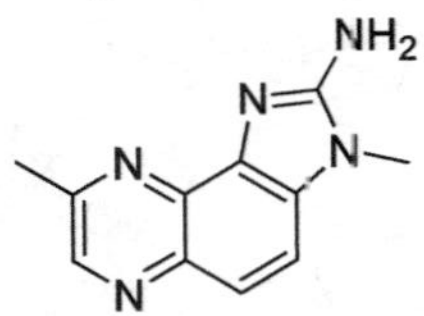 | 2B[2] Reasonably anticipated to be a human carcinogen[3] | 0.7 | 11 | 145 000 | 14 000 |
| DiMeIQx | 2-amino-3,4,8-trimet hylimidazo[4,5-f]qui noxaline | 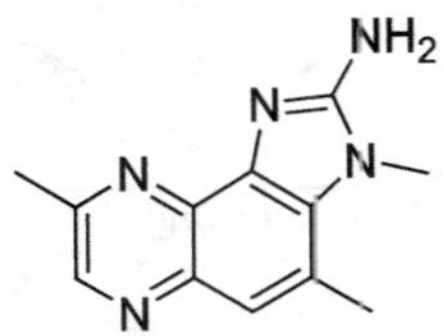 | – | – | – | 183 000 | 8000 |

(*Continued*)

© Burleigh Dodds Science Publishing Limited, 2025. All rights reserved.

**Table 4** (*Continued*)

| Abbreviated name | Chemical name | Structure | Cancer classification | $TD_{50}$ (mg/kg/day)[1] | | Mutagenicity in *S. typhimurium* with S9 mix (Revertants/μg)[1] | |
|---|---|---|---|---|---|---|---|
| | | | | Rats | Mice | TA98 | TA100 |
| PhIP | 2-Amino-1-methyl-6-phenylimidazo[4,5-b] pyridine | 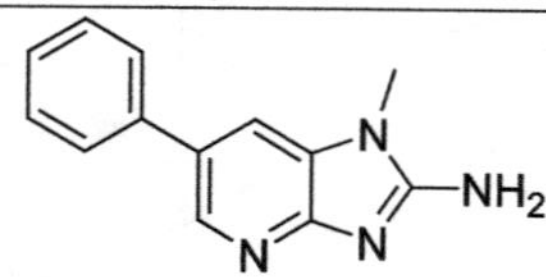 | 2B[2] Reasonably anticipated to be a human carcinogen[3] | 2.2 | 64.6 | 1800 | 120 |
| Pyrolytic HCAs/Non-IQ type/Aminocarbolines | | | | | | | |
| Glu-P-1 | 2-Amino-6-methy ldipyrido[1,2-A:3',2 '-D]imidazole | 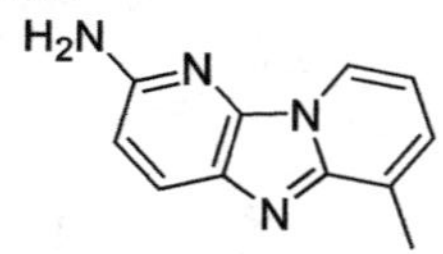 | 2B[2] | – | – | 49 000 | 3200 |
| Glu-P-2 | 2-Aminodipyrido[1,2-A:3',2'-D]imidazole | 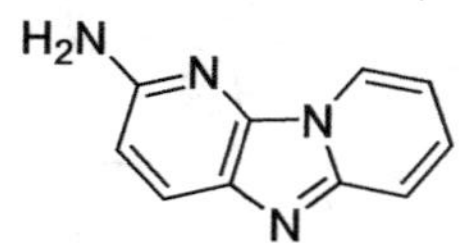 | 2B[2] | – | – | 1900 | 1200 |
| Trp-P-1 | 3-Amino-1,4-dimethyl -5H-pyrido[4,3-b] indole | 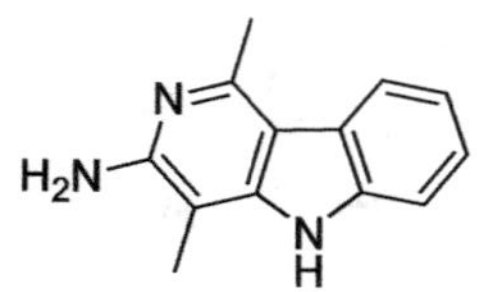 | 2B[2] | – | – | 39 000 | 1700 |
| Trp-P-2 | 3-Amino-1-methyl-5H-pyrido[4,3-b] indole | 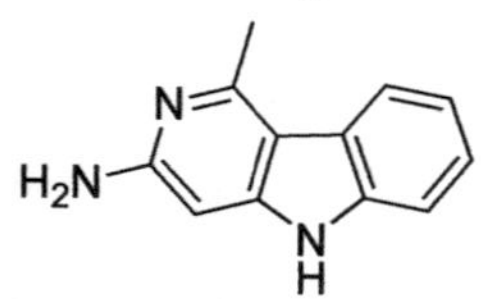 | 2B[2] | – | – | 104 200 | 1800 |

© Burleigh Dodds Science Publishing Limited, 2025. All rights reserved.

| | | | | | | | |
|---|---|---|---|---|---|---|---|
| AαC | 2-Amino-9H-pyrido[2,3-b]indole | 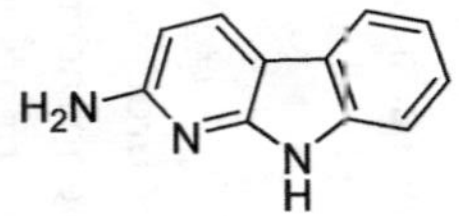 | 2B[2] | – | 15.8 | 300 | 20 |
| MeAαC | 2-Amino-3-methyl-9H-pyrido[2,3-b]indole | 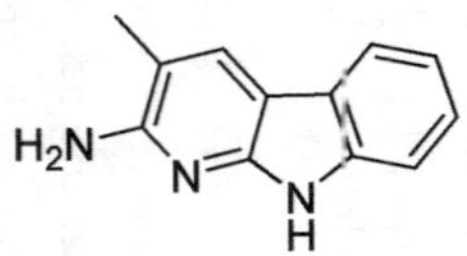 | 2B[2] | 6.4 | 5.8 | 200 | 120 |
| Harman | 1-Methyl-9H-pyrido[3,4-b]indole | 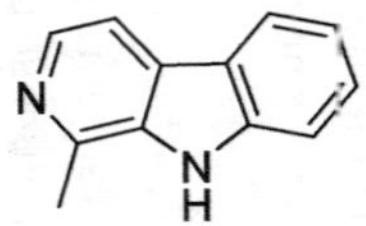 | – | – | – | – | – |
| Norharman | 9H-pyrido[3,4-b]indole | 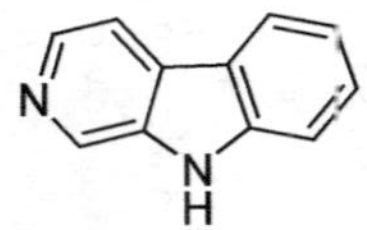 | – | – | – | – | – |

[1]Sugimura et al. (2004).
[2]Classified by International Agency for Research on Cancer (IARC).
[3]Classified by National Toxicology Program (NTP).

© Burleigh Dodds Science Publishing Limited, 2025. All rights reserved.

mechanisms of HCA formation can be classified into four categories based on their structural characteristics: (1) IQ type, (2) IQx type, (3) IP type, and (4) pyrolytic HCAs (Murkovic, 2004a; Dong et al., 2020).

Thermic HCAs are categorized into IQ type (e.g. IQ, MeIQ), IQx type (e.g. IQx, MeIQx), and IP type (e.g. PhIP). The formation of IQ and IQx HCAs is associated with the Maillard reaction, involving amino acids, reducing sugars, and creatine. During this reaction, reducing sugars and amino acids undergo Strecker degradation, producing pyridines and pyrazines, which are critical intermediates for the formation of IQ and IQx types, respectively. Pyridines or pyrazines subsequently react with creatinine and aldehydes – formed during the heating of creatine and amino acids, respectively – via aldol reactions to produce IQ and IQx HCAs (Dong et al., 2020).

In addition, a free radical pathway for IQ and IQx HCAs has been proposed, supported by the detection of free radicals and the inhibitory effects of antioxidants (e.g. polyphenols). While this pathway resembles the Maillard reaction, it is distinct in that pyridines and pyrazines are derived from their respective cation radicals, formed through Strecker degradation involving dicarbonyl compounds and amino acids (Geng et al., 2024; Murkovic, 2004b).

Among IP-type HCAs, the formation of PhIP is well-documented and involves phenylalanine, creatine, and reducing sugars as precursors. Phenylacetaldehyde, produced from phenylalanine via Strecker degradation, reacts with creatinine (generated from creatine under heating conditions) to form an aldol addition intermediate. This intermediate is dehydrated to form a condensation product, ultimately leading to PhIP formation. Although reducing sugars are not directly used as precursors in this mechanism, they play a critical role in modulating PhIP yields from creatine and phenylalanine (Dong et al., 2020).

Among pyrolytic HCAs, the formation of β-carbolines has been extensively studied and is known to occur at lower temperatures. In this process, tryptophan acts as the primary precursor, with glucose serving as an enhancer. Tryptophan reacts with acetaldehyde and formaldehyde via Pictet-Spengler condensation to form tetrahydro-β-carbolines, which are subsequently oxidized to β-carbolines. Additionally, the tryptophan Amadori rearrangement product, formed from the reaction between tryptophan and glucose, undergoes dehydration followed by β-elimination, facilitated by the lone-pair electrons on the oxygen in the aromatic ring. The resulting reactive intermediate stabilizes either through dehydration or carbon-carbon cleavage. These intermediates participate in intramolecular nucleophilic substitution reactions, ultimately forming norharman. In general, pyrolytic HCAs are produced from amino acids and proteins under high-temperature conditions (Murkovic, 2004b; Dong et al., 2020).

© Burleigh Dodds Science Publishing Limited, 2025. All rights reserved.

**Table 5** Contents of heterocyclic amines (HCAs) in cooked foods (adapted from Murkovic (2004a))

| Cooked foods | Thermic HCAs (ng/g) | | | | Pyrolytic HCAs (ng/g) | | | | |
|---|---|---|---|---|---|---|---|---|---|
| | IQ | MeIQx | DiMeIQx | PhIP | AαC | Trp-P-1 | Trp-P-2 | Harman | Norharman |
| Red meat | 0–2 | 0–10 | 0–5 | 0–35 | 0–20 | 0–1 | 0–2 | 0–30 | 0–20 |
| Poultry | 0–1 | 0–3 | 0–3 | 0–330 | 0–1 | 0–2 | 0–1 | 0–13 | 0–12 |
| Fish | 0–1 | 0–2 | 0–14 | 0–10 | 0–10 | 0–1 | n.d.[1] | 0–200 | 0–130 |
| Meat extract, gravy | 0–15 | 0–80 | 0–9 | 0–10 | 0–3 | 0–5 | n.d. | 0–100 | 0–120 |
| Fried bacon | 5 | 8 | 5 | 28 | –[2] | – | – | – | – |
| Grilled bacon | 4 | 2 | 9 | 5 | – | – | – | – | – |

[1]Not determined.
[2]Not reported.

Based on their formation mechanisms, thermic HCAs are primarily found in meat and fish, as these foods provide creatine from animal tissue. In contrast, β-carbolines, a class of pyrolytic HCAs, are commonly observed in a wide range of foods. Their formation requires only tryptophan and reducing sugars and occurs at lower temperatures compared to other pyrolytic HCAs. The concentrations of HCAs vary across different cooked foods and are influenced by the type of meat and cooking methods (Table 5) (Murkovic, 2004a).

Typically, IQ and IQx HCAs are detected at low levels, around 0–10 ng/g, in meat and fish. In comparison, PhIP, harman, and norharman are found in higher concentrations. Among meat types, poultry generates higher levels of PhIP, likely due to its lower glucose content (Murkovic, 2004a; Geng et al., 2024). Heterocyclic amine formation increases significantly in meat extracts, where high precursor concentrations and water evaporation during high-temperature cooking intensify the effect.

Fried bacon produces more HCAs than grilled bacon, as pan-frying concentrates HCAs in the fat. Conversely, grilling allows HCAs to drip away with the fat, reducing their final concentration (Murkovic, 2004b).

In the Korean TDS (Lee et al., 2022), a total of 1232 samples were analyzed for 12 HCAs, and the detected levels showed considerable variation. Among the 12 HCAs, IQ, Glu-P-1, Glu-P-2, and Trp-P-1 were not detected, while MeIQ, Trp-P-2, and MeAαC were each detected in a single sample, with concentrations of 1.20, 0.276, and 0.133 ng/g, respectively. The β-carbolines (harman and norharman) showed the highest levels (harman: n.d.[4]–1353

4 The method detection limit (MDL) ranged from 0.02 to 1.23 ng/g across eleven representative matrices, including porridge, apple juice, radish, corn oil, peanut butter, beef, milk, 20% ethanol, flounder, tuna, and seaweed (MFDS, 2016b). For each sample, the most appropriate representative matrix was selected based on its food composition, and the corresponding MDL was applied.

© Burleigh Dodds Science Publishing Limited, 2025. All rights reserved.

ng/g; norharman: n.d.[5]–4129 ng/g) and detection rates (harman: 53.2%; norharman: 65.3%), followed by PhIP, which had a detection rate of 2.92%, with concentrations ranging from n.d.[6] to 5.98 ng/g. The levels of AαC and MeIQx were n.d.[7]–3.15 ng/g and n.d.[8]–2.51 ng/g, with detection rates of 2.35% and 1.62%, respectively. The highest concentrations of harman (1353 ng/g) and norharman (4129 ng/g) were observed in coffee. Notably, the second-highest levels of these compounds were found in sesame oil, a popular edible oil in Korea and Japan made by pressing roasted sesame seeds (Liu et al., 2022). Harman and norharman in coffee and sesame oil could form through the oxidative decarboxylation of tetrahydro-β-carboline-3-carboxylic acid during the roasting process.

The distribution of β-carbolines in various foods was also reported by Herraiz (2000). Harman was detected in wine (0.8 – 11.7 ng/mL), beer (7.3 – 140 ng/mL), vinegar (0.015–0.73 ng/mL), soy sauce (0.13–0.25 ng/mL), cooked meats (0–20 ng/g), cooked fish (0–130 ng/g), instant coffee and dried coffee (0.02–1.67 ng/g), and ready-to-eat coffee (26.9–39.9 ng/mL). Norharman was found in wine (0.3–0.7 ng/mL), vinegar (0.0019–0.096 ng/mL), cooked meats (0–30 ng/g), cooked fish (0–200 ng/g), instant coffee and dried coffee (0.08–9.34 ng/g), and ready-to-eat coffee (91–165.6 ng/mL).

### 3.2 Exposure and risk characterization

Exposure assessments of HCAs from Korea, the USA, and Switzerland are summarized in Table 6. Both the USA and Switzerland estimated daily intake based on HCA levels in specific food groups (e.g. meat, fish), while Korea included all food groups to estimate the overall dietary intake of HCAs. The estimated daily intakes of IQ, MeIQ, MeIQx, and PhIP, which are commonly generated in meat and fish, are much lower in Korea compared to the other countries. This difference can be attributed to variations in national consumption habits regarding meat and fish, with higher consumption of boiled meat and fish, as well as the practice of eating raw fish, in Korea. Additionally, the Korean TDS used a probabilistic model to estimate population exposure based on individual dietary intake distributions, incorporating personal body weight, food consumption, and HCA concentrations in foods, while the other countries used a deterministic model. To compare dietary exposure levels, the mean values from the Korean TDS were referenced.

In the Korean TDS, the mean dietary exposure to HCAs in the Korean population was as follows: MeIQx (0.042 ng/kg bw/day), PhIP (0.315 ng/kg

5 The MDL ranged from 0.02 to 1.36 ng/g.
6 The MDL ranged from 0.06 to 1.01 ng/g.
7 The MDL ranged from 0.04 to 0.86 ng/g.
8 The MDL ranged from 0.11 to 2.59 ng/g.

© Burleigh Dodds Science Publishing Limited, 2025. All rights reserved.

**Table 6** Estimated dietary exposure to heterocyclic amines (ng/kg bw/day) and margin of exposure (MOE) of PhIP in different countries

| | | | | Heterocyclic amines (ng/kg bw/day) | | | | | | | | | | MOE1 |
|---|---|---|---|---|---|---|---|---|---|---|---|---|---|---|
| Country | Age | Race | Gender | IQ | MeIQ | MeIQx | DiMeIQx | PhIP | Trp-P-2 | AαC | MeAαC | Harman | Norharman | PhIP |
| Korea[2] | 1–2 | Korean | All | 0[3] | 0 | 0.045 | –[4] | 0.381 | 0.003 | 0.295 | 0.002 | 102 | 129 | 1942000* |
| | 3–5 | Korean | All | 0 | 0 | 0.065 | – | 0.482 | 0.001 | 0.271 | 0.007 | 111 | 135 | 1535000* |
| | 6–11 | Korean | All | 0 | 0 | 0.056 | – | 0.476 | 0.000 | 0.249 | 0.004 | 98 | 127 | 1555000* |
| | 12–18 | Korean | All | 0 | 0 | 0.048 | – | 0.372 | 0.000 | 0.223 | 0.003 | 81 | 134 | 1989000* |
| | 19–64 | Korean | All | 0 | 0 | 0.043 | – | 0.315 | 0.001 | 0.190 | 0.002 | 172 | 587 | 2349000* |
| | ≥65 | Korean | All | 0 | 0 | 0.014 | – | 0.119 | 0.001 | 0.076 | 0.001 | 127 | 431 | 6218000* |
| USA[5] | <16 | White | Male | 1.73[6] | – | 1.3 | 0.27 | 7.7 | – | [6] | – | – | – | 96000 |
| | | | Female | 1.82[6] | – | 1.2 | 0.28 | 8.1 | – | [6] | – | – | – | 91000 |
| | | African-American | Male | 2.73[6] | – | 1.8 | 0.67 | 14.7 | – | [6] | – | – | – | 50000 |
| | | | Female | 3.58[6] | – | 1.5 | 0.42 | 10.8 | – | [6] | – | – | – | 69000 |
| | >30 | White | Male | 2.52[6] | – | 1.4 | 0.28 | 9.2 | – | [6] | – | – | – | 80000 |
| | | | Female | 2.72[6] | – | 1.4 | 0.28 | 9.2 | – | [6] | – | – | – | 80,000 |
| | | African-American | Male | 3.23[6] | – | 2.1 | 0.47 | 12.6 | – | [6] | – | – | – | 59000 |
| | | | Female | 3.69[6] | – | 1.6 | 0.41 | 12.7 | – | [6] | – | – | – | 58000 |
| Switzerland[7] | [8] | Swiss | All | 0.3 | 0.3 | 0.6 | 0.3 | 1.1 | – | – | – | – | – | 673000 |

[1]The MOE was calculated using PhIP intake and a $BMDL_{10}$ value of 740 μg/kg bw/day, with the resulting values rounded to approximately 1000 (Carthew, DiNovi & Woodrow Setzer, 2010; Lee et al., 2022). The values marked with * were obtained from the reference, while the others were calculated.
[2]Mean values from Lee et al. (2022).
[3]When the compound was analyzed but not determined in food, the exposure value is expressed as zero.
[4]The compound was not analyzed in this study.
[5]Keating and Bogen (2004)
[6]The values are the sum of IQ and AαC, which are derived from the total exposure levels of five HCAs and the individual levels of MeIQx, DiMeIQx and PhIP.
[7]Zimmerli et al. (2001)

© Burleigh Dodds Science Publishing Limited, 2025. All rights reserved.

bw/day), Trp-P-2 (0.001 ng/kg bw/day), AαC (0.188 ng/kg bw/day), MeAαC (0.002 ng/kg bw/day), harman (150 ng/kg bw/day), and norharman (474 ng/kg bw/day). Meats were the primary contributors to the daily intake of MeIQx, PhIP, AαC, and MeAαC, while beverages and alcoholic drinks were the main contributors to the daily intake of harman and norharman. The top food sources for each compound were as follows: imported beef for AαC (48.6%), pork belly for MeIQx (35.0%), and pork for PhIP (33.4%). MeAαC was detected only in beef, and Trp-P-2 was exclusively found in croaker. Beverages and alcoholic drinks accounted for 50% of the daily intake of harman and 81% of the daily intake of norharman. Specifically, coffee mix was the leading contributor to the estimated daily intake (EDI) of harman (27.6%) and norharman (56.8%), primarily consumed by adults. For children and teenagers, soy sauce and sesame oil were the main contributors to daily intakes of harman and norharman, respectively.

The health risk of PhIP was assessed using the MOE approach among various HCAs, as $BMDL_{10}$ values for PhIP are available (Carthew, DiNovi & Woodrow Setzer, 2010). Three $BMDL_{10}$ values were reported: the most conservative value is 480 μg/kg bw/day, based on prostate ventral carcinoma in male rats; the second most conservative value is 740 μg/kg bw/day, based on mammary carcinoma in female rats; and the least conservative value is 2710 μg/kg bw/day, based on colon tumors in male rats. In this chapter, the second most conservative value (740 μg/kg bw/day) was selected for the risk characterization of PhIP, as the most conservative value was derived from prostate ventral carcinomas in rats, which are not considered relevant to humans (Lee et al., 2022; Hwang & Kwon, 2022). The MOE values of PhIP were calculated and summarized in Table 6. Across all countries, the MOE values of PhIP were significantly higher than 10 000. Specifically, MOE values ranged from 1 535 000 to 6 218 000 in Korea, 673 000 in the EU, and 50 000 to 96 000 in the USA. These results indicate that the health risk of PhIP is considered low and thus deemed safe.

### 3.3 Management

Heterocyclic amines in foods have not been specifically regulated in any country. As processing contaminants, the formation of HCAs can be minimized by managing (1) the raw materials, (2) the processing conditions, and (3) the additives, based on their formation mechanisms (Dong et al., 2020; Geng et al., 2024).

Amino acids (e.g. phenylalanine, tryptophan, leucine, and isoleucine), creatine/creatinine, and sugars are the main precursors of concern for HCAs. The composition of these precursors varies depending on the animal species and meat cuts, which directly affects HCA generation. High concentrations

© Burleigh Dodds Science Publishing Limited, 2025. All rights reserved.

of PhIP have been reported in cooked chicken, which contains relatively low glucose and high tryptophan compared to other meats. Increasing the glucose level has been shown to reduce PhIP formation but increase the production of β-carbolines. Chicken breast tends to produce more HCAs than chicken thighs due to higher levels of precursors such as creatine (Geng et al., 2024).

Cooking temperature and duration play critical roles in HCA generation. To reduce HCA formation, cooking methods involving lower temperatures and shorter cooking times are recommended (MFDS, 2024; Dong et al., 2020; Geng et al., 2024). Additionally, the use of certain ingredients can inhibit HCA formation. Antioxidants such as vitamins and polyphenols are particularly effective due to the involvement of radicals in the formation mechanism. Adding ingredients like onions and berries (e.g. *Rubus coreanus*) to hamburger patties can reduce HCA formation by up to 100%. Marinating chicken with extracts of *Nelumbo nucifera L.* or olive leaves before cooking can reduce HCA generation by 28–44% (MFDS, 2024)

## 4 Polycyclic aromatic hydrocarbons

Polycyclic aromatic hydrocarbons are a group of over 100 compounds that contain two or more fused aromatic rings, produced by the incomplete combustion of carbon-based materials through pyrosynthesis and pyrolysis processes (Edna Hee et al., 2024). The formation of PAHs begins with a primary aromatic ring, followed by reactions like hydrogen abstraction and acetylene addition, which lead to the formation of additional aromatic rings (Edna Hee et al., 2024). PAHs originate from both natural sources, such as wildfires and volcanic activity, and human activities, including fuel combustion, vehicle emissions, and cigarette smoke. Polycyclic aromatic hydrocarbons are categorized by origin: combustion-related (pyrogenic), petroleum-based (petrogenic), and natural (biogenic). Biogenic PAHs come from organisms like microorganisms and algae, pyrogenic PAHs are formed in high-temperature processes involving organic materials (e.g. biomass, fossil fuels), and petrogenic PAHs, abundant in petroleum, are released through activities such as oil storage, transport, and accidental spills (Patel et al., 2020). These compounds accumulate in the environment, contaminating water, air, and soil through processes such as aerial deposition, leaching into groundwater, and absorption by plants. They ultimately enter the food chain, posing potential risks to human health and ecosystems. Furthermore, PAHs are generated during food processing and cooking at high temperatures, making food ingestion the primary exposure pathway for non-smokers.

Representative PAHs, classified as PAH8 by EFSA, are shown in Table 7, detailing their structures, cancer classifications, and toxic equivalency factors (TEFs). Among various PAHs, benzo[a]pyrene (BaP) is well-studied and

© Burleigh Dodds Science Publishing Limited, 2025. All rights reserved.

**Table 7** Classification of polycyclic aromatic hydrocarbons (EFSA, 2008; EPA, 2012; Law et al., 2002; Patel et al., 2020)

| Abbreviated name | Chemical name | Structure | Classification[1] | IARC cancer classification | Toxic equivalency factor (TEF) USFDA[2] | USEPA[3] | Nisbet and LaGoy[4] |
|---|---|---|---|---|---|---|---|
| BaP | Benzo[*a*]pyrene | 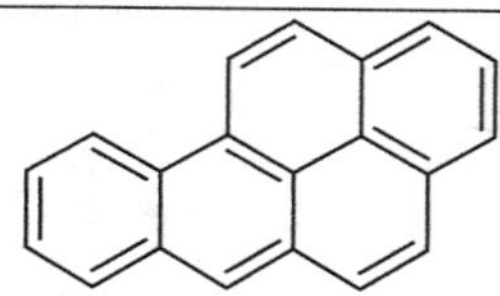 | PAH2 | 1 | 1.00 | 1.00 | 1.00 |
| CHR | Chrysene | 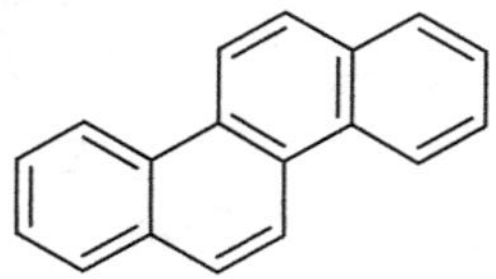 | PAH2 | 2B | 0.013 | 0.01 | 0.01 |
| BaA | Benz[*a*]anthracene | 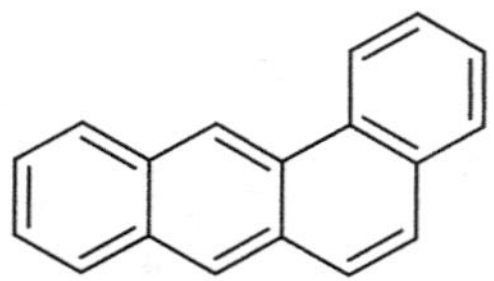 | PAH4 | 2B | 0.014 | 0.1 | 0.1 |
| BbF | Benzo[*b*]fluoranthene | 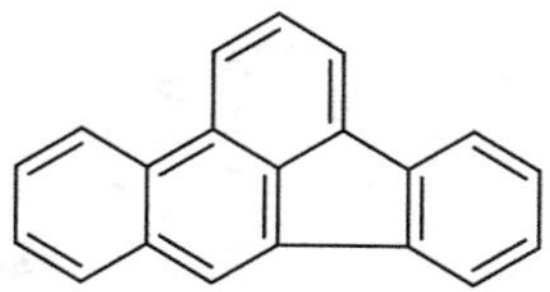 | PAH4 | 2B | 0.11 | 0.1 | 0.1 |
| BkF | Benzo[*k*]fluoranthene | 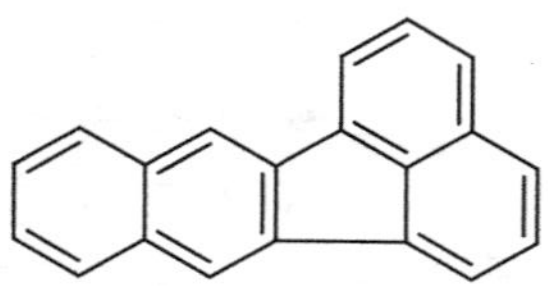 | PAH8 | 2B | 0.07 | 0.1 | 0.1 |

© Burleigh Dodds Science Publishing Limited, 2025. All rights reserved.

| | | | | | | | |
|---|---|---|---|---|---|---|---|
| BghiP | Benzo[*ghi*]perylene | 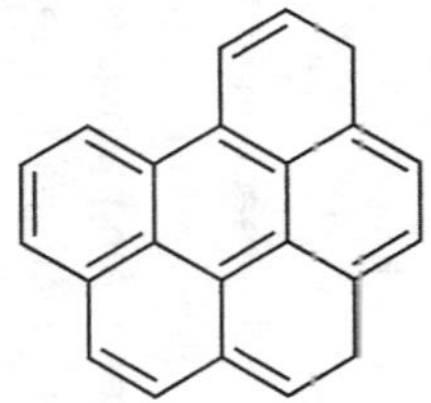 | PAH8 | 3 | 0.03 | 0.01 | 0.01 |
| DBahA | Dibenz[*a,h*] anthracene | 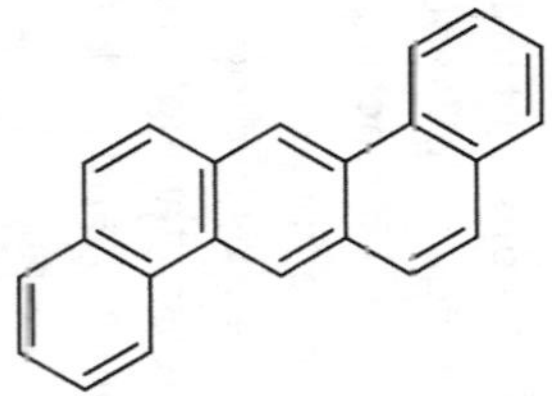 | PAH8 | 2A | 1.05 | 1.00 | 5.00 |
| IP | Indeno[*1,2,3-cd*] pyrene | 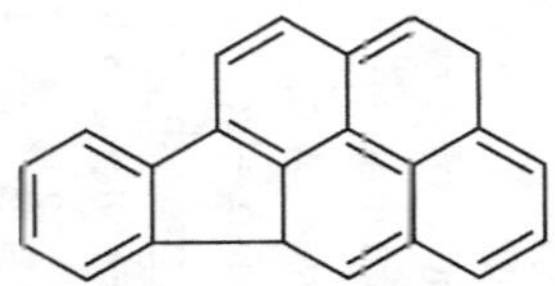 | PAH8 | 2B | 0.25 | 0.1 | 0.1 |

[1]PAH4 and PAH8 include the compounds in PAH2 and PAH4, respectively.
[2]Law et al. (2002).
[3]EPA (2012).
[4]Nisbet and LaGoy (1992).

© Burleigh Dodds Science Publishing Limited, 2025. All rights reserved.

regulated due to its recognition as a human carcinogen and endocrine disruptor. EFSA (2008) has used eight PAHs, individually or in combinations (PAH2, PAH4, and PAH8), to evaluate carcinogenic potency in foods, supported by data on PAH8's oral carcinogenicity. The PAH4 and PAH8 are reliable markers for assessing carcinogenicity in foods (EFSA, 2008). The US EPA designated 16 PAHs, including PAH8 and others like naphthalene, acenaphthene, and phenanthrene, as priority compounds based on their prevalence, exposure risks, persistence, and toxicity (Patel et al., 2020).

Toxic equivalency factors are used in hazard and risk assessments to estimate cancer risk as a single value for chemicals within the same class that share similar structures. The TEFs represent the relative potencies of these chemicals, calculated by multiplying each compound's concentration by its TEF to determine the toxic equivalent (TEQ). Toxic equivalency factor values for PAHs, determined through bioassays assessing individual PAH carcinogenic potency, are presented in Table 7, with PAH carcinogenicity typically represented as a BaP equivalent ($TEQ_{BaP}$). EFSA (2008) did not adopt the TEF approach for PAHs due to several limitations. These included insufficient oral carcinogenicity data for various PAHs, differences in their modes of action, and the limited reliability of current TEF values in predicting the carcinogenic potency in PAH mixtures (Table 8).

### 4.1 Mechanism and occurrence

The formation of the first ring in PAHs, which consist of multiple fused rings, is critical as it serves as the foundation for further growth. Various mechanisms have been proposed for this initial step, including the reaction of a 1,3-butadienyl radical with acetylene, followed by hydrogen elimination, and the reaction of a vinyl radical with 1,3-butadiene, also followed by hydrogen elimination. Once the first ring is formed, PAH growth progresses through three main pathways: acetylene additions, vinylacetylene additions, and radical reactions (Reizer et al., 2022; Edna Hee et al., 2024).

The growth mechanism involving acetylene additions includes the well-known hydrogen abstraction and acetylene or carbon addition (HACA) process and the Diels-Alder mechanism. In the HACA process, two steps are repeated: (1) hydrogen abstraction to activate the aromatic molecule and (2) acetylene addition to the radical site created by hydrogen abstraction. The Diels-Alder mechanism, commonly used in laboratory synthesis of aromatic compounds, involves the cycloaddition of acetylene near the bay region of PAH structures, followed by hydrogen loss. Another growth mechanism involves the addition of vinyl acetylene, known as hydrogen abstraction vinyl acetylene addition. In this pathway, vinyl acetylene adds to a PAH radical, resulting in cyclization.

© Burleigh Dodds Science Publishing Limited, 2025. All rights reserved.

**Table 8** Contents of BaP and PAH4 detected from the Korean monitoring study (adapted by MFDS (2018))

| Food categories | BaP | | | | PAH4 | | | |
|---|---|---|---|---|---|---|---|---|
| | Detection rate (%) | Range (ng/g)[1] | | | Detection rate (%) | Range (ng/g)[1] | | |
| Confectioneries | 9.8 | n.d. | – | 0.580 | 37.7 | n.d. | – | 4.940 |
| Breads | 8.5 | n.d. | – | 0.328 | 20.3 | n.d. | – | 0.600 |
| Rice cakes | – | n.d. | – | n.d. | 20 | n.d. | – | 0.510 |
| Frozen desserts | 1.7 | n.d. | – | 0.080 | 30 | n.d. | – | 0.510 |
| Cocoa products or chocolates | 8.6 | n.d. | – | 0.740 | 74.3 | n.d. | – | 3.030 |
| Saccharides | – | n.d. | – | n.d. | 10 | n.d. | – | 0.290 |
| *Tofu* (Soybean curd) | – | n.d. | – | n.d. | 16.7 | n.d. | – | 0.160 |
| *Muk* (Starch jellies) | 3.3 | n.d. | – | 0.357 | 20 | n.d. | – | 0.610 |
| Edible oils and fats | 29.3 | n.d. | – | 2.000 | 73.3 | n.d. | – | 12.900 |
| Noodles | 6.9 | n.d. | – | 0.270 | 30.6 | n.d. | – | 0.900 |
| Leached tea | 99.5 | n.d. | – | 12.453 | 100 | 0.543 | – | 44.246 |
| Liquid tea | 99.4 | n.d. | – | 0.341 | 100 | 0.083 | – | 1.699 |
| Solid tea | 28.7 | n.d. | – | 3.037 | 95.1 | n.d. | – | 13.442 |
| Coffee | 60 | n.d. | – | 1.870 | 87.5 | n.d. | – | 10.050 |
| Beverages (fruit, vegetables, carbonated and others) | 0.05 | n.d. | – | 0.120 | 0.08 | n.d. | – | 0.041 |
| Foods for special dietary use | 1.9 | n.d. | – | 0.740 | 19.5 | n.d. | – | 1.235 |
| Soy sauces and pastes | 11.6 | n.d. | – | 1.510 | 54.3 | n.d. | – | 11.460 |
| Seasoning | 10.8 | n.d. | – | 5.220 | 36.9 | n.d. | – | 37.400 |
| *Kimchi* and picked food | 0.3 | n.d. | – | 0.268 | 15.6 | n.d. | – | 4.3 |
| Alcoholic beverages | 6.0 | n.d. | – | 0.040 | 20.5 | n.d. | – | 0.090 |
| Processed agricultural foods | 7.0 | n.d. | – | 5.970 | 27.9 | n.d. | – | 25.990 |

*(Continued)*

© Burleigh Dodds Science Publishing Limited, 2025. All rights reserved.

**Table 8** (*Continued*)

| Food categories | BaP | | PAH4 | |
|---|---|---|---|---|
| | Detection rate (%) | Range (ng/g)[1] | Detection rate (%) | Range (ng/g)[1] |
| Processed meat products | 2.1 | n.d. – 0.290 | 33.0 | n.d. – 2.360 |
| Dairy products | 14.9 | n.d. – 0.570 | 32.6 | n.d. – 1.100 |
| Processed fishery foods | 2.2 | n.d. – 0.890 | 40.2 | n.d. – 32.630 |
| Honey | – | n.d. – n.d. | 16.7 | n.d. – 0.750 |
| Ready-to-eat foods | 3.6 | n.d. – 0.600 | 23.6 | n.d. – 1.270 |
| Fruit, vegetable, plant extract and seaweed | 10.8 | n.d. – 15.8 | 29.1 | n.d. – 46.200 |
| Meat and poultry | 4.2 | n.d. – 0.500 | 23.8 | n.d. – 4.500 |
| Seafoods | 2.8 | n.d. – 1.080 | 23.5 | n.d. – 2.200 |

[1]The limit of detection (LOD) ranged from 0.01 to 0.31 μg/kg for BaP, 0.01 to 0.35 μg/kg for BaA, 0.01 to 0.48 μg/kg for CHR, and 0.02 to 0.32 for BbF.

© Burleigh Dodds Science Publishing Limited, 2025. All rights reserved.

Lastly, PAHs can also form through reactions with various radicals, including methyl, ethynyl, and phenyl radicals (Reizer et al., 2022) (Table 9).

The Korea monitoring study (MFDS, 2018) analyzed data from 5450 raw samples across 74 food types collected between 2015 and 2017. The BaP and PAH4 were detected in 14.8% and 39.6% of the total samples, respectively, with significant variation in concentrations. High levels of BaP and PAH4 were notably found in tea, coffee, and edible oils and fats. *Sookjihawang* (processed *Rehmannia* root) and dried adhesive *Rehmannia* roots exceeding the maximum allowable level of BaP (5 μg/kg) were withdrawn from the market. Additionally, two acorn flour products were found to contain 0.534 and 5.970 μg/kg of BaP, which are linked to the detection in *Muk* (starch jellies).

The monitoring study only considered commercial food, failing to account for PAHs produced during final cooking. Therefore, dietary exposure to PAHs based on the monitoring study may have been underestimated, as PAH formation significantly increases at high temperatures. This study confirmed that PAH4 and PAH8 are appropriate markers for PAH evaluation, based on a strong correlation ($R^2 \geq 0.87$) between PAH8 and PAH4, compared to a weaker correlation ($R^2 \geq 0.66$) between PAH8 and PAH2 (Choi et al., 2022)

The Korean TDS (MFDS, 2016c) analyzed PAHs in more than 1200 samples as a pair of food and cooking method. Cooked black pepper had the highest concentrations of BaA, CHR, BbF, and BkF, while the highest levels of BaP, DBahA, BghiP, and IP were detected in *Katsuobushi* (uncooked), *Bulgogi* sauce, bacon, and grilled pork belly, respectively. The BghiP was frequently found in processed meat products like sausage and ham, while IP was primarily detected in meat and fish (MFDS, 2016b) (Table 10).

The study investigated PAH formation during the processing of meat and edible oils (Choi et al., 2022), with cooking methods selected based on the Korean National Health and Nutrition Examination Survey. Among beef samples, the highest PAH8 concentration was found in grilled sirloin (3.8 μg/kg), followed by grilled chitterlings (3.62 μg/kg). The highest BaP equivalent toxicity value was noted in grilled chitterlings (1.35 μg/kg), followed by grilled sirloin (1.3 μg/kg). The high level of saturated fat in chitterlings may contribute to increased PAH production.

Among pork samples, grilled wiener sausage contained the highest levels of PAH8 (5.9 μg/kg) and BaP (1.2 μg/kg), followed by stir-fried wiener sausage (4.74 μg/kg for PAH8 and 1.09 μg/kg for $TEQ_{BaP}$), while uncooked samples contained 4.2 μg/kg of PAH8 and 0.81 μg/kg of $TEQ_{BaP}$. In processed meat products, PAHs can be formed both during cooking and manufacturing.

In poultry, the highest PAH8 levels were found in grilled chicken (2.29 μg/kg) and deep-fried chicken wings (1.81 μg/kg), with BaP equivalents reaching 0.70 μg/kg for both samples (Choi et al., 2022). Air-fried chicken wings and

© Burleigh Dodds Science Publishing Limited, 2025. All rights reserved.

**Table 9** Concentration of BaP, PAH8, and benzo[a]pyrene equivalents ($TEQ_{BaP}$) of PAH8 expressed as ug/kg in selected meat and edible oils from the Korean total diet study (adapted by (Choi et al., 2022))

| Category | Sample | Processing method | BaP[1] | PAH8[2] | $TEQ_{BaP}$[3] |
|---|---|---|---|---|---|
| Beef | Beef (imported) | None | 0.16 | 0.63 | 0.39 |
| | Rib (imported) | Boiling | n.d. | 0.53 | 0.22 |
| | | Grilling | 0.20 | 2.08 | 0.66 |
| | | Pan-frying | n.d. | 1.84 | 0.85 |
| | | Stir-frying | n.d. | 0.92 | 0.21 |
| | Sirloin (imported) | Boiling | n.d. | 0.72 | 0.04 |
| | | Grilling | 0.42 | 3.80 | 1.30 |
| | | Simmering | n.d. | – | – |
| | | Stir-frying | 0.39 | 2.91 | 1.11 |
| | Chitterlings | None | 0.13 | 0.82 | 0.44 |
| | | Boiling | n.d. | 0.35 | 0.24 |
| | | Grilling | 0.38 | 3.62 | 1.35 |
| | | Stir-frying | 0.21 | 1.93 | 0.79 |
| Pork | Wiener vienna sausage | None | 0.30 | 4.20 | 0.81 |
| | | Boiling | 0.19 | 3.61 | 0.67 |
| | | Blanching | n.d. | 0.93 | 0.05 |
| | | Grilling | 0.44 | 5.90 | 1.20 |
| | | Stir-frying | 0.46 | 4.74 | 1.09 |
| | Loin | Boiling | n.d. | 0.23 | 0.01 |
| | | Grilling | 0.23 | 1.75 | 0.55 |
| | | Deep-frying | 0.20 | 2.09 | 0.58 |
| | | Air-frying | n.d. | 1.01 | 0.30 |
| | | Stir-frying | 0.14 | 0.95 | 0.37 |
| | | Pan-frying | 0.12 | 0.91 | 0.34 |
| | | Steaming | n.d. | – | – |
| | | Simmering | n.d. | 0.10 | 0.01 |
| Poultry | Chicken | None | n.d. | 0.33 | 0.13 |
| | | Grilling | 0.23 | 2.29 | 0.70 |
| | | Stir-frying | 0.18 | 1.44 | 0.58 |
| | | Boiling | n.d. | – | – |
| | Chicken thigh | Air-frying | n.d. | 0.75 | 0.20 |
| | | Deep-frying | 0.16 | 1.59 | 0.52 |
| | | Stir-frying | n.d. | 0.64 | 0.24 |
| | Duck meat | None | n.d. | 0.14 | 0.01 |
| | | Boiling | n.d. | 0.34 | 0.15 |
| | | Grilling | 0.16 | 1.21 | 0.42 |

(*Continued*)

© Burleigh Dodds Science Publishing Limited, 2025. All rights reserved.

**Table 9** (*Continued*)

| Category | Sample | Processing method | BaP[1] | PAH8[2] | $TEQ_{BaP}$[3] |
|---|---|---|---|---|---|
| Edible oil | Rice bran oil | None | 0.29 | 2.95 | 0.93 |
| | | Grilling | 0.30 | 3.54 | 1.14 |
| | | Pan-frying | 0.31 | 3.48 | 1.02 |
| | | Deep-frying | 0.36 | 3.68 | 1.12 |
| | | Stir-frying | 0.35 | 3.63 | 1.11 |
| | Perilla oil | None | 0.29 | 3.49 | 0.65 |
| | | Boiling | 0.19 | 2.19 | 0.44 |
| | | Stir-frying | 0.56 | 5.71 | 1.29 |
| | | Pan-frying | 0.43 | 4.67 | 1.01 |
| | Sesame oil | None | 0.26 | 3.17 | 0.53 |
| | | Grilling | 0.44 | 3.81 | 0.87 |
| | | Stir-frying | 0.27 | 3.53 | 0.65 |
| | | Boiling | 0.46 | 2.93 | 0.71 |
| | Soybean oil | None | 0.11 | 1.15 | 0.26 |
| | | Pan-frying | 0.21 | 1.42 | 0.39 |
| | | Deep-frying | 0.16 | 1.51 | 0.36 |
| | | Stir-frying | 0.12 | 1.37 | 0.33 |
| | | Boiling after Stir-frying | 0.11 | 1.03 | 0.23 |
| | | Boiling | 0.10 | 0.97 | 0.23 |

[1]The limit of detection (LOD) of BaP was 0.1167 μg/kg for beef and 0.0844 μg/kg for edible oil.
[2]Sum of eight PAHs in Table 7.
[3]BaP toxic equivalent quantity ($TEQ_{BaP}$) of PAH8. Toxic equivalent factors from Nisbet and LaGoy (1992) was used.

thighs produced fewer PAHs compared to deep-fried samples (Lee et al., 2020; Choi et al., 2022).

Among edible oils, stir-fried perilla oil had the highest PAH levels (5.71 μg/kg) and BaP equivalents (1.29 μg/kg), followed by rice bran oil across various cooking methods, including grilling, deep-frying, stir-frying, and pan-frying. PAH levels generally increased with grilling, stir-frying, pan-frying, deep-frying, and air-frying, while water-based cooking methods like blanching, boiling, simmering, and steaming led to decreased PAH formation (Choi et al., 2022).

The ratio between PAHs can be used to detect the origin of the PAHs. When the ratio of BaA/(BaA + CHR) is below 0.2, the PAHs are considered of petrogenic origin, while a ratio above 0.35 indicates a combustion source. The ratios of IP/(IP + BghiP) are categorized into three sections: <0.2 (petrogenic), 0.2–0.5 (petroleum burning), and >0.5 (combustion).

© Burleigh Dodds Science Publishing Limited, 2025. All rights reserved.

**Table 10** Estimated mean dietary exposure to polycyclic aromatic hydrocarbons (ng/kg bw/day) and their margin of exposure (MOE) in different countries

| | | Polycyclic aromatic hydrocarbons (ng/kg bw/day) | | | | | | | | MOE[1] | | | | |
|---|---|---|---|---|---|---|---|---|---|---|---|---|---|---|
| Country | Age | BaP | CHR | BaA | BbF | PAH2 | PAH4 | PAH8 | $TEQ_{BaP}$ | BaP | PAH2 | PAH4 | PAH8 | Comment |
| Korea[2] | 1–2 | 4.707 | 6.738 | 6.282 | 6.005 | – | 23.733 | – | – | 14900* | – | 14300* | – | Processed foods |
| | 3–6 | 3.905 | 5.728 | 5.094 | 4.991 | – | 19.743 | – | – | 17900* | – | 17200* | – | |
| | 7–12 | 2.306 | 3.696 | 3.169 | 2.995 | – | 12.191 | – | – | 30400* | – | 27900* | – | |
| | 13–19 | 1.836 | 2.763 | 2.274 | 2.288 | – | 9.176 | – | – | 38100* | – | 37100* | – | |
| | 20–64 | 2.779 | 3.321 | 2.391 | 3.075 | – | 11.574 | – | – | 25200* | – | 29400* | – | |
| | ≥65 | 1.389 | 1.923 | 1.443 | 1.631 | – | 6.414 | – | – | 50400* | – | 53000* | – | |
| Korea[3] | 1–2 | 4.037 | 5.050 | 2.610 | 7.198 | 9.088 | 18.896 | 30.375 | – | 17340* | 19000 | 18000 | 16292* | Total diet study |
| | 3–5 | 4.676 | 5.329 | 3.294 | 7.252 | 10.003 | 20.549 | 32.998 | – | 14970* | 17000 | 17000 | 15016* | |
| | 6–11 | 4.258 | 4.502 | 3.064 | 6.141 | 8.759 | 17.965 | 28.954 | – | 16440* | 19000 | 19000 | 17133* | |
| | 12–18 | 3.439 | 3.399 | 2.589 | 4.744 | 6.838 | 14.171 | 22.944 | – | 20355* | 25000 | 24000 | 21450* | |
| | 19–64 | 4.037 | 5.050 | 2.610 | 7.198 | 9.088 | 18.896 | 30.375 | – | 19348* | 19000 | 18000 | 21626* | |
| | ≥65 | 1.685 | 2.437 | 1.429 | 3.396 | 4.122 | 8.947 | 12.956 | – | 41543* | 41000 | 38000 | 37626* | |
| France[4] | Children | 0.304 | 1.053 | 0.466 | 0.439 | – | 2.259 | 3.493 | – | 230000 | – | 150000* | 140000 | Total diet study |
| | Adults | 0.191 | 0.688 | 0.300 | 0.301 | – | 1.478 | 2.281 | – | 366000 | – | 230000* | 215000 | |
| EU[5] | All | 3.1–4.3 | – | – | – | 8.3–12.8 | 15.6–24.2 | 23.6–35.6 | – | 16000–23000 | 13000–20000 | 14000–22000 | 14000–21000 | |
| USA[6] | | 0.7–2.7 | – | – | – | – | – | – | – | 26000–100000 | – | – | – | |
| Nigeria[7] | Children | 1850 | 1960 | 2440 | 1330 | 3820 | 7590 | 14000 | 4120[8] | 38 | 45 | 45 | 35 | Only meat and fish |
| | Adults | 1160 | 1200 | 1670 | 870 | 2360 | 4900 | 8880 | 2560[8] | 60 | 72* | 69* | 55* | |

[1]Margin of exposure (MOE) values were calculated based on the exposure levels and corresponding $BMDL_{10}$ values. The $BMDL_{10}$ values were 0.07 mg/kg bw/day for BaP, 0.17 mg/kg bw/day for PAH2, 0.34 mg/kg bw/day for PAH4, and 0.49 mg/kg bw/day for PAH8 (EFSA, 2008). The values marked with * were obtained from the reference, while the others were calculated.

[2]MFDS (2018), adapted the data of upper bound, where n.d. (not detected) was used as the limit of detection.

[3]MFDS (2016c).

[4]Veyrand et al. (2013).

[5]EFSA (2008). The range of average dietary exposure levels (ng/kg bw/day) were summarized from 16 countries in EU assuming a body weight of 60 kg.

[6]Menzie, Potocki and Santodonato (1992)· The range of average dietary exposure levels (ng/kg bw/day) were calculated assuming a body weight of 60 kg.

[7]Okoye et al. (2021).

[8]Mean dietary exposure to total BaP toxic equivalent quantity ($TEQ_{BaP}$) of PAH16, which includes PAH8, naphthalene, acenapthylene, acenaphthene, fluorene, phenanthrene, anthracene, flouranthene, and pyrene. Toxic equivalent factors from Nisbet and LaGoy (1992) was used.

© Burleigh Dodds Science Publishing Limited, 2025. All rights reserved.

When the origin of PAHs in Korean marine products was determined using a scatter plot that compared the ratio of BaA/(BaA + CHR) to that of IP/(IP + BghiP) (Parmar et al., 2024), PAHs in grilled fish were typically associated with combustion, while those in dried fish were more likely to originate from environmental sources. Additionally, PAH8 in shellfish may have been attributed to environmental contamination, particularly oil spills in the sea (Parmar et al., 2024).

## 4.2 Exposure and risk characterization

In Korea, dietary exposure assessments have been conducted through both a monitoring study (MFDS, 2018) and a TDS (MFDS, 2016b). According to the monitoring study, the average dietary exposure to BaP and PAH4 for the Korean population was 1.395 ng/kg bw/day and 5.470 ng/kg bw/day, respectively. These values are lower than those reported in the TDS, likely due to the significant formation of PAHs during high-temperature processes. The dietary exposure levels from the Korean TDS (mean: 14.7 ng/kg bw/day for PAH4 and 22.6 ng/kg bw/day for PAH8) are comparable to those reported in the EU (15.6–24.2 ng/kg bw/day for PAH4 and 23.6–35.6 ng/kg bw/day for PAH8), but higher than the levels found in the French TDS (1.5–2.3 ng/kg bw/day for PAH4 and 2.3–3.5 ng/kg bw/day for PAH8) (MFDS, 2016b, 2018; EFSA, 2008; Veyrand et al., 2013).

In contrast to developed countries, Nigeria has reported significantly higher PAH concentrations in foods collected from heavily polluted regions. PAH16 concentrations in Nigerian foods range from 7.32–25.4 mg/kg in chicken meat, 7.51–16.7 mg/kg in cow meat, 1.24–9.53 mg/kg in goat meat, 3.27–7.72 mg/kg in fish, and 6.58–8.42 mg/kg in plants. Additionally, soil, fish feed, and chicken feed contained PAHs ranging from 6.33–8.8 mg/kg, 6.48–28.7 mg/kg, and 7.4–15.3 mg/kg, respectively – levels that are comparable to or higher than those found in plants, fish, and chicken. Due to the extremely high PAH concentrations in these foods, the mean dietary exposure to PAH4 and PAH8 in Nigeria is at least 300 times higher than that reported in the Korean TDS (Okoye et al., 2021).

The health risks of PAHs were assessed using the MOE approach, based on the exposure levels and the $BMDL_{10}$ values. The $BMDL_{10}$ for BaP, PAH2, PAH4, and PAH8 was calculated based on total tumor-bearing animals in a 2-year carcinogenicity study of female mice exposed to the coal tar mixtures. As markers for the carcinogenic PAHs in foods, the $BMDL_{10}$ values were determined to be 0.07 mg/kg bw/day for BaP, 0.17 mg/kg bw/day for PAH2, 0.34 mg/kg bw/day for PAH4, and 0.49 mg/kg bw/day for PAH8 (EFSA, 2008). Margin of exposure levels across all countries, except Nigeria, indicated a low risk for PAHs. In contrast, MOE levels in Nigeria were below 75, reflecting a

© Burleigh Dodds Science Publishing Limited, 2025. All rights reserved.

significantly high risk. Therefore, prioritizing mitigation strategies to reduce PAH exposure in Nigeria would be critical.

Given the widespread presence of PAHs in the environment from various sources, including pollution, integrated exposure assessments have gained attention as a comprehensive approach to evaluating the risks PAHs pose to human health.

### 4.3 Management

Polycyclic aromatic hydrocarbons in food are regulated in both the EU and the Republic of Korea, as summarized in Table 11 (MFDS, 2024). The EU has established maximum levels for both BaP and PAH4, while Korea regulates only BaP. For both regions, the maximum allowable level of BaP and PAH4 in foods for infants and young children, including formula, is 1 μg/kg – the most stringent control standard. In the case of oils and fats, the maximum level of BaP is set at 2 μg/kg in both regulations. However, for smoked meats and their products, the maximum BaP level in Korea is 5 μg/kg, which is higher than the EU's limit of 2 μg/kg. Regarding marine products, the maximum BaP level in Europe ranges from 2 to 6 μg/kg, while in Korea, it ranges from 2 to 10 μg/kg, depending on the specific food type.

Additionally, both Europe and Korea have established regulatory limits for dried and plant-based products. The EU regulates BaP and PAH4 in banana chips, dried herbs, dried spices, cocoa beans and their derivatives, plant-based powders for beverages, and food supplements containing botanicals, propolis, royal jelly, and spirulina. In Korea, BaP is regulated in medicinal herbs such as *Sookjihwang* (processed *Rehmannia* root), dried adhesive *Rehmannia* roots, and food supplements like black ginseng.

Significant levels of PAHs have been detected in banana chips, likely due to common preparation methods involving frying in coconut oil or other highly saturated oils such as palm oil. In Korea, PAH contamination has also been reported in herbal medicines, primarily as a result of high-temperature drying processes, smoking methods used during preparation, and the cultivation of plants in contaminated soil.

The formation of PAHs can be minimized by opting for steaming or boiling instead of frying or grilling. Adding antioxidants such as polyphenols or onion extract can help reduce PAHs formation by scavenging the radicals involved in their production. Furthermore, preheating charcoal before grilling is recommended to prevent PAH formation caused by incomplete combustion (MFDS, 2017; Singh, Agarwal & Simal-Gandara, 2023). In Korean, unrefined oils like sesame and perilla oil are high in PAHs and significantly contribute to dietary exposure. In Korea, unrefined oils like sesame and perilla oils are particularly high in PAHs, significantly contributing to dietary exposure. On

© Burleigh Dodds Science Publishing Limited, 2025. All rights reserved.

**Table 11** Regulation for polycyclic aromatic hydrocarbons (PAHs) in food from EU and Korea

| | Maximum level (µg/kg) | | |
|---|---|---|---|
| | EU[1] | | Republic of Korea[2] |
| Food | BaP | PAH4[3] | BaP |
| Banana chips | 2.0 | 20.0 | – |
| Powders of food of plant origin for the preparation of beverages except cocoa products | 10.0 | 50.0 | – |
| Dried herbs | 10.0 | 50.0 | – |
| Cocoa beans and derived products except cocoa fibre products | 5.0 µg/kg fat | 30.0 µg/kg fat | – |
| Cocoa fibre and products derived from cocoa fibre intended for use as an ingredient in food | 3.0 | 15.0 | – |
| Smoked meat and smoked meat products | 2.0 | 12.0 | 5.0 |
| Smoked fishery products except products listed in the next row | 2.0 | 12.0 | 2 for fish, 5 for smoked fish, 10 for smoked dried fish, 10 for shellfish, 5 for mollusks and crustaceans |
| Smoked sprats and canned smoked sprats (*Sprattus sprattus*) Smoked Baltic herring ≤14 cm length and canned smoked Baltic herring ≤14 cm length (*Clupea harengus membras*) Katsuobushi (dried bonito, *Katsuwonus pelamis*) Bivalve molluscs (fresh, chilled or frozen) Heat treated meat and heat treated meat products placed on the market for the final consumer | 5.0 | 30.0 | |
| Smoked bivalve molluscs | 6.0 | 35.0 | |
| Dried spices | 10.0 | 50.0 | – |
| Oils and fats placed on the market for the final consumer or use as an Ingredient in food | 2.0 | 10.0 | 2.0 |
| Coconut oil placed on the market for the final consumer or use as an ingredient in food | 2.0 | 20.0 | 2.0 |
| Infant formula, follow-on formula and young-child formula | 1.0 | 1.0 | 1.0 |
| Baby food and processed cereal-based food for infants and young children | 1.0 | 1.0 | 1.0 |
| Food for special medical purposes intended for infants and young children | 1.0 | 1.0 | 1.0 |
| Food supplements containing botanicals and their preparations Food supplements containing propolis, royal jelly, spirulina or their preparations | 10.0 | 50.0 | 5 for *Sookjihawang* (processed rehmannia root) and dried adhesive rehmannia roots 2 for black ginseng (including its powder) 4 for black ginseng concentrate |

[1]Commission Regulation (EU) 2023/915.
[2]MFDS (2024).
[3]Sum of BaP, CHR, BaA, and BbF.

© Burleigh Dodds Science Publishing Limited, 2025. All rights reserved.

an industrial scale, the reduction of PAHs in sesame oil production has been successfully achieved using supercritical extraction methods.

## 5 Conclusion and future trends

Acrylamide, HCAs, and PAHs are major processing contaminants that form during high-temperature cooking and food processing. Acrylamide, primarily found in carbohydrate-rich foods, is produced through the Maillard reaction, particularly between asparagine and reducing sugars. Heterocyclic amines are formed in various foods through the reaction of amino acids, sugars, or creatine during high-temperature processes, while PAHs are generated from incomplete combustion and contaminate foods during cooking methods such as smoking and grilling.

These compounds may pose significant health risks due to their mutagenic and carcinogenic properties, driving regulatory efforts and research to minimize their presence in food. Future trends are expected to focus on more effective mitigation strategies, including optimizing cooking techniques (e.g. reducing cooking temperatures or adopting alternative methods like steaming), incorporating food additives such as antioxidants to inhibit their formation, and utilizing advanced technologies. Additionally, ongoing research into the molecular mechanisms of contaminant formation will be essential for refining risk assessments and establishing comprehensive regulations to enhance food safety.

## 6 References

Abt, E., et al. (2022). Acrylamide: perspectives from international, national, and regional exposure assessments. *Current Opinion in Food Science*, 47:100891.

Ahmed, Z. A. & Mohammed, N. K. (2024). Investigating influencing factors on acrylamide content in fried potatoes and mitigating measures: a review. *Food Production, Processing and Nutrition*, 6(1):24.

Alaejos, M.S. & Afonso, A.M. (2011). Factors that affect the content of heterocyclic aromatic amines in foods. *Comprehensive Reviews in Food Science and Food Safety*, 10(2):52–108.

Blank, I., et al. (2005). Mechanisms of acrylamide formation: maillard-induced transformation of asparagine. In: Friedman, M. & Mottram, D., (eds.). *Chemistry and Safety of Acrylamide in Food*. Vol. 561. Advances in Experimental Medicine and Biology. Boston, MA: Springer US; 171–189. https://doi.org/10.1007/0-387-24980-X_14.

Carthew, P., et al. (2010). Application of the Margin of Exposure (MOE) approach to substances in food that are genotoxic and carcinogenic: Example: CAS No: 105650-23-5 PhIP (2-amino-1-methyl-6-phenylimidazo[4,5-b]pyridine). *Food and Chemical Toxicology*, 48:S98–S105. https://doi.org/10.1016/j.fct.2009.10.035.

Choi, J., et al. (2022). Risk assessment of polycyclic aromatic hydrocarbons in meat and edible oils: results of a total diet study in South Korea. *Food Science and Biotechnology*, 31(12):1523–1535.

© Burleigh Dodds Science Publishing Limited, 2025. All rights reserved.

Claeys, W.L., et al. (2005). Quantifying the formation of carcinogens during food processing: acrylamide. *Trends in Food Science & Technology*, 16(5):181–193.

Dong, H., et al. (2020). Potential carcinogenic heterocyclic aromatic amines (HAAs) in foodstuffs: formation, extraction, analytical methods, and mitigation strategies. *Comprehensive Reviews in Food Science and Food Safety*, 19(2):365–404.

Edna Hee, P.-T., et al. (2024). Formation mechanisms, detection methods and mitigation strategies of acrylamide, polycyclic aromatic hydrocarbons and heterocyclic amines in food products. *Food Control*, 158:110236.

EFSA (2005). Opinion of the Scientific Committee on a request from EFSA related to A harmonised approach for risk assessment of substances which are both genotoxic and carcinogenic. *EFSA Journal*, 3(10):282.

EFSA (2008). Polycyclic aromatic hydrocarbons in food - scientific opinion of the panel on contaminants in the food chain. *EFSA Journal*, 6(8):724.

EPA (2012). Classification of polycyclic aromatic hydrocarbons. https://epa.tas.gov.au/Documents/Advisory%20Note%20for%20Classification%20of%20PAHs.pdf.

European Union (EU) (2017). Commission Regulation (EU) 2017/2158 of 20 November 2017 establishing mitigation measures and benchmark levels for the reduction of the presence of acrylamide in food (https://eur-lex.europa.eu/eli/reg/2017/2158/oj/eng).

Feng, Y., et al. (2025). Polycyclic aromatic hydrocarbons (PAHs): environmental persistence and human health risks. *Natural Product Comm*. 20(1):1–8.

Filippini, T., et al. (2022). Dietary Acrylamide Exposure and Risk of Site-Specific Cancer: A Systematic Review and Dose-Response Meta-Analysis of Epidemiological Studies. *Front Nutr*. 25(9):875607.

FoodDrinkEurope. (2019). Acrylamide Toolbox 2019. (https://www.fooddrinkeurope.eu/wp-content/uploads/2021/05/FoodDrinkEurope_Acrylamide_Toolbox_2019.pdf).

Geng, Y., et al. (2024). Heterocyclic amines in meat and meat products: occurrence, formation, mitigation, health risks and intervention. *Food Reviews International*, 40(5):1503–1519.

Granvogl, M. & Schieberle, P. (2006). Thermally generated 3-aminopropionamide as a transient intermediate in the formation of acrylamide. *Journal of Agricultural and Food Chemistry*, 54(16):5933–5938.

Herraiz, T. (2000). Tetrahydro-beta-carboline-3-carboxylic acid compounds in fish and meat: possible precursors of co-mutagenic beta-carbolines norharman and harman in cooked foods. *Food Additives & Contaminants*, 17(10):859–866.

Hogervorst, J. et al. (2022). Maternal acrylamide exposure during pregnancy and fetal growth: A systematic review and dose-response meta-analysis of epidemiological studies. *Environ Res*. 213:113705.

Hwang, I. & Kwon, H. (2022). Acrylamide formation in carbohydrate-rich food powders consumed in Korea. *Quality Assurance and Safety of Crops & Foods*, 14(3):43–54.

IARC (1987). *Overall Evaluations of Carcinogenicity: An Updating of IARC Monographs Volumes 1–42*. France: International Agency for Research on Cancer.

IARC (1993). *Some Naturally Occurring Substances: Food Items and Constituents, Heterocyclic Aromatic Amines and Mycotoxins*. France: International Agency for Research on Cancer.

International Agency for Research on Cancer (IACR) Monograph Working Group. (2015). Carcinogenicity of consumption of red and processed meat. *Lancet Oncol*. 16(16):1599–1615.

© Burleigh Dodds Science Publishing Limited, 2025. All rights reserved.

Jeong, H., et al. (2020). Survey for acrylamide in processed foods from Korean market and individual exposure estimation using a non-parametric probabilistic model. *Food Additives & Contaminants: Part A*, 37(6):916–930.

Jin, C., et al. (2013). Relationship between antioxidants and acrylamide formation: a review. *Food Research International*, 51(2):611–620.

Keating, G.A. & Bogen, K.T. (2004). Estimates of heterocyclic amine intake in the US population. *Journal of Chromatography B*, 802(1):127–133. https://doi.org/10.1016/j.jchromb.2003.10.047.

Law, R.J., et al. (2002). Toxic equivalency factors for PAH and their applicability in shellfish pollution monitoring studies. *Journal of Environmental Monitoring*, 4(3):383–388.

Lee, Y., et al. (2020). Validation of analytical methods for heterocyclic amines in seven food matrices using high-performance liquid chromatography-tandem mass spectrometry. *Food Additives & Contaminants: Part A*, 37(3):408–420.

Lee, Y., et al. (2022). Assessment of dietary exposure to heterocyclic amines based on the Korean total diet study. *Food Additives & Contaminants: Part A*, 39(3):429–439.

Liu, W., et al. (2022). Bioactive β-Carbolines Harman and Norharman in Sesame Seed Oils in China. *Molecules*, 27(2):402.

Menzie, C.A., et al. (1992). Exposure to carcinogenic PAHs in the environment. *Environmental Science & Technology*, 26(7):1278–1284.

MFDS (2016a). Risk assessment of acrylamide. http://www.nifds.go.kr/brd/m_271/view.do?seq=10120&srchFr=&srchTo= &srchWord=&srchTp=&itm_seq_1=0&itm_seq_2= 0&multi_itm_seq=0&company_cd=&company_nm= &page=5.

MFDS (2016b). *Risk Assessment of Heterocyclic Amines (HCAs)*. Korea: Ministry of Food and Drug Safety.

MFDS (2016c). *Risk Assessment of Polycyclic Aromatic Hydrocarbons (PAHs)*. Korea: Ministry of Food and Drug Safety.

MFDS (2017). Information of hazard compound - Polycyclic aromatic hydrocarbons. https://www.foodsafetykorea.go.kr/portal/board/boardDetail.do?menu_no=2944&bbs_no=bbsfs062&ntctxt_no=1067609&menu_grp=MENU_NEW04.

MFDS (2018). Risk assessment of polycyclic aromatic hydrocarbons in foods. http://www.nifds.go.kr/brd/m_271/view.do?seq=12537&srchFr=&srchTo=&srchWord=%EB%8B%A4%ED%99%98&srchTp=0&itm_seq_1=0&itm_seq_2=0&multi_itm_seq=0&company_cd=&company_nm=&page=1.

Mottram, D. S., et al. (2002). Acrylamide is formed in the Maillard reaction. *Nature*, 419:448–449.

Murkovic, M. (2004a). Chemistry, formation and occurrence of genotoxic heterocyclic aromatic amines in fried products. *European Journal of Lipid Science and Technology*, 106(11):777–785.

Murkovic, M. (2004b). Formation of heterocyclic aromatic amines in model systems. *Journal of Chromatography B*, 802(1):3–10. https://doi.org/10.1016/j.jchromb.2003.09.026

Nisbet, I.C.T. & LaGoy, P.K. (1992). Toxic equivalency factors (TEFs) for polycyclic aromatic hydrocarbons (PAHs). *Regulatory Toxicology and Pharmacology*, 16(3):290–300.

NTP (2021). *Report on Carcinogens*, 15th ed. National Toxicology Program https://ntp.niehs.nih.gov/ntp/roc/content/profiles/heterocyclicamines.pdf (accessed 23 April 2019).

Okoye, E.A., et al. (2021). Concentrations of polycyclic aromatic hydrocarbons in samples of soil, feed and food collected in the Niger Delta region, Nigeria: a probabilistic human health risk assessment. *Environmental Research*, 202:111619.

© Burleigh Dodds Science Publishing Limited, 2025. All rights reserved.

Parmar, M., et al. (2024). A current perspective on polycyclic aromatic hydrocarbons contamination and their bioremediation aspects. *Environmental Earth Sciences*, 83(9):275.

Patel, A.B., et al. (2020). Polycyclic aromatic hydrocarbons: sources, toxicity, and remediation approaches. *Frontiers in Microbiology*, 11:562813.

Reizer, E., et al. (2022). Formation and growth mechanisms of polycyclic aromatic hydrocarbons: a mini-review. *Chemosphere*, 291:132793.

Rifai, L. & Saleh, F. A. (2020). A review of acrylamide in food: occurrence, toxicity and mitigation strategies. *Int. J. Toxicol*. 39(2):93–102.

Singh, L., et al. (2023). Summarizing minimization of polycyclic aromatic hydrocarbons in thermally processed foods by different strategies. *Food Control*, 146:109514.

Sugimura, T., et al. (2004). Heterocyclic amines: mutagens/carcinogens produced during cooking of meat and fish. *Cancer Science*, 95(4):290–299.

Tareke, E., et al. (2002). Analysis of acrylamide, a carcinogen formed in heated foodstuffs. *J Agric Food Chem*. 50(17):4998–5006.

Veyrand, B., et al. (2013). Human dietary exposure to polycyclic aromatic hydrocarbons: results of the second French Total Diet Study. *Environment International*, 54:11–17.

WHO (2005). *Evaluation of Certain Contaminants in Food: Sixty-Fourth Report of the Joint FAO/WHO Expert Committee on Food Additives*. Rome, Italy: World Health Organization WHO Technical Report Series.

WHO (2011). *Evaluation of Certain Contaminants in Food: Seventy-Second Report of the Joint FAO/WHO Expert Committee on Food Additives*. Rome, Italy: World Health Organization WHO Technical Report Series.

Xiong, K., et al. (2024). Formation and Reduction of Toxic Compounds Derived from the Maillard Reaction During the Thermal Processing of Different Food Matrices. *J Food Prot*. 87(9):100338.

Zimmerli, B, et al. (2001). Occurrence of heterocyclic aromatic amines in the Swiss diet: analytical method, exposure estimation and risk assessment. *Food Additives & Contaminants*, 18(6):533–551.

© Burleigh Dodds Science Publishing Limited, 2025. All rights reserved.

# Chapter 17

## Assessing pyrrolizidine alkaloids in food

*Natalia Casado, Universidad Rey Juan Carlos, Spain; Florian Kaltner and Ludwig Maximilian University of Munich, Germany; Begoña Fernández-Pintor and Sonia Morante-Zarcero, Universidad Rey Juan Carlos, Spain; and Isabel Sierra, Universidad Rey Juan Carlos and Instituto de Investigación de Tecnologías para la Sostenibilidad, Spain*



### 1 Introduction

Pyrrolizidine alkaloids (PAs) are a group of secondary plant metabolites, with more than 660 known compounds to date. It is widely accepted that PAs are produced by certain plants to protect themselves from herbivorous insects. Chemotaxonomic estimates suggest that PAs are present in more than 6000 plant species, accounting for 3% of all flowering plants worldwide. All PAs share a common pyrrolizidine-like core structure (1-azabicyclo[3.3.0]octane), typically featuring a hydroxymethyl group at C1 and a hydroxyl group at C7, collectively referred to as the necine base (Figure 1). This base is usually esterified to one or more polyhydroxylated carboxylic acids, known as necic acids, which allow PAs to be classified as monoesters, open-chained diesters or cyclic diesters. Nearly all PAs can exist in two forms: as tertiary amine PAs and as their more hydrophilic counterparts, PA *N*-oxides. It is estimated that PA *N*-oxides are the preferred storage form in the vacuolated tissues of PA-producing plants, while the more lipophilic tertiary amine PAs tend to accumulate in their seeds (Toppel et al. 1988).

http://dx.doi.org/10.19103/AS.2025.0158.27
© Burleigh Dodds Science Publishing Limited, 2025. All rights reserved.

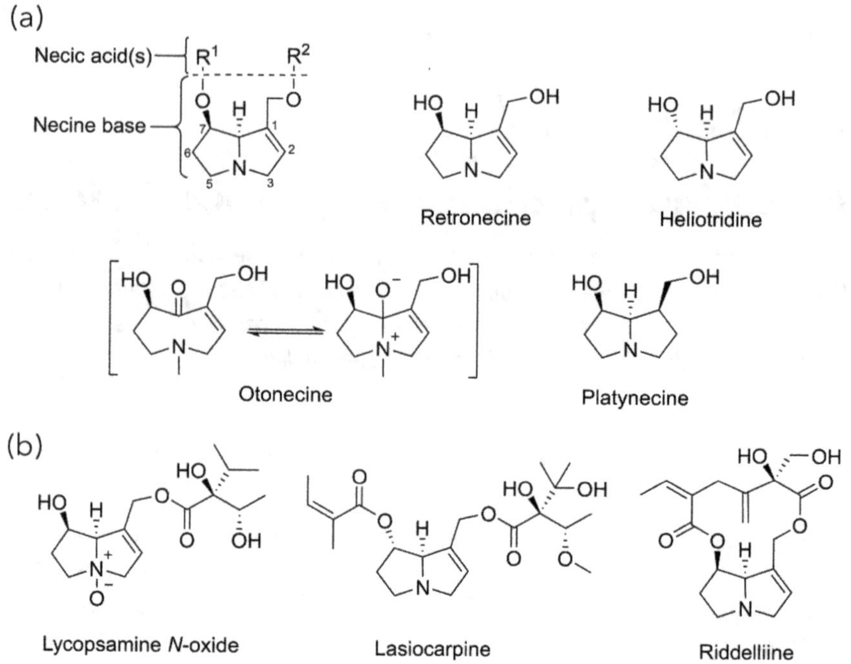


**Figure 1** Overview of important pyrrolizidine alkaloid (PA) chemical structures. (a) PAs consist of a pyrrolizidine core, called necine base, esterified to one or more necic acids. Four types of necine bases are distinguished: the 1,2-unsaturated isomers retronecine and heliotridine, the tautomeric otonecine and the saturated PA platynecine. (b) Examples of different ester types of PAs: the monoester lycopsamine, the open-chained diester lasiocarpine and the cyclic diester riddelliine. Lycopsamine *N*-oxide demonstrates the possible formation of *N*-oxides if the nitrogen atom is not methylated, like it is the case for otonecine-type PAs.

Several decades ago, PA-producing plants (and later PAs) as food contaminants were identified as causes of isolated intoxications in livestock and humans (Bach et al. 1989), as well as large-scale outbreaks of certain diseases. For instance, increased incidences of liver cirrhosis, liver failure or hepatic sinusoidal obstruction syndrome (previously known as veno-occlusive disease) in certain regions were linked to the consumption of PA-producing plants or their parts (Kakar et al. 2010). Subsequent research demonstrated that 1,2-unsaturated PAs undergo metabolic activation in the liver after ingestion. In this process, 1,2-unsaturated PAs are oxidized by hepatic cytochrome P450 monooxygenase enzymes into carbinol amines, which can form highly reactive pyrrole esters. These esters can alkylate cellular nucleophiles or DNA, form protein adducts that inhibit normal cellular functions, or cause DNA cross-links (Fu et al. 2004). PA *N*-oxides exhibit similar behaviour, as they are converted to tertiary amines after ingestion (Wang et al. 2005). In summary, PAs possess

© Burleigh Dodds Science Publishing Limited, 2025. All rights reserved.

hepatotoxic, genotoxic, mutagenic and carcinogenic properties, as shown in long-term animal studies where lasiocarpine or riddelliine were fed to mice or rats (NCI 1978; NTP 2003).

In conclusion, PAs are recognized as plant toxins owing to their toxicological effects on both humans and animals, which can result in severe intoxications. Since these toxins have been detected as contaminants in certain food and feed products, they have attracted significant attention from food safety authorities globally. As a result, the development of sensitive analytical methods for reliably detecting PAs in relevant food matrices has become a key focus for researchers worldwide, aiming to improve food safety and enhance consumer protection. Hence, this chapter aims to provide insight into the issue of PAs as food contaminants, placing special emphasis on the different strategies carried out for their analysis in different food matrices (from 2018 to 2024). As well as this, a section about the effect of food processing on the stability of PAs is also included.

## 2 Contamination pathways for pyrrolizidine alkaloids in food

The main plant families that produce PAs are Boraginaceae (genera *Echium*, *Heliotropium* and *Symphytum*), Asteraceae (tribes Senecioneae and Eupatorieae), Fabaceae (genus *Crotalaria*), Apocynaceae and Orchidaceae. Many of these plant species are used in traditional medicine for the treatment of different diseases. For instance, *Symphytum officinale* (commonly known as comfrey) is traditionally used in rheumatism and gout treatment, while *Tussilago farfara* (commonly known as coltsfoot) is used in the treatment of respiratory conditions such as bronchitis and asthma, among others (Vogl et al. 2013).

On the other hand, most of these plants are common weeds which are considered invasive species that may pose a risk to the environment (D'Alessandro 2024). Accordingly, there are different ways that result in PAs reaching the food supply chain.

One source is the direct intake of PA-producing plants. For instance, borage (*Borago officinalis*) is one of the plants that naturally contains PAs that are part of some European countries' gastronomy, such as Spain, Greece, Italy, France and Germany (Miceli et al. 2015). PAs can be ingested through the consumption of herbal beverages or food supplements prepared from PA-producing plants, owing to the medicinal properties attributed and marketed towards these plants (Lis-Cieplak et al. 2024).

The other source is the intake of food contaminated with PAs (Figure 2). As indicated above, most of these PA-producing plants are invasive plants that appear as weeds in food crop fields. Thus, different parts of these plants and their seeds can contaminate the soil (horizontal natural transfer) or accidentally

© Burleigh Dodds Science Publishing Limited, 2025. All rights reserved.

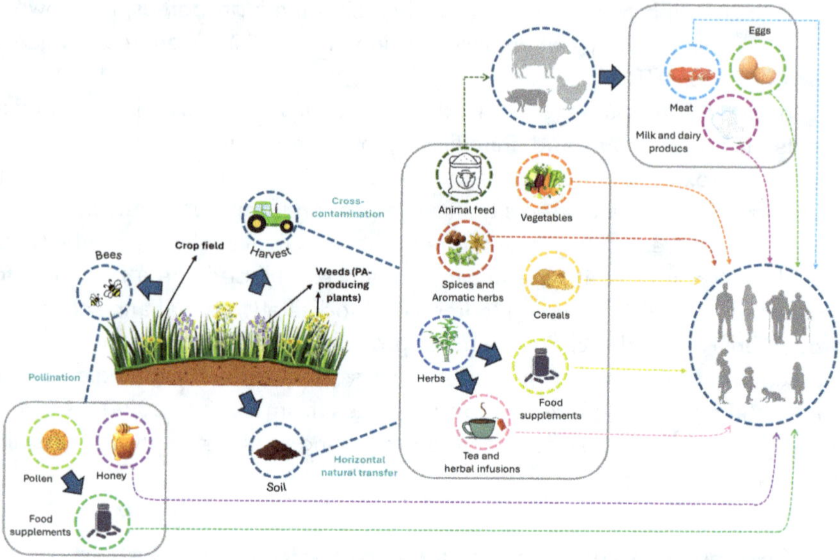


**Figure 2** Overview of the main pathways humans can be exposed to pyrrolizidine alkaloids through contaminated food.

be mixed with the harvested vegetables, cereals or herbs (cross-contamination) (Casado et al. 2022a). For this reason, plant-based food is considered the main source of PAs in the human diet. Nonetheless, in the latter case, contamination takes place in raw materials intended for human consumption, but also contaminates crops for animal feed (Figure 2). In consequence, there may be products of animal origin that also contain PAs in their composition as contaminants (e.g. carry-over of these alkaloids from plant material into eggs and dairy products) (Mulder et al. 2018; Klein et al. 2022). Likewise, honey is also an animal-based product that can be contaminated with PAs (Figure 2). During pollination, the honeybees also harvest pollen and nectar from PA-producing plants, resulting in contaminated honey (He et al. 2020).

Therefore, PAs can be widely present in a large variety of food products, which poses a high risk of consumer exposure to these toxic compounds through their diet (Figure 2). Consequently, owing to the potential adverse health effects of PAs, it is important to carry out good agricultural and harvest practices (e.g. effective weed control and quality control of feed) to mitigate the contamination of raw materials at the primary production level. Accordingly, in 2014, the Codex Alimentarius established an international food standard that outlined the best practices for weed management control, to help prevent and reduce the PA contamination in food and animal feed (CXC 74-2014). However, it is highlighted in this standard that the total eradication of PA-producing plants is not feasible or ecologically desirable. It is also

© Burleigh Dodds Science Publishing Limited, 2025. All rights reserved.

noted that good livestock feeding practices are also important, in addition to weed control management. In 2018, the European Tea Committee (ETC) and the European Herbal Infusion Association (EHIA) published an addendum to this international standard, which is a code of practice to prevent and reduce PA contamination in raw materials for tea and herbal infusions (Tea & Herbal Infusions Europe 2018).

## 3 Regulations governing pyrrolizidine alkaloid levels in food

It is also crucial to monitor the levels of PAs found in marketed food commodities. In this regard, in Europe, the European Food Safety Authority (EFSA) is the international European body in charge of setting up the guidelines to monitor the presence of these alkaloids in food and feed. From 2007 to 2017, the EFSA published different scientific opinions about the occurrence of PAs in food and the consequences related to their intake (EFSA 2007, 2011, 2016, 2017). As a result, the first European regulation to control the occurrence of PAs in food was published in 2020 (Commission Regulation (EU) 2020/2040), which amended the former European regulation on food contaminants 1881/2006. However, in April 2023, this regulation was revised and updated, leading to the current regulation to enforce the control of these alkaloids in Europe (Commission Regulation (EU) 2023/915).

The current regulation sets maximum allowable levels of PAs in a range of products, including: borage leaves; dried herbs such as borage, lovage, marjoram and oregano; teas and herbal infusions or adults, infants and young children; cumin, food supplements; pollen, pollen-based supplements; and other pollen-derived products. These maximum limits range from 1.0 to 1000 µg/kg depending on the type of product. The maximum concentration values refer to the sum of a set of 21 PAs (intermedine, lycopsamine, intermedine-*N*-oxide, lycopsamine-*N*-oxide, senecionine, senecivernine, senecionine-*N*-oxide, senecivernine-*N*-oxide, seneciphylline, seneciphylline-*N*-oxide, retrorsine, retrorsine-*N*-oxide, echimidine, echimidine-*N*-oxide, lasiocarpine, lasiocarpine-*N*-oxide, senkirkine, europine, europine-*N*-oxide, heliotrine and heliotrine-*N*-oxide), which are mandatory to be determined in any analytical methodology designed to monitor the occurrence of alkaloids in food. The regulation also suggests including the monitoring of 14 additional PAs, which are known to co-elute with one or more of the previous 21 PAs, provided they can be individually and separately identified in the analytical method.

Additionally, a new complementary regulation laying down methods for sampling and analysing the control of PAs in food has been published (Commission Implementing Regulation (EU) 2023/2783), which repeals Regulation (EU) 2015/705. This regulation establishes guidelines for

© Burleigh Dodds Science Publishing Limited, 2025. All rights reserved.

developing analytical strategies that aim to monitor the occurrence of PAs in food. Therefore, this document sets the performance criteria for the validation of these analytical methods.

In the United States, the regulation of PAs in food is not as specific as in Europe. The Food and Drug Administration (FDA), together with other public health agencies, is responsible for monitoring and managing the presence of contaminants in the food supply. Unlike the European Union, the FDA has not set maximum concentration limits for PAs in food (Lis-Cieplak et al. 2024). However, it monitors and assesses the risk associated with the presence of PAs and may take action if concerning levels of the compounds are detected in food products. Additionally, the FDA recommends that food and supplement producers follow good agricultural and manufacturing practices to minimize PA contamination. This includes careful selection of raw materials and quality control during processing.

Similarly, in Asian countries, specific limits for PAs in food have not been established. Nonetheless, some countries including China, Japan and India do carry out risk assessments and the implementation of good practices through their different food safety agencies (Lu et al. 2024a).

In the case of African countries, there are also no regulatory guidelines for PAs, so they follow the guidelines established by the World Health Organization (WHO) and the Food and Agriculture Organization of the United Nations (FAO) (Letsyo et al. 2023). Other countries, such as Australia, New Zealand and Switzerland also base their controls on these WHO and FAO guidelines, which are aligned with the European Union legislation (FSANZ; Swiss Federal Food Safety and Veterinary Office). Therefore, it can be concluded that to date, the European Union has the most detailed and specific regulation for the control of PAs in food.

## 4 Assessing levels of contamination of food with pyrrolizidine alkaloids

The Rapid Alert System for Food and Feed (RASFF) is a tool created by the European Union that enables the rapid exchange of information on food risks among the member countries, so that immediate measures can be taken to protect the health of consumers. Some of these measures include withdrawal or recall of the product, re-dispatching to the place of origin; informing the public, etc. This system has an interactive database called RASFF in which all the food alerts notified in the European Union can be consulted (Rapid Alert System for Food and Feed 2024).

According to this database, from 2012 to 2024, a total of 228 notifications have been reported regarding the contamination of different food products with PAs. Figure 3 shows the evolution of the number of these food alerts in

© Burleigh Dodds Science Publishing Limited, 2025. All rights reserved.

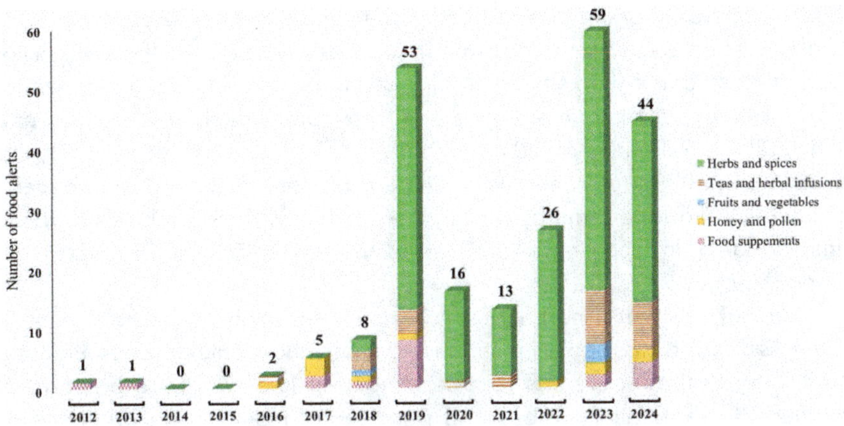


**Figure 3** Evolution of number of food alerts in different food categories related to the occurrence of pyrrolizidine alkaloids that have been notified in the European Rapid Alert System for Food and Feed (RASFF) in the period from 2012 to 2024. Data obtained from the RASFF window portal from 1 January 2012 to 18 September 2024.

this period, including the type of product in which the PAs have been detected. As it can be observed, the year with the highest number of alerts was 2023, followed by 2019 and 2024.

In 2019, there was a significant increase in the number of alerts, just one year before the first European regulation for PAs in food was enforced (Commission Regulation (EU) 2020/2040). Afterwards, the alerts have been more moderate in quantity, but their number has increased significantly again in the last two years. This may be owing to the new update in the regulations of PAs, which requires monitoring them and their compounds in a wider variety of food products (Commission Regulation (EU) 2023/915). On the other hand, it is impprtant to note that the RASFF system predominantly focuses on: (i) contaminants that have established regulatory limits, and (ii) cases where regulatory limits are exceeded. In consequence, the number of alerts related to PAs in this database might have been too low before the regulatory limit for PAs was implemented in the European Union. Moreover, products without a legislated maximum concentration level might not be tested or included in the RASFF. Therefore, it is highly possible that the actual number of food contaminations with PAs is much higher than reported in the RASFF database.

As it can be observed (Figure 3), most of the alerts have been notified for herbs and spices (74%) and followed far by teas and herbal infusions (11%), food supplements (8%), honey and pollen (5%) and fruits and vegetables (2%). This confirms that high levels of these compounds are mainly found as food contaminants in plant-based products, apart from honey.

In the case of herbs and spices, the main food alerts related to their high levels of PAs are reported in oregano and cumin. As previously

© Burleigh Dodds Science Publishing Limited, 2025. All rights reserved.

explained, cross-contamination during cultivation and harvesting will be the contamination pathway most common for these type of products (Figure 2). Nonetheless, some works have highlighted that natural horizontal transfer is also another possible contamination pathway to be considered (Nowak et al. 2016, 2017; Selmar et al. 2019; Chmit et al. 2021; Letsyo et al. 2021; Jiao et al. 2022, 2023). In some cases, PAs are present in these foods because of intentional adulteration by mixing the product with undisclosed herbs for financial gain. In fact, oregano is one of the most adulterated herbs in the world (Maquet et al. 2021).

Similarly, teas and herbal teas are also susceptible to be contaminated with PAs through cross-contamination and natural horizontal transfer (Figure 2). According to the food alerts reported, high levels of PAs have been mainly notified in chamomile, rooibos and peppermint teas, among others (Rapid Alert System for Food and Feed 2024).

Since the first European regulation for PAs was published (Commission Regulation (EU) 2020/2040), there have not been many alerts in the case of food supplements. In the period 2012–2019, high levels of PAs in this product category were mainly found in food supplements based on borage (*Borago officinalis*) and St John's wort (*Hypericum perforatum*). However, since the regulation was updated, some alerts have been notified in *Ginkgo biloba* extracts, gokshura (*Tribulus terrestris*) and psyllium (husk and fibre) (Rapid Alert System for Food and Feed 2024).

On the other hand, among the alerts notified for the high occurrence of PAs in honey and pollen, most of the reports have been from the pollen (Rapid Alert System for Food and Feed 2024). Generally, pollen is more contaminated with toxins compared to honey because it is directly related to the plant materials. Conversely, the contamination of honey with PAs usually occurs because of the deliberate or accidental introduction of pollen in this matrix.

Finally, in fruits and vegetables, only high levels of PAs have been notified in 2018 and 2023 (Figure 3). All of the alerts have been reported for vegetables, specifically borage, chives and lovage (Rapid Alert System for Food and Feed 2024). It is worth noting that borage is a PA-producing plant, and as previously mentioned, is part of the gastronomy of some countries. Nonetheless, despite its intended consumption, its maximum concentration levels of PAs are regulated and must be monitored to ensure the health of consumers (Commission Regulation (EU) 2023/915).

## 5 Analytical methods for the assessment of pyrrolizidine alkaloids in food

As previously mentioned, PAs typically occur as food contaminants within complex matrices and are therefore generally detected at trace levels in

© Burleigh Dodds Science Publishing Limited, 2025. All rights reserved.

food samples. For this reason, the development of analytical methods for the determination of PAs is closely related to a suitable sample preparation procedure combined with sensitive instrumental analysis techniques.

Table 1 collects the main analytical strategies developed from 2018 to 2024 for the assessment of PAs in food samples. As it can be observed, in general, the first step involves the extraction of PAs from the matrix. The most common techniques are solid–liquid extraction (SLE) or liquid–liquid extraction (LLE) depending on the sample nature (solid or liquid food matrix, respectively). Due to the polarity of PAs, acidified aqueous solutions, polar organic solvents or mixtures of both are generally used as extraction solvents.

In some cases, to determine the total content of PAs (expressed as a sum parameter of retronecine equivalents), a solution of zinc and $LiAlH_4$ in TFH is used to reduce PAs. This is performed either to reduce the PA *N*-oxides into their corresponding tertiary amine form (zinc) or to obtain the necine backbones ($LiAH_4$), which are then commonly further derivatised to increase the molecular weight or to increase chromatographic retention (Kowalczyk et al. 2018; Celano et al. 2019; Chmit et al. 2019; Letsyo et al. 2021). These approaches avoid using many reference standards, which tend to be expensive, and provide the sum levels of PAs in a sample. However, they do not allow to obtain of information on the PA pattern present in the samples analysed.

On the other hand, once the sample extracts are obtained, it is usually necessary to purify them prior to instrumental analysis, owing to the high complexity of the food samples. As summarised in Figure 4, to date, among the different purification techniques carried out for the determination of PAs, solid phase extraction (SPE) has been widely employed (56% of methods), followed by the QuEChERS approach (22%). In the case of SPE, the purification of PAs from food samples has primarily employed strong cation-exchange (SCX) sorbents, reversed-phase sorbents with octadecylsilane (C18) ligands, and mixed-mode sorbents (MCX), which combine SCX and reversed-phase interactions (see Table 1). Alternatively, the QuEChERS procedure has been applied with some modifications from its original version to the determination of PAs. This clean-up strategy has been mainly used for the analysis of PAs in honey, aromatic herbs, teas and herbal teas, cereals, legumes, vegetables and food supplements (Table 1). This procedure is very suitable to carry out the multiple extraction of analytes of different nature. For this reason, it has generally been used for the simultaneous analysis of PAs with other compounds, such as other plant alkaloids (tropane, isoquinoline, ergot, opium or glycoalkaloids), pesticides, drugs, mycotoxins, etc. (Casado et al. 2022b).

Likewise, other purification methods have also begun to be used in recent years, such as salting-out liquid-liquid extraction (SALLE, 5%) and the improved miniaturized version of the SPE called µSPEed (3%) (Figure 4). Also, other

© Burleigh Dodds Science Publishing Limited, 2025. All rights reserved.

**Table 1** Analytical procedures for the determination of pyrrolizidine alkaloids and their concentration levels found in food samples (2018–2024).

| Food matrix | No. of PAs | Sample preparation | Analysis | LOD/LOQ | Recoveries (%) | Range of PAs found[a] | Ref. |
|---|---|---|---|---|---|---|---|
| | | | **Plant-based food matrices** | | | | |
| | | | **Teas, herbal teas and others (dry product)** | | | | |
| Teas | 28 | SLE with $H_2SO_4$ (0.05 M), pH adjusted to 6.0–7.0 with ammonia solution followed by SPE-C18 | UHPLC-QTRAP-MS/MS ESI positive ion and MRM mode Column: C18 (40°C) | –/0.010–0.76 µg/kg | 50–120% | – | Chung and Lam (2018) |
| Peppermint tea | 25 | SLE with $H_2SO_4$ (0.05 M) followed by SPE-SCX | HPLC-TQ-MS/MS ESI positive ion and MRM mode Column: polar-reversed phase (30°C) | 0.01–1.60/0.03–5.40 µg/kg | 49–121% | – | Kaltner et al. (2018) |
| Teas and herbal teas | 31 | SLE with MeOH containing 0.1% FA followed by SPE with graphitized non-porous carbon as sorbent | UHPLC-TQ-MS/MS ESI positive ion and MRM mode Column: C18 (45°C) | –/0.1–1 ng/g (plant extracts) | 86–125% | 0.01–187151 µg/kg | Picron et al. (2018) |
| Tea leaves | 27 | QuEChERS (clean-up with PSA and ENVI-Carb sorbents) | UHPLC-TQ-MS/MS ESI positive ion and MRM mode Column: C18 (40°C) | 0.03–0.48/0.10–1.61 µg /kg | 88–107% | 0.9–119 µg/kg | Takatsuji et al. (2018) |
| Teas and herbal teas | 44 | SLE with $H_2SO_4$ (0.05 M) followed by SPE-SCX | HPLC-TQ-MS/MS ESI positive ion and MRM mode Column: C18 (30°C) | 0.1–7.0/0.1–27.9 µg/kg | 52–152% | 0.1–47.9 µg/kg | Kaltner et al. (2019) |
| Herbal tea | 33 | SLE with MeOH/$H_2O$/FA (60/39.6/0.4, v/v) followed by dSPE. | UHPLC-QTRAP-MS/MS ESI positive ion and MRM mode Column: C18 (50°C) | –/0.5–10 µg/kg | 83–117% | – | Dzuman et al. (2020) |
| Black tea and Herbal tea | 21 | SLE with 50% MeOH solution with 0.05 M $H_2SO_4$ followed by SPE-MCX | UHPLC-MS/MS ESI positive ion and MRM mode Column: C18 (40°C) | 0.1–3/0.3–9 µg/kg | 86–101% | 0.1–113.4 µg/kg | Kwon (2022) |

© Burleigh Dodds Science Publishing Limited, 2025. All rights reserved.

| | | | | | | | |
|---|---|---|---|---|---|---|---|
| Teas and weeds | 14 | SLE with $H_2SO_4$ (0.1 M) followed by SPE-MCX. | UHPLC-MS/MS ESI positive ion and MRM mode Column: HSS T3 (40°C) | 0.001–0.4/1–5 µg/kg | 68–110% | 1.9–5017.1 µg/kg | Han et al. (2022a) |
| Teas (black and green tea) | 15 | SLE with $H_2SO_4$ (0.1 M) followed by SPE-MCX | UHPLC-QTRAP-MS/MS ESI positive ion and MRM mode Column: HSS T3(40°C) | –/1–5 µg/kg | 67–112% | – | Han et al. (2022b) |
| Tea and herbal tinctures | 30 | QuEChERS (without clean-up step) | Nano-LC-Orbitrap-MS/MS ESI positive ion and HRMS mode Capillary column: C18 (50°C) | –/0.027–11 µg/kg | – | 5.8–215 µg/kg | Jansons et al. (2022) |
| Teas | 17 | SLE with FA in $H_2O$ (2% v/v) and adjusted the pH to 10 with 25% ammonia solution followed by automated SPE. | UPLC-TQ-MS/MS ESI positive ion and MRM mode Column: BEH C18 (40°C) | >2.0/>5.0 µg/kg | 62–147% | 6.7 µg/kg | Keuth et al. (2022) |
| Teas and herbal teas | 28 | QuEChERS (clean-up with $MgSO_4$, PSA, C18 and GCB sorbents) | LC-HRMS H-ESI positive ion and PRM mode Column; C18 (30°C) | –/5 µg/kg | 87–111% | – | León et al. (2022) |
| Scented teas | 10 | SLE with 10% ammonium hydroxide, addition of MeOH:dichloromethane (1:1), evaporation and reconstitution in MeOH: $H_2O$ (1:1), pH adjusted to neutral with FA, followed by SPME-MIP | UHPLC-QTOF-MS/MS ESI positive ion and MRM mode Column: HSS T3 (35°C) | 0.08–0.54/0.26–1.77 µg/L | 91–108% | 0.1–64.5 µg/kg | Luo et al. (2022) |
| Teas | 14 | SLE with $H_2SO_4$ (0.05 M) followed by SPE-MCX | UHPLC-TQ-MS/MS SIM mode Column: XB-C18 (25°C) | – | – | 0.873–498.5 µg/kg | Girard et al. (2023) |
| Tea (*Camellia sinensis* L.) | 15 | SLE with $H_2SO_4$ (0.1 M), two times, followed by clean-up with GCB, PSA and C18 sorbents | UPLC-MS/MS ESI positive ion MRM mode Column: HSS (40°C) | – | – | 8–66440 µg/kg | Jiao et al. (2023) |

*(Continued)*

© Burleigh Dodds Science Publishing Limited, 2025. All rights reserved.

**Table 1** (*Continued*)

| Food matrix | No. of PAs | Sample preparation | Analysis | LOD/LOQ | Recoveries (%) | Range of PAs found[a] | Ref. |
|---|---|---|---|---|---|---|---|
| Dried teas and herbal infusions | 35 | SLE with $H_2SO_4$ (0.05 M) followed by SPE-MCX | UPLC-TQ-MS/MS ESI positive ion and MRM mode Column: C8 | 0.5/1–5 μg/kg | 70–106% | 5–3410 μg/kg | Peloso et al. (2023) |
| Teas and herbal infusions | 118 | SLE with $H_2SO_4$ (0.05 M) followed by SALLE | UHPLC-Q-Orbitrap-MS/MS ESI positive ion and HRMS mode Column: Polar C18 (40°C) | LOD: 0.6–30 μg/kg | 69–113% | – | Rizzo et al. (2023) |
| Fresh tea leaves | 15 | SLE with $H_2SO_4$ (0.1 M) followed by clean-up with GCB, PSA and C18 sorbents | UPLC-MS/MS ESI positive ion and MRM mode Column: HSS (40°C) | –/0.2 μg/kg | 79–109% | 1.87–123.3 μg/kg | Jiao et al. (2024b) |
| *Camellia sinensis L.* | 2 | SLE with $H_2SO_4$ (0.1 M) followed by SPE-PCX | UPLC-MS/MS ESI positive ion and MRM mode Column: HSS (40°C) | –/1–2 μg/kg | 80–110% | – | Lu et al. (2024b) |
| **Liquid infusions and other beverages** | | | | | | | |
| Herbal tea infusions | 38 | Infusion with boiling water followed by SPE-C18 | UHPLC-TQ-MS/MS ESI positive ion and MRM mode Column: C18 (50°C) | 0.2–3.8/– μg/kg | 30–122% | 0.05–0.16 μg/L | Mulder et al. (2018) |
| Tea infusions and ice-tea beverages | 31 | *Infusion extracts*: infusion of herbal teas with boiling water, basification to pH 9.0–10.0 with 28–30% ammonia followed by SPE-C18 *Ice-tea beverages*: basification to pH 9.0–10.0 followed by SPE-C18 | UHPLC-TQ-MS/MS ESI positive ion and MRM mode Column: C18 (45°C) | –/0.01 ng/mL | 86–125% | 0.01–2106 μg/L | Picron et al. (2018) |
| Bottled tea | 27 | QuEChERS (clean-up with PSA and ENVI-Carb sorbents) | UHPLC-TQ-MS/MS ESI positive ion and MRM mode Column: C18 (40°C) | 2–32/2–88 ng/L | 88–107% | – | Takatsuji et al. (2018) |

© Burleigh Dodds Science Publishing Limited, 2025. All rights reserved.

| | | | | | | | |
|---|---|---|---|---|---|---|---|
| Herbal tea infusions | 70 | Infusion with boiling water. | UHPLC-TQ-MS/MS MRM mode Column: C18 (50°C) | 0.01–0.02/ 0.05 μg/L | 73–107% | 30.7–1120 μg/L | Chen et al. (2019) |
| Liqueurs, elixirs and herbal juices | 30 | LLE with $H_2SO_4$ (0.05 M) and purification by SPE-SCX. | HPLC-QTRAP-MS/MS ESI positive ion and MRM mode Column: C18 (25°C) | – | – | 0.21–3121 μg/kg | Chmit et al. (2019) |
| Edible flower infusions | 21 | Infusion with boiling water followed by μSPEed-C18 purification | UHPLC-IT-MS/MS ESI positive ion and MRM mode Column: Polar C18 (25°C) | 0.1–0.3/0.3–1.0 μg/L | 79–97% | 23–41 μg/L | Casado et al. (2022c) |
| Herbal infusions | 21 | Infusion with boiling water followed by μSPEed-C18 purification | UHPLC-IT-MS/MS ESI positive ion and MRM mode Column: Polar C18 (25°C) | –/– | 76–103% | – | Casado et al. (2022d) |
| Herbal beverage | 7 | QuEChERS (clean-up with PSA sorbent) | UHPLC-IM- QTOF-MS/MS ESI positive ion and HDMS$^E$ mode Column: C18 (50°C) | –/1–20 μg/Kg | 61–120% | | Guo et al. (2022) |
| Teas (black and green tea) | 15 | Infusion with boiling water followed by SPE-MCX | UHPLC-QTRAP-MS/MS ESI positive ion and MRM mode Column: HSS T3 (40°C) | –/0.05–0.1 μg/L | 67–112% | – | Han et al. (2022b) |
| Chamomile tea infusion | 21 | Infusion with boiling water followed by μSPEed-C18 purification | UHPLC-IT-MS/MS ESI positive ion and MRM mode Column: Polar C18 (25°C) | –/– | 76–101% | – | Fernández-Pintor et al. (2023) |
| Teas and herbal infusions | 118 | Infusion with boiling water followed by SALLE | UHPLC-Q-Orbitrap-MS/MS ESI positive ion and HRMS mode Column: Polar C18 at 40°C | 0.6–30/– μg/kg | 69–113% | – | Rizzo et al. (2023) |
| Edible flower infusions | 21 | Infusion with boiling water followed by μSPEed-C18 purification | UHPLC-IT-MS/MS ESI positive ion and MRM mode Column: Polar C18 (30°C) | 0.03–0.07/0.09–0.2 μg/L | 87–97% | 0.16–0.2 μg/L | Fernández-Pintor et al. (2024) |

(Continued)

© Burleigh Dodds Science Publishing Limited, 2025. All rights reserved.

**Table 1** (*Continued*)

| Food matrix | No. of PAs | Sample preparation | Analysis | LOD/LOQ | Recoveries (%) | Range of PAs found[a] | Ref. |
|---|---|---|---|---|---|---|---|
| | | | **Herbs and spices** | | | | |
| Spices | 28 | SLE with $H_2SO_4$ (0.05 M), pH adjusted to 6.0–7.0 with ammonia solution followed by SPE-C18 | UHPLC-QTRAP-MS/MS ESI positive ion and MRM mode Column: C18 (40°C) | –/0.010–0.76 µg/kg | 50–120% | – | Chung et al. (2018) |
| Herbs and spices | 31 | SLE with MeOH containing 0.1% FA followed by SPE with graphitized non-porous carbon as sorbent | UHPLC-TQ-MS/MS ESI positive ion and MRM mode Column: C18 (45°C) | –/0.1–1 ng/g (plant extracts) | 86–125% | 0.01–1770 µg/kg | Picron et al. (2018) |
| Herbs | 12 | SLE with 2% FA | HPLC-TQ-MS/MS ESI positive ion and MRM mode Column: C18 | – | – | 11–1777 µg/kg | Selmar et al. (2019) |
| Oregano | 21 | µ-QuEChERS (clean-up with $MgSO_4$ and PSA sorbent) | UHPLC-IT-MS/MS ESI positive ion and MRM mode Column: Polar C18 (25°C) | 0.1–7.5/0.5–25.0 µg/kg | 77–96% | 334–6375 µg/kg | Izcara et al. (2020) |
| Herbs | 30 | QuEChERS (clean-up with graphene sorbent) | HPLC-QTRAP-MS/MS ESI positive ion and MRM mode Column: C18 (40°C) | –/1 µg/kg | 61–128 % | 8–41 µg/kg | Kaczynski (2020) |
| Spices and culinary herbs | 44 | SLE with $H_2SO_4$ (0.05 M) followed by S-SPE-SCX | HPLC-TQ-MS/MS ESI positive ion and MRM mode Column: C18 (30°C) | <0.1–2.6/ µg/kg | 50–119% | 0.1–24600 µg/kg | Kaltner et al. (2020b) |
| Oregano | 33 | SLE with MeOH/$H_2O$/FA (60/39.6/0.4, v/v) followed by dSPE. | UHPLC-QTRAP-MS/MS ESI positive ion and MRM mode Column: C18 (50°C) | –/0.5–10 µg/kg | 80–104% | – | Dzuman et al. (2020) |
| Aromatic herbs | 21 | µ-QuEChERS (clean-up with $MgSO_4$ and LP-MS-$NH_2$ sorbent) | UHPLC-IT-MS/MS ESI positive ion and MRM mode Column: Polar C18 (25°C) | 0.4–3.0/1.2–9.9 µg/kg | 73–105% | 49–553 µg/kg | Izcara et al. (2022) |

© Burleigh Dodds Science Publishing Limited, 2025. All rights reserved.

| | | | | | | | |
|---|---|---|---|---|---|---|---|
| Dried herbs | 35 | SLE with $H_2SO_4$ (0.05 M) followed by SPE-MCX | UPLC-TQ-MS/MS ESI positive ion and MRM mode Column: C8 | 0.5/1–5 µg/kg | 70–106% | 11–4678 µg/kg | Peloso et al. (2023) |
| Spices | 21 | QuEChERS (clean-up with $MgSO_4$ and PSA sorbent) | UHPLC-HRMS HESI positive mode, FS mode Column: C18 (40°C) | 0.2–0.3/0.4–0.8 µg/kg | 77–111% | 4–725.1 µg/kg | Rollo et al. (2023) |
| Oregano | 28 | SLE with $H_2SO_4$ (0.05 M), two times, followed SPE-SCX | UPLC-TQ-MS/MS ESI positive ion and MRM mode Column: C18 | –/10 µg/kg | – | – | Kucukoglu et al. (2024) |
| **Food supplements** | | | | | | | |
| Herbal dietary supplements | 38 | *Food supplements*: SLE with $H_2SO_4$ (0.05 M), pH adjusted to 6.0–7.0 with ammonia solution followed by SPE-C18. *Oily food supplements*: SLE with $H_2SO_4$ (0.05 M) followed by SPE-SCX | UHPLC-TQ-MS/MS ESI positive ion and MRM mode Column: C18 (50°C) | 0.2–3.8/– µg/kg | 30–122% | 0.05–0.16 µg/L/0.10–2410275 µg/kg | Mulder et al. (2018) |
| Herbal dietary supplements | 10 | Extraction with 10% ammonium hydroxide and dichloromethane:MeOH (1:1), three times. Redissolve with MeOH with 0.1% of FA, followed by SPE-BCX | UPLC-Q-ToF-MS ESI positive ion and MRM mode Column: HSS (35°C) | 0.05–0.54/0.15–1.77 µg/kg | 75–110% | 2.97–45.92 µg/kg | Luo et al. (2019a) |
| Herbal dietary supplements | 10 | Extraction with 10% ammonium hydroxide and dichloromethane:MeOH (1:1), three times. Redissolve with MeOH with 0.1% of FA, followed by SPME-MIP | UPLC-Q-ToF-MS ESI positive ion and MRM mode Column: C18 (30°C) | 0.32–0.60/1.08–2.0 ng/g | 89–105% | 17–178 ng/g | Luo et al. (2019b) |

(*Continued*)

© Burleigh Dodds Science Publishing Limited, 2025. All rights reserved.

**Table 1** (*Continued*)

| Food matrix | No. of PAs | Sample preparation | Analysis | LOD/LOQ | Recoveries (%) | Range of PAs found[a] | Ref. |
|---|---|---|---|---|---|---|---|
| Herbal dietary supplements and bee product-based food supplements (royal jelly, propolis syrup and manuka honey syrup) | 44 | SLE with $H_2SO_4$ (0.05 M) followed by SPE-SCX | HPLC-TQ-MS/MS ESI positive ion and MRM mode Column: C18 (30°C) | – | – | 0.1–105.1 ng/g (solid samples) 0.03–2.20 ng/mL (liquid samples) | Kaltner et al. (2020a) |
| Herbal dietary supplements | 28 | SLE with 50% MeOH with 0.05 M $H_2SO_4$. Followed by SPE-MCX | HPLC-MS/MS ESI positive ion and MRM mode Column: C18 (40°C) | 0.03–2.1/0.1–6.5 µg/kg | 67–151 % | 84.7–120.9 µg/kg | Jeong et al. (2021) |
| Herbal dietary supplements | 6 | SLE with 1% FA followed by LLE with dichloromethane | DART-IT-MS Positive ion mode | 0.55–0.85/1.83–2.82 ng/mL | 89–112 % | 2.13–355 µg/kg | Chen et al. (2021) |
| Dietary supplements | 118 | SLE with $H_2SO_4$ (0.05 M) followed by SALLE | UHPLC-Q-Orbitrap-MS/MS ESI positive ion and HRMS mode Column: Polar C18 (40°C) | 0.6–30/– µg/kg | 69–113% | – | Rizzo et al. (2023) |
| Herbal dietary supplements | 30 | QuEChERS (without clean-up step) | HPLC-TQ-MS/MS ESI positive ion and MRM mode Column: C18 (40°C) | –/0.25–500 µg/kg | 86–119% | 0.62–1097 µg/kg | Berzina et al. (2024) |
| Ginseng | 18 | QuEChERS (clean-up with $MgSO_4$ and nano $TiO_2$ sorbent) | HPLC-TQ-MS/MS ESI positive ion and MRM mode Column: C18 (35°C) | 0.05–1.49/0.17–4.95 µg/kg | 71–111% | 0–139 µg/kg | Jiao et al. (2024a) |

© Burleigh Dodds Science Publishing Limited, 2025. All rights reserved.

| | | | | | | | |
|---|---|---|---|---|---|---|---|
| | | | **Cereals, legumes, vegetables and derivates** | | | | |
| Cereal products | 28 | SLE with $H_2SO_4$ (0.05 M), pH adjusted to 6.0-7.0 with ammonia solution followed by SPE-C18 | UHPLC-QTRAP-MS/MS ESI positive ion and MRM mode Column: C18 (40°C) | –/0.010–0.76 μg/kg | 50–120% | – | Chung and Lam (2018) |
| Salads | 31 | SLE with MeOH containing 0.1% FA followed by SPE with graphitized non-porous carbon as sorbent | UHPLC-TQ-MS/MS ESI positive ion and MRM mode Column: C18 (45°C) | –/0.1–1 ng/g (plant extracts) | 86–125% | 0.01–10.47 μg/kg | Picron et al. (2018) |
| Sorghum | 33 | SLE with MeOH/$H_2O$/FA (60/39.6/0.4, v/v) followed by dSPE | UHPLC-QTRAP-MS/MS ESI positive ion and MRM mode Column: C18 (50°C) | –/0.5–10 μg/kg | 87–115% | – | Dzuman et al. (2020) |
| Maize | 1 | SLE with $H_2SO_4$ (0.05 M), filtration followed by SPE-SCX | HPLC-QTRAP-MS/MS ESI positive ion and MRM mode Column: C12 (40°C) | – | – | 0.9–6.6 μg/kg | Letsyo (2021) |
| Pea, soy, wheatflour and quinoa | 56 | QuEChERS (without clean-up step) | UHPLC-Q-Orbitrap-MS/MS ESI positive and negative ion, HRMS mode Column: C18 (50°C) | –/1 μg/kg | 31–132% | – | Bessaire et al. (2021) |
| Cereal, flours and others | 28 | QuEChERS (clean-up with $MgSO_4$, PSA and C18 sorbents) | HPLC-TQ-MS/MS ESI positive ion and MRM mode Column: C18 | 0.02–0.29/0.5–1.0 μg/kg | 72–119% | 0.6–8.0 μg/kg | García-Juan et al. (2023) |
| Fresh borage leaves | 35 | SLE with $H_2SO_4$ (0.05 M) followed by SPE-MCX for the pollen samples | UPLC-TQ-MS/MS ESI positive ion and MRM mode Column: C8 | 0.5/1–5 μg/kg | 70–106% | 5–3410 μg/kg | Peloso et al. (2023) |
| Cereals | 21 | QuEChERS (clean-up with $MgSO_4$ and PSA sorbent) | UHPLC-HRMS HESI positive and FS mode Column: C18 (40°C) | 0.2–0.3/0.4–0.8 μg/kg | 77–111% | – | Rollo et al. (2023) |

(Continued)

© Burleigh Dodds Science Publishing Limited, 2025. All rights reserved.

**Table 1** (*Continued*)

| Food matrix | No. of PAs | Sample preparation | Analysis | LOD/LOQ | Recoveries (%) | Range of PAs found[a] | Ref. |
|---|---|---|---|---|---|---|---|
| Borage | 18 | SLE with $H_2SO_4$ (0.05 M) two times and adjusted the pH to 6–7 followed by SPE-C18 | UHPLC-Q-TOF ESI positive ion and MRM mode Column: C18 (25°C) | – | 85–112% | 19304–1152214 μg/kg | Sattler (2024) |
| Cereal samples and fruit and vegetables purées | 56 | QuEChERS (clean-up with $MgSO_4$ and C18 sorbent) | UHPLC-Q-Orbitrap-MS/MS ESI positive and negative ion, HRMS mode Column: C18 (50°C) | –/1 μg/kg | 4–78% | 2–26 μg/kg | Bessaire et al. (2024) |
| Borago leaves, flowers and microgreens | 17 | SLE with 0.2% FA and 10% MeOH in $H_2O$ | UHPLC-Q-TOF ESI positive ion and MRM mode Column: C18 (25°C) | – | 13–118% | 144–25518 μg/kg | Sattler et al. (2024) |
| Bakery products | 21 | SLE with HCl (1%) in water followed by SPE-MCX | UHPLC-IT-MS/MS ESI positive ion and MRM mode Column: C18 (30°C) | 0.25–1.63/0.84–5.43 μg/kg | 86–95% | 3.6–11.3 μg/kg | Vera-Baquero et al. (2024) |
| | | | **Animal-based food matrices** | | | | |
| | | | **Honey and bee products** | | | | |
| Honey | 28 | LLE with $H_2SO_4$ (0.05 M), pH adjusted to 6.0–7.0 with ammonia solution followed by SPE-C18 | UHPLC-QTRAP-MS/MS ESI positive ion and MRM mode Column: C18 (40°C) | –/0.010–0.76 μg/kg | 50–120% | – | Chung and Lam (2018) |
| Honey | 25 | LLE with $H_2SO_4$ (0.05 M) followed by SPE-SCX | HPLC-TQ-MS/MS ESI positive ion and MRM mode Column: polar-reversed phase (30°C) | 0.01–1.60/0.03–5.40 μg/kg | 49–121% | – | Kaltner et al. (2018) |

© Burleigh Dodds Science Publishing Limited, 2025. All rights reserved.

| | | | | | | | |
|---|---|---|---|---|---|---|---|
| Honey | 4 | LLE with HCl (0.15 M), addition of zinc, filtration through cellulose filters, followed by SPE-MCX | GC-Q-MS EI mode Column: DB-5 MS (nonpolar phenyl arylene polymer) capillary column | –/1 μg/kg | 73–94% | 1.0–64.1 μg/kg | Kowalczyk et al. (2018) |
| Honey | 9 | Dilution with acidified water (0.25 M acetic acid), addition of zinc and pH adjusted to 9.5 followed by DLLME with chloroform and isopropyl alcohol | UHPLC-QTRAP-MS/MS ESI positive ion and MRM mode Column: Polar C18 (30°C) | –/0.03–0.06 μg/kg | 63–103% | 0.2–17.5 μg/kg | Celano et al. (2019) |
| Honey | 7 | QuEChERS (without clean-up step) | HPLC-QTRAP-MS/MS ESI positive ion and MRM mode Column: C18 | –/8–18 μg/kg | 50–100% | 43–75 μg/kg | Sixto et al. (2019) |
| Honey | 12 | LLE with $H_2SO_4$ (0.05 M) followed by SPE-MCX | HPLC-QTOF-MS/MS ESI positive ion mode and HRMS mode Column: C18 at 40°C | 0.2–0.6/0.5–1.3 μg/kg | 79–104% | 1.4–14.2 μg/kg | Wang et al. (2019) |
| Honey | 25 | LLE with $H_2SO_4$ (0.05 M) followed by SPE-SCX | HPLC-TQ-MS/MS ESI positive ion and MRM mode Column: reversed phase (phenyl)-polar (40°C) | 0.01–0.19/0.03–0.59 μg /kg | 82–121% | 2.4–446 μg/kg | Gottschalk et al. (2020) |
| Honey | 17 | LLE with 0.05 M $H_2SO_4$: MeOH (85:15, v/v) followed by SPE-MCX | HPLC-TQ-MS/MS ESI positive ion and MRM mode Column: C18 | – | – | 0.2–281.1 μg/kg | He et al. (2020) |
| Honey | 2 | LLE with $H_2SO_4$ (0.05 M) followed by SPE-SCX | HPLC-DADλ: 223 nm Column: C18 | – | – | – | Moreira et al. (2020) |
| Honey | 26 | LLE with 6.5 mmol/L $NH_4OH$ | UHPLC-IM- QTOF-MS/MS ESI positive ion and HRMS mode Column: C18 (50°C) | 1–7/10–20 μg/kg | 75–120% | 0–141.8 μg/kg | Bandini and Spisso (2021) |

*(Continued)*

© Burleigh Dodds Science Publishing Limited, 2025. All rights reserved.

**Table 1** (*Continued*)

| Food matrix | No. of PAs | Sample preparation | Analysis | LOD/LOQ | Recoveries (%) | Range of PAs found[a] | Ref. |
|---|---|---|---|---|---|---|---|
| Honey and pollen | 8 | LLE with 70% MeOH in $H_2O$ acidified with 2% FA | HPLC-QTRAP-MS/MS ESI positive ion and MRM mode Column: C18 (30°C) | – | – | 0–623 µg/kg (honey) 24.9–221 mg/kg (pollen) | Valese et al. (2021) |
| Honey | 22 | Dilution (20-fold dilution factor) | UHPLC-IM- QTOF-MS/MS ESI positive ion and HRMS mode Column: C18 (50°C) | –/10–20 µg/kg | 71–105% | 10.4–141.8 µg/kg | Bandini and Spisso (2022) |
| Pollen | 44 | SLE with $H_2SO_4$ (0.05 M) followed by a second extraction followed by SPE-C18 | UHPLC-MS/MS ESI positive ion and dMRM mode Column: CSH C18 (50°C) | 0.09–3.6/0.26–7.9 µg/kg | 63–120% | 0.48–48400 ng/g | Friedle et al. (2022) |
| Honey | 7 | QuEChERS (clean-up with PSA sorbent) | UHPLC-IM- QTOF-MS/MS ESI positive ion and HDMS$^E$ mode Column: C18 (50°C) | –/1–20 µg/kg | 61–120% | 200–5500 µg/kg | Guo et al. (2022) |
| Honey | 30 | QuEChERS (without clean-up step) | Nano-LC-Orbitrap-MS/MS ESI positive ion and HRMS mode Capillary column: C18 (50°C) | –/ 0.027–11 µg/kg | – | 0.14–74 µg/kg | Jansons et al. (2022) |
| Honey | 17 | SLE with FA in water (2% v/v) and adjusted the pH to 10 with 25% ammonia solution followed by automated SPE | UPLC-TQ-MS/MS ESI positive ion and MRM mode Column: BEH C18 (40°C) | >0.5 µg/kg/>1.5 µg/kg | 62–147% | 1.2–103 µg/kg | Keuth et al. (2022) |
| Honey and pollen | 30 | SLE with $H_2SO_4$ (0.05 M) in MeOH followed by SPE-MCX | LC-MS/MS ESI positive ion and MRM mode Column C18 (40°C) | – | – | 0.9–72.6 µg/kg | Kwon et al. (2022) |

© Burleigh Dodds Science Publishing Limited, 2025. All rights reserved.

| | | | | | | | |
|---|---|---|---|---|---|---|---|
| Honey and pollen | 9 | Honey: dilution with $H_2O$ followed by SALLE Pollen: SLE with $H_2SO_4$ (0.05 M) followed by SALLE | UHPLC-QTRAP-MS/MS ESI positive ion and MRM mode Column: Polar C18 (40°C) | 0.02–0.07/0.07–0.22 μg/kg | 50–113% | 0–37.3 μg/kg | Rizzo et al. (2022) |
| Honey | 4 | Dilution with water followed by SPE-HNT-$SO_3H$ | UHPLC-TQ-MS/MS ESI positive ion and MRM mode Column: C18 (40°C) | 3.2–4.8/6.7–12.3 μg/L | 55–92% | – | Schlappack et al. (2022a) |
| Honey | 4 | LLE with FA (0.05 M) followed by SPE with alkyl-sulfonated halloysite nanotube sorbents (HNT-Ph$SO_3H$ or HNT-MPTMS-$SO_3H$) | UHPLC-TQ-MS/MS ESI positive ion and MRM mode Column: C18 (50°C) | 0.6–1.2/1.9–3.6 μg/L | 78–101% | – | Schlappack et al. (2022b) |
| Honey | 32 | SLE with $H_2SO_4$ (0.05 M) followed by SPE-MCX | UHPLC-MS/MS ESI positive ion and MRM mode Column: C18 (30°C) | 0.06–0.25/0.22–0.82 μg/kg | 66–91% | 2.2–207.0 μg/kg | Wu et al. (2023) |
| Honey and pollen | 35 | *Honey*: SLE with $H_2SO_4$ (0.1 M) followed by QuEChERS (without clean-up step) *Pollen*: SLE with $H_2SO_4$ (0.05 M) followed by SPE-MCX | UPLC-TQ-MS/MS ESI positive ion and MRM mode Column: C8 | 0.5/1–5 μg/kg | 70–106% | 1–121.1 μg/kg (honey) 6–10168 μg/kg (pollen) | Peloso et al. (2023) |
| Honey and pollen | 118 | *Honey*: dilution with water followed by SALLE *Pollen*: SLE with $H_2SO_4$ (0.05 M) followed by SALLE | UHPLC-Q-Orbitrap-MS/MS ESI positive ion and HRMS mode Column: Polar C18 (40°C) | LOD: 0.6–30 μg/kg | 69–113% | 0–37.3 μg/kg | Rizzo et al. (2023) |
| Honey | 21 | SLE with $H_2SO_4$ (0.05 M) followed by μSPEed-PS/DVB | UHPLC-IT-MS/MS ESI positive ion and MRM mode Column: Polar C18 (30°C) | 0.12–0.3/0.4–1.0 μg/kg | 72–100% | 35–159 μg/kg | Casado et al. (2024) |

(*Continued*)

© Burleigh Dodds Science Publishing Limited, 2025. All rights reserved.

**Table 1** (*Continued*)

| Food matrix | No. of PAs | Sample preparation | Analysis | LOD/LOQ | Recoveries (%) | Range of PAs found[a] | Ref. |
|---|---|---|---|---|---|---|---|
| Honey | 11 | SLE with FA/MeOH followed by QuEChERS clean-up with $MgSO_4$ and PSA sorbent) | UHPLC-Q-TOF ESI positive ion and MRM mode Column EC-C18 (40°C) | 3–14/11–45 μg/kg | 92–102% | 89–8188 μg/kg | Cvetanoska (2024) |
| Honey | 30 | SLE with $H_2SO_4$ (0.05 M) followed by SPE-SCX | UHPLC-MS/MS ESI positive ion and MRM mode Column XB-C18 (5°C) | – | – | 0.6–85 ng/g | Hungerford (2024) |
| Brazilian bee bread and pollen | 8 | SLE with MeOH:$H_2O$ (70:30 v/v) acidified with 2% of FA followed by a dilution 1:1 with mobile phase | LC-MS/MS ESI positive ion and MRM mode Column: C18 (30°C) | 0.1–1.0/0.2–1.5 μg/kg | – | 268–263849 μg/kg | Seraglio et al. (2024) |
| | | | **Milk and other dairy products milk-based** | | | | |
| Milk and dairy products | 38 | LLE with $H_2O$ and hexane with 0.2% FA. The aqueous phase was adjusted to pH 9.0–10.0 with ammonia solution followed by reversed phase SPE | UHPLC-TQ-MS/MS ESI positive ion and MRM mode Column: C18 (50°C) | 0.03–0.05/ μg/L (milk, yoghurt) 0.05–0.15/ μg/kg (cheese) | 30–122% | 0.11–0.17 μg/L | Mulder et al. (2018) |
| Milk | 28 | SLE with $H_2SO_4$ (0.05 M), pH adjusted to 6.0–7.0 with ammonia solution followed by SPE-C18 | UHPLC-QTRAP-MS/MS ESI positive ion and MRM mode Column: C18 (40°C) | –/0.010–0.76 μg/kg | 50–120% | – | Chung et al. (2018) |
| Milk | 6 | LLE with 0.5% FA followed by LLE with dichloromethane | DART-IT-MS Positive ion mode | 0.55–0.85/1.83–2.82 ng/mL | 89–112 % | – | Chen et al. (2021) |
| Milk | 30 | QuEChERS (without clean-up step) | Nano-LC-Orbitrap-MS/MS ESI positive ion and HRMS mode Capillary column: C18 (50°C) | –/0.027–11 μg/kg | – | – | Jansons et al. (2022) |

© Burleigh Dodds Science Publishing Limited, 2025. All rights reserved.

| | | | | | | | |
|---|---|---|---|---|---|---|---|
| Milk | 51 | LLE with 2% FA in $H_2O$ followed by SPE-PCX | HPLC-TQ-MS/MS ESI positive ion mode Column: C18 (30°C) | 0.005–0.054/0.009–0.123 µg/L | 13–157% | 0.008–0.035 µg/L | Klein et al. (2022) |
| Milk | 54 | LLE with 2% FA in $H_2O$ and n-hexane followed by SPE-PCX | LC-MS/MS ESI positive ion and MRM mode Column: C18 (30°C) | 0.005–0.054/0.009–0.123 µg/L | 64–27% | 0.007–5.6 µg/L | Klein et al. (2024) |
| **Eggs** | | | | | | | |
| Eggs | 28 | LLE with $H_2SO_4$ (0.05 M), pH adjusted to 6.0–7.0 with ammonia solution followed by SPE-C18 | UHPLC-QTRAP-MS/MS ESI positive ion and MRM mode Column: C18 (40°C) | –/0.010–0.76 µg/kg | 50–120% | – | Chung and Lam (2018) |
| Eggs | 38 | LLE with $H_2O$ and hexane with 0.2% FA. The aqueous phase was adjusted to pH 9.0–10.0 with ammonia solution followed by reversed phase SPE | UHPLC-TQ-MS/MS ESI positive ion and MRM mode Column: C18 (50°C) | 0.05–0.15/ µg/kg | 30–122% | 0.12 µg/kg | Mulder et al. (2018) |
| | | | **Meat** | | | | |
| Meat | 28 | SLE with $H_2SO_4$ (0.05 M), pH adjusted to 6.0–7.0 with ammonia solution followed by SPE-C18 | UHPLC-QTRAP-MS/MS ESI positive ion and MRM mode Column: C18 (40°C) | –/0.010–0.76 µg/kg | 50–120% | – | Chung and Lam (2018) |
| Meat and meat products | 38 | SLE with $H_2O$ and hexane with 0.2% FA. The aqueous phase was adjusted to pH 9.0–10.0 with ammonia solution followed by reversed phase SPE | UHPLC-TQ-MS/MS ESI positive ion and MRM mode Column: C18 (50°C) | 0.05–0.15/ µg/kg (chicken, pork meat) 0.1–0.25/ µg/kg (beef, liver) | 30–122% | – | Mulder et al. (2018) |

(*Continued*)

© Burleigh Dodds Science Publishing Limited, 2025. All rights reserved.

**Table 1** (*Continued*)

| Food matrix | No. of PAs | Sample preparation | Analysis | LOD/LOQ | Recoveries (%) | Range of PAs found[a] | Ref. |
|---|---|---|---|---|---|---|---|
| Meat and liver | 14 | Lyophilization, SLE with 1% FA in methanol | HPLC-QTRAP-MS/MS ESI positive ion and MRM mode Column: C18 (30°C) | 0.02–0.54/ μg/kg (muscle) 0.06–0.60/ μg/kg (liver) | – | 5.5–84.9 μg/kg (muscle) 10.4–315 μg/kg (liver) | Taenzer et al. (2025) |
| **Edible insects** | | | | | | | |
| Lesser mealworm | 47 | Lyophilization, SLE with 0.2% FA followed by 10-fold dilution with aqueous methanol (10%) | UHPLC-TQ-MS/MS ESI positive ion and MRM mode Column: C18 (50°C) | –/2.5 μg/kg (larvae) | – | 1.8–897 μg/kg | Mulder et al. (2024) |

[a]Expressed as sum.

BCX: Benzenesulfonyl cation exchange; BEH: Ethylene bridged hybrid; C8: Octyl bonded silica; C12: Dodecyl bonded silica; C18: Octadecyl bonded silica; CSH: Charged surface hybrid; DAD: Diode array detection; DART: Direct analysis in real time; DLLME: Dispersive liquid–liquid microextraction; dMRM: Dynamic multiple reaction monitoring mode; dSPE: Dispersive solid-phase extraction; EC: Endcapping; EI: Electron ionization; ESI: Electrospray ionization; FA: Formic acid; GC: Gas chromatography; GCB: Graphitized carbon black; HDMS$^E$: High definition MS$^E$; H-ESI: Heated-electrospray ion source; HNT-$SO_3H$: Sulfonated halloysite nanotubes; HPLC: High-performance liquid chromatography; HRMS: High-resolution mass spectrometry: HSS T3: High-strength silica reversed phase; IM: Ion mobility; IT: Ion-trap; LC: Liquid chromatography; LLE: Liquid–liquid extraction; LOD: Limit of detection; LOQ: Limit of quantification; LP-MS-$NH_2$: large pore mesostructured silica with amino groups; MCX: Mixed-mode sorbent with non-polar and cation exchange properties; MeOH: Methanol; MIP: Molecularly imprinted polymer; MRM: Multiple reaction monitoring; MS/MS: Tandem mass spectrometry; MS: Mass spectrometry; Nano-$TiO_2$: nanotitanium dioxide; PAs: Pyrrolizidine alkaloids; PCX: Polymeric cation Exchange; PRM: Product reaction monitoring; PS/DVB: Polystyrene/divinylbenzene; PSA: Primary secondary amine; Q: Single quadrupole; QTOF: Quadrupole time-of-flight; QTRAP: Hybrid triple quadrupole-linear ion trap; QuEChERS: Quick, easy, cheap, effective, rugged and safe; SALLE: Salting-out assisted liquid–liquid extraction; SCX: Strong cation Exchange; SIM: Selected ion monitoring; SLE: Solid–liquid extraction; SPE: Solid-phase extraction; SPME: Solid-phase microextraction; TQ: Triple quadrupole; UHPLC: Ultra-high performance liquid chromatography.

© Burleigh Dodds Science Publishing Limited, 2025. All rights reserved.

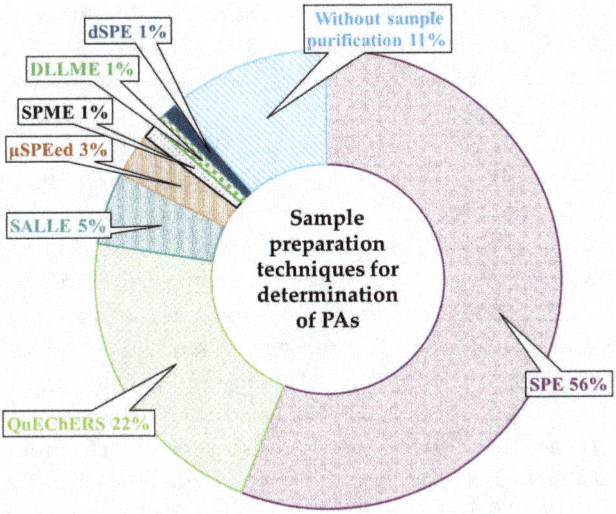


**Figure 4** Overview of the main sample preparation techniques carried out for the determination of pyrrolizidine alkaloids in food over the last 7 years (from 2018 to 2024). Data obtained from Scopus, Web of Science and Google Scholar databases up to October 2024. A total of 103 sample preparation procedures have been considered for this figure. DLLME: dispersive liquid-liquid microextraction; dSPE: dispersive solid-phase extraction, QuEChERS: quick, easy, cheap, effective, rugged and safe; SALLE: salting-out liquidliquid extraction; SPE: solid-phase extraction; SPME: solid-phase microextraction.

miniaturized techniques have been used, such as solid-phase microextraction (SPME) and dispersive liquid–liquid microextraction (DLLME), although they have been used to a lesser extent (1%), like dispersive SPE (dSPE, 1%).

On the other hand, there are some sample preparation procedures proposed that only performed extraction without further purification (11%) (Figure 4). Nonetheless, purification is an important step in the analytical procedure, since it greatly reduces matrix interferences, which improves the sensitivity and selectivity of the method. In addition, it helps to increase the lifetime of the instrumental equipment used for the analysis, as it provides cleaner extracts (Casado et al. 2023a).

For the final analytical step, once the sample extracts have been purified, they are generally analysed by liquid chromatography (LC) or the improved version – ultra-high-performance LC (UHPLC) coupled to mass spectrometry (MS) (Table 1). In fact, regulation (EU) 2023/2783 related to the analysis of PAs in food sets these instrumental techniques as the most suitable to achieve sensitive, selective identification and quantification of these compounds, in addition to gas chromatography coupled to MS (GC-MS). However, GC-MS has been scarcely used because PAs are not volatile, so they need to be subjected to derivatization for their analysis (e.g. with ethyl acetate anhydrous

© Burleigh Dodds Science Publishing Limited, 2025. All rights reserved.

and heptafluorobutyric anhydride) (Kowalczyk 2018). Furthermore, nano-LC coupled with MS, as well as direct analysis in real-time (DART) coupled with MS, have also been employed as analytical techniques for the detection of PAs. (Chen et al. 2021; Jansons et al. 2022).

For the chromatographic separation, C18 stationary phases have mainly been used, followed by polar reversed, HSS T3 (high-strength silica reversed phase) and C12 separation columns, which have been used in the temperature range from 25°C to 50°C (Table 1). Regarding MS detection, electrospray ionization (ESI) in positive ion mode has been commonly used as the ionization source of these compounds. Different MS analysers have been employed, the triple quadrupole (TQ) being the most common. Nonetheless, others such as ion trap (IT), or the combination of quadrupole with ion trap (Q-TRAP or Q-Orbitrap) or time-of-flight (QTOF) have also been explored (Table 1). On the other hand, ultraviolet (UV) detection has also been used (Moreira et al. 2020), but this detection technique is not selective enough as MS because PAs do not have a unique UV spectrum, and they are registered at wavelengths of 200-220 nm, where many compounds exhibit UV absorption. For this reason, MS is the detection technique included and preferred in regulation (EU) 2023/2783.

## *5.1 Analysis of pyrrolizidine alkaloids in teas and herbal teas*

The determination of PAs in (herbal) teas has two different strategies (Table 1). One of them is the analysis of the dry product. Usually in these cases, authors first carry out a SLE with $H_2SO_4$ (0.05–0.1 M) followed by SPE with C18, SCX or MCX sorbents. Alternatively, other works have performed the QuEChERS approach with slight modifications, using different extraction solvents (mainly acidified acetonitrile) and clean-up sorbents, such as PSA, GCB and C18 (Table 1). Dzuman (2020) also tested the dSPE for the simultaneous extraction of PAs and tropane alkaloids from herbal tea samples. The analytes were first extracted from the matrix with a mixture of methanol/water/formic acid (60:39.6:0.4, *v/v*), and the extract obtained was then purified by dSPE using 100 mg of sorbent. On the other hand, Luo et al. (2022) used a miniaturization approach for the selective extraction of 10 PAs from scented teas. For this purpose, they performed SPME with molecularly imprinted fibres modified with cation-exchange active groups.

The other strategy is to directly analyse the prepared infusion (Table 1). This approach is more appropriate to estimate the real exposure of consumers to PAs through these beverages. Otherwise, the estimation of PAs in the dry product can lead to an overestimation of the real intake, since it has been confirmed that the transfer rate of PAs during brewing is not always 100% (Casado et al. 2023b; Fernández-Pintor et al. 2023). In this strategy, once the infusion is prepared in boiling water, it can be directly analysed, or it can be

© Burleigh Dodds Science Publishing Limited, 2025. All rights reserved.

subsequently purified before the chromatographic analysis. In the latter case, the aqueous extracts have been purified by different procedures, such as SPE with MCX sorbents, SALLE and μSPEed (Table 1). The SALLE technique involves a partition equilibrium between two liquid phases (aqueous and organic) assisted by a dehydrating agent that enables the partition of the analytes between them. For instance, Rizzo et al. (2022) used $MgSO_4 \cdot 7H_2O$ and $Na_2SO_4$ as dehydrating agents to extract PAs with acetonitrile from the aqueously prepared infusions. On the other hand, μSPEed is an improved miniaturization of conventional SPE that leads to faster, more efficient and greener extractions at a lower overall cost. This technique has proved its potential for the analysis of PAs in different flower infusions and PA-producing plants using C18 sorbents (Casado et al. 2022c, 2022d; Fernández-Pintor et al. 2023, 2024).

Regarding the levels of PAs found in these works (Table 1), it can be observed that in many cases, the levels estimated in the dry product alarmingly exceed the legislated value for these compounds in the matrices. For instance, Picron et al. (2018), Han et al. (2022a), Jiao et al. (2023) and Peloso et al. (2023) quantified PA levels higher than 5000 μg/kg, when the limit depending on the type of tea or herbal infusion ranges between 150 μg/kg and 400 μg/kg (Commission Regulation (EU) 2023/915). Likewise, Jansons et al. (2022) and Girard et al. (2023) determined significant values that exceeded these limits in some of the samples analysed in their works. Nonetheless, as mentioned above, it is also important to analyse the infusion sample, since the concentration levels can be different from the dry plant because of the transfer rate effectiveness. However, some authors have also quantified high levels of PAs in these beverages (Picron et al. 2018; Chen et al. 2019), but in general, the exposure is lower than if it is estimated from the dry product. Hence, the analysis of the liquid infusion beverage provides a more reliable scenario of consumer exposure to these contaminants. Moreover, sample treatment of the dry product requires more steps and reagents than the analysis of the liquid infusion, which it is simpler (Table 1).

## *5.2 Analysis of pyrrolizidine alkaloids in herbs and spices*

In general, PAs have been extracted from herbs and spices by SLE using aqueous or organic solvents acidified with $H_2SO_4$ or formic acid. Usually, the sample extracts are then purified by SPE using different sorbents (SCX, C18 and graphitized non-porous carbon). However, many other works have also carried out methods based on the QuEChERS approach for the determination of PAs in these products. For this purpose, different clean-up sorbents have been used, including PSA, graphene, C18, zirconia-coated silica and mesostructured silicas (Table 1). Likewise, some authors have performed a miniaturized version of the QueChERS procedure for the determination of PAs in herbs and spices, leading

© Burleigh Dodds Science Publishing Limited, 2025. All rights reserved.

to a more sustainable approach (Izcara et al. 2020, 2022). This modification of the procedure involves reducing up to 10 times the amount of sample, solvents and sorbents. It is worth highlighting that the PA levels quantified in the samples analysed in most of these works (Table 1) significantly exceed the maximum limits established in Commission Regulation (EU) 2023/915 for these products (400–1000 μg/kg, depending on the product). Mainly, these high concentrations (even up to 24600 μg/kg) have been identified in oregano, cumin, marjoram, thyme, parsley, dill, savory and chive (Picron et al. 2018; Selmar et al. 2019; Kaltner 2020b; Izcara et al. 2020, 2022; Peloso et al. 2023; Rollo et al. 2023).

## 5.3 Analysis of pyrrolizidine alkaloids in food supplements

As with other sample matrices, an initial SLE step using $H_2SO_4$ and methanol has been proposed for the determination of PAs in food supplements, followed by the purification of the sample extracts by SPE using C18 or SCX sorbent; or SALLE (Table 1). Some authors have performed the SLE with a basic mixture of dichloromethane/methanol (1/1, *v/v*), followed by SPE, with a cation-exchange sorbent based on benzenesulfonic acids (BCX cartridges) or purification by SPME with molecularly imprinted fibres modified with cation-exchange active groups (Luo et al. 2019a, 2019b). On the other hand, other authors have carried out the QuEChERS strategy for this purpose (Berzina et al. 2024; Jiao et al. 2024a). Nonetheless, it is worth highlighting the method proposed by Chen et al. (2021) using DART-MS without chromatographic separation. First, an SLE with 0.1% formic acid was carried out followed by LLE with dichloromethane. The organic phase was then evaporated and re-dissolved for direct analysis. However, with this instrumental technique, a smaller number of PAs were determined (only 6) in comparison to other methods with chromatographic separation (Mulder et al. 2018; Kaltner et al. 2020a; Jeong et al. 2021; Rizzoet al. 2023; Berzina et al. 2024; Jiao et al. 2024a).

With regards to the concentration levels of PAs found in the food supplement samples (Table 1), generally the values were within the maximum limit established for these products (400 μg/kg, according to Commission Regulation (EU) 2023/915). Nonetheless, it is worth highlighting the data reported in the work of Mulder et al. (2018), in which herbal dietary supplements yielded PA concentrations up to 2410275 μg/kg. It must be noted that some of these supplements analysed were based on material from PA-producing plants. However, in the work of Berzina et al. (2024), some samples without PA-producing plants among their ingredients also exceeded the maximum limit legislated. One of the food supplements based on hawthorn leaves with flowers, dry extract of valerian roots and dry extract of lemon balm leaves

© Burleigh Dodds Science Publishing Limited, 2025. All rights reserved.

showed a total content of 1097 μg/kg, while another supplement marketed to improve liver health containing different herbal products (prickly caper root, chicory seed, black evening primrose seed, terminalia bark, cassia seed, yarrow and gallic tamarind) contained 466 μg/kg of PAs.

## 5.4 Analysis of pyrrolizidine alkaloids in cereals and bakery products

Although no food alerts have been notified regarding high levels of PAs in cereals and bakery products (Figure 3), some works have assessed the determination of PAs in these matrices (Table 1). In many cases, they have been simultaneously analysed with other alkaloids, such as tropane and opium alkaloids (Bessaire et al. 2021; García-Juan et al. 2023; Rollo et al. 2023; Dzuman et al. 2020; Vera-Baquero et al. 2024). For this reason, the QuEChERS strategy has been commonly used for the analysis of PAs in these products, mainly using $MgSO_4$ with PSA and/or C18 as clean-up sorbents, although some authors have avoided the clean-up step. SLE with acidic aqueous solution ($H_2SO_4$ or HCl) followed by SPE with C18, SCX or MCX has also been used as sample preparation procedure in these matrices (Table 1). It should be noted that Commission Regulation (EU) 2023/915 does not set maximum limits established for PAs in these types of products. This is mainly because the occurrence of PAs in these foods do not pose a serious risk to consumers, as seen in Table 1, PAs are found in these products at trace levels or low concentrations (up to 26 μg/kg).

## 5.5 Analysis of pyrrolizidine alkaloids in vegetables

As shown in Figure 3, some food alerts have notified high levels of PAs in certain vegetables. Some analytical methods have been developed for the determination of PAs in these products (Table 1). For instance, PAs were determined in salad samples by using SLE with acidified methanol, followed by SPE with graphitized non-porous carbon as sorbent. Most of the samples showed contamination with PAs, mainly the senecionine-type. However, the levels found did not pose a serious risk to consumers, since the highest concentration was 10.47 μg/kg (Picron et al. 2018). On the other hand, different strategies have been proposed for the analysis of PAs in borage (Peloso et al. 2023; Sattler 2023; Sattler et al. 2024). This is a PA-producing plant, so for this reason some authors have just performed a SLE with acidified water and methanol followed by dilution, owing to the high concentrations of PAs that can be present in this vegetable (Sattler et al. 2024). Other authors have used SLE with $H_2SO_4$ followed by SPE with MCX sorbents (Peloso et al. 2023; Sattler 2023). As expected, high levels of PAs have been detected in borage (Table 1), in some

© Burleigh Dodds Science Publishing Limited, 2025. All rights reserved.

cases exceeding the maximum limit established in Commission Regulation (EU) 2023/915 (750 μg/kg). In fact, Sattler et al. (2024) determined that PA contents are higher in emerging plants and decrease during leaf maturity. So, this should be considered when these plants are intended for human consumption.

## 5.6 Analysis of pyrrolizidine alkaloids in honey and derived products

In the case of this food category, most of the works published mainly addressed the determination of PAs in honey and pollen samples. Only one work analysed other bee-based derived products, such as bee bread (Seraglio et al. 2024). Nonetheless, the sample preparation procedures carried out for the determination of these toxins in these matrices are very similar. For the extraction, the matrices are generally dissolved in an acidic aqueous medium, commonly 0.05 M $H_2SO_4$, since it helps to release the alkaloids from the samples and reduces its viscosity. However, other acids have also been used, such as HCl, formic acid and acetic acid. Likewise, mixtures of acidic aqueous solutions with methanol have been proposed, while other authors have just performed a dilution with water or aqueous ammonium hydroxide solution (Table 1).

Regarding sample purification, SPE has been widely preferred, mainly using SCX and MCX sorbents compared to C18 sorbents. Indeed, some authors have prepared new sorbents with cation-exchange interactions based on sulfonated halloysite nanotubes, and then applied them to the purification and extraction of four PAs from honey samples (Schlappack et al. 2022a, 2022b). On the other hand, modified QuEChERS and SALLE have also been explored as sample preparation procedures for the determination of PAs in honey and pollen samples (Table 1). Some miniaturized procedures have been proposed for the sustainable extraction of PAs from honey samples. For instance, Celano et al. (2019) carried out the DLLME of 9 PAs with chloroform and isopropyl alcohol from an aqueous basic honey solution. Alternatively, μSPEed with a PS/DVB sorbent was used for the simultaneous microextraction of 21 PAs and 2 tropane alkaloids from an aqueous acidic honey solution in only 3 minutes (Casado et al. 2024).

Conversely, in some cases, purification of the honey or pollen sample solutions has been avoided (Bandini and Spisso 2021, 2022; Valese et al. 2021; Seraglio et al. 2024). These non-purified solutions have therefore been injected directly into the instrumental analytical system for their analysis. However, this practice may result in lower sensitivity due to increased matrix interferences, as well as causing high impurities in the ion source and the ion path of the instrument, as mentioned above.

Most of the works published have addressed the determination of PAs in honey rather than in pollen samples (Table 1), yet most of the food alerts

© Burleigh Dodds Science Publishing Limited, 2025. All rights reserved.

notified for high levels of PAs in honey and derived products have been reported in pollen (Rapid Alert System for Food and Feed 2024 as there is a higher concentration. This fact can be confirmed by the data reported in Table 1 relative to the levels of PAs found in these matrices. As it can be observed, PAs have been generally found in a range between 0 and 281 µg/kg in honey, except for a few works where higher concentrations of PAs in this matrix have been quantified (Gottschalk et al. 2020; Valese et al. 2021; Guo et al. 2022; Cvetanoska et al. 2024). For instance, Guo et al. (2022) and Cvetanoska et al. (2024) reported very high concentrations of PAs (up to 5500 and 8188 µg/kg, respectively) in the honey samples analysed. However, it should be noted that maximum concentration limits for PAs in honey are not included in Commission Regulation (EU) 2023/915. In contrast, food supplements containing pollen are limited to a maximum concentration of 500 µg/kg. In this regard, many authors have found very high concentrations of PAs in pollen samples (Table 1), even in the order of mg/kg (Valese et al. 2021; Friedle et al. 2022; Peloso et al. 2023; Seraglio et al. 2024). These results highlight the need for surveillance of products intended for human consumption that contain pollen, owing to their high degree of PA contamination and the risk of exposure they may pose to the population. Other bee-based food supplements, such as royal jelly, propolis syrup and manuka honey syrup have also been analysed (Kaltner et al. 2020a). However, no big concentrations of PAs have been detected in them (Table 1).

## 5.7 Analysis of pyrrolizidine alkaloids in other foods of animal origin (meat, eggs, dairy products)

Although no alerts of PAs in animal-based foods other than honey have been reported in the RASFF system (Figure 3), some studies have investigated the transfer of PAs into such products via feed contaminated with PA-containing plants or plant parts. As a result, sensitive analytical methods capable of detecting trace levels of PAs in these food matrices have been developed (Table 1).

The transfer of PAs into milk, liver, meat and eggs has been well demonstrated. However, detecting PAs in fat-rich foods presents two major challenges during extraction: the high lipid content and the relatively low levels of PAs, necessitating highly sensitive methods. For milk samples, researchers initially performed LLE using weak acidic polar extraction solvents, such as formic acid solution, commonly used in PA analysis, along with *n*-hexane to remove lipids. The aqueous phase was then purified using suitable SPE cartridges, such as C18 or MCX. While MCX sorbents worked directly with 2% formic acid raw extracts without pH adjustment (Klein et al. 2022), using C18 sorbents required a more time-consuming pH adjustment. This process could be expedited by initially using a lower concentration of formic acid (0.2%) (Mulder et al. 2018).

© Burleigh Dodds Science Publishing Limited, 2025. All rights reserved.

The method was successfully applied to Gouda-type cheese and yoghurt (De Nijs et al. 2017).

Instead of using LLE, another strategy to remove lipids from raw extracts was to freeze the samples overnight (Chung and Lam 2018). In all the presented methods, the SPE eluates were dried under nitrogen, reconstituted, and filtered prior to measurement using an MS instrument. An alternative method, which did not require SPE, was applied to analyse PA levels in the muscle meat and liver of dairy cows, goats, and sheep. In this approach, the tissue was minced, lyophilized, milled, extracted, filtered, and then measured (Taenzer et al. 2025).

Insects intended for food or feed are an emerging topic in sustainable food systems. The lesser mealworm (*Alphitobius diaperinus*) is one of four insect species recently deemed safe for human consumption by EFSA (2022), classifying it as food of animal origin. One study explored the potential transfer of PAs from contaminated plant-based substrates to insects. The sample preparation procedure involved a seven-day freeze-drying process, followed by homogenization, extraction, centrifugation and filtration (Mulder et al. 2024).

Regarding the levels of PAs found in animal-based foods, except honey, values are not as high as in plant-based matrices (Table 1), and they do not constitute a concerning source of exposure to these toxins. Nonetheless, it can be noted that higher levels of PAs have been detected in meat products (mainly liver) than in milk and eggs. In contrast, it is worth highlighting, that relevant levels of PAs have been detected in lesser mealworm (up to 897 µg/kg). Therefore, it is important to delve into the analysis of PAs in this food matrix, which has been little explored to date.

## 6 Stability and degradation products of pyrrolizidine alkaloids in food processing

Food processing encompasses a wide range of techniques, varying significantly in complexity and effort – from simple storage of food or food commodities to the production of highly processed foods. During the processing or storage of food, both nutritive ingredients and potential contaminants or residues may undergo reactions that affect their molecular structure. Examples of such reactions include conjugation with other food ingredients, adduct formation, hydrolysis, hydroxylation or demethylation (Bletsou et al. 2015). Currently, little is known about how PAs are affected by these types of reactions, and only a few studies have investigated the impact of processing or storage on PA levels in food (Kaltner et al. 2022).

The issue of PA stability across various food matrices was initially highlighted in honey, where bees can transfer PAs by foraging pollen and nectar from plants containing these compounds. Early studies addressing this question revealed inconsistent results in honey samples over time, particularly

© Burleigh Dodds Science Publishing Limited, 2025. All rights reserved.

owing to decreased PA *N*-oxide levels (Gottschalk et al. 2018). Research on their stability in artificially contaminated honey showed an almost complete reduction (nearly 100%) of spiked PA *N*-oxides within a few days (Kaltner et al. 2018). However, this decrease did not correspond to a proportional increase in the related tertiary amine PAs linked to the diminished *N*-oxides, suggesting that simple chemical reduction was unlikely. The study also demonstrated that the rate of *N*-oxide reduction depended on their structural characteristics.

In contrast, senkirkine, an *N*-methylated tertiary amine PA incapable of forming a corresponding *N*-oxide, remained stable during a six-month storage period. Based on the varying rates of *N*-oxide reduction across the investigated PA groups, bee digestive enzymes were hypothesized to be responsible for the observed losses (Kaltner et al. 2018). Beyond honey, PA *N*-oxide degradation was also noted in bee bread – a honey-based product made from pollen mixed with bee saliva and stored in honeycombs, where it ferments through the action of *Lactobacillus* species. Kast (2019) found that while tertiary amine PAs remained stable in bee bread over a six-month period at 30°C, PA *N*-oxides progressively declined, though they did not completely disappear (Kast 2019). Both studies suggested that bee enzymes might contribute to PA *N*-oxide degradation in honey and bee bread, though no transformation products were identified during the investigations.

PA contamination can typically occur in plant-based foods owing to co-harvesting, affecting products like cereals, spices, culinary herbs and especially herbal teas. While many studies have reported findings on the occurrence of PAs in these matrices, only a limited number have focused on the effects of food processing or storage on PA levels. In one study, peppermint tea artificially contaminated with freeze-dried PA-containing plants was stored for 6 months, demonstrating that PAs are highly persistent in dried plant materials (Kaltner et al. 2018). This finding confirmed earlier results, where PAs were detected in herbarium samples preserved for over a century (Colegate et al. 2014), indicating that these plant toxins remain remarkably stable, even for decades, when stored under dry and proper conditions.

Another study examined the effects of typical microbial sterilization processes on oregano contaminated with PAs. Steam sterilization at 93°C for three minutes converted up to 58% of PA *N*-oxides into their corresponding tertiary amine PAs, while UV or gamma irradiation did not significantly alter PA levels. Ozone treatment of the contaminated oregano reduced PA levels by up to 62%, affecting both tertiary amine PAs and PA *N*-oxides. Some degradation, likely owing to oxidation, was suggested, though no specific degradation products were identified during the study. It was proposed that ozone might attack the 1,2-double bond of PAs, causing oxidative cleavage reactions (Kucukoglu et al. 2024).

© Burleigh Dodds Science Publishing Limited, 2025. All rights reserved.

PAs also seem to withstand very high temperatures typical of various food processes, such as baking. Poisoning case studies have shown that high PA contamination in cereals and flour can still lead to intoxication, even after the contaminated flour has been baked into bread (Mohabbat et al. 1976; Kakar et al. 2010). These findings suggest that standard bread-baking processes and temperatures are insufficient to reliably eliminate PAs, demonstrating their stability even at elevated temperatures. However, slight degradation of certain PAs has been observed during tea preparation. In one study, chamomile tea was spiked with high concentrations of 21 different PAs (ranging from 200 to 10,000 μg/kg of dry material) and then brewed with hot or boiling water (75°C or 100°C). Degradation values ranged from 0% to 38%, with the highest average degradation of 33% reported for europine (Fernández-Pintor et al. 2023).

The transfer of PAs from animal feed into the milk of dairy cows and other milk-producing animals, such as goats or sheep, has been demonstrated. Consequently, the persistence of plant toxins during the processing of contaminated milk has been the focus of some research studies, though only a limited number are available to date. One study examined the effects of Gouda-type cheese production and yogurt fermentation on PA levels. Using typical microorganisms for yogurt fermentation (*Streptococcus thermophiles* and *Lactobacillus delbrueckii* subsp. *bulgaricus*), a 27% reduction in PA levels was observed after six hours of fermentation at 42°C. In cheese, the total PA content decreased by 38% after production and 6 weeks of ripening. It was suggested that microbial activity may contribute to the reduction of PAs in fermented dairy products, although no transformation products were identified (De Nijs et al. 2017).

In summary, the number of studies investigating PA stability in food is limited. A key finding from these studies is that PAs exhibit remarkable persistence during food processing and storage. Based on current knowledge, only fermentative processes, such as yoghurt production, seem to have a significant effect, with the potential to convert PA *N*-oxides to their corresponding tertiary amine PAs to some extent. However, as the total levels of these analytes before and after fermentation were unbalanced, the formation of unknown transformation products cannot be ruled out. In terms of storage, the significant loss of PA *N*-oxides in certain honeys highlights an important area for future research. To date, the exact fate of these toxins, as well as any potential degradation or transformation products, remains unknown and awaits further investigation using advanced analytical strategies (Kaltner 2022). Researching the transformation of PAs into potentially toxic degradation products is of high interest, so that the risk exposure of consumers to these compounds can be more reliably determined. Therefore, it is crucial not only to identify these transformation products but also to assess their toxic

© Burleigh Dodds Science Publishing Limited, 2025. All rights reserved.

properties to ensure a comprehensive risk assessment. However, this research line is very scarce and results are in many cases inconclusive. There are still many challenges to be solved, therefore future studies should focus on PA stability and the identification of potential transformation products to enhance consumer protection against these plant toxins.

## 7 Conclusion and future trends

The ingestion of PAs can result in severe intoxication in both humans and animals. For this reason, these alkaloids are currently in the spotlight of food safety, due to their widespread presence in many foods and animal feeds. Consequently, the development of sensitive analytical methods for reliably detecting PAs in relevant food matrices has become a key focus for researchers worldwide. As a result, the analytical strategies published for the assessment of PAs within the last seven years show that this research topic is rapidly increasing within the scientific community. The analytical approaches for the determination of PAs in different food matrices mainly involve a first LLE or SLE step with acidified aqueous/organic solvent mixtures followed by purification with SPE or QuEChERS and final analysis by UHPLC-MS/MS. Nonetheless, other sample preparation techniques are emerging for the determination of these compounds, such as the application of SALLE or μSPEed. This last technique, together with other micro-extractive techniques (e.g. SPME and DLLME), allows the development of more sustainable and cost-effective methodologies. However, they have scarcely been applied to the determination of PAs in food samples. The analysis of PAs has mainly focused on plant-based matrices, while animal-based foods have been scarcely studied, except for honey. The main concerns involve plant-based products (mainly teas and herbal teas, aromatic herbs and pollen) as that is where there are higher traces of PAs, rather than in animal-based matrices. However, it is also important to evaluate the transfer of these alkaloids from feed to animal-based products and determine the occurrence of PA metabolites in them. The evaluation of PA degradation products resulting from food processing is an important field to explore, so that the actual exposure of consumers to these compounds can be reliably assessed. Also, the current trend of including new insect-based products in the diet may represent a new route of exposure to PAs, which has been practically unconsidered until now. Therefore, these issues constitute different research lines of great interest in the future assessment of PAs.

## 8 References

Bach, N., et al. (1989), 'Comfrey herb tea-induced hepatic veno-occlusive disease', *Am. J. Med.*, 87 (1), 97–99.

© Burleigh Dodds Science Publishing Limited, 2025. All rights reserved.

Bandini, T. B. and Spisso, B. F. (2021), 'Development and validation of an LC-HRMS method for the determination of pyrrolizidine alkaloids and quinolones in honey employing a simple alkaline sample dilution', J. *Food Meas. Charact.*, 15 (5), 4758–4770.

Bandini, T. B. and Spisso, B. F. (2022), 'Detection, dietary exposure assessment and risk evaluation of quinolones and pyrrolizidine alkaloids in commercial honey from Brazil', *Food Addit. Contam. Part B*, 15 (2), 89–97.

Berzina, Z., et al. (2024), 'Mycotoxins and pyrrolizidine alkaloids in herbal dietary supplements', *Food Addit. Contam. Part B*, 17 (2), 180–192.

Bessaire, T., et al. (2021), 'High resolution mass spectrometry workflow for the analysis of food contaminants: Application to plant toxins, mycotoxins and phytoestrogens in plant-based ingredients', *Food Addit. Contam. Part A*, 38 (6), 978–996.

Bessaire, T., et al. (2024), 'Enhanced surveillance of> 1100 pesticides and natural toxins in food: harnessing the capabilities of LC-HRMS for reliable identification and quantification', *Foods*, 13 (19), 3040.

Bletsou, A. A., et al. (2015), 'Targeted and non-targeted liquid chromatography-mass spectrometric workflows for identification of transformation products of emerging pollutants in the aquatic environment', *TrAC*, 66, 32–44.

Casado, N., et al. (2022a), 'The concerning food safety issue of pyrrolizidine alkaloids: An overview', *Trends Food Sci. Technol.*, 120, 123–139.

Casado, N., et al. (2022b), 'Application of the QuEChERS strategy as a useful sample preparation tool for the multiresidue determination of pyrrolizidine alkaloids in food and feed samples: a critical overview', *Appl. Sci.*, 12 (9), 4325.

Casado, N., et al. (2022c), 'Quick and green microextraction of pyrrolizidine alkaloids from infusions of mallow, calendula, and hibiscus flowers using ultrahigh-performance liquid chromatography coupled to tandem mass spectrometry analysis', *J. Agric. and Food Chem.*, 70 (25), 7826–7841.

Casado, N., et al. (2022d), 'ePrep application note: µSPEed extraction and UHPLC-MS/MS analysis of pyrrolizidine alkaloids in herbal infusions (Pub No. 98-35030 Rev 01)'. (https://eprep-analytical.com/wp-content/uploads/2023/05/98-35030-01-%CE%BCSPEed-pyrrolizidine-alkaloids-in-herbals.pdf); accessed October 14th, 2024.

Casado, N., et al. (2023a), 'Green sorbent-based microextraction techniques in sample preparation for food analysis', *Comprehen. Anal. Chem.* (in press). doi:10.1016/bs.coac.2023.09.001

Casado, N., et al. (2023b), 'Insight into the impact of food processing and culinary preparations on the stability and content of plant alkaloids considered as natural food contaminants', *Appl. Sci.*, 13 (3), 1704.

Casado, N., et al. (2024), 'Miniaturized analytical strategy based on µ-SPEed for monitoring the occurrence of pyrrolizidine and tropane alkaloids in honey', *J. Agric. and Food Chem.*, 72 (1), 819–832.

Celano, R., et al. (2019), 'Determination of selected pyrrolizidine alkaloids in honey by dispersive liquid–liquid microextraction and ultrahigh-performance liquid chromatography–tandem mass spectrometry', *J. Agric. and Food Chem.*, 67, 8689–8699.

Chen, L., et al. (2019), 'Risk assessment of intake of pyrrolizidine alkaloids from herbal teas and medicines following realistic exposure scenarios', *Food Chem. Toxicol.*, 130, 142–153.

© Burleigh Dodds Science Publishing Limited, 2025. All rights reserved.

Chen, Y., et al. (2021), 'Rapid identification and determination of pyrrolizidine alkaloids in herbal and food samples via direct analysis in real-time mass spectrometry', *Food Chem.*, 334, 127472.

Chmit, M. S., et al. (2019), 'Quantitative and qualitative analysis of pyrrolizidine alkaloids in liqueurs, elixirs and herbal juices', *Fitoterapia*, 136, 104172.

Chmit, M. S., et al. (2021), 'Pyrrolizidine alkaloids in the food chain: is horizontal transfer of natural products of relevance?', *Foods*, 10 (8), 1827.

Chung, S. W. C. and Lam, C. H. (2018), 'Development of an analytical method for analyzing pyrrolizidine alkaloids in different groups of food by UPLC-MS/MS', *J. Agric. Food Chem.*, 66 (11), 3009–3018.

Codex Alimentarius CXC 74-2014, Code of practice for weed control to prevent and reduce pyrrolizidine alkaloid contamination in food and feed. (https://www.fao.org/fao-who-codexalimentarius/codex-texts/codes-of-practice/en/); accessed October 14th, 2024.

Colegate, S. M., et al. (2014), 'Profiling of dehydropyrrolizidine alkaloids and their N-oxides in herbarium-preserved specimens of amsinckia species using HPLC-esi(+)MS', *J. Agric. Food Chem.*, 62 (30), 7382–7392.

Commission Implementing Regulation (EU) 2023/2783 of 14 December 2023 laying down the methods of sampling and analysis for the control of the levels of plant toxins in food and repealing Regulation (EU) 2015/705. (https://eur-lex.europa.eu/eli/reg_impl/2023/2783); accessed October 14th, 2024.

Commission Regulation (EU) 2020/2040 of 11 December 2020, amending Regulation (EC) No 1881/2006 as regards maximum levels of pyrrolizidine alkaloids in certain foodstuffs. (https://eur-lex.europa.eu/legal-content/EN/TXT/?uri=CELEX%3A32020R2040); accessed October 14th, 2024.

Commission Regulation (EU) 2023/915 of 25 April 2023 on maximum levels for certain contaminants in food and repealing Regulation (EC) No 1881/2006. (https://eur-lex.europa.eu/eli/reg/2023/915/oj); accessed October 14th, 2024.

Cvetanoska, M., et al. (2024), 'UHPLC-Q-TOF analysis of pyrrolizidine alkaloids in North-Macedonian honey', *Food Addit. Contam. Part B*, 17 (1), 5–15.

D'Alessandro, R., et al. (2024), 'The spread of invasive and poisonous plants: a lesson from alkaloids', *Appl. Sci.*, 14 (17), 8058.

De Nijs, M., et al. (2017), 'Fate of pyrrolizidine alkaloids during processing of milk of cows treated with ragwort', *Food Addit. Contam. Part A*, 34 (12), 2212–2219.

Dzuman, Z., et al. (2020), 'Development of a new LC-MS method for accurate and sensitive determination of 33 pyrrolizidine and 21 tropane alkaloids in plant-based food matrices', *Anal. Bioanal. Chem.*, 412, 7155–7167.

EFSA-European Food Safety Authority. (2007), 'Opinion of the Panel on contaminants in the food chain [CONTAM] related to pyrrolizidine alkaloids as undesirable substances in animal feed', *EFSA J.*, 447, 1–51.

EFSA-European Food Safety Authority. (2011), 'Scientific Opinion on Pyrrolizidine alkaloids in food and feed', *EFSA J.*, 9, 1–134.

EFSA-European Food Safety Authority. (2016), 'Dietary exposure assessment to pyrrolizidine alkaloids in the European population', *EFSA J.*, 14, 4572.

EFSA-European Food Safety Authority. (2017), 'Risks for human health related to the presence of pyrrolizidine alkaloids in honey, tea, herbal infusions and food supplements', *EFSA J.*, 15, 4908.

© Burleigh Dodds Science Publishing Limited, 2025. All rights reserved.

EFSA-European Food Safety Authority. (2022), 'Safety of frozen and freeze-dried formulations of the lesser mealworm (Alphitobius diaperinus larva) as a Novel food pursuant to Regulation (EU) 2015/2283', *EFSA J.*, 20 (4), e07325.

Federal Food Safety and Veterinary Office. (https://www.blv.admin.ch/blv/en/home.html); accessed October 16th, 2024.

Fernández-Pintor, B., et al. (2023), 'Evaluation of the thermal stability and transfer rate of pyrrolizidine alkaloids during the brewing of herbal infusions contaminated with Echium vulgare and Senecio vulgaris weeds', *Food Control*, 153, 109926.

Fernández-Pintor, B., et al. (2024), 'Simultaneous determination of 23 pyrrolizidine and tropane alkaloids in infusions from dry edible flowers using optimized µSPEed® microextraction prior to their analysis by UHPLC-IT-MS/MS', *Foods*, 13 (11), 1740.

Food Standards Australia and New Zealand (FSANZ), 'Pyrrolizidine alkaloids in foods'. (https://www.foodstandards.gov.au/consumer/chemicals/Pyrrolizidine-alkaloids-in-foods); accessed October 16th, 2024.

Friedle, C., et al. (2022), 'High abundance of pyrrolizidine alkaloids in bee pollen collected in July 2019 from Southern Germany', *Environ. Monit. Assess.*, 194 (4), 250.

Fu, P. P., et al. (2004), 'Pyrrolizidine alkaloids - genotoxicity, metabolism enzymes, metabolic activation, and mechanisms', *Drug Metab. Rev.*, 36 (1), 1–55.

García-Juan, A., et al. (2023), 'Development and validation of an analytical method for the simultaneous determination of 12 ergot, 2 tropane, and 28 pyrrolizidine alkaloids in cereal-based food by LC-MS/MS', *Food Res. Int.*, 174, 113614.

Girard, M. F. C., et al. (2023), 'Vacuum differential mobility spectrometry combined with column-switching liquid chromatography-mass spectrometry for the analysis of pyrrolizidine alkaloids in tea samples', *J. Chromatogr. A*, 1705, 464174.

Gottschalk, C., et al. (2018), 'Uncertainties in the determination of pyrrolizidine alkaloid levels in naturally contaminated honeys and comparison of results obtained by different analytical approaches', *Food Addit. Contam. Part A*, 35 (7), 1366–1383.

Gottschalk, C., et al. (2020), 'Spread of Jacobaea vulgaris and occurrence of pyrrolizidine alkaloids in regionally produced honeys from Northern Germany: inter-and intra-site variations and risk assessment for special consumer groups', *Toxins*, 12 (7), 441.

Guo, Q., et al. (2022), 'Screening for plant toxins in honey and herbal beverage by ultrahigh-performance liquid chromatography-ion mobility-quadrupole time of flight mass spectrometry', *Am. J. Anal. Chem.*, 13 (3), 108–134.

Han, H., et al. (2022a), 'Development, optimization, validation and application of ultra high performance liquid chromatography tandem mass spectrometry for the analysis of pyrrolizidine alkaloids and pyrrolizidine alkaloid N-oxides in teas and weeds', *Food Control*, 132, 108518.

Han, H., et al. (2022b), 'Dissipation pattern and conversion of pyrrolizidine alkaloids (PAs) and pyrrolizidine alkaloid N-oxides (PANOs) during tea manufacturing and brewing', *Food Chem.*, 390, 133183.

He, Y., et al. (2020), 'Comprehensive investigation and risk study on pyrrolizidine alkaloid contamination in Chinese retail Honey', *Environ. Pollut.*, 267, 115542.

Hungerford, N. L., et al. (2024), 'Analysis of Pyrrolizidine Alkaloids in Stingless Bee Honey and Identification of a Botanical Source as Ageratum conyzoides', *Toxins*, 16 (1), 40.

Izcara, S., et al. (2020), 'A miniaturized QuEChERS method combined with ultrahigh liquid chromatography coupled to tandem mass spectrometry for the analysis of pyrrolizidine alkaloids in oregano samples', *Foods*, 9, 1319.


© Burleigh Dodds Science Publishing Limited, 2025. All rights reserved.

Izcara, S., et al. (2022), 'Miniaturized and modified QuEChERS method with mesostructured silica as clean-up sorbent for pyrrolizidine alkaloids determination in aromatic herbs', *Food Chem.*, 380, 132189.

Jansons, M., et al. (2022), 'Nanoflow liquid chromatography mass spectrometry method for quantitative analysis and target ion screening of pyrrolizidine alkaloids in honey, tea, herbal tinctures, and milk', *J. Chromatogr. A*, 1676, 463269.

Jeong, S. H., et al. (2021), 'LC-ESI-MS/MS simultaneous analysis method coupled with cation-exchange solid-phase extraction for determination of pyrrolizidine alkaloids on five kinds of herbal medicines', *J. AOAC Int.*, 104 (6), 1514–1525.

Jiao, W., et al. (2022). "Source and route of pyrrolizidine alkaloid contamination in tea simples", *J. Vis. Exp. JoVE*, 187, e64375.

Jiao, W., et al. (2023), 'Pyrrolizidine alkaloids in tea (Camellia sinensis L.) from weeds through weed–soil–tea transfer and risk assessment of tea intake', *J. Agric. Food Chem.*, 71 (48), 19045–19053.

Jiao, R., et al. (2024a), 'The single adsorbent nano-TiO2 modified QuEChERS for simultaneous determination of 18 pyrrolizidine alkaloids in ginseng samples via LC-MS/MS', *Microchem. J.*, 203, 110804.

Jiao, W., et al. (2024b), 'Simultaneous determination of 15 pyrrolizidine alkaloids and their N-oxides in weeds, soil, fresh tea leaves, and tea: exploring the pollution source of pyrrolizidine alkaloids in tea', *Food Chem.*, 434, 137305.

Kaczynski, P. and Łozowicka, B. (2020), 'A novel approach for fast and simple determination pyrrolizidine alkaloids in herbs by ultrasound-assisted dispersive solid phase extraction method coupled to liquid chromatography–tandem mass spectrometry', *J. Pharm. Biomed. Anal.*, 187, 113351.

Kakar, F., et al. (2010), 'An outbreak of hepatic veno-occlusive disease in Western Afghanistan associated with exposure to wheat flour contaminated with pyrrolizidine alkaloids', *J. Toxicol.*, 2010, 313280.

Kaltner, F., et al. (2018), 'Influence of storage on the stability of toxic pyrrolizidine alkaloids and their N-oxides in peppermint tea, hay, and honey', *J. Agric. Food Chem.*, 66 (20), 5221–5228.

Kaltner, F., et al. (2019), 'Development of a sensitive analytical method for determining 44 pyrrolizidine alkaloids in teas and herbal teas via LC-ESI-MS/MS', *Anal. Bioanal. Chem.*, 411, 7233–7249.

Kaltner, F., et al. (2020a), 'Screening of food supplements for toxic pyrrolizidine alkaloids', *J. Consum. Protect. Food Safety*, 15 (3), 237–243.

Kaltner, F., et al. (2020b), 'Occurrence and risk assessment of pyrrolizidine alkaloids in spices and culinary herbs from various geographical origins', *Toxins*, 12, 155.

Kaltner, F. (2022), 'Fate of food-relevant toxic plant alkaloids during food processing or storing and analytical strategies to unveil potential transformation products', *J. Agric. Food Chem.*, 70 (20), 5975–5981.

Kast, C.,et al. (2019), 'Pyrrolizidine alkaloids: The botanical origin of pollen collected during the flowering period of Echium vulgare and the stability of pyrrolizidine alkaloids in bee bread', *Molecules*, 24 ( 12), 2214.

Keuth, O., et al. (2022), 'Determination of pyrrolizidine alkaloids in tea and honey with automated SPE clean-up and ultra-performance liquid chromatography/tandem mass spectrometry', *Food Addit. Contam. Part A*, 39 (1), 149–157.


© Burleigh Dodds Science Publishing Limited, 2025. All rights reserved.

Klein, L. M., et al. (2022), 'A sensitive LC–MS/MS method for isomer separation and quantitative determination of 51 pyrrolizidine alkaloids and two tropane alkaloids in cow's milk', *Anal. Bioanal. Chem.*, 414 (28), 8107–8124.

Klein, L. M., et al. (2024), 'Pyrrolizidine alkaloids and tropane alkaloids in milk samples from individual dairy farms of the German federal states of Bavaria and Schleswig-Holstein', *Food Addit. Contam. Part A*, 41 (6), 629–647.

Kowalczyk, E., et al. (2018), 'Determination of pyrrolizidine alkaloids in honey with sensitive gas chromatography-mass spectrometry method', *Food Anal. Methods*, 11, 1345–1355.

Kucukoglu, A. S., et al. (2024), 'Effects of thermal and nonthermal treatments on microorganisms, pyrrolizidine alkaloids and volatile compounds in oregano (Origanum vulgare L.)', *Food Chem.*, 440, 138235.

Kwon, Y., et al. (2022), 'Determination of pyrrolizidine alkaloids in teas using liquid chromatography–tandem mass spectrometry combined with rapid-easy extraction', *Foods*, 10 (10), 2250.

León, N., et al. (2022), 'A green analytical method for the simultaneous determination of 30 tropane and pyrrolizidine alkaloids and their N-oxides in teas and herbs for infusions by LC-Q-Orbitrap HRMS', *J. Chromatogr. A*, 1666, 462835.

Letsyo, E., et al. (2021), 'Uptake and accumulation of pyrrolizidine alkaloids in the tissues of maize (Zea mays L.) plants from the soil of a 4-year-old Chromolaena odorata dominated fallow farmland', *Chemosphere*, 270, 128669.

Letsyo, E., et al. (2023), 'Hepatic phytotoxins in herbal medicines: a review of Africa's pyrrolizidine alkaloid-containing plants, their traditional uses, contamination pathways and associated health risks', *South Afric. J. Bot.*, 161, 78–89.

Lis-Cieplak, A., et al. (2024), 'Pyrrolizidine alkaloids as hazardous toxins in natural products: current analytical methods and latest legal regulations', *Molecules*, 29 (14), 3269.

Lu, Y. S., et al. (2024a), 'Levels, toxic effects, and risk assessment of pyrrolizidine alkaloids in foods: a review', *Foods*, 13 (4), 536.

Lu, Y. S., et al. (2024b), 'Uptake, accumulation, translocation and transformation of seneciphylline (Sp) and seneciphylline-N-oxide (SpNO) by Camellia sinensis L.', *Environ. Int.*, 188, 108765.

Luo, Z., et al. (2019a), 'Development of UPLC-Q-TOF-MS coupled with cation-exchange solid-phase extraction method for the determination of ten pyrrolizidine alkaloids in herbal medicines', *Anal. Sci.*, 35 (12), 1317–1325.

Luo, Z., et al. (2019b), 'Molecularly imprinted polymer solid-phase microextraction coupled with ultra high performance liquid chromatography and tandem mass spectrometry for rapid analysis of pyrrolizidine alkaloids in herbal medicine', *J. Sep. Sci.*, 42 (21), 3352–3362.

Luo, Z., et al. (2022), 'Multi-template imprinted solid-phase microextraction coupled with UPLC-Q-TOF-MS for simultaneous monitoring of ten hepatotoxic pyrrolizidine alkaloids in scented tea', *Front. Chem.*, 10, 1048467.

Maquet, A., et al. (2021), *Results of an EU wide coordinated control plan to establish the prevalence of fraudulent practices in the marketing of herbs and spices, EUR 30877 EN*, Luxembourg: Publications Office of the European Union, 978-92-76-42979-1.

Miceli, A., et al. (2015), 'The influence of addition of Borago officinalis with antibacterial activity on the sensory quality of fresh pasta', *Int. J. Gastro. Food Sci.*, 2 (2), 93–97.

Mohabbat, O., et al. (1976), 'An outbreak of hepatic veno-occlusive disease in North-Western Afghanistan', *The Lancet*, 308 (7980), 269–271.

© Burleigh Dodds Science Publishing Limited, 2025. All rights reserved.

Moreira, R., et al. (2020), 'Echium plantagineum L. honey: search of pyrrolizidine alkaloids and polyphenols, antiinflammatory potential and cytotoxicity', *Food Chem.*, 328, 127169.

Mulder, P. P. J., et al. (2018), 'Occurrence of pyrrolizidine alkaloids in animal-and plant-derived food: results of a survey across Europe', *Food Addit. Contam. Part A*, 35, 118–133.

Mulder, P. P. J., et al. (2024), 'Effects of dietary exposure to plant toxins on bioaccumulation, survival, and growth of black soldier fly (*Hermetia illucens*) larvae and lesser mealworm (*Alphitobius diaperinus*) ', *Heliyon*, 10 (4), e26523.

NCI. (1978), *Bioassay of lasiocarpine for possible carcinogenicity*, National Cancer Institute carcinogenesis technical report series, 39, pp. 1–66.

Nowak, M., et al. (2016), 'Interspecific transfer of pyrrolizidine alkaloids: an unconsidered source of contaminations of phytopharmaceuticals and plant derived commodities', *Food Chem.*, 213, 163–168.

Nowak, M., et al. (2017), 'Horizontal natural product transfer: a so far unconsidered source of contamination of medicinal plants', *Med. Plants Environ. Challenges*, 215–225. doi:10.1007/978-3-319-68717-9_12

NTP. (2003), *Toxicology and carcinogenesis studies of riddelliine*, National Toxicology Program Technical Report, 508.

Peloso, M., et al. (2023), 'Pyrrolizidine alkaloids in food on the Italian market', *Molecules*, 28 (14), 5346.

Picron, J. F., et al. (2018), 'Analytical strategies for the determination of pyrrolizidine alkaloids in plant based food and examination of the transfer rate during the infusion process', *Food Chem.*, 266, 514–523.

Rapid Alert System for Food and Feed. (2024), 'Rapid alert system for food and feed (RASFF) window'. (https://webgate.ec.europa.eu/rasff-window/screen/search); accessed October 16th, 2024.

Rizzo, S., et al. (2022), 'Salting-out assisted liquid-liquid extraction for the rapid and simple simultaneous analysis of pyrrolizidine alkaloids and related N-oxides in honey and pollen', *J. Food Comp. Anal.*, 108, 104457.

Rizzo, S., et al. (2023), 'Target screening method for the quantitative determination of 118 pyrrolizidine alkaloids in food supplements, herbal infusions, honey and teas by liquid chromatography coupled to quadrupole orbitrap mass spectrometry', *Food Chem.*, 423, 136306.

Rollo, E., et al. (2023), 'QuEChERS method combined to liquid chromatography high-resolution mass spectrometry for the accurate and sensitive simultaneous determination of pyrrolizidine and tropane alkaloids in cereals and spices', *J. Mass Spectr.*, 58 (10), e4969.

Sattler, M., et al. (2024), 'Pyrrolizidine alkaloid contents and profiles in Borago officinalis leaves, flowers and microgreens: Implications for safety', *Food Control*, 168, 110930.

Schlappack, T., et al. (2022a), 'Effective solid phase extraction of toxic pyrrolizidine alkaloids from honey with reusable organosilyl-sulfonated halloysite nanotubes', *Separations*, 9 (10), 270.

Schlappack, T., et al. (2022b), 'Sulfonated halloysite nanotubes as a novel cation exchange material for solid phase extraction of toxic pyrrolizidine alkaloids', *Anal. Met.*, 14 (27), 2689–2697.


© Burleigh Dodds Science Publishing Limited, 2025. All rights reserved.

Selmar, D., et al. (2019), 'Transfer of pyrrolizidine alkaloids between living plants: a disregarded source of contaminations', *Environ. Pollut.*, 248, 456–461.

Seraglio, S. K. T., et al. (2024), 'LC–MS/MS analysis of pyrrolizidine alkaloids in bee bread and commercial pollen from Brazil', *Eur. Food Res. Technol.*, 6, e03593.

Sixto, A., et al. (2019), 'Straightforward determination of pyrrolizidine alkaloids in honey through simplified methanol extraction (QuPPE) and LC-MS/MS modes', *ACS Omega*, 4 (27), 22632–22637.

Taenzer, J., et al. (2025), 'Selective and low transfer of pyrrolizidine alkaloids from Jacobaea vulgaris Gaertn. into muscle and liver of dairy cattle, goat and sheep', *Food Control*, 167, 110766.

Takatsuji, Y., et al. (2018), 'A novel method for the detection of pyrrolizidine alkaloids in bottled tea and tea leaves by LC-MS/MS', *Jap. J. Food Chem. Saf.*, 25 (2), 97–104.

Tea & Herbal Infusions Europe. European Tea Committee (ETC) and European Herbal Infusions Association (EHIA), (2018), "Code of practice to prevent and reduce pyrrolizidine alkaloid contamination in raw materials for tea and herbal infusions". (https://thie-online.eu/files/thie/docs/2018-07-12_THIE_Code_of_Practice_PA_in_TEA-HFI_ISSUE_1.pdf); accessed October 16th, 2024.

Toppel, G., et al. (1988), 'N-Oxidation and degradation of pyrrolizidine alkaloids during germination of Crotalaria scassellatii', *Phytochem.*, 27 (12), 3757–3760.

Valese, A. C., et al. (2021), 'Quantification of pyrrolizidine alkaloids in Senecio brasiliensis, beehive pollen, and honey by LC-MS/MS', *J. Environ. Sci. Health Part B*, 56 (7), 685–694.

Vera-Baquero, F. L., et al. (2024), 'Improving the food safety of bakery products by simultaneously monitoring the occurrence of pyrrolizidine, tropane and opium alkaloids', *Food Chem.*, 460, 140769.

Vogl, S., et al. (2013), 'Ethnopharmacological in vitro studies on Austria's folk medicine–An unexplored lore in vitro anti-inflammatory activities of 71 Austrian traditional herbal drugs', *J. ethnopharm.*, 149 (3), 750–771.

Wang, T., et al. (2019), 'Pyrrolizidine alkaloids in honey: Quantification with and without standards', *Food Control*, 98, 227–237.

Wang, Y.-P., et al. (2005), 'Human liver microsomal reduction of pyrrolizidine alkaloid N-oxides to form the corresponding carcinogenic parent alkaloid', *Toxicol. Letters*, 155 (3), 411–420.

Wu, H., et al. (2023), 'Development and validation of an UHPLC–Ms/Ms method for the determination of 32 pyrrolizidine alkaloids in Chinese wild honey', *J. AOAC Int.*, 106 (1), 56–64.

© Burleigh Dodds Science Publishing Limited, 2025. All rights reserved.

# Chapter 18

# Tropane alkaloids in food

*Inês Parente, University of Coimbra, Portugal; Ana Rita Soares Mateus, University of Coimbra, Portugal/National Institute for Agricultural and Veterinary Research (INIAV), Portugal/Centre for Animal Science Studies (CECA), ICETA, Portugal; Sílvia Cruz Barros, National Institute for Agricultural and Veterinary Research (INIAV), Portugal; Angelina Pena, University of Coimbra, Portugal; and Ana Sanches Silva, University of Coimbra, Portugal/Centre for Animal Science Studies (CECA), ICETA, Portugal/Associate Laboratory for Animal and Veterinary Sciences (AL4AnimalS), Portugal*



## 1 Introduction

Tropane alkaloids (TAs) are naturally occurring compounds present in numerous species of plants, being the most prevalent in the *Solanaceae* family (González-Gómez et al., 2022a). Over 200 types of TAs are known. Still, the most researched ones are atropine and scopolamine due to their pharmacological importance through their non-selective inhibition of the muscarinic acetylcholine receptors (Kohnen-Johannsen and Kayser, 2019). Moreover, due to its pharmaceutical status and economic interest, the global demand for these TAs has skyrocketed, either through the cultivation of TA-containing plants (such as Australian *Duboisia*), through the synthetic transformation of an intermediate compound (tropinone), or through biotechnological approaches (including plant cell cultures and microbial production) (Kohnen-Johannsen and Kayser, 2019; Srinivasan and Smolke, 2021).

TAs are highly potent substances that are simple to distribute globally, which has contributed to their use in traditional medicine for centuries.

http://dx.doi.org/10.19103/AS.2025.0158.35
© Burleigh Dodds Science Publishing Limited, 2025. All rights reserved.

Unfortunately, plants that carry atropine and scopolamine can easily contaminate crops of different cereals like buckwheat, millet, and sorghum, they can also contaminate herbal teas and infusions and honey, causing their accidental ingestion (Kohnen-Johannsen and Kayser, 2019).

In recent years, the cases of TAs contamination and consequently health risks have increased significantly, forcing the EU to take action and establish a maximum level allowed for atropine and scopolamine for specific foodstuffs, to control the global trade of the foods containing these compounds.

Several extraction and determination methods have been used over the years. The most frequently used is liquid chromatography (LC), more specifically high-performance LC (HPLC) coupled with mass spectrometry (MS). Furthermore, recent techniques such as ultra-high-performance LC (UHPLC) and high-resolution MS (HRMS) have had an increase in interest.

The present chapter evaluates the presence of atropine and scopolamine in food and the main methods used for their extraction, detection, and quantification, as well as the hazards they may cause to the health of the population. Decontamination methods of TAs in food will also be addressed.

## 2 Tropane alkaloids

TAs are naturally produced secondary metabolites in various plant families, like *Brassicaceae* (mustard family), *Solanaceae* (nightshade or potato family), *Erythroxylaceae* (coca family), *Convolvulaceae* (the bindweed or morning glory family), *Euphorbiaceae* (spurge family), *Proteaceae,* and *Rhizophoraceae* (mangrove family).

Though TAs are found more predominately in *Solanaceae*, especially in plants like *Atropa belladonna*, commonly designated as simply belladonna or deadly nightshade, and *Datura stramonium* known as jimsonweed (de Nijs et al., 2023; EFSA CONTAM Panel, 2013). *D. stramonium* is an invasive weed across temperate and tropical regions in the world, but they have spread in European countries due to climate change. This plant can grow between 50 cm and 100 cm high, with dark green leaves that can vary in ovate sizes and produce singular white bell trumpet-shaped flowers. The fruits from the flowers are green spinous capsules that are densely populated with small black/brown seeds, as showed in Fig. 1 (Abia et al., 2021).

TAs can be distributed over all parts of the plant, including seeds, leaves, flowers, pericarp, and roots. TAs are synthesized in the roots and subsequently transported to the aerial parts of the plant via the xylem, ensuring their distribution throughout the organism (Kohnen-Johannsen and Kayser, 2019). Nevertheless, seeds have the highest concentration of TAs, and they are usually the preferred plant tissue to study (FAO/WHO, 2020; Marín-Sáez et al., 2017).

© Burleigh Dodds Science Publishing Limited, 2025. All rights reserved.

**Figure 1** Plants from Solanaceae family: *Datura stramonium* and *Atropa belladonna*.

## *2.1 Atropine*

Atropine is widely used in medicine mostly as an injectable solution, having other dosage forms such as tablets, ophthalmic solution, and ophthalmic ointment. The World Health Organization (WHO) named atropine as one of the Essential Medicines.

This TA is used as preoperative medication and sedation for short-term procedures since it has a pre-anesthetic effect. It is used as an antidote for intoxication with anticholinesterases (e.g. organophosphate insecticides) and the treatment of bradycardia. Also, it is used in ophthalmology as a diagnostic agent and for mydriatic treatment due to their ability to dilate pupils. Additionally, homatropine has been used in ophthalmology to produce a less paralyzing and quicker acting action compared to atropine (Kohnen-Johannsen and Kayser, 2019). Atropine, as a medical drug, can be used in different doses depending on the intended effect.

Atropine is the racemic mixture of (−)-hyoscyamine and (+)-hyoscyamine, of which only the enantiomer (−)-hyoscyamine is pharmacologically active (Cirlini et al., 2018). In addition, the analytical procedures to determine TAs cannot distinguish between the two enantiomers; thus, the racemic mixture, called atropine, is the preferable compound used in research studies (Castilla-Fernández et al., 2021).

© Burleigh Dodds Science Publishing Limited, 2025. All rights reserved.

## 2.2 Scopolamine

Scopolamine is an important drug for the treatment of nausea and vomiting associated with anesthesia. This TA was also used as a transdermal system for the treatment of motion sickness, but it also shows hallucinogenic side effects and even death at higher doses (Lusthof et al., 2017).

Furthermore, scopolamine is formulated as N-butylscopolamine (hyoscine butylbromide) and marketed as Buscopan® (Kohnen-Johannsen and Kayser, 2019). This drug has been marketed for decades, and it is used for the relief of abdominal pain and/or discomfort associated with spasms of the gastrointestinal tract. Scopolamine is administered as an N-butyl derivate because, despite its anticholinergic and antispasmodic effects, scopolamine also has a potent sedative side effect, being a much safer option (Ribeiro et al., 2022). Similar to atropine, butylscopolamine is available as an injectable solution but also as a tablet.

## 2.3 Key factors contributing to food contamination by tropane alkaloids

Even though the intentional ingestion of TAs has happened for centuries through the use of parts of plants for medicinal or recreational purposes at low doses, that is not what most concerns the international authorities (Kohnen-Johannsen and Kayser, 2019; Marín-Sáez et al., 2017). The real danger is when, for unintentional reasons, TAs make their way into our food chain.

TA-producing plants are distributed throughout tropical and temperate climates, being indigenous to North and South America, Europe, Africa, and Asia (EFSA CONTAM Panel, 2013; Kowalczyk and Kwiatek, 2022).

As mentioned above, *D. stramonium* and *A. belladonna* are well known for their content in TAs, with *D. stramonium* currently being the primary source of TA contamination (Boros et al., 2010). But how does the contamination happen? Well, general belief says that certain parts of this plant (usually seeds, due to their similar size to cereals) are accidentally mixed with crops like millet, buckwheat, linseed, soybean, sorghum, oats, etc. during harvest or processing (Fig. 2). Consequently, foodstuffs mainly made with these cereals are showing considerable levels of TAs in foods like pasta, bread, breakfast cereals, biscuits, and baby food. Nevertheless, herbal infusions, vegetables, and honey have also registered, to a lower extent, considerable levels of TAs (Baslé et al., 2020; Chan, 2017; Cirlini et al., 2018; EFSA CONTAM Panel, 2013; Kowalczyk and Kwiatek, 2022; Marín-Sáez et al., 2017).

Jimsonweed *(D. stramonium)*, as the name suggests, is a weed, meaning it is a species that thrives at high temperatures and grows incredibly fast, having the ability to negatively impact the habitat where it settles. Climate change

© Burleigh Dodds Science Publishing Limited, 2025. All rights reserved.

**Figure 2** Contamination of buckwheat with tropane alkaloids from *Datura* seeds.

influences the vulnerability of ecosystems, making them more susceptible to the growth of these invasive plants. Due to the increase in global temperature, $CO_2$ concentration in the atmosphere, and changes in precipitation patterns, the growth of Jimsonweed may increase severely, leading to the invasion of new crops and an intensification of TA contamination (Jank and Rath, 2021; Kowalczyk and Kwiatek, 2022).

Moreover, celiac disease is a serious health condition that requires caution with the food the patient ingests. For this reason, the search for gluten-free options in supermarkets has intensified, leading to an increased number of buckwheat-based products available for consumption. This cereal is free of gluten and sold not only as flour but also in processed foods such as pasta, bread, cakes, biscuits, and baby food. Therefore, a lot more crops of buckwheat have been cultivated to keep up with the demand. Attention is required in the monitoring of these fields to prevent unwanted contamination with TAs (Cirlini et al., 2018).

## 2.4 Toxicity and mechanisms of action of tropane alkaloids

Numerous classes of alkaloids exist; the one that possesses a tropane skeleton is called TA (Fig. 3). Besides this structure, TAs also share the fact that they are derived from tropine and organic acids (Castilla-Fernández et al., 2021; Kohnen-Johannsen and Kayser, 2019).

Atropine and scopolamine have a high level of penetration into the central nervous system and act as non-selective inhibitors of the muscarinic

© Burleigh Dodds Science Publishing Limited, 2025. All rights reserved.

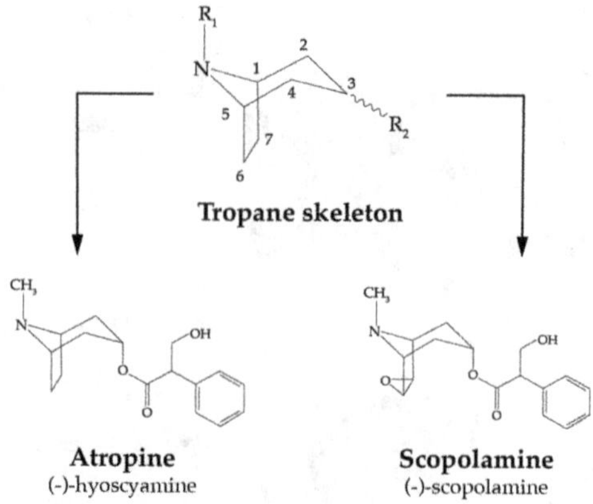


**Figure 3** Tropane skeleton and two main tropane alkaloids in foodstuffs: atropine and scopolamine.

acetylcholine receptors. These are odorless and tasteless substances and heat-stable (Jank and Rath, 2021; Ribeiro et al., 2022).

The administration of low doses of TAs can be helpful for medical treatment, as previously mentioned. Nevertheless, TAs have a dose-dependent response and the intake of large concentrations of these compounds can cause adverse effects (Adamse et al., 2014; Jank and Rath, 2021). Frequently reported adverse effects of atropine and scopolamine are attributed to their anticholinergic action; they induce dizziness, dry mouth and difficulty in swallowing, mydriasis, muscle spasms, clouded consciousness, tachycardia, occasionally vomiting, and when consumed in higher doses, delirium and hallucinations. Fortunately, there are a few antidote options for these toxic effects, such as benzodiazepines, physostigmine, and activated charcoal (Adamse et al., 2014).

Infants and young children should have close monitoring, not only due to their higher sensitivity to the adverse effects of anticholinergics, but also due to the fact that they are exposed to a greater concentration of TAs per kg of body weight (Jank and Rath, 2021).

Cases of toxic effects as a consequence of the intake of TAs in food have long been disclosed. However, no chronic or genotoxic toxicity was exhibited. Since these compounds do not accumulate in the human body, the side effects reported originate from acute exposure. Until this day, no long-term adverse effects have been reported (EFSA CONTAM Panel, 2013; Kohnen-Johannsen and Kayser, 2019).

The metabolic course of scopolamine and atropine in the human system differs from one another. Both TAs are rapidly absorbed through the gastrointestinal system (between 30 min and 60 min) and distributed throughout the body (with 1–2 min of half-life in the plasma). Atropine has great oral bioavailability, while scopolamine does not. About 75% of atropine

© Burleigh Dodds Science Publishing Limited, 2025. All rights reserved.

is excreted in urine in active form within 24 hours, while only approximately 3% of the pharmacologically active form of scopolamine is excreted in urine (about 22% is excreted as Phase II metabolites), suggesting a significant first-pass effect. Cytochrome P450 enzymes seem to be involved in the metabolism of TAs (Kohnen-Johannsen and Kayser, 2019; Mulder et al., 2016).

## 2.5 Tropane alkaloids legislation

The EU established maximum levels for contaminants in foodstuffs, dictating the importance of keeping contaminants at levels toxicologically acceptable for the protection of public health, but also following the ALARA principle (as low as reasonably achievable), which means that the permitted concentration of contaminants must not compromise the availability of the main food nutrients. TAs are one of the contaminants present in Regulation No. 2023/915 and its amendments which establish the maximum levels for certain contaminants in food (European Commission, 2023a). However, TAs have only recently been considered in the legal framework for contaminants.

The first reference to TAs in EU legislation was in 2016 where maximum levels for atropine and scopolamine were established by Commission Regulation (EU) 2016/239 in processed cereal-based foods and baby foods for infants and young children, containing millet, sorghum, buckwheat, or their derived products (1.0 μg/Kg for each TA) (European Commission, 2016). This first maximum level for TAs was established based on the European Food and Safety Authority (EFSA) opinion on this category of alkaloids for food and feed. The Scientific Panel on Contaminants in the Food Chain (CONTAM) of EFSA established an acute reference dose (ARfD) of 0.016 μg/Kg of body weight, expressed as the sum of atropine and scopolamine and assuming potency equivalence (EFSA CONTAM Panel, 2013). Based on the limited information available, the dietary exposure of toddlers could significantly exceed the ARfD due to the high levels of TAs found in cereal-based foods and the individual consumption per day.

After two years, the EFSA presented a scientific report that showed that the ARfD had been exceeded in different population groups, for several acute exposure estimates, which raised concerns about the health effects of TAs (Arcella et al., 2018). As a result, in 2021, the EU found it necessary to expand the previous legislation and included maximum levels for the sum of atropine and scopolamine for unprocessed or processed millet, sorghum, maize, maize for popping, and buckwheat and for herbal infusions. The summary of current maximum levels for TAs established by the Commission Regulation (EC) No. 2023/915 is shown in Table 1. These maximum levels in food took effect on the 1 September 2022.

© Burleigh Dodds Science Publishing Limited, 2025. All rights reserved.

**Table 1** Maximum levels (μg/Kg) established for TAs in foodstuffs in the European Union since 2023

| Annex I | Foodstuff | MLs (μg/Kg) | |
|---|---|---|---|
| 2.2.1 | Baby food and processed cereal-based food for infants and young children, containing millet, sorghum, buckwheat, maize, or their derived products | Atropine | Scopolamine |
| | | 1.0 | 1.0 |
| | | **Sum of Atropine and Scopolamine** | |
| 2.2.2<br>2.2.5 | • Unprocessed millet and sorghum grains;<br>• Maize for popping;<br>• Millet, sorghum, and maize placed on the market for the final consumer;<br>• Milling products of millet, sorghum, and maize. | 5.0 | |
| 2.2.4<br>2.2.6 | • Unprocessed buckwheat grains;<br>• Buckwheat placed on the market for the final consumer;<br>• Milling products of buckwheat. | 10 | |
| 2.2.3 | Unprocessed maize grains, except for unprocessed maize intended to be processed by wet milling and unprocessed maize grains for popping. | 15 | |
| 2.2.7 | Herbal infusions (dried product) except anise seeds. | 25 | |
| 2.2.8 | Herbal infusions (dried product) of anise seeds. | 50 | |
| 2.2.9 | Herbal infusions (liquid product). | 0.20 | |

Adapted from the Commission Regulation (EC) No. 2023/915 and its amendments establishing the maximum levels for certain contaminants in food (European Commission, 2023a).

From the numbers displayed, there is a clear difference between TAs in cereal-based foods and in herbal infusion; the latter has a higher limit in the dried form, probably due to possible matrix effects of these products. The family of TAs does not only comprise atropine and scopolamine but also homatropine and anisodamine, among others. Despite the EC recommending the detection of other TAs, only the first two are legislated (Cirlini et al., 2019). Furthermore, certain foodstuffs with a high volume of consumption, like honey and frozen vegetables, are also not mentioned in the legislation.

The reviewed literature revealed no existing legislation outside of Europe that sets limits for TAs in food or animal feed. But the Joint FAO/WHO Expert Committee on Food Additives has recommended data collection on the occurrence of TAs in food, recognizing the potential health risks, especially for staple foods prone to contamination, like grains. Currently, there is call open for data on TAs to discuss the need for MLs in the Codex General Standard for Contaminants and Toxins in Food and Feed.

© Burleigh Dodds Science Publishing Limited, 2025. All rights reserved.

# 3 Analytical methodologies to determine tropane alkaloids

Several analytical techniques have been described for the determination of TAs in various samples. Since these compounds are found in complex matrices and are naturally present in low concentrations, only a limited number of techniques can determine such analytes. The analysis of TAs requires different steps, beginning with sample collection and its homogenization, followed by the extraction process and clean-up (if necessary) and finally, the detection and quantification of the analytes.

Atropine and scopolamine are the only legislated TAs in the EU and consequently the ones that most methods focus on. Despite this, they are not the only existing tropane compounds; some methods determine more than 10 TAs, including, for example, homatropine, anisodamine, pseudotropine, and tropine (Han et al., 2024; Romera-Torres et al., 2020). Homatropine and anisodamine are also reoccurring TAs present in foodstuffs capable of causing health complications (EFSA CONTAM Panel, 2013); however, there is no established limit for their concentration in food or feed. Moreover, some methodologies were developed to determine not only TAs but also other plant toxins, such as pyrrolizidine alkaloids (PAs), ergot alkaloids (EAs), and opium alkaloids (OAs), given the possible co-occurrence of these natural contaminants (García-Juan et al., 2023; Kowalczyk and Kwiatek, 2022; León et al., 2022; Vera-Baquero et al., 2024).

## 3.1 Sampling

Sampling plays an important role in the determination of TAs, so it is as precise as possible. The challenge in this process is in obtaining a representative sample, since the contamination with TAs leads to a heterogeneous dispersion of these compounds in the crops, culminating in discrete contamination in the food. To mitigate this heterogeneity, the protocols mentioned in Commission Regulation No. 2023/2783 for sampling methods for the official control of the levels of plant toxins in food (European Commission, 2023b).

Most of the research articles compiled in Table 2 focus on cereals, like wheat, buckwheat, millet, sorghum, rye, and rice as the primary sample. Although tea and herbal infusions, baby food, baked goods, and honey also belong to the set of samples.

According to the EU, baby food is defined as "*food intended to fulfill the particular requirements of infants in good health while they are being weaned, and of young children in good health as a supplement to their diet and/or for their progressive adaptation to ordinary food, excluding processed cereal-based food, and milk-based drinks and similar products intended*

© Burleigh Dodds Science Publishing Limited, 2025. All rights reserved.

**Table 2** Extraction methods for the determination of TAs in different food matrices

| Sample | TAs Analysed | Extraction Method | Extraction Conditions | Ref. |
|---|---|---|---|---|
| **Cereals, Pseudocereals, and Cereal-Based Food Products** | | | | |
| Bakery products (bread, slice bread, bread toasts, breadsticks, crackers, and biscuits) made with different flours, spices, aromatic herbs, and/or seeds) | Atropine; scopolamine 21 PAs and 6 OAs | SLE + SPE | 0.5 g + 5 mL HCl 1% OASIS® MCX cartridges | (Vera-Baquero et al., 2024) |
| Cereal food samples, including raw (millet, barley, sorghum, buckwheat, rice, corn, and wheat) and processed cereals (baby food, confectionery, bread, and rice cake) | 15 TAs, including anisodamine, atropine, homatropine, pseudotropine, scopine, scopolamine, scopoline, tropine and tropinone | SLE + SPE | 2 g + 20 mL 0.4% FA in MeOH/water (75:25, v/v) Oasis MCX 150 mg cation exchange | (Han et al., 2024) |
| Buckwheat flour and grains | Atropine; Scopolamine | SLE, with low temperature partitioning | 1 g + 5 mL H2O + 15 mL ACN 1 g $MgSO_4$, followed by freeze 12 h | (Caminhas et al., 2024) |
| Cereal based-food, including wheat and rye flours and breadcrumbs, corn flour, and two multigrain baby foods | Atropine; Scopolamine 12 EAs and 28 PAs | QuEChERS + d-SPE | 2.5 g + 30 mL ACN 75 % (v/v), with 0.1 % FA; + 1 g of sodium citrate, 0.5 g of sodium hydrogenocitrate sesquihydrate, 4 g of $MgSO_4$, and 1 g of NaCl; + 150 mg PSA, 150 mg C18, and 900 mg $MgSO_4$ | (García-Juan et al., 2023) |
| Bread (with wheat, rye, wheat-rye and multigrain) | 6 TAs, including atropine, scopolamine and 20 EAs | SLE | 4 g + 40 mL MeOH 60% w/0.4% FA | (Veršilovskis et al., 2020) |
| Cereals, including wheat, corn, rice, oat, millet, and cereal-based products | Atropine and scopolamine | QuEChERS | 5 g + 10 mL $H_2O$ + 10 mL ACN w/5% acetic acid 4 g of $MgSO_4$ and 1 g NaCl + 5 mL *n*-hexane | (Baslé et al., 2020) |

© Burleigh Dodds Science Publishing Limited, 2025. All rights reserved.

| | | | | |
|---|---|---|---|---|
| **Tea, Herbs, Spices, and Vegetables** | | | | |
| Herbal teas, including black tea, green tea, mixed tea, chamomile, sage, linden, fennel and rosehips, thyme, and peppermint | Atropine; Scopolamine and 25 PAs | SLE | 50 mg + 2.5 mL MeOH (x 4) | (Gumus, 2023) |
| Tea and herbal tea infusions | Atropine; Scopolamine | µSPE | 2 g + 100 mL of boiling water, 5 min | (González-Gómez et al., 2022b) |
| Teas and herbs for infusions, including, for example, rooibos, chamomile, and mixture of herbs | Atropine; Scopolamine 28 PAs | QuEChERS + d-SPE | 1 g + 30 mL ACN 75% w/0.5% FA; + 6 g $MgSO_4$, 1.5 g sodium acetate; + 200 mg MgSO4, 400 mg PSA, 400 mg C18, and 400 mg GCB | (León et al., 2022) |
| Spinaches (frozen and derived based infant products) | Atropine and scopolamine | QuEChERS + d-SPE | 10 g + 10 mL ACN + 4 g anhydrous $MgSO_4$, 1 g NaCl, 1 g trisodium citrate dihydrate, and 0.5 g sodium hydrogen citrate sesquihydrate + 750 mg anhydrous $MgSO_4$ + 250 mg PSA | (Castilla-Fernández et al., 2021) |
| **Honey** | | | | |
| Honey | Atropine; Scopolamine15 PAs | SPE | heat 40–45°C 10 g + 20 mL of 0.05 M sulphuric acid 1 g of zinc dust (24 h at room temperature) Oasis MCX cation exchange cartridges – 500 mg alkaloids eluted 10 mL ethyl acetate, methanol, acetonitrile, ammonia solution, and triethylamine (8:1:1:0.3:0.1, v/v) | (Kowalczyk and Kwiatek, 2022) |
| Honey | Atropine and scopolamine | SALLE | 5 g + 10 mL concentrated sodium acetate solution+ 10 mL ACN | (Thompson et al., 2020) |
| Honey | 11 TAs, including atropine and scopolamine | SLE | 2.5 g + 10 mL of MeOH 75 % w/0.4% FA 300 mg of $MgSO_4$ + 50 mg of GBC | (Romera-Torres et al., 2020) |

ACN – Acetonitrile; C18 – Octadecyl Silica; d-SPE – dispersive Solid Phase Extraction; EAs – Ergot Alkaloids; FA – Formic acid; GCB – Graphitized Carbon Black; HCl – Hydrochloric acid; MeOH- methanol; $MgSO_4$ – Magnesium Sulfate; NaCl – Sodium Chloride; OAs – Opium Alkaloids; PAs – Pyrrolizidine alkaloids; PSA – Primary Secondary Amine; QuEChERS – Quick, Easy, Cheap, Effective, Rugged and Safe; SALLE – Salting-out assisted liquid-liquid extraction; SLE – Solid-Liquid Extraction; SPE – Solid-Phase Extraction; TAs – Tropane Alkaloids.

© Burleigh Dodds Science Publishing Limited, 2025. All rights reserved.

*for young children"* (European Commission, 2013). This is one of the most important groups of foods to be analyzed for the presence of TAs, because, as mentioned before, children have a higher predisposition to the hazards of these compounds.

## 3.2 Extraction and clean-up

Sample preparation plays a crucial role in the analysis of food. By cleaning up the matrix components that may interfere with the analytical method, it enables a lower detection limit while also providing good sensitivity (González-Gómez et al., 2022b). This procedure depends on the chemical properties of the compounds to be analyzed, the nature of the sample matrix, and the method of detection that will be performed. The most frequently used methods for the analysis of TAs are solid-liquid extraction (SLE) and QuEChERS (Table 2).

QuEChERS stands for Quick, Easy, Cheap, Effective, Rugged, and Safe. This approach was first introduced in 2002 with the aim of recovering residues of pesticides from vegetables and fruits. Since then, it has gained popularity and has been used for different food matrices and analytes (Perestrelo et al., 2019). QuEChERS is seen as a green, user-friendly, and cheap extraction method that requires small amounts of organic solvents, samples, as well as laboratory glassware. This fast and simple technique encompasses two steps. The primary phase is the extraction, where water is removed from the organic layer via a salting-out effect with acetonitrile (ACN). In the second phase, the clean-up mostly uses dispersive solid-phase extraction (d-SPE), with different sorbents and salts that remove interferents from the sample matrix (Bruzzoniti et al., 2014; Chen et al., 2017).

Alternatively, SLE is one of the most widely used extraction methods for foodstuffs (Table 1). This technique is based on dilution and osmosis, meaning that a solid sample is immersed in a liquid capable of extracting its compounds, and this liquid begins to be enriched with the analytes that were removed from the solid sample (Gumus, 2023; Han et al., 2024; Romera-Torres et al., 2020). TAs are highly polar compounds, so in SLE they are usually extracted with either methanol (MeOH) or ACN as the organic phase. Additionally, the acidification of the solvent has shown great results in extraction efficiency (Baslé et al., 2020).

Additionally, to achieve optimal results and maintain the integrity of the equipment, SLE requires an extra clean-up step. This is where SPE comes into play – a reliable, efficient, and reproducible technique used to further eliminate interferences during the clean-up process. The SPE procedure includes the extraction of TAs dissolved in the sample matrix into solid support with a sorbent. Afterwards, the compounds go through a clean-up step, and finally, the matrix interferences are eluted. It is important to mention that, depending on the type of sorbent, the mechanism of the extraction can vary. Regardless of

© Burleigh Dodds Science Publishing Limited, 2025. All rights reserved.

the vast list of advantages, SPE also has some drawbacks; it is a time-consuming method that requires several steps, and the sample needs to be in a liquid stage (Han et al., 2024; Li et al., 2022; Vera-Baquero et al., 2024).

Other modern extraction techniques have been applied for determination of these natural contaminants. For example, Thompson et al. (2020) used the salting-out assisted liquid-liquid extraction (SALLE) for honey samples, in which the addition of salts to a sample solution composed of a miscible mixture of water and an organic solvent causes the separation of two phases. Likewise, González-Gómez et al. (2022b) used the micro-solid-phase extraction (μSPE) for herbal infusions. This microextraction technique uses cartridges containing sorbents with a smaller particle size, providing a higher surface area and more efficient separations with smaller volumes, i.e. faster, more environmentally friendly, and cheaper methods compared to conventional SPE.

### 3.3 Chromatographic techniques

The chromatographic step separates analytes either by gas chromatography (GC) or by LC, being the latter the most widely used technique for determination of TAs in different food matrices. Although GC has a good resolution between TAs and the matrix, the tropane compounds suffer thermolysis. By using LC, that problem is overcome (Chen et al., 2017; Marín-Sáez et al., 2017).

For quantitative determination of plant toxins, LC with tandem mass spectrometry (LC-MS/MS) is the preferable chromatographic method, given its high precision, high sensitivity, and low detection limits. Regarding the mobile phases, both MeOH and ACN were equally used during all the experiments (Table 3). Different types of analytical columns can be used in LC; they differ in several parameters such as length, inner diameter, and particle size, but the most commonly used is the C18 column (Han et al., 2024).

Meanwhile, Hydrophilic Interaction Chromatography (HILIC) has become an interesting choice for hydrophilic, polar, and ionized compounds in foods (Kartsova et al., 2019; Marrubini et al., 2018). HILIC is more complex; it combines features of reversed-phase (polar mobile phase), normal-phase (polar stationary phase), and ion exchange. HILIC can be applied to highly polar uncharged and amphiphilic analytes that are not able to be separated by reverse-phase or ion exchange chromatography (Kartsova et al., 2019). For instance, Dzuman et al. (2020) applied a complementary HILIC to solve the problem of co-elution of multiple groups of isomeric alkaloids, improving the selectivity. Additionally, Marín-Sáez et al. (2017) coupled a HILIC column with C18 (second column) to overcome the problem of low molecular weight substances, such as tropine and pseudotropine, which had a low retention time on the C18 column and could be eluted together with the compounds present in the matrix. With these two columns coupled, the authors obtained better

© Burleigh Dodds Science Publishing Limited, 2025. All rights reserved.

**Table 3** Compilation of techniques used for the determination of TAs in foodstuff

| Sample | TAs Analysed | Analytical Techinque | LC/MS Conditions | LOQ* (μg/kg) | Ref. |
|---|---|---|---|---|---|
| | | | **Cereals, Pseudocereals, and Cereal-Based Food Products** | | |
| Bakery products | Atropine; Scopolamine 21 PAs and 6 OAs | UHPLC-MS/MS Ion Trap analyser | Luna Omega Polar C18 column (100 mm × 2.1 mm, 1.6 μm) 30°C; 0.3 mL/min; ESI + 0.2% FA and 0.2% NH3•H2O in MeOH | At: 0.38 Sc: 0.59 | (Vera-Baquero et al., 2024) |
| Cereal food samples and processed cereals | 15 TAs, including atropine and scopolamine | UHPLC – MS/MS Xevo TQ-XS | BEH C18 column (150 mm × 2.1 mm, 1.7 μm) 50°C; 0.4 mL/min 10 mM ammonium carbonate in water (A) and ACN (B) ESI + MRM | 0.20–36.07 | (Han et al., 2024) |
| Buckwheat flour and grains | Atropine; Scopolamine | LC – UHPLC – MS/MS QqQ MS | OASIS HLB reverse-phase column (2.1 × 30 mm; 20 μm); 40°C ACQUITY BEH C18 (50 × 2.1 mm; 1.7 um); ESI | 2.5 | (Caminhas et al., 2024) |
| Cereal based-food | Atropina; Scopolamine 12 EAs and 28 PAs | HPLC-MS/MS QqQ MS | Luna 3u C18 column (100 × 2.0 mm, 3 μm) 0.1% formic acid (A) and methanol with 0.1% formic acid (B) 0.3 mL/min ESI+; MRM | 0.5 | (García-Juan et al., 2023) |
| Bread | 6 TAs, including atropine, scopolamine and 20 EAs | UHPLC – MS/MS Xevo TQ-S MS | Acquity BEH C18 (150 × 2.1 mm; 1.7 μm) 50°C; 0.4 mL/min 10mM ammonium carbonate (A) and ACN (B) ESI+ | At: 1.0 Sc: 0.8 | (Veršilovskis et al., 2020) |
| Cereals, including wheat, corn, rice, oat, millet, and cereal-based products | Atropine and Scopolamine | HPLC-MS/MS QqQ MS | Acquity UPLC BEH C18 column (2.1 × 100 mm, 1.7 μm), 50°C; 0.4 mL/min 1.5 mL/L FA and 10 mM ammonium formate (A) and 0.5 mL/L FA in MeOH (B); ESI+; MRM | 0.5 | (Baslé et al., 2020) |
| | | | **Tea, herbs, spices and vegetables** | | |
| Teas and Herbal infusions | Atropine; Scopolamine and 25 PAs | HPLC-Q-ToF-MS/MS | Agilent Poroshell 120 EC-C18 (2.1 × 150 mm, 2.7 μm) 40°C; 0.27 mL/min; ESI + water with 0.1% FA (A) and acetonitrile with 0.1% FA (B) | At: 0.54 Sc: 0.59 | (Gumus, 2023) |

© Burleigh Dodds Science Publishing Limited, 2025. All rights reserved.

| | | | | | |
|---|---|---|---|---|---|
| Tea and herbal infusions | Atropine; Scopolamine | HPLC – MS/MS QqQ MS | C18 Kromaphase 10 (150 mm × 2.0 mm, 3.5 μm); 30°C; 0.25 mL/min ACN w/0.1% FA (A) and H2O w/0.1% FA (B); ESI+; MRM | At: 0.09Sc: 0.06 | (González-Gómez et al., 2022b) |
| Tea and herbal infusions | Atropine; Scopolamine 28 PAs | UHPLC-Q-Orbitrap- MS/MS H-ESI | Hypersil GOLD C18 (100 mm × 2.1 mm, 1.9 μm); 40°C; 300 μL/min 0.1% of FA (A) and 0.1% FA in MeOH (B); ESI +; PRM | 5 | (León et al., 2022) |
| Spinaches (frozen and derived based infant products) | Atropine and Scopolamine | UHPLC -MS/MS QqQ MS | C18 column Zorbax Eclipse Plus RRHD (2.1 mm × 50 mm, 1.8 μm) 25°C; 0.5 mL/min water (A) and MeOH (B), both containing 0.1% FA and 5 mM ammonium formateESI+; MRM | At: 0.018 Sc: 0.015 | (Castilla-Fernández et al., 2021) |
| **Honey** | | | | | |
| Honey | Atropine; Scopolamine 15 PAs | HPLC - MS | Kinetex C18 Column (100 × 4.6 mm, 2.6 μm), 30°C; 0.6 mL/min0.3% FA (A) and MeOH with ACN 2:1 (v/v) (B) ESI + | At: 0.36 Sc: 0.49 | (Kowalczyk and Kwiatek, 2022) |
| Honey | Atropine and Scopolamine | UHPLC–MS/MS TQ-XS tandem quadrupole | Kinetex HILIC core–shell column (100 × 2.1 mm, 1.7 μm) 40°C; 0.5 mL/min A = 5 mM ammonium formate in 0.1% FA (A) and 5 mM ammonium formate in 0.1% FA in ACN 95% (B) ESI +; SRM | 0.01 μg/kg | (Thompson et al., 2020) |
| Honey | 11 TAs, including atropine and scopolamine. | HPLC - MS/MS Exactive Orbitrap MS | ACE HILIC-A column (250 × 4.6 mm, 3 μm); 25°C; 0.3 mL/min 5 mM of ammonium formate and 0.1% FA (A) and MeOH (B) | At: 20 Sc: 20 | (Romera-Torres et al., 2020) |

* The LOQ correspond for atropine and scopolamine.
ACN - Acetonitrile; At – Atropine; ESI - Electrospray ionization; HESI-II – heated electrospray ionization; HILIC – Hydrophilic Interaction Chromatography; HPLC – High Performance Liquid Chromatography; LOQ – Limit of Quantification; MeOH – Methanol; MS – Mass Spectrometry; MS/MS – Tandem Mass Spectrometry; PAs – Pyrrolizidine alkaloids; PRM – Product Reaction Monitoring; QqQ – Triple Quadrupole; Sc – Scopolamine; SRM – Selected Reaction Monitoring; TA – Tropane Alkaloid; ToF – Time of Flight; UHPLC – Ultra-High-Performance Liquid Chromatography.

© Burleigh Dodds Science Publishing Limited, 2025. All rights reserved.

results, which means a better peak shape and elution time. Nevertheless, in all the studies, the most frequently used column is C18 columns. The lowest LOD and/or LOQ are generally achieved when RP-C18 columns are used, confirming that RP continues to be the most effective mode in LC for food analysis (Marrubini et al., 2018).

Furthermore, ultra-high liquid pressure chromatography (UHPLC) has started to get attention in the academic field, since it also provides good resolution, better sensitivity, and consequently lower LOD values. UHPLC achieves broader linearity, the run time is decreased by more than half, and it has a higher separating efficiency due to the decrease of the particle size (<2 μm) (Vargas Medina et al., 2021). One of the most differentiating factors in this technique is its green perspective, because of its shorter run time and small volume of organic solvents needed, which do not compromise the sensitivity and resolution of determination (Cielecka-Piontek et al., 2013). In a time when climate change is a hot topic, the green aspects of UHPLC bring a high advantage when compared to other chromatographic methods.

After the chromatographic analysis, it is imperative to conduct a MS analysis, being the most common method the triple quadrupole (QqQ), where the first quadrupole isolates the precursor (parent) ion, the second quadrupole acts as a collision cell, and the third quadrupole selects a specific fragment of the precursor ion, which allows low detection and quantification limits (Baslé et al., 2020; Castilla-Fernández et al., 2021; García-Juan et al., 2023).

However, high-resolution mass spectrometry (HRMS) has gained interest, competing strongly with the QqQ technique. HRMS, such as Time-of-Flight (ToF) and Orbitrap (orbital ion trap) or hybrid detectors such as Q-ToF (quadrupole-ToF) or Q-orbitrap (quadrupole-orbital ion trap), provides the highest precision in measuring the mass-to-charge ratio (m/z) of analytes (Gumus, 2023; León et al., 2022; Romera-Torres et al., 2020).

Orbital ion trap, also known as Orbitrap, has been considered one of the most popular HRMS techniques. This method has superior performance due to its high resolution, meaning improved selectivity to recognize the targeted ions from the background interferences, culminating in greater sensitivity and mass accuracy that facilitates the reliable identification of the ion of interest. Nevertheless, one of the greatest advantages of Orbitrap is its ability to identify not only targeted ions but also unknown compounds, resulting in the perfect method for non-targeted approaches (Dong et al., 2021; Marín-Sáez et al., 2019a; Yan et al., 2022). Orbitrap-MS is an interesting approach because it provides better insight into the sample composition through the full-scan MS; it enables a retrospective data analysis and, as previously mentioned, the search for unknown compounds. However, this technique still needs to improve its linear range, high throughput, and analytical speed parameters (Dong et al., 2021). Despite the advantages for non-targeted analyses, HRMS

© Burleigh Dodds Science Publishing Limited, 2025. All rights reserved.

instrumentation tends to be expensive and generates large amounts of highly complex data, which is not necessary for routine, focused analyses to determine a set of contaminants (Géhin and Holman, 2021).

Out of all the studies compiled in Table 2 and 3, the lowest levels of LOQ were determined by Castilla-Fernández et al. (2021) who analyzed spinach-based infant food products using UHPLC-QqQ-MS/MS as the chromatographic and detection step and QuEChERS extraction. By using this method, they were able to obtain a LOQ of 0.014 μg/Kg for atropine and 0.018 μg/Kg for scopolamine. The EU recommends the level of LOQ to be "preferably below 1 μg/Kg for processed cereal-based foods for infants and young children," which means that Castilla-Fernández et al. were able to obtain LOQ considerably lower than the recommended values (European Commission, 2023b). Also, Han et al. (2024) achieve a lower LOQ (0.30 μg/Kg for atropine and 0.35 μg/Kg for scopolamine) with SLE followed by UHPLC-MS/MS, meeting the recommendation of the EU Commission of LOQ below 2 μg/Kg. On the other hand, Caminhas et al. (2024) had LOQ of 2.5 μg/Kg for both TAs, higher than the recommendation of LOQ below 2 μg/Kg for cereals and cereal products.

Tea, herbal infusions, and honey are more complex matrices than pseudocereals and consequently have a higher matrix effect, meaning the influence of compound(s) present in the sample can disrupt the detector response, possibly causing an increase in the LOQ (León et al., 2022; Romera-Torres et al., 2020).

## 4 Occurrence of TAs in food

In the EU, Rapid Alert System for Food and Feed (RASFF) is an important tool with great value for food safety. This tool facilitates the rapid notification and response in case of possible risk to human health related to food and feed (European Commission, 2002). Since 2020, 41 notifications regarding food contaminated with TAs have been reported, summarized in Table 4 (European Commission, 2024). About 73% of notifications are cereals and bakery products, including teff, millet, maize, sorghum, and buckwheat (Fig. 4). Nonetheless, foods like soybean, aromatic plants, and vegetables have also been mentioned in RASFF with the presence of TAs at levels exceeding the MLs.

The highest levels of TAs were registered recently in France, when in May 2024 a concentration of sum of atropine and scopolamine of 1 240 000 μg/Kg was found in dried herbs for infusions from Albania. The herbal infusion was recalled from consumers in France, but 3 persons were affected, showing symptoms of TA poisoning (RASFF, 2024a).

Previously, an outbreak with more than 100 people complaining of dizziness, nausea, dry mouth, and mydriasis occurred in Slovakia due to the consumption of spinach puree deeply frozen contaminated with atropine (850–3446 μg/

© Burleigh Dodds Science Publishing Limited, 2025. All rights reserved.

**Table 4** RASFF notifications due to TAs contamination from 2020 to 2024 (adapted from RASFF portal)

| Date | Food Product | Notifying Country | Origin Country | Level (μg/Kg) |
|---|---|---|---|---|
| 12-07-2024 | Teff flour | Switzerland | Not indicated | Not indicated |
| 27-05-2024 | Herbal teas | France | Albania | 1240000* |
| 02-05-2024 | Infant formula | Germany | Germany | 5.2* |
| 11-04-2024 | Teff | Netherlands | Spain | 42* |
| 11-04-2024 | Teff | Netherlands | Germany,Italy | 1416* |
| 18-12-2023 | Dried whole raspberry leaves | Germany | Albania | Atropine: 376<br>Scopolamine: 27.8 |
| 11-12-2023 | Datura in organic corn flour | France | France | Atropine: 57.2<br>Scopolamine: 5.3 |
| 03-10-2023 | Bulk corn (conventional) | France | France | Atropine: 52<br>Scopolamine: 14 |
| 29-09-2023 | Sorghum flour | Norway | Germany | Atropine: 11.1<br>Scopolamine: 3.3 |
| 18-09-2023 | Buckwheat flour | France | France | Atropine: 7.4<br>Scopolamine: 7.1 |
| 13-09-2023 | Sorghum flour (organic) | Norway | Italy | Atropine: 15.9<br>Scopolamine: 5 |
| 08-09-2023 | Maize flour, white | Portugal | France | 355.6 |
| 06-09-2023 | Teff flour, whole grain (organic) | Germany | Germany | Atropine: 463.9<br>Scopolamine: 158.6 |
| 02-08-2023 | Sourdough powder | Netherlands | France | 40* |
| 02-08-2023 | Biscuits made with corn flour | France | France | 10.3* |
| 10-07-2023 | Millet, whole brown (organic) | Italy | Italy | Atropine: 119.8 |
| 30-06-2023 | Savory | Germany | Germany,Spain | Atropine: 1001<br>Scopolamine: 1068 |

© Burleigh Dodds Science Publishing Limited, 2025. All rights reserved.

| | | | | |
|---|---|---|---|---|
| 28-06-2023 | Gluten-free chocolate cookies | Netherlands | France | 35* |
| 08-06-2023 | Teff flour | Sweden | Spain, Sweden | 107.84* |
| 25-05-2023 | Teff flour (organic) | Germany | Bulgaria,South Africa | Scopolamine: 21 |
| 11-05-2023 | Breads with teff flour | Germany | Spain | Atropine: 174.2–1236.4 |
| 30-03-2023 | Corn (organic) | Belgium | France | Atropine: 160<br>Scopolamine: 20 |
| 24-03-2023 | Teff flour, dark (organic) | Germany | Germany, Netherlands, Spain | Atropine: 105–526<br>Scopolamine: 31.6–300 |
| 16-03-2023 | Teff flour (organic) | Germany | Germany | Atropine: 8.83–190.4<br>Scopolamine: 3.11–60.2 |
| 16-03-2023 | Popcorn | Germany | Germany | Atropine: 16.7<br>Scopolamine: 2.3 |
| 06-01-2023 | Hulled millet (organic) | Denmark | Poland,Sweden | Atropine: 29<br>Scopolamine: 23 |
| 01-07-2022 | Corn tortilla chips and flour | Germany | Belgium,France,Spain | Not indicated |
| 06-05-2022 | Savory | Germany | Austria | 106* |
| 07-04-2022 | Soybean meal | Germany | Austria | Atropine: 56 |
| 09-11-2021 | Infusion | Belgium | Netherlands | Atropine: 184<br>Scopolamine: 21 |
| 08-11-2021 | Flaxseed meal (organic) | Germany | Hungary | Atropine: 238 |
| 05-11-2021 | Corn flour (organic) | Belgium | France | Atropine: 25.12<br>Scopolamine: 6.5 |
| 12-08-2021 | Buckwheat flour | Germany | Austria,Germany | Atropine: 240<br>Scopolamine: 87 |
| 17-03-2021 | Deep frozen spinach puree | Slovakia | Slovakia | Atropine: 850–3446<br>Scopolamine: 1033–3860 |

*(Continued)*

© Burleigh Dodds Science Publishing Limited, 2025. All rights reserved.

**Table 4** (*Continued*)

| Date | Food Product | Notifying Country | Origin Country | Level (µg/Kg) |
|---|---|---|---|---|
| 18-01-2021 | Popcorn maize | Germany | Spain | Atropine: 16–35<br>Scopolamine: 1–1.9 |
| 11-12-2020 | Ground wild brown millet | Germany | Germany | Atropine: 24.2<br>Scopolamine: 13.3 |
| 27-11-2020 | Popcorn maize | Belgium | Belgium,Netherlands | Scopolamine: 14.6 |
| 04-11-2020 | Peppermint | Germany | Türkiye | Atropine: 66.7 |
| 04-09-2020 | Ground wild brown millet | Germany | Germany | Atropine: 36<br>Scopolamine: 22.4 |
| 25-05-2020 | Blackberry leaves (organic) | Germany | Bulgaria | Atropine: 543.1<br>Scopolamine: 31.4 |
| 24-01-2020 | Soy flakes (organic) | Germany | Austria,Italy | Atropine: 12.1–20.2<br>Scopolamine: 4.0–6.9 |

*Sum of atropine and scopolamine.

© Burleigh Dodds Science Publishing Limited, 2025. All rights reserved.

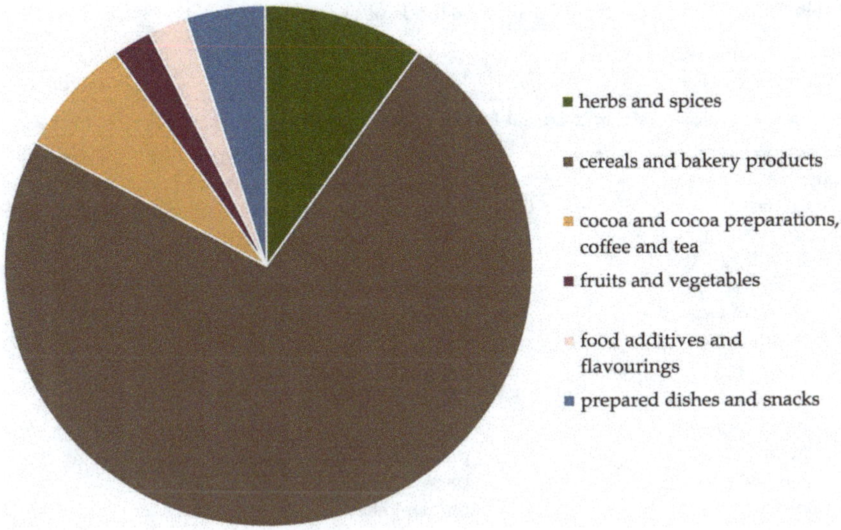


**Figure 4** Pie chart with number of notifications for TAs per food category (2020–2024) (European Commission, 2024).

Kg) and scopolamine (1033–3860 μg/Kg) (RASFF, 2021). The EU has not yet established a limit for TAs in vegetables. In 2022, 4 cases of intoxications were reported in Italy related to the consumption of leafy vegetables contaminated with TAs, with concentrations of atropine and scopolamine up to 4642 μg/kg and 8158 μg/kg, possibly due to cross contamination with leaves of *Datura* species (Caprai et al., 2024). More recently, another food poisoning related to the presence of TAs in fresh spinach was reported in Italy, affecting six people (RASFF, 2024b), alerting to the need to control and establish maximum limits for this food group.

Another food poisoning outbreak with nearly 200 illnesses was linked to the consumption of type of cornbread in Portugal. After identification of raw materials used in the manufacture of the bread, a white maize flour, a concentration of 355.6 μg/Kg for the sum of atropine and scopolamine was found (RASFF, 2023).

The origins of these contaminations are all from Europe; it is important to guarantee that the good agricultural and collecting practices are being fulfilled. All of these notifications emphasize the importance of close monitoring of these compounds and the relevance of TAs in food safety.

Moreover, analytical researchers worldwide develop the methodologies for the determination of TAs in different food matrices (as mentioned in the previous section), and they also analyze samples, mostly cereals, herbs for infusions, and honey (Table 5).

© Burleigh Dodds Science Publishing Limited, 2025. All rights reserved.

**Table 5** Occurrence of tropane alkaloids in foodstuffs worldwide

| Sample (Number of Samples) | TAs Analysed | Main Conclusions | Reference |
|---|---|---|---|
| **Cereals, Pseudocereals, and Cereal-Based Food Products** | | | |
| Bakery products (bread, slice bread, bread toasts, breadsticks, crackers and biscuits) made with different flours, spices, aromatic herbs and/or seeds) (n = 15) | Atropine; Scopolamine 21 PAs and 6 OAs | 80% of the samples contained TAs, with concentrations ranging from 0.2 μg/kg to 10 μg/kg. Atropine was detected in all the samples, while scopolamine was only presented in five samples. The positive samples were mainly made of buckwheat flour, millet, corn, teff, and/or honey. | (Vera-Baquero et al., 2024) |
| Cereal food samples, including raw (millet, barley, sorghum, buckwheat, rice, corn and wheat) and processed cereals (baby food, confectionery, bread, and rice cake) (n = 71) | 15 TAs | Six TAs were detected in at least one sample, ranging from 0.12 μg/kg to 88.10 μg/kg. TAs were presented only in millet and barley. Four samples of baby were contaminated with atropine (0.21 μg/kg), scopolamine (0.12 μg/kg), pseudotropine (10.71 μg/kg), and 6-hydroxytropinone (2.78 μg/kg), where millet was an ingredient. | (Han et al., 2024) |
| Buckwheat flour and grains (n = 18) | Atropine; Scopolamine | No contamination above the LOQ was found in any samples of buckwheat. | (Caminhas et al., 2024) |
| Bread (with wheat, rye, wheat-rye and multigrain) (n = 40) | Six TAs, including atropine, scopolamine and 20 EAs | None of the 6 TAs were not detected. In one wheat-rye bread a trace (0.22 μg/kg) of scopolamine was detected, but below LOQ 18 EAs were detected, and total levels varied between <LOQ and 335 μg/kg. | (Veršilovskis et al., 2020) |
| Cereals, including wheat, corn, rice, oat, millet and cereal-based products (n = 95) | Atropine and scopolamine | All samples were found free of the two targeted TAs, with the exception of a rice-based product in which both atropine and scopolamine were quantified at 9.6 μg/kg and 2.6 μg/kg, respectively. | (Baslé et al., 2020) |
| **Tea, Herbs, Spices, and Vegetables** | | | |
| Tea and herbal teas (n = 44) | Atropine; scopolamine and 25 PAs | TAs were found in 4 herbal teas. Scopolamine was found in a mixed tea sample and in the rosehip tea in amounts of 35.1 ng/g and 244.5 ng/g, respectively, while atropine was found in one flavored tea (29.3 ng/g) and in the green tea (482.3 ng/g) and chamomile tea (559.2 ng/g) samples. | (Gumus, 2023) |

*(Continued)*

© Burleigh Dodds Science Publishing Limited, 2025. All rights reserved.

**Table 5** *(Continued)*

| | | | |
|---|---|---|---|
| Tea and herbal tea infusions (n = 17) | Atropine; Scopolamine | Atropine was found in 16 infusions, and with scopolamine in 5 infusions. The highest concentrations were found in peppermint (1.86 ng/g for atropine and 1.09 ng/g for scopolamine) and green tea (3.88 ng/g for atropine and 1.13 ng/g for scopolamine). | (González-Gómez et al., 2022b) |
| Teas and herbs for infusions (n = 11) | Atropine; Scopolamine 28 PAs | No alkaloids were found above LOQ levels, meaning they follow European regulations. | (León et al., 2022) |
| Spinaches (frozen and derived based infant products) (n = 66) | Atropine and scopolamine | Scopolamine was found in 21% of the samples and at higher concentrations than atropine in most cases. Atropine was found in 3 samples of infant food while scopolamine was found in one sample of infant food, but below the maximum level set for children. Only one frozen spinach sample exceeded the ML, with 4.5 µg/kg of atropine and 8.2 µg/kg of scopolamine. | (Castilla-Fernández et al., 2021) |
| **Honey** | | | |
| Honey (n = 23) | Atropine and scopolamine | Only two samples were above LOQ for both atropine and scopolamine, both at 0.012 µg/kg. | (Thompson et al., 2020) |
| Honey(n = 19) | 11 TAs, including atropine and scopolamine | 52% of the samples showed the occurrence of TAs. Scopolamine was detected in a multifloral honey sample at 27 µg/kg. Physoperuvine and tropine were detected in 26% of samples, but at concentrations below LOQ. | (Romera-Torres et al., 2020) |

The most contaminated sample was found by Gumus (2023) when analyzing teas and herbal infusions. Samples of chamomile infusions were contaminated with atropine up to 559.2 µg/Kg and a level of 244.5 µg/Kg of scopolamine was found in a sample of rosehip infusion. This number severely exceeds its legislated limit set at 25 µg/Kg for the sum of atropine and scopolamine.

Regarding cereals and pseudocereals, the positive samples for TAs were mainly made up of buckwheat, millet, barley, corn, and teff. In the EU, the MLs are only set for millet, sorghum, maize, and buckwheat. Due to the high frequency of occurrence of TAs in teff, MLs should be set for the presence of toxins in this cereal. Also, a rice-based product analyzed by Baslé et al. (2020) was contaminated with TAs, in which both atropine and scopolamine were

© Burleigh Dodds Science Publishing Limited, 2025. All rights reserved.

quantified at 9.6 μg/kg and 2.6 μg/kg, respectively. Although no ML is defined for rice-based products, the sample exceeds the ML of 10 μg/kg for buckwheat, the highest ML for products placed on the market for the final consumer.

Honey is not mentioned in the EU legislation; however, some studies have shown the presence of TAs in this food. For example, a level of 27 μg/kg of scopolamine was found by Romera-Torres et al. (2020). The presence of TAs has also been confirmed in the floral nectar of *Datura* species. Thus, when bees collect nectar from flowers of these plants, they can transfer these compounds into the honey during the bees' nectar processing and storage (Boros et al., 2010). Although the values differentiate among them, further research studies should focus on this matrix and, if a risk to human health is confirmed, an amendment to the legislation should be carried out to include a maximum permitted level of TAs in honey (Kowalczyk and Kwiatek, 2022; Martinello et al., 2017; Romera-Torres et al., 2020; Thompson et al., 2020).

Furthermore, other alkaloids could be present; for example, Vera-Baquero et al. (2024) detected three different alkaloid families, including PA, tropane, and OAs, in two bakery products, confirming the importance of simultaneously monitoring these alkaloids in food. Besides atropine and scopolamine, the only TAs with MLs in the EU, the presence of other TAs has been reported in cereals and honey, such as physoperuvine, tropine, pseudotropine, and 6-hydroxytropinone (Han et al., 2024; Romera-Torres et al., 2020).

As previously mentioned, babies and children have a higher predisposition to the consequences of the ingestion of TAs. In studies present in Table 5, TAs were found in some baby food samples, but below the ML (Castilla-Fernández et al., 2021; Han et al., 2024). However, in previous studies by Marín-Sáez et al. (2019a) and Mulder et al. (2015) high TAs' concentrations were found, exceeding the established limit of 1 μg/kg for each TA. However, Mulder and co-authors monitored for four years the presence of TAs in cereal-based products for children and found that the samples collected in 2011 had a higher TAs contamination (4.6 μg/kg) than those collected in 2014 (0.5 μg/kg). Even though all the sampling occurred before the legislation was set in place, it means that good agricultural practices were followed during the latter year or that the level of contamination of the harvests varied greatly throughout those years (Mulder et al., 2015).

## 4 Contamination mitigation

The contamination with TAs generally occurs during harvest and processing. As a preventive step, the implementation of good agricultural and collection practices can greatly minimize the contamination of crops (Chan, 2017; European Commission, 2023a). In case of contamination, sorting and cleaning cereals can be an effective method to remove the contaminating seeds. This

© Burleigh Dodds Science Publishing Limited, 2025. All rights reserved.

process is based on visual inspection by differentiating the cereal from the seed through density, color, and size (EFSA CONTAM Panel, 2013; European Commission, 2023a; FAO/WHO, 2020). Unfortunately, despite all this effort during pre- and post-harvest processing and control, some seeds may go undetected to further phases of the food process, leading to unavoidable contamination of foodstuff with TAs (Cirlini et al., 2018).

In recent years, studies relating the effect of food processing on the decrease of TA concentrations have emerged (Marín-Sáez et al., 2019b, 2019c; Torrents-Masoliver et al., 2024). For example, Marín-Sáez et al. evaluated the effect of tea making, boiling, and banking processes in the degradation of TAs, determining atropine, scopolamine, and their main metabolites (comprising 17 compounds). Samples of bread and homemade fresh pasta made with buckwheat and green tea were intentionally contaminated with a known amount of *Datura*. The baking (190°C for 40 min) process leads to a decrease in atropine and scopolamine concentrations of 73% and 84%, respectively. The boiling (100°C for 10 min) process also leads to a decrease approximately 32% of the main TAs. The degradation in tea samples is significantly lower, probably due to the shorter time at lower temperature when compared to the conditions for pasta and bread samples (Marín-Sáez et al., 2019b, 2019c).

However, the results of a small number studies have been considered inconsistencies, and it appears that the percentage of TAs degraded is small; besides, there is a lot of uncertainty regarding which degradation products of TAs are generated during the cooking process, the influence of the food matrix, and specific food processing techniques (Adamse et al., 2014; FAO/WHO, 2020). Since TAs are heat-stable and in order to protect consumers' health, the expert meeting organized by the Food and Agricultural Organization of the United Nations and WHO in 2020 agreed, as a default, to consider that no baking or boiling process impacts the concentration of TAs in food (FAO/WHO, 2020; Jank and Rath, 2021).

Some detoxification techniques have been reported for other alkaloids, like EA, PA, and aconitum alkaloids (Jiang et al., 2022; Kaltner, 2022; Lai et al., 2019).

Decontamination processes were studied for various plants, one of them is Fuzi, a plant root containing aconitum alkaloids, used in Chinese medicine for over 2000 years. Procedures like pressure-steaming, oven roasting, microwave, and microbial fermentation have been used as detoxification methods. However, they require high pressure and/or temperature, which consumes large amounts of energy (Jiang et al., 2022). Recently, the use of deep eutectic solvents (DES) could affect the detoxification of these alkaloids and the hydrolysis of their toxic compounds (Lai et al., 2019).

Nevertheless, even in alkaloids other than tropane, a lot of questions remain. The mitigation of toxic alkaloids cannot be reclaimed without

© Burleigh Dodds Science Publishing Limited, 2025. All rights reserved.

confirming that their transformation products are less toxic than the parent alkaloid itself (Kaltner, 2022).

In closing, further research on alkaloid transformation products and subsequently decontamination methods is necessary, so that – once proven effective – such methods can offer an alternative solution other than destroying batches in the event of serious contamination (Kaltner, 2022).

## 5 Conclusion and future trends

Atropine and scopolamine are used as medical drugs all over the globe. Despite this, these naturally occurring compounds can reach our food chain and cause a serious public health threat for consumers. That led to the establishment of MLs in certain food products by the EU.

The RASFF platform mentions various cases of contamination with TAs; some are more serious than others. Nevertheless, the cease of contamination notifications does not seem probable. This way, the implementation of good agricultural and harvest practices is fundamental as a preventive step against the contamination of crops with TAs. Besides, the monitorization of these substances in foods is of great importance, especially in baby and infant foods, since this group of people is more susceptible to the adverse effects of TAs. The concern about climate changes can induce tremendous consequences regarding TA contamination, since the rise of $CO_2$ and extreme temperatures increases the growth of contaminating plants.

Adequate sample treatment is essential for an analytical procedure. So, analysts are consistently being challenged to develop or enhance greener, faster, more accurate, and more precise methods that are capable of guaranteeing good quality, safety, authenticity, and traceability of the targeted analytes in several different matrices. In this way, research suggests SLE as the preferable extraction mode.

Regarding the detection and quantification methods, the UHPLC method coupled with MS and a C18 column was the one that demonstrated the most benefit by having a smaller run time, broader linearity, higher separating efficiency, and significant reduction in particle size. HRMS techniques like Orbitrap, despite their cost, are being used due to their unequivocal advantages such as information on exact mass and identification of unknown compounds.

Nevertheless, an update of the current legislation should be considered, since quite a few TAs, like anisodamine and homatropine – though not as well-studied as atropine and scopolamine – may also pose health risks. Therefore, establishing specific MLs for foodstuffs like honey and spinach would be beneficial for ensuring human health.

Finally, being this a hot topic, due to not only the environmental crises but also the variety of diets (vegetarian, vegan, gluten-free, etc.), investments should

© Burleigh Dodds Science Publishing Limited, 2025. All rights reserved.

be made in preventive and decontamination measures. Although the recent interest in research about TAs, limited information exists in this field, making the removal process of TAs from foodstuff nearly impossible. This culminates in food waste or, in the worst-case scenario, public health consequences.

## 6 References

Abia, W.A., et al., 2021. Tropane alkaloid contamination of agricultural commodities and food products in relation to consumer health: learnings from the 2019 Uganda food aid outbreak. *Comprehensive Reviews in Food Science and Food Safety* 20, 501–525. https://doi.org/10.1111/1541-4337.12664

Adamse, P., et al., 2014. Tropane alkaloids in food: poisoning incidents. *Quality Assurance and Safety of Crops & Foods* 6, 15–24. https://doi.org/10.3920/QAS2013.0314

Baslé, Q., et al., 2020. Application of a streamlined LC-MS/MS methodology for the determination of atropine and scopolamine in cereals from Asian and African countries. Food Additives & Contaminants: Part A 37(11), 1744–1754. https://doi.org/10.1080/19440049.2020.1800828

Boros, B., et al., 2010. LC-MS quantitative determination of atropine and scopolamine in the floral nectar of *Datura* Species. *Chromatographia* 71, 43–49. https://doi.org/10.1365/s10337-010-1524-y

Bruzzoniti, M.C., et al., 2014. QuEChERS sample preparation for the determination of pesticides and other organic residues in environmental matrices: a critical review. *Analytical and Bioanalytical Chemistry* 406, 4089–4116. https://doi.org/10.1007/S00216-014-7798-4

Caminhas, L.D., et al., 2024. Assessment of tropane alkaloid levels in Brazilian buckwheat flour products: a novel LC-UHPLC-MS/MS approach using solid-liquid extraction at low temperature. *Food Chemistry* 438, 137004. https://doi.org/10.1016/j.foodchem.2023.138010

Caprai, E., et al., 2024. Case reports of tropane alkaloid contamination in spinach from Italy and its potential implications for consumer health. *Food Control* 160, Article 110334. https://doi.org/10.1016/j.foodcont.2024.110334

Castilla-Fernández, D., et al., 2021. Determination of atropine and scopolamine in spinach-based products contaminated with genus datura by UHPLC–MS/MS. *Food Chemistry* 347, 129020. https://doi.org/10.1016/j.foodchem.2021.129020

Chan, T.Y.K., 2017. Worldwide occurrence and investigations of contamination of herbal medicines by Tropane Alkaloids. *Toxins (Basel)* 9, 284. https://doi.org/10.3390/toxins9090284

Chen, H., et al., 2017. Simultaneous determination of atropine and scopolamine in buckwheat and related products using modified QuEChERS and liquid chromatography tandem mass spectrometry. *Food Chemistry* 218, 173–180. https://doi.org/10.1016/j.foodchem.2016.09.075

Cielecka-Piontek, J., et al., 2013. UHPLC: the greening face of liquid chromatography. *Chromatographia* 76, 1429–1437. https://doi.org/10.1007/S10337-013-2434-6

Cirlini, M., et al., 2018. Are tropane alkaloids present in organic foods? Detection of scopolamine and atropine in organic buckwheat (Fagopyron esculentum L.) products by UHPLC–MS/MS. *Food Chemistry* 239, 141–147. https://doi.org/10.1016/j.foodchem.2017.06.028

© Burleigh Dodds Science Publishing Limited, 2025. All rights reserved.

Cirlini, M., et al., 2019. A sensitive UHPLC-ESI-MS/MS method for the determination of tropane alkaloids in herbal teas and extracts. *Food Control* 105, 285–291. https://doi.org/10.1016/j.foodcont.2019.05.030

de Nijs, M., et al., 2023. Emerging issues on tropane alkaloid contamination of food in Europe. *Toxins (Basel)* 15(2), 98. https://doi.org/10.3390/toxins15020098

Dong, H., et al., 2021. Advances in analysis of contaminants in foodstuffs on the basis of orbitrap mass spectrometry: a review. *Food Analytical Methods* 15(3), 803–819. https://doi.org/10.1007/S12161-021-02168-0

Dzuman, Z., et al., 2020. Development of a new LC-MS method for accurate and sensitive determination of 33 pyrrolizidine and 21 tropane alkaloids in plant-based food matrices. *Analytical and Bioanalytical Chemistry* 412, 7155–7167. https://doi.org/10.1007/S00216-020-02848-6/TABLES/2

EFSA CONTAM Panel, 2013. Scientific opinion on tropane alkaloids in food and feed. *EFSA Journal* 11(10), 3386. https://doi.org/10.2903/j.efsa.2013.3386

European Commission, 2002. RASFF - European Commission [WWW Document]. https://food.ec.europa.eu/safety/rasff_en#Learn (accessed 9.12.24).

European Commission, 2013. Regulation (EU) No 609/2013 of the European Parliament and of the Council of 12 June 2013 on food intended for infants and young children, food for special medical purposes, and total diet replacement for weight control and repealing Council Directive 92/52/EEC, Commission Directives 96/8/EC, 1999/21/EC, 2006/125/EC and 2006/141/EC, Directive 2009/39/EC of the European Parliament and of the Council and Commission Regulations (EC) No 41/2009 and (EC) No 953/2009 Text with EEA relevance. *Official Journal of the European Union* 181, 35–56.

European Commission, 2016. Commission Regulation (EU) 2016/239 of 19 February 2016 amending Regulation (EC) No 1881/2006 as regards maximum levels of tropane alkaloids in certain cereal-based foods for infants and young children. *Official Journal of the European Union* 45, 3–5.

European Commission, 2023a. Commission Regulation (EU) 2023/915 of 25 April 2023 on maximum levels for certain contaminants in food and repealing Regulation (EC) No 1881/2006. *Official Journal of the European Union* 119, 103–157.

European Commission, 2023b. Commission Implementing Regulation (EU) 2023/2783 of 14 December 2023 laying down the methods of sampling and analysis for the control of the levels of plant toxins in food and repealing Regulation (EU) 2015/705. *Official Journal of the European Union* 2023/2783, 1–95.

European Commission, 2024. RASFF Window (Notifications on natural toxins from Jan 2020 to Aug 2024) [WWW Document]. https://webgate.ec.europa.eu/rasff-window/screen/search (accessed 9.12.24).

FAO/WHO, 2020. *Joint FAO/WHO Expert Meeting on Tropane Alkaloids, Food Safety and Quality Series*. FAO. https://doi.org/10.4060/cb1857en

Arcella, D., et al., 2018. Human acute exposure assessment to tropane alkaloids. *EFSA Journal* 16, e05160. https://doi.org/10.2903/j.efsa.2018.5160

García-Juan, A., et al., 2023. Development and validation of an analytical method for the simultaneous determination of 12 ergot, 2 tropane, and 28 pyrrolizidine alkaloids in cereal-based food by LC-MS/MS. *Food Research International* 174, 113614. https://doi.org/10.1016/j.foodres.2023.113614

© Burleigh Dodds Science Publishing Limited, 2025. All rights reserved.

Géhin, C. and Holman, S.W., 2021. Advances in high-resolution mass spectrometry applied to pharmaceuticals in 2020: a whole new age of information. *Analytical Science Advances* 2, 142. https://doi.org/10.1002/ANSA.202000149

González-Gómez, L., et al., 2022a. Occurrence and chemistry of tropane alkaloids in foods, with a focus on sample analysis methods: a review on recent trends and technological advances. *Foods* 11, 407. https://doi.org/10.3390/FOODS11030407/S1

González-Gómez, L., et al., 2022b. Green extraction approach based on μSPEed® followed by HPLC-MS/MS for the determination of atropine and scopolamine in tea and herbal tea infusions. *Food Chemistry* 394, 133512. https://doi.org/10.1016/j.foodchem.2022.133512

Gumus, Z.P., 2023. Assessment of toxic pyrrolizidine and tropane alkaloids in herbal teas and culinary herbs using LC-Q-ToF/MS. *Foods* 12, 3572. https://doi.org/10.3390/foods12193572

Han, S., et al., 2024. Occurrence and health risk assessment of tropane alkaloids in cereal foods consumed in Korea. *Food and Chemical Toxicology* 186, 114589. https://doi.org/10.1016/j.fct.2024.114589

Jank, B. and Rath, J., 2021. Emerging tropane alkaloid contaminations under climate change. Trends in Plant Science 26(11), 1101–1103. https://doi.org/10.1016/j.tplants.2021.08.001

Jiang, Z.M., et al., 2022. Deep eutectic solvents in the efficient detoxification and extraction of alkaloids of aconiti lateralis radix praeparata. *ACS Sustainable Chemistry & Engineering* 10, 5674–5682. https://doi.org/10.1021/ACSSUSCHEMENG.2C00754/ASSET/IMAGES/LARGE/SC2C00754_0008.JPEG

Kaltner, F., 2022. Fate of food-relevant toxic plant alkaloids during food processing or storing and analytical strategies to unveil potential transformation products. *Journal of Agricultural and Food Chemistry* 70, 5975–5981. https://doi.org/10.1021/ACS.JAFC.2C01489/ASSET/IMAGES/MEDIUM/JF2C01489_0003.GIF

Kartsova, L.A., et al., 2019. Hydrophilic interaction chromatography. *Journal of Analytical Chemistry* 74(5), 415–424. https://doi.org/10.1134/S1061934819050058

Kohnen-Johannsen, K.L. and Kayser, O., 2019. Tropane alkaloids: chemistry, pharmacology, biosynthesis and production. *Molecules* 24, 796. https://doi.org/10.3390/molecules24040796

Kowalczyk, E. and Kwiatek, K., 2022. Simultaneous determination of pyrrolizidine and tropane alkaloids in honey by liquid chromatography-mass spectrometry. *Journal of Veterinary Research* 66, 235–243. https://doi.org/10.2478/JVETRES-2022-0032

Lai, Y.C., et al., 2019. Quantification and simplified detoxification investigation on Fuzi, Root of aconitum carmichaelii. Natural Product Communications 14(10), Article 1934578X1988154. https://doi.org/10.1177/1934578X19881548/ASSET/IMAGES/LARGE/10.1177_1934578X19881548-FIG4.JPEG

León, N., et al., 2022. A green analytical method for the simultaneous determination of 30 tropane and pyrrolizidine alkaloids and their N-oxides in teas and herbs for infusions by LC-Q-Orbitrap HRMS. Journal of Chromatography A 1666, Article 462835. https://doi.org/10.1016/j.chroma.2022.462835

Li, P., et al., 2022. Solid-phase extraction methods for nucleic acid separation: a review. *Journal of Separation Science* 45, 172–184. https://doi.org/10.1002/JSSC.202100295


© Burleigh Dodds Science Publishing Limited, 2025. All rights reserved.

Lusthof, K.J., et al., 2017. Toxicological results in a fatal and two non-fatal cases of scopolamine-facilitated robberies. *Forensic Science International* 274, 79–82. https://doi.org/10.1016/j.forsciint.2017.01.024

Marín-Sáez, J., et al., 2019a. Reliable determination of tropane alkaloids in cereal based baby foods coupling on-line spe to mass spectrometry avoiding chromatographic step. *Food Chemistry* 275, 746–753. https://doi.org/10.1016/j.foodchem.2018.09.137

Marín-Sáez, J., et al., 2019b. Effect of tea making and boiling processes on the degradation of tropane alkaloids in tea and pasta samples contaminated with Solanaceae seeds and coca leaf. *Food Chemistry* 287, 265–272. https://doi.org/10.1016/j.foodchem.2019.02.091

Marín-Sáez, J., et al., 2019c. Degradation of tropane alkaloids in baked bread samples contaminated with Solanaceae seeds. *Food Research International* 122, 585–592. https://doi.org/10.1016/j.foodres.2019.01.027

Marín-Sáez, J., et al., 2017. Multi-analysis determination of tropane alkaloids in cereals and solanaceaes seeds by liquid chromatography coupled to single stage exactive-orbitrap. *Journal of Chromatography A* 1518, 46–58. https://doi.org/10.1016/j.chroma.2017.08.052

Marrubini, G., et al., 2018. Hydrophilic interaction chromatography in food matrices analysis: an updated review. *Food Chemistry* 257, 53–66. https://doi.org/10.1016/J.FOODCHEM.2018.03.008

Martinello, M., et al., 2017. Development and validation of a QuEChERS method coupled to liquid chromatography and high resolution mass spectrometry to determine pyrrolizidine and tropane alkaloids in honey. *Food Chemistry* 234, 295–302. https://doi.org/10.1016/j.foodchem.2017.04.186

Mulder, P.P.J., et al., 2015. Tropane and ergot alkaloids in grain-based products for infants and young children in the Netherlands in 2011–2014. *Food Additives and Contaminants: Part B Surveillance* 8, 284–290. https://doi.org/10.1080/19393210.2015.1089947

Mulder, P.P.J., et al., 2016. *Occurrence of Tropane Alkaloids in Food*. EFSA Supporting Publications 13. https://doi.org/10.2903/sp.efsa.2016.en-1140

Perestrelo, R., et al., 2019. QuEChERS - Fundamentals, relevant improvements, applications and future trends. *Analytica Chimica Acta* 1070, 1–28. https://doi.org/10.1016/j.aca.2019.02.036

RASFF, 2021. Notification 2021.1390 - Foodborne outbreak in deep frozen spinach puree from Slovakia [WWW Document]. https://webgate.ec.europa.eu/rasff-window/screen/notification/470490 (accessed 9.12.24).

RASFF, 2023. Notification 2023.6126 - Exceedance of tropane alkaloids limit [WWW Document]. https://webgate.ec.europa.eu/rasff-window/screen/notification/631525 (accessed 9.12.24).

RASFF, 2024a. Notification 2024.4104 - Tropane alkaloids in herbal teas [WWW Document]. https://webgate.ec.europa.eu/rasff-window/screen/notification/687425 (accessed 9.12.24).

RASFF, 2024b. Notification 2024.5947 - Food poisoning from tropane alkaloids in product "Fresh SPINACH in bag" [WWW Document]. https://webgate.ec.europa.eu/rasff-window/screen/notification/701553 (accessed 9.12.24).

© Burleigh Dodds Science Publishing Limited, 2025. All rights reserved.

Ribeiro, M.M.A.C., et al., 2022. Simultaneous determination of scopolamine and butylscopolamine in pharmaceutical and beverage samples by capillary zone electrophoresis. *Microchemical Journal* 172, 106985. https://doi.org/10.1016/J.MICROC.2021.106985

Romera-Torres, A., et al., 2020. Comprehensive tropane alkaloids analysis and retrospective screening of contaminants in honey samples using liquid chromatography-high resolution mass spectrometry (Orbitrap). Food Research International 133, Article 109130. https://doi.org/10.1016/J.FOODRES.2020.109130

Srinivasan, P. and Smolke, C.D., 2021. Engineering cellular metabolite transport for biosynthesis of computationally predicted tropane alkaloid derivatives in yeast. *Proceedings of the National Academy of Sciences* 118, e2104460118. https://doi.org/10.1073/pnas.2104460118

Thompson, T.S., et al., 2020. Hyoscyamine and scopolamine in honey by HILIC–ESI-MS/MS. *Chromatographia* 83, 683–689. https://doi.org/10.1007/S10337-020-03880-5/TABLES/3

Torrents-Masoliver, B., et al., 2024. Effect of pH and temperature on tropane alkaloids within a processing strategy to provide safe infant cereal-based food. *Food Research International* 188, Article 114439. https://doi.org/10.1016/j.foodres.2024.114439

Vargas Medina, D.A., et al., 2021. Current role of modern chromatography and mass spectrometry in the analysis of mycotoxins in food. *TrAC Trends in Analytical Chemistry* 135, 116156. https://doi.org/10.1016/j.trac.2020.116156

Vera-Baquero, F.L., et al., 2024. Improving the food safety of bakery products by simultaneously monitoring the occurrence of pyrrolizidine, tropane and opium alkaloids. *Food Chemistry* 460, 140769. https://doi.org/10.1016/j.foodchem.2024.140769

Veršilovskis, A., et al., 2020. Simultaneous quantification of ergot and tropane alkaloids in bread in the Netherlands by LC-MS/MS. *Food Additives & Contaminants: Part B* 13, 215–223. https://doi.org/10.1080/19393210.2020.1771777

Yan, X., et al., 2022. Technical overview of orbitrap high resolution mass spectrometry and its application to the detection of small molecules in food (Update since 2012). *Critical Reviews in Analytical Chemistry* 52, 593–626. https://doi.org/10.1080/10408347.2020.1815168


© Burleigh Dodds Science Publishing Limited, 2025. All rights reserved.

# Index



© Burleigh Dodds Science Publishing Limited, 2025. All rights reserved.

© Burleigh Dodds Science Publishing Limited, 2025. All rights reserved.

© Burleigh Dodds Science Publishing Limited, 2025. All rights reserved.

© Burleigh Dodds Science Publishing Limited, 2025. All rights reserved.

© Burleigh Dodds Science Publishing Limited, 2025. All rights reserved.

© Burleigh Dodds Science Publishing Limited, 2025. All rights reserved.

© Burleigh Dodds Science Publishing Limited, 2025. All rights reserved.

© Burleigh Dodds Science Publishing Limited, 2025. All rights reserved.

© Burleigh Dodds Science Publishing Limited, 2025. All rights reserved.

© Burleigh Dodds Science Publishing Limited, 2025. All rights reserved.

www.ingramcontent.com/pod-product-compliance
Lightning Source LLC
LaVergne TN
LVHW010620300926
849162LV00001B/10

* 9 7 8 1 8 3 5 4 5 0 5 0 5 *